Elastic structures, conceived as slender bodies able to transmit loads, have been studied by scientists and engineers for centuries. By the seventeenth century, several useful theories of elastic structures had emerged, with applications to civil and mechanical engineering problems. In recent years improved mathematical tools have extended applications into new areas such as geomechanics and biomechanics.

This book offers a critically filtered collection of the most significant theories dealing with elastic slender bodies, as well as the mathematical models, that have been developed over time. The book also shows how these models are used to solve practical problems involving elastic structures with particular emphasis on nonlinear problems. Chapters progress from simpler to more complicated structures, including rods, strings, membranes, plates, and shells.

As a collection of interesting and important problems in elastic structures, this book will appeal to a broad range of scientists and engineers working in the area of structural mechanics.

MATHEMATICAL MODELS FOR ELASTIC STRUCTURES

MATHEMATICAL MODELS FOR ELASTIC STRUCTURES

PIERO VILLAGGIO
Università di Pisa

CAMBRIDGE UNIVERSITY PRESS
Cambridge, New York, Melbourne, Madrid, Cape Town, Singapore, São Paulo

Cambridge University Press
The Edinburgh Building, Cambridge CB2 2RU, UK

Published in the United States of America by Cambridge University Press, New York

www.cambridge.org
Information on this title: www.cambridge.org/9780521573245

First published 1997
This digitally printed first paperback version 2005

A catalogue record for this publication is available from the British Library

Library of Congress Cataloguing in Publication data

Villaggio, Piero.
Mathematical models for elastic structures/Piero Villaggio.
p. cm.
Includes bibliographical references and indexes.
ISBN 0 521 57324 6 (hardcover)
1. Elastic analysis (Engineering) 2. Nonlinear mechanics-
Mathematical models. 3. Structural analysis (Engineering)
I. Title.
TA653.V55 1997 96-44172
624.1'71 DC20 CIP

ISBN-13 978-0-521-57324-5 hardback
ISBN-10 0-521-57324-6 hardback

ISBN-13 978-0-521-01798-5 paperback
ISBN-10 0-521-01798-X paperback

Contents

Preface

Few words are used with so many different meanings as the term "model." In everyday language the word "model" can be applied in a moral, fashion, economic, linguistic, or scientific context; in each case it means something completely different. Even if we restrict ourselves to the category of scientific models, the notion is ambiguous, because it could signify the reproduction in miniature of a certain physical phenomenon, and at the same time present a theoretical description of its nature that preserves the broad outline of its behavior. It is the theoretical aspect of models that we wish to consider; in order to emphasize this, we describe this type of model as "mathematical" (Tarski 1953). Formulating a mathematical model is a logical operation consisting in: (i) making a selection of variables relevant to the problem; (ii) postulating statements of a general law in precise mathematical form, establishing relations between some variables said to be data and others unknown; and (iii) carrying out the treatment of the mathematical problem to make the connections between these variables explicit.

The motivations underlying the use of mathematical models are of different types. Sometimes a model is the passage from a lesser known theoretical domain to another for which the theory is well established, as, for example, when we describe neurological processes by means of network theory. In other cases a model is simply a bridge between theory and observation (Aris 1978). The word "model" must be distinguished from "simulation." The simulation of a phenomenon increases in usefulness with the quantity of specific details incorporated, as, for example, in trying to predict the circumstances under which an epidemic propagates. The mathematical model should instead include as few details as possible, but preserve the essential outline of the problem. The "simulation' is concretely descriptive, but applies to only one case; the "mathematical model" is abstract and universal. Another special property of a good mathematical model is that it can isolate only some aspects of the physical fact, but not all. The merit of such a model is not of finding what is common to two groups of observed facts, but rather of indicating their diversities. A long-debated and important question is that of how to formulate a model in its most useful form. The answer is clearly not unique, because there are examples in which the same observed phenomenon can be described equally well by two completely different models. However, in order to initiate the formulation of a good model, six precepts have been proposed (Hammersley 1973): (i) notation should be clarified; (ii) suitable units should be chosen; (iii) the number of variables should be reduced, whenever possible; (iv) rough sketches should be made and particular cases examined; (v) rigor should be avoided at this stage; and (vi) equations should be adjusted to have roughly the same number of terms on each side. Among these rules the third is by far the most important. The essential elements needed to describe a physical phenomenon should initially be isolated. The description is not improved by adding new terms to them. All the theories that have contributed decisively to the progress of physics are remarkable for their simplicity. If we look at the history of mechanics we see that the most important advances conform to the Baconian criterion of *dissecare naturam*, that is to strive to retain only those ingredients in a model that give the answer we require from a specific physical event.

The theory of elastic structures is, by definition, the collection of all reasonable models, proposed during almost three centuries, concerned with simplifying the solutions of problems involving elastic bodies. The equations describing the motion and equilibrium of a three-dimensional elastic body were formulated in full generality during the first half of the nineteenth century, but their solutions are known only in a few cases. From the beginning of the theory of elasticity the interest of many scientists was focused on the solution of the problems of the bending of a beam, the vibrations of bars and plates, and the stability of columns. Later, other problems were formulated and solved, as, for example, those concerning the torsion of a beam, the equilibrium and vibrations of thin shells, and the longitudinal impact of rods. The progress of the theory is not uniform, because we can find frequent retrogressions on the part of experimentalists in adopting hypotheses that had not been properly established, or on the part of mathematicians in using approximate methods beyond the limits of their validity. However, in the case of linear elasticity, a satisfactory degree of knowledge about the range of applicability of each of the theories has been achieved.

Nevertheless, things are radically different when the hypothesis of small strains and small displacements in elastic materials is removed. We may think that the equations of elastic structures with large deformations can be simply obtained from those valid with small strains and displacements by replacing the linear constitutive equations of classical elasticity with the constitutive equations of nonlinear elasticity, while referring the data and the unknown to the reference configuration. This is what is customarily done in deriving the equations of three-dimensional finite elasticity. In the theory of structures, however, a new aspect arises. When formulating the model of a structure, such as a rod or a plate, we try to describe it without using the three-dimensional equations. The aim is to find certain reduced equations capable of representing the essential features of the state of strain and stress in simplified form. To do this we make some conjecture about the form of the solutions, as, for example, the conservation of the planarity of the cross-sections in a bent beam, and we write certain averaged equations for the simplified unknowns. The procedure is perfectly rigorous, provided that the hypotheses are coherent and physically plausible. Unfortunately, the range of applicability of assumptions of this kind is very narrow. It may happen that it succeeds if the strains are infinitesimal, although not when there are large strains, and vice versa. One of the reasons why the theory of structures is exposed to serious criticisms is that, very often, a set of equations valid under certain restrictions has been illegitimately extended to different situations.

In the history of mechanics of structures the use of models incapable of giving results for suitably large extensions is very frequent. Theories perfectly sound for thin rods have been applied to thick rods or to thin-walled beams; the theory of Kirchhoff's plate has been extrapolated to thick and sandwich plates; and methods of solution well established for curved membranes and shells in small deformations have been employed in the presence of large deformations, with arbitrary adjustments that invalidate the results. Those who have proposed these generalizations have often realized the weakness of some arguments. However, instead of setting up new, consistent, formulations of problems, they have limited themselves to adding small corrections to the old models, with the consequence that they have proposed dubious theories from both physical and logical aspects.

This kind of attitude has been common for several decades, save for a few praiseworthy exceptions, although remarkable change has occurred in the last twenty years. Though the scientific literature is still abundant in old-style articles, we can now find a number of novel contributions. These are important in two respects: they pay more attention to the formulation and to the treatment of the mathematical problem, drawing on developments in modern mathematics; and they tackle new problems arising in the fields of technology or everyday life. An example of the first type of advance is seen in the application of bifurcation theory. This allows the characterization of solutions of a nonlinear problem even when they are very far from those of the corresponding linearized problem. An example of the second development comes in the extended application of the theory of structures to new fields, such as those of geomechanics, biology, and medicine.

The purpose of this book is to give an account of the advances in the theory of structures in both directions. The number of papers actually published on subjects involving the theory is so vast that it is physically impossible to enumerate them. The selection of the most significant contributions is naturally conditioned by the taste and prejudices of the author. It is therefore natural to ask what criterion has been used in selecting or rejecting papers. The papers that were rejected were judged on their feature of offering only "slight generalizations." Very often the essential features of an important mechanical problem were isolated and formulated in mathematical equations more than a century ago. Since then much work has been put into adding small corrections to improve the physical validity of the system. The result has been an immediate complicating of the mathematics. In general, there are two types of development in a certain field. Sometimes the problem is an old one, but treated now by a new mathematical technique that allows the properties of solutions to be illustrated. In other cases an unsatisfactory analytical treatment of an important mechanical problem has prompted questions normally considered outside the domain of engineering. There are also papers that have the distinction of encompassing both these aspects, resulting in new, original, problems treated with elegant mathematics.

A book of this kind is the result of many direct and indirect contributions from friends, correspondents, and the authors I have read. However, one person deserves special mention. Ernest Wilkes has been a continual source of support throughout the preparation of this book, and the extraordinary care with which he has offered suggestions for improvements to the entire manuscript has been decisive. I wish to express my warmest thanks to him.

Piero Villaggio
Università di Pisa

Introduction

What does it mean to solve a problem in classical elasticity? The question may appear trivial, but, if we ask scholars working in the field, we receive surprisingly different answers. Let us assume, in order to make the subject more explicit, that the problem concerns the impact between two elastic bodies. For an experimental physicist solving the problem means interpreting those crucial experiments that make it possible to decide which are the important variables in the phenomenon: in this specific case, the densities and elastic moduli of the materials are important, but the temperature and atmospheric pressure, for instance, are not. A theoretical physicist will say instead that the solution consists in formulating the general equations of the problem, having inserted all the significant variables. For a mathematician it will be obvious that solving the problem means finding an existence, uniqueness, and possibly regularity, theorem for the equations of elastic impact. Yet another answer will be given by an engineer, who will require an explicit formula giving the stress components within the two bodies at each point and at each instant.

Confronted with such a variety of answers, a typical student feels disoriented, being immediately aware of the basic ambiguity in the way in which the question itself has been posed. Ludwig Wittgenstein would say in explanation that the confusion arises from the vague use of the verb "to solve." For the same word has been used in different contexts with different meanings.

Though dissimilar, the four answers have a common characteristic. They represent four attempts at describing the same phenomenon by abstraction, setting aside the unessential details, with the purpose not merely of illustrating but also of predicting. We say that they propose four models for the elastic impact. At this point we immediately ask whether there is a rational criterion for deciding which model is preferable, provided that all satisfy the three necessary requisites of being realistic, logically coherent, and simple. It is evident that a model must not be in obvious conflict with the physical data, nor must it be self-contradictory or too complicated. However, unfortunately, there is no incontrovertible way of establishing that one model is better than another. Setting up a model means creating conceptual conditions suitable for posing a particular question about the problem. The choice of these conditions then depends on the kind of answer we wish to obtain. For instance, if the question we ask ourselves about the elastic impact is one concerning the qualitative interdependence of some quantities, then the rough tests of the experimentalist represent the right method; if, alternatively, we want to know the detailed distribution of stresses on the surface of contact of the two colliding bodies, then the formulae proposed by the engineer work better.

However, even if a rigorous science for formulating good models is lacking we have at least the comfort of returning to the traditional analysis of those models that have been most deeply involved in the development of continuum mechanics, as, for example, the Euler equations of motion of perfect fluids, or the Navier equations of classical elasticity. Besides the three necessary properties of realistic representation, coherence, and simplicity, these models possess two other qualities. The first is that there is a surprising harmony between the originality of the physical problem that must be described and the novelty of the mathematical method by which the problem is treated. Newton's second law is a model for explaining the motions of heavenly bodies, but, at the same time, it requires the solution of ordinary differential equations of second order, a problem which was just at that time beginning to be studied. On the other hand, there are several examples of physically interesting problems treated with primitive mathematics, or of insignificant problems accompanied by a brilliant mathematical manipulation. Using an adjective coined by Nietzsche, this first quality may be called the "Apollonian" attribute of a good model. But this is not all. The great models of classical mathematical physics are characterized by a second quality which may be called "reproductive," because we devise from them the methodological example for tackling new problems. A typical instance of this is found in the classical theory of elasticity, which, as Love (1927) says in the Historical Introduction to his treatise, is not only important for its contribution to the material advance of mankind, but also, and more importantly, for the light thrown on other branches of the physics of other areas, such as optics and geomechanics.

It would appear that, at this point, the question of finding a good model does not necessarily depend on these terms. We cannot hope to find an infallible recipe for suitably formulating and treating all the problems proposed by physics, but, at least, we have inherited from earlier work a significant number of examples suggesting means of dealing with those that interest us. However, in spite of this, it is hardly likely that these examples will provide satisfactory solutions to fresh problems. We must also consider that the great men of the past, when unable to find a general closed solution, have often reformulated their problem in more manageable terms, even at the cost of convoluted approximations. This is the so-called path of "reasonable" approximations. When confronted with a problem of too great a complexity, simplification is necessary; we can achieve this, for instance, by eliminating the residual non-linear terms, which jeopardizes the application of classical procedures, or by changing the shape of the original domain into one of more regular form. All this implies the substitution of the primitive model with a new one, and means that the new model must be reconsidered *ab initio* from the point of view of physical completeness and mathematical consistency.

Models thus generate submodels, thereby creating the possibility of multiplicity, and deciding on their suitability can cause confusion. But here again examples from the past give us encouragement. Workers such as Kelvin (1848), Hertz (1881) and others, have passed into history for their contributions to the general principles of mechanics, but are not always appreciated for their formidable capacity in making compromises. In order to evaluate the effect of a force acting at a point in an indefinitely extended solid, Kelvin considered the case in which body forces act within a finite volume T and vanish outside it. He constructed the solution of the problem and then passed to a limit by diminishing T indefinitely and supposing that the resultant force with T has a finite limit. The procedure seems tortuous, but the final solution is very simple. Hertz, in his solution of the problem of the pressure between two bodies in contact, assumed that the compressed area, common to the two bodies, is an ellipse, and that, provided this area is small, each body may be regarded as a half-space loaded over the bounding plane.

We are tempted to believe that, in mechanics, reductions like those of Stokes and Rayleigh are the outcome of long reflections, but it has not always been so. To be more precise, it was so until the end of the last century, but after that, those working in mechanics, pressed by the need to achieve manageable solutions, have preferred to accept the technical artifices, without submitting them under critical scrutiny. The treatises of the nineteenth century, such as those of Clebsch (1862), Kelvin and Tait (1867), and Love (1927), are full of mathematical development, but still maintain a critical grasp whenever any new approximations are introduced. More recent books, such as those by Sommerfeld (1944) and Landau and Lifschitz (1971), only occasionally state the postulates introduced during the process of deriving the equations. Books written for engineers are even less punctilious and all this leaves the reader in a state of permanent disorientation. There are, of course, some brilliant exceptions: for instance, a paradigm of deduction of the linearized equations of the motion of a string, with exemplary justification of all assumptions, can be found in the book by Weinberger (1965), which is not a book on elasticity, but a course on partial differential equations!

Thus a sort of no-man's land has been created between the principles of elasticity and their applications. And this terrain is shaky, either because one fears having somewhere violated the basic principles of mechanics and thermodynamics in conjecturing some simplifying property of a solution, or, because, given the arbitraries of choice, one is constantly afflicted by the doubt that another much simpler but more elegant model ought to have been used instead. However, in spite of possible misapplications, the technical literature is rich in contributions proposing expedient methods of treating more general problems. There are theories of plates and shells derived from the three-dimensional equations of elasticity through expansions of the displacements as functions of the first, second and third order of the distance from the middle surface; theories of finite deformations with unjustified linearizations; and semi-inverse solutions in elastostatics for domains having noncylindrical forms. A great number of these attempts are mediocre, but some have contributed greatly to the development of technical mechanics. On the other hand, books on general mechanics are reluctant to supply the general equations with some practical application, as if the analysis of the motion of a toothed wheel or of a bicycle would compromise the elegance of the book. An attempt to fill the gap has been made by Szabó (1963, 1964), whose work ranges from celestial mechanics to plasticity theory with surprising versatility, and above all with a taste for applications. But the work is too fragmentary: it does not develop a method, but rather offers only a collection of smart artifices in solving particular problems.

The divisions in an intellectual discipline are always pernicious. They damage the foundations of mechanics because the objects being systematized, whenever they have lost any connection with experience, become the "golden mountains" of which Russel (1903) speaks. In applications, they are even lethal. For historical reasons, in continuum mechanics subdivisions have been created between, roughly, the mechanics of incompressible fluids, of compressible fluids, and of solids. This division is necessary because the specific problems of each sector are so particular that an *ad hoc* procedure must be employed in their solution, as, for instance, in the theory of surface waves or in that of elastic structures. But these methods are in some cases so cogent that they might well be employed in related sectors, where they are unknown.

An objection made to this observation is that nowadays the use of numerical calculus has overcome the difficulty of describing, in the limit, small technical detail by solving the general equations. Unfortunately, however, even precise numerical solutions are too restrictive and do give no idea of how solutions behave as data vary.

Many ambiguities arise from three tacit premises: that classical elasticity is a logical science; that logic can be raised to the level of awareness; and that thinking in mechanical terms can be refined by its intelligent application.

The logic of mechanics is not a formal logic of deductive inference having the symmetric structure of Aristotelian syllogism, nor is it an inductive logic, like that of John Stuart Mill. Using a word introduced by Peirce (1931), it is a process of "abduction," that is, of formulating a hypothesis and of deducting what would be the case if the hypothesis were true. Usually the most important advances have been achieved by workers who had a particular problem in mind, the solution of which did not obviously follow by applying the general equations. They have then started to manipulate those equations and to extract some new ones, until the solution furnished by the latter conform to their expectations. The success of the method has been ensured by two ingredients: a prior intuition of how things should evolve and, consequently, of how the solutions must be; and a sort of esthetic taste in choosing the clearest route in working through the intermediate stages.

In these processes, as there are no definite prescriptions for reaching the result, it is easier to try to identify the common ways in which errors have been made. Fischer (1969), in his brilliant essay on the logic of historical thought, has endeavored to indicate the most frequent occasion of mistake. Essentially, these are the fallacies resulting from excessive motivation; that is, the neglect of relevant quantities simply because they disturb the desired result. On the other hand, there are the fallacies of overgeneralization, which arise from the belief that a model improves, the greater the number of its constituent entities. It is clearly forgotten in this fallacy that a good model can never be exhaustive, and must be simple to be effective. There are several examples of bad models, some of which originated with illustrious authors as, for instance, the celebrated equation of Föppl–v. Kármán in describing the large transverse displacements of a thin plate subjected to pressure. The defects of the theory have been pointed out by Truesdell (1978), whose criticisms can be summoned up in the following terms: unnecessary geometric approximations, and unjustified assumptions about the way in which the stress varies over a cross-section.

In the history of mechanics the separation between theory and application has never really existed. Workers in the nineteenth century have simply followed their seventeenth and eighteenth century predecessors, who could pass from abstract mathematics to technical solutions with extraordinary ease. It is known that James Bernoulli (1694) formulated the problem of the elastic line or *elastica* of a beam, and that Euler (1744) proposed a simple theory to explain the onset of buckling in a thin strut. But they are not exceptions, since eclecticism is a characteristic of scientists less famous than J. Bernoulli and Euler. For instance, in 1823, Lamé and Clapeyron were called on to assess the stability of the dome of the cathedral of St Isaac in St Petersburg, and on which occasion they virtually re-invented the technique of slicing the dome into lunes, each of which has to have an independent equilibrium.

At this point it might by naïvely asked which have been the most creative ideas in the development of the branch of classical elasticity oriented towards applications. The answer is clearly complicated because, in general, an ingenious result is not the outcome of a single author but the result of many refinements by a sequence of authors over a long period of time. Nevertheless, let us suppose that one is required to construct a list of outstanding results produced in the last three centuries, solely with the purpose of providing a picture of the evolution of interests over time. A tentative list might be as follows: James Bernoulli's (1705) equation of the *elastica*; Euler's (1744) integration of the equation of the *elastica* and Euler's (1757) definition of buckling load; Navier's (1827) equations of three-dimensional elasticity; Cauchy's (1829) definition of elastic material; Green's (1839, 1842) definition of hyperelastic material; Saint-Venant's theory (1855) of the torsion of prisms and the invention of the semi-inverse method of solution; Kirchhoff's (1850) formulation of the boundary conditions at the edge of a plate; Kelvin's (1848) solution for the elastic displacement due to a point force in an indefinite medium; Boussinesq's (1885) solution for the half-plane normally loaded by a point force; Hertz's (1881) theory of the contact between two bodies; Michell's (1904) theory of trusses; Geckeler's (1926) approximate theory for evaluating edge effects in shells; Kolosov's (1909, 1914) solution of the biharmonic equation in two variables; the three functions *Ansatz* of Boussinesq, Papkovič, and Neuber (Papkovič 1932; Neuber 1934); Vlasov's (1958) theory of thin-walled beams; Ericksen and Truesdell's (1958) director theory of rods; and Ericksen's (1973) concept of a loading device.

The list may suggest that all efforts to convert the problems created by practical applications into simple and precise terms have succeeded. But this impression is illusory, as there have also been numerous failures. For instance, Euler's (1766) theory of axisymmetric shells is not accepable; Coulomb's (1787) theory of twisting is wrong; Cauchy's (1828) theory of "rari-constants" has been contradicted by experience and thermodynamics; and Greenhill's (1881) solution for the buckling of a heavy column is incorrect, because the equation he solved neglects terms of the same order of magnitude as those maintained in the equation.

It is instructive to recognize that faults occur in the works of those who have decisively contributed to the progress of continuum mechanics: this means that mechanics is a perfectible science, and what seems a definite conquest today may be demolished tomorrow. An unknown Italian philosopher, Giovanni Vailati

(1863–1909), pointed out the importance of error in the development of science. Vailati maintained that an error is worth as much as an ingenious discovery, if it contributes equally in orienting our intellectual faculties: "Every error is a reef to be avoided, while each discovery doesn't always indicate the path to follow."

The present-day literature on mechanics abounds with papers proposing modifications of the classical models, either by insignificant extension or by eliminating terms that generate difficulty. These works are not wrong. They are, of course, absolutely correct, but as mathematical works, they are deplorably inelegant.

Chapter I
Basic Concepts

1. Density, Motion, and Temperature

Mechanics studies the conditions for the motion and equilibrium of natural objects. In order to treat these in a sufficiently general form, mechanics introduces the notion of the *body*, which is a mathematical concept designed to give an abstract representation of the most important properties of these physical objects in order to describe their mechanical behavior. By the term "body" we mean a regularly open set $\mathscr{B}$ in some topological space. The elements of a body are called *particles*, or *substantial points* to avoid confusion with the term "particle" as used in physics (Truesdell 1991, Ch. I, p.3). Bodies are available in their *configurations*, which are the regions $\chi(\mathscr{B})$ that they may occupy in Euclidean space at some time. It is commonly assumed that configurations are regularly open sets in Euclidean space and there is a one-to-one mapping between the particles of the body and a possible configuration. It is often convenient to select one particular configuration $\kappa(\mathscr{B})$ and to identify each particle of $\kappa(\mathscr{B})$ by its position in this configuration, called the *reference configuration*. If, in the reference configuration, we choose Cartesian coordinates, the particle can be designated X_A or X, Y, Z.

At the time t varies the particles change their positions in space, and consequently the position of a particle X_A at the instant t is characterized by a vector function x_i, or x, y, z, of the form

$$x_i = \chi_i(X_A, t), \tag{1.1}$$

which is called the *motion* of the body. The motion determines the shape of the body at each instant.

The inertia of a particle is determined by a scalar function $\rho = \rho(X_A, t)$, called the *density*; and the hotness of a particle is determined by another scalar function $\theta = \theta(X_A, t)$, called *temperature*.

Motion, density, and temperature are regarded as *primitive concepts*: that is, they need not be defined in terms of other known quantities, and must be measurable, at least in principle. The aim of mechanics is to determine the fields associated with χ_i, ρ, and θ. However, these quantities are not completely unrestricted, because both density and temperature must be positive-valued functions, and the functions χ_i must be continuous and invertible so as to exclude the possibility of two particles assuming the same position.

At time t the particle X_A has the position x_i, and we may use this position, instead of X_A, to identify the particle. Thus, provided that the mapping (1.1) is continuous and invertible, carrying a regularly open set into regularly open sets, it is possible to represent the fields of mechanics in the form $x_i = x_i(t)$, $\rho = \rho(x_i, t)$, $\theta = \theta(x_i, t)$, which is called the *spatial description*. The representation in terms of X_A and t as independent variables is said to be the *material description*. In the spatial description we regard as the object of our investigation a knowledge of the density and temperature at all points occupied by the body, at all instants in time; in the material description we seek the history of every particle.

In continuum mechanics Green's theorem is of central importance in establishing mechanical relations in terms of equations. It is thus necessary that all the possible configurations of bodies are regular enough for the application of Green's formula, whenever the fields are sufficiently smooth. As the assumption made about the configurations is too weak, it is convenient to impose the condition that they are *fit regions* (Noll and Virga 1988) that is, regular open sets, bounded by a finite perimeter, and with negligible boundary. The restriction may appear too strong, since it would exclude the treatment of infinite regions. However, including infinite regions in the class of fit regions can be done with some further specification (Truesdell 1991, Ch. II, p. 1).

Besides density, temperature, and motion, in classical mechanics there are two other primitive quantities: namely, *force* and *heat*.

Forces are vectors introduced to describe the purely mechanical actions exerted on the parts $\mathscr{P}$ of a body $\mathscr{B}$ in a configuration $\chi(\mathscr{B})$. There are two kinds of forces: the *body forces* f_i, acting in the interior of $\chi(\mathscr{P})$; and *tractions* t_i, acting on the surface of $\chi(\mathscr{P})$. Body forces and tractions depend, in general, on x_i and t, and also on $\mathscr{B}$ and $\mathscr{P}$, but in classical mechanics attention is restricted to body forces that are unaffected by the presence or absence of other bodies in space. These forces therefore have the form

$$f_i = f_i(x_i t), \tag{1.2}$$

and are commonly given per unit mass. Tractions t_i at any place and time have a common value for all parts of the surface of $\chi(\mathscr{P})$ having a common tangent plane and lying on the same side of it. This implies that

$$t_i = t_i(x_i, t, e_i), \tag{1.3}$$

where e_i is the outward normal to the surface of $\chi(\mathscr{P})$. The assumption embodied in (1.3) is called the *Cauchy stress principle*, which characterizes *simple bodies*.

Heat supply is a scalar quantity representing the thermal actions exerted on the parts $\mathscr{P}$ of $\mathscr{B}$. There are two kinds of heating: the *internal production* r, having the form

$$r = r(x_i, t), \tag{1.4}$$

given per unit mass; and the *normal heat flux* h across the surface of $\chi(\mathscr{P})$. An assumption like Cauchy's principle is also made for h. That is, h depends on x_i, t, and e_i, the exterior normal to the boundary of $\chi(\mathscr{P})$, in the form:

$$h = h(x_i, t, e_i). \tag{1.5}$$

This equation is typical of *thermally simple bodies*.

In contrast to density, motion, and temperature, which are unknown, body forces and heat production are taken as known. The body force f_i is specified at every point of $\chi(\mathscr{B})$; tractions t_i are known at the boundary of $\chi(\mathscr{B})$, but not in its interior. Heat production r is considered to be given; the normal heat flux h is known on the boundary of $\chi(\mathscr{B})$, but unknown in its interior. This distinction means that we assume a knowledge of how the body $\mathscr{B}$ interacts with its exterior, but we do not know how the parts of $\mathscr{B}$ interact with one another. Although the assumption may appear very natural and useful, because it permits us to study a body as isolated, it is open to question because, in many cases, the quantities we consider as known must be found by considering $\mathscr{B}$ as enveloped by a wider body called the *environment* of $\mathscr{B}$ and the interaction between these two bodies must be evaluated. (An attempt at making a rational definition of these interactions has been made by Ericksen (1973).)

Once the primitive quantities have been introduced, other quantities, known as *derived quantities*, can be defined. For instance, the *velocity* v_i or u, v, w of a particle X_A is defined as the rate of change of its position:

$$v_i = \frac{\partial}{\partial t}\chi_i(X_A, t) = \dot{\chi}_i. \tag{1.6}$$

The rate of change of the velocity

$$a_i = \frac{\partial}{\partial t}v_i(X_A, t) = \dot{v}_i = \ddot{\chi}_i \tag{1.7}$$

is called the *acceleration* of the particle X_A.

Formula (1.7) gives the acceleration whenever we have adopted the material description of the velocity. But, if we consider the spatial description of the velocity field, then v_i must be considered as a function of x_i, t instead of X_A, t. Applications of the chain rule thus yields

$$a_i = \frac{\partial v_i}{\partial t} + \frac{\partial v_i}{\partial x_j}\frac{\partial \chi_j(X_A, t)}{\partial t} = \frac{\partial v_i}{\partial t} + \frac{\partial v_i}{\partial x_j}v_j. \tag{1.8}$$

Here, and in the remainder of the text, summation is implied by repeated subscripts. If x_i is fixed, the acceleration reduces to $\partial v_i/\partial t$, which is called the spatial time derivative of the velocity.

In addition to the time derivatives of the motion, we can also introduce its partial derivatives with respect to the variables X_A, and define the matrix with coefficients

$$F_{iA} = \frac{\partial}{\partial X_A}\chi(X_A, t) \tag{1.9}$$

as the *deformation gradient*. This matrix must be nonsingular; that is, its determinant $J = \det(F_{iA})$ must always be positive in order to ensure that the mapping (1.1) is invertible for all X_A and t. This requirement is said to be the axiom of *permanence of matter* (Truesdell and Toupin 1960, sec. 16).

The deformation gradient is a measure of deformation since it permits us to evaluate the distance vector dx_i between two neighboring points in the present configuration $\chi(\mathscr{B})$ in terms of their distance vectors dX_A in the reference configuration through the formula

$$dx_i = \frac{\partial}{\partial X_{\rm A}} \chi(X_{\rm A}, t) dX_{\rm A} = F_{i{\rm A}}\, dX_{\rm A}. \tag{1.10}$$

But, if $F_{i\rm A}$ is known, it is possible to evaluate both the change in length of $dX_{\rm A}$ and its rotation. A theorem, known as polar decomposition (Ericksen 1960, p. 840), determines these two contributions to the deformation. The theorem states that there are two, and only two, decompositions of $F_{i\rm A}$:

$$F_{i\rm A} = R_{i\rm B} U_{\rm BA} \text{ and } F_{i\rm A} = V_{ij} R_{j\rm A}, \tag{1.11}$$

where $R_{i\rm B}$ is an orthogonal matrix, and $U_{\rm BA}$ and V_{ij} are two symmetric positive-definite matrices. The theorem of polar decomposition has a simple geometric interpretation: $U_{\rm BA}$ and V_{ij} characterize the pure changes in length along the three orthogonal axes, while $R_{i\rm B}$, as an orthogonal matrix, represents a pure rotation; thus the passage of $dX_{\rm A}$ to $dx_{\rm i}$ can be obtained either by a stretch followed by a rotation, as indicated by the first equation (1.11), or starting with a pure rotation of $dX_{\rm A}$ followed by pure stretching in the directions of the axes, as indicated by the second equation (1.11).

The set of nine components $F_{i\rm A}$ can also be represented by the single matrix symbol $\mathbf{F}$. The transpose of $\mathbf{F}$ is denoted by $\mathbf{F}^{\rm T}$ and the inverse by $\mathbf{F}^{-1}$. The generic component of $\mathbf{F}^{\rm T}$ has the form

$$(\mathbf{F}^{\rm T})_{i\rm A} = F_{{\rm A}i}, \tag{1.12}$$

and that of $\mathbf{F}^{-1}$ can be written as

$$(\mathbf{F}^{-1})_{i\rm A} = \frac{1}{J}\left(\frac{1}{2}\varepsilon_{ijk}\varepsilon_{\rm ABC} F_{j\rm B} F_{k\rm C}\right), \tag{1.13}$$

where ε_{ijk} and $\varepsilon_{\rm ABC}$ are two matrices which are antisymmetric with respect to the permutation of any two indices so that $\varepsilon_{\rm rss} = 0$ and ε_{123} is equal to unity, with $J = \det \mathbf{F} = \varepsilon_{ijk} F_{i1} F_{j2} F_{k3}$.

From the definition of the deformation gradient it is also possible to express the resulting change in an element of area consequent upon the motion. In fact, two distance vectors $dX_{\rm B}^{(1)}$, $dX_{\rm C}^{(2)}$ in the reference configuration span an element of area the size and direction of which are represented by the vector $dA_{\rm A} = \varepsilon_{\rm ABC}\, dX_{\rm B}^{(1)}\, dX_{\rm C}^{(2)}$. In the present configuration this element is defined by the vector

$$da_k = \varepsilon_{kij}\, dx_i^{(1)}\, dx_j^{(2)}, \tag{1.14}$$

where $dx_i^{(1)} = F_{i\rm B}\, dX_{\rm B}^{(1)}$ and $dx_j^{(2)} = F_{j\rm C}\, dX_{\rm C}^{(2)}$. Thus (1.14) becomes

$$da_k = \varepsilon_{kij} F_{i\rm B} F_{j\rm C}\, dX_{\rm B}^{(1)}\, dX_{\rm C}^{(2)}.$$

Multiplying this formula by $\partial x_k / \partial X_{\rm A} = F_{k\rm A}$, we obtain

$$F_{k\rm A} da_k = \varepsilon_{kij} F_{k\rm A} F_{i\rm B} F_{j\rm C}\, dX_{\rm B}^{(1)}\, dX_{\rm C}^{(2)},$$

which is equivalent to

$$da_k = J(\mathbf{F}^{-1})_{Ak}\, dA_{\rm A}, \tag{1.15}$$

which is called Nanson's formula (Nanson 1878).

If we consider three vectors of the type $dX_{\rm A}^{(1)}$, $dX_{\rm B}^{(2)}$, and $dX_{\rm C}^{(3)}$ in the reference configurations, the volume of the parallelepiped spanned by them is

$$dV = |\varepsilon_{\rm ABC}\, dX_{\rm A}^{(1)}\, dX_{\rm B}^{(2)}\, dX_{\rm C}^{(3)}|. \tag{1.16}$$

In the present configuration the volume of the element is given by

$$dv = |\varepsilon_{ijk} F_{i\rm A} F_{j\rm B} F_{k\rm C}\, dX_{\rm A}^{(1)}\, dX_{\rm B}^{(2)}\, dX_{\rm C}^{(3)}| = |\det \mathbf{F}|\, dV. \tag{1.17}$$

Since (1.1) is invertible, $J = \det \mathbf{F}$ is of one sign for all $X_{\rm A}$ and t in a given reference configuration.

A motion for which $J = 1$, that is, for which the volume element is preserved, is said to be *isochoric*.

So far we have not imposed any restriction on the fundamental quantities density, motion, or temperature, except the requirements that density and temperature are strictly positive and the map representing the motion is invertible. There is, however, a large class of bodies having the characteristic that the smallest portions into which we can conceive them to be divided have the same properties as those of the substance in bulk. For these bodies the fundamental quantities are *fields*, which are functions of $X_{\rm A}$ and t, varying smoothly over portions of space and time. It is, of course, clear that abrupt changes of the fields are allowed, but they are localized on surfaces propagating in the medium called *waves*. However, on each side of the wave the fields are smooth.

The assumption is peculiar to classical continuum mechanics, but it is extremely restrictive because it excludes the treatment of discrete systems, the particles of which, of small but finite size, are incapable of further subdivision without changing their essential physical properties.

The question of deciding between which of the descriptions of the matter is preferable is less drastic than may at first appear. There is, in fact, a wide set of problems in which the discrete approach is necessary, because in such problems the extensions of the bodies under examination are so small as to be comparable with the dimensions of their molecules. If, on the other hand, we need to treat portions of matter appreciably larger in relation to the extent of its molecules, then the continuous approach is more appropriate for at least two reasons (Truesdell and Noll 1965, sec. 3): the first is that we can employ a theory valid for classes of bodies and not for single bodies; and the second is that the assumption of continuity permits the application of classical differential and integral calculus.

2. Balance Equations

We have seen that in continuum mechanics we try to find the density, motion, and temperature of each particle of a body in terms of other quantities such as body forces, surface tractions, heat supply, and surface heat flux, which are customarily considered as given.

The mathematical relations between known and unknown quantities take the name *balance equations*, because they express the common requirement that during the motion certain quantities are conserved. The balance equations are valid for any kind of body, irrespective of the physical nature of the material. For instance, the same equations govern the motion of a piece of rubber as for a mass of water.

Since balance equations have a common structure, deriving them can be done once and for all in general form, whence the particular balance equations are derived by specifying those quantities to be conserved in a particular model.

Let Ψ be an *additive* quantity defined on an arbitrary part $\mathscr{P}$ of $\mathscr{B}$, occupying at the instant t the region $\chi(\mathscr{P})$, denoted for simplicity, by V. The function Ψ can also be written in the form

$$\Psi = \int_V \psi \, dv, \tag{2.1}$$

where ψ is called the *mathematical density* of Ψ, in order to avoid confusion with ρ. We now consider the time variation of ψ and we *postulate* that it originates from two sources. The first is the outflux Φ across the surface of V, denoted by $S(V)$. In this case the function Φ can also be written as

$$\Phi = \int_{S(V)} \varphi_i \, da_i, \tag{2.2}$$

where φ_i are components of the *surface flux density* of the outflow. The second is the bulk production Σ inside V, which again can be put in the form

$$\Sigma = \int_V \sigma \, dv, \tag{2.3}$$

having denoted by σ its (mathematical) *volume density*.[1]

The general balance equation states that Ψ, the time variation of Ψ, is equal to the sum of $-\Phi$, the inflow, and Σ:

$$\dot{\overline{\left(\int_V \psi \, dv\right)}} = -\int_{S(V)} \varphi_i \, da_i + \int_V \sigma \, dv. \tag{2.4}$$

But the left-hand side of (2.4) can be expressed in a more suitable form by observing that, if V_0 in $\kappa(\mathscr{P})$ is the image of $\chi(\mathscr{P})$ in the reference configuration, then by the formula for change of variables for triple integrals, we obtain the relation

$$\int_V \psi \, dv = \int_{V_0} \psi J \, dV, \tag{2.5}$$

where dV is the volume element in the reference configuration.

By differentiation with respect to the time we obtain

[1] Müller (1972, Kap. II, p. 22) distinguishes between "intrinsic" production and production "at a distance," in order to point out that one part of σ is created inside and another is due to the infuence of surrounding bodies.

$$\overline{\left(\int_V \psi \, dv\right)}^{\cdot} = \int_V (\dot{\psi} J + \psi \dot{J}) \, dV. \tag{2.6}$$

The time derivative of J is given by the following identity of Euler:

$$\dot{J} = \frac{d}{dt}\left(\det \frac{\partial x_i}{\partial X_A}\right) = \frac{\partial v_i}{\partial X_A} \cdot \left(\text{cofactor of } \frac{\partial x_i}{\partial X_A}\right)$$

$$= \frac{\partial v_i}{\partial x_j}\frac{\partial x_j}{\partial X_A} \cdot \left(\text{cofactor of } \frac{\partial x_i}{\partial X_A}\right) = \frac{\partial v_i}{\partial x_i} J, \tag{2.7}$$

and so (2.6) becomes

$$\overline{\left(\int_V \psi \, dv\right)}^{\cdot} = \int_{V_0} \left(\dot{\psi} J + \psi \frac{\partial v_i}{\partial x_i} J\right) dV = \int_V \left(\frac{\partial \psi}{\partial t} + \frac{\partial}{\partial x_i}(\psi v_i)\right) dv, \tag{2.8}$$

or, by application of the divergence theorem to the second integral on the right-hand side

$$\overline{\left(\int_V \psi \, dv\right)}^{\cdot} = \int_V \frac{\partial \psi}{\partial t} \, dv + \int_{S(V)} \psi v_i \, da_i, \tag{2.9}$$

which is called the *transport theorem* for a volume.

In deriving the transport theorem we have assumed that ψ is continuous and differentiable everywhere in V. But, it may happen that V is separated into two parts, V^+ and V^-, by a surface s on which ψ is discontinuous (Figure 2.1). The singular surface moves with its own independent velocity u_i, not necessarily equal to v_i. Let us denote the part of $S(V)$ belonging to V^+ by $S^+(V)$ and that belonging to V^- by $S^-(V)$. The piece of s contained inside V is again called s. The values assumed

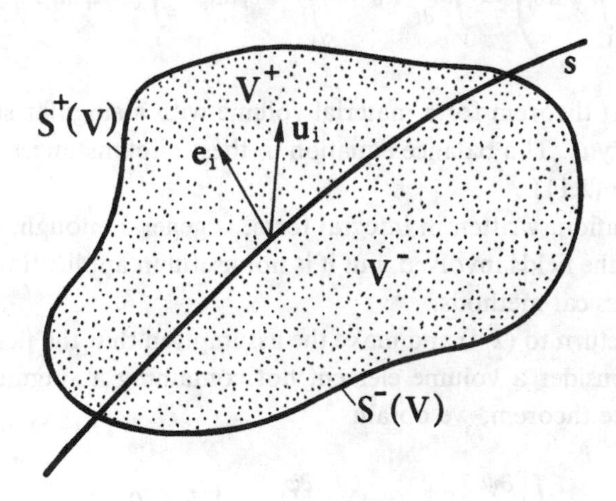

Fig. 2.1

by ψ on the singular surface when one considers the limits from V^+ or from V^- are denoted by ψ^+ and ψ^-, and $[\psi]$ represents the jump $\psi^+ - \psi^-$.

Since ψ is additive, we can write

$$\overline{\left(\int_V \psi\, dv\right)}^{\cdot} = \overline{\left(\int_{V^+} \psi\, dv\right)}^{\cdot} + \overline{\left(\int_{V^-} \psi\, dv\right)}^{\cdot},$$

and hence by applying the transport theorem to each integral separately we obtain

$$\overline{\left(\int_V \psi\, dv\right)}^{\cdot} = \int_{V^+} \left(\frac{\partial \psi}{\partial t} + \frac{\partial}{\partial x_i}(\psi v_i)\right) dv + \int_{V^-} \left(\frac{\partial \psi}{\partial t} + \frac{\partial}{\partial x_i}(\psi v_i)\right) dv. \tag{2.10}$$

We now again apply the divergence theorem, recalling that the velocity of $\partial\chi(\mathscr{P})$ is v_i and that of s is u_i:

$$\begin{aligned} \overline{\left(\int_V \psi\, dv\right)}^{\cdot} = & \int_{V^+} \frac{\partial \psi}{\partial t}\, dv + \int_{S^+} \psi v_i\, da_i - \int_s \psi^+ u_i\, da_i \\ & + \int_{V^-} \frac{\partial \psi}{\partial t}\, dv + \int_{S^-} \psi v_i\, da_i + \int_s \psi^- u_i\, da_i, \end{aligned}$$

where the two different signs of the surface integrals over s are due to the fact that the normal unit vector e_i points into V^+ and out of V^-. By combining the integrals in V^+ and V^-, and in S^+ and S^-, the last expression becomes

$$\overline{\left(\int_V \psi\, dv\right)}^{\cdot} = \int_V \frac{\partial \psi}{\partial t}\, dv + \int_{S(V)} \psi v_i\, da_i - \int_s [\psi] u_i\, da_i. \tag{2.11}$$

This is the transport theorem for a material volume with a singular surface moving with its own velocity u_i. The balance equation in these circumstances is obtained by replacing (2.4) with (2.11).

The balance equation, written in integral form, is general enough, requiring only the integrability of the fields involved, but it is not useful in applications because it is not solvable by classical methods.

However, if we return to (2.4) and make the assumption that the fields are smooth enough, and we consider a volume element not containing a singular surface. On using the divergence theorem, we obtain

$$\int_V \left[\frac{\partial \psi}{\partial t} + \frac{\partial}{\partial x_i}(\psi v_i) + \frac{\partial \varphi_i}{\partial x_i} - \sigma\right] dv = 0. \tag{2.12a}$$

This equation must hold for every measurable domain in the Euclidean space. Thus, under the assumption that the integrand is continuous, we conclude that the equation

$$\frac{\partial \psi}{\partial t} + \frac{\partial}{\partial x_i}(\psi v_i) + \frac{\partial \varphi_i}{\partial x_i} = \sigma \tag{2.12b}$$

is a necessary and sufficient condition for validity of (2.12a). Equation (2.12b) is called the *local form* of balance for ψ.

The same device can be applied for obtaining the local form of the balance equation on a surface of discontinuity. Let us consider an infinitesimal cylindrical region, intersecting s in S_G and bounded by two end faces S_G^+ and S_G^- parallel to the surface s, having lateral surface S_L constituted by the elements of the normal to s connecting the boundaries of the faces (Figure 2.2). Let us now apply the balance equation (2.4), recalling that its left-hand side is given by (2.11), and let S_L tend to zero by progressively diminishing the length of the cylindrical element. If all quantities are smooth enough, the contributions of the volume integral vanish and the surface integrals can be extended over S_G, the intersection of the cylindrical box with s. From (2.4) and (2.11), we thus obtain the equation

$$[\varphi_i e_i] + [\psi(v_i - u_i)e_i] = 0, \tag{2.13}$$

where e_i is the unit vector normal to s, pointing, from V^- to V^+. Equation (2.13) is known as *Kotchine's theorem* (Kotchine 1926).

In order to derive (2.13) we have made the assumption that all fields are smooth, with the consequence that (2.13) have a very simple form. But, if this is not the case, we must also consider the contributions of the singular terms. For instance, it may happen that, as S_L tends to zero with S_G constant, the quantity

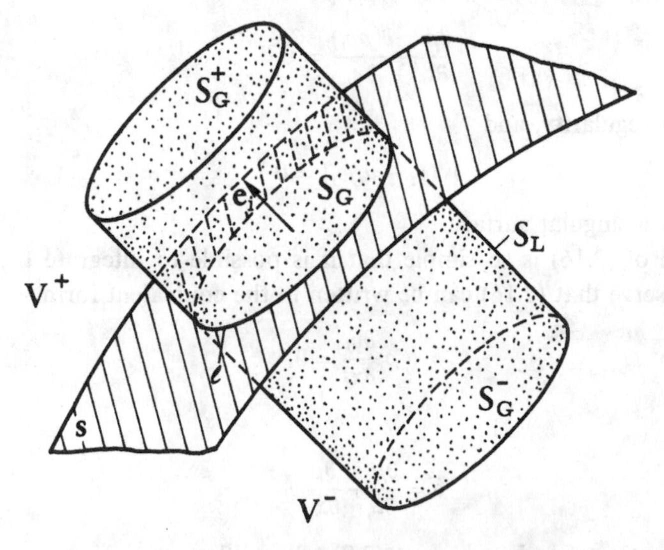

Fig. 2.2

$$\lim_{S_L \to 0} \int_{S_L} \varphi_i \, da_i$$

is finite. The value of this limit can be expressed as an integral of the form $\int_\ell \phi_i^s \, d\ell_i$, where $\ell = S_L \cap s$ (Figure 2.2), ϕ_i^s is a vector field defined on s, and $d\ell_i$ gives the magnitude and direction of an element of ℓ. But, if we apply Stokes's theorem, we can write

$$\int_\ell \phi_i^s \, d\ell_i = \int_{S_G} \varepsilon_{ijk} \frac{\partial}{\partial x_j} \phi_k^s \, da_i,$$

and therefore, instead of (2.13) we obtain the jump condition

$$[\varphi_i e_i] + [\psi(v_i - u_i) e_i] = -\varepsilon_{ijk} \frac{\partial}{\partial x_j} \phi_k^s e_i. \tag{2.14}$$

Situations do occur for which this equation is necessary (Müller 1972, Kap. II, p. 26).

The result we have obtained so far is the formal structure of a balance equation without specifying the physical quantities to be conserved. The balance equations that are of interest in continuum mechanics are the result of experience, which proves that certain quantities, called mass, momentum, and energy, are additive quantities, and are preserved during the motion of the body.

The first balance equation states that the mass of every subbody V of $\chi(\mathscr{B})$, which can be written as

$$\Psi = \mathscr{M}(V) = \int_V \rho \, dv, \tag{2.15}$$

is maintained even if its outflux φ_i across a material surface is zero, and its internal production r is also zero, since mass cannot be expelled across $S(V)$ or created inside V.

Consequently, the local version of the equation of balance of mass has the form

$$\frac{\partial \rho}{\partial t} + \frac{\partial(\rho v_i)}{\partial x_i} = 0 \tag{2.16}$$

at each point of regularity, and

$$[\rho(v_i - u_i) e_i] = 0 \tag{2.17}$$

at each point of a singular surface.

The structure of (2.16) is so simple that it is possible to integrate the equation directly. We observe that (2.16) can be written in the equivalent form

$$\dot{\rho} + \rho \frac{\partial v_i}{\partial x_i} = 0, \tag{2.18}$$

where

$$\dot{\rho} = \frac{\partial \rho}{\partial t} + \frac{\partial \rho}{\partial x_i} v_i \tag{2.19}$$

is the total derivative of ρ. In addition, (2.7) tells us that

$$\frac{\partial v_i}{\partial x_i} = \frac{1}{J}\dot{J},$$

and therefore (2.18) becomes

$$\frac{\dot{\rho}}{\rho} + \frac{\dot{J}}{J} = 0,$$

and this equation can be integrated at once, recalling that, for $t = 0$, in the reference configuration, $J = 1$ and $\rho = \rho_\kappa$ is the density of the material in this configuration. The result is

$$\rho = \rho_\kappa \frac{1}{J}, \tag{2.20}$$

which means that, once we know ρ_κ, the density ρ at each instant is a function of J.

The second balance equation is a natural extension of Newton's second law of motion. It requires that the quantity

$$\Psi_i = \int_V \rho v_i \, dv$$

which is the momentum of the subbody V, satisfies an equation of the form

$$\overline{\left(\int_V \rho v_i \, dv\right)}^{\boldsymbol{\cdot}} = \int_{S(V)} \sigma_{ij} \, da_j + \int_V \rho f_i \, dv, \tag{2.21}$$

where $-\sigma_{ij}$ is a second-order tensor playing the role of φ_i in the general equation (2.4), and ρf_i is the density of the body force per unit volume. Equation (2.21) is a vector equation, but in stating (2.4), we have not made the restriction that ψ is a scalar. The only difference now is that, if ψ is a vector, then the associated outflux φ_i is a second-order tensor. The tensor σ_{ij} occurring in (2.21) is called the *Cauchy stress tensor*. The Cauchy stress tensor is related to t_i, the surface traction over $S(V)$, by a linear equation of the form

$$\sigma_{ij} e_j = t_i, \tag{2.22}$$

where e_i is the unit normal to $S(V)$, directed outwards, and it can be proved that σ_{ij} is independent of e_i. The proof of this relationship between the stress tensor and the vector traction is a consequence of the balance equation (2.21), obtained by applying it to a tetrahedron, three faces of which are mutually perpendicular. This result is known as *Cauchy's fundamental theorem*.

The local version of the balance of momentum now reads

$$\frac{\partial}{\partial t}(\rho v_i) + \frac{\partial}{\partial x_j}(\rho v_i v_j - \sigma_{ij}) = \rho f_i, \tag{2.23}$$

or, by use of (2.19),

$$\rho \dot{v}_i - \frac{\partial \sigma_{ij}}{\partial x_j} = \rho f_i. \tag{2.24}$$

These two equations hold at the regular points of V. If instead we consider a point on a surface of discontinuity s, we can again repeat the argument employed to derive (2.14), the only difference being that now ψ is the vector ρv_i and φ_i is the second-order tensor $-\sigma_{ij}$, and hence the jump condition becomes

$$-[\sigma_{ij}e_j] + [\rho v_i(v_j - u_j)e_j] = 0. \tag{2.25}$$

The condition of discontinuity (2.25) holds whenever the limit

$$\int_{S_L} \sigma_{ij}\, da_j$$

is zero, but in some cases this limit is finite and expressible in the form of a curvilinear integral along $\ell = S_L \cap s$:

$$\lim_{S_L \to 0} \int_{S_L} \sigma_{ij}\, da_j = \int_{\ell} T^s_{ij}\, d\ell_j, \tag{2.26}$$

T^s_{ij} being a tensor field on s. Since, here again, the integral along ℓ can be transformed into an integral over S_G, the piece of s inside ℓ, by applying Stokes's theorem we can arrive at the modified jump condition:

$$-[\sigma_{ij}e_j] + [\rho v_i(v_j - u_j)e_j] = \varepsilon_{hjk} \frac{\partial}{\partial x_j} T^s_{ik} e_h, \tag{2.27}$$

which is important when the singular surface s is not simply a diaphragm separating two discontinuous fields, but itself transmits non-negligible stresses.

The equation of balance of momentum has two important corollaries.

The first is obtained by multiplying (2.23) by $\varepsilon_{k\ell i}x_\ell$ and rearranging the terms to give

$$\frac{\partial}{\partial t}(\rho \varepsilon_{k\ell i} x_\ell v_i) + \frac{\partial}{\partial x_j}(\rho \varepsilon_{k\ell i} x_\ell v_i v_j - \varepsilon_{k\ell i} x_\ell \sigma_{ij})$$

$$= \varepsilon_{kij}\sigma_{ij} + \varepsilon_{k\ell i} x_\ell \rho f_i, \tag{2.28}$$

which is called the equation of balance for the moment of momentum, or the balance of angular momentum. The term $\varepsilon_{k\ell i} x_\ell \rho v_i$ has the physical meaning of density of angular momentum, $-\varepsilon_{k\ell i} x_\ell \sigma_{ij}$ is the flux of angular momentum, $\varepsilon_{kij}\sigma_{ij}$ is the production of angular momentum, and $\varepsilon_{k\ell i} x_\ell \rho f_i$ is the torque of the body force. In contrast to the linear momentum ρv_i, the moment of momentum is not a conservative quantity, because, even in the absence of body forces, there is a production density $\varepsilon_{kij}\sigma_{ij}$ on the right-hand side of (2.28). However, this production density is zero when the stress tensor is symmetric.

Alternatively, the second consequence of the balance of momentum can be derived by multiplying (2.23) by v_i and summing with respect the repeated index i. Putting $v^2 = v_i v_i$, after some rearrangement we obtain

$$\frac{\partial}{\partial t}\left(\frac{1}{2}\rho v^2\right) + \frac{\partial}{\partial x_j}\left(\frac{1}{2}\rho v^2 v_j - \sigma_{ij} v_i\right) = -\sigma_{ij}\frac{\partial v_i}{\partial x_j} + \rho f_i v_i, \tag{2.29}$$

which is known as equation of balance of kinetic energy. The term $\frac{1}{2}\rho v^2$ is the density of kinetic energy, $-\sigma_{ij}v_j$ is its flux, $-\sigma_{ij}(\partial v_i/\partial x_j)$ its production, and $\rho f_i v_i$ is the power of the body force. The kinetic energy is not a conservative quantity either, since there is a production even if the body force vanishes.

The last balance equation of significance in classical continuum mechanics is motivated by the realization that kinetic energy is not the total energy of a body, but that there is an additional part called heat or internal energy, the density of which is denoted by $\rho\varepsilon$. Thus the total energy density is

$$\rho e = \rho\varepsilon + \frac{1}{2}\rho v^2,$$

and if we now let J_j be its flux and ρp its supply, then the energy balance equation reads

$$\frac{\partial}{\partial t}(\rho e) + \frac{\partial}{\partial x_j}(\rho e v_j + J_j) = \rho p. \tag{2.30}$$

In practice, it is more convenient to introduce the quantity

$$q_i = J_i + \sigma_{ij}v_j, \tag{2.31}$$

which is called the *heat flux*, and the quantity

$$\rho r = \rho p - \rho f_i v_i, \tag{2.32}$$

which is called the *heat supply*. Then, replacing the terms ρe, J_j, and ρp, in (2.30) and subtracting from it the balance equation (2.29), we obtain

$$\frac{\partial(\rho\varepsilon)}{\partial t} + \frac{\partial}{\partial x_i}(\rho\varepsilon v_i + q_i) = \sigma_{ij}\frac{\partial v_i}{\partial x_j} + \rho r, \tag{2.33}$$

where $\sigma_{ij}(\partial v_i/\partial x_j)$ represents the *density production* of internal energy.

Equation (2.33) is the *First Law of Thermodynamics.*

The balance of internal energy across a singular surface s, can be derived again, with the same argument which led to (2.13), where now φ_i is $q_i - \sigma_{ij}v_j$ and ψ is $[\rho\varepsilon + (\rho/2)v^2]$. We thus have

$$[(q_i - \sigma_{ij}v_j)e_i] + [\rho(\varepsilon + \frac{1}{2}v^2)(v_i - u_i)e_i] = 0. \tag{2.34}$$

This equation holds when the quantity

$$\lim_{S_L \to 0} \int_{S_L} (-q_i + \sigma_{ij}v_j)\, da_i$$

is zero. If, instead, the limit is finite, an additional term will appear on the right-hand side of (2.34), when we are considering the intrinsic contribution of s to the energy balance.

The singular surface often represents a thin wall for which the tangential components of velocity at the two sides of s vanish so that $u_j = (u_i e_i)e_j = u_n e_j$, the velocity being purely normal. In this case, the equations of balance of mass, momentum, and energy reduce to

$$[\rho(v_i - u_n e_i)e_i] = 0,$$

$$-[\sigma_{ij}e_j] + [\rho v_i(v_j - u_n e_j)e_j] = 0,$$

$$[(q_i - \sigma_{ij}v_j)e_i] + [\rho(\varepsilon + \frac{1}{2}v^2)(v_i - u_n e_i)e_i] = 0.$$

By virtue of the first of these equations, the other two can be written in the equivalent forms

$$-[\sigma_{ij}e_j] + [\rho(v_i - u_n e_i)(v_j - u_n e_j)e_j] = 0,$$

$$[q_i e_i] - [\sigma_{ij}(v_j - u_n e_j)e_i] + [\rho(\varepsilon + \frac{1}{2}(v - u_n)^2)(v_i - u_n e_i)e_i] = 0.$$

In particular, if the wall is impermeable so that $v_i e_i = u_n$ and if the tangential component of v_i vanishes on either side of the wall, the last equation reduces to

$$[q_i e_i] = 0, \tag{2.35}$$

which proves that the normal component of the heat flux is continuous.

3. Constitutive Equations

The balance equations discussed so far should give the necessary and sufficient mathematical relations for determining density, motion, and temperature as functions of the data, body force, and heat supply. Unfortunately, the balance equations also contain the stress tensor σ_{ij}, the internal energy ε, and the heat flux q_i, which are unknown. The system of balance equations is thus unable to produce the required fields ρ, x_i, and θ, unless other relations between the unknown fields are introduced. These additional equations are called *constitutive* in order to point out that, unlike balance equations, they characterize the nature of the materials that make up the bodies.

In mechanics, constitutive equations are relations expressing stress, heat flux, and internal energy in terms of density, motion, and temperature, in a form which depends on the material. We may object that external body forces and heat supply might also depend on density, motion, and temperature, and, in effect, this dependence is not negligible in some circumstances. But the assumption that the body force and heat supply are external means that we restrict our attention to the situation in which these quantities are practically unaffected by the other fields.

Although in the past examples of constitutive equations have been proposed by James Bernoulli, Euler, Cauchy, Green, Fourier, and Stokes, only in recent years have all these attempts been critically revised by Noll (1958) and integrated into a unified theory. Previously, the notion of a constitutive equation has not always been clear, sometimes being confused with that of a balance equation. This misunderstanding has been particularly troublesome in the treatment of the last balance equation, which involves internal energy, heat flux, and heat supply.[2]

[2] See Truesdell's (1980) account of the debate on the constitutive equations in thermodynamics.

Noll has proposed a general theory of constitutive equations that is capable of relating their essential properties to those of experiment and of specifying the fundamental restrictions that they must obey to be physically meaningful. Noll's theory starts with three axioms, called axioms of *determinism*, *local action*, and *material frame indifference*, reflecting three properties of constitutive equations, in agreement with experiment.

The axiom, or principle, of determinism states that stress, heat flux, and specific internal energy at a point X_A at time t are determined by the history of ρ, χ_i, and θ in the whole body up to the time t. In other words, the concept of determinism represents the common observation that natural materials exhibit memory of their past experiences, sometimes long after the experiences took place.

The principle of determinism allows σ_{ij}, q_i, and ε to depend on the history of ρ, χ_i, and θ at points of the body that lie far away from X_A. However, because this action at a distance is practically negligible, in many cases we can assume a second constitutive axiom requiring that the values of density, motion, and temperature at a finite distance from X_A may be disregarded when calculating stress, heat flux, and energy at X_A. This axiom, characterizes *simple* materials. The values of a constitutive function in the neighborhood of X_A are given by its value in X_A and the value of its gradient there. However, by (2.20), we have a finite relation between ρ and ρ_κ, the density in the reference configuration. Thus ρ can be dropped from the list of variables. The same can be done for its gradient, because the density gradients represent the second gradients of the motion, and these have been excluded. It turns out that the only variables are the histories of motion, temperature, and their gradients.

The principle of material frame indifference, or objectivity, expresses the property of invariance of the constitutive equations with respect to the Euclidean transformations. The position of a particle is given by three Cartesian coordinates x_i in a system at rest in an inertial frame. A change to a Cartesian coordinate system at rest in a new noninertial frame is characterized by the transformation

$$x_i^* = 0_{ij}(t)x_j + b_i(t), \tag{3.1}$$

where $0_{ij}(t)$ is an orthogonal matrix and $b_i(t)$ are three arbitrary functions. If (3.1) holds, we say that x_i^* and x_i are related by Euclidean transformation. Under Euclidean transformations, σ_{ij}, q_i, and ε change according to the laws of tensor, vector, and scalar quantities, respectively, that is,

$$\sigma_{ij}^* = 0_{ih}0_{jk}\sigma_{hk}, \quad q_i^* = 0_{ij}q_j, \quad \varepsilon^* = \varepsilon; \tag{3.2}$$

whereas the independent variables transform as

$$x_i^* = 0_{ij}(t)x_j + b_i(t), \quad \theta^* = \theta; \tag{3.3}$$

and their spatial gradients as

$$F_{iA}^* = \frac{\partial x_i^*}{\partial x_j}\frac{\partial x_j}{\partial X_A} = 0_{ij}F_{jA}, \quad \theta_{,B}^* = \theta_{,B}. \tag{3.4}$$

The principle of material frame indifference states simply that, under an Euclidean change of frame, the constitutive fields and their variables are modified according to

(3.2), and (3.3) and (3.4), respectively, but the *structure* of the constitutive function remains unchanged. A simple consequence is that the constitutive quantities must be independent of the history of motion. But this is not all: A second consequence is that the constitutive quantities are not dependent on the history of F_{jA}, but only on the history of its stretch U_{AB}, while the history of the rotational part of the deformation does not appear. These reductions are necessary consequences of the principle of indifference, but it can be proved that they are also sufficient, in the sense that no other restrictions on the constitutive equations can be obtained.

So far we have considered materials for which the constitutive quantities at a given time t are determined by the whole history of the independent fields, so that the constitutive equations are necessarily functional equations. However, this level of generality is often superfluous, because in most bodies the influence of memory is so weak that it can be disregarded with no appreciable consequence. For these materials without memory, instead of the histories of U_{AB}, θ, and $\theta_{,A}$, we can insert their actual values at the time t into the constitutive equations, which consequently become ordinary functions and not functional equations. Materials of this type are called simple *thermoelastic* materials.

By applying the restrictions imposed by the principle of material frame indifference to simple thermoelastic bodies, it is possible to prove that their constitutive equations are functions of the forms

$$\sigma_{ij} = R_{iC}\mathcal{T}_{CD}(U_{AB}, \theta, \theta_{,A})R_{Dj}, \tag{3.5a}$$

$$q_i = R_{iC}\mathcal{L}_C(U_{AB}, \theta, \theta_{,A}), \tag{3.5b}$$

$$\varepsilon = \mathcal{E}(U_{AB}, \theta, \theta_{,A}), \tag{3.5c}$$

where R_{iC} is the rotational part of the deformation gradient F_{iA}.

But there are alternative versions of these relations. For instance, we may introduce the *material stress* and the *material heat flux*, as defined by

$$T_{AB} = J(\mathbf{F}^{-1})_{iA}\sigma_{ij}(\mathbf{F}^{-1})_{jB}, \tag{3.6a}$$

$$Q_A = J(\mathbf{F}^{-1})_{iA}q_i, \tag{3.6b}$$

and we write (3.5) after having replaced σ_{ij} and q_i with T_{AD} and Q_A. The result of the substitution is

$$T_{AB} = J(\mathbf{F}^{-1})_{iA}R_{iC}\mathcal{T}_{CD}R_{Dj}(\mathbf{F}^{-1})_{jB},$$

$$Q_A = J(\mathbf{F}^{-1})_{iA}R_{iC}\mathcal{L}_C.$$

This means that, if we recall the polar decomposition (1.11), T_{AB} and Q_A are only functions of the stretch U_{AB}. Often, instead of U_{AB}, it is more convenient to introduce the symmetric tensor

$$C_{AB} = U_{AC}U_{CB} = F_{iA}F_{iB}, \tag{3.7}$$

called the *right Cauchy–Green tensor*, so that, after redefinition of the constitutive functions, (3.5) can be written as

$$T_{AB} = \overset{\circ}{\mathscr{T}}_{AB}(C_{CB}, \theta, \theta_{,C}),$$
$$Q_A = \overset{\circ}{\mathscr{L}}_A(C_{CD}, \theta, \theta_{,C}), \tag{3.8}$$
$$\varepsilon = \overset{\circ}{\mathscr{E}}(C_{CD}, \theta, \theta_{,C}).$$

It might seem that only simple thermoelastic bodies admit reduced constitutive equations as simple as (3.8), but this is not so. All simple materials with memory have reduced constitutive equations like (3.8), with the difference that $\overset{\circ}{\mathscr{T}}_{AB}$, $\overset{\circ}{\mathscr{L}}_A$, and $\overset{\circ}{\mathscr{E}}$ are functionals of the history of C_{AB}, θ, and $\theta_{,C}$.

So far we have not emphasized the fact that the form of the constitutive equations depends on the particular choice of the reference configuration $\boldsymbol{\kappa}(\mathscr{B})$. In this connection, let us imagine that, in the distant past, a particle was in the reference configuration $\boldsymbol{\kappa}(\mathscr{B})$ and was subjected to a prescribed deformation and thermal history. At time t its stress, heat flux, and internal energy will have certain values determined by the constitutive functionals. If, instead, the particle was initially in another reference configuration $\hat{\boldsymbol{\kappa}}(\mathscr{B})$ and was subsequently subjected to the same history, we expect different values of the constitutive quantities, since the response functionals depend on the reference configuration. However, if it happens that the constitutive quantities assume the same values as before, we say that the particle is *isotropic* with respect to the transformation

$$\hat{X}_A = \hat{X}_A(X_B) \tag{3.9}$$

relating $\hat{\boldsymbol{\kappa}}(\mathscr{B})$ with $\boldsymbol{\kappa}(\mathscr{B})$, the gradient of which is $P_{AB} = (\partial \hat{X}_A/\partial X_B)$.

Experiment shows that, in order to preserve isotropy, the transformation (3.9) must be unimodular, that is, with $|\det P_{AB}| = 1$, because there is no natural material for which constitutive equations remain unaltered when there are variations in density.

Isotropy reflects symmetries in the molecular structure of matter. For example, if a particle consists of cubic lattice cells, it is isotropic with respect to rotations of 90° around the edges of the cells. But it is clear that, if the particle is deformed, it loses its original symmetry and, consequently, does not exhibit the isotropy observed in the initial configuration. This means that isotropy is a property depending on the choice of the reference configuration. The configuration having the largest number of isotropic transformations is called *undistorted.*

Among unimodular transformations there are those with $\det P_{AB} = -1$, which represent mirror reflections. In classical continuum mechanics all materials are isotropic under reflections. Rotations are also unimodular transformations. A material with a configuration the isotropic transformations of which contain all rotations is said to be an *isotropic* material.

Both unimodular and orthogonal transformations form a group. There are materials for which the isotropy group with respect to the undistorted configuration contains all the orthogonal transformations, and there are materials whose isotropy group is a subgroup of the orthogonal group; and, finally, there are materials whose isotropy group is the full unimodular group. The first are called *isotropic solids,* the second *crystalline solids,* and the third *fluids.* The isotropy group reflects the crystallographic symmetrics of a material. Neumann (1885) propounded the principle that any crystallographic symmetry is also a physical symmetry.[3] A question which arises immediately is whether there are materials whose isotropy group is both contained by the unimodular group and also contains the orthogonal group. The answer is no, because, according to a theorem of group theory, the orthogonal group is maximal in the unimodular group, and therefore there is no intermediate group between them.

The consequence of isotropy is a drastic reduction in the form of the constitutive equation of a material. For simplicity, let us look at the reduction resulting from isotropy in the constitutive equation of a simple thermoelastic material; clearly the argument also applies, with no conceptual changes, to more general materials.

We have seen that the general form of the constitutive equation of a simple thermoelastic material can be of the type (3.8). Let us consider a transformation like (3.9) from a configuration $\kappa(\mathscr{B})$. The deformation gradient and the temperature gradient in the new configuration are

$$\hat{F}_{iA} = \frac{\partial \chi_i}{\partial \hat{X}_A} = \frac{\partial \chi_i}{\partial X_B}\frac{\partial X_B}{\partial \hat{X}_A} = F_{iB}(\mathbf{P}^{-1})_{BA}, \tag{3.10}$$

$$\hat{\theta}_{,A} = \frac{\partial \theta}{\partial \hat{X}_A} = \frac{\partial \theta}{\partial X_B}\frac{\partial X_B}{\partial \hat{X}_A} = \theta_{,B}(\mathbf{P}^{-1})_{BA}, \tag{3.11}$$

and, consequently, the Cauchy–Green tensor $\hat{C}_{AB}$ in $\hat{\kappa}(\mathscr{B})$ is given by

$$\hat{C}_{AB} = C_{CD}(\mathbf{P}^{-1})_{CA}(\mathbf{P}^{-1})_{DB}. \tag{3.12}$$

Now, if the particle is isotropic, P_{AB} can be an arbitrary orthogonal matrix, say Q_{AB}. Thus, for an orthogonal transformation characterized by the property that $(\mathbf{Q}^{-1})_{AB} = Q_{BA}$, C_{AB} and $\theta_{,A}$ are replaced by $C_{CD}Q_{AC}Q_{BD}$ and $\theta_{,B}Q_{AB}$, respectively. With this replacement the material stress T_{AB} changes to $T_{CD}Q_{CA}Q_{DB}$, and the heat flux to $q_A = q_B Q_{BA}$. Therefore, the constitutive functionals $\mathring{\mathscr{T}}_{AB}$, $\mathring{q}_A$, and $\mathring{\varepsilon}$ must satisfy three restrictions of the forms:

$$\begin{aligned}
\mathring{\mathscr{T}}_{AB}(C_{RS}Q_{CR}Q_{DS}, \theta, \theta_{,R}Q_{CR}) &= Q_{AC}Q_{BD}\mathring{\mathscr{T}}_{CD}(C_{RS}, \theta, \theta_{,R}),\\
\mathring{q}_A(C_{RS}Q_{CR}Q_{DS}, \theta, \theta_{,R}Q_{CR}) &= Q_{AB}\mathring{q}_B(C_{RS}, \theta, \theta_{,R}),\\
\mathring{\varepsilon}(C_{RS}Q_{CR}Q_{DS}, \theta, \theta_{,R}Q_{CR}) &= \mathring{\varepsilon}(C_{RS}, \theta, \theta_{,R}).
\end{aligned} \tag{3.13}$$

Functions satisfying conditions of this type are called *isotropic functions*.

There is, in addition, a theorem of representation for tensor, vector, and scalar functions, which specifies the forms of $\mathring{\mathscr{T}}_{AB}$, $\mathring{q}_A$, and $\mathring{\varepsilon}$, obeying the isotropy condition (3.13). For simple thermoelastic solids the representation theorem requires the constitutive functions to have the following forms:

$$\begin{aligned}
\mathring{\mathscr{T}}_{AB} &= T_0\delta_{AB} + T_1C_{AB} + T_2(\mathbf{C}^2)_{AB} + T_3\theta_{,A}\theta_{,B} + T_4\theta_{,C}C_{C(A}\theta_{,C)} + T_5\theta_{,C}(\mathbf{C}^2)_{C(A}\theta_{,C)}\\
\mathring{q}_A &= Q_0\theta_{,A} + Q_1C_{AB}\theta_{,B} + Q_2(\mathbf{C}^2)_{AB}\theta_{,B},\\
\mathring{\varepsilon} &= \varepsilon[\theta;\ C_{AA};\ (\mathbf{C}^2)_{AA};\ (\mathbf{C}^3)_{AA};\ \theta_{,A}\theta_{,A};\ \theta_{,A}C_{AB}\theta_{,B};\ \theta_{,A}(\mathbf{C}^2)_{AB}\theta_{,B}],
\end{aligned} \tag{3.14}$$

[3] Note that a crystal may possess physical symmetries which are not possessed by the crystallographic form.

where δ_{AB} is equal to unity for $A=B$ and zero for $A \neq B$, $\mathbf{C}$ is the Cauchy-Green tensor, $C_{C(A}\theta_{,C)}$ denotes the symmetric part of the matrix $C_{C(A}\theta_{,C)}$, and all the coefficients $T_0, \ldots, Q_2$ may be functions of the seven scalars on which ε depends.

The general representation forms (3.14) take into account the influence of strain C_{AB} and temperature gradient $\theta_{,A}$ simultaneously. But in many practical situations we have to deal with processes close to the equilibrium, that is to say, processes in which the gradients of velocity and temperature are so small that they can be dropped from the constitutive equations. Since velocity gradients do not occur in (3.14), only terms containing $\theta_{,A}$ must be eliminated so that (3.14) become:

$$\overset{\circ}{\mathscr{T}}_{AB}\big|_E = T_0\big|_E \delta_{AB} + T_1\big|_E C_{AB} + T_2\big|_E (\mathbf{C}^2)_{AB}, \tag{3.15a}$$

$$\overset{\circ}{Q}_A\big|_E = 0, \tag{3.15b}$$

$$\overset{\circ}{\varepsilon}\big|_E = \varepsilon[\theta;\ C_{AA};\ (\mathbf{C}^2)_{AA};\ (\mathbf{C}^3)_{AA}], \tag{3.15c}$$

where the coefficients T_0, T_1, and T_2 now depend on the same scalars as ε. Of course, on recalling (3.6a), we can derive the constitutive equation at the equilibrium for the Cauchy stress σ_{ij}:

$$\sigma_{ij}\big|_E = JF_{iA}F_{jB}T_{AB}\big|_E, \tag{3.16}$$

which, after replacing (3.15a) and introducing the symmetric tensor $B_{ij} = F_{iA}F_{jA}$, called *left Cauchy–Green tensor*, becomes (Müller 1972, Kap. III, p. 72)

$$\sigma_{ij}\big|_E = T_2\big|_E J\delta_{ij} + J^{-1}(T_0\big|_E - T_2\big|_E I_2)B_{ij} + J^{-1}(T_1\big|_E + T_2\big|_E I_1)(\mathbf{B}^2)_{ij}, \tag{3.17}$$

where I_1 and I_2 are the first and second invariants of B_{ij}. Sometimes it is more expedient to write the last equation as

$$\sigma_{ij}\big|_E = a_0\delta_{ij} + a_1 B_{ij} + a_2(\mathbf{B}^2)_{ij}, \tag{3.18}$$

where again a_0, a_1, and a_2 are functions of θ, B_{ij}, $(\mathbf{B}^2)_{ij}$, and $(\mathbf{B}^3)_{ij}$.

So far, we have derived the constitutive equations of a thermoelastic material without any restriction on the magnitude of the deformation gradient F_{iA} or of the Cauchy–Green tensors, which are defined in terms of F_{iA}. But there is a wide class of materials that, even under large stresses, do not have appreciable deformation gradients. To render this behavior more precise, let us assume that the reference configuration is free from stress, so that $\sigma_{ij} = 0$ for $F_{iA} = \delta_{iA}$. The passage from X_A to x_i is thus characterized by the vector

$$u_i = x_i - \delta_{iA}X_A, \tag{3.19}$$

called the *displacement vector*, and by its gradient

$$H_{iA} = \frac{\partial u_i}{\partial X_A}, \tag{3.20}$$

called the *displacement gradient*. Now, if the instantaneous configuration does not differ appreciably from the reference configuration, the displacements are small with respect the linear dimensions of the reference configuration and all components of H_{iA} are quantities negligible with respect to unity. It turns out that squares and products of these components can be neglected and C_{AB} can be represented approximately as

$$C_{AB} = F_{iA}F_{iB} = (\delta_{iA} + H_{iA})(\delta_{iB} + H_{iB}) \approx \delta_{AB} + (\delta_{iA}H_{iB} + \delta_{iB}H_{iA}). \tag{3.21}$$

Moreover, by applying the same approximation to the polar decomposition $F_{iA} = O_{iB}U_{BA}$ of F_{iA}, it is apparent that we can write

$$F_{iA} = \delta_{iA} + H_{iA} = (\delta_{iB} + \tilde{R}_{iB})(\delta_{BA} + \tilde{E}_{BA}), \tag{3.22}$$

where $\tilde{R}_{iB}$ is antisymmetric and $\tilde{E}_{BA}$ is symmetric. Since

$$H_{iA} = \delta_{BA}\tilde{R}_{iB} + \delta_{iB}\tilde{E}_{BA},$$

it follows that we necessarily have

$$\tilde{R}_{iA} = \frac{1}{2}(H_{iA} - H_{Ai}), \qquad \delta_{iB}\tilde{E}_{BA} = \frac{1}{2}(H_{iA} + H_{Ai}). \tag{3.23}$$

$\tilde{R}_{iA}$ and $\tilde{E}_{BA}$ are called the *tensors of infinitesimal rotation* and *strain*.

For B_{ij} and $(\mathbf{B}^2)_{ij}$ we also have the approximate expressions

$$B_{ij} = F_{iA}F_{jA} \approx \delta_{ij} + (\delta_{iA}H_{jA} + \delta_{jA}H_{iA}),$$
$$(\mathbf{B}^2)_{ij} = B_{ik}B_{kj} \approx \delta_{ij} + 2(\delta_{iA}H_{jA} + \delta_{jA}H_{iA}),$$

and consequently (3.18) can be written as

$$\sigma_{ij}\big|_E \approx (a_0 + a_1 + a_2)\delta_{ij} + (a_1 + 2a_2)(\delta_{iA}H_{jA} + \delta_{jA}H_{iA}),$$

where, if only terms of first order are retained, $(a_0 + a_1 + a_2)$ is a linear homogeneous function of trace $(\mathbf{H})$ and $a_1 + 2a_2$ is independent of H_{iA}. On introducing the *Lamé moduli* λ and μ, defined by

$$(a_0 + a_1 + a_2) = \lambda \operatorname{trace}(\mathbf{H}),$$
$$a_1 + 2a_2 = \mu,$$

we have

$$\sigma_{ij}\big|_E \approx \lambda\delta_{ij} \operatorname{trace}(\mathbf{H}) + \mu(\delta_{iA}H_{jA} + \delta_{jA}H_{iA}). \tag{3.24}$$

The simple form of (3.24) is due to the assumption that the material admits a natural configuration such that $\sigma_{ij} = 0$ for $F_{iA} = \delta_{iA}$. If, on the other hand, we want to see the form of the constitutive equation for infinitesimal deformations superimposed on a stressed state, characterized by a Cauchy stress σ^0_{ij}, we must start from the general constitutive equation (3.5a) and assume that $\mathscr{T}_{CD}$ is continuously differentiable at $U_{AB} = \delta_{AB}$. The function $\mathscr{T}_{CD}$ in the neighborhood of δ_{AB} can be approximated by its Taylor expansion up to the first order:

$$\mathscr{T}_{CD}(U_{AB}) \approx \mathscr{T}_{CD}(\delta_{AB}) + \frac{\partial \mathscr{T}_{CD}}{\partial U_{AB}}\bigg|_{U_{AB}=\delta_{AB}} \frac{1}{2}(\delta_{iA}H_{iB} + \delta_{iB}H_{iA}).$$

The derivative

$$\left.\frac{\partial \mathscr{T}_{\mathrm{CD}}}{\partial U_{\mathrm{AB}}}\right|_{U_{\mathrm{AB}}=\delta_{\mathrm{AB}}}$$

is a fourth-order tensor which is commonly denoted by C_{CDAB} and its coefficient $\frac{1}{2}(\delta_{i\mathrm{A}}H_{i\mathrm{B}}+\delta_{i\mathrm{B}}H_{i\mathrm{A}})$ is just the infinitesimal strain $\tilde{E}_{\mathrm{AB}}$. On returning to (3.5a), the approximate expression for the Cauchy stress is now

$$\sigma_{ij}=(\delta_{i\mathrm{C}}+\tilde{R}_{i\mathrm{C}})[\mathscr{T}_{\mathrm{CD}}(\delta_{\mathrm{AB}})+C_{\mathrm{CDAB}}\tilde{E}_{\mathrm{AB}}](\delta_{\mathrm{D}j}+\tilde{R}_{\mathrm{D}j}).$$

Since $\tilde{R}_{\mathrm{D}j}$ is antisymmetric, it may be replaced by $-\tilde{R}_{j\mathrm{D}}$. On exploiting the small magnitudes of $\tilde{R}_{i\mathrm{C}}$, $\tilde{E}_{\mathrm{AB}}$, and $\tilde{R}_{j\mathrm{D}}$ we arrive at the following approximate form of σ_{ij}:

$$\sigma_{ij}\approx\sigma_{ij}^{0}+\tilde{R}_{i\mathrm{C}}\mathscr{T}_{\mathrm{CD}}\delta_{\mathrm{D}j}-\tilde{R}_{j\mathrm{D}}\mathscr{T}_{\mathrm{CD}}\delta_{i\mathrm{C}}+\delta_{i\mathrm{C}}\delta_{\mathrm{D}j}C_{\mathrm{CDAB}}\tilde{E}_{\mathrm{AB}}, \tag{3.25}$$

where $\sigma_{ij}^{0}=\delta_{i\mathrm{C}}\delta_{\mathrm{D}j}\mathscr{T}_{\mathrm{CD}}(\delta_{\mathrm{AB}})$ is the Cauchy stress, coinciding with the material stress, in the reference configuration.

Equations (3.25) were derived by Cauchy (1828), but by a procedure criticized by Todhunter and Pearson (1886, Vol. 2, pp. 84–85). However, their practical importance is considerable. Initial states of stress may be induced during the manufacturing process of a material, or by the action of pre-existing body forces. In cast iron, initial stresses arise from the unequal rates of cooling throughout the body. A body in equilibrium under the mutual gravitation of its parts is in a state of stress, which sometimes can be enormous, as happens, for instance, in the interior of the Earth. Another class of problems in which Cauchy's relations are useful is that of the elastic stability of bodies subject to a state of a too severe exterior pressure; a small change in the data results in an unpredictable change in the qualitative character of the set of solutions.

We have noted that the concept of elasticity embodied in equations (3.5) is due to Cauchy, but a more restricted concept of elasticity has been proposed by Green, (1839, 1842), using a different argument. According to Green, elasticity means the existence of a scalar function W of the material coordinates X_{A} and the deformation gradient $F_{i\mathrm{A}}$ such that the rate of work done in deforming the interior of the body is given by

$$\frac{\rho}{\rho_{\kappa}}\dot{W}=\sigma_{ij}d_{ij}, \tag{3.26}$$

where $d_{ij}=(\partial v_i/\partial x_j)$, is the symmetric part of the velocity gradient, which is called the *stretching tensor*, and the factor ρ/ρ_{κ} is introduced purely for convenience.

W is given the name *strain energy*, and a material endowed with strain energy is said to be *hyperelastic*.

$W(F_{i\mathrm{A}})$ is subject to the principle of material frame indifference, since the strain energy must be the same for all observers. It thus follows that W does not depend on $F_{i\mathrm{A}}$, but rather on the symmetric part U_{AB} of the multiplicative decomposition of $F_{i\mathrm{A}}$. This means that W depends on C_{AB}, and (3.26) can be written as

$$\frac{\rho}{\rho_{\kappa}}\frac{\partial W}{\partial C_{\mathrm{AB}}}\dot{C}_{\mathrm{AB}}=\sigma_{ij}d_{ij}. \tag{3.27}$$

Since

$$C_{AB} = F_{iA}F_{iB},$$

by differentiation with respect to time we obtain

$$\dot{C}_{AB} = \dot{F}_{iA}F_{iB} + F_{iA}\dot{F}_{iB} = \frac{\partial v_i}{\partial x_j} F_{jA}F_{iB} + \frac{\partial v_i}{\partial x_j} F_{jB}F_{iA} = 2d_{ij}F_{jA}F_{iB}. \tag{3.28}$$

If we substitute $\dot{C}_{AB}$ from (3.28) in (3.27) we have an identity which must hold for any d_{ij}. Consequently, the Cauchy stress σ_{ij} can be expressed in terms of W according to the law

$$\sigma_{ij} = \frac{2\rho}{\rho_\kappa} F_{jA}F_{iB} \frac{\partial W}{\partial C_{AB}}. \tag{3.29}$$

There are two remarkable variants of this stress–strain relation. If we exploit (3.6a) we immediately find the equation

$$T_{AB} = 2J \frac{\partial W}{\partial C_{AB}}, \tag{3.30}$$

a formula obtained by Kelvin in 1863 (see Kelvin and Tait 1867). Alternatively, we may introduce the so-called *Piola–Kirchhoff stress tensor*, which is related to σ_{ij} as follows:

$$\sigma_{ij} = \frac{\rho}{\rho_\kappa} T_{iA}F_{jA}. \tag{3.31}$$

T_{iA} is the stress per unit area of the reference configuration. In order to justify relation (3.31), it is sufficient to observe that $\sigma_{ij}da_j$, the contact force acting on the area element da_j, must be equal to a vector $T_{iA}\,dA_A$ acting on the element dA_A, which corresponds to da_j in the reference configuration. By using (1.15) we can thus write

$$T_{iA}\,dA_A = J\sigma_{ij}(\mathbf{F}^{-1})_{Aj}\,dA_A, \tag{3.32}$$

which proves (3.31). It is worth noting that, as a consequence of the symmetry of σ_{ij}, T_{iA} is not symmetric but satisfies the relation

$$T_{iA}F_{jA} = T_{jA}F_{iA}. \tag{3.33}$$

Having introduced the Piola–Kirchhoff stress, the alternative expression (3.30) becomes

$$T_{iA} = \rho_\kappa \frac{\partial W}{\partial F_{iA}}. \tag{3.34}$$

In hyperelasticity, a material is isotropic if its strain energy is an isotropic function of C_{AB}. As a consequence of the representation theorem for isotropic scalar functions of a symmetric tensor, it turns out that W is a certain function of the principal invariants of C_{AB}:

$$W = W(I_1, I_2, I_3). \tag{3.35}$$

The notion of hyperelasticity does not restrict the magnitude of the strains, and may therefore be exploited in treating problems involving large deformations. If, however, we deal with infinitesimal deformations, then we can specify the form of W, since this must be a homogeneous quadratic form in the infinitesimal strains $\tilde{E}_{AB}$. In order to follow the traditional notation of classical elasticity, it is convenient to put

$$\tilde{E}_{11} = e_{11},\ \tilde{E}_{22} = e_{22},\ \tilde{E}_{33} = e_{33},$$

$$2\tilde{E}_{23} = e_{23},\ 2\tilde{E}_{31} = e_{31},\ 2\tilde{E}_{12} = e_{12},$$

and write the expression for $2W$ in the form

$$\begin{aligned}2W = C_{11}e_{11}^2 &+ 2c_{12}e_{11}e_{22} + 2c_{13}e_{11}e_{33} + 2c_{14}e_{11}e_{23} + 2c_{15}e_{11}e_{31} + 2c_{16}e_{11}e_{12}\\ &+ c_{22}e_{22}^2 + 2c_{23}e_{22}e_{33} + 2c_{24}e_{22}e_{23} + 2c_{25}e_{22}e_{31} + 2c_{26}e_{22}e_{12}\\ &+ c_{33}e_{33}^3 + 2c_{34}e_{33}e_{23} + 2c_{35}e_{33}e_{31} + 2c_{36}e_{33}e_{12}\\ &+ c_{44}e_{23}^2 + 2c_{45}e_{23}e_{31} + 2c_{46}e_{23}e_{12}\\ &+ c_{55}e_{31}^2 + 2c_{56}e_{31}e_{12}\\ &+ c_{66}^2e_{12}^2,\end{aligned} \tag{3.36}$$

where the 21 coefficients c_{rs} $(r, s = 1, \ldots, 6)$ are the Cauchy *elastic moduli* of the material.

If the material possesses some plane or some axis of symmetry, the number of independent moduli is less than 21 and can be 13, 9, 7, 6, 5, 3, or 2, each of these numbers being the index of a crystallographic class. The class with only two independent moduli characterizes isotropic materials. For these materials every plane is a plane of symmetry, and every axis is an axis of symmetry for a rotation of unlimited amount. The strain energy then reduces to the form

$$\begin{aligned}2W = c_{11}(e_{11}^2 + e_{22}^2 + e_{33}^2) &+ 2c_{12}(e_{22}e_{33} + e_{33}e_{11} + e_{11}e_{22})\\ &+ \frac{1}{2}(c_{11} - c_{12})(e_{23}^2 + e_{31}^2 + e_{12}^2).\end{aligned} \tag{3.37}$$

For linear hyperelastic materials the stress–strain relationship can be simply expressed by the equations

$$\sigma_{11} = \frac{\partial W}{\partial e_{11}}, \ldots; \qquad \sigma_{23} = \frac{\partial W}{\partial e_{23}}, \ldots . \tag{3.38}$$

In the case of an isotropic material we can write

$$\sigma_{11} = c_{11}e_{11} + c_{12}(e_{22} + e_{33}), \ldots$$

$$\sigma_{23} = \frac{1}{2}(c_{11} - c_{12})e_{23}, \ldots ,$$

that is to say, by comparing this with (3.24), we recover the Lamé equations, provided that we put

$$\mu = \frac{1}{2}(c_{11} - c_{12}), \qquad \lambda = c_{12}. \tag{3.39}$$

Many attempts have been made in the past to justify the concept of hyperelasticity in the physical context. These attempts at adaptation usually proceed as follows. They consider a particular process, for instance an adiabatic or an isothermal change, and apply the thermodynamic laws in order to infer that a strain energy function exists out of necessity. But, as Truesdell (1966) has pointed out, these "demonstrations" are questionable for two reasons: they consider only particular processes; and they appeal to principles, like those of thermodynamics, which are strictly extraneous to elasticity. For instance, thermodynamics excludes perpetual motion, but in linear elasticity, or even in rigid-body mechanics, this is perfectly admissible. The pendulum without friction is an example.

A more important question is that of finding a program of experiments which relate to the concepts of stress and strain and so to a strain energy function. A first difficulty arises from the fact that the components of stress and strain can never be measured directly. Their values can be found only by a process of inference from measurements of quantities that are not stresses or strains. Instruments, in general, measure displacements and loads, and the corresponding strains and stresses are derived by averaging the measured displacements over finite lengths and averaging loads over finite portions of surface. It turns out that the experiments which lead to the characterization of the form of the strain energy do not constitute a proof of its existence. But the mathematical consequences that can be deduced by postulating the existence of a strain energy are sometimes capable of experimental verification. These consequences have been investigated in the case of small strain and constant temperature, that is to say when the strain energy function W is a homogeneous quadratic form of the strain components. For instance, the fact that all solid bodies can be thrown into a state of isochronous vibrations was recognized by Stokes (1845) as a decisive proof of the existence of a strain energy quadratic in the strains. Other examples come from the theory of structures, where the measured displacements of certain points under given loads are almost exactly equal to those predicted by the theory.

When we want to apply the linearized theory it is necessary to pay attention to the way in which loads are applied. In most experiments the load that is increased, or diminished, represents only a part of the external forces. This is because there are pre-existing forces, for example the weight of the body, which give rise to a state of prestressing. The final stress thus consists of two systems: the stresses in the initial state, and a stress induced by the additional loads. Likewise, the overall strain is composed of two components: one induced from the unstressed to the initial state, and one from the initial to the actual state. Of course, experiments measure the relationship between the second stress system and the second strain. The general statement of a law of proportionality between stress and strain implies that the stress in the initial state is also proportional to the strain in that state. Now this is true if the initial strains and the strains assumed under the load are linearly related or are sufficiently small to be calculated according to the linearized theory. Without linear stress–strain relations superposition is impossible and nothing can be deduced about the relation between the components of total stress and the total strain.

The elastic deformations of a body are usually calculated and measured by neglecting changes in temperature. If, however, strong variations in temperature occur, it is necessary to modify the stress–strain relations in order to include the influence of temperature. If the material is isotropic and the strains are small, an increase in temperature of $(\theta^* - \theta)$ gives rise to a dilatation by an amount proportional to $(\theta^* - \theta)$, produced without any change of pressure. This implies that each linear element is extended by $\frac{1}{3}c(\theta^* - \theta)$, where c is a constant called the *coefficient of expansion.*

For linear isotropic solids the equations relating strain to stress can easily be obtained by solving the Lamé equations (3.24) in terms of the infinitesimal strains:

$$e_{11} = H_{11}, \qquad e_{22} = H_{22}, \qquad e_{33} = H_{33},$$
$$e_{23} = H_{23} + H_{32}, \qquad e_{31} = H_{31} + H_{13}, \qquad e_{12} = H_{12} + H_{21}.$$

If we introduce Young's modulus

$$E = \frac{\mu(3\lambda + 2\mu)}{\lambda + \mu}, \tag{3.40}$$

and Poisson's ratio

$$\sigma = \frac{\lambda}{2(\lambda + \mu)}, \tag{3.41}$$

the inverted Lamé equations assume the form

$$\begin{aligned} e_{11} &= \frac{1}{E}[\sigma_{11} - \sigma(\sigma_{22} + \sigma_{33})], \ldots \\ e_{23} &= \frac{2(1+\sigma)}{E}\sigma_{23}, \ \ldots . \end{aligned} \tag{3.42}$$

If an additional dilatation due to a change in temperature occurs, instead of (3.42) we must use three equations of the type

$$e_{11} = \frac{1}{3}c(\theta^* - \theta) + \frac{1}{E}[\sigma_{11} - \sigma(\sigma_{22} + \sigma_{33})], \ \ldots \tag{3.43}$$

and three of the type

$$e_{23} = \frac{2(1+\sigma)}{E}\sigma_{23}, \ \ldots .$$

The equations derived above are known as the constitutive equations of *linear thermoelasticity*. Their range of application is, however, not very wide, firstly because the changes in temperature may be so great that the dilation is not proportional to $(\theta^* - \theta)$, and secondly because the elastic moduli themselves are functions of the temperature and, in general, diminish with a rise in temperature.

4. The Search for Solutions

The constitutive equations are a necessary ingredient in the process of solving field equations, which would be otherwise undetermined since the number of unknowns is larger than that of the equations. Nevertheless, even after the introduction of constitutive equations, finding a solution of the field equations in their full generality is an overwhelming task usually beyond the capacity of mathematics, even if one is content with finding only certain general properties rather than explicit solutions. If no further restriction on the constitutive equations is made, the possibility of finding a solution might still be in doubt.

In order to have a more precise idea of what is known in the mathematical treatment of field equations, let us start from a particular situation, with the warning that the more generality we add the greater the difficulties we must face. The case we wish to discuss is that of the motion and the equilibrium of a simple hyperelastic isotropic body at constant temperature. This example isolates the purely mechanical aspects of the phenomenon. Since the equation of balance of mass can be integrated directly, the density ρ is no longer an unknown, but a function of ρ_κ through the relation (2.20). Since temperature does not appear in an explicit form in the momentum balance equations, we may try to solve these equations separately, in the hope of determining the motion.

The original equations (2.24) are of no use since they are concerned with the present configuration, which is unknown. For this reason it is more convenient to repeat the argument of localization of the momentum balance equations, considering not the volume V in the present configuration, but its image V_0 in the reference configuration. On recalling the meaning of the Piola–Kirchhoff stress tensor, the balance equation is

$$\overline{\left(\int_{V_0} \rho_\kappa v_i \, dV\right)}^{\cdot} = \int_{S(V_0)} T_{iB} \, dA_B + \int_{V_0} \rho_\kappa f_i, \tag{4.1}$$

where dV is the volume element of V_0. Since V_0 is arbitrary, the local form of (4.1) has the form

$$\rho_\kappa \dot{v}_i = \frac{\partial T_{iB}}{\partial X_B} + \rho_\kappa f_i. \tag{4.2}$$

In these equations X_A are the independent variables, and therefore v_i and f_i, in constrast to (2.24), must be regarded as functions of X_A. It is clear that, with this replacement, we have

$$\dot{v}_i(X_A, t) = \frac{\partial^2}{\partial t^2} x_i(X_A, t), \tag{4.3}$$

and the total derivative with respect to t coincides with the partial derivative.

These equations hold at the points X_A inside $\kappa(\mathscr{B})$ and must be completed by the information we have on the boundary of $\kappa(\mathscr{B})$. It may be that some parts of the boundary remain fixed during the motion. For the points of these parts we must impose the boundary condition

$$\chi_i(X_A, t) = \delta_{iA} X_A, \tag{4.4}$$

which is said to be the condition of *place*. Alternatively, we may know that other parts of the boundary are free of traction or subjected to prescribed tractions during the motion. In this case we should require that

$$t_i = \sigma_{ij} e_j \tag{4.5}$$

must have prescribed value for all times. But a condition of this type is useless, since the configuration $\chi(\mathscr{B})$ itself is unknown. On the other hand, if we merely specify the traction on the boundary of $\kappa(\mathscr{B})$, the deformation produced by the prescribed traction will move its point of application and deform the surface, with the consequence that again the latter is unknown. In order to avoid this difficulty, the *boundary condition of traction* customarily imposed is that

$$t_i^0 = T_{iB} e_B^0, \tag{4.6}$$

where e_B^0, the unit outward normal to the boundary of $\kappa(\mathcal{B})$, has a given value. Since t_i^0 is parallel to t_i, the boundary condition of traction means that the surface traction keeps its direction during the deformation but adjusts its magnitude in the ratio of the element of area dA_A to the element da_i. In this way, we assign the force per unit initial area.

Of course there are other realistic ways in which boundary tractions may appear. One is *surface pressure*, which occurs when the surface traction remains normal to each element of the boundary, whatever the deformation. This type of surface load must be considered when we wish to study the large deformations of a submerged elastic body.

In a problem involving motion, in addition to the boundary conditions, the initial conditions must also be supplied, which prescribe the position and velocity of each point of the body at a given instant. If this instant is the time $t = 0$, the initial data are said to *Cauchy's initial conditions.*

Notwithstanding the existence of many partial results in particular cases, as was implied at the beginning of this chapter, the general problem of finite elastodynamics is far from being solved. The only known result is the following (Ciarlet 1988, Sec. 5.10). Suppose that the material is hyperelastic and we introduce the *elasticity tensor*

$$a_{iAjB}(\mathbf{F}) = \frac{\partial T_{iA}}{\partial F_{jB}}, \tag{4.7}$$

then the following strong ellipticity condition[4]

$$a_{iAjB}(\mathbf{F}) b_i b_j c_A c_B > 0, \tag{4.8}$$

for all vectors b_i and c_A, is a sufficient condition for ensuring the existence in the small, that is for short times, of solutions of the pure displacement problem.

For the restricted case of pure elastostatic problems, that is for zero inertial forces and the body in equilibrium, more can be said about solutions.

However, before attempting to solve the problem, it is necessary to know some of the properties that are required of the solutions in order for them to be physically admissible. A first requirement is that the mapping from $\kappa(\mathcal{B})$ to $\chi(\mathcal{B})$ is continuous in terms of its first partial derivatives and is locally invertible for all points $X_A \in \kappa(\mathcal{B})$. This implies the condition $\det F_{iA} = J > 0$. But local invertibility does not entail global invertibility, as is known from the standard textbooks of mathematical analysis. It is thus necessary to find the conditions that ensure global injectivity. Two useful sufficient conditions answering this requirement are:[5] that $|\det u_{i,A}|$ is sufficiently small; and that $J > 0$, but the boundary conditions are purely spatial and the mapping from $\partial\kappa(\mathcal{B})$, the boundary of $\kappa(\mathcal{B})$, to $\partial\chi(\mathcal{B})$ is globally invertible.

[4] This is also known as the strong Legendre–Hadamard condition.
[5] For the proof, see Ciarlet (1988, Sec. 5.5).

However, when the boundary conditions concern tractions, an additional condition is needed in order to exclude interpenetration of the material, as illustrated in Figure 4.1. Here, as a result of deformation, two distinct regions of $\kappa(\mathscr{B})$ (the hatched areas in Figure 4.1) would coalesce in the same region of $\chi(\mathscr{B})$. A condition which excludes situations of this kind has been proved by Ciarlet and Nečas (1987) and requires that the two following inequalities be satisfied simultaneously:

$$J > 0 \text{ in } \kappa(\mathscr{B}),$$

and

$$\int_{\kappa(\mathscr{B})} J\,dV \leq \text{vol}\,[\chi(\mathscr{B})]. \tag{4.9}$$

Another striking property of solutions of finite elastostatics is their lack of uniqueness, because, if the mathematical model constituted by the equations of finite elastostatics is adequate, it must predict the occurrences of nonuniqueness observed in physical experiments, such as the eversion of a cylindrical tube or simply the buckling of a plate.

Boundary value problems frequently figure in finite elastostatics, and much effort has been expended in attempts to deal with such problems.

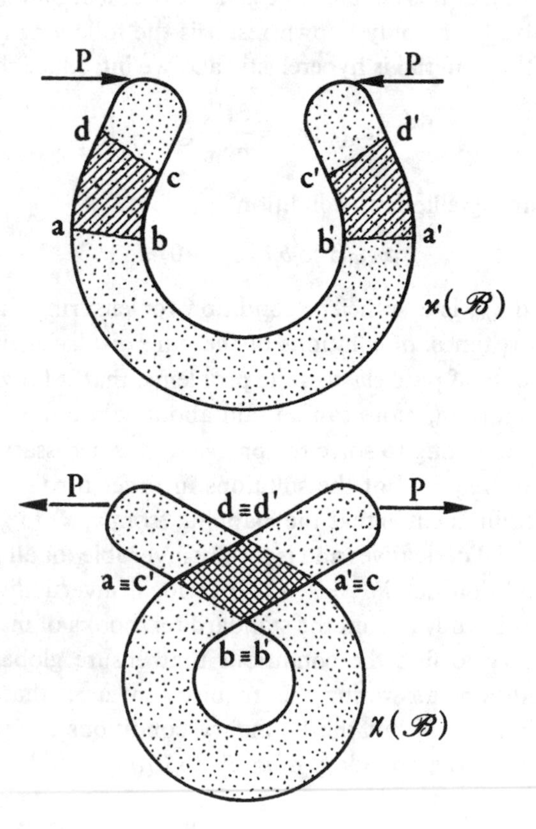

Fig. 4.1

The first method was proposed by Signorini (1930, 1949) who considered the pure traction problems:

$$\frac{\partial T_{iB}}{\partial X_B} = \rho_\kappa f_i \text{ in } \kappa(\mathscr{B}),$$
$$T_{iB} = t_i^0 \text{ on } \partial\kappa(\mathscr{B}). \tag{4.10}$$

Since only forces are applied to the body, they cannot be arbitrary but must be balanced, which implies that

$$\int_{\kappa(\mathscr{B})} \rho_\kappa f_i \, dV + \int_{\partial\kappa(\mathscr{B})} T_{iB} \, dA_B = 0, \tag{4.11}$$

and that for any fixed point x_i^0 in $\chi(\mathscr{B})$ we have

$$\int_{\partial\kappa(\mathscr{B})} \varepsilon_{ijk}(x_j - x_j^0) T_{kB} \, dA_B + \int_{\kappa(\mathscr{B})} \varepsilon_{ijk}(x_j - x_j^0) \rho_\kappa f_k \, dV = 0. \tag{4.12}$$

Since $x_i = \delta_{iA} X_A + u_i$, and u_i is unknown, this condition cannot be used directly as a criterion for the assigned loads t_k^0 and f_κ.

Signorini assumed that t_i^0 and f_i are analytic functions of a parameter ε:

$$t_i^0 = \sum_{n=1}^{\infty} \varepsilon^n (t_i^0)_n,$$
$$f_i = \sum_{n=1}^{\infty} \varepsilon^n (f_i)_n, \tag{4.13}$$

and he tried to find a solution of the form

$$u_i = \sum_{n=1}^{\infty} \varepsilon^n (u_i)_n. \tag{4.14}$$

The coefficients $(u_i)_n$ of the series are unknown and must be calculated by substituting the expansions (4.13) and (4.14) in the system (4.10) and equating powers of ε. Since T_{iA} also depends on u_i, it is necessary to assume that T_{iA} is an analytic function of F_{iA}. For each n we thus obtain a system which is linear in $(u_i)_n$ and contains all the coefficients $(u_i)_1, \ldots, (u_i)_{n-1}$, evaluated in the preceding $n-1$ stages of the process. The loads of each system of order n are not only $(t_i^0)_n$ and $(f_i)_n$, but also combinations of terms involving $(u_i)_1, \ldots, (u_i)_{n-1}$. Thus, in order that each system has a solution $(u_i)_n$ at all, it is necessary that a condition of compatibility should be satisfied by $(u_i)_{n-1}$ found in the previous stage. Signorini's method gives a theorem of compatibility and uniqueness for each iterative system, but does not prove the existence of the solution.

A second, more systematic, solution rests on the application of the implicit function theorem, or its corollary, that is, the local inversion theorem to the boundary value problem concerned purely with spatial location

$$\frac{\partial T_{iB}}{\partial X_B} = \rho_\kappa f_i \text{ in } \boldsymbol{\kappa}(\mathscr{B}), \qquad u_i = 0 \quad \text{on } \partial\boldsymbol{\kappa}(\mathscr{B}), \tag{4.15}$$

in a neighborhood of the null solution $u_i = 0$ corresponding to $f_i = 0$. It can be seen that the theorem works only for the displacement problem, and this restriction is essential (Ciarlet 1988, Sec. 6.4). The theorem states that, if the boundary is of class $\mathscr{C}^2$, and the force $\rho_\kappa f_i$ is of class L^p with $p > 3$, then there is a neighborhood F^p of the origin of the space L^p and a neighborhood U^p of the origin in the subspace of $W^{2,p}$ with $u_i = 0$ on $\delta\boldsymbol{\kappa}(\mathscr{B})$, such that, for each $\rho_\kappa f_i \in F^p$, the problem (4.15) has exactly one solution $u_i \in U^p$.

The successful application of the implicit function theorem to the system (4.15) depends on the proof that, if $u_i \in W^{2,p}$ with $u_i = 0$ on $\partial\boldsymbol{\kappa}(\mathscr{B})$, then $(\partial T_{iB}/\partial X_B) \in L^p$ for $p > 3$. For this reason, the extension of the method to boundary conditions different from those of place has failed.

A third method is of the variational type and applies only to a hyperelastic material (Ball 1977). However, it works for the more general mixed boundary value problem

$$\frac{\partial T_{iB}}{\partial X_B} = \rho_\kappa f_i \text{ in } \boldsymbol{\kappa}(\mathscr{B}), \qquad u_i = 0 \quad \text{on } \partial_1\boldsymbol{\kappa}(\mathscr{B}), \qquad T_{iB}\overset{0}{e}_B = \overset{0}{t}_i \quad \text{on } \partial_2\boldsymbol{\kappa}(\mathscr{B}), \tag{4.16}$$

with $\overline{\partial_1\boldsymbol{\kappa}(\mathscr{B})} \cup \partial_2\boldsymbol{\kappa}(\mathscr{B}) = \partial\boldsymbol{\kappa}(\mathscr{B})$, provided that exterior loads are *dead*, that is to say that they are independent of u_i.

In a hyperelastic body the strain energy stored in $\boldsymbol{\kappa}(\mathscr{B})$ is represented by the integral

$$\mathscr{W} = \int_{\boldsymbol{\kappa}(\mathscr{B})} W(F_{iA})\, dV,$$

and the work of exterior loads is given by

$$\mathscr{L} = \int_{\boldsymbol{\kappa}(\mathscr{B})} \rho_\kappa f_i u_i\, dV + \int_{\partial_2\boldsymbol{\kappa}(\mathscr{B})} T_{iA} u_i\, dA_A.$$

Solving the boundary value problem (4.16) is equivalent to finding a minimum of the total potential energy $\mathscr{E} = \mathscr{W} - \mathscr{L}$, over the set of admissible functions to be defined later.

The question of the existence of a solution is therefore reduced to proving the existence of a minimum of the functional $\mathscr{E}$. If loads are dead and of class L^p, the term $\mathscr{L}$ is a linear continuous functional of u_i and, as such, can be handled easily. However, the treatment of the functional $\mathscr{W}$ is more difficult, because the function $W(F_{iA})$ is, in general, nonconvex in F_{iA} and not defined for det $F_{iA} \leq 0$. In order to overcome these difficulties, it is necessary to replace the simple convexity condition on F_{iA} with a weaker condition known as *polyconvexity*. Polyconvexity requires the existence of a convex function W^* of F_{iA}, $(\mathrm{cof}\,\mathbf{F})_{iA} = \frac{1}{2}\varepsilon_{ijk}\varepsilon_{ABC}F_{jB}F_{kC}$ and det F_{iA} such that

$$W^*(F_{iA}, (\mathrm{cof}\,\mathbf{F})_{iA}, \det F_{iA}) = W(F_{iA}), \tag{4.17}$$

for all F_{iA} with det $F_{iA} > 0$. A simple indication of the advantages of the notion of polyconvexity may be inferred by the following example (Ciarlet 1988, Sec. 4.9). If W has the form $W = a \det F_{iA}$, where a is a constant, and we consider the matrices

$$F_{iA} = \lambda\,\mathrm{diag}(2, 1, 1) + (1 - \lambda)\,\mathrm{diag}(1, 2, 1), \qquad 0 \le \lambda \le 1,$$

we can easily calculate $a \det F_{iA} = a(2 + \lambda - \lambda^2)$, which is not convex in λ. However, it is polyconvex, since $W(\delta) = a\delta$ is convex.

Besides polyconvexity, W must satisfy a *coerciveness inequality* of the type

$$W(F_{iA}) \ge \alpha\{\|F_{iA}\|^p + \|(\mathrm{cof}\,\mathbf{F})_{iA}\|^q + (\det F_{iA})^r\} + \beta, \tag{4.18}$$

for all F_{iA} with det $F_{iA} > 0$, where $\alpha > 0$, β is a real number, and $\|\|$ denote the Euclidean norms of the matrices F_{iA} and $(\mathrm{cof}\,\mathbf{F})_{iA}$. In addition, the exponents p, q, and r must satisfy the inequalities

$$p \ge 2, q \geqslant \frac{p}{p-1}, \quad r > 1. \tag{4.19}$$

In order to preserve the property of conservation of matter the functional $W(F_{iA})$ must also tend to infinity as det $F_{iA} \to 0^+$.

As for the set of admissible functions, the functional can be specified as follows: $u_i \in W^{1,p}$, $(\mathrm{cof}\,F)_{iA} \in L^q$, det $F_{iA} \in L^r$, $u_i = 0$ almost everywhere on $\overline{\partial_1\kappa(\mathscr{B})}$ and det $F_{iA} > 0$ almost everywhere in $\kappa(\mathscr{B})$.

Now, provided that the set of admissible functions is nonempty, and that $\mathscr{E}$ is not identically $+\infty$ in this set, then there is at least one function u_i for which $\mathscr{W}$ attains a minimum.

The proof of the theorem of existence is rather involved, but its logical steps are simple: we must first prove that the functional $\mathscr{E}$ is well defined in the set of admissible functions; that we can find a lower bound for $\mathscr{E}$; that there is a minimizing sequence u_i^k which weakly converges to a function u_i; that u_i belongs to the set of admissible functions; and, finally, that $\mathscr{E}$ attains a minimum at u_i.

These are the principal techniques employed to obtain an existence theorem for problems in finite deformations. If, on the other hand, we consider the case of infinitesimal deformations, the existence theory is much easier because the equations are linear and the properties of linear operators are better known. For simplicity, we concentrate our attention on linear homogeneous isotropic materials, characterized by Lamé's constants λ and μ, and assume that the reference configuration is a natural state.

In the linearized theory, under certain assumptions of regularity the material stress, also called the *second Piola–Kirchhoff stress tensor*, can be simply approximated by the expressions

$$T_{11} = \lambda(e_{11} + e_{22} + e_{33}) + 2\mu e_{11}, \ldots \tag{4.20}$$

for the normal stress components, and

$$T_{23} = \mu e_{23}, \ldots \tag{4.21}$$

for the tangential stress components. Expressed in terms of the displacement components u_i, the mixed boundary value problem of linear elastodynamics takes the following form:

$$\begin{aligned} \mu\nabla^2 u_i + (\lambda+\mu)\frac{\partial}{\partial x_i}(u_{j,j}) + \rho f_i = \rho\frac{\partial^2 u_i}{\partial t^2} \quad &\text{in} \quad \boldsymbol{\kappa}(\mathscr{B}), \\ u_i = 0 \quad &\text{on} \quad \overline{\partial_1\boldsymbol{\kappa}(\mathscr{B})}, \\ \sigma_{ij}e_j = t_i^0 \quad &\text{on} \quad \partial_2\boldsymbol{\kappa}(\mathscr{B}), \end{aligned} \tag{4.22}$$

where now we have denoted the independent variables by x_i and the components of the Cauchy stress tensor by σ_{ij}, these correspond to the second and first Piola–Kirchhoff stress tensors, respectively. Note that also ρ and ρ_κ are practically the same. Of course, apart from the boundary conditions we need two initial conditions

$$u_i = u_i^0, \quad \frac{\partial u_i}{\partial t} = v_i^0, \tag{4.23}$$

where u_i^0 and v_i^0 are two functions vanishing over $\overline{\partial_1\boldsymbol{\kappa}(\mathscr{B})}$.

The key to proving the existence theorem is the observation that, if we introduce the subspace V_0 of functions $v_i \in H^1$ such that $v_i = 0$ on $\overline{\partial_1\boldsymbol{\kappa}(\mathscr{B})}$, problem (4.22) is equivalent to finding a solution u_i of the integrodifferential equation

$$\left(\frac{\partial^2 u_i}{\partial t^2}, v_i\right) + a(u_i, v_i) = L(v_i) \qquad \forall v_i \in V_0, \tag{4.24}$$

where

$$\left(\frac{\partial^2 u_i}{\partial t^2}, v_i\right) = \int_{\boldsymbol{\kappa}(\mathscr{B})} \rho\frac{\partial^2 u_i}{\partial t^2} v_i \, dV, \tag{4.25}$$

$$a(u_i, v_i) = \int_{\boldsymbol{\kappa}(\mathscr{B})} \{\lambda e_{hh}(u_i)e_{kk}(v_i) + 2\mu e_{hk}(u_i)e_{hk}(v_i)\}\, dV, \tag{4.26}$$

$$L(v_i) = \int_{\boldsymbol{\kappa}(\mathscr{B})} \rho f_i v_i \, dV + \int_{\partial_2\boldsymbol{\kappa}(\mathscr{B})} \sigma_{ij}e_j v_i \, dA_j. \tag{4.27}$$

This is called the *weak* formulation of problem (4.22).

It is possible to prove that the bilinear form (4.26) is *coercive*, that is to say that satisfies an inequality of the type

$$a(v_i, v_i) \geq \alpha\left\{\int_{\boldsymbol{\kappa}(\mathscr{B})} [v_i v_i + v_{i,j}v_{i,j}]\, dV\right\}^{1/2} = \alpha\|v_i\|_1, \tag{4.28}$$

for all $v_i \in V_0$, and with $\alpha > 0$. Another property of $a(u_i, v_i)$ is its *continuity*:

$$a(u_i, v_i) \leqslant B\|u_i\|_1 \|v_i\|_1, \tag{4.29}$$

for all $u_i, v_i \in V_0$, and with $B > 0$.

After these preliminaries, we may choose a sequence of functions $w_i^{(1)} w_i^{(2)}, \ldots,$ such that, for every m, $w_i^{(1)}, \ldots, w_i^{(m)}$ are linearly independent in V_0 and such that the linear independent combinations

$$u_i^{(m)} = a_1(t)w_i^{(1)} + a_2(t)w_i^{(2)} + \ldots + a_m(t)w_i^{(m)} \tag{4.30}$$

are dense in V_0, and where $w_i^{(1)} = u_i^0$ (if $u_i^0 \neq 0$). The combination is taken as an approximate solution and, because $w_i^{(1)}, \ldots, w_i^{(2)}, \ldots,$ are known, the problem is completely determined once the coefficients $a_1(t)$, $a_2(t)$, ..., are determined. For this purpose we write the system of m ordinary differential equations of second order:

$$\left(\frac{\partial^2}{\partial t^2} u_i^{(m)}, v_i\right) + a\left(u_i^{(m)}, v_i\right) = L(v_i) \qquad \forall v_i = [w_i^{(1)}, \ldots, w_i^{(m)}]$$

$$u_i^{(m)}\Big|_{t=0} = u_i^0, \tag{4.31}$$

$$\frac{\partial u_i^{(m)}}{\partial t}\Big|_{t=0} = v_i^{0(m)}, \quad v_i^{0(m)} \in [w_i^{(1)}, \ldots, w_i^{(m)}].$$

Since the equations are linear and the system is not singular because $w_i^{(1)}$ and $w_i^{(m)}$ are linearly independent, the functions $u_i^{(m)}$ are uniquely determined.

In this way we have constructed the approximation $u_i^{(m)}$ of order m. Let us now let m tend to infinity. Under the assumptions made on the form of $a(u_i, v_i)$ and with other mild restrictions on the data, it is not difficult to prove that $u_i^{(m)}$ and its first time derivatives remain in two bounded subsets of $L^\infty(0, T; V_0)$ and $L^\infty(0, T; H)$,[6] and therefore we can derive a subsequence $u_i^{(\mu)}$ which is weakly convergent with its first time derivative, to a function u_i. A further step in the demonstration proves that u_i is a solution of the original system. The final result is that the solution is unique, and a remarkable feature of the method is that if offers a *constructive procedure* for finding approximate solutions.[7]

When we pass to treating the elastic equilibrium of a body under mixed type boundary conditions, we have to solve the following system of partial differential equations:

$$\mu\nabla^2 u_i + (\lambda + \mu)\frac{\partial}{\partial x_i}(u_{j,j}) + \rho f_i = 0 \qquad \text{in} \qquad \kappa(\mathscr{B}),$$

$$u_i = 0 \qquad \text{on} \qquad \overline{\partial_1 \kappa(\mathscr{B})}, \tag{4.32}$$

$$\sigma_{ij} e_j = t_i^0 \text{ on } \quad \partial_2 \kappa(\mathscr{B}).$$

The system of linear elasticity has been the object of much important research during the last 150 years. The methods of solution may be classified into three categories.

[6] $L^\infty(0, T; V_0)$ is the space of functions from $[0, T]$ to V_0 such that the norm $(\int_0^T \|u_i\|_1^2\, dt)^{1/2} < \infty$.
[7] For this reason, the above technique for proving the existence of a solution in elastodynamics is called the *Faedo–Galerkin method* (see Lions and Magenes 1968).

The first, in chronological order, is a method based on the equivalence of finding the solution of the system (4.33) and determining a minimum of the total potential energy:

$$\mathscr{E} = \mathscr{W} - \mathscr{L} = \frac{1}{2}a(u_i, u_i) - L(u_i), \tag{4.33}$$

where $a(u_i, u_i)$ and $L(u_i)$ are defined by (4.26) and (4.27), with u_i belonging to the subspace V_0 introduced above.

Since $a(u_i, u_i)$ is continuous and coercive and $L(u_i)$ is linear, the function $\mathscr{E}$ is continuous and bounded below, since

$$\mathscr{E} \geqslant \frac{\alpha}{2}\|u_i\|_1^2 - M\|u_i\|_1 \tag{4.34}$$

and tends to $+\infty$ as $\|u_i\|_1^2 \to \infty$. It thus admits a unique minimum in the subspace V_0, and the corresponding minimum value of u_i is the solution of the boundary value problem (4.33).

A second method of proving the existence theorem of linear elastostatics rests on the idea of transforming the boundary problem into one concerned with a system of linear integral equations.

The adaptation of the method of integral equations to elasticity was first done by Betti (1872), and was subsequently modified by Lauricella (1906) and Korn (1907, 1908). Betti's idea is that of expressing the dilation $\Delta = (e_{11} + e_{22} + e_{33})$ and the rotations

$$\tilde{\omega}_1 = \frac{1}{2}\left(\frac{\partial u_3}{\partial x_2} - \frac{\partial u_2}{\partial x_3}\right), \quad \tilde{\omega}_2 = \frac{1}{2}\left(\frac{\partial u_1}{\partial x_3} - \frac{\partial u_3}{\partial x_1}\right), \quad \tilde{\omega}_3 = \frac{1}{2}\left(\frac{\partial u_2}{\partial x_1} - \frac{\partial u_1}{\partial x_2}\right)$$

by means of formulae that explicitly contain the surface displacements and the surface tractions. These formulae can be obtained by exploiting the property that $\Delta, \tilde{\omega}_1, \tilde{\omega}_2$, and $\tilde{\omega}_3$ are harmonic functions when the body forces are absent. In this case the field equations of elasticity can be written in the form

$$\nabla^2\left[u_i + \frac{1}{2}\left(1 + \frac{\lambda}{\mu}\right)x_i\Delta\right] = 0, \qquad i = 1, 2, 3, \tag{4.35}$$

and thus determining u_i, when Δ is known and the surface values of u_i over the *whole* of $\partial\kappa(\mathscr{B})$ are prescribed, is reduced to the Dirichlet problem for harmonic functions. If, on the other hand, the surface tractions t_i are prescribed over the *whole* of $\partial\kappa(\mathscr{B})$, the normal derivatives $\partial u_i/\partial \nu$ at the boundary can be expressed as

$$\frac{\partial u_1}{\partial \nu} = \frac{1}{2\mu}t_1 - \frac{\lambda}{2\mu}\Delta e_1 + \tilde{\omega}_2 e_3 - \tilde{\omega}_3 e_2, \ldots, \tag{4.36}$$

so that, when $\Delta, \tilde{\omega}_1, \tilde{\omega}_2$, and $\tilde{\omega}_3$ are known, the normal surface derivatives $\partial u_i/\partial \nu$ are also known, and the problem of determining u_i is reduced to Neumann problem for harmonic functions.

The consequence is that we can solve the elastic problem for pure boundary displacements or pure surface tractions whenever we know the values of the dilatation and of the rotations. The problem is thus reduced to finding these quantities in terms of the prescribed surface displacements or surface tractions by solving some linear integral equations. These equations can be obtained using a suitable application of Betti's reciprocal theorem, which states that when no body forces are applied,

$$\int_{\partial\kappa(\mathscr{B})} \sigma_{ij} u'_i \, dA_j = \int_{\partial\kappa(\mathscr{B})} \sigma'_{ij} u_i \, dA_j, \tag{4.37}$$

where u_j is a displacement field satisfying the equilibrium equations, and $\sigma_{ij} e_j = t_i$ are the corresponding surface tractions; u'_j is another displacement field and $\sigma'_{ij} e_j = t'_i$ are the associated surface tractions. For u'_i we may take the expressions

$$u'_i = \frac{\partial}{\partial x_i}\left(\frac{1}{r}\right), \tag{4.38}$$

where $r = \sqrt{x_i x_i}$ represents the distance of a generic point of $\kappa(\mathscr{B})$ from the origin, which is a point inside $\kappa(\mathscr{B})$. Since the function $1/r$ is harmonic, it is immediately verified that (4.38) is a solution of the equations of elasticity in the absence of body forces. However, the solution (4.38) is singular at the origin, and the region in which Betti's theorem can be applied is not the entire $\kappa(\mathscr{B})$, but $\kappa(\mathscr{B})$ bounded internally by a closed surface surrounding the origin, for instance a sphere of surface Σ whose radius tends to zero.

Since the values of e_i on Σ are $-(x_i/r)$, the contribution of Σ to the left-hand side of (4.37) is

$$\int_{\Sigma} \sigma_{ij} \frac{\partial}{\partial x_i}\left(\frac{1}{r}\right)\left(-\frac{x_j}{r}\right) d\Sigma, \tag{4.39}$$

where we have put $dA_j = e_j \, d\Sigma$, $d\Sigma$ being the area element of Σ. But, by using the constitutive equations (4.9) and (4.10), with σ_{ij} instead of T_{AB}, the integral (4.39) becomes

$$\int_{\Sigma} \left[\lambda \frac{\Delta}{r^2} + 2\mu\left(\frac{x_1^2}{r^4} e_{11} + \frac{x_2^2}{r^4} e_{22} + \frac{x_3^2}{r^4} e_{33}\right) + 2\mu\left(\frac{x_2 x_3}{r^4} e_{23} + \frac{x_3 x_1}{r^4} e_{13} + \frac{x_1 x_2}{r^4} e_{12}\right)\right] d\Sigma. \tag{4.40}$$

All integrals of the type $\int_{\Sigma} x_2 x_3 \, d\Sigma$ vanish, and each of the type $\int_{\Sigma} x_1^2 d\Sigma$ is equal to $\frac{4}{3}\pi r^4$. Therefore, in the limit when r tends to zero, the value of (4.40) is $4\pi(\lambda + \frac{2}{3}\mu)\Delta_0$, where Δ_0 denotes the value of Δ at the origin.

The contribution of Σ to the right-hand side of (4.37) can be evaluated by observing that

$$t'_1 = \sigma'_{1j} e_j = (2\mu e'_{11} + \lambda \Delta')\left(-\frac{x_1}{r}\right) + \mu e'_{12}\left(-\frac{x_2}{r}\right) + \mu e'_{13}\left(-\frac{x_3}{r}\right), \tag{4.41}$$

and analogous expressions hold for t'_2 and t'_3. In the last formula, the dilatation Δ' is obviously zero, and t'_1 may be written as

$$t'_1 = 2\mu\left(-\frac{x_1}{r}\frac{\partial}{\partial x_1} - \frac{x_2}{r}\frac{\partial}{\partial x_2} - \frac{x_3}{r}\frac{\partial}{\partial x_3}\right)\frac{\partial}{\partial x_1}\left(\frac{1}{r}\right).$$

The contribution of Σ thus becomes

$$2\mu \int_{\Sigma} \left(-2\frac{u_i x_i}{r^4}\right) d\Sigma.$$

In order to evaluate the last integral, we expand the functions u_i in the neighborhood of the origin

$$u_i = (u_i)_0 + x_j \left(\frac{\partial u_i}{\partial x_j}\right)_0 + \dots,$$

retaining the first powers in x_j. In the limit, when the radius of Σ tends to zero, the value of the integral is $-\frac{16}{3}\pi\mu\Delta_0$.

Equation (4.37) thus yields

$$4\pi(\lambda + 2\mu)\Delta_0 = \int_{\partial\kappa(\mathscr{B})} (\sigma'_{ij} u_i - \sigma_{ij} u'_i)\, dA_j. \tag{4.42}$$

The same conceptual procedure can be applied in order to find the rotations at an internal point of $\kappa(\mathscr{B})$, taken as the origin of the coordinate system. Let Σ denote the surface of a small sphere surrounding the origin, and then again apply Betti's theorem to the region $\kappa(\mathscr{B})$ deprived of the sphere having Σ as boundary. The procedure now differs from the one above. As an auxiliary state of displacement we take the state

$$(u''_1, u''_2, u''_3) = \left[0, \frac{\partial}{\partial x_3}\left(\frac{1}{r}\right), -\frac{\partial}{\partial x_2}\left(\frac{1}{r}\right)\right],$$

and the corresponding surface tractions t''_1, expressed in terms of e''_{1j} by the formulae (4.41). By repeating the calculation of the integrals over Σ in the reciprocal theorem (4.37), we should find that the contribution of the left-hand side vanishes and that of the right-hand side is $8\pi\mu(\tilde{\omega}_1)_0$, where $(\tilde{\omega}_1)_0$ denotes the value of $\tilde{\omega}_1$ at the origin. We therefore arrive at the formula

$$8\pi\mu(\tilde{\omega}_1)_0 = \int_{\partial\kappa(\mathscr{B})} (\sigma_{ij} u''_i - \sigma''_{ij} u_i)\, dA_j, \tag{4.43}$$

and corresponding results of the same type can be written for $\tilde{\omega}_2$, and $\tilde{\omega}_3$.

Since the choice of the origin inside $\kappa(\mathscr{B})$ is arbitrary, (4.42) and (4.43) give the dilatation Δ and the rotations $\tilde{\omega}_1$, $\tilde{\omega}_2$ and $\tilde{\omega}_3$ at each point in the interior of $\kappa(\mathscr{B})$ in terms of the surface values of the displacements and the tractions. In general, the surface values of both these quantities cannot be prescribed simultaneously, and thus the above formulae are of no use in solving the elastic problem, unless we are able to eliminate the unknown terms on the right-hand side of (4.42) and (4.43) with the aid of particular subsidiary solutions.

To get an idea of the procedure, let us consider (4.42) when the surface displacements u_i are given, but the surface tractions t_i are not. In this case we seek a displacement which is continuous in $\kappa(\mathscr{B})$, satisfies the equations of elasticity in the absence of body forces, and has the same values of u'_i on the boundary $\partial\kappa(\mathscr{B})$. Let us denote these displacements by $\bar{u}'_i$ and the corresponding stresses by $\bar{\sigma}'_{ij}$. We then apply the reciprocal theorem to the displacements u_i and $\bar{u}_i$, which have no singularities inside $\partial\kappa(\mathscr{B})$, and obtain

$$\int\limits_{\partial\kappa(\mathscr{B})} \bar{\sigma}'_{ij}u_i\,dA_j = \int\limits_{\partial\kappa(\mathscr{B})} \sigma_{ij}\bar{u}'_i\,dA_j = \int\limits_{\partial\kappa(\mathscr{B})} \sigma_{ij}u'_i\,dA_j.$$

We can, therefore, write (4.42) in the form

$$4\pi(\lambda+2\mu)\Delta_0 = \int\limits_{\partial\kappa(\mathscr{B})} (\sigma'_{ij}-\bar{\sigma}'_{ij})u_i\,dA_j. \tag{4.44}$$

With the introduction of the auxiliary state $\bar{u}'_i$ we have constructed a *Green's function* for the dilatation.

Finally, there is a third method that permits us to prove the existence of a solution of the displacement boundary value problem by means of linear combinations of given functions (Cosserat and Cosserat 1898, 1901).

Denoting $(\lambda+\mu)/\mu$ by τ, the equations of equilibrium have the form

$$\nabla^2 u_i + \tau\frac{\partial}{\partial x_i}\Delta = 0 \quad \text{on} \quad \kappa(\mathscr{B}). \tag{4.45}$$

We may suppose that there exists a sequence of real numbers $\tau_1, \tau_2, \ldots$, such that the system

$$\nabla^2 U_i^{(k)} + \tau_k\frac{\partial}{\partial x_i}\Delta^{(k)} = 0 \quad \text{on} \quad \kappa(\mathscr{B}),$$

for $k = 1, 2, \ldots$, allows nontrivial solutions $U_i^{(k)}$ which vanish at the surface. The dilatations $\Delta^{(k)}$ are harmonic functions, and we may verify that, if k' is different from k, $\Delta^{(k)}$ and $\Delta^{(k')}$ satisfy the following condition of orthogonality:

$$\int\limits_{\kappa(\mathscr{B})} \Delta^{(k)}\Delta^{(k')}\,dV = 0. \tag{4.46}$$

Having assumed the existence of the functions $U_i^{(k)}$ and the numbers τ_k, we can try to solve problem (4.45) by assuming for u_i an expansion of the form

$$u_i = u_i^0 + \sum_{k=1}^{\infty} B_k U_i^{(k)}, \tag{4.47}$$

where the B_k are constants to be determined and u_i^0 are harmonic functions. The dilatation corresponding to u_i^0 is $\Delta_0 = u_{i,i}^0$, and this is also a harmonic function. Substituting (4.47) in (4.45), we obtain

$$\sum_{k=1}^{\infty} B_k\nabla^2 U_1^{(k)} + \tau\frac{\partial}{\partial x_i}\left(\Delta_0 + \sum_{k=1}^{\infty} B_k\Delta^{(k)}\right) = 0,$$

which is equivalent to

$$-\sum_{k=1}^{\infty} \tau_k B_k\frac{\partial}{\partial x_i}\Delta^{(k)} + \tau\frac{\partial}{\partial x_i}\left(\Delta_0 + \sum_{k=1}^{\infty} B_k\Delta^{(k)}\right) = 0.$$

These equations are clearly satisfied provided that

$$\sum_{k=1}^{\infty}(\tau - \tau_k)B_k\Delta^{(k)} = \tau\Delta_0,$$

whence the constants B_k may be derived explicitly by multiplying the last equation by $\Delta^{(k')}(k' = 1, 2, \ldots)$ and integrating over $\kappa(\mathscr{B})$. Since (4.46) holds, we obtain the constants B_k using the formula

$$(\tau - \tau_k)B_k \int_{\kappa(\mathscr{B})} (\Delta^{(k)})^2 \, dV = \tau \int_{\kappa(\mathscr{B})} \Delta_0\Delta^{(k)} \, dV. \tag{4.48}$$

This method of determining the solution u_i is constructive because it gives an explicit representation of the solution, but it is more involved than it may at first appear, because not only must we find the coefficients B_k, but we must also show that the limits of the series satisfy the differential equations and the boundary conditions. In order to prove these results the series must converge uniformly together with its term-by-term partial derivatives up to second order. These proofs are intricate.

5. The Notion of a Loading Device[8]

The boundary conditions commonly used in continuum mechanics are those of place or traction in a purely mechanical problem and those of temperature or heat flux in purely thermal problem. There are, however, many situations in which the boundary conditions are of a different type. For instance, exterior tractions may not have a fixed direction at each point on the surface, but may follow, in a prescribed manner, the deformation of the surface element upon which they are acting; these are called "follower" tractions, to illustrate their particular behavior. More general than these are "configuration-dependent loadings," which are defined by the property that the load acting on a material point on the boundary can be calculated from an assigned function of the displacement at the point and within its neighborhood (Sewell 1967). As a further generalization, assume that the applied surface tractions are functions of the displacement, velocity, acceleration, and their gradients of first and higher order with respect to the position of a particle in the reference configuration. These loads are termed "motion-dependent" in order to indicate that the body interacts locally with the environment (Nemat-Nasser 1970).

An elegant way of treating all these situations is to introduce the concept of the "loading device," which is another body in contact with the body under consideration along a common piece of surface. Let us call the reference configuration of this latter body V and the reference configuration of the loading device R (Figure 5.1); and let C be the common boundary of the two regions. The boundary conditions on C account to some extent for the deformation about R on introducing some quantities defined on the interface C by suitably averaging the stresses and strains of R.

[8] The term "loading device" was introduced by Ericksen in his Baltimore lectures, but the detailed theory of its application is due to Batra (1972).

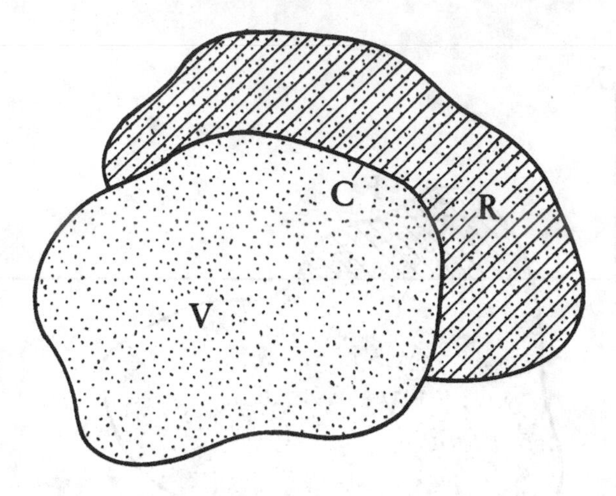

Fig. 5.1

For example, assigning zero surface displacements at the boundary surface of V is equivalent to saying that V is permanently glued to R, which is a rigid body incapable of suffering overall movement. An approximate but systematic procedure for deriving the boundary conditions on C from the effect of the overall deformation of R consists in averaging the field equations valid in the interior of R and transforming them into another set of equations defined on C, but equivalent in certain respects to the previous ones. A technique of this kind is customarily used to obtain balance equations for rods and shells, starting from the three-dimensional theory. The details are different, however, since in the present case it is not possible to take advantage of the particular shape of the body.

In order to avoid unnecessary complications, let us assume that C is a simply connected plane domain contained in the plane $X_3 = 0$, and that the region V is located in the half-space $X_3 < 0$ and region R is in the half-space $X_3 > 0$ (Figure 5.2). This assumption is less restrictive than may at first appear, because we can transform other problems into these terms by making a suitable choice of the curvilinear coordinates. We further assume that the straight lines $X_1 =$ constant, $X_2 =$ constant intersect $F \equiv \partial R - C$ at one point only.

For simplicity, we assume that F is *isolated*, that is to say there is no mechanical work or flux of energy through it. The contact between R and V is called *intimate* if the following continuity conditions hold at the common surface C:

$$\left.\begin{aligned} [x_i(X_A, t)] &= 0, \\ [t_i(X_A, t)] &= 0, \\ [\theta(X_A, t)] &= 0, \\ [q_i(X_A, t)] &= 0, \end{aligned}\right\} \quad \forall (X_1, X_2) \in C,\ t > 0, \tag{5.1}$$

where t_i and q_i are the surface tractions and the heat flux per unit coordinate area, respectively. In thermomechanical terms, conditions (5.1) state that there is no slip at the interface, that the contiguous part of the body and the loading device are at the

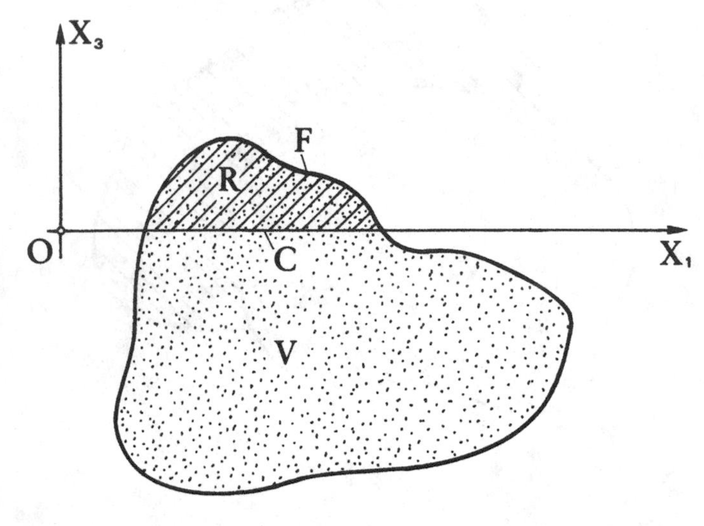

Fig. 5.2

same temperature, and that the surface tractions and the heat flux are continuous across C, respectively. These contact loading devices do not embrace all conceivable kinds of surface contrivance. For example, during the deformation process, different parts of R might come into contact with ∂V, as happens in a deformable roller bearing pushed against a rigid plane.

The idea governing the notion of a loading device is that of replacing the influence of R on V by a set of quantities only over C, which consequently play the same role as classical boundary conditions prescribed on the surface of V, which is regarded as isolated.

In order to illustrate this procedure of averaging, we set out the details for the case when the loading device is a homogeneous, isotropic, linear elastic half-space occupying the entire region $X_3 > 0$. C is then glued to R along a plane domain C which is a proper subset of the bounding plane ∂R of R. The reference configurations for R and V are those occupied by them at the instant of gluing. The reference configuration for R is assumed to be natural, that is, without initial stresses and with constant initial temperature, but no assumption is made for either the constitutive equations for V or for its reference configuration.

We discuss a purely mechanical static problem in which the deformation of the half-space R is time independent and its temperature remains constant. If the loading device is assumed to be free of body forces and the part $\partial R - C$ of its boundary is isolated, R can be deformed either by applying surface tractions on that part of the boundary of V not glued to R or by varying the body force field in V. The elastic displacements in R are determined by the Navier equations (4.32) with zero body forces:

$$\mu \nabla^2 u_i + (\lambda + \mu) \frac{\partial}{\partial X_i}(u_{j,j}) = 0, \text{ in } R,$$

together with the boundary conditions

$$t_i = 0 \quad \text{on } \partial R - C, \tag{5.2}$$

$$t_i = {}^{\ell}t_i \quad \text{on } C, \tag{5.3}$$

where ${}^{\ell}t_i$ represent the surface tractions exerted by V on R. In addition, we must prescribe a decay law for the tractions at infinity; this is of the form

$$|t_i| = {}_0(1) \text{ as } |X_i| \to \infty. \tag{5.4}$$

In classical problems the surface tractions are prescribed, but here they are quantities that must be determined.

For this purpose, we temporarily suppose that ${}^{\ell}t_i$ are known. Thus the displacements in R permit the explicit representation[9]

$$u_\alpha = \frac{1}{4\pi\mu}\int_C \left[\frac{\delta_{i\alpha}}{r} + (\delta_{i\beta}(X_\beta - X'_\beta) + \delta_{i3}X_3)\frac{1}{r^3}(X_\alpha - X'_\alpha) + \frac{\mu}{\lambda+\mu}\frac{\partial}{\partial X_i}\left(\frac{x_\alpha - X'_\alpha}{X_3 + r}\right)\right]{}^{\ell}t_i\, dA(X'_\alpha),$$

$$u_3 = \frac{1}{4\pi\mu}\int_C \left[\frac{\delta_{i3}}{r} + (\delta_{i\beta}(X_\beta - X'_\beta) + \delta_{i3}X_3)\frac{1}{r^3}X_3 + \frac{\mu}{\lambda+\mu}\frac{\partial}{\partial X_i}\ell n(r + X_3)\right]{}^{\ell}t_i\, dA(X'_\alpha), \tag{5.5}$$

where

$$r^2 = (X_\alpha - X'_\alpha)(X_\alpha - X'_\alpha) + X_3^2,$$

and $\alpha = 1$ or 2, $i = 1, 2$ or 3. Since these kernels have singularities of the order l/r, which are integrable, we can calculate the surface displacements taking the limits

$${}^{\ell}u_i(\bar{X}_1, \bar{X}_2, 0) = \lim_{\substack{X_\alpha \to \bar{X}_\alpha \\ X_3 \to 0}} u_i(X_1, X_2, X_3),$$

and obtain

$${}^{\ell}u_\alpha = \frac{1}{4\pi\mu}\int_C \left[\frac{\lambda+2\mu}{\lambda+\mu}\frac{\delta_{i\alpha}}{r_0} - \delta_{i3}\frac{\mu}{\lambda+\mu}\frac{(\bar{X}_\alpha - X'_\alpha)}{r_0^2} + \delta_{i\beta}\frac{\lambda}{\lambda+\mu}(\bar{X}_\alpha - X'_\alpha)(\bar{X}_\beta - X'_\beta)\frac{1}{r_0^3}\right]{}^{\ell}t_i\, dA(X'_\alpha)$$

$${}^{\ell}u_3 = \frac{1}{4\pi\mu}\int_C \left[\frac{\lambda+2\mu}{\lambda+\mu}\frac{\delta_{i3}}{r_0} + \delta_{i\alpha}\frac{\mu}{\lambda+\mu}(\bar{X}_\alpha - X'_\alpha)\frac{1}{r_0^2}\right]{}^{\ell}t_i\, dA(X'_\alpha), \tag{5.6}$$

where

$$r_0^2 = (\bar{X}_\alpha - X'_\alpha)(\bar{X}_\alpha - X'_\alpha).$$

[9] Known as the Boussinesq–Cerruti solution (see Love 1927, Art. 134, 166).

Since R is assumed to be permanently glued to V, the first two conditions (5.1) hold at the contact surface C, and, consequently, from (5.6) we can derive the boundary conditions for the part C of ∂R. But, in order to write these conditions explicitly, we should invert (5.6). However, the inversion is not simple, and we therefore modify the boundary conditions of our original problem by imposing (instead of (5.2), (5.3) and (5.4)) the following conditions:

$$u_i = 0 \quad \text{on } \partial R - C, \tag{5.7}$$

$$u_i = {}^{\ell}u_i \quad \text{on } C, \tag{5.8}$$

$$|u_i| = 0(1) \quad \text{for } |X_i| \to \infty. \tag{5.9}$$

The solution to this problem, which is simpler than (5.5), can be written explicitly:[10]

$$u_i = \frac{1}{2\pi}\int_C \left[-\frac{\partial}{\partial X_3}\left(\frac{1}{r}\right)\delta_{ik} + \bar{\sigma}X_3\frac{\partial^2}{\partial X_i \partial X_k}\left(\frac{1}{r}\right)\right]{}^{\ell}u_k\, dA(X'_\alpha), \tag{5.10}$$

where

$$\bar{\sigma} = \frac{\lambda + \mu}{\lambda + 3\mu}.$$

The kernel has a singularity of the type $1/r^2$, but the displacements in the interior of R, where $X_3 > 0$, are well defined. As we need to know the tractions on C, we must proceed to evaluate the gradients of u_i at points of C. The tangential derivatives $u_{i,\alpha}$ can be calculated directly from the boundary data (5.8). Calculation of the normal derivatives $u_{i,3}$ is more difficult, however. We start by observing that, by virtue of (5.7), the region of integration in (5.10) can be replaced by ∂R. In addition, by exploiting the transformation formulae for double integrals from Cartesian to polar coordinates, we arrive at the results

$$\begin{aligned}
&-\frac{1}{2\pi}\int_{\partial R}\frac{\partial}{\partial X_3}\left(\frac{1}{r}\right)dA(X'_\alpha) = 1,\\
&\int_{\delta R}(X_\alpha - X'_\alpha)\frac{\partial}{\partial X_3}\left(\frac{1}{r}\right)dA\,(X'_\alpha) = 0,\\
&\int_{\partial R}\frac{\partial^2}{\partial X_i \partial X_k}\left(\frac{1}{r}\right)dA\,(X'_\alpha) = 0,\\
&\frac{1}{2\pi}\int_{\partial R}(X_\alpha - X'_\alpha)\frac{\partial^2}{\partial X_i \partial X_k}\left(\frac{1}{r}\right)dA\,(X'_\alpha) = 2\delta_{3(i}\delta_{k)\alpha},
\end{aligned} \tag{5.11}$$

where $\delta_{3(i}\delta_{k)\alpha}$ denotes symmetrization with respect the indices i, k.

By definition, ${}^{\ell}u_{i,3}$ is the limit

$${}^{\ell}u_{i,3} = \lim_{X^3 \to 0^+}\frac{u_i(X_\alpha, X_3) - u_i(X_\alpha, 0)}{X_3},$$

[10] In this case too the result is due to Boussinesq (Love 1927, Art. 134, 166).

and, therefore, from (5.10) we get

$$ {}^{\ell}u_{i,3} = \lim_{X^3\to 0^+} \frac{1}{X_3}\left\{\frac{1}{2\pi}\int_{\partial R}\left[-\frac{\partial}{\partial X_3}\left(\frac{1}{r}\right)\delta_{ik} + \bar{\sigma}X_3\frac{\partial^2}{\partial X_i \partial X_k}\left(\frac{1}{r}\right)\right]{}^{\ell}u_k\, dA(X'_\alpha) - u_i(X_\alpha, 0)\right\}. \tag{5.12} $$

Assuming the ${}^{\ell}u_i$ is differentiable almost everywhere in ∂R, we consider the truncated Taylor expansion

$$ {}^{\ell}u_i(X_\alpha X'_\alpha) = {}^{\ell}u_i(X'_\alpha, 0) - {}^{\ell}u_i(X_\alpha, 0) - (X_\beta - X'_\beta){}^{\ell}u_{i,\beta}(X_\alpha, 0). $$

Now, making use of the relations (5.11), we can rewrite (5.12) as

$$ \begin{aligned} {}^{\ell}u_{i,3}(X_\alpha, 0) &= \lim_{X^3\to 0^+} \frac{1}{X_3}\left\{\frac{1}{2\pi}\int_{\partial R}\left[-\frac{\partial}{\partial X_3}\left(\frac{1}{r}\right)\delta_{ik} + \bar{\sigma}X_3\frac{\partial^2}{\partial X_1 \partial X_k}\left(\frac{1}{r}\right)\right]{}^{\ell}u_k dA(X'_\alpha) \right. \\ &\quad \left. + 2\bar{\sigma}X_3\frac{\partial}{dX_\alpha}\,{}^{\ell}u_k(x_\alpha, 0)\delta_{3(i}\delta_{k)\alpha}\right\} \\ &= 2\bar{\sigma}\frac{\partial}{\partial X_\alpha}{}^{\ell}u_k\,(X_\alpha, 0)\delta_{3(i}\delta_{k)\alpha} + \int_{\partial R} K_{ij}(X_\alpha X'_\alpha){}^{\ell}u_j(X_\alpha X'_\alpha)\, dA(X'_\alpha), \end{aligned} \tag{5.13} $$

where

$$ K_{ij}(X_\alpha X'_\alpha) = \frac{1}{2\pi}\left[\frac{\delta_{ij}}{r_0^3} + \bar{\sigma}\delta_{i\beta}\frac{\partial^2}{\partial X_\beta \partial X_\alpha}\left(\frac{1}{r_0}\right)\delta_{\alpha j}\right]. \tag{5.14} $$

In deriving (5.14) from (5.13) we have interchanged the limit and integration, which is permitted because the singularity of the integrand has been reduced to l/r. With the definitions

$$ E_{i3j}(\bar{X}_\alpha, 0) = -\int_{\partial R - C} K_{ij}(\bar{X}_\alpha, X'_\alpha)\, dA(X'_\alpha), $$

$$ G_{i3j\alpha}(\bar{X}_\alpha, 0) = 2\bar{\sigma}\delta_{3(i}\delta_{j)\alpha} - \int_{\partial R - C} (\bar{X}_\alpha - X'_\alpha)K_{ij}(\bar{X}_\alpha, X'_\alpha)\, dA(X'_\alpha), $$

we can write (5.13) as

$$ {}^{\ell}u_{i,3}(\bar{X}_\alpha, 0) = E_{i3k}{}^{\ell}u_k + G_{i3k\alpha}{}^{\ell}u_{k,\alpha} + \int_C K_{ij}(\bar{X}_\alpha, X'_\alpha){}^{\ell}u_j(\bar{X}_\alpha, X'_\alpha)\, dA(X'_\alpha). \tag{5.15} $$

Note that E_{i3k} and $G_{i3j\alpha}$ are well defined at all interior points of C, although $G_{i3j\alpha}$ may not be well defined for some infinite regions. Remembering that the outer unit normal to C has the components $(0, 0, -1)$, and using the constitutive equations

$$ \sigma_{3\alpha} = \mu(u_{3,\alpha} + u_{\alpha,3}), \qquad \alpha = 1, 2, $$

$$ \sigma_{33} = 2\mu u_{3,3} + \lambda(u_{\alpha,\alpha} + u_{3,3}), $$

we obtain the surface tractions on C:

$$
\begin{aligned}
{}^{\ell}t_\alpha(\bar{X}_\alpha, 0) = &- \mu[{}^{\ell}u_{3,\alpha} + E_{\alpha 3k}{}^{\ell}u_k + G_{\alpha 3k\beta}{}^{\ell}u_{k,\beta}] \\
&- \mu \int_C K_{\alpha j}(\bar{X}_\alpha, X'_\alpha){}^{\ell}u_j(\bar{X}_\alpha, X'_\alpha)\, dA(X'_\alpha), \\
{}^{\ell}u_3(\bar{X}_\alpha, 0) = &- (2\mu + \lambda)(E_{33k}{}^{\ell}u_k + G_{33k\alpha}{}^{\ell}u_{k,\alpha}) \\
&- \frac{\lambda}{2\pi} \int_C \frac{1}{r_0^3}{}^{\ell}u_3(X_\alpha, X'_\alpha)\, dA(X'_\alpha).
\end{aligned} \tag{5.16}
$$

Since the continuity (5.1) holds on C, we can replace ${}^{\ell}u_i, {}^{\ell}u_{i,\alpha}, \ldots,$ in the above expressions by ${}^{v}u_i, {}^{v}u_{i,\alpha}, \ldots,$ which are the displacements and their tangential gradients evaluated for the body V. This substitution yields:

$$
\begin{aligned}
{}^{v}t_\alpha(\bar{X}_\alpha, 0) =& \mu[{}^{v}u_{3,\alpha} + E_{\alpha 3k}{}^{v}u_k + G_{\alpha 3k\beta}{}^{v}u_{k,\beta}] + \mu \int_C K_{\alpha j}(\bar{X}_\alpha, X'_\alpha){}^{v}\bar{u}_j(\bar{X}_\alpha, X'_\alpha)\, dA(X'_\alpha) \\
{}^{v}t_3(\bar{X}_\alpha, 0) =& (2\mu + \lambda)[E_{33k}{}^{v}u_k + G_{33k\beta}{}^{v}u_{k\beta}] + \frac{\lambda}{2\pi}\int_C \frac{1}{r_0^3}{}^{v}\bar{u}_3(X_\alpha, X'_\alpha)\, dA(X'_\alpha).
\end{aligned} \tag{5.17}
$$

These boundary conditions show that, when the body V is loaded by a deformable loading device, the mechanical boundary conditions are a mixture of those of traction and those of place. The loading device interacts nonlocally, in the sense that the surface tractions at a point on C depend on the deformation of all points on C and not just on the deformation of its neighborhood. This form of the surface tractions is in sharp contrast to the customary way of prescribing the loads in classical problems. Furthermore, the form of (5.17) reveals that the surface tractions at C depend on the field of displacements defined on C and not on the normal derivatives of the displacements. Another unexpected property of ${}^{v}t_i$ is that they depend on the elastic constants λ and μ of the loading device and not on those of the body.

In the particular example discussed so far we have succeeded in obtaining an explicit expression for the elastic solution in the half-space R while the associated stresses inside R necessarily satisfy the equilibrium equations. But, if R has a more general shape, the construction of the solution may be complicated or impossible. In these situations we can again apply the procedure leading to the definition of the surface traction ${}^{v}t_i$, provided that we make the following approximation.[11] We consider a material tube $T \subset R$, with one end consisting of the points $\sigma \in C$ and the lateral surface ∂T_0 consisting of the points $\{X_\alpha \in R, (X_\alpha, 0) \in \partial\sigma\}$. Denoting $T \cap F$ by σ_0, the boundary of T is given by

$$\partial T = \sigma \cup \partial T_0 \cup \sigma_0. \tag{5.18}$$

We now write the balance equations, which in the present case are the equilibrium equations for the stresses, over an arbitrary tube T, exploiting the circumstance that the integrals of the form $\int_T \sigma_{ij,i}\, dV$ can be written as $\int_\sigma d\sigma \int_0^{X_3(X_\alpha)} \sigma_{ij,i}\, dV$, where $X_3 = X_3(X_\alpha)$ denotes the X_3 coordinate of the points of σ_0. Since we have assumed

[11] Called "averaging" (see Batra 1972).

that the straight lines parallel to the X_3 axis intersect F at one point only, the function $X_3 = X_3(X_\alpha)$ is uniquely defined and assumed to be continuous. The balance equations for the stresses are thus

$$\int_\sigma d\sigma \left[\int_0^{X_3} \sigma_{\alpha j,\alpha}\, dX_3 + [\sigma_{3j}]_0^{X_3(X_\alpha)} \right] = 0,$$

or putting (5.19)

$$\Phi_{j\alpha} = \int_0^{X_3} \sigma_{\alpha j}\, dX_3, \qquad q_j = [\sigma_{3j}]_0^{X_3(X_\alpha)},$$

they can be written

$$\int_\sigma [\phi_{j\alpha,\alpha} + q_j]\, d\sigma = 0.$$

With the postulate that these equations hold for every element σ and the integrand is continuous, we arrive at the local averaged equations

$$\phi_{j\alpha,\alpha} + q_j = 0. \tag{5.20}$$

When the exact solution for R is difficult to obtain, we may content ourselves with finding a state of stress in R which satisfies equations (5.19) instead of finding the stresses associated with three-dimensional elasticity. In the example treated above, the loading device was assumed to be a semi-infinite body. The other extreme case would be that of a very thin shell glued to V along the surface C. A loading device of this type is commonly called a *reinforcement*.

6. An Example of Alternative Modeling

The description of a mechanical phenomenon is never unique, since we are always at liberty to select some aspects of the experimental data and to disregard others. The crucial phase in setting up a model, which is both explanatory and fits the experimental facts, is that of deciding on the significant mathematical variables. The choice of the appropriate variables depends on the nature of the result we wish to extract from the model. A crude representation may not be adequate to offer the information we need, but a sophisticated theory may give too many unessential details. A good model must be a compromise between economy and accuracy.

In order to illustrate the different ways in which a problem of mechanics may be formulated, let us analyse the impact of a heavy ball falling onto the upper end of a straight rod, which is fixed at the lower end and free at the upper end, as indicated in Figure 6.1. Let us denote the mass of the ball by M, the initial distance of the center of mass of the ball from the upper end of the bar by h, and the length of the bar by H. The bar is cylindrical with a cross-sectional area A, and is made of an elastic material of Young's modulus E, and has a density per unit length equal to ρ.

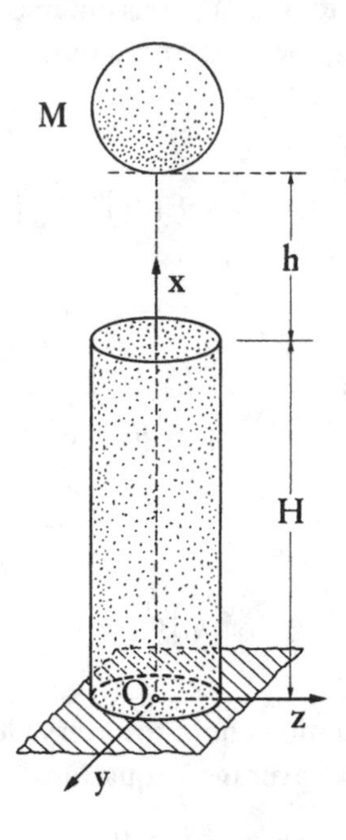

Fig. 6.1

A crude model of the idealized experiment of letting the ball fall onto the top of the bar rests on the assumption that the bar is so slightly deformable that we may regard it as perfectly rigid. The consequence of this is that the upper end of the bar behaves like a rigid wall against which the ball collides. Provided that the duration of the impact is large compared with the longest period of free vibration of the impacting sphere, all points on the sphere have approximately the same velocity. The motion is thus completely known when we have determined this vertical velocity $v(t)$ before and after the impact. Let v_0 $(=\sqrt{2gh})$ be the velocity of the sphere at the instant of the impact and $P(t)$ be the thrust between the bar and the sphere during compression $(v \geqslant 0)$. The equation governing the motion during the period of compression is

$$\frac{d}{dt}(Mv) = P(t),$$

which, when integrated over the interval of compression $(0, \tau_1)$, gives

$$[Mv]_0^{\tau_1} = -Mv_0 = \int_0^{\tau_1} P(t)\,dt = I_1. \qquad (6.1)$$

The following interval of restitution is (τ_1, τ_2) during which time the sphere reaches its final velocity of rebound v_2. If Q is the thrust exerted on the sphere during this second period, the balance of momentum requires that

$$[Mv]_{\tau_1}^{\tau_2} = Mv_2 = \int_{\tau_1}^{\tau_2} Q(t)\, dt = I_2. \tag{6.2}$$

We cannot obtain v_2 from (6.1) and (6.2) unless we impose a constitutive relationship between I_1 and I_2. For instance, if we want to characterize a fully elastic impact we must assume that $I_1 = I_2$, and consequently we obtain $v_2 = v_0$. It is useful to observe that the model analyzed so far is adequate for determining the velocity of rebound v_2, but is too rough for determining the duration of the impact and the thrust during contact. In order to overcome this indeterminacy, let us remove the hypothesis of rigidity of the bar by assuming the longitudinal stress σ_x to be proportional to the longitudinal strain ε_x according to the law $\sigma_x = E\varepsilon_x$, where E is the Young's modulus. If we neglect inertia and suppose that the vertical displacement of the rod at $t = \tau_1$ has the form $u = (x/H)u_1$, u_1 being the displacement of the top of the bar, the equation of conservation of energy, for times $t = 0$ and $t = \tau_1$, is

$$\frac{1}{2}Mv_0^2 = \frac{1}{2}EAH\left(\frac{u_1}{H}\right)^2. \tag{6.3}$$

From (6.3) we can immediately determine not only the largest contraction of the bar during compression, but also the stress σ_x at $t = \tau_1$. In particular, we can write

$$\frac{u_1}{H} = \frac{v_0}{c}\sqrt{\frac{M}{M_b}}, \tag{6.4}$$

where $c = \sqrt{E/\rho}$ is the velocity of sound in the bar and $M_b = \rho AH$ is the mass of the bar.

A more refined model can be introduced for considering the inertia of the material constituting the bar, provided that we assume the velocities of its points to vary linearly along the longitudinal x axis and to be zero at the bottom. In order to evaluate the common velocity of the mass of the ball and the top of the bar after first contact, it is sufficient to apply the principle of conservation of momentum immediately before and after the time $t = 0$. We denote the velocity of the mass by $\bar{u}$ when the velocity of any point of the bar is $\bar{u}(x/H)$ (see Szabó 1963, Sec. 24). The equation of the balance of momentum thus gives

$$Mv_0 = (M\bar{u} + \int_0^H \rho A\bar{u}\frac{x}{H}\, dx) = (M + \frac{1}{2}\rho AH)\bar{u} = (M + \frac{1}{2}M_b)\bar{u}, \tag{6.5}$$

where M_b is the mass of the bar. At the time τ_1, the end of the period of compression, the kinetic energy is totally converted into strain energy stored along the bar. Hence the equation of the conservation energy becomes

$$\frac{1}{2}\left[M\bar{u}^2 + \int_0^H \rho A\bar{u}\left(\frac{x}{H}\right)^2 dx\right] = \frac{1}{2}EAH\left(\frac{u_1'}{H}\right)^2, \tag{6.6}$$

u_1' being the new value of the displacement of the top of the bar. By substituting the value of $\bar{u}$ given by (6.5) and replacing ρAH with M_b, we obtain

$$\left(M + \frac{1}{3}M_b\right)\frac{M^2 v_0^2}{(M + \frac{1}{2}M_b)^2} = c^2 M_b \left(\frac{u_1'}{H}\right)^2,$$

with $c^2 = E/\rho$, and formula (6.4) must be replaced by[12]

$$\frac{u'}{H} = \frac{v_0}{c}\sqrt{\frac{M}{M_b}\frac{1 + \frac{1}{3}\frac{M_b}{M}}{\left(1 + \frac{1}{2}\frac{M_b}{M}\right)^2}}. \tag{6.7}$$

If, however, we need a description of the impact instant by instant and we want to know the displacements of the points of the bar avoiding the unjustified assumption that they vary linearly along the x axis, we must start from the equation of momentum written for each cross-section provided that the planes $x =$ constant move together in the x direction.[13] Let $u(x, t)$ be the displacement at time t of a cross-section the equilibrium position of which is at x. Since the elongation per unit length $\varepsilon_x = \partial u/\partial x$, the normal force T at x is simply

$$T = \sigma_x A = EA\,\frac{\partial u}{\partial x},$$

and the equation of the balance of momentum becomes

$$c^2\,\frac{\partial^2 u}{\partial x^2} = \frac{\partial^2 u}{\partial t^2}. \tag{6.8}$$

with $c^2 = E/\rho$. The terminal condition at $x = 0$ is $u = 0$; the condition at $x = H$ is an equation of balance between the normal force T and the inertial force exerted by the striking mass, and is valid for the whole interval of time during which the mass remains in contact with the section $x = H$. This condition reads

$$-T = -EA\,\frac{\partial u}{\partial x} = M\,\frac{\partial^2 u}{\partial t^2}.$$

Alternatively, by introducing $M_b = \rho AH$, and the ratio $m = M/M_b$,

$$\frac{\partial^2 u}{\partial t^2} = -\frac{c^2}{mH}\frac{\partial u}{\partial x}. \tag{6.9}$$

For the initial conditions at $t = 0$,

$$u(x, 0) \equiv 0, \tag{6.10}$$

and the initial velocity is also zero for all values of x between 0 and H. At $x = H$ we have

$$\lim_{t \to 0^+} \frac{\partial u}{\partial t}(H, t) = -v_0, \tag{6.11}$$

since the velocities of the ball and end of the bar are identical at the moment of impact.

[12] A slightly different formula was obtained by Szabó (1963, Sec. 24).

[13] This theory is due to Saint-Venant, but, according to Love (1927, Art. 281), a new and powerful method of solution was derived by Bromwich.

To obtain a solution it is necessary to write the general integral of (6.8) in the form of d'Alembert (1747),

$$u(x, t) = p(x + ct) + q(x - ct), \tag{6.12}$$

where p and q are any differentiable functions of the single variables $\xi = x + ct$, $\eta = x - ct$, respectively. In order that the second partial derivatives appearing in (6.8) exist, p and q must be twice-differentiable functions, except at the points $x = 0$ and $x = H$, where the velocity may be discontinuous. In order to solve the initial boundary-value problem, we must find the functions p and q. A first condition is immediately obtained by setting $t = 0$ in (6.12). The second condition is obtained by differentiating (6.12) using the chain rule and setting $t = 0$. The initial conditions then become

$$p(x) + q(x) = 0 \quad \text{for} \quad 0 < x \leqslant H, \tag{6.13a}$$

$$cp'(x) - cq'(x) = 0 \quad \text{for} \quad 0 < x < H. \tag{6.13b}$$

After differentiating the first of these equations and eliminating $q'(x)$, we find that

$$2cp'(x) = 0,$$

and hence, by integration,

$$p(\xi) = K \quad \text{for} \quad 0 < \xi < H, \tag{6.14}$$

where K is a constant. From (6.13a) with $x = H$, we then have

$$q(\eta) = -K \quad \text{for} \quad 0 < \eta < H. \tag{6.15}$$

Noting that K cancels when $p(\xi)$ and $q(\eta)$ are added, we may put $K = 0$. Therefore the general solution $u(x, t)$ can be written as

$$u(x, t) = 0,$$

provided that $0 \leqslant x + ct < H$ and $0 < x - ct < H$. Then, in the triangular region

$$t < \frac{x}{c}, \; t < \frac{H - x}{c}, \quad t \geqslant 0,$$

the solution is uniquely determined by the initial data and is precisely zero for $0 < x < H$.

In order to construct the solution for longer times, we write the boundary conditions at $x = 0$ and H:

$$p(ct) + q(-ct) = 0 \quad \text{for} \quad t \geqslant 0,$$

$$p''(H + ct) + q''(H - ct) = -\frac{1}{mH}[p''(H - ct)] \quad \text{for} \quad t \geqslant 0.$$

On setting $\zeta = -ct$, the first of these conditions becomes

$$q(\zeta) = -p(-\zeta) \quad \text{for} \quad \zeta < 0. \tag{6.16}$$

At the other end, if we let $\zeta = H + ct$, the condition at $x = H$ becomes

$$p''(\zeta) + \frac{1}{mH}p'(\zeta) = -[q''(2H - \zeta) + \frac{1}{mH}q'(2H - \zeta)] \quad \text{for} \quad \zeta > H. \tag{6.17}$$

We already know the solution (6.14) for $0 < \xi < H$. Thus, from (6.16) we obtain $q(\eta)$ in the interval $-H < \eta < 0$, by observing that $0 < -\eta < H$, and using (6.14) with $\eta = \zeta$:

$$q(\eta) = 0 \quad \text{for} \quad -H < \eta < 0. \tag{6.18}$$

In order to exploit the boundary condition (6.17), we first obtain $p'(\zeta)$ by integration

$$p'(\zeta) = C\,e^{-\zeta/mH} - e^{-\zeta/mH}\int_0^{\zeta} e^{\bar{\xi}/mH}\left[q''(2H-\bar{\xi}) + \frac{1}{mH}q'(2H-\bar{\xi})\right]d\bar{\xi} \quad (\zeta \geqslant H), \tag{6.19}$$

where C is a constant.

We now have $q(\eta) = 0$ for $-H < \eta < H$. Therefore, if $H < \xi < 3H$, the expression under the integral vanishes and we can put $\xi = \zeta$ in (6.19) to obtain

$$p'(\xi) = C\,e^{-\xi/mH}, \quad H < \xi < 3H. \tag{6.20}$$

Now the condition (6.11), written for $t = 0$ and $x = H$, that is for $\xi = \eta = H$, gives

$$cp'(H) = cC\,e^{-\frac{1}{m}} = -v_0, \quad \text{or} \quad C\,e^{-1/m} = -\frac{v_0}{c}. \tag{6.21}$$

Hence we have the expression

$$p'(\xi) = -\frac{v_0}{c}\,e^{-(\xi-H)/mH}, \quad \text{for} \quad H < \xi < 3H.$$

We note that $p'(\xi)$ is discontinuous at $\xi = H$. Conversely, $p(\xi)$ is continuous at $\xi = H$, and vanishes. Therefore, by integration, we find

$$p(\xi) = -\frac{mHv_0}{c}(1 - e^{-(\xi-H)/mH}), \quad \text{for} \quad H < \xi < 3H. \tag{6.22}$$

Since we now have $p(\xi)$ for $0 < \xi < 3H$, with $\zeta = \eta$ (6.16) gives $q(\eta)$ for $-3H < \eta < 0$. In particular $q(\eta)$ vanishes for $-H < \eta < 0$, and so we obtain

$$q(\eta) = \frac{mHv_0}{c}\left[1 - e^{(\eta+H)/mH}\right], \quad \text{for} \quad -3H < \eta < -H. \tag{6.23}$$

Hence we find

$$q''(2H-\eta) + \frac{1}{mH}q'(2H-\eta) = -2\frac{v_0}{cmH}\,e^{(3H-\eta)} \quad \text{for} \quad -3H < \eta < H. \tag{6.24}$$

Again, having $q(\eta)$ in the interval $-3H < \eta < -H$, equation (6.19) permits us to determine $p'(\xi)$for $3H < \xi < 5H$. The result of the integration is

$$\begin{aligned} p'(\xi) &= C_1 e^{-\xi/mH} + \frac{2v_0}{cmH}\,e^{-\xi/mH}\int_{3H}^{\xi} e^{\bar{\xi}/mH}\,e^{(3H-\bar{\xi})/mH}\,d\bar{\xi} \\ &= C_1\,e^{-\xi/mH} + \frac{2v_0}{cmH}(\xi - 3H)\,e^{-(\xi-3H)/mH}, \end{aligned}$$

where C_1 is a new constant. The condition of continuity of velocity at $x - H$ and $t = 2H/c$, that is for $\xi = 3H$ and $\eta = -H$, gives

$$-v_0\,e^{-2/m} = cC_1\,e^{-3/m} + v_0,$$

and hence

$$C_1 = -\frac{v_0}{c}(e^{3/m} + e^{1/m}).$$

This process can be repeated to give $q(\eta)$ for any $\eta < 0$ and $p(\xi)$ for any $\xi > 0$, and the solution is determined for all $0 < x < H$ and $t > 0$. The constants like $C, C_1, \ldots,$ which appear in the formulae giving $p'(\xi)$ in the intervals $5H < \xi < 7H$, etc., can be determined by imposing continuity of velocity at $x = H$, $t = 4H/c$, $t = 6H/c$, etc., that is for $\xi = 5H, 7H, \ldots$ and $\eta = -3H$, $-5H, \ldots$.

The function $p(\xi)$ is found by integrating $p'(\xi)$, and the constant of integration is determined by the condition that there is no sudden change in the displacement at $x = H$.

The solution represents the intuitive structure of the mechanism of the impact. At the instant of impact a compression wave leaves the upper end of the bar and travels towards the fixed end where it is reflected, generating a steady series of waves traveling back and forth along the rod. In constructing the solution the assumption has been permanently maintained that the striking mass is attached to the rod for all the time after impact. But there is also the possibility of abandoning this assumption and determining the time at which detachment occurs, when the velocity of the mass at the struck end becomes positive. In the time interval $0 < t < 2H/c$, the velocity is equal to

$$\frac{\partial u}{\partial t}(H, t) = cp'(H + ct) - cq'(H - ct) = -v_0 e^{-ct/mH},$$

which is negative. The mass might then detach itself during the interval $2H/c < t < 4H/c$, when the velocity is

$$\begin{aligned}\frac{\partial u}{\partial t}(H, t) &= cC_1\,e^{-(H+ct)/mH} + \frac{2v_0}{mH}(ct - 2H)\,e^{-(ct-2H)/mH} - v_0\,e^{-(ct-2H)/mH}\\ &= -\,v_0\,e^{-ct/mH}\left[1 + 2e^{2/m}\left(1 - \frac{ct - 2H}{mH}\right)\right].\end{aligned}$$

This expression vanishes when $2ct/mH = (4/m) + 2 + e^{-2/m}$, and the above equation admits a root in the interval $2H < ct < 4H$, provided that $2 + e^{-2/m} < 4/m$, or, equivalently, provided that $m < 1.73$. If $m > 1.73$ we must see whether or not the impact ceases during the subsequent interval $4H/c < t < 6H/c$, and so on.

Saint-Venant's theory is adequate for describing the motion of the cross-sections of the rod after the first impact. It is, however, incapable of accounting for the influence of the inertial motion of the lateral cross-sections when, extended or contracted in their own planes. This effect may be relevant if the transverse dimensions of the section are not negligible compared to the length of the bar. In order to account for the lateral motion, we may assume that the longitudinal displacement $u(x, t)$ is accompanied by two components of displacement[14]

[14] This extension of the theory was proposed by Pochhammer (1876), and Chree (1886).

$$v = -\sigma y u_x, \quad w = -\sigma z u_x,$$

where y and z are the coordinates of any point of the cross-section, referred to a system of yz axes with the center of mass of the section as the origin, and σ is the Poisson ratio for the material. In terms of these displacements the kinetic energy per unit length is

$$T = \frac{1}{2}\rho A(u_t^2 + \sigma^2 K^2 u_{xt}^2),$$

K being the radius of gyration of the cross-section about the x axis. The potential energy per unit length is

$$W = \frac{1}{2} EAu_x^2.$$

The equation of motion can be deduced from the Hamilton principle, which states that the variation of the integral

$$\int_{t_0}^{t_1} dt \int_0^H (T - W)\, dx,$$

taken between an initial and a final value t_0 and t_1 must be zero for all variations of u vanishing at t_0 and t_1, and satisfying the boundary condition $u = 0$ at $x = 0$. If we form the variation of u, integrate by parts, and apply the fundamental lemma of the calculus of variations, we arrive at the equation

$$\rho(u_{tt} - \sigma^2 K^2 u_{xxtt}) = Eu_{xx} \quad \text{for} \quad 0 < x < H. \tag{6.25}$$

This new equation, being of fourth order, requires additional boundary conditions for a unique solution to be obtained. In addition to Saint-Venant's conditions, we now require that $u_x(0, t) = 0$ at $x = 0$, which limits the lateral displacements and tangential stresses at $x = H$ to zero (*i.e.* $u_{xt}(H, 0) = 0$), together with $u_{xt}(x, 0) = 0$ for $0 < x < H$, making the lateral velocities zero. No explicit solution to this problem is known. The only existing results (Love 1927, Art. 278) concern the relation between the speed of propagation of the longitudinal waves generated and the lateral motion of the particles.

The model of the longitudinal impact on a rod can be extended by specifying the displacements in cylindrical polar form:

$$u = u(x, t), \quad u_r = r\Phi(x, t), \quad u_\theta = 0, \tag{6.26}$$

the x axis being along the cylindrical axis, with r, θ being in the plane of cross-section, where $\Phi(x, t)$ is an unknown function. This is a generalization of the case proposed by Pochhammer and Chree who used the case $\Phi(x, t) = -\sigma u_x$.

In order to derive the field equation of motion we again apply Hamilton's principle by considering the kinetic energy per unit length

$$T = \frac{1}{2}\rho A(u_t^2 + K^2 \Phi_t^2), \tag{6.27}$$

and the strain energy per unit length

$$W = \frac{1}{2} \iint_A \left[(2\mu + \lambda)(\varepsilon_x + \varepsilon_r + \varepsilon_\theta)^2 + 4\mu(\varepsilon_{r\theta}^2 - \varepsilon_x\varepsilon_r - \varepsilon_r\varepsilon_\theta - \varepsilon_\theta\varepsilon_x) \right] dA$$

where

$$\varepsilon_x = u_x, \ \varepsilon_r = \varepsilon_\theta = \Phi, \ \varepsilon_{xr} = \frac{1}{2} r\Phi_x,$$

are the strains associated with the displacements (6.26). The explicit form of W is

$$W = \frac{1}{2} A[(2\mu + \lambda)u_x^2 + 4(\mu + \lambda)\Phi^2 + 4\lambda u_x \Phi + \mu K^2 \Phi_x^2]. \tag{6.28}$$

The equations resulting from the application of Hamilton's principle are, therefore,

$$\begin{aligned} (2\mu + \lambda)u_{xx} + 2\lambda\Phi_x &= \rho u_{tt}, \\ \mu K^2 \Phi_{xx} - 4(\mu + \lambda)\Phi - 2\lambda u_x &= \rho K^2 \Phi_{tt}, \end{aligned} \tag{6.29}$$

for $0 < x < H$ and $t > 0$. These are known as the Herrmann–Mindlin equations (Mindlin and Herrmann 1952). If the rod is initially at rest except at the terminal section where the velocity has a jump $-v_0$, the initial conditions are

$$u(x, 0) = \Phi(x, 0) = 0, \ u_t(x, 0) = \Phi_t(x, 0) = 0 \text{ for } 0 < x < H, \tag{6.30}$$

plus a condition like (6.11) for the axial velocity u_t. As the rod is fixed at the lower end, the boundary conditions at $x = 0$ are

$$u(0, t) = \Phi(0, t) = 0; \tag{6.31}$$

at the $x = H$ end, the corresponding boundary conditions become

$$\begin{aligned} -T = -\iint_A \sigma_x \, dA = -A[(2\mu + \lambda)u_x + 2\lambda\phi] &= M u_{tt}, \\ \Phi_x(H, t) &= 0, \end{aligned} \tag{6.32}$$

where the first term represents the balance between the axial force and the inertia of the striking body and the second term denotes the absence of tangential stresses on the upper face.

The integration of the system of equations derived above cannot be performed with the elegance that characterizes Saint-Venant's solution. But it is likely that the kind of motion described by system (6.29) is much more accurate when the transverse dimensions of the cross-section have an appreciable magnitude.

A natural question is which of the different formulations of the problem of the extensional motion of a rod by a localized shock is the best. However, the answer to this is not as simple as it may appear at first sight. All the models proposed so far are correct in the sense that they respect the principles of mechanics and originate well-posed mathematical problems, although of different levels of difficulty. In this respect they may be considered as good models. The choice of one model or another is suggested by the particular aspect of the phenomenon we want to model and predict. For instance, if we want to know the rebound velocity of the striking body after the impact, the first description is the simplest and most elegant; if, on the other hand, we want to follow the motion of the points of the rod for all instants subsequent to the first shock and consider also the lateral components of displacement, then the two last descriptions are necessary. Once a particular model has been chosen, comparison must be made between its predictions and the results of experiment. However, the confirmation of a model must only be done *a posteriori*, since the model has its own integrity with regard to the relevance and clarity of the hypotheses and the rigor and appropriateness of its development.

Chapter II

Rod Theories: Three-dimensional Approach

7. Geometrical Properties of Rods

From the beginnings of human civilization, rods have been used in making rudimentary tools and weapons. In the theory of the elasticity, rods have been used to model a range of physical structures such as beams, columns, and shafts, the characteristics of which is slenderness that is of having a longitudinal dimension much greater than either of the transverse dimensions. The term "bar" is sometimes used synonymously with rod, although in general speech it need not imply that one dimension is very much greater than others. Occasionally, when modeling a physical object, it is possible to identify it immediately as a rod; for example, the mast of a boat. On other occasions a physical structure can be regarded as an assembly of rods; for example, certain buildings or bridges. Rods are important because they offer the simplest but nontrivial model to which the theory of elasticity can be applied and verified experimentally. The earliest theories of rods formulated at the beginning of the seventeenth century arose as a consequence of the research of Beeckman and Mersenne (Truesdell 1960, Prologue, Sec. 3.4), but the most significant progress made in the theory was due to James Bernoulli, Euler, Saint-Venant, Kirchhoff, and Michell. Implicit in using rod models is the facility of being able to work with a single independent variable representing the arc measured along a certain curve. In slender bodies this avoids the mathematical obstacles that arise from having the three independent variables of general continuum theory.

This then requires the construction of a rational scheme for approximating the system of field equations of general continua by a system containing one spatial independent variable and, ultimately, the time. However, in order to be acceptable, the approximation procedure must explain at each stage the relationship between the approximate solutions and those of the original three-dimensional problem; in addition, it must contain, at least in principle, the method of estimating the approximation error; finally, the approximate quantities must have a significant mechanical interpretation in the form of stress averages over certain sections of the body.

The first question we must answer is that of defining in a precise but sufficiently general way the possible geometric configurations of the bodies that can be treated as rods. The most obvious definition of rod is that of a body which in the unstressed state is circularly cylindrical or prismatic, with the length of the generators being much greater than the dimensions of the cross-sections. This definition, however, is too narrow because it excludes bodies such as arches and springs, which have always been treated using the theory of rods. Another objection is that slender bodies that are not strictly tubular, such as the blades of a propeller or the wings of an airplane, would also be excluded because their cross-sections are not constant.

In order to overcome these difficulties, a more general definition of a rod is required (Antman 1972, Sec. 4). A rod is a connected, but not necessarily a simply connected, solid body ($\mathscr{B}$) with a smooth invertible mapping $X_A = X_A(Z^p)$ relating the rectangular Cartesian coordinates to the curvilinear coordinates $Z^p (p = 1, 2, 3)$ such that the (Z^1, Z^2) coordinates range within bounded sets of real numbers, and such that the Jacobian determinant of the matrix $\partial X_A / \partial Z^p$ is bounded. These requirements ensure that rods have a finite thickness. According to this definition, nonslender bodies can also be included in the class of rods. The surfaces $Z^3 =$ constant are usually called the *cross-sections* of the rod, but the Z^3 axis is not, in general, the curve of the centroids of the cross-sections, because, although for prisms or cylinders the curve of the centroids is easily identifiable, for arbitrary bodies it is not obvious how this curve should be determined unambiguously. In addition, it is not easy to construct a curve of the centroids from the geometrical structure of the body.

The position of a material point with curvilinear coordinates Z^p $(p = 1, 2, 3)$ is denoted by the vector $\mathbf{r}(\mathbf{Z}, t)$, and the vectors

$$\mathbf{g}_p = \frac{\partial \mathbf{r}}{\partial Z^p},$$

represent the base vectors. Their duals $\mathbf{g}^p$ are defined by the relations

$$\mathbf{g}^p \cdot \mathbf{g}_q = \delta^p_q.$$

Any vector $\mathbf{v}$ can be written as a linear combination of the base vectors:

$$\mathbf{v} = v^p \mathbf{g}_p = v_p \mathbf{g}^p,$$

where $v^p = \mathbf{v} \cdot \mathbf{g}^p$ and $v_p = \mathbf{v} \cdot \mathbf{g}_p$ are the contravariant and covariant components of $\mathbf{v}$, respectively.

The products $g_{pq} = \mathbf{g}_p \cdot \mathbf{g}_q$ define the components of the metric tensor, and the scalar $g = \det(g_{pq})$ represents the square root of the volume of a parallelepiped having the vectors $\mathbf{g}_p$ as edges. Thus we have the identity

$$\sqrt{g} = (\mathbf{g}_1 \times \mathbf{g}_2) \cdot \mathbf{g}_3.$$

Note that the deformation gradient $F_{iA} = \partial x_i / \partial X_A$, in Z^p coordinates, assumes the form

$$\frac{\partial Z^p}{\partial X_A} \frac{\partial}{\partial Z^p} \left(\frac{\partial x_i}{dX^q} x^q \right),$$

where x^q are the contravariant components of $\mathbf{x}$.

The coordinate Z^3 is often denoted by S in order to indicate that this line may not be a Cartesian axis, but rather a suitably chosen curvilinear abscissa. If S ranges within a bounded interval $S_1 \leq S \leq S_2$, the sections $S = S_1$ and $S = S_2$ are called the *ends* of the rod, but either S_1, S_2, or both, may be infinite. As a natural generalization of the traditional treatment of cylindrical bars we introduce the set

$$\mathscr{S} = \partial\mathbf{Z}(\mathscr{B}) \cap \{\mathbf{Z} : S_1 < S < S_2\}, \tag{7.1}$$

and the set

$$\mathscr{A} = \mathbf{Z}(\mathscr{B}) \cap \{\mathbf{Z} : S = T\}. \tag{7.2}$$

The region $\mathbf{r}(\mathscr{S}, t)$ is the *lateral surface* of the rod at time t, and the region $\mathbf{r}(\mathscr{A}(T), t)$ is the *cross-sectional surface* at $S = T$. At the points where $\mathbf{r}(\mathscr{S})$ is sufficiently smooth, it is possible to introduce two surface parameters U and V such that the parametric equations of the lateral surface have the form

$$Z^1 = Z^1(U, V), \quad Z^2 = Z^2(U, V), \quad Z^3 = S = S(V), \tag{7.3}$$

and these equations determine the unit vector normal to the surface as well the area element through the ordinary formulae of differential geometry. If $\boldsymbol{\nu}$ denotes the unit outward normal to $\mathbf{r}(\mathscr{S})$, then the condition

$$\nu_\alpha \frac{\partial Z^\alpha}{\partial U} = 0, \quad \alpha = 1, 2, \tag{7.4}$$

ensures that $\boldsymbol{\nu}$ is perpendicular to the tangent vector to the lines V = constant, and the condition

$$\nu_p \frac{\partial Z^p}{\partial V} = 0, \quad p = 1, 2, 3, \tag{7.5}$$

expresses the orthogonality of $\boldsymbol{\nu}$ to the lines U = constant.

The element of area on $\mathbf{r}(\mathscr{S}, t)$ can be represented by the formula

$$dA = \frac{\sqrt{G}}{1 - \nu_3\nu^3}\left(\frac{\partial Z^2}{\partial U}\nu^1 - \frac{\partial Z^1}{\partial U}\nu^2\right) dU\, dV, \tag{7.6}$$

where G denotes the value of g in the reference configuration, for instance at $t = 0$, and ν^p represent the contravariant components of $\boldsymbol{\nu}$. Other variations of (7.6) can be found in standard texts on tensor analysis.

The approximate theories of rods try to determine the distribution of the stresses and strains inside slender bodies by taking into account the geometrical requirements specified above. They are based on the idea of seeing whether the particular shape of the body offers any conjecture about the nature of the solutions, which might lead to a simplifying procedure for constructing them. However, the question of whether or not this can be done is not properly posed unless, in addition to shape of the body, some restrictions are also made on the nature of the exterior loads. No definite form for the solution can be expected if the loads are too irregular. For this reason, all rigorous theories rest on the assumption that the lateral surface is either free or subjected to a load that is uniformly distributed along the longitudinal axis.

Under these assumptions we might hope to obtain, without solving the full three-dimensional problem, some partial information about the solutions. By "partial information" we mean, for example, a knowledge of the deformation of the curve corresponding to the X_3 axis in the reference configuration, or knowledge of certain averages of the stresses resulting from the three-dimensional deformation of the body. Since what we seek is more restrictive, the corresponding theory should be simpler.

The problem of establishing a coherent simplified theory for rods can be solved in two ways. The first is that of starting from the three-dimensional equations, and taking as the unknowns the stresses and strains at any point on the body which appear in these equations. This method consists in assuming in advance that the stress components take a particular analytical form. This form is suggested by intuition or prompted by comparison with other cases for which the established solution is exact. Having anticipated the form of the solution, it cannot be expected to satisfy the three-dimensional equations exactly; however, we might average these equations over the cross-section of the rod and try to adjust the coefficients of the tentative solution so that the averaged equations are satisfied. This procedure is clearly open to objections. The choice of the form of the solution is largely free, but it is almost always conditioned on tacit assumptions, which are often undesirable. In addition, the method of averaging provides only an approximate solution to the three-dimensional equations, and, in addition, where approximations are iterative, the convergence of the method must be proved and a procedure for estimating the error devised.

The second type of rod theory adopts another approach. Theories in this category endeavor to replace the continuum that constitutes the body with another continuum. Such models study the behavior of a flexible curve of particles of the material identified as geometrical points and of vectors associated with these points – called *directors*. This approach is used so that when trying to represent the main features of the original body with sufficient precision, the number of factors available in the one-dimensional case is wide enough.

The directors can rotate and elongate independently of the deformation of the curve calculated in the classical way.[15] At first sight the director theory might seem to avoid the criticisms made about the approach derived from three-dimensional theory, because its equations can be formulated directly without the need to justify them or estimate their degree of approximation. The equations are intrinsically correct provided that they obey the principles of mechanics. But, in effect, they too pose difficulties when we try to derive stresses and strains in the corresponding three-dimensional continuum starting from a knowledge of the deformation of the longitudinal axis and its directors. The result is that neither of the two theories is decisively superior to the other.

8. Approximation of Three-dimensional Equations

The three-dimensional theory of rods can be developed in full generality by considering the mechanical and thermal effects simultaneously. We assume the position $\mathbf{r}(Z^p, t)$ and the temperature $\theta(Z^p, t)$ as the fundamental dependent variables (Antman 1972, Sec. 5). However, here we prefer to ignore the influence of the

[15] The concept of directors was introduced by Duhem (1893), but the first systematic theory of directed, or polar, media was proposed by Cosserat and Cosserat (1907, 1909). However, the Cosserats considered only media characterized by a set of mutually rigid vectors. Extension of the theory to continua endowed at any point with n stretchable directors was made by Ericksen and Truesdell (1958).

temperature and limit ourselves to a purely mechanical description of the problem, as the mechanical equations are simpler.

The aim of the attempted three-dimensional theory is to approximate $\mathbf{r}(Z^p, t)$ by a representation of the form

$$\mathbf{r}(Z^p, t) \sim \mathbf{b}(\mathscr{R}_N(S, t), Z^p, t), \tag{8.1}$$

where $\mathscr{R}_N(S, t)$ denotes an ordered set of $N+1$ vectors $\{\mathscr{R}_k, k = 0, \ldots, N\}$ which is to be determined, and $\mathbf{b}$ is a given vector function on $\mathscr{R}^{3(N+1)} \times \mathbf{Z}(\mathscr{B}) \times \mathscr{R}$. The representation (8.1) may not only be merely an approximation for $\mathbf{r}(\mathbf{Z}, t)$, but also an exact solution of some special problem. The essential advantage of replacing $\mathbf{r}$ by $\mathbf{b}$ is that the new unknowns $\mathscr{R}_N$ depend only on the variables S and t.

In the following we will employ the symbol

$$\mathbf{B}^k = \frac{\partial \mathbf{b}}{\partial \mathscr{R}_k}, \tag{8.2}$$

to denote the second-order tensor, the components of which are $(\mathbf{B}^k)_i = \dfrac{\partial b_i}{\partial(\mathscr{R}_k)}$.

As a consequence of (8.2) the base vectors $\mathbf{g}_p = \mathbf{r}_{,p}$ can be written in the form

$$\mathbf{g}_p = \mathbf{r}_{,p} \sim \mathbf{b}_{,p} = \delta_p^3 \sum_{k=1}^{N} \mathbf{B}^k \mathscr{R}_{k,s} + \frac{\partial \mathbf{b}}{\partial Z^p}, \tag{8.3}$$

and, consequently, the Cauchy–Green tensor $\mathbf{C}$ is represented by

$$(\mathbf{C})_{pq} = \mathbf{r}_{,p} \cdot \mathbf{r}_{,q} \sim \mathbf{b}_{,p} \cdot \mathbf{b}_{,q}. \tag{8.4}$$

The condition that an element of (positive) volume cannot be compressed into an element of zero volume, can be represented in the approximate form as

$$(\mathbf{b}_{,1} \times \mathbf{b}_{,2}) \cdot \mathbf{b}_{,3} > 0; \tag{8.5}$$

whereas the stronger requirement that no fiber of positive length can be compressed into one of zero length is equivalent to the positive-definiteness of $\mathbf{C}$, namely

$$(\mathbf{C})_{pq} v_p v_q > 0, \tag{8.6}$$

for all vectors $v_p \neq 0$. Accordingly, the approximate version of (8.6) is

$$\mathbf{b}_{,p} \cdot \mathbf{b}_{,q} v_p v_q > 0. \tag{8.7}$$

Thus far we have simply presented an alternative, approximate method for defining the deformed state. We now consider the state of stress in the body, which is known once we have determined the Cauchy stress tensor through its contravariant components

$$t^{pq} = \sigma_{ij} \frac{\partial Z^p}{\partial x_i} \frac{\partial Z^q}{\partial x_j}.$$

The stress vector acting across the surface X^p = constant can be represented by

$$\mathbf{t}^p = t^{pq} \mathbf{g}_q. \tag{8.8}$$

If we introduce the so-called *convected* Piola–Kirchhoff stress vector $G^{-\frac{1}{2}}\boldsymbol{\tau}^p$ and the stress tensor $G^{-\frac{1}{2}}\mathbf{T}$ through the relations

$$\boldsymbol{\tau}^p = g^{\frac{1}{2}}\mathbf{t}^p, \quad T^{pq} = g^{\frac{1}{2}}t^{pq}, \tag{8.9}$$

we find the following representation for $\boldsymbol{\tau}^p$ in terms of T^{pq},

$$\boldsymbol{\tau}^p = T^{pq}\mathbf{g}_q. \tag{8.10}$$

Let us now introduce a sequence $\{\mathbf{A}^k(\mathscr{R}_N, \mathbf{Z}, t)\}$ of independent differentiable second-order tensors defined on $\mathscr{R}^{3(N+1)} \times \mathbf{Z}(\mathscr{B}) \times \mathscr{R}$, and define the following moments of the dependent constitutive variables:

$$\boldsymbol{\sigma}^k(S, t) = \int_{\mathscr{A}(s)} \mathbf{A}^k(\mathscr{R}_N, \mathbf{Z}, t)\boldsymbol{\tau}^3(\mathbf{Z}, t)\, dZ^1\, dZ^2, \tag{8.11}$$

$$\bar{\boldsymbol{\sigma}}^k(S, t) = \int_{\mathscr{A}(s)} \mathbf{A}^k_{,p}(\mathscr{R}_N, \mathbf{Z}, t)\boldsymbol{\tau}^p(\mathbf{Z}, t)\, dZ^1\, dZ^2. \tag{8.12}$$

The choice of tensors $\mathbf{A}^k$ is arbitrary. However, we now take

$$\mathbf{A}^k = (\mathbf{B}^k)^T, \tag{8.13}$$

which is the most usual choice for $\mathbf{A}^k$, although this restriction is not necessary.

Having introduced the Piola–Kirchhoff stress vector $G^{-\frac{1}{2}}\boldsymbol{\tau}^p$ we can obtain the expression for the force for the element of area across $\mathbf{r}(\mathscr{S}, t)$. If $\boldsymbol{\nu}$ is the unit normal to $\mathbf{r}(\mathscr{S}, t)$ this force is given by

$$G^{-\frac{1}{2}}\boldsymbol{\tau}^p \nu_p\, dA = G^{-\frac{1}{2}}\boldsymbol{\tau}_{(\nu)}, dA, \tag{8.14}$$

where we have set $\boldsymbol{\tau}_{(\nu)} = \boldsymbol{\tau}^p \nu_p$. Thus, by expressing $\nu_p\, dA$ in terms of the surface coordinates U and V, we can write

$$G^{-\frac{1}{2}}\boldsymbol{\tau}_{(\nu)}\, dA = \left[\boldsymbol{\tau}^1 \frac{\partial Z^2}{\partial U} - \boldsymbol{\tau}^2 \frac{\partial Z^1}{\partial U} + \boldsymbol{\tau}^3 \frac{\nu_3}{1 - \nu_3\nu^3}\left(\frac{\partial Z^2}{\partial U}\nu^1 - \frac{\partial Z^1}{\partial U}\nu^2\right)\right] dU\, dS. \tag{8.15}$$

The body force per unit volume in the reference configuration is $\rho_0\sqrt{G}$, and if we let $\mathbf{f}$ be the body force per unit mass, then following the procedure for stresses we can write down the contributions of the moment per unit area of S and define the tractions on the lateral surface:

$$\begin{aligned}\mathbf{f}^k &= \int_{\mathscr{A}(s)} \mathbf{A}^k\mathbf{f}\rho_0\sqrt{G}\, dZ^1\, dZ^2 \\ &\quad + \oint_{\partial\mathscr{A}(s)} \mathbf{A}^k[\boldsymbol{\tau}^1\, dZ^2 - \boldsymbol{\tau}^2\, dZ^1 + \boldsymbol{\tau}^3\nu_3(1 - \nu_3\nu^3)^{-1}(\nu^1\, dZ^2 - \nu^2\, dZ^1)] \\ &= \int_{\mathscr{A}(s)} \mathbf{A}^k\mathbf{f}\rho_0\sqrt{G}\, dZ^1\, dZ^2 + \oint_{\partial\mathscr{A}(s)} (1 - \nu_3\nu^3)^{-1}\mathbf{A}^k\boldsymbol{\tau}_{(\nu)}(\nu^1\, dZ^2 - \nu^2\, dZ^1).\end{aligned} \tag{8.16}$$

If the traction boundary conditions are not prescribed over the entire lateral surface, the $\mathbf{f}^k$ may not be well defined. This difficulty can be circumvented by judicious choice of the $\{\mathbf{A}^k\}$ (Antman 1972, Sec. 6), but the situation in which the lateral surface is either free or subjected to prescribed surface tractions is by far the most important.

To obtain the relationship between σ^k, $\bar{\sigma}^k$, and $\mathbf{f}^k$ we come back to the equations of balance of momentum, which, after having assumed the Z^p to be the independent variables, are

$$\boldsymbol{\tau}^p_{,p} + \rho_0\sqrt{G}\mathbf{f} = \rho_0\sqrt{G}\ddot{\mathbf{r}}, \tag{8.17}$$

where the comma denotes differentiation with respect to Z^p and the dot denotes differentiation with respect to time. We multiply (8.17) by $\mathbf{A}^k$ and integrate the resulting expression over $\mathscr{A}$. Since we have

$$\sigma^k_{,3} = \int_{\mathscr{A}} (\mathbf{A}^k\boldsymbol{\tau}^3)_{,3}\, dZ^1\, dZ^2 - \oint_{\partial\mathscr{A}} \mathbf{A}^k\boldsymbol{\tau}^3\nu_3(1-\nu_3\nu^3)^{-1}(\nu^1\, dZ^2 - \nu^2\, dZ^1), \tag{8.18}$$

then, by applying Stokes's theorem for the plane, we find that the weighted integrals of (8.17) can be written in the form

$$\sigma^k_{,s} - \bar{\sigma}^k + \mathbf{f}^k = \int_{\mathscr{A}} \mathbf{A}^k\ddot{\mathbf{r}}\rho_0\sqrt{G}\, dZ^1\, dZ^2. \tag{8.19}$$

For the purpose of verification, let us set $\mathbf{A}^k = 1$ in (8.19). Then from (8.11) we immediately find that

$$\sigma^k = \int_{\mathscr{A}} \boldsymbol{\tau}^3\, dZ^1\, dZ^2 = \mathbf{n}, \tag{8.20}$$

where $\mathbf{n}$ is the resultant stress vector, and from (8.12) we see that $\bar{\sigma}^k$ vanishes. Thus, if we denote the value of $\mathbf{f}^k$ by $\mathbf{f}_0$ when $\mathbf{A}^k = 1$, we obtain

$$\mathbf{n}_{,s} + \mathbf{f}_0 = \int_{\mathscr{A}} \ddot{\mathbf{r}}\rho_0\sqrt{G}\, dZ^1\, dZ^2, \tag{8.21}$$

which is known as the balance equation of linear momentum for rods.

Let us now assume instead that $\mathbf{A}^k = (\mathbf{r}-\mathbf{p})\times$, where $\mathbf{p} = \mathbf{r}(0, 0, S, t)$ and put

$$\begin{aligned} \boldsymbol{\ell} = &\int_{\mathscr{A}} (\mathbf{r}-\mathbf{p})\times\mathbf{f}\rho_0\sqrt{G}\, dZ^1\, dZ^2 \\ &+ \int_{\mathscr{A}} (\mathbf{r}-\mathbf{p})\times[\boldsymbol{\tau}^1\, dZ^2 - \boldsymbol{\tau}^2\, dZ^1 + \boldsymbol{\tau}^3\nu_3(1-\nu_3\nu^3)^{-1}(\nu^1\, dZ^2 - \nu^2\, dZ^1)], \end{aligned} \tag{8.22}$$

which represents the moment of the exterior loads with respect to the point $(0, 0, S)$. Let us also define the resultant couple vector as

$$\mathbf{m} = \int_{\mathscr{A}} (\mathbf{r}-\mathbf{p})\times\boldsymbol{\tau}^3\, dZ^1\, dZ^2. \tag{8.23}$$

Then the integration of the angular momentum balance equation

$$(\mathbf{r}\times\boldsymbol{\tau}^p)_{,p} + \rho_0\sqrt{G}\mathbf{r}\times\mathbf{f} = \rho_0\sqrt{G}\mathbf{r}\times\ddot{\mathbf{r}} \tag{8.24}$$

over $\mathscr{A}$, with the use of Stokes's theorem, yields

$$\mathbf{m}_{,s} + \mathbf{p}_{,s} \times \mathbf{n} + \boldsymbol{\ell} = \int_{\mathscr{A}} (\mathbf{r} - \mathbf{p}) \times \ddot{\mathbf{r}} \rho_0 \sqrt{G}\, dZ^1\, dZ^2. \tag{8.25}$$

This is known as the balance equation of angular momentum for rods.

Any attempt to solve equations like (8.21) and (8.25) is evidently impossible, because the unknown $\mathbf{r}$, appears in the definition of the quantities $\boldsymbol{\ell}$ and $\mathbf{p}_{,s} \times \mathbf{n}$, and must be known at any point in $\mathscr{A}$. In order to overcome this difficulty, we return to the approximation formula (8.1) and its consequences, which have the advantage of expressing $\mathbf{r}(Z^p, t)$ as a function of the variable S only. Thus we replace $\mathbf{r}$ by $\mathbf{b}$ in (8.19) and with this substitution the acceleration terms on the right-hand side are approximated by

$$\int_{\mathscr{A}} \mathbf{A}^k \ddot{\mathbf{r}} \rho_0 \sqrt{G}\, dZ^1 dZ^2 \sim \mathbf{a}^k(\mathscr{R}_N, \dot{\mathscr{R}}_N, \ddot{\mathscr{R}}_N, S, t) \equiv \int_{\mathscr{A}} \mathbf{A}^k \ddot{\mathbf{b}} \rho_0 \sqrt{G}\, dZ^1\, dZ^2. \tag{8.26}$$

But $\mathbf{f}^k$ also changes when we replace $\mathbf{r}$ by $\mathbf{b}$, although $\mathbf{r}$ does not appear explicitly in the right-hand side of (8.16). This happens in general, since the body forces $\mathbf{f}$ depend on the position $\mathbf{r}$. If we now introduce the notation

$$\mathbf{f}^k \sim \hat{\mathbf{f}}^k(\mathscr{R}_N, S, t) \tag{8.27}$$

to represent the approximation of the term containing the body forces in (8.19), we obtain the approximate equation of motion

$$\sigma^k_{,s} - \bar{\sigma}^k + \hat{\mathbf{f}}^k(\mathscr{R}_N, S, t) \sim \mathbf{a}^k(\mathscr{R}_N, \dot{\mathscr{R}}_N, \ddot{\mathscr{R}}_N, S, t). \tag{8.28}$$

In order to solve these equations it is necessary to derive an averaged form for the constitutive equations. Although this might be done for the constitutive relations of general materials (Antman 1972, Sec. 8), here we restrict the deduction to simple hyperelastic bodies. For materials with other stress–strain relations the procedure is conceptually the same.

The three-dimensional constitutive equations for hyperelastic materials in unreduced form are given using a scalar function $\check{\psi}(\mathbf{r}_{,p}, \mathbf{Z})$ such that

$$\boldsymbol{\tau}^p(Z, t) = \rho_0 \sqrt{G} \frac{\partial}{\partial \mathbf{r}_{,p}} \check{\psi}(\mathbf{r}_{,q}, \mathbf{Z}). \tag{8.29}$$

This is for equations to which the principle of frame indifference has not been applied. The function $\check{\psi}$ must satisfy the condition of symmetry of the tensor T^{pq} in (8.10). This condition reads

$$\mathbf{g}_p \times \boldsymbol{\tau}^p = 0,$$

and hence we have the equation

$$\mathbf{g}_p \times \frac{\partial \check{\psi}}{\partial \mathbf{r}_{,p}(\mathbf{r}_{,q}, Z) = 0.} \tag{8.30}$$

Once we know $\check{\psi}$, we can define the function $\tilde{\psi}$ as

$$\tilde{\psi}(\mathscr{R}_N, \mathscr{R}_{N,S}, \mathbf{Z}, t) = \check{\psi}(\mathbf{b}_{,p}, \mathbf{Z}). \tag{8.31}$$

From (8.2) we also obtain

$$\frac{\partial \tilde{\psi}}{\partial \mathscr{R}_{k,S}} = (\mathbf{B}^K)^T \frac{\partial \check{\psi}}{\partial \mathbf{r}_{,S}}, \quad \frac{\partial \tilde{\psi}}{\partial \mathscr{R}_k} = \left(\mathbf{B}^k_{,p}\right)^T \frac{\partial \check{\psi}}{\partial \mathbf{r}_{,p}}. \tag{8.32}$$

The constitutive relations for the averaged quantities (8.11) and (8.12) assume the approximate form

$$\sigma^k \sim \check{\sigma}^k(\mathscr{R}_N, \mathscr{R}_{N,S}, S, t) \equiv \int_{\mathscr{A}} \mathbf{A}^k \tau^3(\mathbf{b}_{,p}, \mathbf{Z})\, dZ^1\, dZ^2, \tag{8.33}$$

$$\bar{\sigma}^k \sim \check{\bar{\sigma}}^k(\mathscr{R}_N, \mathscr{R}_{N,S}, S, t) \equiv \int_{\mathscr{A}} \mathbf{A}^k_{,p} \tau^p(\mathbf{b}_{,p}, \mathbf{Z})\, dZ^1\, dZ^2, \tag{8.34}$$

for $k = 0, \ldots, N$.

Boundary value problems are governed by equations approximated to order N. These equations are obtained by replacing the approximation symbol ($\sim$) by the equality symbol ($=$) whenever it appears in the above formulae. We can thus write

$$\sigma^k_{,S} - \bar{\sigma}^k + \hat{\mathbf{f}}^k(\mathscr{R}_N, S, t) = \mathbf{a}^k(\mathscr{R}_N, \dot{\mathscr{R}}_N, \ddot{\mathscr{R}}_N, S, t), \quad (k = 0, \ldots, N) \tag{8.35}$$

where the functions $\hat{\mathbf{f}}^k$ and $\mathbf{a}^k$ are prescribed. These equations can be written in component form by taking their inner products using a suitable set of independent vectors. The base vectors $\mathbf{g}_p$ evaluated at $Z^1 = Z^2 = 0$ suggest the most natural independent set. However, in practice an appropriate system of orthogonal unit vectors proves more useful.

In order to solve (8.35) with respect to $\mathscr{R}_N$ it is necessary to have boundary and initial conditions in terms of $\mathscr{R}_N$. It is clear that the boundary conditions on the lateral surface require no attention as they can be regarded as having been incorporated in equations (8.35). Thus only the boundary conditions at $S = S_1, S_2$ are important. If the rod is a ring, namely if the ends coincide, we assume that all the given functions of S (reference configuration, loads, and constitutive functions) can be smoothly extended to the entire S axis with period $S_2 - S_1$ in S. We then require the unknown $\mathscr{R}_k$ to satisfy the periodicity conditions

$$\mathscr{R}_k(S_1, t) = \mathscr{R}_k(S_2, t) \quad (k = 0, \ldots, N).$$

If position conditions are prescribed at the ends, we must simply choose $\mathscr{R}_k$ so that $\mathbf{b}(\mathscr{R}_k, Z^1, Z^2, S, t)$ gives the best approximation to $\mathbf{r}(S, t)$ at $S = S_1$ and $S = S_2$. If traction conditions are prescribed at the ends, we must find the moments σ^k by averaging the stresses at the ends. As far as the initial conditions are concerned, they are the values of $\mathscr{R}_k$ and $\dot{\mathscr{R}}_k$ at the time $t = 0$, calculated by averaging the initial values of $\mathbf{r}(\mathbf{Z}, t)$and $\dot{\mathbf{r}}(\mathbf{Z}, t)$. However, the construction of the boundary and initial conditions in terms of $\mathscr{R}_k$ is never unique because these depend on the choice of the vector $\mathbf{b}$ and the tensors $\mathbf{A}^k$.

9. Rod Problems in Linear Elasticity

Many classical theories of rods have been derived from the three-dimensional equations of linear elasticity by averaging the equations over the cross-section. The first example is Saint-Venant's (1855) solution to the problem of a cylinder loaded at the ends and free from tractions over its lateral surface. The rod is made from linearly elastic, homogeneous, isotropic material and its reference configuration, in a system of Cartesian axes $\mathbf{Z} = (X, Y, S)$, occupies the region defined by the cylinder $\Phi(X, Y) = 0$ bounded by the two planes $S = 0$ and $S = L$. The S axis coincides with the straight line of the centroids of the cross-sections, and the X and Y axes are the principal axes of inertia of the section $S = 0$. As a simplification of what was postulated before, we now assume the cross-section to be simply connected. Under the assumption of a linearly elastic material, we can apply the theory of small strains. In addition, we suppose that the cylinder is not too thin, such that the relative displacements of its parts are small, with the consequence that we can identify the reference configuration with the present configuration and also assume that the Piola–Kirchhoff and Cauchy stress tensors are identical.

Let $\mathbf{e}_1$, $\mathbf{e}_2$, and $\mathbf{e}_3$ be the unit vectors applied at the origin and oriented along the X, Y, and S axes, respectively. These vectors form coincident covariant and contravariant bases. Since the lateral surface is free, the boundary data reduce to two fields of tractions defined over $S = 0$ and $S = L$ of the form $\mathbf{t}^3 = \boldsymbol{\tau}^3$ $(g^{\frac{1}{2}} = 1)$. In order to ensure global equilibrium, the resultant force and moment of the surface tractions must be zero.

We first consider the case in which the resultant of the tractions $\mathbf{t}^3$ acting on the face $S = L$ is statically equivalent to a single force $N\mathbf{e}_3 = \mathbf{n}$ applied at the center of gravity of the section, and the resultant moment $\mathbf{m}$ of $\mathbf{t}_3$ with respect to the same point is zero. This kind of loading characterizes the simple tension of the rod.

Within a rigid motion, we conjecture a position vector which may be represented by a formula like (8.1), or, more precisely, by its linear version

$$\mathbf{r}(X, Y, S) = \sum_{k=0}^{1} \mathbf{B}^K(X, Y)\mathscr{R}_K(S), \tag{9.1}$$

where

$$\mathbf{B}^0 = 1, \quad \mathbf{B}^1 = \begin{pmatrix} X & Y & 0 \\ Y & -X & 0 \\ 0 & 0 & X \end{pmatrix}, \tag{9.2}$$

and

$$\mathscr{R}_0 = \begin{pmatrix} 0 \\ 0 \\ \dfrac{N}{EA}S \end{pmatrix}, \quad \mathscr{R}_1 = -\begin{pmatrix} \sigma\dfrac{N}{EA} \\ 0 \\ 0 \end{pmatrix}, \tag{9.3}$$

A being the cross-sectional area, and E and σ being the elastic moduli. We thus get

$$\mathbf{r}(X, Y, S) = \frac{N}{EA}S\,\mathbf{e}_3 - \sigma\frac{N}{EA}(X\mathbf{e}_1 + Y\mathbf{e}_2). \tag{9.4}$$

The next case is that in which the resultant of the tractions over the terminal section $S = L$ is zero, and its moment with respect to the centroid has the form $\mathbf{m} = M\mathbf{e}_2$, that is to say it is a couple M about the Y axis. This example represents the pure bending of a bar.

Here again the position vector may be expressed as a linear combination of the form

$$\mathbf{r}(X, Y, S) = \sum_{k=0}^{2} \mathbf{B}^k(X, Y)\mathcal{R}_k(S), \tag{9.5}$$

where $\mathbf{B}^0$ and $\mathbf{B}^1$ have the same form as given in (9.2), but $\mathbf{B}^2$ is now

$$\mathbf{B}^2 = \begin{pmatrix} \frac{1}{2}(X^2 - Y^2) & XY & 0 \\ XY & -\frac{1}{2}(X^2 - Y^2) & 0 \\ 0 & 0 & Y \end{pmatrix}, \tag{9.6}$$

the terms $\mathcal{R}_k(S)$ having the forms

$$\mathcal{R}_0 = \begin{pmatrix} \frac{1}{2}\frac{M_2}{EI_2}S^2 \\ 0 \\ 0 \end{pmatrix}, \quad \mathcal{R}_1 = -\begin{pmatrix} 0 \\ 0 \\ \frac{M_2}{EI_2}S \end{pmatrix},$$

$$\mathcal{R}_2 = \begin{pmatrix} \sigma\frac{M_2}{EI_2} \\ 0 \\ 0 \end{pmatrix}, \tag{9.7}$$

where I_2 is the moment of inertia of the cross-section about the Y axis. On substituting these quantities (9.5) we obtain

$$\mathbf{r}(X, Y, S) = \frac{M_2}{EI_2}\left(\frac{1}{2}S^2\mathbf{e}_1 - SX\mathbf{e}_3 + \frac{1}{2}\sigma(X^2 - Y^2)\mathbf{e}_1 + \sigma XY\mathbf{e}_2\right), \tag{9.8}$$

which is a well-known formula of Saint-Venant (Love 1927, Art. 87).

Another case which presents itself is that of pure torsion, in which the tractions over the face $S = L$ are statically equivalent to a couple M_3 directed along the S axis. In this case the position vector may be written as

$$\mathbf{r}(X, Y, S) = \sum_{k=0}^{3} \mathbf{B}^k(X, Y)\mathcal{R}_k(S), \tag{9.9}$$

where $\mathbf{B}^0$, $\mathbf{B}^1$, and $\mathbf{B}^2$ are the same as above, and $\mathbf{B}^3$ is the matrix

$$\mathbf{B}^3 = \begin{pmatrix} 0 & 0 & 0 \\ 0 & 0 & 0 \\ \Phi_1(X, Y) & 0 & \Phi_3(X, Y) \end{pmatrix}. \tag{9.10}$$

Φ_1 and Φ_2 are harmonic functions, the normal derivatives of which at the boundary points of the section have prescribed values and are solutions of two Neumann problems, which will be specified later. The terms $\mathcal{R}_k(S)$ now take the form

$$\mathcal{R}_0 = \mathcal{R}_2 = 0, \quad \mathcal{R}_1 = -\begin{pmatrix} 0 \\ \dfrac{M_3}{\mu D} S \\ 0 \end{pmatrix}, \quad \mathcal{R}_3 = \begin{pmatrix} 0 \\ 0 \\ \dfrac{M_3}{\mu D} \end{pmatrix}, \tag{9.11}$$

D being given by the integral $\int_{\mathcal{A}} (X^2 + Y^2 + X\Phi_{3,Y} - Y\Phi_{3,X})\, dX\, dY$. The quantity μD is called the *torsional rigidity* of the section.

The position vector is thus represented by the formula

$$\mathbf{r}(X, Y, S) = \mathbf{B}^1 \mathcal{R}_1 + \mathbf{B}^3 \mathcal{R}_3 = \frac{M_3}{\mu D}(-YS\mathbf{e}_1 + XS\mathbf{e}_2 + \Phi_3(X, Y)\mathbf{e}_3), \tag{9.12}$$

and this is again a classical solution due to Saint-Venant.

As a last example we consider the case in which the surface tractions over the face $S = L$ are statically equivalent to a single force Q_1 directed along the X axis and applied at the centroid of the section. This implies that in a typical cross-section at a distance $(L - S)$ from the terminal end, the stress resultant is composed of the force Q_1 and a couple $M_2 = Q_1(L - S)$ about the Y axis. Let us assume the representation (9.9) for the position vector, but choose the vectors $\mathcal{R}_k$ to have the following form (Antman 1972, Sec. 11):

$$\mathcal{R}_0 = \begin{pmatrix} \dfrac{1}{2EI_2}\left[Q_1\left(L - \dfrac{S}{3}\right) + M_2\right]S^2 \\ 0 \\ 0 \end{pmatrix}, \quad \mathcal{R}_1 = -\begin{pmatrix} 0 \\ \dfrac{1}{ED_1} Q_1 S \\ \dfrac{1}{EI_2}\left[Q_1\left(L - \dfrac{S}{2}\right) + M_2\right]S \end{pmatrix}$$

$$\mathcal{R}_2 = \begin{pmatrix} \dfrac{\sigma}{EI_2}[Q_1(L - S) + M_2] \\ 0 \\ 0 \end{pmatrix}, \quad \mathcal{R}_3 = \begin{pmatrix} \dfrac{1}{EI_2} Q_1 \\ 0 \\ \dfrac{1}{ED_1} Q_1 \end{pmatrix}. \tag{9.13}$$

Here D_1 is a geometric constant equal to

$$DI_2\left[\int_{\mathcal{A}} \left(Y\Phi_{1,X} - X\Phi_{1,Y} + (1 - \frac{1}{2}\sigma)Y^3 - (2 + \frac{1}{2}\sigma)X^2 Y\right) dX\, dY\right]^{-1}.$$

The position vector may be calculated using (9.9), and this result coincides with the solution found by Saint-Venant.

To use this method of solution formally would require the following sequence of steps in each of the above four cases: calculation of the stresses as functions of position; determining the averaged quantities σ^k and $\bar{\sigma}^k$; and checking that these quantities are solutions of the averaged equations of equilibrium. However this operation is unnecessary provided that we choose the harmonic functions

$\Phi_1(X, Y)$ and $\Phi_3(X, Y)$ in such a manner that their normal derivatives at the boundary of the cross-section satisfy the conditions:

$$\frac{\partial \Phi_1}{\partial \nu} = -\left\{\frac{1}{2}\sigma X^2 + \left(1 - \frac{1}{2}\sigma\right) Y^2\right\} \cos(X, \nu) - (2 + \sigma) XY \cos(Y, \nu)\}, \quad (9.14)$$

$$\frac{\partial \Phi_3}{\partial \nu} = Y \cos(X, \nu) - X \cos(Y, \nu), \quad (9.15)$$

where $\cos(X, \nu)$ and $\cos(Y, \nu)$ denote the direction cosines of the outward normal to the bounding curve. In fact the stresses corresponding to this choice of Φ_1 and Φ_3 give exact solutions to the local equations of equilibrium at the interior points of the cylinder and zero traction at each point of the lateral surface. The only points of the surface at which the boundary conditions are not locally satisfied are the terminal sections, on which we can only require the stresses to be statically equipollent to the stress resultants N, M_2, M_3, and Q_2. This equivalence can be verified. Of course, if the tractions applied over the terminal faces are adjusted so that they coincide with the surface tractions associated with Saint-Venant's representation of the displacements, then we have the full, unique, solution to the elastic problem. But the practical utility of Saint-Venant's solution rests on the principle first enunciated by Saint-Venant and known as the principle of the "elastic equivalence of statically equipollent systems of loads." According to this principle, if two sets of loadings are statically equivalent at each end of the cylinder, then the difference in the stress fields is negligible, except possibly near the ends. We thus infer that, when the length of the bar is large compared with any linear measure of the bar over its cross-section, the stresses and strains set up in its interior depend, in practice, only on the resultants N, M_2, etc. This applies at all points of the bar, except in comparatively small regions near its ends.[16]

The fact that Saint-Venant's solution satisfies all the equations of the problem exactly, except those on the terminal face, makes it superior to other approximate solutions containing only four terms in the expansion (9.9). Therefore, it is natural to look at the underlying features that distinguish Saint-Venant's solution from its competitors. The question may be approached in various ways. For instance, Clebsch (1862) deduced all four cases of Saint-Venant's solution from the single unifying hypothesis that the traction vector on any material plane normal to the cross-section of the cylinder is parallel to its generators. Alternatively, Voigt showed that Saint-Venant's results for the case of extension, bending, or torsion are necessary consequences of the assumption that the stress field is independent of the axial coordinate; the solution of the flexure case becomes determinate if the stress field is allowed to involve the axial coordinate linearly at most (Voigt 1887; Love 1927, Art. 236, 237, 238).

Another characterization of Saint-Venant's solution may be obtained in terms of certain minimum properties of the strain energy (Sternberg and Knowles 1966). It is possible to show specifically that his extension and bending solutions are uniquely

[16] Saint-Venant (1855) enunciated the principle as a conjecture for a cylindrical body; the extension of the principle to the general case of any body loaded on a small part of its surface is due to Boussinesq (1885).

determined by the property that they render the total strain energy an absolute minimum in a certain subclass of solutions. This class requires the same resultant load or the same bending couple, as long as the shearing tractions vanish pointwise on the ends of the cylinder. In addition, Saint-Venant's solution for torsion minimizes the total strain energy among all other solutions, which, for a fixed torque, have normal tractions vanishing everywhere on the end cross-sections. Although it might seem natural to believe that an analogous minimizing principle also holds for Saint-Venant's flexure problem, this is not so. However, it is true if the Poisson ratio of the material is zero.

By the way of demonstrating these properties, let us consider the case of pure extension of a cylinder. This state of extension is defined by the following boundary conditions of mixed–mixed type:

$$u_3 = 0, \ \sigma_{31} = \sigma_{32} \text{ on the face } S = 0,$$

$$u_3 = \frac{NL}{EA}, \ \sigma_{31} = \sigma_{32} = 0 \text{ on the face } S = L,$$

where u_3 denotes the displacement component along the S axis. In addition, the lateral surface is free, body forces are absent, and the resultant of the stress component σ_{33} over the face $S = L$ is statically equivalent to a force N applied at the point $(0, 0, L)$. The solution of Saint-Venant's problem in this case should coincide with the minimum of the functional

$$\mathscr{E} = \mathscr{W} - \mathscr{L} = \frac{1}{4\mu}\int_0^L dS \int_{\mathscr{A}(S)} \left[\hat{\sigma}_{ij}\hat{\sigma}_{ij} - \frac{\sigma}{1+\sigma}\hat{\sigma}_{ii}\hat{\sigma}_{jj}\right] dX\,dY - \int_{\mathscr{A}(L)} \hat{\sigma}_{33}u_3\,dX\,dY, \tag{9.16}$$

over all *statically admissible stress fields*. These are stresses $\hat{\sigma}_{ij}$ such that their partial derivatives $\hat{\sigma}_{ij,j}$ are square integrable and satisfy the field equations of equilibrium and all boundary conditions involving stresses. As $u_3(X, Y, L)$ is constant, we may write

$$\mathscr{E} = \mathscr{W} - \int_{\mathscr{A}(L)} \hat{\sigma}_{33}u_3\,dX\,dY = \mathscr{W} - \frac{NL}{EA}\int_{\mathscr{A}(L)} \hat{\sigma}_{33}\,dX\,dY \tag{9.17}$$

whence, because of the condition

$$\int_{\mathscr{A}(L)} \hat{\sigma}_{33}\,dX\,dY = N,$$

we obtain

$$\mathscr{E} = \mathscr{W} - \frac{N^2L}{EA}, \tag{9.18}$$

or, alternatively,

$$\mathscr{W} = \mathscr{E} + \frac{N^2L}{EA}. \tag{9.19}$$

Hence $\mathscr{E}$ and $\mathscr{W}$ assume their respective minima for the same $\hat{\sigma}_{ij}$ and correspond to the minimum of the strain energy in Saint-Venant's state, which minimizes the integrand in (9.16).

This type of demonstration can be repeated, using the corresponding changes, for bending, torsion, and flexure. However, in this last case we do not obtain a formula analogous to (9.19), since the second term on the right-hand side is no longer positive-definite, unless Poisson's ratio σ is zero (Sternberg and Knowles 1966).

Once a criterion for the superiority of Saint-Venant's solutions has been established, it is easy to understand the popularity of the method of comparing other approximate solutions with this as a test for accuracy (Novozilov 1948). Of course, it is still debatable how the notion of closeness between solutions should be guaged. For example, should we require them to be close in energy, or in quadratic mean, or simply to fit at a finite number of points? Unfortunately, this question has not been answered satisfactorily. The problem is particularly important if we wish to estimate the adequacy of rough approximations in technical applications to rods for which the displacements are linear functions X and Y.

The attempts at characterizing the Saint-Venant solution as minimizers of the strain energy do not end here. For instance, it is possible to formulate a boundary value problem for a beam where the displacement is prescribed to be zero at one end and to have the form of an infinitesimal rigid displacement at the other. The corresponding solution is clearly a minimizer for the strain energy in the class of functions satisfying the nonhomogeneous boundary conditions on the displacements. If we ask which of Saint-Venant's solutions satisfies such boundary conditions, we see that a few do but most do not (Maisonnève 1971).

However, in practice, there are good reasons for selecting intuitively some plausible solutions suggested by physical considerations. In fact, very often, the symmetry and the nature of the loading are such that, if we were given the load distribution on one end, the load distribution on the opposite end would be the same, so that the actual distribution will conform to the equation

$$\sigma_{i3}(x_\alpha, L) = \sigma_{i3}(x_\alpha, 0). \tag{9.20}$$

This condition is hardly appropriate for problems for flexure, but it is reasonable for problems of pure traction, bending, and twisting.

Now it has been noted that a common prerogative of Saint-Venant's solutions is that, if we know the displacements $u_i(x_\alpha, 0)$, the displacements at any point of the cylinder have the form (Ericksen 1980)

$$u_i(x_\alpha, x_3) = u_i(x_\alpha, 0) + x_3(a_i + b_i x_3 + \varepsilon_{ijk} c_j x_k), \tag{9.21}$$

where $u_i(x_\alpha, 0)$ are arbitrary functions and a_i, b_i, and c_j are constants. In order to represent Saint-Venant's solutions the constants must be related by the conditions

$$2b_1 + c_2 = 2b_2 - c_1 = b_3 = 0. \tag{9.22}$$

Putting

$$\alpha_i = (a_i + b_i L)L, \quad \beta_i = c_i L,$$

we define

$$v_i = \alpha_i + \varepsilon_{ijk}\beta_j x_k, \tag{9.23}$$

which is a rigid motion, so that

$$u_i(x_\alpha, L) = u_i(x_\alpha, 0) + (\alpha_i + \varepsilon_{ijk}\beta_j x_k)_{x_3=L}. \tag{9.24}$$

Let overbars to denote any solution compatible with (9.20), and giving rise to the same resultant force and moment as a solution of the Saint-Venant type, with displacement u_i. Then

$$\bar{\bar{u}}_i = \bar{u}_i - u_i \tag{9.25}$$

will give a solution with zero resultant force and moment.

On calculating the total strain energy, we have, in obvious notation,

$$\bar{\mathscr{W}} = \mathscr{W} + \bar{\bar{\mathscr{W}}} + \int\limits_V \bar{\bar{\sigma}}_{ij} u_{i,j}\, dV. \tag{9.26}$$

By integrating by parts and using the fact that

$$\bar{\bar{\sigma}}_{ij,j} = 0 \text{ in } V,$$

and $\sigma_{ij}\nu_j = 0$ on the lateral surface, together with (9.20) and (9.24) we obtain the result

$$\int\limits_V \bar{\bar{\sigma}}_{ij} u_{i,j}\, dV = \int\limits_{\mathscr{A}(S)} \bar{\bar{\sigma}}_{i3} u_i\, dS \,\Big|_0^L = \alpha_i \int\limits_{\mathscr{A}(L)} \bar{\bar{\sigma}}_{i3}(x_\alpha, L)\, dS + \beta_i \int\limits_{\mathscr{A}(L)} \varepsilon_{ijhx_j} \bar{\bar{\sigma}}_{h3}(x_\alpha, L)\, dS.$$

However, since $\bar{\bar{u}}_i$ is a solution with zero resultant force and moment, these integrals vanish, so we have

$$\bar{\mathscr{W}} - \mathscr{W} = \bar{\bar{\mathscr{W}}} \geqslant 0, \tag{9.27}$$

which shows that the Saint-Venant solutions do minimize the energy in the solutions that conform to (9.20) and have matching resultant forces and moments.

10. Uniformly Loaded Beams (Michell's Theory)

We shall suppose that there is a body force, specified by a component f_1 parallel to the X axis and surface tractions on the lateral surface, specified similarly by t_1 parallel to the X axis, and that both these quantities are independent of S (Michell 1901).

In order to find a solution to the field and boundary equations we start from a tentative representation of the stresses and strains in the form of quadratic functions of S. For example

$$\sigma_{ij} = \sigma_{ij}^{(0)} + \sigma_{ij}^{(1)} S + \sigma_{ij}^{(2)} S^2, \tag{10.1a}$$

$$e_{ij} = e_{ij}^{(0)} + e_{ij}^{(1)} S + e_{ij}^{(2)} S^2. \tag{10.1b}$$

The equations of equilibrium take the form

$$\sigma_{ij,j} + \rho f_i = 0, \quad (f_2 = f_3 = 0), \tag{10.2}$$

where the stress components $\sigma_{ij}^{(0)}$ and $\sigma_{ij}^{(2)}$ must satisfy the homogeneous equations of equilibrium and the $\sigma_{ij}^{(1)}$ terms must satisfy the same equations containing the body forces.

Let us try to give a tentative representation of the solution in terms of strains. Selecting the terms that contain S^2, we suppose that they have the form

$$\begin{aligned} e_{33}^{(2)} &= \varepsilon_2 - \kappa_2 X, \\ e_{11}^{(2)} &= e_{22}^{(2)} = -\sigma e_{33}^{(2)}, \quad e_{12}^{(2)} = 0 \\ e_{31}^{(2)} &= \tau_2\left(\frac{\partial \Phi_3}{\partial X} - Y\right), \quad e_{32}^{(2)} = \tau_2\left(\frac{\partial \Phi_3}{\partial Y} + X\right), \end{aligned} \tag{10.3}$$

where ε_2, κ_2, and τ_2 are constants and Φ_3 is a harmonic function satisfying a boundary condition like (9.15). A choice of strain components like (10.3) is equivalent to the assumption that the strains $e_{ij}^{(2)}$ are produced by an extensional force, a couple about the Y axis, and a pure torsion, but not by a shear force.

Again, selecting the terms of first order in S, we may show that τ_2 and ε_2 must vanish, and that we may put

$$\begin{aligned} e_{33}^{(1)} &= \varepsilon_1 - \kappa_1 X, \\ e_{11}^{(1)} &= e_{22}^{(1)} = -\sigma e_{33}^{(1)}, \quad e_{12}^{(1)} = 0, \\ e_{31}^{(1)} &= \tau_1\left(\frac{\partial \Phi_3}{\partial X} - Y\right) + 2\kappa_2\left[\frac{\partial \Phi_1}{\partial X} + \frac{1}{2}\sigma X^2 + (1 - \frac{1}{2}\sigma)Y^2\right], \\ e_{32}^{(1)} &= \tau_1\left(\frac{\partial \Phi_3}{\partial Y} + X\right) + 2\kappa_2\left[\frac{\partial \Phi_1}{\partial Y} + (2+\sigma)XY\right], \end{aligned} \tag{10.4}$$

where ε_1, κ_1, and τ_1 are other constants and Φ_1 is a harmonic function satisfying a boundary condition like (9.14). Unlike the solution of order two, the first-order solution contains, in addition to the other stress resultants, a term due to flexure.

As for the terms of zero order, let us assume the following forms for the strain components representing $e_{33}^{(0)}$, $e_{31}^{(0)}$, and $e_{32}^{(0)}$:

$$\begin{aligned} e_{33}^{(0)} &= \varepsilon_0 - \kappa_0 X + 2\kappa_2(\Phi_1 + XY^2) + \tau_1\varphi, \\ e_{31}^{(0)} &= \tau_0\left(\frac{\partial \Phi_3}{\partial X} - Y\right) + \kappa_1\left[\frac{\partial \Phi_1}{\partial X} + \frac{1}{2}\sigma X^2 + \left(1 - \frac{1}{2}\sigma\right)Y^2\right], \\ e_{32}^{(0)} &= \tau_0\left(\frac{\partial \Phi_3}{\partial Y} + X\right) + \kappa_1\left[\frac{\partial \Phi_1}{\partial Y} + (2+\sigma)XY\right], \end{aligned} \tag{10.5}$$

where ε_0, κ_0, and τ_0 are constants, and Φ_3 and Φ_1 are the functions introduced previously. Otherwise, we leave the remaining components indeterminate for the moment.

As we now have the complete expressions for the strains specified in (10.1b), we may further obtain the stresses from the ordinary stress–strain relations. These stresses must be solutions of the equilibrium equations and the strains must obey the six compatibility conditions (known as the Saint-Venant–Beltrami equations):

$$\frac{\partial^2 e_{22}}{dX^2}+\frac{\partial^2 e_{11}}{\partial Y^2}=\frac{\partial^2 e_{12}}{\partial X\,\partial Y},\ \ldots$$
$$\frac{\partial^2 e_{11}}{\partial Y\,\partial S}=\frac{\partial}{\partial X}\left(-\frac{\partial e_{23}}{\partial X}+\frac{\partial e_{31}}{\partial Y}+\frac{\partial e_{12}}{\partial S}\right),\ \ldots, \tag{10.6}$$

where the dots indicate that the other four equations can be obtained from the two written above by circular permutation of the variables. The result of the substitution is that the equilibrium equations are incompatible unless the constant ε_1 vanishes, and then the compatibility equations are identically satisfied.

At this point, Michell's solution is still complicated, because it requires the solution of an auxiliary problem of plane strain. However, if we limit ourselves to the knowledge of some particular components of strain and stress, namely e_{33}, e_{31}, and e_{32}, and σ_{31} and σ_{32}, then the solution may be achieved by imposing simple equilibrium conditions on the stresses averaged over the cross-section.

Let q_1 denote the component parallel to the X axis of the uniform load per unit of length along the longitudinal S axis, so that we have

$$q_1=\int_{\mathscr{A}(S)}\rho f_1\,dX\,dY+\oint_{\partial\mathscr{A}(S)} t_1\,ds=\text{constant}.$$

The load acting on a slice of beam of length dS must be balanced by the resultants of tangential stresses σ_{31} acting on the faces of the slice, so that we have:

$$\int_{\mathscr{A}(S)}\frac{\partial\sigma_{31}}{\partial S}\,dX\,dY+q_1=0.$$

Since q_1 is independent of S, only the term $\sigma_{31}^{(1)}$ can appear in the equation:

$$q_1=-\int_{\mathscr{A}(S)}\sigma_{31}^{(1)}\,dX\,dY. \tag{10.7}$$

Let us now write the equation

$$\begin{aligned}\int_{\mathscr{A}(S)}\sigma_{31}\,dX\,dY&=\int_{\mathscr{A}(S)}\left[\frac{\partial}{\partial X}(X\sigma_{31})+\frac{\partial}{\partial Y}(X\sigma_{32})-X\left(\frac{\partial\sigma_{31}}{\partial X}+\frac{\partial\sigma_{32}}{\partial Y}\right)\right]dX\,dY\\&=\oint_{\partial\mathscr{A}(S)}X\{\sigma_{31}\cos(X,\nu)+\sigma_{32}\cos(Y,\nu)\}\,ds-\int_{\mathscr{A}(S)}X\frac{\partial\sigma_{33}}{\partial S}\,dX\,dY,\end{aligned} \tag{10.8}$$

where we have used the third equation of equilibrium in the second integral of the last equation. If we examine the right-hand side of (10.8) in more detail, we find that, because we have taken t_3 to be

$$t_3=\sigma_{13}\cos(X,\nu)+\sigma_{23}\cos(Y,\nu)=0,$$

the first integral vanishes. On the other hand, we have

$$\frac{\partial\sigma_{33}}{\partial S}=\sigma_{33}^{(1)}+2S\sigma_{33}^{(2)},$$

whence, from the tentative formulae (10.3) and (10.4), we easily obtain

$$\int_{\mathscr{A}(S)} X\left(\sigma_{33}^{(1)} + 2S\sigma_{33}^{(2)}\right) dX\, dY = -EI_2(\kappa_1 + 2S\kappa_2).$$

Hence we find

$$2EI_2\kappa_2 = q_1. \tag{10.9}$$

Thus the constant κ_2 is determined.

If the body forces and the surface tractions on the cylindrical bounding surface give rise to a couple about the S axis, the moment of this couple is

$$M_3 = \int_{\mathscr{A}(S)} \rho(Xf_2 - Yf_1)\, dX\, dY + \oint_{\partial\mathscr{A}(S)} (Xt_2 - Yt_1)\, ds = \text{constant}.$$

Here again, as M_3 does not depend on S, we find that it is related to the tangential stresses $\sigma_{31}^{(1)}$ and $\sigma_{31}^{(1)}$ by the equation

$$M_3 = -\int_{\mathscr{A}(S)} \left(X\sigma_{31}^{(1)} - Y\sigma_{31}^{(1)}\right) dX\, dY. \tag{10.10}$$

On putting $\sigma_{31}^{(1)} = \mu e_{31}^{(1)}$ and $\sigma_{32}^{(1)} = \mu e_{32}^{(1)}$, and using the expressions for $e_{31}^{(1)}$ and $e_{32}^{(1)}$, we have an equation for determining τ_1. When the twisting couple is zero and the section is symmetrical with respect to the X axis, τ_1 also vanishes.

We have thus determined all the constants except $\varepsilon_0, \kappa_0, \kappa_1$, and τ_0. These constants are equivalent to those which give the solution for an elastic cylinder loaded in the terminal faces only. These constants therefore depend on the force – and couple – resultants of the tractions applied to the terminal sections. In order to evaluate these we must observe that the terms containing κ_2 and τ_1 involve the presence of tractions on the normal sections. Thus the force – and couple – resultants on a terminal section are the result of superimposing the contributions of the terms in κ_2 and τ_1 on those due to the terms in $\varepsilon_0, \kappa_0, \kappa_1$, and τ_0. Since the resultants at the ends are given, we may choose one set of them and impose the condition that they must be equal to the resultants of the tractions on the same face. We have just four conditions to determine $\varepsilon_0, \kappa_0, \kappa_1$, and τ_0.

It remains for us to find the functions Φ_3 and Φ_1. This is a geometrical problem involving the solution of two Neumann problems for the cross-sectional surfaces.

As we have seen, Michell's solution is not complete, because we must still solve a problem of plane strain. However, three of the components of this strain are determined without us needing to solve the auxiliary problem, namely that of finding e_{33}, e_{31}, and e_{32}. However, even with just this partial knowledge of the full solution, we are able to illustrate some special aspects of the state of stress and the deformation of the beam.

The constant ε_0 is the extension of the central line. From the expression for e_{33} we note that this quantity is not proportional to the resultant longitudinal tension, as shown by the following. The curvature of the central line in the XS plane is the value of $\partial^2 u_3/\partial S^2$ for $X = Y = 0$. This curvature may be evaluated in terms of e_{33} and e_{31} through the identity

$$\frac{\partial^2 u_3}{\partial S^2} = \frac{\partial e_{31}}{\partial S} - \frac{\partial e_{33}}{\partial X},$$

from which we derive the formula

$$\frac{\partial^3 u_3}{\partial S^2}(0, 0, S) = \kappa_0 + \kappa_1 S + \kappa_2 S^2.$$

If we now insert the curvature in the expression for e_{33}, we see that the longitudinal extension and curvature of the central line are not simply proportional.

Of the stress components only σ_{31} and σ_{32} can be determined without solving the problem of plane strain. The stress component σ_{33} is not equal to Ee_{33} because the stress components σ_{11} and σ_{22} are not zero. If, however, we calculate the resultant longitudinal tension

$$N = \int_{\mathscr{A}(S)} \sigma_{33}\, dX\, dY,$$

and the bending moment

$$M_2 = -\int_{\mathscr{A}(S)} X\sigma_{33}\, dX\, dY,$$

we find that they can be expressed in terms of the constants of the solution without solving the problem of plane strain (Love 1927, Art. 241). It is also possible to express the bending moment in terms of the curvature but their relationship is no longer simply proportional.

In order to complete the solution we must determine the stress components σ_{11}, σ_{22}, and σ_{12}, which clearly coincide with the components $\sigma_{11}^{(0)}, \sigma_{22}^{(0)}$, and $\sigma_{12}^{(0)}$. To determine the latter, we have the equations of equilibrium:

$$\frac{\partial \sigma_{11}^{(0)}}{\partial X} + \frac{\partial \sigma_{12}^{(0)}}{\partial Y} + \sigma_{13}^{(1)} + \rho f_1 = 0, \tag{10.11a}$$

$$\frac{\partial \sigma_{12}^{(0)}}{\partial X} + \frac{\partial \sigma_{22}^{(0)}}{\partial Y} + \sigma_{23}^{(1)} = 0, \tag{10.11b}$$

$$\frac{\partial \sigma_{13}^{(0)}}{\partial X} + \frac{\partial \sigma_{23}^{(0)}}{\partial Y} + \sigma_{33}^{(1)} = 0, \tag{10.11c}$$

and the boundary conditions

$$\sigma_{11}^{(0)} \cos(X, \nu) + \sigma_{12}^{(0)} \cos(Y, \nu) - t_1 = 0, \tag{10.12a}$$

$$\sigma_{12}^{(0)} \cos(X, \nu) + \sigma_{22}^{(0)} \cos(Y, \nu) = 0, \tag{10.12b}$$

$$\sigma_{13}^{(0)} \cos(X, \nu) + \sigma_{23}^{(0)} \cos(Y, \nu) = 0. \tag{10.12c}$$

Equations (10.11c) and (10.12c) are incompatible unless the constant ε_1 in (10.4) vanishes, thus confirming a result mentioned earlier, but then without proof. It remains to determine $\sigma_{11}^{(0)}, \sigma_{22}^{(0)}$, and $\sigma_{12}^{(0)}$ from (10.11a) and (10.11b) and (10.12a) and (10.12b). This is a standard problem in plane elasticity which may be solved by introducing a stress function (Love 1927, Art. 144). For some particular forms of the cross-section, the stress function can be determined explicitly.

In order to illustrate the theory further, let us consider the case of a beam of rectangular section loaded uniformly along its upper surface, in the absence of body forces. Let $2a$ be the height of the beam, $2b$ its width, and L its length. Let us take the S axis coinciding with the horizontal central line, and the X axis coinciding with the median of the section parallel to the height and directed downwards. Finally, let us assume that the left end $S = 0$ is fixed. If the width is sufficiently small with respect to the other dimensions, we may characterize the solution solely by the stress components $\bar{\sigma}_{11}$, $\bar{\sigma}_{33}$, and $\bar{\sigma}_{13}$ averaged over the thickness, defined by

$$\bar{\sigma}_{11} = \frac{1}{2b}\int_{-b}^{b} \sigma_{11}\, dY, \quad \bar{\sigma}_{33} = \frac{1}{2b}\int_{-b}^{b} \sigma_{33}\, dY, \quad \bar{\sigma}_{13} = \frac{1}{2b}\int_{-b}^{b} \sigma_{13}\, dY.$$

We suppose that the normal traction σ_{22} vanishes throughout the beam; on the other hand, the average values $\bar{\sigma}_{21}$ and $\bar{\sigma}_{23}$ vanish, as σ_{21} and σ_{23} are odd functions of Y on the lateral faces. Stresses of this kind are termed states of *generalized plane stress*.

Now, if $t_1 = p =$ constant is the surface load on the upper face of the beam, the averaged stresses can be expressed as follows (Worch 1967, Vol. II, p. 61):

$$\begin{aligned}
\bar{\sigma}_{11} &= \frac{p}{4a^3}(2a^3 + 3a^2X - X^3),\\
\bar{\sigma}_{33} &= -\frac{3p}{4a^3}(L-S)^2X - \frac{p}{10a^3}(3a^2 - 5X^2)X,\\
\bar{\sigma}_{31} &= \frac{3}{4}\frac{p}{a^3}(L-S)(a^2 - X^2),
\end{aligned} \tag{10.13}$$

which are associated with a bending moment M_2 and a shear force Q_1 given by:

$$M_2 = pb(L-S)^2, \quad Q_1 = 2pb(L-S).$$

Since, by the stress–strain law, we know that

$$\bar{e}_{33} = \frac{1}{E}(\bar{\sigma}_{33} - \sigma\bar{\sigma}_{11}), \quad \bar{e}_{31} = \frac{2(1+\sigma)}{E}\bar{\sigma}_{31},$$

we can immediately find the curvature of the axis by using the identity above:

$$\begin{aligned}
\frac{\partial^2 u_3}{\partial S^2}(0,0,S) &= \frac{\partial \bar{e}_{31}}{\partial S}(0,0,S) - \frac{\partial \bar{e}_{33}}{\partial X}(0,0,S)\\
&= \frac{3}{4}\frac{p}{Ea^2}\left[(L-S)^2 - \left(\frac{8}{5} + \sigma\right)a^2\right].
\end{aligned}$$

The term containing a^2 gives the correction to the curvature due to the uniform load.

The extension of the central line is simply

$$\bar{e}_{33}(0,0,S) = -\frac{\sigma p}{2E}.$$

11. Timoshenko's Correction

Let us consider a cylindrical beam as illustrated in Figure 11.1 (Timoshenko 1921). The S axis is taken to coincide with the line of the centroids of the cross-sections. To

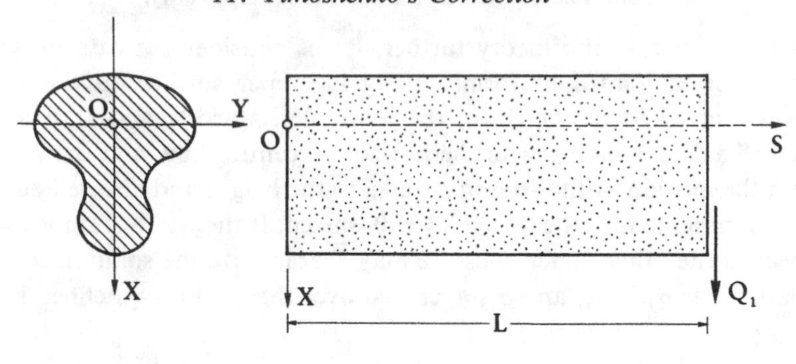

Fig. 11.1

exclude complications arising from the presence of torsional deflections, we assume that the cross-section and the applied loads are symmetrical with respect to the XS plane, which is therefore the plane of deflection of the beam. The beam is loaded by body forces ρf_1 and ρf_2 along the X and Y axes, respectively, with surface tractions t_1 and t_2 along the same axes. For simplicity, we will exclude the S components of the body forces and surface tractions.

If we restrict ourselves to the infinitesimal theory, the problem of determining stresses and strains at any point can be solved exactly only when the exterior forces are applied at the ends or when the beam is loaded uniformly along its length. Of course the solutions to both these problems satisfy the boundary conditions at the ends only in the mean.

In the case of a beam like the one represented in the Figure 11.1, fixed at the left and loaded by forces at the right end, which are statically equivalent to a vertical load Q_1 acting in a line through the centroid of the section, the state of stress is completely given by two harmonic functions Φ_1 and Φ_3 satisfying the boundary conditions (9.14) and (9.15). From the stresses we may calculate the strains, and in particular, the shearing component e_{31}. We may then evaluate this strain component at any point on the central line. Denoting this strain by s_0, we have

$$s_0 = -\frac{Q_1}{EI_2}\frac{\partial \Phi_1}{\partial X}(0, 0).$$

It thus follows that the total deflection of the central line, obtained by the superposition of bending and shearing effects, is

$$u_1(0, 0, S) = \frac{Q_1}{EI_2}\left(\frac{1}{2}S^2 L - \frac{1}{6}S^3\right) + s_0 S,$$

the second term on the right-hand side being known as *Timoshenko's correction.*

In order to extend the above solutions to more complex load conditions, Timoshenko proposed an approximation method for evaluating the transverse displacement u_1 of the centroid of the current section. Instead of trying to solve the problem exactly and then specifying the values of the function $u_1(X, Y, S)$ at points on the S axis, it may occur that, if the dimensions of the cross-section are small with respect the length L, an averaage value of u_1 taken over the cross-section, might, in practice, be almost as useful as the actual value of u_1 at the centroid. We thus define this averaged quantity as the mean deflection of the cross-section.[17]

$$\bar{u}_1 = \frac{1}{A} \int_{\mathscr{A}(S)} u_1 \, dX \, dY, \tag{11.1}$$

where the integration extends over the cross-section of area A.

The equation of balance in the X direction is

$$\frac{\partial \sigma_{11}}{\partial X} + \frac{\partial \sigma_{21}}{\partial Y} + \frac{\partial \sigma_{31}}{\partial S} + \rho f_1 = \rho \frac{\partial^2 u_1}{\partial t^2}, \tag{11.2}$$

where generality has been preserved with the inclusion of inertial terms. Let us now integrate (11.2) over the cross-section, recalling that $\mathscr{A}(S) =$ constant:

$$\int_{\mathscr{A}(S)} \left(\frac{\partial \sigma_{11}}{\partial X} + \frac{\partial \sigma_{21}}{\partial Y} + \rho f_1 \right) dX \, dY + \frac{\partial}{\partial S} \int_{\mathscr{A}(S)} \sigma_{31} \, dX \, dY = \rho \frac{\partial^2}{\partial t^2} \int_{\mathscr{A}(S)} u_1 \, dX \, dY. \tag{11.3}$$

In terms of shearing force Q_1 on the cross-section

$$Q_1 = \int_{\mathscr{A}(S)} \sigma_{31} \, dX \, dY,$$

and the total transverse load p per unit length

$$\begin{aligned} p &= \int_{\mathscr{A}(S)} \left(\frac{\partial \sigma_{11}}{\partial X} + \frac{\partial \sigma_{21}}{\partial Y} + \rho f_1 \right) dX \, dY \\ &= \oint_{\partial \mathscr{A}(S)} [\sigma_{11} \cos(X, \nu) + \sigma_{21} \cos(Y, \nu)] \, ds + \int_{\mathscr{A}(S)} \rho f_1 \, dX \, dY, \end{aligned}$$

equation (11.3) may be written in the form

$$\frac{\partial Q_1}{\partial S} + p = \rho A \frac{\partial^2 \bar{u}_1}{\partial t^2}. \tag{11.4}$$

This is the first equation of Timoshenko's beam theory.

Let us now consider the balance equation in the S direction:

$$\frac{\partial \sigma_{13}}{\partial X} + \frac{\partial \sigma_{23}}{\partial Y} + \frac{\partial \sigma_{33}}{\partial S} = \rho \frac{\partial^2 u_3}{\partial t^2}.$$

Multiplication of this equation by X and integration over $\mathscr{A}(S)$ yields

$$\int_{\mathscr{A}(S)} X \left(\frac{\partial \sigma_{13}}{\partial X} + \frac{\partial \sigma_{23}}{\partial Y} \right) dX \, dY + \frac{\partial}{\partial S} \int_{\mathscr{A}(S)} X \sigma_{33} \, dX \, dY = \rho \frac{\partial^2}{\partial t^2} \int_{\mathscr{A}(S)} X u_3 \, dX \, dY. \tag{11.5}$$

[17] This deduction of Timoshenko's equations is due to Cowper (1966).

Let us introduce the bending moment about the Y axis

$$M_2 = -\int_{\mathscr{A}(S)} X\sigma_{33}\, dX\, dY,$$

and transform the first integral of (11.3) using integration by parts and the divergence theorem. The result is

$$\oint_{\partial\mathscr{A}(S)} X(\sigma_{13}\cos(X,\nu) + \sigma_{23}\cos(Y,\nu))\, dX\, dY - \int_{\mathscr{A}(S)} \sigma_{13}\, dX\, dY - \frac{\partial}{\partial S} M_2$$
$$= \rho\frac{\partial^2}{\partial t^2}\int_{\mathscr{A}(S)} Xu_3\, dX\, dY. \qquad (11.6)$$

In this formula the first integral vanishes as a consequence of the assumption that the surface tractions have no components along the S axis; the second integral is equal to Q_1; and the integral on the right-hand side may be written as

$$\int_{\mathscr{A}(S)} Xu_3\, dX\, dY = -\frac{1}{I_2}\bar{\varphi},$$

where I_2 is the moment of inertia of the section about the Y axis, and $\bar{\varphi}$ represents the mean angle of rotation of the section about the same axis. As, under bending, cross-sections do not remain plane and u_3 is not proportional to X, we take into account the warping from (11.6) in the form

$$\frac{\partial M_2}{\partial S} + Q_1 = \rho I_2\frac{\partial^2\bar{\varphi}}{\partial t^2}. \qquad (11.7)$$

This is the second equation of Timoshenko's beam theory.

In order to derive the relations between the stress resultants M_2 and Q_1, the average displacement $\bar{u}_1$, and rotation $\bar{\varphi}$, we express the displacements u_1 and u_3 in the following form:

$$u_1 = \bar{u}_1 + v_1, \quad u_{23} = \bar{u}_3 - X\bar{\varphi} + v_3, \qquad (11.8)$$

where

$$\bar{u}_3 = \frac{1}{A}\int_{\mathscr{A}(S)} u_3\, dX\, dY$$

is the mean displacement in the S direction. From (11.8) we immediately see that

$$\int_{\mathscr{A}(S)} v_1\, dX\, dY = 0, \quad \int_{\mathscr{A}(S)} v_3\, dX\, dY = 0, \quad \int_{\mathscr{A}(S)} Xv_3\, dX\, dY = 0.$$

Using (11.8), the stress–strain relation

$$\frac{\partial u_1}{\partial S} + \frac{\partial u_3}{\partial X} = \frac{\sigma_{13}}{\mu}$$

may now be written as

$$\frac{\partial \bar{u}_1}{\partial S} - \bar{\varphi} = \frac{\sigma_{13}}{\mu} - \frac{\partial v_3}{\partial X} - \frac{\partial v_1}{\partial S}.$$

If we integrate this equation over $\mathscr{A}(S)$, we obtain

$$A\left(\frac{\partial \bar{u}_1}{\partial S} - \bar{\varphi}\right) = \frac{1}{\mu} \int_{\mathscr{A}(S)} \left(\sigma_{13} - \mu \frac{\partial v_3}{\partial X}\right) dX\, dY. \tag{11.9}$$

Another equation can be obtained from the stress–strain relation:

$$E \frac{\partial u_3}{\partial S} = \sigma_{33} - \sigma(\sigma_{11} + \sigma_{22}).$$

Multiplication of this equation by X and integration over $\mathscr{A}(S)$ yields

$$E \frac{\partial}{\partial S} \int_{\mathscr{A}(S)} u_3\, dX\, dY = \int_{\mathscr{A}(S)} X\sigma_{33}\, dX\, dY - \sigma \int_{\mathscr{A}(S)} X(\sigma_{11} + \sigma_{22})\, dX\, dY,$$

or

$$EI_2 \frac{\partial \bar{\varphi}}{\partial S} = M_2 + \sigma \int_{\mathscr{A}(S)} X(\sigma_{11} + \sigma_{22})\, dX\, dY. \tag{11.10}$$

To sum up the results obtained so far, the problem is reduced to the integration of two balance equations, (11.2) and (11.7), and two averaged constitutive equations, (11.9) and (11.10).

Up to now no approximation has been introduced beyond the assumption that strains are small and that the material is linearly elastic according to Lamé's relations. However, in order to solve the equations, two additional assumptions about the distribution of stress and strain within the beam are needed. The first assumption is that the contribution of the stresses σ_{11} and σ_{22} is negligible. The relation between couple and curvature (11.10) then becomes

$$M_2 = EI_2 \frac{\partial \bar{\varphi}}{\partial S}. \tag{11.11}$$

The second assumption concerns the distribution of shear stresses within each cross-section, knowledge of which would allow us to evaluate the integral in (11.9). If we exclude two small portions near the ends, we know the exact solutions in at least two particular cases of loading: a cantilever beam under a single transverse load at its tip, given by (9.13); and a uniformly loaded beam (Michell's solution). It is natural to adopt one or the other of these distributions yielding a constant shear or one varying linearly along the longitudinal axis respectively. These form the basis of the approximations for more general load conditions, including dynamic loading, provided that the shear force does not vary too rapidly along the length. The cross-section is symmetric with respect to the Y axis, and thus, as there is no twist, the stress components σ_{31} and σ_{32} have a simpler form than that obtainable from (9.13) using stress–strain relations. Their forms are now (Love 1927, Art. 229):

$$\sigma_{31} = -\frac{Q_1}{2(1+\sigma)I_2}\left\{\frac{\partial \Phi_1}{\partial X} + \frac{1}{2}\sigma X^2 + (1 - \frac{1}{2}\sigma)Y^2\right\}, \tag{11.12}$$

$$\sigma_{32} = -\frac{Q_1}{2(1+\sigma)I_2}\left\{\frac{\partial \Phi_1}{\partial Y} + (2+\sigma)XY\right\},$$

where Φ_1 is a harmonic function satisfying the boundary condition (9.14). In order to check the preceding formulae, let us consider the case of a single load Q_1 applied at the right-hand end of the beam, so that the shear force Q_1 is constant. The bending moment at the cross-section a distance S from the fixed end is $Q_1(L-S)$. If we assume that the tension on any element of this section is given by the equation

$$\sigma_{33} = -\frac{Q_1}{I_2}(L-S)X,$$

the third equation of equilibrium becomes

$$\frac{\partial \sigma_{31}}{\partial X} + \frac{\partial \sigma_{32}}{\partial Y} + \frac{Q_1 X}{I_2} = 0, \tag{11.13}$$

and the condition that the cylindrical lateral surface is free from tractions is

$$\sigma_{31}\cos(X,\nu) + \sigma_{32}\cos(Y,\nu) = 0. \tag{11.14}$$

Now by (11.13) and (11.14), we may write

$$\int_{\mathscr{A}(S)} \sigma_{31}\, dX\, dY = \int_{\mathscr{A}(S)} \left[\sigma_{31} + X\left(\frac{\partial \sigma_{31}}{\partial X} + \frac{\partial \sigma_{32}}{\partial Y}\right) + \frac{Q_1 X^2}{I_2}\right] dX\, dY$$

$$= Q_1 + \oint_{\partial\mathscr{A}(S)} X[\sigma_{31}\cos(X,\nu) + \sigma_{32}\cos(Y,\nu)]\, ds = Q_1. \tag{11.15}$$

In a similar manner, observing that $\int_{\mathscr{A}(S)} XY\, dX\, dY$ vanishes, we may prove the result

$$\int_{\mathscr{A}(S)} \sigma_{32}\, dX\, dY = 0. \tag{11.16}$$

The expression for u_3 is given in the form

$$u_3 = Xf(S) - \frac{Q_1}{EI_2}(\Phi_1 + XY^2), \tag{11.17}$$

either for the case of constant shear Q_1 or that of Q_1 linearly variable along the axis, where $f(S)$ is a polynomial in S. For instance, when $Q_1 = \text{constant}$, $f(S)$ is given by (Love 1927, Art. 230)

$$f(S) = -\frac{Q_1}{EI_2}(LS - \frac{1}{2}S^2),$$

although in our work here the exact form of $f(S)$ need not be known.

Assuming that the shear stresses have the form (11.12), and the longitudinal displacement has the form (11.17), we find the expression for $\bar{u}_3$:

$$\bar{u}_3 = -\frac{1}{A}\frac{Q_1}{EI_2}\int_{\mathscr{A}(S)} (\Phi_1 + XY^2)\, dX\, dY,$$

and hence

$$v_3 = \frac{Q_1}{EI_2}\left[-(\Phi_1 + XY^2) + \frac{1}{A}\int_{\mathscr{A}(S)} (\Phi_1 + XY^2)\, dX\, dY + \frac{X}{I_2}\int_{\mathscr{A}(S)} X(\Phi_1 + XY^2)\, dX\, dY \right]. \tag{11.18}$$

We may now evaluate the integral

$$\int_{\mathscr{A}(S)} \left(\sigma_{31} - \frac{\partial v_3}{\partial X}\right) dX\, dY = \frac{Q_1}{2(1+\sigma)I_2}\left[\frac{\sigma}{2}(I_1 - I_2) - \frac{A}{I_2}\int_{\mathscr{A}(S)} X(\Phi_1 + XY^2)\, dX\, dY\right],$$

where

$$I_1 = \int_{\mathscr{A}(S)} Y^2\, dX\, dY.$$

If we put

$$K = 2(1+\sigma)I_2\left[\frac{\sigma}{2}(I_1 - I_2) - \frac{A}{I_2}\int_{\mathscr{A}(S)} X(\Phi_1 + XY^2)\, dX\, dY\right], \tag{11.19}$$

we may write equation (11.9) as

$$\frac{\partial \bar{u}_1}{\partial S} - \bar{\varphi} = \frac{Q_1}{KA\mu}. \tag{11.20}$$

The final form of the equations of Timoshenko's beam theory is then

$$\frac{\partial Q_1}{\partial S} + p = \rho A \frac{\partial^2 \bar{u}_3}{\partial t^2}, \tag{11.21a}$$

$$\frac{\partial M_2}{\partial S} + Q_1 = \rho I_2 \frac{\partial^2 \bar{\varphi}}{\partial t^2}, \tag{11.21b}$$

$$EI_2 \frac{\partial \bar{\varphi}}{\partial S} = M_2, \tag{11.21c}$$

$$\frac{\partial u_1}{\partial S} - \bar{\varphi} = \frac{Q_1}{KA\mu}. \tag{11.21d}$$

For particular forms of the cross-section, the constant K can be evaluated exactly. For example, if $\mathscr{A}(S)$ is a circle of radius a, the function Φ_1 is given by (Love 1927, Art. 231)

$$\Phi_1 = -\left(\frac{3}{4} + \frac{1}{2}\sigma\right)a^2 X + \frac{1}{4}(X^3 - 3XY^2),$$

and the value of K is (Cowper 1966)

$$K = \frac{6(1+\sigma)}{7+6\sigma}. \tag{11.22}$$

There has been a diversity of opinion as to whether it is preferable to work with an integrated expression of the shear displacement or with the pointwise value, as given in Timoshenko's first treatment (Stephen 1981). The answer is that, as in every technical theory of beams integrated stress quantities such as bending moments and shearing forces are used, consistency requires the use of integrated displacement quantities.

Notwithstanding the plausibility of the assumptions, Cowper's derivation of the factor K may be improved further if, in (11.10), we retain the stresses σ_{11} and σ_{22} and we try to obtain a more precise relationship between the moment and the curvature. The improvement can be introduced only in the presence of uniform loads, because the normal stresses σ_{11} and σ_{22} vanish when the beam is loaded only on the terminal faces and Saint-Venant's theory applies.

In order to find the improved value of K without solving the problem of plane strain required by Michell's solution, we consider again the case of the cantilever beam (Figure 11.1). Let us define an "integrated" curvature by the formula (Stephen 1980)

$$\frac{\partial^2 \bar{u}_1}{\partial S^2} = \frac{1}{A}\int_{\mathscr{A}(S)} \left(\frac{\partial e_{31}}{\partial S} - \frac{\partial e_{33}}{\partial X}\right) dX\, dY.$$

If we use (10.1), and give $e_{ij}^{(0)}$, $e_{ij}^{(1)}$ and $e_{ij}^{(2)}$ their values, we obtain

$$e_{33} = \varepsilon_0 - (\kappa_0 + \kappa_1 S + \kappa_2 S^2)X + 2\kappa_2(\Phi_1 + XY^2),$$

$$e_{31} = (\kappa_1 + 2\kappa_2 S)\left[\frac{\partial \Phi_1}{\partial X} + \frac{1}{2}\sigma X^2 + \left(1 - \frac{1}{2}\sigma\right)Y^2\right],$$

where we have exploited the fact that the constants τ_2 and ε_2 must vanish, as also does τ_1 due to the absence of twist. The curvature may then be expressed as

$$\frac{\partial^2 u_1}{\partial S^2} = \frac{\partial e_{31}}{\partial S} - \frac{\partial e_{33}}{\partial X} = (\kappa_0 + \kappa_1 S + \kappa_2 S^2) + \kappa_2 \sigma (X^2 - Y^2),$$

and the integrated curvature becomes

$$\frac{\partial^2 \bar{u}_1}{\partial S^2} = (\kappa_0 + \kappa_1 S + \kappa_2 S^2) + \kappa_2 \sigma \frac{1}{A}(I_1 - I_2). \tag{11.23}$$

From the formula giving the bending moment

$$M_2 = -\int_{\mathscr{A}(S)} X\sigma_{33}\, dX\, dY,$$

we have

$$\frac{\partial M_2}{\partial S} = -\int_{\mathscr{A}(S)} X\left(\sigma_{33}^{(1)} + 2S\sigma_{33}^{(2)}\right) dX\, dY = EI_2(\kappa_1 + 2\kappa_2 S),$$

and this equation shows that M_2 is expressible in the form

$$M_2 = EI_2(\kappa_0 + \kappa_1 S + \kappa_2 S^2) + \text{constant}, \tag{11.24}$$

where the constant added to the right-hand side of (11.24) does not, in general, vanish. To determine this constant and the other constants κ_0, κ_1 and κ_2, let us consider the case in which one $S = 0$ end is held fixed, the $S = L$ end is free from tractions, and the load is statically equivalent to a force p per unit length acting at the centroid of the cross-section in the direction of the X axis. The bending moment is then given by

$$M_2 = \frac{1}{2}p(L - S)^2,$$

and, on comparing this with (11.24), we obtain the values

$$\kappa_1 = -\frac{pL}{EI_2}, \quad \kappa_2 = \frac{1}{2}\frac{p}{EI_2}. \tag{11.25}$$

To determine the other constants we observe that the value of M_2 at $S = 0$ is

$$M_2(0) = EI_2\kappa_0 + \text{constant} = -\int_{\mathscr{A}(S)} X\left[E\varepsilon_{33}^{(0)} + \sigma\left(\sigma_{11}^{(0)} + \sigma_{22}^{(0)}\right)\right] dX\,dY \tag{11.26}$$

and hence

$$M_2 = EI_2(\kappa_0 + \kappa_1 S + \kappa_2 S^2) - \int_{\mathscr{A}(S)} X\left[E\left(\varepsilon_{33}^{(0)} + \kappa_0 X\right) + \sigma\left(\sigma_{11}^{(0)} + \sigma_{22}^{(0)}\right)\right] dX\,dY. \tag{11.27}$$

But we also have the identities (Love 1927, Art. 242)

$$\begin{aligned}
&\int_{\mathscr{A}(S)} X\left(\sigma_{11}^{(0)} + \sigma_{22}^{(0)}\right) dX\,dY \\
&= \int_{\mathscr{A}(S)} \left\{\frac{\partial}{\partial X}\left[\frac{1}{2}(X^2 - Y^2)\sigma_{11}^{(0)} + XY\sigma_{21}^{(0)}\right] + \frac{\partial}{\partial Y}\left[\frac{1}{2}(X^2 - Y^2)\sigma_{12}^{(0)} + XY\sigma_{22}^{(0)}\right]\right. \\
&\quad \left. - \left[\frac{1}{2}(X^2 - Y^2)\left(\frac{\partial\sigma_{11}^{(0)}}{\partial X} + \frac{\partial\sigma_{21}^{(0)}}{\partial Y}\right)\right] + XY\left(\frac{\partial\sigma_{21}^{(0)}}{\partial X} + \frac{\partial\sigma_{22}^{(0)}}{\partial Y}\right)\right\} dX\,dY \\
&= \oint_{\partial\mathscr{A}(S)} \left[\frac{1}{2}(X^2 - Y^2)t_1^{(0)}\right] ds + \int_{\mathscr{A}(S)} \left[\frac{1}{2}(X - Y^2)(\rho f_1 + \sigma_{31}^{(0)}) + XY\sigma_{23}^{(1)}\right] dX\,dY,
\end{aligned} \tag{11.28}$$

where we have used (10.11) and (10.12) to obtain the third term. Hence we have the result

$$\begin{aligned}
M_2 = {} & EI_2(\kappa_0 + \kappa_1 S + \kappa_2 S^2) - \int_{\mathscr{A}(S)} XE(\varepsilon_{33}^{(0)} + \kappa_0 X) - \sigma \oint_{\partial\mathscr{A}(S)} \frac{1}{2}(X^2 - Y^2)t_1^{(0)}\,ds \\
& - \sigma \int_{\mathscr{A}(S)} \left[\frac{1}{2}(X^2 - Y^2)(\rho f_1 + \sigma_{31}^{(1)}) + XY\sigma_{23}^{(1)}\right] dX\,dY.
\end{aligned} \tag{11.29}$$

In this equation the stress and strain components $\sigma_{31}^{(1)}$, $\sigma_{32}^{(1)}$, and $e_{33}^{(0)}$ are given by

$$\sigma_{31}^{(1)} = 2\mu\kappa_2\left[\frac{\partial\Phi_1}{\partial X} + \frac{1}{2}\sigma X^2 + (1 - \frac{1}{2}\sigma)Y^2\right],$$
$$\sigma_{32}^{(1)} = 2\mu\kappa_2\left[\frac{\partial\Phi_1}{\partial Y} + (2 + \sigma)XY\right],$$
$$e_{33}^{(0)} = \varepsilon_0 - \kappa_0 X + 2\kappa_2(\Phi_1 + XY^2).$$

The bending moment is therefore expressed by the equation

$$M_2 = EI_2(\kappa_0 + \kappa_1 S + \kappa_2 S^2) - 2E\kappa_2 \int_{\mathscr{A}(S)} X(\Phi_1 + XY^2)\,dX\,dY$$
$$- \sigma \oint_{\partial\mathscr{A}(S)} \frac{1}{2}(X^2 - Y^2)t_1^{(0)}\,ds$$
$$- \sigma \int_{\mathscr{A}(S)} \frac{1}{2}(X^2 - Y^2)\rho f_1\,dX\,dY - \sigma \int_{\mathscr{A}(S)} \left[\frac{1}{2}(X^2 - Y^2)\sigma_{31}^{(1)} + XY\sigma_{32}^{(1)}\right] dX\,dY. \tag{11.30}$$

As M_2 is equal to $\frac{1}{2}p(L - S)^2$, this equation determines the constant κ_0. The explicit expression for κ_0 is, however, not necessary, since from (11.21c) and (11.21d) we obtain

$$M_2 = EI_2\left(\frac{\partial^2 \bar{u}_1}{\partial S^2} - \frac{1}{KA\mu}\frac{\partial Q_1}{\partial S}\right). \tag{11.31}$$

Alternatively, recalling (11.23), M_2 can be written as

$$M_2 = EI_2(\kappa_0 + \kappa_1 S + \kappa_2 S^2) + \sigma\frac{EI_2}{A}\kappa_2(I_1 - I_2) - \frac{EI_2}{KA\mu}\frac{\partial Q_1}{\partial S}. \tag{11.32}$$

Finally, by comparing (11.30) and (11.32) we arrive at an equation that allows us to determine K. The surprising result is that the new value of K may differ from the one given by (11.19).

It is important to note that a sequence of evaluations to improve the accuracy of the factor K does not necessary improve our knowledge of $\bar{u}_1$, as might well be the case when body forces and surface tractions are not constant (Nicholson and Simmonds 1977).

By way of a particular counterexample, valid only in the special case described here,[18] let us take the same beam shown in Figure 11.1, clamped at the $S = 0$ end and loaded in a manner to be specified shortly. Let us assume that the cross-section is rectangular, of height $2H$ and sufficiently thin for plane stress theory to apply. For simplicity, the thickness of the rectangle is taken to be unity. Thus, with reference to the XS axes in Figure 11.1, the equilibrium equations become

$$\frac{\partial\sigma_{11}}{\partial X} + \frac{\partial\sigma_{31}}{\partial S} + \rho f_1 = 0, \quad \frac{\partial\sigma_{13}}{\partial X} + \frac{\partial\sigma_{33}}{\partial S} + \rho f_3 = 0, \tag{11.33}$$

[18] Cautions on the validity of the counterexample have been made by van der Heijden et al. (1977, pp. 357–360) and by Christensen et al. (1977, pp. 797–799).

and the Lamé equations have the form

$$\sigma_{11} = \frac{E}{(1-\sigma^2)}\left(\frac{\partial u_1}{\partial X} + \sigma\frac{\partial u_3}{\partial S}\right), \quad \sigma_{33} = \frac{E}{(1-\sigma^2)}\left(\frac{\partial u_3}{\partial S} + \sigma\frac{\partial u_1}{\partial X}\right), \tag{11.34}$$

$$\sigma_{31} = \sigma_{13} = \frac{E}{2(1+\sigma)}\left(\frac{\partial u_1}{\partial X} + \frac{\partial u_3}{\partial S}\right).$$

The bending moment M_2 and the shear force Q_1 are related to the mean rotation $\bar{\varphi}$ and to the mean displacement $\bar{u}_1$ by two equations, of the type

$$M_2 = \frac{2}{3}EH^3\frac{\partial\bar{\varphi}}{\partial S}, \quad \bar{K}Q_1 = EH\left(\frac{\partial\bar{u}_1}{\partial S} - \bar{\varphi}\right), \tag{11.35}$$

where $\bar{K}$ is a constant easily expressible in terms of K.

To compare the accuracy of Timoshenko's theory with that of elementary beam theory, we must relate the local displacement $u_1(X, S)$ with the mean displacement $\bar{u}_1(S)$. However, we may define $\bar{u}_1$, not simply as the mean displacement given by (11.1), but in other forms. For instance, it may be defined as the displacement of the center line $u_1(0, S)$, or as the weighted displacement

$$w_1(S) = \frac{3}{4}H\int_{-H}^{H}\left(1 - \frac{X^2}{H^2}\right)u_1(X, S)\,dX; \tag{11.36}$$

or as the "work-consistent" displacement (Nicholson and Simmonds 1977)

$$\delta_1(s) = \frac{1}{Q_1}\int_{-H}^{H}\sigma_{31}(X, S)u_1(X, S)\,dX. \tag{11.37}$$

For the subsequent discussion it is convenient to introduce dimensionless quantities by setting

$$\varepsilon = \frac{H}{L}, \quad S = L\tilde{S}, X = H\tilde{X}, \quad (u_1, u_3) = \frac{\sigma_0 L^2}{EH}(\tilde{u}_1, \varepsilon\tilde{u}_3),$$

$$(\sigma_{11}, \sigma_{31}, \sigma_{33}) = \sigma_0(\tilde{\sigma}_{11}, \tilde{\sigma}_{31}, \tilde{\sigma}_{33}),$$

$$(\rho f_1, \rho f_3) = \frac{\sigma_0}{L}(\rho\tilde{f}_1, \rho\tilde{f}_3), \quad Q_1 = \sigma_0 H\varepsilon\tilde{Q}_1, \quad M = \sigma_0 H^2\tilde{M},$$

$$\bar{u}_1 = \frac{\sigma_0 L^2}{EH}\tilde{\bar{u}}_1, \quad \bar{\varphi}_1 = \frac{\sigma_0 L}{EH}\tilde{\bar{\varphi}}_1,$$

where σ_0 is a reference stress. However, here we drop the tildes, and adopt the tacit convention that we work with the dimensionless quantities.

Let us now take a (dimensionless) state of displacement of the form

$$u_3(X, S) = \beta(S)X - \frac{\varepsilon^2}{3}g'(S)\left(X^3 - \frac{3}{5}X\right), \tag{11.38}$$

$$u_1(X, S) = \eta(S) + \varepsilon^2 g(S)\left(X^2 - \frac{1}{5}\right), \tag{11.39}$$

where, for now, β, g, and η are arbitrary, but regular, functions which satisfy the initial conditions

$$\beta(0) = g(0) = g'(0) = \eta(0) = 0,$$

expressing the clamped nature of the left end of the beam. Once the displacements are given, the associated stresses become

$$\begin{aligned}
(1-\sigma^2)\sigma_{33} &= [\beta'(S) + 2\sigma g(S)]X - \frac{\varepsilon^2}{3} g''(S)\left(X^3 - \frac{3}{5}X\right),\\
(1-\sigma^2)\sigma_{11} &= \varepsilon^{-2}[2g(S) + \sigma\beta'(S)]X - \frac{\sigma}{3} g''(S)\left(X^3 - \frac{3}{5}X\right),\\
2(1+\sigma)\sigma_{31} &= \varepsilon^{-2}[\beta(S) + \eta'(S)].
\end{aligned} \tag{11.40}$$

These stresses must fulfil the equilibrium equations, and, as the body forces have not been specified, it is always possible to determine f_1 and f_2 so that the equilibrium is satisfied.

The stress resultants $Q_1(S)$ and $M_2(S)$ are given by the usual relations:

$$(1+\sigma)Q_1(S) = (1+\sigma)\int_{-1}^{1} \sigma_{31}\, dX = \varepsilon^{-2}[\beta(S) + \eta'(S)],$$

$$(1-\sigma^2)M_2(S) = -(1-\sigma^2)\int_{-1}^{1} X\sigma_{33}\, dX = -\frac{2}{3}[\beta'(S) + 2\sigma g(S)].$$

From these, by using the constitutive relations (11.35), and the boundary conditions $\bar{u}_1(0) = \bar{\varphi}(0) = 0$, we obtain

$$(1-\sigma^2)\bar{\varphi} = -\left[\beta(S) + 2\sigma\int_0^S g(\xi)\, d\xi\right], \tag{11.41}$$

$$(1-\sigma^2)\bar{u}_1 = -\int_0^S \left[\beta(\xi) + 2\sigma(S-\xi)g(\xi)\right] d\xi - \bar{K}(1-\sigma)\int_0^S \left[\beta(\xi) + \eta'(\xi)\right] d\xi. \tag{11.42}$$

In order to discuss the result, we set

$$\beta(S) + \eta'(S) = \varepsilon^2\gamma(S), \quad 2g(S) + \sigma\beta'(S) = \varepsilon^2\zeta(S),$$

where the functions γ, ζ, and η are arbitrary, although they must satisfy the boundary conditions

$$\gamma(0) = \gamma'(0) = \gamma''(0) = 0, \quad \zeta(0) = \zeta'(0) = 0, \quad \eta(0) = \eta'(0) = \eta''(0) = \eta'''(0) = 0,$$

to ensure the requirement that

$$u_1(0, S) = u_3(0, S) = 0.$$

In terms of γ, ζ, and η the vertical deflection calculated according to (11.42) has the form

$$\bar{u}_1 = \eta(S) - \varepsilon^2(1-\sigma^2)^{-1}\int_0^S \{(1-\sigma)[\bar{K}+(1+\sigma)]\gamma(\xi) + \sigma(S-\xi)\zeta(\xi)\}\,d\xi. \quad (11.43)$$

Now the other measures of gross vertical deflection can be calculated from (11.39). Because γ, ζ, and η are arbitrary functions, save for the boundary conditions at $S = 0$, it is clear that, in two-dimensional theory, there is no value of the constant $\bar{K}$ such that there is at least one corresponding expression for $\bar{u}_1$ from (11.43), except possibly for some particular distribution of the body forces. Timoshenko's theory thus predicts a value of $\bar{u}_1$ that is no better, or worse, than that predicted by the elementary theory, in which $\bar{K}$ vanishes.

12. Vlasov's Theory for Thin-walled Beams

Saint-Venant's theory for linearly elastic rods is not applicable to a thin-walled beam the cross-section of which may be described geometrically by means of its median line, the edges of its lateral sides, and its thickness (Vlasov 1958, Kap. XI). We shall take the thickness to be constant (although this assumption is not necessary), and describe this by saying that any normal to the midline cuts the upper and lower sides at two points at distance h from the midline on opposite sides of it. The case in which the midline is open, so that there are edges (Figure 12.1), is much more important than the case of at closed middle curve, because an open section can be deformed into an appreciably different shape without producing in it strains that are too large to be dealt with by using linearized theory.

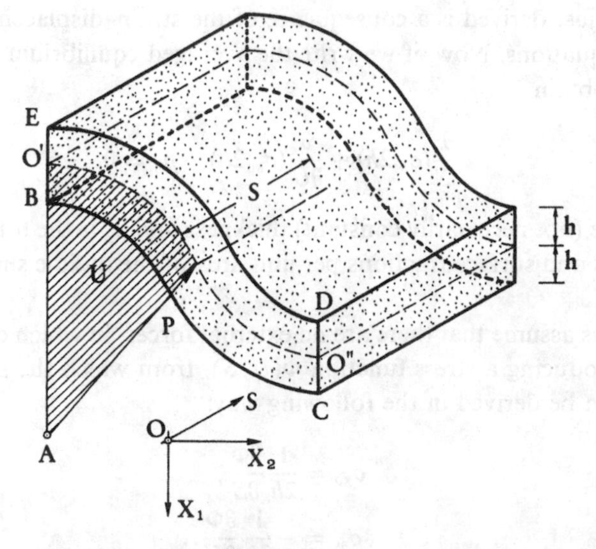

Fig. 12.1

Let us refer the points of the middle cylindrical surface to a system of curvilinear coordinates U, S, where S represents the distance of a current section from a given section, for instance B, C, D, and E, measured along the generators, and U is the arc length along the midline evaluated from a given point on the midline. Let $u(U, S)$ represent the displacement of any point of the middle surface in the S direction, and $v(U, S)$ be the displacement in the direction of the tangent to the line $S =$ constant, which is positive in the direction of increasing U.

We shall assume that the state of strain in the beam is characterized by only two quantities, namely:

$$e_{ss} = \frac{\partial u}{\partial S}, \quad e_{su} = \frac{\partial u}{\partial U} + \frac{\partial v}{\partial S}; \tag{12.1}$$

while elongations of the type $\partial v/\partial U$ in the direction of the tangent to the coordinate lines $S =$ constant are disregarded.

Let σ_{uu} and σ_{ss} denote the normal stresses parallel to U and S lines; the stresses are considered positive if they are tensile and negative if compressive. The tangential stress component σ_{su} is considered positive if, on the area elements the normals of which are directed along the positive S axis, it is directed in the sense of increasing U. These stress components are related to the strains e_{ss} and e_{su} by Lamé's equations:

$$\sigma_{ss} = E\frac{\partial u}{\partial S}, \tag{12.2a}$$

$$\sigma_{su} = \mu\left(\frac{\partial u}{\partial U} + \frac{\partial v}{\partial S}\right). \tag{12.2b}$$

On differentiating (12.2b) with respect to S and (12.2a) with respect U, and then eliminating the mixed second-order partial derivative of u, we obtain

$$-\frac{1}{E}\frac{\partial \sigma_{ss}}{\partial U} + \frac{1}{\mu}\frac{\partial \sigma_{su}}{\partial S} - \frac{\partial^2 v}{\partial S^2} = 0. \tag{12.3}$$

The equation just derived is a consequence of the strain–displacement relationships and Lamé's equations. Now, if we write the averaged equilibrium equation in the S direction we obtain

$$\frac{\partial}{\partial S}(\sigma_{ss}2h) + \frac{\partial}{\partial U}(\sigma_{su}2h) + p_3 = 0, \tag{12.4}$$

where $2h$ is the (not necessarily constant) thickness and p_3 is the total force, inclusive of body forces and surface tractions per unit area of the middle surface in its underformed state.

Firstly, let us assume that there are no external forces, in which case (12.4) may be solved by introducing a stress function $\Phi(U, S)$, from which the stress components σ_{ss} and σ_{su} can be derived in the following way:

$$\sigma_{su} = \frac{1}{2h}\frac{\partial \Phi}{\partial S}, \tag{12.5a}$$

$$\sigma_{ss} = -\frac{1}{2h}\frac{\partial \Phi}{\partial U}. \tag{12.5b}$$

On expressing the stresses in (12.3) by means of the function Φ, and restricting our initial assumption supposing that the thickness depends only on the U coordinate, that is to say that the thickness is constant along the generators, we may write

$$\frac{1}{\mu h(U)}\frac{\partial^2\Phi}{\partial S^2}+\frac{1}{E}\frac{\partial}{\partial U}\left(\frac{1}{h(U)}\frac{\partial\Phi}{\partial U}\right)-\frac{\partial^2 v}{\partial S^2}=0. \tag{12.6}$$

In order to solve this equation we make an assumption about the nature of the displacement v in each cross-section, known as *Vlasov's hypothesis* (Vlasov 1958, Kap. XI, p. 4). Let us take a fixed point in the plane of the section $S=0$, and consider a system of Cartesian axes with the origin at a point such that the S axis is parallel to the generators and the X_1 and X_2 axes are orthogonal axes placed in the plane of the section $S=0$ (Figure 12.1). With reference to the system, let us call $X_1(U)$ and $X_2(U)$ the Cartesian coordinates of a generic point of the middle curve of the initial section. Vlasov's hypothesis states that the displacement component $v(U,S)$ of each point of the central line of a generic section may be expressed in the form

$$v(U,S)=\xi(S)X'_1(U)+\eta(S)X'_2(U)+\theta(S)\omega'(U), \tag{12.7}$$

where ξ, η, and θ are functions to be determined and $\omega'(U)$ is a known function, given as the derivative with respect to U of the so-called *sectorial area*, which has an arbitrary pole at A in the plane $S=0$. This area is double that of the area of the sector demarcated by the arc O′P and the two straight sides AO′ and AP which connect the ends of this arc to the pole A (see Figure 12.1).

If we substitute the expression (12.7) for v in (12.6) we are still not yet able to find Φ because ξ, η, and θ are unknown functions. But there are still three equilibrium equations that apply for any slice of beam of unit width in the S direction. Let $p_1(U,S)$ and $p_2(U,S)$ be the exterior loads per unit area of the middle surface, acting along the X_1 and X_2 axis, respectively. The resultants of these surface loads acting on a slice of beam of unit width are given by

$$q_1(S)=\int_0^{\bar U}p_1(U,S)\,dU,\quad q_2(S)=\int_0^{\bar U}p_2(U,S)\,dU, \tag{12.8}$$

where $\bar U$ is the curvilinear abscissa of the right end-point of the middle arc O′O″ in Figure 12.1 measured from the left end-point O′. These represent the total forces acting on the entire transverse strip per unit width along the generators; accordingly,

$$m_3(S)=\int_0^{\bar U}(p_1X_2-p_2X_1)\,dU, \tag{12.9}$$

is the moment of these forces about an axis through O parallel to the generators in the S direction.

Now, for the global equilibrium of a strip of unit length we must have

$$\int_0^{\bar{U}} \frac{\partial}{\partial S}(\sigma_{su}2h)X_1'(U)\,dU + q_1(S) = 0,$$
$$\int_0^{\bar{U}} \frac{\partial}{\partial S}(\sigma_{su}2h)X_2'(U)\,dU + q_2(S) = 0, \qquad (12.10)$$
$$\int_0^{\bar{U}} \frac{\partial}{\partial S}(\sigma_{su}2h)\omega'(U)\,dU + m_3(S) = 0.$$

On expressing the unknown shearing force $\sigma_{su}2h$ in terms of the stress function Φ in (12.10) and then adding (12.6), we obtain the system

$$\frac{\partial^2\Phi}{\partial S^2} + \frac{\mu}{E}h\frac{\partial}{\partial U}\left(\frac{1}{h}\frac{\partial\Phi}{\partial U}\right) - \mu h(\xi''X_1' + \eta''X_2' + \theta''\omega') = 0, \qquad (12.11a)$$

$$\int_0^{\bar{U}} \frac{\partial^2\Phi}{\partial S^2} X_1'\,dU + q_1 = 0, \qquad (12.11b)$$

$$\int_0^{\bar{U}} \frac{\partial^2\Phi}{\partial S^2} X_2'\,dU + q_2 = 0, \qquad (12.11c)$$

$$\int_0^{\bar{U}} \frac{\partial^2\Phi}{\partial S^2} \omega'\,dU + m_3 = 0. \qquad (12.11d)$$

These equations govern the equilibrium of a thin-walled beam with variable thickness $2h$ along the curves $S = \text{constant}$. They must be completed by the boundary conditions for the function $\Phi(S, U)$ on the longitudinal and transverse edges, and for the functions $\xi(S)$, $\eta(S)$, and $\theta(S)$ along the transverse edges only. In the case of open profiles with longitudinal edges free of shearing forces, and with transverse edges free of normal forces, and prevented from moving in the planes of these edges, the boundary conditions require $\partial\Phi/\partial S$ to be zero along the longitudinal edges, and $\partial\Phi/\partial U$ to be zero along the curvilinear edges; in addition, ξ, η, and θ must correspondingly vanish on these curvilinear edges. On the other hand, when the profile is closed with only one contour line, the function $\Phi(U, S)$ must be periodic in U.

A simple solution to the integrodifferential system (12.11) may be obtained for an open profile if we neglect the shearing deformation; that is, when the product μh in the first equation tends to infinity. We may thus replace this equation by its approximation

$$\frac{\partial\Phi}{\partial U} = Eh[\xi''X_1 + \eta''X_2 + \theta''\omega - \zeta'(U)] \qquad (12.12)$$

where $\zeta'(U)$ is an arbitrary function, which will be left undetermined for the moment.

From (12.12), recalling (12.5b) we find that the expression for the normal longitudinal stress is

$$\sigma_{ss} = \frac{1}{2}E[\zeta'(U) - \xi''X_1 - \eta''X_2 - \theta''\omega]. \tag{12.13}$$

This means that, as an extension of Saint-Venant's theory, the σ_{ss} stress is linear in ω, in addition to X_1 and X_2.

Considering now the other equations (12.11), after an integration by parts we may write

$$\begin{aligned}
&\left[\frac{\partial^2\Phi}{\partial S^2}X_1\right]_0^{\bar{U}} - \int_0^{\bar{U}} X_1\frac{\partial^3\Phi}{\partial S^2\partial U}dU + q_1 = 0,\\
&\left[\frac{\partial^2\Phi}{\partial S^2}X_2\right]_0^{\bar{U}} - \int_0^{\bar{U}} X_2\frac{\partial^3\Phi}{\partial S^2\partial U}dU + q_2 = 0,\\
&\left[\frac{\partial^2\Phi}{\partial S^2}\omega\right]_0^{\bar{U}} - \int_0^{\bar{U}} \omega\frac{\partial^3\Phi}{\partial S^2\partial U}dU + m_3 = 0,
\end{aligned} \tag{12.14}$$

where, if the tangential stress σ_{su} vanishes along the longitudinal sides, the terms within the brackets are necessarily zero. The terms under the sign of integration can be expressed as functions of ξ, η, and θ on observing the equality

$$\frac{\partial^3\Phi}{\partial S^2\partial U} = Eh\left(\xi^{\mathrm{IV}}X_1 + \eta^{\mathrm{IV}}X_2 + \theta^{\mathrm{IV}}\omega\right),$$

and we consequently obtain three ordinary differential equations in ξ, η, and θ. The resulting system may be simplified further by choosing the origin O of the X_1 and X_2 axes and the pole A (which have been arbitrary so far), so that the following orthogonality relations hold:

$$\int_0^{\bar{U}} h(U)X_1\,dU = \int_0^{\bar{U}} h(U)X_2\,dU = \int_0^{\bar{U}} h(U)X_1X_2\,dU = 0,$$

$$\int_0^{\bar{U}} h(U)\omega\,dU = \int_0^{\bar{U}} h(U)\omega X_1\,dU = \int_0^{\bar{U}} h(U)\omega X_2\,dU = 0.$$

We thus arrive at the system of uncoupled equations:

$$\begin{aligned}
&-EI_1\xi^{\mathrm{IV}} + q_1 = 0,\\
&-EI_2\eta^{\mathrm{IV}} + q_2 = 0,\\
&-EI_\omega\theta^{\mathrm{IV}} + m_3 = 0,
\end{aligned} \tag{12.15}$$

where I_1 and I_2 are the moments of inertia of the cross-section with respect to the principal axes, and I_ω is the *sectorial moment of inertia.*

We must still determine the arbitrary function $\zeta(U)$, but we may write an averaged version of the equilibrium equation in the S direction, which, in the absence of longitudinal surface loads, becomes

$$\int_0^{\bar{U}} \frac{\partial}{\partial S}(\sigma_{ss}2h)\, dU + \left[\sigma_{su}2h\right]_0^{\bar{U}} = 0, \tag{12.16}$$

where the term in brackets again vanishes because the lateral sides are free. If we now use (12.13) to replace σ_{ss} under the integral sign in (12.16), and recall the orthogonality conditions, we immediately derive the equation

$$EA\zeta'' = 0, \tag{12.17}$$

where A is the area of the cross-section.

Once we know the conditions that the four functions ξ, η, θ and ζ must satisfy at the faces $S = 0$ and $S = L$, L being the length of the beam, the problem is completely determined.

Vlasov's theory can be extended in several directions. A first attempt (Vlasov 1958, Kap. XI, p. 3) at extending the theory was to remove the hypothesis of plane deformation of each cross-section. This may be done in a relatively simple way by the following argument. Let us assume for the moment that, in a generic slice of beam of unit length, distance S from the origin (Figure 12.2), we fix the face at $U = 0$, and let us consider only the effect of the tangential forces $\partial(\sigma_{su}2h)/\partial S$, neglecting the influence of the normal forces $\partial(\sigma_{ss}2h)/\partial S$, on the deformation of the strip in the plane $S =$ constant. By regarding the transverse strip as a plane beam, we may determine the tangential displacement of any point on abscissa U of its central filament produced by a unit force applied at a point on the abscissa V of the mean filament and directed along the tangent. If we introduce a contour coordinate W, this displacement can be represented in the form

$$K(U, V) = \int_0^{\bar{U}} M_u(W) M_v(W) \frac{dW}{EI}, \tag{12.18}$$

where $M_u(W)$ is the bending moment at W generated by a unit tangential force applied at U and $M_v(W)$ is the bending moment at W due to a unit tangential

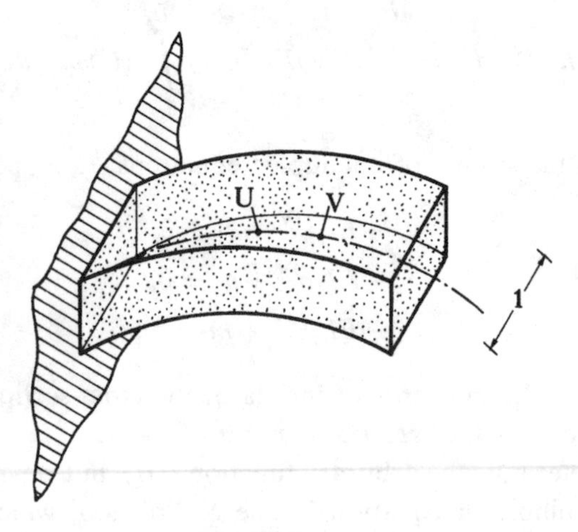

Fig. 12.2

force applied at V; EI, with $I = (1/12)h^3(W)$, is a flexural rigidity of the strip of unit width. Incidentally, note that $K(U, V)$ is symmetric with respect to its variables, as a consequence of Maxwell's theorem of reciprocity between elastic displacements.

In the unit strip considered here, the tangential load will include the force $\partial(\sigma_{su}2h)/\partial S$ directed along the tangent to the mean line. The effect of the entire tangential load, due to these distributed forces, on the tangential displacement of a point on the abscissa U of the central line of a section S = constant, is expressed by the formula

$$v_t(U, S) = \int_0^{\bar{U}} K(U, V) \frac{\partial(\sigma_{su}2h)}{\partial S}\, dU.$$

If we replace $\sigma_{su}2H$ by $\partial\Phi/\partial S$, we obtain

$$v_t(U, S) = \int_0^{\bar{U}} K(U, V) \frac{\partial^2\Phi}{\partial S^2}\, dU. \tag{12.19}$$

In this way we have calculated the additional tangential displacement of a unit strip under the effect of the internal tangential stresses acting on its faces perpendicular to the longitudinal S axis. This analysis could be extended to include exterior loads in the tangential displacement of the central line by using a suitably defined $K(U, V)$.

The result is that the effective displacement of the central line of a cross-section is the superposition of (12.7) and (12.19), and of other terms should the problem require them. Thus (12.11a) must be replaced by the new equation

$$\frac{\partial^2\Phi}{\partial S^2} + \frac{\mu h}{E}\frac{\partial}{\partial U}\left(\frac{1}{h}\frac{\partial\Phi}{\partial U}\right) - \mu h\left(\xi''X_1' + \eta''X_2' + \theta''\omega' + \int_0^{\bar{U}} \frac{\partial^4\Phi}{\partial S^4} K(U, V)dV\right) = 0. \tag{12.20}$$

However, unlike the earlier equations, this is no longer a differential equation, but an integrodifferential equation; the elimination of the function Φ from (12.11b)–(12.11d) is not simple.

Another refinement arising naturally from Vlasov's theory is that of increasing the number of unknown functions in the formulae representing the displacements. Formula (12.7) contains three terms for the v-component of displacement, whereas the u component has been left undetermined with the possibility, at least in principle, of finding it after having solved the problem fully. In a more sophisticated approach we might take

$$\begin{aligned} u(U, S) &= \sum_{i=1}^{m} U_i(S)\varphi_i(U), \\ v(V, S) &= \sum_{i=1}^{n} V_i(S)\psi_i(U), \end{aligned} \tag{12.21}$$

where the functions U_i and V_i are unknown, while the functions φ_i and ψ_i are given. It is clear that, provided the functions ϕ_i and ψ_i are linearly independent in the interval $(0, \bar{U})$, it is not necessary that the choice should be unique. However, as we have seen in the choice of the functions X_1, X_2, and ω, it is useful that they should also be orthogonal, because in the final equations the terms of highest order become uncoupled.

However, despite its popularity, Vlasov's theory is not satisfactory – a criticism applicable to many other theories in solid mechanics.[19] The assumption that some stress components may be neglected and others can be retained is not justified; representing the v displacement in the form (12.7) is acceptable for small overall displacements and rotations of the cross-sections, but these displacements are often large, even though the strains are small. Finally, the "improvements" to the theory that take account of the deformability of the cross-sections in their own planes are themselves only partial, and do no more than open the route to an infinite sequence of improvements, which is reminiscent of Aristotle's argument of the third man.

As a final comment, we should like to suggest that a worthwhile undertaking would be a complete revision of Vlasov's theory extended to large strain and displacements using Antman's method of approximation described in Section 8.

13. The Influence of Concentrated Loads

The state of stress produced in the interior of a beam is given with sufficient accuracy by Saint-Venant's solution for considerable distances from a concentrated load or from a support exerting a concentrated reaction. However, in the neighborhood of concentrated loads or reactions the actual distribution of stresses will be very different from that predicted by Saint-Venant's theory, and the full three-dimensional theory of elasticity must be applied in order to determine the stresses. This problem was tackled by Stokes[20] in its two-dimensional form. Stokes considered a beam of span $2a$ (Figure 13.1), height b, and thickness $2h$ loaded with a concentrated force W applied at the midpoint A of the upper edge, and supported by two rollers at the end-points C and B of its lower edge. If the thickness of the beam is small with respect to the other dimensions, the state of stress in the interior may be regarded as one of generalized plane stress, and is characterized completely by only three average stress components σ_{xx}, σ_{yy}, and σ_{xy} with respect to a suitably placed Cartesian system of reference, having its x,y coordinate plane in the mean plane of the thickness. Taking A as the origin, the x axis is directed along AB$'$ and the y axis along AD.

Stokes's approximation consists in assuming that the force W generates in the neighborhood of A a purely radial distribution of stresses directed towards A of the type

$$\sigma_r = -\frac{2W\cos\theta}{\pi r}, \tag{13.1}$$

[19] For instance, the plate theory for moderately large deformations, proposed by Föppl (1907) and von Kármán (1910).

[20] According to Love (1927, Art. 245) Stokes's theory had been confirmed experimentally by Carus Wilson in 1891.

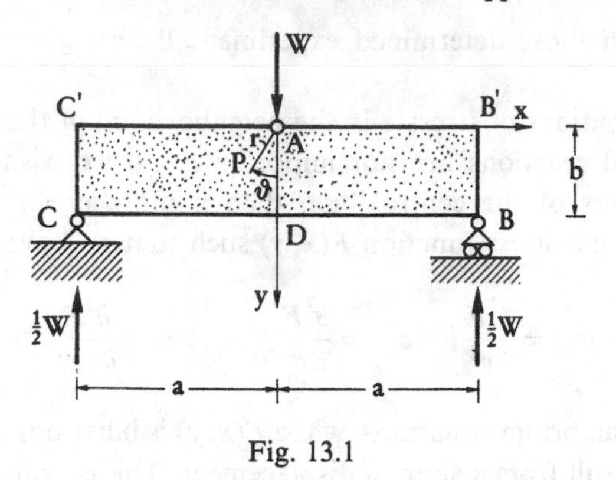

Fig. 13.1

where r is the distance of a generic point P from A, and θ is the angle that AP forms with the y axis. Formula (13.1) is known as Flamant's solution in plane elasticity. In like manner, the reactions $\frac{1}{2}W$ at B and C will give rise to two systems of radial compressive stresses emanating from B and C, respectively. In order to find the actual state of stress in the beam, it is necessary to superimpose a second system of stresses in order to satisfy the boundary conditions that require the sides of the beam to be free from tractions. The complete solution cannot be determined explicitly, but it is still possible to find an additional state of stress that satisfies certain anticipated properties of the stress along the length of AD. If we neglect the influence of the stresses due to the reactions at B and C, we may assume that the fundamental state of stress along AD is simply given by (13.1) and has the form

$$\sigma_y^{(0)} = -\frac{2W}{\pi y}. \tag{13.2}$$

The additional stress along AD is, instead, taken to be of the form

$$\sigma_x^{(1)} = A + By, \quad \sigma_y^{(1)} = Cy, \tag{13.3}$$

where A, B, and C are constants. These constants are determined by requiring that the vertical pressure calculated from the two systems vanishes at D and that calculated from the second system vanishes at A. In addition, the resultant moment of the horizontal tensions about A must obviously be equal to $\frac{1}{2}aW$. From these conditions we obtain the following values for the stress components at any point on AD:

$$\sigma_x = \frac{W}{b}\left(\frac{4}{\pi} - \frac{3a}{b}\right) + \frac{6W}{b^2}\left(\frac{a}{b} - \frac{1}{\pi}\right)y, \tag{13.4}$$

$$\sigma_y = -\frac{2W}{\pi}\left(\frac{1}{y} - \frac{y}{b^2}\right).$$

As the solution is approximate, it is necessary to give a criterion for deciding the restrictions on the ratio a/b within which Stokes's assumptions are legitimate. To this end, Stokes proposed that the ratio a/b must be large enough for there to be points in the interval AD for which $\sigma_x = \sigma_y$. If we impose this condition on solution (13.4), we find that these points are real provided that $6a/b \geqslant 40/\pi$, or approximately $a/b \geqslant 4.25$. For strict inequality there are two such points and their calculated positions agree with those determined experimentally using a method employing polarized light.

If the local distributions of stresses in the neighborhood of the concentrated load and of concentrated reactions are not important, and we wish to determine the vertical displacements of the central line, then a more precise solution may be found by introducing a stress function $F(x, y)$ such that we have

$$\sigma_{xx} = \frac{\partial^2 F}{\partial y^2}, \quad \sigma_{yy} = \frac{\partial^2 F}{\partial x^2}, \quad \sigma_{xy} = -\frac{\partial^2 F}{\partial x \partial y}, \tag{13.5}$$

which satisfy the equilibrium equations, while $F(x, y)$ is biharmonic ensuring that the associated strains result from a state of displacement. This condition is related to the equations known as "Beltrami's equations" (Love 1927, Art. 17).

By exploiting the symmetry, the deflection of the beam considered in Figure 13.1 may be studied considering a half-beam of length a, rigidly clamped along its left vertical edge, and loaded with an unspecified distribution of tangential stresses along the right vertical edge, the resultant of which is $\frac{1}{2}W$ (Figure 13.2).

On taking the x axis to coincide with the longitudinal central-line of the rectangle, and the y axis upwardly directed, it is easy to verify that a function $F(x, y)$ of the form

$$F(x, y) = \frac{W}{2b^2}(b - x)(3b^3 - y^3)y, \tag{13.6}$$

is biharmonic, and the associated stresses

$$\begin{aligned} \sigma_{xx} &= -\frac{6W}{b^3}(a - x)y, \\ \sigma_{yy} &= 0, \\ \sigma_{xy} &= \frac{3W}{b^3}\left(\frac{b^2}{4} - y^2\right), \end{aligned} \tag{13.7}$$

satisfy the boundary conditions $\sigma_{yy} = \sigma_{xy} = 0$ along the edges $y = \pm(b/2)$. The boundary conditions along the edges $x = 0$ and $x = a$ can be satisfied only in the mean. More precisely, the resultant of the tangential stresses at $x = a$ and $-(b/2) \leqslant y \leqslant (b/2)$ is exactly $\frac{1}{2}W$; moreover, the displacements u and v parallel to the x and y axes, have the form (Worch 1967, p. 61)

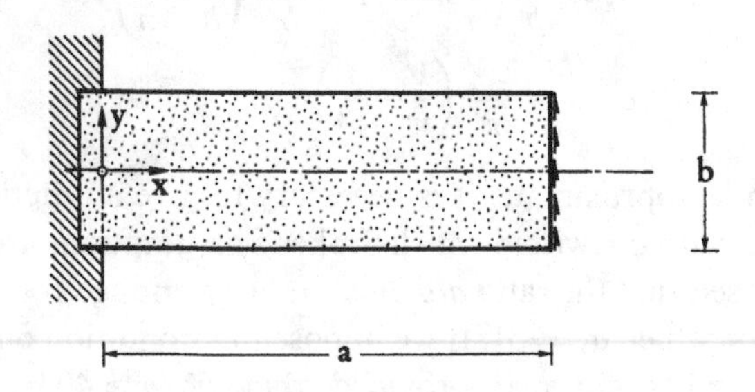

Fig. 13.2

$$Eu = -\frac{6}{b^3}(2a - x)xy + \frac{2}{b^3}(2 + \sigma)\left(\frac{b^2}{4} - y^2\right)y, \tag{13.8a}$$

$$Ev = \frac{2}{b^3}(3a - x)x^2 + \frac{12}{5b}(1 + \sigma)x - \frac{2}{5b^3}\left[(4 - \sigma)\frac{b^2}{2}x - 15\sigma(a - x)y^2\right], \tag{13.8b}$$

where the constants of integration, which must be introduced when constructing the displacements from the strains associated with (13.7), have been determined by the conditions $u = v = 0$ for $x = y = 0$ and $u = 0$ for $x = 0$, $y = b/2$. It is worth noting that the choice of these boundary conditions is not unique (Love 1927, Art. 230).

From (13.8b) we can calculate the vertical displacement of the midpoint of the terminal section $x = a$:

$$v(a, 0) = \frac{4a^3}{Eb^3} + \frac{12}{5b}\frac{(1 + \sigma)}{E}a - \frac{(4 - \sigma)}{5bE}a,$$

which can be put in the equivalent form

$$v(a, 0) = v_0\left[1 + \frac{b^2}{a^2}\left(\frac{3}{10}\frac{E}{\mu} + \frac{\sigma}{20} - \frac{1}{5}\right)\right], \tag{13.9}$$

where v_0 is the displacement of the same point, evaluated using elementary beam theory. A formula that is similar, but uses a slightly different numerical coefficient for the ratio b^2/a^2, has been given by Geckeler (1928, Ziff. 23).

The influence of the concentrated loads has been investigated in a more precise manner by Lamb (1909). He considered the case of an infinite strip of height $2b$ (Figure 13.3) loaded with a vertical body force the value of which, per unit length of the midline $y = 0$, is $H \cos mx$. The displacements induced by this load can be calculated by solving a problem of linear elasticity, and are expressed as:

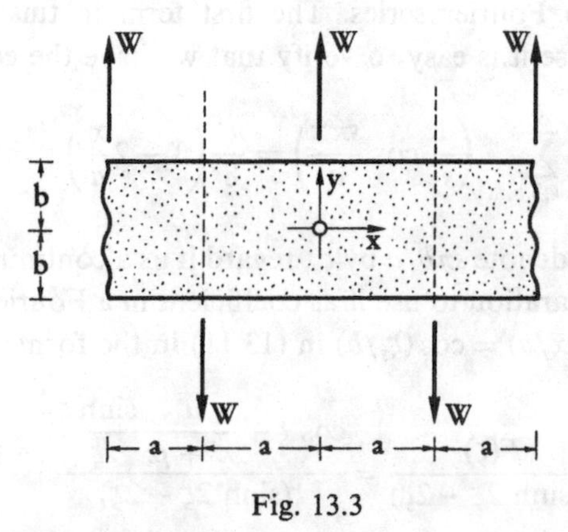

Fig. 13.3

$$u = -\frac{H}{2\mu m}\left[\cosh mb \sinh my - \frac{\sinh mb}{mb}\left(my \cosh my + \frac{\lambda' + 2\mu}{\lambda' + \mu}\sinh my\right)\right]$$
$$\times \frac{\sin mx}{\sinh 2mb - 2mb},$$
$$v = \frac{H}{2\mu m^2 b}\cos mx + \frac{H}{2\mu m}\left[\cosh mb \cosh my\right.$$
$$\left. - \frac{\sinh mb}{mb}\left(my \sinh my - \frac{\mu}{\lambda' + \mu}\cos my\right)\right]\frac{\cos mx}{\sinh 2mb - 2mb}, \tag{13.10}$$

where

$$\lambda' = \frac{2\lambda\mu}{\lambda + 2\mu}.$$

The corresponding stresses can be determined using the stress–strain relations, and it is easy to see that they satisfy the equilibrium equations when the body force is $H \cos mx$.

To calculate the effect of a system of forces concentrated in the sections at a distance a from the origin (Figure 13.3), we must write

$$H = \frac{2W}{ha}, \quad m = \frac{s\pi}{a},$$

where h denotes the width of the beam, and take the sum with respect to s on the right-hand side of (13.10) for all odd values of s. Thus for the vertical displacement of the middle line $y = 0$ we have

$$v(x, 0) = \frac{Wa}{\pi^2 \mu bh}\sum_{s=1,3,5,\ldots}^{\infty}\frac{1}{s^2}\cos\frac{s\pi x}{a}$$
$$+ \frac{Wb}{\mu ha}\sum_{s=1,3,5,\ldots}^{\infty}\frac{\cosh\dfrac{s\pi b}{a} + \dfrac{\mu}{\lambda' + \mu}\dfrac{\sinh\dfrac{s\pi b}{a}}{\dfrac{s\pi b}{a}}}{\dfrac{s\pi b}{a}\left(\sinh\dfrac{2s\pi b}{a} - \dfrac{2s\pi b}{a}\right)}\cos\frac{s\pi x}{a}. \tag{13.11}$$

This is the sum of two Fourier series. The first term in this expression can be evaluated exactly, because it is easy to verify that we have the equation

$$\sum_{s=1,3,5,\ldots}^{\infty}\left(\frac{1}{s^2}\cos\frac{s\pi x}{a}\right) = \frac{\pi^2}{8}\left(1 - 2\frac{x}{a}\right). \tag{13.12}$$

For the second term, we denote $s\pi b/a$ by ζ_s, regard it as a continuous variable ζ, and, following Lamb, in preparation to use it as coefficient in a Fourier series, we express the coefficient of $\cos(s\pi x/a) = \cos(\zeta_s/b)$ in (13.11) in the form

$$\frac{F(\zeta)}{\zeta(\sinh 2\zeta - 2\zeta)} = \frac{\cos\zeta + \dfrac{\mu}{\lambda' + \mu}\dfrac{\sinh\zeta}{\zeta}}{\zeta(\sinh 2\zeta - 2\zeta)}, \tag{13.13}$$

where $F(\zeta)$ is an even function.

The left-hand side is singular at $\zeta = 0$, and thus its Laurent expansion, the coefficients of which are the residues of a meromorphic function of a complex variable, can be written as

$$\frac{F(\zeta)}{\zeta(\sinh 2\zeta - 2\zeta)} = \frac{A}{\zeta^4} + \frac{B}{\zeta^2} + f(\zeta), \tag{13.14}$$

where $f(0)$ is finite for $\zeta = 0$. On finding the residues we obtain

$$A = \frac{3(\lambda' + 2\mu)}{4(\lambda' + \mu)}, \qquad B = \frac{9\lambda' + 8\mu}{\lambda' + \mu}.$$

To complete the evaluation of the displacement $v(x, 0)$ in (13.11) we need the sum

$$\sum_{s=1,3,5,\ldots}^{\infty} \frac{F(\zeta_s)}{\zeta_s(\sinh 2\zeta_s - 2\zeta_s)} \cos \frac{s\pi x}{a},$$

which can now be expressed in the form

$$\frac{Aa^4}{\pi^4 h^4} \sum_{s=1,3,5,\ldots}^{\infty} \frac{1}{s^4} \cos \frac{s\pi x}{a} + \frac{Ba^2}{\pi^2 h^2} \sum_{s=1,3,5,\ldots}^{\infty} \frac{1}{s^2} \cos \frac{s\pi x}{a} + \text{regular terms.}$$

The sum

$$\sum_{s=1,3,5,\ldots}^{\infty} \frac{1}{s^2} \cos \frac{s\pi x}{a}$$

has already been derived in (13.12), and a further application of Fourier theory gives

$$\sum_{s=1,3,5,\ldots}^{\infty} \frac{1}{s^4} \cos \frac{s\pi x}{a} = \frac{\pi^4}{96}\left(1 - \frac{6x^2}{a^2} + \frac{4x^3}{a^3}\right).$$

It then follows that the sum, with $\zeta_s = (s\pi b)/a$, takes the form

$$\sum_{s=1,3,5,\ldots} \frac{F\left(\dfrac{s\pi b}{a}\right)}{\dfrac{s\pi b}{a}\left(\sinh \dfrac{2s\pi b}{a} - \dfrac{2s\pi b}{a}\right)} \cos \frac{s\pi x}{a}$$

$$= \frac{A}{96}\frac{a^4}{h}\left(1 - \frac{6x^2}{a^2} + \frac{4x^3}{a^3}\right) + \frac{B}{8}\frac{a^2}{h^2}\left(1 - \frac{2x}{a}\right) + \text{regular terms,} \tag{13.15}$$

so that (13.11) may be written in the form

$$v(x, 0) = \frac{\lambda' + 2\mu}{128(\lambda' + \mu)\mu} \frac{Wa^3}{bh^3}\left(1 - 6\frac{x^2}{a^2} + 4\frac{x^3}{a^3}\right)$$

$$+ \frac{49\lambda' + 48\mu}{320(\lambda' + \mu)\mu} \frac{Wa}{bh}\left(1 - 2\frac{x}{a}\right) + \text{regular terms.} \tag{13.16}$$

The first part of this expression is identical to the deflection as given by the ordinary Euler–Bernoulli theory, that is

$$v(x,0)=\frac{Wa^3}{48EI}\left(1-6\frac{x^2}{a^2}+4\frac{x^3}{a^3}\right), \tag{13.17}$$

where I is the moment of inertia of the cross-section with respect to a transverse axis through its center, and E is Young's modulus. We have in fact that

$$I=\frac{2}{3}bh^3,\quad E=\frac{\mu(2\mu+3\lambda)}{\lambda+u}=\frac{4\mu(\lambda'+\mu)}{\lambda'+2\mu}.$$

The additional deflection represented by the second term in (13.16) is of the order of h^2/a^2 compared with the former. It indicates that this deflection is a zig-zag line the successive straight portions of which make very obtuse angles with one another at the points of loading. It is also possible to prove that, if the ratio h/a is moderately small, the remaining terms in (13.16) are negligible, except in the immediate neighborhood of the loads.

14. The Influence of Curvature

One of the most important problems in the three-dimensional theory of rods is that of deciding on a suitable form for the local position vector in order to obtain the averaged equations in the simplest form without disregarding certain essential kinematic aspects of the deformation. In prescribing the form of the position we try to impose certain inherent constraints on the nature of the deformation, an operation which is acceptable in some cases but not others.

In order to obtain more definite expressions for the strains, we consider a type of rod more specific than the one discussed in Section 7. The rod is now defined as the solid body generated by a plane domain the centroid of which describes a tortuous curve, and the principal axes of this domain make angles with the principal normals to this curve, which vary from point to point on the curve. In contrast to the definition of a rod given before, we suppose that the line of the centroids of each section is well defined and that each section may be obtained from another by a rigid shift along the central line, each section remaining perpendicular to this curve.

The hypothesis that governs the deformation of sufficiently slender rods is that the cross-sections remain planar and at right angles to the central line, and undergo no strain in these planes. This characterization of the kinematic structure of the deformation of a rod is known as *Kirchhoff's hypothesis*. This theory has the advantage of taking into account the fact that the relative displacements of the parts of a long thin rod may not be small, although the strains that occur in any part of the rod are, in general, small enough to justify adopting the infinitesimal theory. If, in addition, the displacements are also small, then the resulting equations may be simplified further.

A detailed analysis of the state of strain in a bent and twisted rod was made by Love (1927, Art. 256, 259) under more general assumptions than those postulated in Kirchhoff's theory. With its greater generality, Love's treatment is able to consider, in addition, large strains and the possible curvature and twist of the rod on its initial unstressed state.

In order to explain Love's theory in an accessible form, let us assume that, in the reference configuration, at any point on the central line there is a tangent which may be associated with a perpendicular cross-section. If the rod is bent or twisted, the curvature of the longitudinal fibres changes. The line elements that are initially linear in the plane of each cross-section and are initially perpendicular to one another, do not, in general, continue to be at right angles or remain perpendicular to the tangent to the strained central line. As we will see, this theory, in contrast to Kirchhoff's theory, allows for the axial extension and shear of the cross-section with respect to the axis.

In order to construct a theory with sufficient geometric structure to include the flexure, torsion, axial extension, and shear of cross-sections, we assume, in the reference configuration, a system of X, Y, S coordinates, such that S denotes the length along the central line, measured from a given fixed point, and X and Y denote the distances along lines orthogonal to the reference line S.

The position of a particle of the rod in the unstrained configuration is thus identified by the vector

$$\mathbf{R} = \mathbf{R}(X, Y, S),$$

and the vector

$$\mathbf{R}_0 = \mathbf{R}(0, 0, \mathbf{S})$$

represents the position of a particle P situated on the central line (Figure 14.1(a)). It thus follows that, at any point on the reference configuration, we can define the base vectors

$$\mathbf{G}_p = \frac{\partial \mathbf{R}}{\partial X^p},$$

where X^p $(p = 1, 2, 3)$ stands for the triad (X,Y,S). In particular, the vectors $\mathbf{G}_p$ may be defined for the points on the central line, and in this case we put

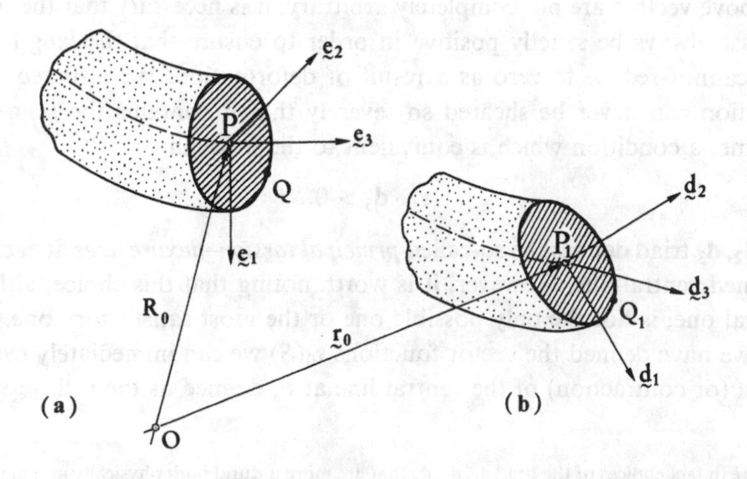

Fig. 14.1

$$\mathbf{G}_p(0, 0, S) = \mathbf{e}_p,$$

and we observe that $\mathbf{e}_p$ are orthogonal unit vectors since the coordinate $X^p = (X, Y, S)$ are arc lengths along orthogonal lines.

The position vector of a particle in the reference configuration may also be written as

$$\mathbf{R}(X, Y, S) = \mathbf{R}_0 + X\mathbf{e}_1 + Y\mathbf{e}_2. \tag{14.1}$$

In the deformed state, each particle that occupied an initial position $\mathbf{R}$ is now located by a vector

$$\mathbf{r} = \mathbf{r}(X, Y, S),$$

and the point P of the strained central line is located by the vector (Figure 14.1(b))

$$\mathbf{r}_0 = r(0, 0, S).$$

In the strained configuration the base vectors are given by

$$\mathbf{g}_p = \frac{\partial \mathbf{r}}{\partial X^p},$$

and their values along the central line, that is to say for $X = Y = 0$, represent the transformed elements of the triad $\mathbf{e}_1, \mathbf{e}_2, \mathbf{e}_3$. However, in contrast to this triad, the new vectors are no longer orthogonal or unitary, the only conserved property being that $\mathbf{g}_3$ is still a tangent to the strained central line. In order to be able to use an orthonormal triad also at the points of the strained central line, we introduce a new triad $\mathbf{d}_1, \mathbf{d}_2, \mathbf{d}_3$, defined as follows: $\mathbf{d}_3$ coincides with $\mathbf{g}_3$, but has unit length; $\mathbf{d}_1$ is the unit vector perpendicular to $\mathbf{d}_3$ and contained in the $\mathbf{d}_3\mathbf{g}_1$ plane; $\mathbf{d}_2$ is the unit vector perpendicular to $\mathbf{d}_1\mathbf{d}_3$ and directed so as to make the triad $\mathbf{d}_1, \mathbf{d}_2, \mathbf{d}_3$ right-handed.

In the deformed state the vector function $\mathbf{r}_0$ and the unit vectors $\mathbf{d}_1, \mathbf{d}_2$, and $\mathbf{d}_3$ depend on the variable S, the arc length parameter of the central line in the reference configuration, with the condition $0 \leqslant S \leqslant L$. We also require derivatives with respect to S for which, in this section, the primed notation will be used, for example, $\mathbf{r}_0' = \mathrm{d}\mathbf{r}_0/\mathrm{d}S$.

The above vectors are not completely arbitrary; it is necessary that the magnitude of $\mathbf{r}_0'$ must always be strictly positive in order to ensure that the length of any S interval cannot reduce to zero as a result of deformation. We also require that a cross-section can never be sheared so severely that it contains the tangent to the central line, a condition which is equivalent to the inequality

$$\mathbf{r}_0' \cdot \mathbf{d}_3 > 0. \tag{14.2}$$

The $\mathbf{d}_1, \mathbf{d}_2, \mathbf{d}_3$ triad defines the so-called *principal torsion–flexure axes* at each point of the strained central line. However, it is worth noting that this choice, although the traditional one, is not the only possible one or the most satisfactory one.[21]

Once we have defined the vector functions $\mathbf{r}_0(S)$ we can immediately evaluate the extension (or contraction) of the central line at P, defined as the ratio

[21] There are in fact choices of the triad $\mathbf{d}_1, \mathbf{d}_2, \mathbf{d}_3$ that are more natural both physically and mathematically (see Antman 1974).

$$\varepsilon = \frac{ds - dS}{dS}, \tag{14.3}$$

where ds is the arc element of the strained central line given by the formula

$$ds = |\mathbf{r}_0'(S)|\, dS = \sqrt{x'^2 + y'^2 + z'^2}\, dS, \tag{14.4}$$

where x, y, and z are the coordinates of P_1 with respect the basis $\mathbf{e}_1, \mathbf{e}_2, \mathbf{e}_3$. In addition to the extension, we can also calculate the other quantities that characterize the local form of the central line after deformation, namely the curvatures κ and κ' and the twist τ. These quantities represent the rigid rotation about the axes of $\mathbf{d}_1$, $\mathbf{d}_2$,and $\mathbf{d}_3$ when passing from the point to a point P_1' in the neighborhood. From analytical mechanics, the quantities can be expressed in the form

$$\kappa = \frac{d\mathbf{d}_2}{ds} \cdot \mathbf{d}_3, \quad \kappa' = \frac{d\mathbf{d}_3}{ds} \cdot \mathbf{d}_1, \quad \tau = \frac{d\mathbf{d}_1}{ds} \cdot \mathbf{d}_2. \tag{14.5}$$

So far we have not restricted the magnitude of ε, except for the fact that $|\mathbf{r}_0'|$ must be greater than zero. However, for any application of the classical theory to be possible, we must assume ε to be small and of the order of magnitude required by the theory. An immediate consequence of the smallness of ε is that differentiation with respect s in (14.5) may be replaced by differentiation with respect S, provided that we multiply κ, κ', and τ by $(1 + \varepsilon)$ and neglect quantities such as $\varepsilon\kappa$, $\varepsilon\kappa'$, and $\varepsilon\tau$. It turns out that the quantities κ, κ', and τ may then be estimated by the formulae

$$\kappa = \frac{d\mathbf{d}_2}{dS} \cdot \mathbf{d}_3 = \mathbf{d}_2' \cdot \mathbf{d}_3, \tag{14.6a}$$

$$\kappa' = \frac{d\mathbf{d}_3}{dS} \cdot \mathbf{d}_1 = \mathbf{d}_3' \cdot \mathbf{d}_1, \tag{14.6b}$$

$$\tau = \frac{d\mathbf{d}_1}{dS} \cdot \mathbf{d}_2 = \mathbf{d}_1' \cdot \mathbf{d}_2. \tag{14.6c}$$

Equations (14.6b) and (14.6c), together with the condition $\mathbf{d}_1 \cdot \mathbf{d}_1' = 0$ can be used to invert the system, giving

$$\mathbf{d}_1' = -\kappa'\mathbf{d}_3 + \tau\mathbf{d}_2, \quad \mathbf{d}_2' = \kappa\mathbf{d}_3 - \tau\mathbf{d}_1, \quad \mathbf{d}_3' = \kappa'\mathbf{d}_1 - \kappa\mathbf{d}_2. \tag{14.7}$$

In a similar manner, the initial curvature and twist are given by

$$\kappa_0 = \mathbf{e}_2' \cdot \mathbf{e}_3, \quad \kappa_0' = \mathbf{e}_3' \cdot \mathbf{e}_1, \quad \tau = \mathbf{e}_1' \cdot \mathbf{e}_2, \tag{14.8}$$

and hence the derivatives of $\mathbf{e}_1$, $\mathbf{e}_2$, and $\mathbf{e}_2$ with respect to S become

$$\mathbf{e}_1' = -\kappa_0'\mathbf{e}_3 + \tau_0\mathbf{e}_2, \quad \mathbf{e}_2' = \kappa_0\mathbf{e}_3 - \tau_0\mathbf{e}_1, \quad \mathbf{e}_3' = \kappa_0'\mathbf{e}_1 - \kappa_0\mathbf{e}_2. \tag{14.9}$$

In accordance with this result we also have

$$\mathbf{G}_1 = \mathbf{e}_1, \tag{14.10a}$$

$$\mathbf{G}_2 = \mathbf{e}_2, \tag{14.10b}$$

$$\begin{aligned} \mathbf{G}_3 &= \mathbf{R}_{0,3} + X\mathbf{e}_1' + Y\mathbf{e}_2' \\ &= (1 - \kappa_0'X + \kappa_0 Y)\mathbf{e}_3 - \tau_0 Y\mathbf{e}_1 + \tau_0 X\mathbf{e}_2. \end{aligned} \tag{14.10c}$$

It is worth noting that $\mathbf{G}_3$ is not a unit vector and is not normal to the $\mathbf{e}_1\mathbf{e}_2$ plane when τ_0 is nonzero.

We now require the strains at any point on the rod originating from an additional deformation which takes any vector $\mathbf{R}(X, Y, S)$ into a new vector $\mathbf{r}(X, Y, S)$. In general, the triad $\mathbf{g}_p(0, 0, s)$, the transform of $\mathbf{e}_1, \mathbf{e}_2, \mathbf{e}_3$, above, is no longer orthonormal. However, the orthonormal triad $\mathbf{d}_1, \mathbf{d}_2, \mathbf{d}_3$, which we introduced above, can be obtained from $\mathbf{e}_1, \mathbf{e}_2, \mathbf{e}_3$ by a rigid motion.

If we know the vectors $\mathbf{r}(X, Y, S)$ and $\mathbf{R}(X, Y, S)$, the strain components are defined as (see, for example, Wempner 1981, 8.4)

$$e_{pp} = \frac{1}{2}\left(\frac{\partial \mathbf{r}}{\partial S_p}\cdot\frac{\partial \mathbf{r}}{\partial S_p} - \frac{\partial \mathbf{R}}{\partial S_p}\cdot\frac{\partial \mathbf{R}}{\partial S_p}\right),$$

$$e_{pq} = \left(\frac{\partial \mathbf{r}}{\partial S_p}\cdot\frac{\partial \mathbf{r}}{\partial S_q} - \frac{\partial \mathbf{R}}{\partial S_p}\cdot\frac{\partial \mathbf{R}}{\partial S_q}\right) \quad (p \neq q), \tag{14.11}$$

where S_p is the length along the undeformed curvilinear coordinate axes. In the present case we have $S_1 = X, S_2 = Y$, but $dS_3 = \sqrt{G_{33}}dS$, since $\mathbf{G}_3$ is not a unit vector. Therefore the strains have the form

$$e_{11} = \frac{1}{2}(\mathbf{g}_1\cdot\mathbf{g}_1 - 1), \quad e_{22} = \frac{1}{2}(\mathbf{g}_2\cdot\mathbf{g}_2 - 1),$$

$$e_{33} = \frac{1}{2}\left(\mathbf{g}_3\cdot\mathbf{g}_3\frac{1}{G_{33}} - 1\right),$$

$$e_{23} = \left(\mathbf{g}_2\cdot\mathbf{g}_3\frac{1}{\sqrt{G_{33}}} - \mathbf{G}_2\cdot\mathbf{G}_3\frac{1}{\sqrt{G_{33}}}\right), \quad e_{13} = \left(\mathbf{g}_1\cdot\mathbf{g}_3\frac{1}{\sqrt{G_{33}}} - \mathbf{G}_1\cdot\mathbf{G}_3\frac{1}{\sqrt{G_{33}}}\right),$$

$$e_{12} = \mathbf{g}_1\cdot\mathbf{g}_2. \tag{14.12}$$

Once we know $\mathbf{r}(X, Y, S)$ and, of course, the shape of the rod before the deformation, formulae (14.12) give the strains. However, it is often useful to write the strains in terms of displacements, that is to say by introducing the displacement vector $\boldsymbol{\xi} = (\xi, \eta, \zeta)$ such that

$$\mathbf{r}(X, Y, S) = \mathbf{R}(X, Y, S) + \boldsymbol{\xi}(X, Y, S). \tag{14.13}$$

It thus follows that a point Q, which before the deformation had the position vector

$$\mathbf{R} = \mathbf{R}_0 + X\mathbf{e}_1 + Y\mathbf{e}_2, \tag{14.14}$$

moves into a point Q_1 at the position

$$\mathbf{r} = \mathbf{r}_0 + (X + \xi)\mathbf{d}_1 + (Y + \eta)\mathbf{d}_2 + \zeta\mathbf{d}_3, \tag{14.15}$$

provided we assume that the unit vectors $\mathbf{e}_1, \mathbf{e}_2$, and $\mathbf{e}_3$ are displaced bodily into the triad $\mathbf{d}_1, \mathbf{d}_2, \mathbf{d}_3$, instead of into the triad $\mathbf{g}_1(0, 0, S)$, $\mathbf{g}_2(0, 0, S)$, $\mathbf{g}_3(0, 0, S)$. If the difference between these two triads is small, (14.15) constitutes a good approximation for the vector $\mathbf{r}$ in terms of $\mathbf{d}_1, \mathbf{d}_2, \mathbf{d}_3$.[22]

[22] This approximation is tacit, but not declared, in Love's treatment (1927, Art. 256).

From (14.15) we can easily derive the following:

$$\begin{aligned}
\mathbf{g}_1 &\sim \left(1+\frac{\partial\xi}{\partial X}\right)\mathbf{d}_1 + \frac{\partial\eta}{\partial X}\mathbf{d}_2 + \frac{\partial\zeta}{\partial X}\mathbf{d}_3, \\
\mathbf{g}_2 &\sim \frac{\partial\xi}{\partial Y}\mathbf{d}_1 + \left(1+\frac{\partial\eta}{\partial Y}\right)\mathbf{d}_2 + \frac{\partial\zeta}{\partial Y}\mathbf{d}_3, \qquad (14.16) \\
\mathbf{g}_3 &\sim \mathbf{d}_3 + \frac{\partial\xi}{\partial S}\mathbf{d}_1 + \frac{\partial\eta}{\partial S}\mathbf{d}_2 + \frac{\partial\eta}{\partial S}\mathbf{d}_3 + (X+\xi)\mathbf{d}_1' + (Y+\eta)\mathbf{d}_2' + \zeta\mathbf{d}_3'.
\end{aligned}$$

Since, from (14.12), we know the vectors $\mathbf{G}_1$, $\mathbf{G}_2$, and $\mathbf{G}_3$, the application of (14.12) permits us to determine all the strain components at any point on the rod in terms of the coordinates X, Y, the displacements ξ, η, and ζ, and the curvatures. The formulae obtained may be simplified further by neglecting the products of terms that are small relative to the others, and this operation gives us the freedom of choice over several variants concerning the definition of strains in rods.

When the central line is rectilinear in the unstrained state and we neglect all terms of higher order in the expression of the strains, formula (14.12) gives

$$\begin{aligned}
e_{11} &= \frac{\partial\xi}{\partial X}, \quad e_{22} = \frac{\partial\eta}{\partial Y}, \quad e_{12} = \frac{\partial\xi}{\partial Y} + \frac{\partial\eta}{\partial X}, \\
e_{13} &= \frac{\partial\zeta}{\partial X} + \frac{\partial\xi}{\partial S} - \tau Y, \quad e_{23} = \frac{\partial\zeta}{\partial Y} + \frac{\partial\eta}{\partial S} + \tau X, \qquad (14.17) \\
e_{33} &= \varepsilon - \kappa' X + \kappa Y + \frac{\partial\zeta}{\partial S},
\end{aligned}$$

where ε is the linear dilatation defined by (14.3).

A second step in the simplification of the formulae for strains comes from the fact that quantities like $\partial\xi/\partial S$, $\partial\eta/\partial S$, and $\partial\zeta/\partial S$ are, in general, small compared with the partial derivatives of ξ, η, and ζ with respect to X and Y. Therefore on omitting $\partial\xi/\partial S$, $\partial\eta/\partial S$, and $\partial\zeta/\partial S$ from (14.17), we obtain the formulae

$$e_{13} = \frac{\partial\zeta}{\partial X} - \tau Y, \quad e_{23} = \frac{\partial\zeta}{\partial Y} + \tau X, \quad e_{33} = \varepsilon - \kappa' X + \kappa Y. \qquad (14.18)$$

If, alternatively, the rod in its unstrained state possesses curvature and twist, and we neglect the derivatives (as above) and the displacements ξ, η, and ζ compared with respect to the coordinates X, Y, we obtain the relations

$$\begin{aligned}
\mathbf{g}_1 &= \left(1+\frac{\partial\xi}{\partial X}\right)\mathbf{d}_1 + \frac{\partial\eta}{\partial X}\mathbf{d}_2 + \frac{\partial\zeta}{\partial X}\mathbf{d}_3, \\
\mathbf{g}_2 &= \frac{\partial\xi}{\partial Y}\mathbf{d}_1 + \left(1+\frac{\partial\eta}{\partial Y}\right)\mathbf{d}_2 + \frac{\partial\zeta}{\partial Y}\mathbf{d}_3, \\
\mathbf{g}_3 &= \mathbf{d}_3 + X\mathbf{d}_1' + Y\mathbf{d}_2' = (1-\kappa' X + \kappa Y)\mathbf{d}_3 - \tau Y\mathbf{d}_1 + \tau X\mathbf{d}_2,
\end{aligned}$$

while $\mathbf{G}_1$, $\mathbf{G}_2$, and $\mathbf{G}_3$ are given by (14.10). It follows that, on retaining only the terms of first order, the strains e_{13}, e_{23}, and e_{33}, defined by (14.12), assume the forms

$$e_{13} = \frac{\partial\zeta}{\partial X} - (\tau-\tau_0)Y, \quad e_{23} = \frac{\partial\zeta}{\partial Y} + (\tau-\tau_0)X, \quad e_{33} = (\kappa-\kappa_0)Y - (\kappa'-\kappa_0')X. \qquad (14.19)$$

These formulae were derived by Clebsch (1862, Sec. 55).

However, when the transverse dimensions of the cross-section are not small compared to the extent of the longitudinal axis, the formulae for the strains derived above are no longer valid. This is because the cross-sections, initially perpendicular to the unstrained central line, undergo strains and shears that cannot be described analytically by a vector function such as (14.15). The problem arises when we wish to calculate the strains and the stresses in a thick beam for which the radius of the initial curvature of the central axis has a magnitude comparable with the diameter of the cross-section. A particular example is the problem of determining the strain and stress in a beam the original form of which is annular, with an exterior radius b, an interior radius $a\,(b > a)$, a rectangular cross-section of height $(b - a)$, and a thickness h (Figure 14.2). If h is small compared to the height, the state of stress in the beam can be reasonably regarded as a plane generalized state, which is tractable using the method employing Airy's stress function.

Let us assume that the bar is bent in the plane by two equal and opposite couples $\bar{M}$ applied at the ends, the circular boundaries being free from stress. If we take a system of plane polar coordinates with its origin O at the center and an angular argument φ measured from a horizontal straight line drawn from the origin, the midsection of the beam, perpendicular to the thickness, is defined by the domain

$$a \leqslant r \leqslant b, \quad -\alpha \leqslant \varphi \leqslant \alpha,$$

where 2α represents the angular opening of the annular sector.

In polar coordinates, a state of stress that satisfies the boundary conditions $\sigma_r = \tau_{r\varphi} = 0$ along the circular sides, and such that

$$\int_a^b \sigma_\varphi \left[\frac{1}{2}(b + a) - r\right] dr = \bar{M}, \tag{14.20}$$

is derived from the following expression of the stress function (Worch 1967, III):

$$F = -\frac{2}{n}\bar{M}\left[\frac{b^2}{b^2 - a^2}r^2 - 2\frac{a^2b^2}{b^2 - a^2}\ln r - \frac{r^2}{\ln\dfrac{b}{a}}\left(\ln\frac{r}{a} - \frac{1}{2}\right)\right], \tag{14.21}$$

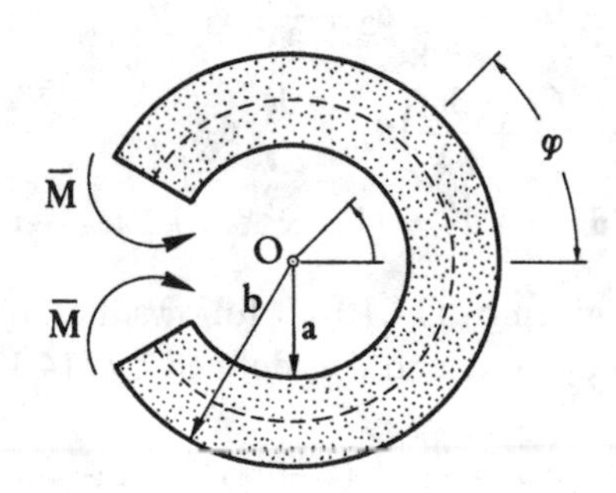

Fig. 14.2

where n is the constant

$$n = \frac{b^2 - a^2}{\ln\frac{b}{a}} - 4\frac{a^2b^2}{b^2 - a^2}\ln\frac{b}{a}. \tag{14.22}$$

From F we can easily obtain the stresses

$$\sigma_r = -\frac{4\bar{M}}{n}\left[\frac{b^2}{b^2 - a^2}\frac{r^2 - a^2}{r^2} - \frac{\ln\frac{r}{a}}{\ln\frac{b}{a}}\right],$$

$$\sigma_\varphi = -\frac{4\bar{M}}{n}\left[\frac{b^2}{b^2 - a^2}\frac{r^2 + a^2}{r^2} - \frac{\ln\frac{r}{a}}{\ln\frac{b}{a}} - \frac{1}{\ln\frac{b}{a}}\right], \tag{14.23}$$

$$\tau_{r\varphi} = 0;$$

and the displacements

$$u_r = -\frac{4\bar{M}}{En}\left\{\frac{1}{\ln\frac{b}{a}} + \frac{b^2}{b^2 - a^2}\left[(1-\sigma) + (1+\sigma)\left(\frac{a}{r}\right)^2\right] - (1-\sigma)\frac{\ln\frac{r}{a}}{\ln\frac{b}{a}}\right\}r,$$

$$u_\varphi = \frac{8\bar{M}}{n}\frac{1}{\ln\frac{b}{a}}r\varphi. \tag{14.24}$$

This stress and displacement distribution represent the exact solution of the elastic problem, provided that the stresses σ_φ on the faces $\varphi = \pm\alpha$ have the form prescribed by the second of (14.23). If the forces giving $\bar{M}$ are distributed over the ends in some other manner, the solution (14.21) is only approximate. In any case, the law of distribution of σ_φ along each section $\varphi =$ constant is never linear in r. However, if $(b - a)$ is small compared to the radius of the central axis, the difference from linearity is not relevant. Finally, it should be observed that the longitudinal normal stress σ_φ is associated with a transverse normal stress σ_r, which is always zero in elementary beam theory.

The solution for a beam having the form of an annular sector, although more accurate than the elementary solution, is not applicable to profiles in which the stress-sustaining part is not an annulus but a circular segment bounded by a circular arc $\widehat{\text{OA}\Omega}$ and a rectilinear segment OBΩ (Figure 14.3). The chord BA, which divides the circular segment into two symmetric parts, is the cross-section of the beam the axis of which is a straight line passing through O and Ω. The region OAΩB is the midsection of a thin plate of thickness $2h$, with a generalized plane stress generated by two equal and opposite horizontal forces Q applied at the vertices O and Ω. A region of this shape may be regarded as a reasonable model for studying the distribution of stresses within the central part of the hook of a railway carriage, provided that we are essentially interested in evaluating the stresses along the chord BA.

In order to solve the elastic problem described above, we may apply the following procedure (Fillunger 1930). Let us consider for the moment an infinite plate the

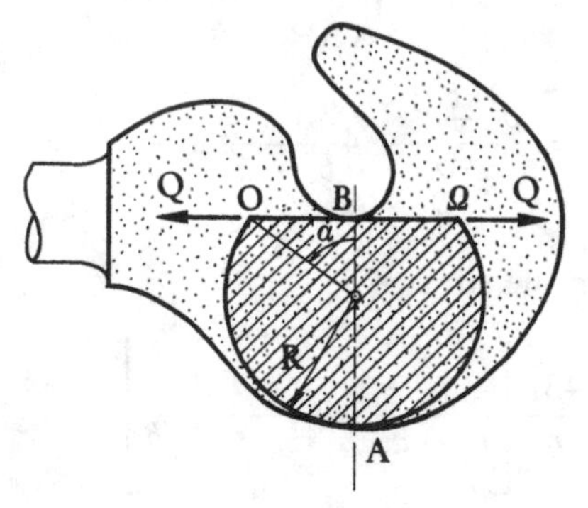

Fig.14.3

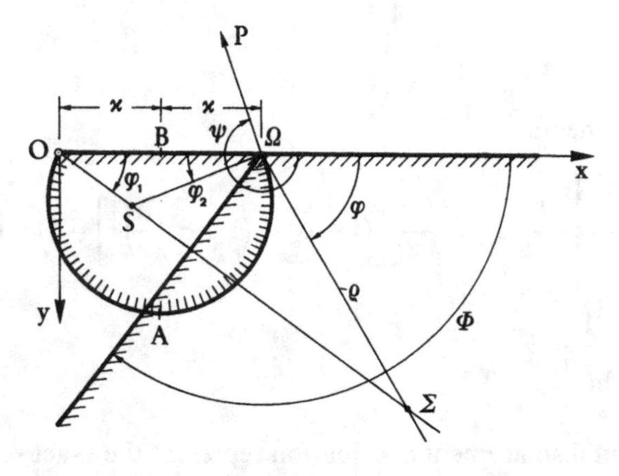

Fig. 14.4

boundary of which is made up of two straight edges meeting at the origin Ω (Figure 14.4) and forming an angle Φ measured clockwise from the horizontal edge. If a concentrated force P is applied at Ω and forms an angle ψ with the horizontal edge, we know that the state of stress within the wedge is characterized by a stress function of the form

$$F = \rho(C_2\varphi \sin\varphi + C_3\varphi \cos\varphi), \tag{14.25}$$

where ρ and φ are the polar coordinates in a system with its origin at Ω, and angles measured clockwise from the horizontal part of the boundary. The constants C_2 and C_3 have the values[23]

$$C_2 = \frac{P}{\Phi^2 - \sin^2\Phi}[\sin\psi \sin^2\Phi - \cos\psi(\Phi - \cos\Phi \sin\Phi)], \tag{14.26}$$

$$C_3 = \frac{P}{\Phi^2 - \sin^2\Phi}[\sin\psi(\Phi + \sin\Phi\cos\Phi) - \cos\psi \sin^2\Phi], \tag{14.27}$$

[23] Calculated by Michell (see Love 1927, Art. 151).

and the stresses have the form

$$\sigma_r = \frac{2}{\rho}(C_2 \cos\varphi - C_3 \sin\varphi), \quad \sigma_\varphi = \tau_{r\varphi} = 0,$$

which is described by Michell as a *simple radial distribution.*

Let us now introduce a change in the independent variables, termed *central inversion*, with respect a point O situated on the straight line $\varphi = 0$ at a distance 2κ from Ω. With the inversion, a point whose polar coordinates with respect to Ω were ρ and φ, and whose Cartesian coordinates were ξ and η, takes new Cartesian coordinates, with respect to a system at O, of the form

$$4\kappa^2 \frac{x}{r^2} = \xi + 2\kappa = \rho\cos\varphi + 2\kappa, \quad 4\kappa^2 \frac{y}{r^2} = \eta = \rho\sin\varphi. \tag{14.28}$$

This transformation maps the interior covered by varying the angle Φ with a vertex at Ω into the interior of a circular segment with vertices at Ω and O. The segment ΩO is the image of the half-line $\varphi = 0$; the image of the half-line $\varphi = \Phi$ is a circular arc $\widehat{\mathrm{OA}\Omega}$ with a central angle $2\pi - 2\alpha$, where $\alpha = \pi - \Phi$, and a radius $R = (\kappa/\sin\alpha)$. If we put

$$\overline{\mathrm{OS}} = r_1, \quad \overline{\Omega\mathrm{S}} = r_2, \quad \Omega\hat{\mathrm{O}}\mathrm{S} = \varphi_1, \quad \mathrm{O}\hat{\Omega}\mathrm{S} = \varphi_2,$$

and we observe that

$$\varphi = \varphi_1 + \varphi_2,$$

we may write (14.28) in the simpler form

$$x = r_1\cos\varphi_1 = 2\kappa - r_2\cos\varphi_2, \quad y = r_1\sin\varphi_1 = r_2\sin\varphi_2. \tag{14.29}$$

We have already found the stress function F at the generic point Σ on the wedge due to a single force P at Ω. Now the function

$$F_1 = r_1^2 F \tag{14.30}$$

is a new stress function for the transformed point S, having the property of being singular at O and Ω in the new domain. However, despite the singularity at the vertices, the boundary of the transformed domain is not free, because it is subject to normal traction. This traction has the same value at all points on the transformed boundary and may be eliminated by superposition of another additional uniform state of stress that can create uniform tractions on the boundary, but of opposite sign to the first.[24]

Under the change of variables (14.29), the stress function F_1 for the transformed domain becomes

[24] This peculiar property of the transformation by inversion was pointed out by Michell.

$$F_1 = r_1^2(C_2\varphi\rho\sin\varphi + C_3\varphi\rho\cos\varphi) = r_1^2(C_2\varphi\eta + C_3\varphi\xi)$$
$$= 4\kappa^2\left[C_2\varphi y + C_3\varphi\left(x - \frac{r_1^2}{2\kappa}\right)\right]$$
$$= 4\kappa^2(\varphi_1 + \varphi_2)\left[C_2 r_1 \sin\varphi_1 + C_3\frac{r_1 r_2}{2\kappa}\cos(\varphi_1 + \varphi_2)\right]. \tag{14.31}$$

The associated stresses are expressed as follows:

$$\sigma_x = \frac{\partial^2 F_1}{\partial y^2} = 4\kappa^2 C_2\left[2\left(\frac{\cos\varphi_1}{r_1} + \frac{\cos\varphi_2}{r^2}\right) - r_1\sin\varphi_1\left(\frac{\sin 2\varphi_1}{r_1^2} + \frac{\sin 2\varphi_2}{r_2^2}\right)\right]$$
$$+ 2\kappa C_3\left[\sin 2(\varphi_1 + \varphi_2) - 2r_1 r_2\left(\frac{\cos\varphi_1}{r_1} + \frac{\cos\varphi_2}{r_2}\right)^2 \sin(\varphi_1 + \varphi_2) - 2(\varphi_1 + \varphi_2)\right],$$
$$\sigma_y = \frac{\partial^2 F_1}{\partial x^2} = 4\kappa^2 C_2 r_1 \sin\varphi_1\left(\frac{\sin 2\varphi_1}{r_1^2} + \frac{\sin 2\varphi_2}{r_2^2}\right)$$
$$+ 2\kappa C_3\left[\sin 2(\varphi_2 + \varphi_2) - 2r_1 r_2\left(-\frac{\sin\varphi_1}{r_1} + \frac{\sin\varphi_2}{r_2}\right)^2 \sin(\varphi_1 + \varphi_2) - 2(\varphi_1 + \varphi_2)\right],$$
$$\tau_{xy} = -\frac{\partial^2 F_1}{\partial x\,\partial y} = 4\kappa^2 C_2\left[\frac{\sin\varphi_1}{r_1} - \frac{\sin\varphi_2}{r^2} + r_1\sin\varphi_1\left(\frac{\cos 2\varphi_1}{r_1^2} - \frac{\cos 2\varphi_2}{r_2^2}\right)\right]$$
$$+ 2\kappa C_3 \sin(\varphi_1 + \varphi_2)\left[-\frac{r^2}{r_1}\sin 2\varphi_1 + \frac{r_1}{r_2}\sin 2\varphi_2 - 2\sin(\varphi_2 - \varphi_1)\right]. \tag{14.32}$$

Since we are particularly interested in finding the stresses along the section BA, characterized by the properties $r_1 = r_2 = r$ and $\varphi_1 = \varphi_2 = \varphi$, we can obtain from the preceding formulae the stresses on these points:

$$\sigma_x = \frac{\partial^2 F_1}{\partial y^2}(0, y) = \bar{\sigma}_x = 16\kappa C_2\cos^4\varphi - 8\kappa C_3(\varphi + \sin\varphi\cos\varphi + 2\sin\varphi\cos^3\varphi),$$
$$\sigma_y = \frac{\partial^2 F_1}{\partial x^2}(0, y) = \bar{\sigma}_y = 16\kappa C_2\sin^2\varphi\cos^2\varphi - 8\kappa C_3(\varphi + \sin\varphi\cos\varphi - 2\sin\varphi\cos^3\varphi),$$
$$\tau_{xy} = -\frac{\partial^2 F_1}{\partial y\,\partial x}(0, y) = \bar{\tau}_{xy} = 0. \tag{14.33}$$

Let us now consider the particular, but important case in which the force P acting at the vertex Ω of the wedge forms an angle $\psi = \pi$ with the edge $\varphi = 0$. After the inversion, the transformed domain is loaded by two horizontal opposite forces Q applied at the vertices Ω and O (Figure 14.5). The magnitude Q of these two forces is no longer P, but must be determined as a function of P. In general, the transformed domain is not free from tractions along its boundary, but in this particular case we find that along the rectilinear piece $OB\Omega$ of the boundary where $\varphi_1 = \varphi_2 = 0$, the stress components σ_y and τ_{xy}, given by (14.32), vanish. It is also possible to prove, however, that the tractions vanish on the circular part $\widehat{OA\Omega}$ of the boundary. Since, in the present case, the angle ψ is equal to π and we have seen that Φ is $\pi - \alpha$, the constants C_2 and C_3 given in (14.26) and (14.27) have the values

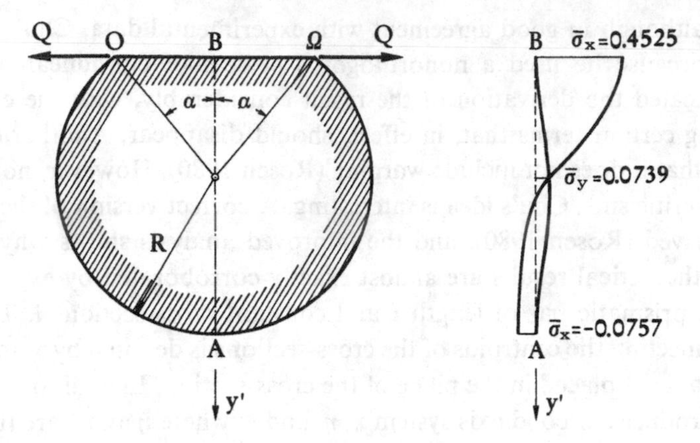

Fig. 14.5

$$C_2 = P\frac{(\pi - \alpha) + \sin\alpha\cos\alpha}{(\pi - \alpha)^2 - \sin^2\alpha},$$

$$C_3 = P\frac{\sin^2\alpha}{(\pi - \alpha)^2 - \sin^2\alpha}.$$

In order to evaluate Q, we use the equilibrium condition

$$\int_B^A \bar{\sigma}_x\, dy' = \int_0^{\frac{\pi-\alpha}{2}} \bar{\sigma}_x \kappa \frac{d\varphi}{\cos^2\varphi} = 4\kappa^2 P = Q,$$

whereas we have the condition

$$\int_A^B \bar{\sigma}_x y'\, dy' = \int_0^{\frac{\pi-\alpha}{2}} \bar{\sigma}_x \kappa^2 tg\varphi \frac{d\varphi}{\cos^2\varphi} = 0.$$

From the first of these equations we obtain $P = Q/(4\kappa^2)$, and thus the constants C_2 and C_3 are completely determined. Figure 14.5 shows the distribution of the stresses $\bar{\sigma}_x$ and $\bar{\sigma}_y$ in the middle section AB for $Q = 1$.

15. The Influence of Initial Twist

When a prismatic bar possesses an initial twist, its torsional rigidity will be greater than that of the same bar with no initial twist. The relative increase in torsional rigidity depends on the initial twist, the shape of the cross-section, and the elastic modulus of the material. This property was first studied by Chu (1951), who calculated the increase in torsional rigidity and confirmed the result with a series of experiments on initially twisted rectangular steel strips. The improvement in the stiffness of a bar by pretwisting is used in aircraft propellers and turbine blades.

However, although in good agreement with experimental data, Chu's theory was incomplete, because he used a nonorthogonal system of curvilinear coordinates, which complicated the derivation of the result considerably, with the consequence of introducing certain terms that, in effect, should disappear. Another criticism of the theory is that it does not include warping (Rosen 1980). However, notwithstanding these two criticisms, Chu's idea is interesting. A correct version of the theory can be simply derived (Rosen 1980), and the improved analysis shows why, in certain cases, Chu's theoretical results are almost exactly corroborated by experiment. Let us consider a prismatic bar of length ℓ and constant cross-section A. The longitudinal axis connecting the centroids of the cross-sections is denoted by x and y, z these are orthogonal axes placed in the plane of the cross-section. In addition to the y and z axes, we introduce a second axis system x, η, and ζ, where η and ζ are the principal axes of the cross-section that may rotate with the cross-section along x.

The initial twist is represented by a rotation β of the cross-section about the x axis. The y and z axes are fixed and the η and ζ axes are movable, and β is the angle between axes η and y. The angle β is a function of x, but we consider the special case in which β is linear in x, so that

$$k = \frac{d\beta}{dx} \tag{15.1}$$

is a constant. It is important to emphasize that the initial configuration, including the initial twist, is free from stress. In other words, the initial twist is a pure geometrical modification of a prismatic beam, ideally obtained by cutting the latter into an infinite number of small transverse slices and then superimposing these slices in such a manner that each of them is rotated by a small angle with respect to the preceding one.

Let us now suppose that, at the ends of the beam, two equal and opposite torques of magnitude M_t are applied by means of tangential tractions exerted on the terminal sections and distributed over these according to specific laws that must be determined *a posteriori*, the only restriction being that they are equivalent to a couple, say, M_t at the section $x = \ell$, and a couple $-M_t$ at the section $x = 0$.

If we denote the displacement components along the fixed axes by u, v, and w, respectively, we assume that they have the form

$$\begin{aligned} u &= u_1 + \theta\psi, \\ v &= -\theta xz, \\ w &= \theta xy, \end{aligned} \tag{15.2}$$

where $\psi = \psi(x, y, z)$ is a function, called the *warping function*, θx is the angle of elastic rotation of a section relative to the section $x = 0$, and $u = u_1(x)$ is another function, unknown at this stage. If the bar is prismatic, then u_1 is, identically, zero and (15.2) coincides with Saint-Venant's solution, provided that ψ depends on y and z but not on x. The function ψ is harmonic and satisfies the condition.

$$\frac{\partial\psi}{\partial\nu} = z\cos(y, \nu) - y\cos(z, \nu),$$

at all points on the bounding curve of each cross-section. Alternatively, as a consequence of (15.2), the strain components that do not vanish are

$$\begin{aligned} e_{xx} &= \frac{\partial u}{\partial x} = \frac{\partial u_1}{\partial x} + \theta\,\frac{\partial \psi}{\partial x}, \\ e_{xy} &= \frac{\partial u}{\partial y} + \frac{\partial v}{\partial x} = \theta\left(-z + \frac{\partial \psi}{\partial y}\right), \\ e_{xz} &= \frac{\partial u}{\partial z} + \frac{\partial w}{\partial x} = \theta\left(y + \frac{\partial \psi}{\partial z}\right). \end{aligned} \tag{15.3}$$

The strain energy per unit volume has the form

$$W = \frac{E}{2}\,e_{xx}^2 + \frac{G}{2}\,(e_{xy}^2 + e_{xz}^2),$$

where E and G are (constant) elastic moduli. The total potential energy of the beam is thus

$$\begin{aligned} W - L = \int_0^{\ell} dx \left\{ \iint_A \left[\frac{E}{2}\left(\frac{\partial u_1}{\partial x} + \theta\,\frac{\partial \psi}{\partial x}\right)^2 + \frac{G}{2}\,\theta^2\left(-z + \frac{\partial \psi}{\partial y}\right)^2 + \frac{G}{2}\,\theta^2\left(y + \frac{\partial \psi}{\partial z}\right)\right] dy\,dz \right\} \\ - M_{\mathrm{t}} \int_0^{\ell} \theta\,dx, \end{aligned} \tag{15.4}$$

which may be written in the simpler form

$$W - L = \int_0^{\ell} \left\{ \frac{E}{2}\left[A\left(\frac{\partial u_1}{\partial x}\right)^2 + 2\,\frac{\partial u_1}{\partial x}\,S\theta + \theta^2 K \right] + \frac{G}{2}\,J_{\mathrm{S}}\theta^2 - M_{\mathrm{t}}\theta \right\} dx, \tag{15.5}$$

where

$$\begin{aligned} J_{\mathrm{S}} &= \iint_A \left[\left(-z + \frac{\partial \psi}{\partial y}\right)^2 + \left(y + \frac{\partial \psi}{\partial z}\right)^2 \right] dy\,dz, \\ S &= \iint_A \frac{\partial \psi}{\partial x}\,dy\,dz, \\ K &= \iint_A \left(\frac{\partial \psi}{\partial x}\right)^2 dy\,dz. \end{aligned} \tag{15.6}$$

It can be seen immediately that GJ_{S} is the torsional rigidity of Saint-Venant's theory, while S and K are integrals associated with the dependence of ψ on x.

If the beam is in equilibrium, the functional $W - L$ attains a stationary point, and the Euler equations that correspond to the vanishing of the first variation of the functional with respect to u_1 and θ are

$$A\,\frac{\partial u_1}{\partial x} + S\theta = 0, \tag{15.7}$$

$$GJ_{\mathrm{S}}\theta + E\left(S\,\frac{\partial u_1}{\partial x} + K\theta\right) = M_{\mathrm{t}}, \tag{15.8}$$

so that elimination of $\partial u_1/\partial x$ yields

$$G\left[J_S + \frac{E}{G}\left(K - \frac{S^2}{A}\right)\right]\theta = M_t. \tag{15.9}$$

This means that the initial twist produces a variation in the torsional rigidity GJ_S, calculated according to Saint-Venant's theory, to give a new value

$$GJ = G\left[J_S + \frac{E}{G}\left(K - \frac{S^2}{A}\right)\right], \tag{15.10}$$

and it is clear that we have $J \geqslant J_S$ as a consequence of Schwarz's inequality.

A useful application of the theory is offered by the case, very common in practice, in which the cross-section is a slender profile that is symmetrical with respect to its principal moving axes η and ζ (Figure 15.1) (Rosen 1980). Since all cross-sections are identical except for the initial rotation β, ψ will not be a function of x, if represented by the coordinates η, ζ. Therefore,

$$\frac{\partial\psi}{\partial x} = \frac{\partial\psi}{\partial\eta}\frac{\partial\eta}{\partial x} + \frac{\partial\psi}{\partial\zeta}\frac{\partial\zeta}{\partial x}, \tag{15.11}$$

with $\partial\psi/\partial\eta$ and $\partial\psi/\partial\zeta$ independent of x. On the other hand, by differentiating the relations (see Figure 15.1)

$$\eta = y\cos\beta + z\sin\beta,$$
$$\zeta = -y\sin\beta + z\cos\beta,$$

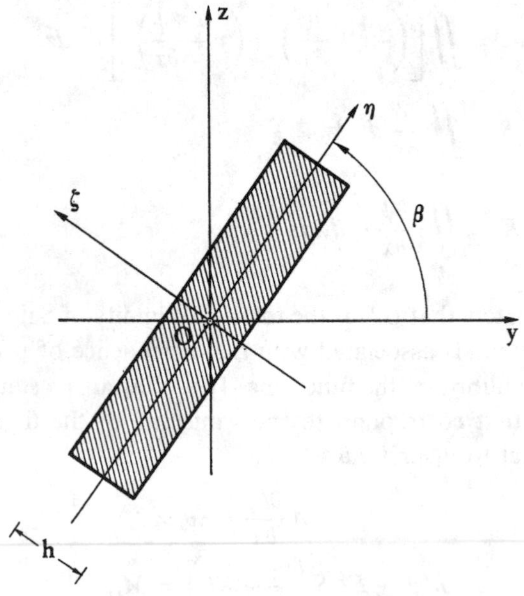

Fig. 15.1

we obtain

$$\frac{\partial \eta}{\partial x} = (-y \sin \beta + z \cos \beta) \frac{d\beta}{dx} = \zeta k,$$
$$\frac{\partial \zeta}{\partial x} = (-y \cos \beta - z \sin \beta) \frac{\partial \beta}{\partial x} = -\eta k.$$

Substitution of these equations in (15.11) and then in (15.6) gives

$$S = k \iint_A \left(\frac{\partial \psi}{\partial \eta} \zeta - \frac{\partial \psi}{\partial \zeta} \eta \right) d\eta \, d\zeta, \tag{15.12}$$

$$K = k^2 \iint_A \left(\frac{\partial \psi}{\partial \eta} \zeta - \frac{\partial \psi}{\partial \zeta} \eta \right)^2 d\eta \, d\zeta. \tag{15.13}$$

Since now the function ψ, considered as a function of η and ζ, is the torsion function according to Saint-Venant's theory, it may be written as

$$\frac{\partial \psi}{\partial \eta} = \zeta + \frac{\partial \phi}{\partial \zeta}, \quad \frac{\partial \psi}{\partial \zeta} = -\eta - \frac{\partial \phi}{\partial \eta}, \tag{15.14}$$

where ϕ is another harmonic function, called *Prandl's function.* It is known that ϕ satisfies the equation

$$\frac{\partial^2 \phi}{\partial \eta^2} + \frac{\partial^2 \phi}{\partial \zeta^2} = -2 \text{ in } A,$$

and vanishes on the boundary. For a thin-walled cross-section the stress function ϕ may be approximated as

$$\phi \simeq \left(\frac{h}{2} \right)^2 - \zeta^2,$$

where h is the thickness of the cross-section measured perpendicular to the η axis. If we substitute this expression for ϕ in (15.14) and then in (15.12) and (15.13), we obtain

$$S = k \iint_A \eta^2 \left(1 - \frac{\zeta^2}{\eta^2} \right) d\eta \, d\zeta,$$

$$K = k^2 \iint_A \eta^4 \left(1 - \frac{\zeta^2}{\eta^2} \right)^2 d\eta \, d\zeta.$$

It is clear that the term ζ^2/η^2 does not contribute very much to the values of S and K, and it is thus sometimes neglected.

The influence of bending effects in a pretwisted section is a very important problem in the analysis of turbine blades, because, when the beam is bent, the bending stiffnesses of a pretwisted cross-section are very different from those of the same section in an untwisted beam. An approximate analysis of the problem can be made under certain strong assumptions about the magnitude of the pretwist (Downs 1979), which must be small.

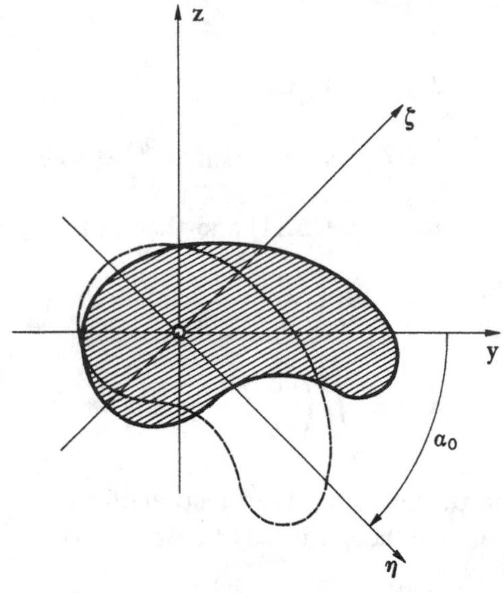

Fig. 15-2

Let us consider a pretwisted beam the cross-section of which (see Figure 15.2) undergoes a uniform initial twist in the direction of the x axis, perpendicular to the plane of the figure and directed inwards. The initial cross-section, at $x = 0$, is referred to a system of fixed Cartesian axes y and z that coincide with the principal axes of the section. Another section, at a distance x from the former is identified by the angle α_0, representing the rigid rotation of the section about the x axis with respect to the fixed axes y and z. A point of coordinates y, z in the fixed section is at a distance

$$d = (y^2 + z^2)^{\frac{1}{2}}$$

from the origin. A corresponding point in a section at unit distance from the fixed section, is connected to the first point by a helical fiber of length

$$\ell = (1 + \alpha_0^2 d^2)^{\frac{1}{2}}. \tag{15.15}$$

By imposing a further elastic twist θ per unit length ($\theta \ll \alpha_0$) in the direction of the pretwist, while allowing no axial movement or bending, we produce a strain

$$e_\theta = \{[1 + (\alpha_0 + \theta)^2 d^2]^{\frac{1}{2}} - (1 + \alpha_0^2 d^2)^{\frac{1}{2}}\}\frac{1}{(1 + \alpha_0^2 d^2)^{\frac{1}{2}}} \simeq \theta(\alpha_0 d^2 - \alpha_0^3 d^4), \tag{15.16}$$

where, in deriving the last formula, we have neglected the terms involving θ^2 and higher powers of θ, and the terms involving α_0^5 and higher powers of α_0. The corresponding stress acting along the helical filament is

$$\sigma_\theta \simeq E\theta(\alpha_0 d^2 - \alpha_0^3 d^4). \tag{15.17}$$

We now impose an axial extension a per unit length, without allowing any other movement, and produce the strain

$$e_a = \{[(1+a)^2 + \alpha_0^2 d^2]^{\frac{1}{2}} - (1+\alpha_0^2 d^2)^{\frac{1}{2}}\}\frac{1}{(1+\alpha_0^2 d^2)^{\frac{1}{2}}} \simeq a(1-\alpha_0^2 d^2), \qquad (15.18)$$

where we have ignored the terms in a^2 and higher powers of a, and the terms in α_0^4 and higher powers of α_0. The corresponding filament stress is

$$\sigma_a \simeq Ea(1-\alpha_0^2 d^2). \qquad (15.19)$$

Through its length the filament is inclined with respect the x axis at a helical angle $\beta = \arctan(\alpha_0 d)$, so that the normal cross-sectional area of the filament associated with the elementary cross-sectional area of the beam dA, is $dA \cos\beta$. It follows that the force dF acting along the filament is $\sigma\, dA \cos\beta$, where σ is the normal stress along the fiber. Since β is small, we may write

$$dF \simeq \sigma\left(1 - \frac{1}{2}\alpha_0^2 d^2\right)dA, \qquad (15.20)$$

and the axial component of this elementary force is

$$dF_{\mathrm{a}} = dF\cos\beta \simeq \sigma(1-\alpha_0^2 d^2)dA, \qquad (15.21)$$

and the tangential component is

$$dF_{\mathrm{t}} = dF\sin\beta \simeq \sigma(\alpha_0 d - \alpha_0^3 d^3)dA. \qquad (15.22)$$

The examination of the terms discarded indicates that the accuracy remains within one percent for a maximum helix angle of 16°, and within two percent for a maximum helix angle of 23°.

Let us now apply a unit relative translation, along the x axis, between two sections unit distance apart; and let us apply three unit relative rotations between the same sections, about the x, y, and z axes, respectively. The displacement generates strains and stresses along the helical fibers, from which we can find the stress resultants N, M_x, M_y, and M_z on the basis of elementary beam theory. Let us call ε the extension, κ and κ' the curvatures, and τ the torsion. We then obtain four constitutive equations of the type

$$\begin{aligned} N &= \kappa_{11}\varepsilon + \kappa_{12}\kappa + \kappa_{13}\kappa' + \kappa_{14}\tau, \\ -M_z &= \kappa_{21}\varepsilon + \kappa_{22}\kappa + \kappa_{23}\kappa' + \kappa_{24}\tau, \\ M_y &= \kappa_{31}\varepsilon + \kappa_{32}\kappa + \kappa_{33}\kappa' + \kappa_{34}\tau, \\ M_x &= \kappa_{41}\varepsilon + \kappa_{42}\kappa + \kappa_{43}\kappa' + \kappa_{44}\tau, \end{aligned} \qquad (15.23)$$

where

$$
\begin{aligned}
\kappa_{11} &= E \iint_a (1-\alpha_0^2 d^2)^2\, dA,\\
\kappa_{12} = \kappa_{21} &= -E \iint_A y(1-\alpha_0^2 d^2)^2\, dA,\\
\kappa_{13} = \kappa_{31} &= -E \iint_A z(1-\alpha_0^2 d^2)^2\, dA,\\
\kappa_{14} = \kappa_{41} &= E \iint_A (\alpha_0 d^2 - \alpha_0^3 d^4)(1-\alpha_0^2 d^2)\, dA,\\
\kappa_{22} &= E \iint_A y^2(1-\alpha_0^2 d^2)^2\, dA,\\
\kappa_{23} = \kappa_{32} &= E \iint_A yz(1-\alpha_0^2 d^2)^2\, dA,\\
\kappa_{24} = \kappa_{42} &= -E \iint_A y(\alpha_0 d^2 - \alpha_0^3 d^4)(1-\alpha_0^2 d^2)\, dA,\\
\kappa_{33} &= E \iint_A z^2(1-\alpha_0^2 d^2)^2\, dA,\\
\kappa_{34} = \kappa_{43} &= -E \iint_A z(\alpha_0 d^2 - \alpha_0^3 d^4)(1-\alpha_0^2 d^2)\, dA,\\
\kappa_{44} &= E \iint_A (\alpha_0 d^2 - \alpha_0^3 d^4)^2\, dA + GT_0,
\end{aligned}
\tag{15.24}
$$

GT_0 being the torsional rigidity of the section according to Saint-Venant's theory.

The formulae for the stress resultant show that there is an intimate coupling between them, in the sense that each stress resultant depends on all the strain parameters. There is the immediate question of whether (15.23) can be simplified; this can be done by introducing the notion of a neutral axis. Let us consider the axial force and the moment induced by imposing a small rotation of a pretwisted section about a line parallel to the neutral axis. Then, if n is the distance of the neutral axis from the line, we must have the equation

$$n \iint_A dF_a = \iint_A s\, dF_a, \tag{15.25}$$

where dF_a is given by (15.19) and (15.21), and s is the distance of the element of force dF_a from the given line. Thus from the last equation we obtain

$$n = \iint_A s(1-\alpha_0^2 d^2)^2\, dA \times \left[\iint_A (1-\alpha_0^2 d^2) \right]^{-1},$$

and, calculating the coefficients of equations (15.23) with y and z measured from the neutral axis, we find that κ_{12}, κ_{13}, and κ_{23} vanish, thereby permitting some simplification.

16. The Energy of an Elastic Rod

When an elastic rod is viewed as a three-dimensional body it is always necessary, at least in principle, to solve a system of partial differential equations in order to determine the state of strain and stress in the body. Solving this system is very difficult even when using the assumptions of the linear theory of elasticity, but practically impossible when large strains or even large displacements are involved. In the absence of a systematic method for tackling the nonlinear problem, efforts have been made towards reducing the original problem to one of a single dimension through the averaging procedure, which we saw in Section 8. However, sometimes these methods are also difficult, so the procedure for solving the averaged equations needs further reduction. Instead of finding a complete solution, we content ourselves with characterizing certain properties of the solutions, or evaluating some useful quantities related to them. The techniques involved in this are almost always of the variational type.

The simplest application of such methods is illustrated by the case of a prismatic beam, of rectangular cross-section and sufficiently thin for plane stress theory to apply (Rychter 1988a, 1988b). The beam is bent in the xz plane, which coincides with the middle plane of the rectangular prism. The x axis coincides with the center line of the prism, the y axis passes across the width, and the z axis points downwards; y and z are the principal axes of inertia of the section. The end surfaces of the beam are defined by $x = \pm\ell$, the side surfaces are described by $y = \pm b$, and the upper and lower surfaces are described by $z = \pm h$.

The beam is bent in the xz plane by a surface load $p(x)$ distributed over the upper face $z = -h$; the lower face and both side faces are free. The boundary conditions for the end faces are specified in the following form: the left face $x = -\ell$ experiences tractions $\sigma_x = \sigma_x^*$ and $\tau_{xz} = \tau_{xz}^*$, and the right face $x = \ell$ experiences displacements $u = u^*$ and $w = w^*$. The body forces are zero.

In order to solve the problem we must know the constitutive equations of the material. In the present case, we assume that the material is homogeneous, linearly elastic, and orthotropic, and is characterized by a stress–strain relation of the type

$$\sigma_x = Eu,_z + D\sigma_z, \quad \tau_{xz} = G(u,_z + w,_z),$$
$$\sigma_z = E'(w,_z + Du,_x)\frac{1}{1-\sigma D}, \tag{16.1}$$

where E and E' are the Young's moduli in the x and z directions, respectively, G is the shear modulus, and $D = \sigma(E/E')$.

As it is difficult to determine an exact solution to the problem, we consider two approximate fields. The first is a kinematically admissible displacement field, that is to say a field of displacements $(\hat{u}, \hat{v}, \hat{w})$ that is sufficiently regular, in particular with piecewise continuous first partial derivatives, and satisfies the kinematic boundary conditions on the face $x = \ell$. For reasons that will be clarified later, it is convenient to split the field into two parts

$$(\hat{u}, \hat{v}, \hat{w}) = (\hat{u}^i, \hat{v}^i, \hat{w}^i) + (\hat{u}^\ell, \hat{v}^\ell, \hat{w}^\ell), \tag{16.2}$$

where $(\hat{u}^i, \hat{v}^i, \hat{w}^i)$, the interior field, need not necessarily satisfy the boundary conditions on the face $x = \ell$, whereas $(\hat{u}^\ell, \hat{v}^\ell, \hat{w}^\ell)$, the exterior field, must satisfy the boundary conditions

$$\hat{u}^\ell = u^* - \hat{u}^i, \quad \hat{w}^\ell = w^* - \hat{w}^i, \quad \text{at } x = \ell. \tag{16.3}$$

Once we have the displacements, we can derive the associated stresses $(\hat{\sigma}_x, \hat{\sigma}_z, \hat{\tau}_{xz})$ through the constitutive equations, and decompose them into the sum

$$(\hat{\sigma}_x, \hat{\sigma}_z, \hat{\tau}_{xz}) = (\hat{\sigma}^i_x, \hat{\sigma}^i_z, \hat{\tau}^i_{xz}) + (\hat{\sigma}^\ell_x, \hat{\sigma}^\ell_z, \hat{\tau}^\ell_{xz}). \tag{16.4}$$

The second field is a statically admissible stress field $(\tilde{\sigma}_x, \tilde{\sigma}_z, \tilde{\sigma}_{xz})$ which satisfies the equilibrium equations in the interior and all the boundary conditions of statical type. Again, it is useful to write

$$(\tilde{\sigma}_x, \tilde{\sigma}_z, \tilde{\tau}_{xz}) = (\tilde{\sigma}^i_x, \tilde{\sigma}^i_z, \tilde{\tau}^i_{xz}) + (\tilde{\sigma}^\ell_x, \tilde{\sigma}^\ell_z, \tilde{\tau}^\ell_{xz}), \tag{16.5}$$

where $(\tilde{\sigma}^i_x, \tilde{\sigma}^i_z, \tau^i_{xz})$ satisfy the equilibrium equations and the traction boundary conditions on the lateral surface, while $(\tilde{\sigma}^\ell_x, \tilde{\sigma}^\ell_z, \tilde{\tau}^\ell_{xz})$ satisfy the equilibrium equations, the boundary condition on the lateral surface, but with $p(x) = 0$, and the end conditions

$$\tilde{\sigma}^\ell_x = \sigma^*_x - \tilde{\sigma}^i_x, \quad \tilde{\tau}^\ell_{xz} = \tau^*_{xz} - \tilde{\tau}^i_{xz}, \quad \text{at } x = -\ell. \tag{16.6}$$

Let us now suppose that the two fields $(\hat{u}, \hat{v}, \hat{w})$ and $(\tilde{\sigma}_x, \tilde{\sigma}_z, \tilde{\tau}_{xz})$ have been determined. In terms of these fields we may define

$$\bar{w}(x) = \frac{1}{2h} \int_{-h}^{h} \hat{w}(x, z)\, dz,$$

$$b(x) = \frac{1}{2h} \int_{-h}^{h} \hat{u}_{,z}(x, z)\, dz = \frac{1}{2h} [\hat{u}(x, h) - \hat{u}(x - h)], \tag{16.7}$$

where $\bar{w}$ is a mean lateral deflection, and b is a rotation represented by the difference between the axial displacement of the upper and lower edges. At the same time we introduce the conventional bending moment M and shear force Q, defined as follows:

$$M(x) = \int_{-h}^{h} z\hat{\sigma}_x(x, z)\, dz, \quad Q(x) = \int_{-h}^{h} \hat{\tau}_{xz}(x, z)\, dz. \tag{16.8}$$

From the field equations we derive the equilibrium condition

$$M'(x) = Q, \tag{16.9a}$$

$$Q'(x) + p = 0 \tag{16.9b}$$

and the constitutive relations

$$M = \frac{2}{3}h^3 Eb' + \left(7D - \frac{E}{G}\right)\frac{h^2 p}{15}, \tag{16.10a}$$

$$Q = 2Gh(b + \bar{w}'). \tag{16.10b}$$

At the ends we must have

$$M = M^*, \quad Q = Q^* \text{ for } x = \ell; \quad b = b^*, \quad \bar{w} = \bar{w}^*, \text{ for } x = -\ell, \tag{16.11}$$

where the starred quantities are prescribed.

To account for the elongation or contraction of the beam, we introduce the average axial displacement

$$\bar{u}(x) = \frac{1}{2h}\int_{-h}^{h} \hat{u}(x, z)\, dz, \tag{16.12}$$

and the axial force

$$N(x) = \int_{-h}^{h} \hat{\sigma}_x(x, z)\, dz.$$

For the equilibrium we also need the conditions

$$N'(x) = 0, \tag{16.13}$$

and

$$N = N^* \text{ at } x = \ell, \quad \bar{v} = v^* \text{ at } x = -\ell,$$

where, again, the starred quantities are known. In addition we have the constitutive relation

$$N = 2hE\bar{u}' + hDp. \tag{16.14}$$

The equations just derived represent the averaged version of the original two-dimensional problem. Our task is to exploit the one-dimensional solution in order to obtain estimates of the two-dimensional solution. This may be done by constructing a displacement field, satisfying only the interior conditions, having the form

$$\begin{aligned} \hat{u}^i &= \bar{u}(x) + thb(x) + (t^3 - t)f(x), \\ \hat{w}^i &= \bar{w}(x) + ts(x) + (3t^2 - 1)g(x) + \left(t^4 - 2t^2 + \tfrac{7}{15}\right)r(x) - \left(t^4 - 6t^2 - 8t + \tfrac{9}{5}\right)q(x), \end{aligned} \tag{16.15}$$

where

$$t = \frac{z}{h}, \tag{16.16a}$$

$$s = -hD\bar{u}', \tag{16.16b}$$

$$g = -h^2 D \frac{b'}{6}, \tag{16.16c}$$

$$r = -hD\frac{f'}{4}, \tag{16.16d}$$

$$q = hp(1 - \sigma D)\frac{1}{16E'}, \tag{16.16e}$$

$$f = -hg' - \frac{Q}{4G}. \tag{16.16f}$$

The corresponding stress field becomes

$$\begin{aligned} \hat{\sigma}_x^i &= E[\bar{u}' + thb' + (t^3 - t)f'] + D\hat{\sigma}_z^i, \\ \hat{\tau}_{xz}^i &= 3(1 - t^2)\frac{Q}{4h} + G[ts' + (t^4 - 2t^2 + \tfrac{7}{15})r'] - G(t^4 - 6t^2 - 8t + \tfrac{9}{5})q', \\ \hat{\sigma}_x^i &= p(2 + 3t - t^3)\frac{1}{4}. \end{aligned} \tag{16.17}$$

The interior statically admissible stress field is given by

$$\begin{aligned} \tilde{\sigma}_x^i &= \frac{1}{2h}N + \frac{3t}{2h^2}M + \frac{1}{20}(5t^3 - 3t)\left(\frac{E}{G} - 2D\right)p, \\ \tilde{\tau}_{xy}^i &= \frac{3}{4h}(1 - t^2)Q - \frac{1}{80}(5t^4 - 6t^2 + 1)\left(\frac{E}{G} - 2D\right)hp', \\ \tilde{\sigma}_z^i &= \frac{p}{4}(2 + 3t - t^3) + \frac{1}{80}(t^5 - 2t^3 + t)\left(\frac{E}{G} - 2D\right)h^2p''. \end{aligned} \tag{16.18}$$

In these equations, we may express f' in terms of the load p, by combining (16.16f) with (16.9b) and (16.16c) with (16.10a). The result is

$$f' = \left(\frac{1}{G} - \frac{D}{E}\right)\frac{p}{4} + O\left(\frac{h^2p''}{E}\right). \tag{16.19}$$

Similarly, from (16.16b) and the two equations written for N, we derive

$$s' = \frac{hD^2}{2E}p'. \tag{16.20}$$

Since we already have formulae (16.17) and (16.18) for the two states of stress, we may write

$$\begin{aligned} &\tilde{\sigma}_x^i - \hat{\sigma}_x^i = O(h^2p''), \quad \tilde{\tau}_{xz}^i - \hat{\tau}_{xz}^i = O(hp'), \\ &\tilde{\sigma}_z^i - \hat{\sigma}_z^i = O(h^2p''). \end{aligned} \tag{16.21}$$

A first conclusion to be drawn from (16.21) is that the two stress states coincide for a constant load p on the upper surface and, consequently, the exact two-dimensional solution is then reached, provided that the end conditions are just that of the state $(\tilde{\sigma}_x^i, \tilde{\sigma}_z^i, \tilde{\tau}_{xz}^i)$ at $x = \ell$ and that of the state $(\hat{\sigma}_x^i, \hat{\sigma}_z^i, \hat{\tau}_{xz}^i)$ at $x = -\ell$. In the general case of a load varying with x we may impose an energy bound in the linear elasticity,[24] which allows the strain energy of an elastic body to be bounded by means of a kinematically and statically admissible state such as those constructed above. Let $(\sigma_x, \sigma_z, \tau_{xz})$ be the exact solution, and let U be a quadratic norm associated with this solution having the form

$$U(\sigma_x, \sigma_z, \tau_{xz}) = \int_{-h}^{h} \int_{-\ell}^{\ell} \left[\sigma_x^2 + (\sigma_z^2 - 2\sigma\sigma_x\sigma_z)\frac{E}{E'} + \tau_{xy}^2 \frac{E}{G} \right] dx\, dz.$$

Then Synge and Prager's theorem states that the following equality holds:

$$\begin{aligned} &U[\sigma_x - \tfrac{1}{2}(\hat{\sigma}_x^i + \tilde{\sigma}_x^i), \quad \sigma_z - \tfrac{1}{2}(\hat{\sigma}_z^i + \tilde{\sigma}_z^i), \quad \tau_{xz} - \tfrac{1}{2}(\hat{\tau}_{xz}^i + \tilde{\tau}_{xz}^i)] \\ &= U(\hat{\sigma}_x^i - \tilde{\sigma}_x^i, \hat{\sigma}_z^i - \tilde{\sigma}_z^i, \hat{\tau}_{xz}^i - \tilde{\tau}_{xz}^i). \end{aligned} \tag{16.22}$$

We can now use (16.21) to obtain an asymptotic estimate of the right-hand side of (16.22).

So far we have treated the case in which the boundary data at the terminal faces have two special distributions such that the other external fields vanish. If, however, the boundary conditions at the ends are different, having forms that allow the states $(\hat{\sigma}_x^\ell, \hat{\sigma}_z^\ell, \hat{\tau}_{xz}^\ell)$ and $(\tilde{\sigma}_x^\ell, \tilde{\sigma}_z^\ell, \tilde{\tau}_{xz}^\ell)$ to be constructed without excessive difficulty, then a formula like (16.22) still holds, and we obtain an estimate for the complete solution.

The level of accuracy of an estimate like that derivable from (16.22) can be improved further through a process of successive generation of approximate two-dimensional fields of any order of accuracy, provided that, at the ends of the beam, we demand no more detail than the shear stress resulting from the average vertical displacement and the stress couple or the gross rotation (Duva and Simmonds 1990).

In order to get an idea of the method, let us consider a rectangular beam which, when undeformed, occupies the region $0 \leqslant x \leqslant L$, $|y| \leqslant B$, $|z| \leqslant H$. We assume that the beam is clamped at $x = 0$ and is stress free at $x = L$, that there are no body forces, and that the faces of the beam are under the tractions $\sigma_z(x, \pm H) = \pm\frac{1}{2}p_0 p(x/L)$ and $\tau_{xy}(x, \pm H) = 0$, where p_0 is a dimensionless normal load.

On assuming that the material is linearly elastic, orthotropic, and that the thickness B is small, the stress in the beam may be considered as a generalized plane state. A statically admissible stress field may be derived from a stress function $F(x, z)$ through the relations

$$\tilde{\sigma}_x = F_{,zz}, \quad \tilde{\sigma}_z = F_{,xx}, \quad \tilde{\tau}_{xz} = -F_{,xz}, \tag{16.23}$$

where $F(x, z)$ must also satisfy the face traction conditions

$$F_{,xx}(x, \pm H) = \pm\frac{1}{2}p_0 p\left(\frac{x}{L}\right), \quad F_{,xz}(x, \pm H) = 0, \tag{16.24}$$

and the condition that the normal and shear stresses at the end $x = L$ have a zero resultant and zero couple with respect to the y axis. A kinematically admissible state is simply generated by assigning the axial and normal displacements U and W, with

[24] Known as Synge and Prager's method (Prager and Synge 1947).

the only restriction that the averaged vertical deflection and the averaged vertical rotation are zero at $x = 0$.

Once we have the two admissible states we can calculate the differences between the strain components

$$\Delta e_z = \hat{e}_z - \tilde{e}_z = W_{,z} - \frac{1}{E'}(F_{,xx} - \sigma F_{,zz}),$$
$$\Delta e_{xz} = \hat{e}_{xz} - \tilde{e}_{xz} = U_{,z} + W_{,x} + \frac{1}{G}F_{,xz}, \qquad (16.25)$$
$$\Delta e_x = \hat{e}_x - \tilde{e}_x = U_{,x} - \frac{1}{E}(F_{,zz} - \sigma F_{,xx}),$$

where E, E', and G are elastic moduli, and σ is Poisson's ratio.

The actual strain field is kinematically and statically admissible and the actual stress function must satisfy the compatibility equation

$$F_{,zzzz} + 2\tilde{E}F_{,xxzz} + \tilde{E}F_{,xxxx} = 0, \qquad (16.26)$$

where

$$\tilde{E} = \frac{1}{2}\frac{E}{G} - \sigma\frac{E}{E'}.$$

In order to construct a statically admissible field, we introduce the dimensionless variables

$$\xi = \frac{x}{L}, \quad \zeta = \frac{z}{H}, \quad f = \frac{F}{p_0 L^2}, \qquad (16.27)$$

and the parameter $\varepsilon = H/L$. It thus follows that (16.26) may be rewritten as

$$f_{,\zeta\zeta\zeta\zeta} = -(2\varepsilon^2\tilde{E}f_{,\zeta\zeta\xi\xi} + \varepsilon^4\tilde{E}_{,\xi\xi\xi\xi}), \qquad (16.28)$$

and the face traction conditions (16.24) as

$$f_{,\xi\xi}(\xi, \pm 1) = \pm\tfrac{1}{2}p(\xi), \quad f_{,\xi\zeta}(\xi, \pm 1) = 0. \qquad (16.29)$$

Because the stress resultant and the couple at the right end of the beam vanish, we may integrate (16.29) with respect to ξ and replace them by the conditions[25]

$$f(\xi, \pm 1) = \pm\tfrac{1}{2}m(\xi), \quad f_{,\zeta}(\xi, \pm 1) = 0 \qquad (16.30)$$

where

$$m(\xi) = \int_{\xi}^{1}(\xi - t)p(t)\,dt,$$

is the dimensionless bending moment along the longitudinal axis of the beam.

We assume that the prescribed stresses and displacements at the ends are such that σ_x and U are odd in z, while τ_{xy} and W are even in z. It thus follows that f must also be odd in ζ. Hence, on integrating (16.28) four times with respect to ζ, we obtain

[25] This modification of the boundary conditions in plane elasticity is known as *Michell's transformation*.

$$f = A(\xi)\zeta + B(\xi)\zeta^3 - 2\varepsilon^2\tilde{E}\left[\int_0^\zeta (\zeta - t)f(\xi, t)\,dt\right]_{,\xi\xi}$$

$$- \tfrac{1}{6}\varepsilon^4\tilde{E}\left[\int_0^\zeta (\zeta - t)^3 f(\xi, t)\,dt\right]_{,\xi\xi\xi\xi}, \tag{16.31}$$

where A and B are unknown functions of ξ. Imposing the boundary conditions (16.30) yields

$$f = f^{(0)} + \varepsilon^2\tilde{E}IF_{,\xi\xi} + \varepsilon^4\tilde{E}Jf_{,\xi\xi\xi\xi}, \tag{16.32}$$

where

$$f^{(0)} = \tfrac{1}{4}m(\xi)\zeta(\zeta^2 - 3), \tag{16.33a}$$

$$If = \zeta\int_0^1 (\zeta^2 t + 2 - 3t)f(\xi, t)\,dt - 2\int_0^\zeta (\zeta - t)f(\xi, t)\,dt, \tag{16.33b}$$

$$Jf = \tfrac{1}{12}\zeta\int_0^1 [\zeta^2(2 + t) - 3t](1 - t^2)f(\xi, t)\,dt$$

$$- \tfrac{1}{6}\int_0^\zeta (\zeta - t)^3 f(\xi, t)\,dt. \tag{16.33c}$$

The integrodifferential equation (16.32) may be solved by using a perturbation method, on putting

$$f = f^{(0)} + \varepsilon^2 f^{(1)} + \ldots, \tag{16.34}$$

and equating the terms of equal order in ε. We already have $f^{(0)}$, and thus $f^{(1)}$ becomes

$$f^{(1)} = -\tfrac{1}{40}\tilde{E}m''(\xi)\zeta(1 - \zeta^2)^2 = \tfrac{1}{40}\tilde{E}p(\xi)\zeta(1 - \zeta^2)^2, \tag{16.35}$$

which may then be followed by the calculation of $f^{(2)}$. The interesting fact is that the approximation $f^{(0)} + \varepsilon^2 f^{(1)}$ leads to the statically admissible stress field obtained by Rychter.

To approximate the displacements we put

$$U = \frac{L^3 p_0}{EH^2}u, \quad W = \frac{L^4 p_0}{E'H^3}w. \tag{16.36}$$

Then (16.25) takes the form

$$\begin{aligned}
\Delta e_z &= \frac{p_0}{E}\varepsilon^{-4}(w_{,\zeta} + \varepsilon^2\sigma\tilde{E}f_{,\zeta\zeta} - \varepsilon^4\tilde{E}f_{,\xi\xi}),\\
\Delta e_{xz} &= \frac{p_0}{E}\varepsilon^{-3}\left(u_{,\zeta} + w_{,\xi} + \varepsilon^2\frac{E}{G}f_{,\xi\xi}\right),\\
\Delta e_x &= \frac{p_0}{E}\varepsilon^{-2}(u_{,\xi} - f_{,\zeta\zeta} + \varepsilon^2\sigma\tilde{E}f_{,\xi\xi}).
\end{aligned} \tag{16.37}$$

In these equations the function f is supposed known in the expansion (16.34). We now assume similar expansions for u and w

$$u = u^{(0)} + \varepsilon^2 u^{(1)} + \dots,$$
$$w = w^{(0)} + \varepsilon^2 w^{(1)} + \dots, \tag{16.38}$$

and choose the coefficients in the expansion so that

$$w^{(n-1)}_{,\zeta} = -\sigma \tilde{E} f^{(n-2)}_{,\zeta\zeta} + \tilde{E} \tilde{f}^{(n-3)}_{,\xi\xi}, \tag{16.39a}$$

$$u^{(n-1)}_{,\zeta} = -w^{(n-1)}_{,\xi} - \frac{E}{G} f^{(n-2)}_{,\xi\xi}, \tag{16.39b}$$

$$u^{(n-1)}_{,\xi} = f^{(n-1)}_{,\zeta\zeta} - \sigma \tilde{E} f^{(n-2)}_{,\xi\xi}, \tag{16.39c}$$

with the convention that the terms of order less than zero vanish. Equations (16.39) may then be solved in sequence. On integrating the first with respect to ζ we obtain

$$w^{(n-1)} = v_{n-1}(\xi) - \sigma \tilde{E} f^{(n-2)}_{,\zeta} + \tilde{E} \left(\int_0^\zeta f^{(n-3)}\, dt \right)_{,\xi\xi}, \tag{16.40}$$

where $v_{n-1}(\xi)$ is an arbitrary function. On inserting (16.40) in (16.39b) and integrating with respect to ζ, we have

$$u^{(n-1)} = -v'_{n-1}(\xi) + \left(\sigma \tilde{E} - \frac{E}{G} \right) f^{(n-2)}_{,\xi} - \tilde{E} \left[\int_0^\zeta (\zeta - t) f^{(n-3)}\, dt \right]_{,\xi\xi\xi}, \tag{16.41}$$

without any function of integration, as u must be odd in ζ. The v_n terms are determined by substituting (16.41) in (16.39b). In particular, we find from (16.33a) the relation

$$\zeta v''_0(\xi) = -f^{(0)}_{,\zeta\zeta} = -\tfrac{3}{2} m(\xi) \zeta,$$

that is, the value

$$v''_0(\xi) = -\tfrac{3}{2} m(\xi), \tag{16.42}$$

which is just the moment–curvature relationship of the elementary theory.

The two methods presented above allow us to estimate the energy of the strains and stresses in a thin rectangular beam, by means of constructing two or more admissible fields. Both methods lead to the conclusion that, if we restrict ourselves to satisfying static or geometric boundary conditions on the ends solely in the mean, then the approximation can be progressively improved, whatever the pressure distribution over the upper face of the rectangular beam may be.

However, it often happens that even this procedure presents difficulties in general circumstances, when, for instance, the cross-section of the beam is not a thin rectangle, or the longitudinal axis is naturally curved. In these situations we can still apply a variational argument, but using a different technique (Berdichevskii 1980) and with a different purpose. Instead of trying to compare the strains and stresses calculated using two-dimensional theory with those obtained using elementary beam theory, we content ourselves with evaluating certain averaged elastic moduli of the stiffness of an equivalent one-dimensional continuum, which approximates a given three-dimensional tubular elastic body. In other words, the problem consists of determining the bending and torsional stiffnesses of the first continuum by averaging in some way the elastic moduli of the second.

The one-dimensional continuum is regarded as a curve Γ endowed with an orthonormal triad $\mathbf{d}_1, \mathbf{d}_2, \mathbf{d}_3$, the last of which is tangential to Γ. The points of Γ are characterized by a radius vector $\mathbf{r}_0(S)$, where S is a parameter in Γ, and the derivative

$$\mathbf{g}_{,3} = \frac{d\mathbf{r}_0}{dS} = \mathbf{r}_0'(S)$$

is a vector tangential to Γ, but not, in general, of unit length.

In the unstrained state the curve Γ occupies the position Γ_0, determined by a radius vector $\mathbf{R}_0(S)$ and by an orthonormal triad $\mathbf{e}_1, \mathbf{e}_2, \mathbf{e}_3$, the last of these being tangential to Γ_0 at the point $\mathbf{R}_0(S)$, and the first two oriented, for instance, along the principal axes of inertia of the cross-section, although this choice is not necessary.

As a result of the deformation, the vector $\mathbf{R}_0(S)$ moves to the vector $\mathbf{r}_0(S)$, and the ratio

$$\gamma = \frac{1}{2}\left(\frac{|d\mathbf{r}_0|^2 - |d\mathbf{R}_0|^2}{|d\mathbf{R}_0|^2}\right)$$

is a measure (one among many) of the elongation of the rod. If the parameter S is the arc length in Γ_0, then we have $|d\mathbf{R}_0| = dS$ and the elongation assumes the form

$$\gamma = \tfrac{1}{2}(|\mathbf{r}_0'|^2 - 1). \tag{16.43}$$

In addition, let us introduce the quantities κ, κ', τ, and κ_0, κ_0', τ_0 defined by (14.6) and (14.8) and assume, as measures of the change in curvature, the differences

$$\begin{aligned} \Omega_1 &= (1+2\gamma)^{\frac{1}{2}}\kappa - \kappa_0, \\ \Omega_2 &= (1+2\gamma)^{\frac{1}{2}}\kappa' - \kappa_0', \\ \Omega &= (1+2\gamma)^{\frac{1}{2}}\tau - \tau_0. \end{aligned} \tag{16.44}$$

Our purpose then is to represent the total energy of the three-dimensional elastic body by means of a functional of the form

$$W - L = \int_{\Gamma_0} \Phi(\gamma, \Omega_\alpha, \Omega)\, ds - L(S) \quad (\alpha = 1, 2), \tag{16.45}$$

where Φ and L are obtained by averaging the elastic moduli of the material over the cross-section. The three-dimensional undeformed configuration of the rod is a domain generated by the motion along the space curve Γ_0 of a plane figure A, intersecting Γ_0 at its centroid and perpendicular to the tangent to this curve at each point of Γ_0. Let us introduce a system of curvilinear coordinates ξ^α, S by means of the formula

$$\mathbf{R}(\xi^\alpha, S) = \mathbf{R}_0(S) + \mathbf{e}_\alpha \xi^\alpha. \tag{16.46}$$

The point with coordinates $\xi^\alpha = 0$ is the centroid of the cross-section. We assume that A is centrally symmetric; this means that, in addition to every point with coordinates ξ^α there is the point $-\xi^\alpha$. The section is, in general, nonhomogeneous, but the inhomogeneities are also assumed symmetric, that is to say the Young's modulus $E(\xi^\alpha)$ and the shear modulus $\mu(\xi^\alpha)$ are even functions of the coordinates ξ^α, and Poisson's ratio σ is constant.

As a consequence of the deformation, the new position of a point becomes $\mathbf{r}(\xi^\alpha, S)$, and the components of the strain tensor are determined by the formula

$$2\varepsilon_{ab} = x^i_{,a} x_{i,b} - \overset{\circ}{g}_{ab} \tag{16.47}$$

where x^i and x_i are the contravariant and covariant components of $\mathbf{r}$, the new position, and $\overset{\circ}{g}_{ab}$ are the components of the metric tensor in the unstrained configuration. The strain energy of the three-dimensional solid assumes the following form:

$$W = \int_{\Gamma_0} ds \iint_A \Lambda \sqrt{\overset{\circ}{g}}\, d\xi^\alpha\, d\xi^\beta, \tag{16.48}$$

with

$$\Lambda = \tfrac{1}{2}[\lambda(\overset{\circ}{g}{}^{ab}\varepsilon_{ab})^2 + 2\mu \overset{\circ}{g}{}^{ab}\overset{\circ}{g}{}^{cd}\varepsilon_{ac}\varepsilon_{bd}], \tag{16.49}$$
$$\overset{\circ}{g} = \det \|\overset{\circ}{g}{}^{ab}\|;$$

and the external work L may be written as

$$L = \int_{\Gamma_0} ds \left[\iint_A F_i x^i \sqrt{\overset{\circ}{g}}\, d\xi^\alpha\, d\xi^\beta + \int_{\partial A} P_i x^i\, ds \right], \tag{16.50}$$

where F_i are the body forces and P_i are the surface forces. In these formulae, the components of the metric tensor in the reference configuration can be derived from (16.46) and their detailed expressions are

$$\begin{aligned}
&\overset{\circ}{g}_{33} = (1 + \overset{\circ}{\kappa}\xi^1 + \overset{\circ}{\kappa}{}'\xi^2)^2 + \overset{\circ}{\tau}{}^2\xi^\alpha\xi^\alpha, \quad \overset{\circ}{g}{}^{33} = (1 + \overset{\circ}{\kappa}\xi^1 + \overset{\circ}{\kappa}{}'\xi^2)^{-2},\\
&\overset{\circ}{g}_{13} = -\overset{\circ}{\tau}\xi^2, \quad \overset{\circ}{g}_{23} = \overset{\circ}{\tau}\xi^1,\\
&\overset{\circ}{g}{}^{13} = -\overset{\circ}{\tau}\xi^2(1 + \overset{\circ}{\kappa}\xi^1 + \overset{\circ}{\kappa}{}'\xi^2)^{-2},\\
&\overset{\circ}{g}{}^{23} = \overset{\circ}{\tau}\xi^1(1 + \overset{\circ}{\kappa}\xi^1 + \overset{\circ}{\kappa}{}'\xi^2)^{-2},\\
&\overset{\circ}{g}_{\alpha\beta} = \delta_{\alpha\beta}, \quad \overset{\circ}{g}{}^{\alpha\beta} = \delta^{\alpha\beta} + \overset{\circ}{\tau}{}^2(\delta^{\alpha\beta}\xi^\gamma\xi^\gamma - \xi^\alpha\xi^\beta)(1 + \kappa\xi^1 + \kappa'\xi^2)^{-2},\\
&\overset{\circ}{g} = \det\|g_{ab}\| = (1 + \overset{\circ}{\kappa}\xi^1 + \overset{\circ}{\kappa}{}'\xi^2)^{-2}.
\end{aligned} \tag{16.51}$$

The strains $\varepsilon_{\alpha\beta}$ may thus be written in explicit form when we recall that the position $\mathbf{r}(\xi^\alpha, S)$ admits the two-fold representation

$$\mathbf{r} = \mathbf{r}_0(S) + \mathbf{d}_\alpha\xi^\alpha = x_i\mathbf{g}^i = x^i\mathbf{g}_i.$$

Now, from (16.47) and (16.51), we can derive the complete expression for the strain energy density $\Lambda\sqrt{\overset{\circ}{g}}$. We now take into account the fact that terms in ξ^α are of the order of h, the diameter of the cross-section. This is small compared with the length L of the beam, making terms of the order of $(h/L)^2$ negligible, and so the strain energy can be written in the form

$$W = \int_{\Gamma_0} \Phi(\gamma, \Omega_\alpha, \Omega)\, ds,$$

where

$$2\Phi = \gamma^2 \left(\iint_A E\, d\xi^\alpha\, d\xi^\beta \right) + \left(\iint_A E \xi^\alpha \xi^\beta\, d\xi^\gamma\, d\xi^\delta \right) \Omega_\alpha \Omega_\beta + C\Omega^2, \tag{16.52}$$

E being Young's modulus of the section and C its torsional rigidity.

This result rests on the assumption that the triad $\mathbf{d}_1, \mathbf{d}_2, \mathbf{d}_3$ is orthogonal, as assumed in the so-called *Kirchhoff hypothesis* for beams. However, the method may be generalized further to the *Timoshenko theory*, in which the vectors $\mathbf{d}_1$ and $\mathbf{d}_2$ are of unit length and mutually orthogonal, but not orthogonal to the vector $\mathbf{d}_3$ (Berdichevskii and Starosel'skii 1983) (see Section 14).

Yet another extension of the method lies in the treatment of naturally twisted rods (Berdichevskii and Starosel'skii 1985), a particular class of curvilinear rods generated as follows. We consider a segment $0 \leqslant S \leqslant L$ located on the x^3 axis of a Cartesian system of coordinates and take a two-dimensional domain A in the plane $x_3 = 0$ and displace it along the x_3 axis, while simultaneously rotating it through an angle $\varphi = \omega x^3$, where ω is a constant, about the x^3 axis. For naturally twisted rods it is also possible to define an averaged strain energy like (16.52).

A problem that can be dealt with by means of a variational argument is that of the so-called *generalized torsion* of a cylinder (proposed by Truesdell 1959, 1978). This problem may be formulated as follows. Saint-Venant's theory of torsion of an isotropic linear elastic cylinder shows that between the torque M and the twist τ the relation

$$M = \mu R \tau \tag{16.53}$$

holds, where R is a geometric quantity. The product μR is the torsional rigidity. However, (16.53) holds only for the exact Saint-Venant's solution corresponding to a very special distribution of tractions on the plane ends of the cylinder, and the question thus arises of deciding whether (16.53) is valid in general for solutions that are not exact. The answer is that (16.53) holds if the twist τ is defined as the constant associated with Saint-Venant's torsion field which approximates a mean-square solution in the strain (Day 1981) instead of the exact solution.

In order to prove the result, let us consider a cylinder V of height $2h$, the cross-section A of which is a domain in the xy plane and the generators of which are parallel to the z axis, where x, y, and z are rectangular Cartesian coordinates.

The vector field

$$\begin{aligned} u_0 &= -zy, \\ v_0 &= zx, \\ w_0 &= \varphi(x, y), \end{aligned} \tag{16.54}$$

where $\varphi(x, y)$ is a harmonic function in A and satisfies the Neumann condition

$$\frac{\partial \varphi}{\partial \nu} = y \cos(x, \nu) - x \cos(y, \nu), \tag{16.55}$$

is called a *torsion field* and corresponds to unit twist.

If (u, v, w) is any continuous differentiable field, we define the *generalized twist* τ, associated with (u, v, w), as the unique number that minimizes the function

$$f(\alpha) = \iiint_V \left\{ \sum_{i,j=1}^{3} [e_{ij}(u, v, w) - \alpha e_{ij}(u_0, v_0, w_0)]^2 \right\} dx\, dy\, dz \tag{16.56}$$

where e_{ij} denote the strains. On squaring the integrand and carrying out a short calculation we obtain

$$\tau = \iiint_V \left\{ \sum_{i,j=1}^{3} e_{ij}(u_0, v_0, w_0) e_{ij}(u, v, w) \right\} dx\, dy\, dz \times \left\{ \iiint_V \left(\sum_{i,j=1}^{3} e_{ij}^2(u_0, v_0, w_0) \right) dx\, dy\, dz \right\}^{-1}. \tag{16.57}$$

In this expression we put

$$\iiint_V \left(\sum_{i,j=1}^{3} e_{ij}^2(u_0, v_0, w_0) \right) dx\, dy\, dz = 2h \iint_A [(\varphi_{,x} - y)^2 + (\varphi_{,y} + x)^2]\, dx\, dy = 2hR, \tag{16.58}$$

and we transform the numerator in (16.57) in the following manner:

$$\begin{aligned}
&\iiint_V \left[\sum_{i,j=1}^{3} e_{ij}(u, v, w) e_{ij}(u_0, v_0, w_0) \right] dx\, dy\, dz \\
&= \iiint_V [(u_{,z} + w_{,x})(\varphi_{,x} - y) + (v_{,z} + w_{,y})(\varphi_{,y} + x)]\, dx\, dy\, dz \\
&= \iiint_V \left\{ \left[\frac{\partial}{\partial z} (u(\varphi_{,x} - y) + v(\varphi_{,y} + x)) \right] + \frac{\partial}{\partial x} [w(\varphi_{,x} - y)] + \frac{\partial}{\partial y} [w(\varphi_{,y} + x)] \right\} dx\, dy\, dz \\
&= \iiint_V \left\{ \frac{\partial}{\partial z} [u(\varphi_{,x} - y) + v(\varphi_{,y} + x)] \right\} dx\, dy\, dz \\
&\quad + \int_{-h}^{h} dz \left\{ \int_{\partial A} [w(\varphi_{,x} - y) \cos(x, \nu) + w(\varphi_{,y} + x) \cos(y, \nu)] \right\} ds \\
&= \iiint_V \left\{ \frac{\partial}{\partial z} [u(\varphi_{,x} - y) + v(\varphi_{,y} + x)] \right\} dx\, dy\, dz,
\end{aligned} \tag{16.59}$$

to obtain the result

$$\tau = \frac{1}{2hR} \iiint_V \left\{ \frac{\partial}{\partial z} [u(\varphi_{,x} - y) + v(\varphi_{,y} + x)] \right\} dx\, dy\, dz. \tag{16.60}$$

Let us now suppose that (u, v, w) is a solution of the torsion problem, in the sense that the stresses associated with these displacements satisfy the equilibrium equations in V, the tractions on the lateral surface vanish, and the tractions on the plane ends of the cylinder are purely tangential. It is clear that the field (u_0, v_0, w_0) is a solution in this sense. Another consequence of the definition is that the torque is given by

$$M = \iint_A [\sigma_{zy}(x, y, h)x - \sigma_{zx}(w, y, h)y]\, dx\, dy$$

$$= -\iint_A [\sigma_{zy}(x, y, -h)x - \sigma_{zx}(x, y, -h)y]\, dx\, dy,$$

where σ_{zx} and σ_{zy} are the tangential tractions on the end faces.

If we have two solutions of the torsion problem, (u, v, w) and $(\tilde{u}, \tilde{v}, \tilde{w})$, then, by Betti's reciprocal term, we may write

$$\iint_A [\sigma_{zx}(w, y, h)\tilde{u} + \sigma_{zy}(x, y, h)\tilde{v}]\, dx\, dy + \iint_A [\sigma_{zx}(x, y, -h)\tilde{u} + \sigma_{zy}(x, y, -h)\tilde{v}]\, dx\, dy$$

$$= \iint_A [\tilde{\sigma}_{zx}(x, y, h)u + \tilde{\sigma}_{zy}(x, y, h)v]\, dx\, dy$$

$$+ \iint_A [\tilde{\sigma}_{zx}(x, y, -h)u + \tilde{\sigma}_{zy}(x, y, -h)v]\, dx\, dy. \tag{16.61}$$

In particular, if we chose $(\tilde{u}, \tilde{v}, \tilde{w}) = (u_0, v_0, w_0)$, the tractions over the plane ends are

$$\tilde{\sigma}_{zx}(x, y, h) = -\tilde{\sigma}_{zx}(x, y, -h) = \mu(\varphi_{,x} - y),$$
$$\tilde{\sigma}_{zy}(x, y, h) = -\tilde{\sigma}_{zy}(x, y, -h) = \mu(\varphi_{,y} + x),$$

and the right-hand side of (16.61), by virtue of the divergence theorem and (16.60), reduces to

$$\mu \iint_A [(\varphi_{,x} - y)u(x, y, h) + (\varphi_{,y} + x)v(x, y, h))\, dx\, dy$$

$$+ \mu \iint_A (\varphi_{,x} - y)u(x, y, -h) + (\varphi_{,y} + x)v(x, y, -h)\, dx\, dy$$

$$= \mu \iiint_V \frac{\partial}{\partial z}[(\varphi_{,x} - y)u + (\varphi_{,y} + x)v]\, dx\, dy\, dz = 2\mu h R\tau. \tag{16.62}$$

On the other hand, the left-hand side of (16.61), with $(\tilde{u}, \tilde{v}, \tilde{w}) = (u_0, v_0, w_0)$, becomes

$$h \iint_A [-\sigma_{zx}(x, y, h)y + \sigma_{zy}(x, y, h)x]\, dx\, dy - h \iint_A [-\sigma_{zx}(x, y, -h)y + \sigma_{zy}(x, y, -h)x]\, dx\, dy = 2hM. \tag{16.63}$$

Thus, by comparing the last two equations, we arrive at the equality $M = \mu R\tau$, for any solution of the torsion problem.

17. The Influence of Large Strains

The systematic application of three-dimensional theory to large deformations leads to equations of considerable complexity, the solutions of which exist for only a few special cases. For practical applications, it is useful to write the equilibrium equations relating the displacements to certain averages of the exterior loads taken over the cross-section. However, these equations must be such that they allow the generation of a sequence of models at increasing orders of approximation. Classical theories should be included in the early stages (Parker 1979) as, for example, in the case of small strains and displacements, Saint-Venant's solution should result as a first approximation. On the other hand, if we allow for small strains but large displacements and rotations, Kirchhoff's theory of rods should naturally emerge.

In order to elaborate the approach, let us suppose, for simplicity, that the rod is straight in its unstressed reference configuration. This will be a cylindrical domain V, with cross-section A, referred to a Cartesian orthogonal system, such that the X_3 axis coincides with the center line and the X_1 and X_2 axes are parallel to the principal axes of inertia of the cross-section. The length of the rod is ℓ, such that $0 \leqslant X_3 \leqslant \ell$, and the diameter of A is assumed to be small compared with ℓ. We recall that we are concerned with deformations in which each cross-section remains nearly plane, although its orientation and displacement may vary appreciably, and so introduce the new variable $X = \varepsilon X_3$, where ε is a parameter, and let $L = \varepsilon\ell$ be the length of the beam.

We now associate with each X a triad of vectors $\mathbf{d}_1(X), \mathbf{d}_2(X), \mathbf{d}_3(X)$ the components of which along the X_1, X_2, and X_3 axes are the columns of an orthogonal matrix $\mathbf{H} = \mathbf{H}(X)$. Thus each position $\mathbf{r}(X_1, X_2, X_3)$ of the deformed configuration may be represented as

$$\mathbf{r}(X_\alpha, X_3) = \mathbf{r}_0(X_3) + u_\alpha \mathbf{d}_\alpha + u_3\mathbf{d}_3 = \mathbf{r}_0(X_3) + \mathbf{H}\mathbf{u} \quad (\alpha = 1, 2), \tag{17.1}$$

where $\mathbf{r}_0(X_3)$ denotes the deformed position of the line of the centroids and $(u_\alpha, u_3) = \mathbf{u}$ denotes the position of each point of the cross-section relative to its centroid. The unit vector $\mathbf{d}_3$ is tangential to the deformed center line, and we choose $\mathbf{d}_1$ and $\mathbf{d}_2$ such that

$$u_1 = X_1 + \varepsilon\hat{u}_1, \quad u_2 = X_2 + \varepsilon\hat{u}_2, \quad u_3 = \varepsilon\hat{u}_3.$$

Equation (17.1) then becomes

$$\mathbf{r}(X_\alpha, X_3) = \mathbf{r}_0(X_3) + X_1\mathbf{d}_1 + X_2\mathbf{d}_2 + \varepsilon(\hat{u}_\alpha\mathbf{d}_\alpha + \hat{u}_3\mathbf{d}_3). \tag{17.2}$$

This means that the deformation of the cross-section can be split into a term representing the motion of the centroid, a rotation which aligns the X_1 and X_2 axes along $\mathbf{d}_1$ and $\mathbf{d}_2$, and a superimposed small displacement $\varepsilon(\hat{u}_1, \hat{u}_2, \hat{u}_3)$ with components parallel to $\mathbf{d}_\alpha$ and $\mathbf{d}_3$. The rotation of this triad is associated with a skew-symmetric matrix $\mathbf{\Omega}$, for which

$$\mathbf{H}'(X) = \mathbf{H}(X)\mathbf{\Omega}(X) \quad (\mathbf{\Omega} + \mathbf{\Omega}^{\mathrm{T}} = 0), \tag{17.3}$$

where $\boldsymbol{\Omega}$ has the form

$$\boldsymbol{\Omega} = \begin{pmatrix} 0 & \tau & -\kappa' \\ -\tau & 0 & \kappa \\ \kappa' & -\kappa & 0 \end{pmatrix}, \tag{17.4}$$

κ and κ' being the curvatures and τ being the twist. Formula (17.3) is exactly equivalent to equations (14.7), which give the derivatives of $(\mathbf{d}_\alpha, \mathbf{d}_3)$ in terms of the curvatures.

The extension of the central line is defined in terms of $|\mathbf{r}_0'(X_3)| \equiv a(X)$, and, as $a(X)$ is close to unity, we write

$$\mathbf{r}_0'(X_3) = [1 + \varepsilon\bar{a}(X) + \varepsilon^2\hat{a}(X)]\mathbf{d}_3.$$

The position of each point on the rod is determined from (17.2), and we know the terms composing its right-hand side, or, at least, an approximation of it. To determine these terms we set

$$(\hat{u}_1, \hat{u}_2, \hat{u}_3) = \hat{\mathbf{u}} = \bar{\mathbf{u}} + \epsilon\bar{\mathbf{v}} + \ldots \tag{17.5}$$

where $\bar{\mathbf{u}}$ has the form

$$\bar{\mathbf{u}} = \kappa\mathbf{w}^{(1)}(X_\alpha) + \kappa'\mathbf{w}^{(2)}(X_\alpha) + \tau\boldsymbol{\varphi}(X_\alpha) + \bar{a}\mathbf{c}(X_\alpha). \tag{17.6}$$

The displacements $\mathbf{w}^{(1)}$ and $\mathbf{w}^{(2)}$ correspond to pure bending about axes parallel to $\mathbf{d}_1$ and $\mathbf{d}_2$, respectively; $\boldsymbol{\varphi}$ is the warping associated with torsion about the $\mathbf{d}_3$ axis; and $\mathbf{c}$ describes the transverse displacement produced by the longitudinal elongation.

Many classical solutions of beam theory can be derived from (17.1), by specifying the form of $\mathbf{r}_0$, $\mathbf{H}$, and $\mathbf{u}$. For instance, let us assume

$$\mathbf{H} = \mathbf{H}_0 \exp(\boldsymbol{\Omega}^{\mathrm{T}} X), \tag{17.7}$$

where $\mathbf{H}_0$ is a constant vector and $\boldsymbol{\Omega}$ is given by (17.4) but with $\kappa = \kappa' = 0$. Using the characteristic equation

$$\boldsymbol{\Omega}^3 + \tau^2\boldsymbol{\Omega} = 0,$$

we can write

$$\exp(\boldsymbol{\Omega}^{\mathrm{T}} X) = [1 + \tau^{-1}\boldsymbol{\Omega}^{\mathrm{T}} \sin\tau X + \tau^{-2}(\boldsymbol{\Omega}^{\mathrm{T}})^2(1 - \cos\tau X)] = \begin{pmatrix} \cos\tau X & -\sin\tau X & 0 \\ \sin\tau X & \cos\tau X & 0 \\ 0 & 0 & 1 \end{pmatrix},$$

and we approximate $\mathbf{r}_0'(X_3)$ by taking

$$\mathbf{r}_0'(X_3) \simeq \mathbf{d}_3 = \mathbf{H}_0 \begin{pmatrix} 0 \\ 0 \\ 1 \end{pmatrix},$$

with a displacement $\bar{\mathbf{u}} = [0, 0, \tau\phi(X_\alpha)]$. It then follows that, correct to terms of first order in ε, the position may be written as

$$\mathbf{r} = \mathbf{r}_0(X_3) + \mathbf{H}_0 \exp(\boldsymbol{\Omega}^{\mathrm{T}} X) \begin{pmatrix} X_1 \\ X_2 \\ X_3 + \varepsilon\tau\varphi(X_\alpha) \end{pmatrix}, \tag{17.8}$$

which, in the special case $\mathbf{r}_0 = 0$ and $\mathbf{H}_0 = 1$ reduces to

$$x_1 = X_1 \cos(\varepsilon\tau X_3) - X_2 \sin(\varepsilon\tau X_3),$$
$$x_2 = X_1 \sin(\varepsilon\tau X_3) + X_2 \cos(\varepsilon\tau X_3), \qquad (17.9)$$
$$x_3 = X_3 + \varepsilon\tau\varphi(X_\alpha).$$

These formulae generalize, for arbitrary X_3, the standard solution for torsion

$$x_1 = X_1 - \varepsilon\tau X_2 X_3,$$
$$x_2 = X_2 + \varepsilon\tau X_1 X_3, \qquad (17.10)$$
$$x_3 = X_3 + \varepsilon\tau\varphi(X_\alpha).$$

Let us now consider the case in which $\kappa = \tau = 0$, with $\mathbf{d}_2$ fixed. We choose

$$\exp(\mathbf{\Omega}^{\mathrm{T}} X) = \begin{pmatrix} \cos\kappa' X & 0 & \sin\kappa' X \\ 0 & 1 & 0 \\ -\sin\kappa' X & 0 & \cos\kappa' X \end{pmatrix},$$

and

$$\mathbf{u} = \sigma\kappa' \begin{pmatrix} \frac{1}{2}(X_1^2 - X_2^2) \\ X_1 X_2 \\ 0 \end{pmatrix},$$

so that

$$\mathbf{r} = \mathbf{r}_0(X_3) + \mathbf{H}_0 \begin{pmatrix} X_1 \cos(\varepsilon\kappa' X_3) \\ X_2 \\ -X_1 \sin(\varepsilon\kappa' X_3) \end{pmatrix} + (\varepsilon\kappa')^{-1}\mathbf{H}_0 \begin{pmatrix} 1 - \cos(\varepsilon\kappa' X_3) \\ 0 \\ \sin(\varepsilon\kappa' X_3) \end{pmatrix}$$
$$+ \varepsilon\kappa'\sigma\mathbf{H}_0 \begin{pmatrix} \frac{1}{2}(X_1^2 - X_2^2)\cos(\varepsilon\kappa' X_3) \\ X_1 X_2 \\ -\frac{1}{2}(X_1^2 - X_2^2)\sin(\varepsilon\kappa' X_3) \end{pmatrix}.$$

In the special case for which $\mathbf{r}_0 = 0$ and $\mathbf{H}_0 = 1$, we derive

$$x_1 = X_1 \cos(\varepsilon\kappa' X_3) + (\varepsilon\kappa')^{-1}[1 - \cos(\varepsilon\kappa' X_3)] + \tfrac{1}{2}\varepsilon\kappa'\sigma(X_1^2 - X_2^2)\cos(\varepsilon\kappa' X_3),$$
$$x_2 = X_2 + \varepsilon\kappa'\sigma X_1 X_2, \qquad (17.11)$$
$$x_3 = [(\varepsilon\kappa')^{-1} - X_1 - \tfrac{1}{2}\varepsilon\kappa'\sigma(X_1^2 - X_2^2)]\sin(\varepsilon\kappa' X_3),$$

and the standard solution for small displacements can be recovered simply by linearizing the functions of $\sin(\varepsilon\kappa' X_3)$ and $\cos(\varepsilon\kappa' X_3)$:

$$x_1 = X_1 + \tfrac{1}{2}\varepsilon\kappa' X_3^2 + \tfrac{1}{2}\varepsilon\kappa'\sigma(X_1^2 - X_2^2),$$
$$x_2 = X_2 + \varepsilon\kappa'\sigma X_1 X_2, \qquad (17.12)$$
$$x_3 = X_3 - \varepsilon\kappa' X_1 X_3.$$

Another interesting solution may be obtained by putting $\kappa = 0$, $\kappa' = b\bar{\kappa}'$, $\tau = b\bar{\tau}$, with $(\bar{\kappa}')^2 + \bar{\tau}^2 = 1$, so that the position vector $\mathbf{r}$ can be approximated as

$$\mathbf{r} = \mathbf{r}_0(X_3) + \mathbf{H}_0(\varepsilon b)^{-1} \begin{pmatrix} \bar{\kappa}'[\cos(bX) - 1] \\ \bar{\kappa}'\bar{\tau}[bX - \sin(bX)] \\ \bar{\tau}^2 bX + \kappa'^2 \sin(bX) \end{pmatrix} + \mathbf{H}_0 \exp(\mathbf{\Omega}^{\mathrm{T}} X)\mathbf{u},$$

where

$$\exp(\mathbf{\Omega}^{\mathrm{T}}X) = \begin{pmatrix} \cos bX & \bar{\tau}\sin bX & -\bar{\kappa}'\sin bX \\ \bar{\tau}\sin bX & \bar{\kappa}'^2 + \bar{\tau}^2\cos bX & \bar{\kappa}'\bar{\tau}[1-\cos(bX)] \\ -\bar{\kappa}'\sin bX & \bar{\kappa}'\bar{\tau}[1-\cos(bX)] & \bar{\tau}^2 + \bar{\kappa}'^2\cos bX \end{pmatrix},$$

and

$$\mathbf{u} = \begin{pmatrix} X_1 + \frac{1}{2}\varepsilon\kappa'\sigma(X_1^2 - X_2^2) \\ X_1 + \varepsilon\kappa'\sigma X_1 X_2 \\ \varepsilon\tau\varphi(X_\alpha) \end{pmatrix}, \quad X = \varepsilon X_3.$$

The orthogonal matrix $\mathbf{H}$ is periodic in X, and the fibers $X_\alpha =$ constant are coaxial helices with axis parallel to $\mathbf{H}\boldsymbol{\kappa}_3 = \mathbf{H}_0\boldsymbol{\kappa}_3$, where $\boldsymbol{\kappa}_3 = (0, \bar{\kappa}', \bar{\tau})^{\mathrm{T}}$ is the unit null vector of $\mathbf{\Omega}$. These solutions are known as Ericksen's *helical deformations* of prisms (Ericksen 1977).

In some cases the purpose of the theory is that of obtaining a set of equations which, while still preserving the relevant nonlinear terms, are sufficiently simple that they can be treated mathematically or even be solved in some particular cases. A theory of slender straight beams undergoing small strains and moderate rotations, which is useful for describing the behavior of rotor and turbine blades, can be formulated as follows (Rosen and Friedmann 1979).

The, initially straight, longitudinal axis of the beam is the line connecting the centroids of the cross-sections. This line is taken as the Z axis of a Cartesian system of coordinates, with the origin O placed at the centroid of the left face of the cylinder; the X and Y axes coincide with the principal axes of inertia of this face and are also the axes of symmetry of the cross-section. Now, we introduce a triad of orthonormal unit vectors $\mathbf{e}_1, \mathbf{e}_2, \mathbf{e}_3$ placed along the coordinate axes so that the position vector of any material point before the deformation is represented by

$$\mathbf{R}(X, Y, Z) = X\mathbf{e}_1 + Y\mathbf{e}_2 + Z\mathbf{e}_3. \tag{17.13}$$

The vector equation of the central axis is then the straight line

$$\mathbf{R}_0(Z) = \mathbf{R}(0, 0, Z) = Z\mathbf{e}_3.$$

After the deformation, the curve $\mathbf{R}_0(z)$ is displaced to a new curved line $\mathbf{r}_0(Z)$, and the tangential vector to this line is the, nonunit, vector

$$\mathbf{g}_3 = \mathbf{r}_0'(Z). \tag{17.14}$$

The unit vectors drawn from the centroid of the generic cross-section parallel to the X and Y axes are changed to a pair of nonunit vectors $\mathbf{g}_1$ and $\mathbf{g}_2$. Knowledge of the triad $\mathbf{g}_1, \mathbf{g}_2, \mathbf{g}_3$ characterizes the deformation of the cross-section. However, in order to explain the process of deformation, it is often convenient to define another orthonormal triad $\mathbf{d}_1, \mathbf{d}_2, \mathbf{d}_3$, such that $\mathbf{d}_3$ coincides with $\mathbf{g}_3$, but is of unit length, $\mathbf{d}_1$ is a unit vector perpendicular to $\mathbf{d}_3$ and contained in $\mathbf{d}_3\mathbf{g}_1$ plane, and $\mathbf{d}_2$ is a unit vector perpendicular to $\mathbf{d}_1$ and $\mathbf{d}_3$. The triad $\mathbf{d}_1, \mathbf{d}_2, \mathbf{d}_3$ defines the principal *torsion–flexure axes*. The introduction of the principal torsion–flexure axes helps in understanding how the deformation of the cross-section can be split into successive stages. First, the triad $\mathbf{e}_1, \mathbf{e}_2, \mathbf{e}_3$ is rigidly translated and rotated so as to coincide with the triad $\mathbf{d}_1, \mathbf{d}_2, \mathbf{d}_3$; next, this triad is stretched and sheared so that it is superimposed on the triad $\mathbf{g}_1, \mathbf{g}_2, \mathbf{g}_3$. In linear theory it is commonly assumed that the two deformations are both infinitesimal, but sometimes it is necessary to remove the hypothesis that the rigid rotation is small. This means that the position of the principal torsion–flexure axes cannot be described by a skew–symmetric matrix like $\mathbf{\Omega}$, but must be treated by means of at least three finite angles. The canonical choice of these angles is the Euler angles of rigid body dynamics. According to Euler's method, the passage from $\mathbf{e}_1, \mathbf{e}_2, \mathbf{e}_3$ to $\mathbf{d}_1, \mathbf{d}_2, \mathbf{d}_3$ may be done by means of successive rotations φ, θ, and ψ, which yield (Love 1927, Art. 253)

$$\begin{aligned}
\mathbf{d}_1 &= (-\sin\psi\sin\varphi + \cos\psi\cos\varphi\cos\theta)\mathbf{e}_1 + (\cos\psi\sin\varphi + \sin\psi\cos\varphi\cos\theta)\mathbf{e}_2 \\
&\quad - \cos\varphi\sin\theta\mathbf{e}_3, \\
\mathbf{d}_2 &= (-\sin\psi\cos\varphi - \cos\psi\sin\varphi\cos\theta)\mathbf{e}_1 + (\cos\psi\cos\varphi - \sin\psi\sin\varphi\cos\theta)\mathbf{e}_2 \\
&\quad + \sin\varphi\sin\theta\mathbf{e}_3, \\
\mathbf{d}_3 &= \cos\psi\sin\theta\mathbf{e}_1 + \sin\psi\sin\theta\mathbf{e}_2 + \cos\theta\mathbf{e}_3,
\end{aligned} \tag{17.15}$$

where θ is restricted to the interval $[0, \pi]$.

The strains and shears which take the triads $\mathbf{d}_1, \mathbf{d}_2, \mathbf{d}_3$ into the triad $\mathbf{g}_1, \mathbf{g}_2, \mathbf{g}_3$ are infinitesimal. An approximate method for describing them is as follows. Let us denote the displacements of a point on the central axis by u, v, w, in the directions $\mathbf{e}_1$, $\mathbf{e}_2$, and $\mathbf{e}_3$, respectively. The derivative $w'(z)$ represents the elongation in the direction of the axis, but w' may be neglected compared to unity; the derivatives $u'(z)$ and $v'(z)$ are now approximate measures of the angle formed by $\mathbf{d}_3$ and $\mathbf{e}_3$, obtained by means of a rotation about $\mathbf{e}_2$ and one about $\mathbf{e}_1$ (Figure 17.1). The Euler angle φ may be taken as an approximate measure of the angle formed by $\mathbf{d}_1$ with $\mathbf{e}_2$, or the one formed by $\mathbf{d}_2$ and $\mathbf{e}_1$.

The introduction of the new quantities u', v', and φ permits us to replace Euler's equations (17.15) by linearized versions: which read[26]

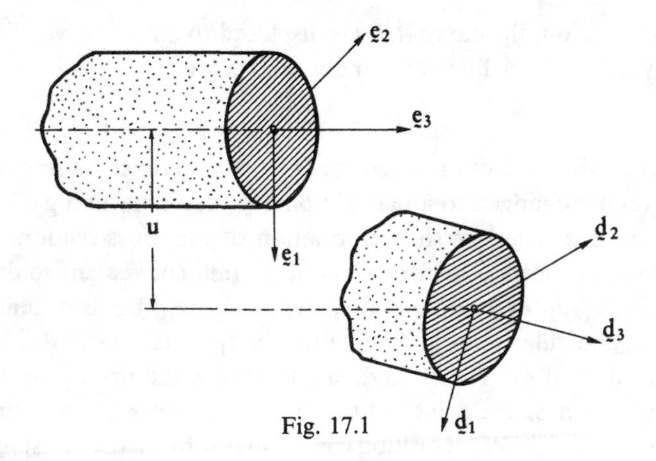

Fig. 17.1

[26] Rosen and Friedmann (1979); however, some terms of second order are retained.

$$\mathbf{d}_1 = \mathbf{e}_1 + \varphi\mathbf{e}_2 - u'\mathbf{e}_3,$$
$$\mathbf{d}_2 = -\varphi\mathbf{e}_1 + \mathbf{e}_2 - v'\mathbf{e}_3, \tag{17.16}$$
$$\mathbf{d}_3 = u'\mathbf{e}_1 + v'\mathbf{e}_2 + \mathbf{e}_3.$$

These expressions for the rigid rotation of the triad $\mathbf{e}_1$, $\mathbf{e}_2$, $\mathbf{e}_3$ are useful for describing the case of small strains accompanied by *moderate* rotations.

Curvature and twist follow from (17.16) through (14.6):

$$\kappa = \mathbf{d}_2' \cdot \mathbf{d}_3 = -\mathbf{d}_2 \cdot \mathbf{d}_3' = v'' + u''\varphi,$$
$$\kappa' = \mathbf{d}_3' \cdot \mathbf{d}_1 = u'' + v''\varphi, \tag{17.17}$$
$$\tau = \mathbf{d}_1' \cdot \mathbf{d}_2 = \varphi' + u''v'.$$

If the strains within the cross-section are neglected, the relevant strains can be derived from (14.17), and they assume the simplified form

$$e_{33} = -\kappa' X + \kappa Y, \quad e_{31} = -\tau Y, \quad e_{32} = \tau X. \tag{17.18}$$

The bending moments about the $\mathbf{d}_2$ and $\mathbf{d}_3$ axes and the torque about the $\mathbf{d}_3$ axis can thus be written as

$$M_2 = \iint_A \sigma_{33} X\, dA = -EI_2\kappa', \quad M_1 = \iint_A \sigma_{33} Y\, dA = EI_1\kappa,$$
$$M_3 = \iint_A (\sigma_{32}X - \sigma_{31}Y)\, dA = GJ\tau. \tag{17.19}$$

These couples are related to one another by the equilibrium equations and the resulting couple is a vector $\mathbf{m}$, which may be expressed as the sum

$$\mathbf{m} = M_1\mathbf{d}_1 + M_2\mathbf{d}_2 + M_3\mathbf{d}_3, \tag{17.20}$$

or, alternatively, written in terms of the fixed triad,

$$\mathbf{m} = M_1(\mathbf{e}_1 + \varphi\mathbf{e}_2 - u'\mathbf{e}_3) + M_2(-\varphi\mathbf{e}_1 + \mathbf{e}_2 - v'\mathbf{e}_3) + M_3(u'\mathbf{e}_1 + v'\mathbf{e}_2 + \mathbf{e}_3). \tag{17.21}$$

The vector couple $\mathbf{m}$ must satisfy the equilibrium equation

$$\mathbf{m}_{,z} + \mathbf{r}_0' \times \mathbf{n} = \mathbf{0}, \tag{17.22}$$

where $\mathbf{n}$ is the resultant of the shear forces and the axial force acting on the cross-section. In the absence of a load distributed along the longitudinal axis the equilibrium equation for $\mathbf{n}$ becomes

$$\mathbf{n}_{,z} = \mathbf{0},$$

whence we immediately derive the result

$$\mathbf{n} = \text{constant vector} = \mathbf{n}(L), \tag{17.23}$$

where L is the value of Z at the right-hand end of the beam. By using the last result we can integrate (17.22):

$$\mathbf{m} + \mathbf{r}_0 \times \mathbf{n}(L) = \text{constant vector}. \tag{17.24}$$

In this vector equation $\mathbf{n}(L)$ is known, $\mathbf{m}$ is an explicit function of φ, u, and v as a result of (17.21) and (17.19). The vector $\mathbf{r}_0$ is also representable, as the sum

$$\mathbf{r}_0 = \mathbf{R}_0 + u\mathbf{e}_1 + v\mathbf{e}_2 + w\mathbf{e}_3,$$

or also by its approximation

$$\mathbf{r}_0 \simeq \mathbf{R}_0 + u\mathbf{e}_1 + v\mathbf{e}_2. \tag{17.25}$$

It turns out that (17.24) is equivalent to a nonlinear system of three ordinary differential equations in u, v, and φ. Despite the difficulty involved in effecting a solution, a remarkable simplification often occurs for beams having a fairly thin cross-section, so that the rigidity factor, say EI_2, is much greater than EI_1 and GJ, and the terms containing EI_1 and GJ may thus be neglected without appreciable error (Rosen and Friedmann 1979).

In some other circumstances, as, for example, in the design of tailored helicopter blades, the nonlinear effects occur in naturally curved and twisted beams. However, their treatment may be simplified if we assume that the cross-section is undeformable in its own plane and that the initial curvatures of the beam are small (Bauchau and Hong 1988).

In its undeformed state, the beam is defined by it axis, which is the line of centroids of the cross-sections, and by a plane perpendicular at each point to the central line, called the plane of the section. The position vector of a particle before the deformation is a vector function of the form

$$\mathbf{R} = \mathbf{R}(X_1, X_2, X_3), \tag{17.26}$$

where X_α and X_3 are curvilinear coordinates chosen in such a way that

$$\mathbf{R}_0 = \mathbf{R}(0, 0, X_3) = \mathbf{R}_0(X_3) \tag{17.27}$$

is the equation of the central line. The direction of the central line and the position of the cross-section for each X_3 are defined by an orthonormal triad $\mathbf{e}_1, \mathbf{e}_2, \mathbf{e}_3$, where the unit vector $\mathbf{e}_3$ coincides with the tangent, and the unit vectors $\mathbf{e}_1$ and $\mathbf{e}_2$ coincide with the principal axes of inertia of the cross-section.

The derivates of this triad with respect to X_3, denoted by $\mathbf{e}_1', \mathbf{e}_2', \mathbf{e}_3'$, are related to the initial curvatures and twist by a formula like (14.9). If, in particular, the coordinates X_1 and X_2 are Cartesian coordinates in the plane of the cross-section, the position vector can be expressed as

$$\mathbf{R}(X_1, X_2, X_3) = \mathbf{R}_0(X_3) + \mathbf{e}_1 X_1 + \mathbf{e}_2 X_2, \tag{17.28}$$

and the base vectors, still recalling (14.10), become

$$\mathbf{G}_1 = \mathbf{R}_{,1} = \mathbf{e}_1, \tag{17.29a}$$

$$\mathbf{G}_2 = \mathbf{R}_{,2} = \mathbf{e}_2, \tag{17.29b}$$

$$\mathbf{G}_3 = \mathbf{R}_{,3} = \mathbf{e}_3 = (\mathbf{R}_0' - \kappa_0' X_1 \mathbf{e}_3 + \kappa_0 X_2 \mathbf{e}_3) - \tau_0 X_2 \mathbf{e}_1 + \tau_0 X_1 \mathbf{e}_2. \tag{17.29c}$$

The vector $\mathbf{R}_0'$ is parallel to $\mathbf{e}_3$, but not of unit length. However, by using the relation

$$\frac{\mathbf{G}_{33}}{\sqrt{G_{33}}} = \frac{\mathbf{G}_{33}}{\sqrt{\mathbf{G}_3 \cdot \mathbf{G}_3}} = \mathbf{e}_3,$$

(17.29c) assumes the form

$$\mathbf{G}_3 = (\sqrt{G_{33}} - \kappa_0' X_1 + \kappa_0 X_2)\mathbf{e}_3 - \tau_0 X_2 \mathbf{e}_1 + \tau_0 X_1 \mathbf{e}_2. \tag{17.30}$$

The geometric structure of the beam after the deformation is defined by a position vector

$$\mathbf{r} = \mathbf{r}(X_1, X_2, X_3), \tag{17.31}$$

the values of which for $X_1 = X_2 = 0$ define the central line

$$\mathbf{r}_0 = \mathbf{r}(0, 0, X_3) = \mathbf{r}_0(X_3). \tag{17.32}$$

In addition, for the strained central line we may introduce an orthonormal triad $\mathbf{d}_1, \mathbf{d}_2, \mathbf{d}_3$, which usually coincides with the principal torsion–flexure axes, such that the derivatives with respect to X_3, denoted by $\mathbf{d}_1', \mathbf{d}_2', \mathbf{d}_3'$, are related to the triad by formulae like (14.7), where κ, κ', and τ are the curvatures and the twist of the deformed axis. If we assume that the cross-section does not deform in its own plane, the position vector in the strained configuration can be written as (Bauchau and Hong 1988)

$$\begin{aligned}\mathbf{r}(X_1, X_2, X_3) = \mathbf{r}_0(X_3) + \mathbf{d}_1 X_1 + \mathbf{d}_2 X_2 + [\delta_3(X_3) w_3(X_1, X_2) \\ + \hat{e}_{13} w_1(X_1, X_2) + \hat{e}_{23} w_2(X_1, X_2)]\mathbf{d}_3,\end{aligned} \tag{17.33}$$

where w_1, w_2, and w_3 are three functions representing the warping of the cross-section, the forms of which are considered as given;[27] $\delta_3(X_3)$ is an unknown function characterizing the magnitude of the torsional warping w_3; and $\hat{e}_{13}$ and $\hat{e}_{23}$ are two unknown averaged shearing strains characterizing the magnitude of the two other bending components of warping.

From (17.33) we can derive the base vectors in the strained configuration:

$$\begin{aligned}\mathbf{g}_1 &= \mathbf{r}_{,1} = \mathbf{d}_1 + [\delta_3 w_{3,1} + \hat{e}_{13} w_{1,1} + \hat{e}_{23} w_{2,1}]\mathbf{d}_3, \\ \mathbf{g}_2 &= \mathbf{r}_{,2} = \mathbf{d}_2 + [\delta_3 w_{3,2} + \hat{e}_{13} w_{1,2} + \hat{e}_{23} w_{2,2}]\mathbf{d}_3, \\ \mathbf{g}_3 &= \mathbf{r}_{,3} = \mathbf{r}_{0,3} + \mathbf{d}_1' X_1 + \mathbf{d}_2' X_2 + [\delta_3' w_3 + \hat{e}_{13}' w_1 + \hat{e}_{23}' w_2]\mathbf{d}_3, \\ &\quad + [\delta_3 w_3 + \hat{e}_{13} w_1 + \hat{e}_{23} w_2]\mathbf{d}_3',\end{aligned} \tag{17.34}$$

where again $\mathbf{d}_1', \mathbf{d}_2', \mathbf{d}_3'$ may be expressed in terms of $\mathbf{d}_1, \mathbf{d}_2, \mathbf{d}_3$ through the curvatures κ and κ' and the twist τ. The vector $\mathbf{r}_{0,3}$ is still parallel to $\mathbf{d}_3$, but not of unit length. It may be written as

$$\mathbf{r}_{0,3} = (1 + \varepsilon)\mathbf{d}_3,$$

where ε is the elongation of the central line.

In order to reduce the number of unknowns, let us introduce the displacement $\mathbf{u}(X_3)$ of a particle of the central line, so that

$$\mathbf{r}_0(X_3) = \mathbf{R}_0(X_3) + \mathbf{u}(X_3).$$

By differentiating this equation with respect to X_3, and recalling the connection[28]

$$\mathbf{R}_{0,3} = \sqrt{G_{33}}\mathbf{e}_{33},$$

we obtain

[27] In fact w_1 is the torsion function φ for the section, and w_2 and w_3 are the flexure functions, usually denoted by χ and χ' (Love 1927, Art. 229).

[28] We note that, in contrast to (14.10c), $\mathbf{R}_{0,3}$ is not equal to $\mathbf{e}_3$ because X_3 is not necessarily the arc length along the axis.

$$(1+\varepsilon)\mathbf{d}_3 = \sqrt{G_{33}}\mathbf{e}_3 + \mathbf{u}'. \tag{17.35}$$

This equation can be exploited in two ways. First, we may express the elongation ε in terms of $\mathbf{u}'$; then we may find a relationship between $\hat{e}_{13}$, $\hat{e}_{23}$, and $\mathbf{u}'$ by observing that, by definition, we have the values

$$\hat{e}_{13} = \frac{\mathbf{g}_1}{\sqrt{g_{11}}}(0, 0, X_3) \cdot \mathbf{d}_3,$$
$$\hat{e}_{23} = \frac{\mathbf{g}_2}{\sqrt{g_{22}}}(0, 0, X_3) \cdot \mathbf{d}_3. \tag{17.36}$$

On applying the general formulae (14.11), we thus obtain the strains in terms of the seven unknowns $(u_1, u_2, u_3) = \mathbf{u}$, κ, κ', τ, and the warping amplitude δ_3. The resulting expressions for the strains are still very complicated, but they can be simplified by neglecting terms that are presumably small with respect to the others, a process of simplification which is not rigorous in general.

It remains for us to discuss the solution of the equations, because there are seven unknowns, whereas the balance equations contain only six unknowns. This indeterminacy can be overcome by recourse to a direct variational method.

The full nonlinear treatment of a prismatic slender beam may be performed using the method outlined in Section 9 (Dökmeci 1972). The beam is referred to a system of Cartesian coordinates, such that the X_3 axis is the locus of centroids of the cross-sections and the X_1 and X_2 axes are the principal axes of the cross-section. In order to ensure that the solutions are sufficiently smooth, say continuous with their partial derivatives up to second order, we assume that the cross-section is bounded by a simply connected Jordan curve C, that is to say, a non-self-intersecting and continuously differentiable curve.

The displacement components of a generic point are represented by

$$u_k(X_1, X_2, X_3, t) \sim \sum_{n+m=0}^{\infty} P_m^{(k)}(X_1)Q_n^{(k)}(X_2)u_k^{(m,n)}(X_3, t), \tag{17.37}$$

where $P_m^{(k)}(X_1)$ and $Q_n^{(k)}(X_2)$ are given functions and $u_k^{(m,n)}$ are functions to be determined. There is a certain freedom of choice in the form of the functions $P_m^{(k)}(X_1)$ and $Q_m^{(k)}(X_2)$, provided that they make a complete system in the sense of the convergence in the mean, and satisfy the conditions

$$P_0^{(k)}(X_1) = Q_0^{(k)}(X_2) = 1. \tag{17.38}$$

For example, we may take

$$P_m^{(k)}(X_1) = X_1^m, \quad Q_n^{(k)}(X_2) = X_2^n, \tag{17.39}$$

or some other system of functions, such as Legendre or Jacobi polynomials.

In terms of the displacement components, the strain is defined by

$$\gamma_{k\ell} = \tfrac{1}{2}(u_{k,\ell} + u_{\ell,k} + u_{r,k}u_{r,\ell}),$$

or by

$$\gamma_{k\ell} = \varepsilon_{k\ell} + \tfrac{1}{2}(\varepsilon_{rk} + \tilde{\omega}_{rk})(\varepsilon_{r\ell} + \tilde{\omega}_{r\ell}), \tag{17.40}$$

where

$$\varepsilon_{k\ell} = \varepsilon_{\ell k} = \tfrac{1}{2}(u_{k,\ell} + u_{\ell,k}), \quad \tilde{\omega}_{k\ell} = -\tilde{\omega}_{\ell k} = \tfrac{1}{2}(u_{k,\ell} - u_{\ell,k}). \tag{17.41}$$

On using (17.37) and (17.39) the strains take the form

$$\gamma_{k\ell} = (X_1, X_2, X_3, t) \sim \sum_{n+m=0}^{\infty} X_1^m X_2^n \gamma_{k\ell}^{(m,n)}(X_3, t), \tag{17.42}$$

where

$$\gamma_{k\ell}^{(m,n)} = \varepsilon_{k\ell}^{(m,n)} + \frac{1}{2}\sum_{p+q=0}^{m+n} [\varepsilon_{rk}^{(m-p,n-q)} + \tilde{\omega}_{rk}^{(m-p,n-q)}][\varepsilon_{rl}^{(p,q)} + \tilde{\omega}_{r\ell}^{(n,q)}], \tag{17.43}$$

with

$$\begin{aligned}
\varepsilon_{\alpha\beta}^{(m,n)} &= \tfrac{1}{2}\{(m+1)[\delta_{1\alpha}u_\beta^{(m+1,n)} + \delta_{1\beta}u_\alpha^{(m+1,n)}] + (n+1)[\delta_{2\alpha}u_\beta^{(m,n+1)} + \delta_{2\beta}u_\alpha^{(m,n+1)}]\},\\
\varepsilon_{\alpha 3}^{(m,n)} &= \tfrac{1}{2}[u_{\alpha,3}^{(m,n)} + (m+1)\delta_{1\alpha}u_3^{(m+1,n)} + (n+1)\delta_{2\alpha}u_3^{(m,n+1)}],\\
\varepsilon_{33}^{(m,n)} &= u_{3,3}^{(m,n)}, \quad \tilde{\omega}_{3,3}^{(m,n)} = 0,\\
\tilde{\omega}_{\alpha\beta}^{(m,n)} &= \tfrac{1}{2}\{(m+1)(\delta_{1\beta}u_\alpha^{(m+1,n)} - \delta_{1\alpha}u_\beta^{(m+1,n)}) + (n+1)(\delta_{2\beta}u_\alpha^{(m,n+1)} - \delta_{2\alpha}u_\beta^{(m,n+1)})\},\\
\tilde{\omega}_{\alpha 3}^{(m,n)} &= \tfrac{1}{2}[u_{\alpha,3}^{(m,n)} - (m+1)\delta_{1\alpha}u_3^{(m+1,n)} - (n+1)\delta_{2\beta}u_3^{(m,n+1)}].
\end{aligned} \tag{17.44}$$

The stresses are derived from the momentum balance equation

$$\boldsymbol{\tau}_{,p}^{p} + \rho_0\sqrt{G}\mathbf{f} = \rho_0\sqrt{G}\,\frac{\partial^2 \mathbf{u}}{\partial t^2}, \tag{17.45}$$

where ρ_0 is the density of the undeformed body, G is the determinant of the metric tensor in the reference configuration, and $G^{-\frac{1}{2}}\boldsymbol{\tau}^p$ is the Piola–Kirchhoff stress vector. This is related to the Cauchy stress vector $\mathbf{t}^p$ by the equation (Antman 1972, Sec. 3)

$$\boldsymbol{\tau}^p = \sqrt{g}\mathbf{t}^p, \tag{17.46}$$

$\sqrt{g}$ being the determinant of the metric tensor in the deformed configuration. The vector $\boldsymbol{\tau}^p$ can be expressed in terms of a stress tensor $G^{-\frac{1}{2}}\mathbf{T}$, with components T^{pq} ($T^{pq} = \sqrt{g}t^{pq}$) by means of the relation

$$\boldsymbol{\tau}^p = T^{pq}\mathbf{g}_q. \tag{17.47}$$

For the problem of the cylinder, the reference coordinates X_1, X_2, X_3 are orthogonal Cartesian coordinates, and hence the metric tensor in the reference state is a unit tensor whereas the base vectors in the actual state may be written as

$$\mathbf{g}_q = \mathbf{r}_{,q} = (\mathbf{R} + \mathbf{u})_{,q} = \mathbf{e}_q + \mathbf{u}_{,q}.$$

It then follows that the balance equations, written in terms of components, become

$$[T^{pr}(\delta_{rq} + u_{q,r})]_{,p} + \rho_0 f_q = \rho_0\,\frac{\partial^2 u_q}{\partial t^2}. \tag{17.48}$$

In general, surface tractions or displacements on the surface of the cylinder are known, as are the displacements and velocities at the instant $t = 0$. However, in order to eliminate unessential terms, let us assume that the lateral surface and the end faces are free from tractions, so that only body forces $\rho_0\mathbf{f}$ and inertial forces $\rho_0(\partial^2\mathbf{u}/\partial t^2)$ act on the cylinder. For each section we may define a resultant body force of order (m, n)

$$f_k^{(m,n)} = \int_{\mathscr{A}} \rho_0 f_k X_1^m X_2^n \, dX_1 \, dX_2, \tag{17.49}$$

and a stress resultant of order (m, n)

$$T_{pq}^{(m,n)} = \int_{\mathscr{A}} T^{pq} X_1^m X_2^n \, dX_1 \, dX_2. \tag{17.50}$$

Similarly, we may introduce an average displacement of order (m, n)

$$U_k^{(m,n)} = \sum_{p+q=0}^{\infty} I^{(m+p,n+q)} u_k^{(p,q)}, \tag{17.51}$$

where $I^{(m+p,n+q)}$ represents a moment of inertia of order $(m+p, n+q)$

$$I^{(m+p,n+q)} = \int_{\mathscr{A}} X_1^{m+p} X_2^{n+q} \, dX_1, dX_2. \tag{17.52}$$

As X_1 and X_2 are the principal axes of inertia, the above definition yields

$$I^{(0,0)} = A, \quad I^{(1,0)} = I^{(0,1)} = I^{(1,1)} = 0$$
$$I^{(2,0)} = I_1, \quad I^{(0,2)} = I_2.$$

Moreover, the following relations hold for a symmetric cross-section with respect to X_1:

$$I^{(m,n)} = \delta_{nn_0} I_{(m,n)},$$

where n_0 stands for any even integer and $I_{(m,n)}$ is a centrifugal moment of order (m, n). It is clear that for a symmetric section with respect to X_2 we have

$$I^{(m,n)} = \delta_{mm_0} I_{(m,n)},$$

with m_0 an even integer. For a section that is symmetric with respect to X_1 and X_2, $I^{(m,n)}$ becomes

$$I^{(m,n)} = \delta_{mm_0} \delta_{nn_0} I_{(m,n)}.$$

Stresses and strains must be related by constitutive equations. In the case of a hyperelastic material, the stress–strain law has the form

$$T^{pq} = \frac{1}{2}\left(\frac{\partial W}{\partial \gamma_{pq}} + \frac{\partial W}{\partial \gamma_{qp}}\right), \tag{17.53}$$

where W is the strain energy function measured per unit volume of the undeformed body.

By using (17.42) and (17.50) we can derive the constitutive relations for the stress resultants:

$$T_{pq}^{(m,n)} = \frac{1}{2}\left(\frac{\partial \Sigma}{\partial \gamma_{pq}^{(m,n)}} + \frac{\partial \Sigma}{\partial \gamma_{qp}^{(n,m)}}\right), \tag{17.54}$$

where

$$\Sigma = \int_{\mathscr{A}} W \, dX_1 \, dX_2$$

represents the strain energy function measured per unit length of the undeformed beam.

We now proceed to construct the balance equations involving the stress resultants of order (m, n) with the purpose of transforming the original system of partial differential equations into a sequence of systems involving the partial derivatives with respect to X_3 and those with respect to time. We start from (17.48), where the displacement components u_q are replaced by their formal expansions

$$u_q \sim \sum_{i+k=0}^{\infty} X_1^i X_2^k u_q^{(i,k)}.$$

The balance equations thus become

$$T_{,p}^{pq} + \left[T^{pr}\left(\sum_{i+k=0}^{\infty} X_1^i X_2^k u_q^{(i,k)}\right)_{,r}\right]_{,p} + \rho_0 f_q = \rho_0 \frac{\partial^2}{\partial t^2}\left(\sum_{i+k=0}^{\infty} X_1^i X_2^k u_q^{(i,k)}\right). \tag{17.55}$$

Let us now multiply both sides of these equations by $X_1^m X_2^n$ and integrate the resulting expressions over $\mathscr{A}$. The term $T_{,1}^{1q} X_1^m X_2^n$ can be integrated by parts, and becomes

$$\int_{\mathscr{A}} T_{,1}^{1q} X_1^m X_2^n \, dX_1 \, dX_2 = \oint_{\partial\mathscr{A}} T^{1q} \nu_1 X_1^m X_2^n \, ds - \int_{\mathscr{A}} m T^{1q} X_1^{m-1} X_2^n \, dX_1 \, dX_2,$$

and a similar expression holds for the term with $p = 2$. The term

$$\int_{\mathscr{A}} \left[T^{1r}\left(\sum_{i+k=0}^{\infty} X_1^i X_2^k u_q^{(i,k)}\right)_{,r}\right]_{,1} X_1^m X_2^n \, dX_1 \, dX_2,$$

after integration by parts becomes

$$\oint_{\partial\mathscr{A}} T^{1r}\left(\sum_{i+k=0}^{\infty} X_1^i X_2^k u_q^{(i,k)}\right)_{,r} X_1^m X_2^n \nu_1 \, ds$$

$$- \int_{\mathscr{A}} T^{1r}\left(\sum_{i+k=0}^{\infty} X_1^i X_2^k u_q^{(i,k)}\right)_{,r} m X_1^{m-1} X_2^n \, dX_1 \, dX_2,$$

where the second term can again be integrated by parts when $r = 1, 2$. A similar expression may be derived for $p = 2$.

Now, for each pair (m, n), the sum of integrals over $\mathscr{A}$ gives an equation of the form

$$T^{(m,n)}_{3q,3} - mT^{(m-1,n)}_{1q} - nT^{(m,n-1)}_{2q} + N^{(m,n)}_{q} = \rho_0 \frac{\partial^2}{\partial t^2} U^{(m,n)}_{q}, \tag{17.56}$$

where $N^{(m,n)}_q$ is the collection of all remaining integrals over $\mathscr{A}$. On the other hand, the sum of the integrals through $\partial\mathscr{A}$ imposes a further condition, which may be written symbolically as

$$P^{(m,n)}_q + R^{(m,n)}_q = 0. \tag{17.57}$$

We have supposed that the boundary conditions on the end faces are free, although this is not an essential restriction. The free state, however, requires, say for $X_3 = 0$ and $X_3 = L$, the equation

$$T^{(m,n)}_{3q} = 0.$$

The integrals $N^{(m,n)}_q$, $R^{(m,n)}_q$ in (17.56) and (17.57) have been isolated because they contain all the nonlinear terms of the problem, and so may be neglected in a linearized theory. For this case, in addition to neglecting $N^{(m,n)}_q$ and $R^{(m,n)}_q$, the constitutive equations may also be linearized using the Lamé equations, for example. The known theories of beams can be recovered by particular choices of the functions $u^{(m,n)}_q$ in (17.37). For instance, let us take

$$\begin{aligned}
u_1 &\sim u^{(2,0)}_1 X_1^2 + u^{(0.2)}_1 X_2^2 + u^{(0,0)}_1, \\
u_2 &\sim u^{(1,1)}_2 X_1 X_2, \\
u_3 &\sim u^{(1,0)}_3 X_1,
\end{aligned}$$

where

$$\begin{aligned}
&u^{(2,0)}_1 = \tfrac{1}{2} R^{-1}\sigma, \quad u^{(0,2)}_1 = -\tfrac{1}{2} R^{-1}\sigma, \quad u^{(0,0)}_1 = \tfrac{1}{2} R^{-1} X_3^2, \\
&u^{(1,1)}_2 = R^{-1}\sigma, \quad u^{(1,0)}_3 = -R^{-1} X_3.
\end{aligned}$$

In this case, the displacements are exactly those corresponding to Saint-Venant's theory of a bar bent by two terminal couples about the X_2 axis (Love 1927, Art. 87).

The method of expanding the solutions in powers of the distances of a point on the cross-section from the centroid may be extended to spatially curved beams (Gordienko 1979). The beam is defined geometrically by a curve in three-dimensional Euclidean space, the points of the curve being specified by the arc length S from a given point on the axis, and a cross-section $\mathscr{A}(S)$, which in general varies with $S (0 \leqslant S \leqslant 1)$. The material is supposed linearly elastic, homogeneous, isotropic and the loading conditions are assumed known.

Let P_0 be a point on the axis in the unstrained state and let $\mathscr{A}(S)$ be the cross-section drawn through P_0 perpendicular to the axis. If $\mathbf{t}$ is the unit tangent vector, and $\mathbf{n}$ and $\mathbf{b}$ are the unit vectors of the normal and binormal to the axial line, we may choose a system of reference axes coinciding with these three unit vectors, and denote the coordinates of a point P of $\mathscr{A}(S)$ with respect $\mathbf{n}$ and $\mathbf{b}$ by η and ζ, respectively. If $\mathbf{R}_0$ is the radius vector that identifies P_0, the point P is specified by the vector

$$\mathbf{R} = \mathbf{R}_0 + \eta\mathbf{n} + \zeta\mathbf{b}. \tag{17.58}$$

After the deformation, the point P moves to another point P_1 and occupies a new position

$$\mathbf{r} = \mathbf{R} + \mathbf{u}, \tag{17.59}$$

where $\mathbf{u}$ denotes the displacement, also representable as the sum of its projections along the directions $\mathbf{t}$, $\mathbf{n}$, and $\mathbf{b}$:

$$\mathbf{u} = \mathbf{t}u_1 + \mathbf{n}u_2 + \mathbf{b}u_3. \tag{17.60}$$

On differentiating (17.59) and utilizing the Frenet–Serret formulae,[29] we have

$$\begin{aligned} d\mathbf{r} &= \frac{\partial \mathbf{R}}{\partial S}\,dS + \frac{\partial \mathbf{R}}{\partial \eta}\,d\eta + \frac{\partial \mathbf{R}}{\partial \zeta}\,d\zeta + \frac{\partial \mathbf{u}}{\partial S}\,dS + \frac{\partial \mathbf{u}}{\partial \eta}\,d\eta + \frac{\partial \mathbf{u}}{\partial \zeta}\,d\zeta \\ &= [(1-\kappa_0\eta)\mathbf{t} - \tau_0\zeta\mathbf{n} + \tau_0\eta\mathbf{b}]\,dS + \mathbf{n}\,d\eta + \mathbf{b}\,d\zeta \\ &\quad + du_1\,\mathbf{t} + du_2\,\mathbf{n} + du_3\,\mathbf{b} + [u_1\kappa_0\mathbf{n} - u_2(\kappa_0\mathbf{t} + \tau_0\mathbf{b}) + u_3\tau_0\mathbf{n}]\,dS. \end{aligned} \tag{17.61}$$

In the unstrained state the base vectors are

$$\mathbf{G}_1 = \frac{\partial \mathbf{R}}{\partial S} = (1-\kappa_0\eta)\mathbf{t} + \tau_0\eta\mathbf{n} - \tau_0\eta\mathbf{b}, \quad \mathbf{G}_2 = \mathbf{n}, \quad \mathbf{G}_3 = \mathbf{b}.$$

Let us denote the derivatives of $\mathbf{u}$ with respect to S, η, and ζ by $\mathbf{u}_{,i}$ $(i = 1, 2, 3)$ and write, still using the Frenet–Serret formulae

$$\begin{aligned} \mathbf{u}_{,1} &= (u_{1,1} - \kappa_0 u_2)\mathbf{t} + (u_{2,1} + \kappa_0 u_1 - \tau_0 u_3)\mathbf{n} + (u_{3,1} + \tau_0 u_2)\mathbf{b}, \\ \mathbf{u}_{,2} &= u_{1,2}\mathbf{t} + u_{2,2}\mathbf{n} + u_{3,2}\mathbf{b}, \\ \mathbf{u}_{,3} &= u_{1,3}\mathbf{t} + u_{2,3}\mathbf{n} + u_{3,3}\mathbf{b}. \end{aligned}$$

It is also convenient to put

$$\begin{aligned} &e_{11} = u_{1,1} - \kappa_0 u_2, \quad e_{12} = u_{2,1} + \kappa_0 u_1 + \tau_0 u_3, \quad e_{13} = u_{3,1} - \tau_0 u_2, \\ &e_{ij} = u_{j,i} \quad (i = 2, 3;\ j = 1, 2, 3). \end{aligned}$$

As we already know the metric tensor in the unstrained state, the metric tensor in the deformed body becomes

$$\begin{aligned} g_{ij} &= \mathbf{g}_i \cdot \mathbf{g}_j = (\mathbf{G}_i + \mathbf{u}_{,i}) \cdot (\mathbf{G}_j + \mathbf{u}_{,j}) \\ &= G_{ij} + 2\varepsilon_{ij}, \end{aligned}$$

where the quantities ε_{ij} are expressed as

$$2\varepsilon_{ij} = e_{ij} + e_{ji} + e_{is}e_{js} - (1 + \delta_i^j)a_{ij}(\eta, \zeta), \tag{17.62}$$

[29] The Frenet–Serret formulae are:

$$\mathbf{t} = \frac{\partial \mathbf{R}_0}{\partial S}, \quad \frac{\partial \mathbf{t}}{\partial S} = \kappa_0\mathbf{n}, \quad \frac{\partial \mathbf{b}}{\partial S} = \tau_0\mathbf{n}$$

$$\frac{\partial \mathbf{n}}{\partial S} = -\kappa_0\mathbf{t} - \tau_0\mathbf{b},$$

where κ_0 and τ_0 are the curvature and the torsion of the unstrained axis, respectively.

and

$$a_{1j} = (\kappa_0 e_{j1} - \tau_0 e_{j3})\eta + \kappa_0 e_{j2}\zeta \quad (j = 1, 2, 3),$$
$$a_{21} = a_{12}, \quad a_{31} = a_{13}, \quad a_{22} = a_{33} = a_{23} = a_{32} = 0.$$

We now expand the functions u_j, e_{ij}, and ε_{ij} in power series in η and ζ, and obtain

$$(u_j; e_{ij}; \varepsilon_{ij}) = (u_{j,pq}; e_{ij,pq}; \varepsilon_{ij,pq})\eta^p \zeta^q, \tag{17.63}$$

having used the following relations

$$e_{11,pq} = u_{1,pq} - \kappa_0 u_{2,pq}, \quad e_{13,pq} = u_{3,pq} + \tau_0 u_{2,pq},$$
$$e_{12,pq} = u_{2,pq} + \kappa_0 u_{1,pq} - \tau_0 u_{3,pq},$$
$$e_{2j,pq} = (p+1)u_{j,(p+1)q}, \quad e_{3j,pq} = (q+1)u_{j,p(q+1)} \quad (j = 1, 2, 3),$$
$$2\varepsilon_{ij,pq} = e_{ij,pq} + e_{ji,pq} + e_{is,k\ell} e_{js,(p-k)(q-\ell)} - (1 + \delta^j_i)a_{ij,pq},$$
$$a_{ij,pq} = (\kappa_0 e_{ji,(p-1)q} - \tau_0 e_{j3,(p-1)q}) + \tau_0 e_{j2,p(q-1)},$$
$$a_{21,pq} = a_{12,pq}, \quad a_{31,pq} = a_{13,pq}, \quad a_{22,pq} = a_{33,pq} = a_{23,pq} = a_{32,pq} = 0.$$

Here the indices p and q denote the summation from zero to infinity, and k and ℓ denote the summation from zero to p and q.

Since the material obeys the Lamé constitutive equations, with moduli E and σ, the Cauchy stresses associated with the strains ε_{ij} have the form

$$\sigma_{ij} = E\mathring{\varepsilon}_{ij} = \frac{E}{(1+\sigma)}(\varepsilon_{ij} + \delta^j_i B\theta),$$

where $\theta = \varepsilon_{SS}$ and $B = \sigma/(1 - 2\sigma)$. The quantities $\mathring{\varepsilon}_{ij}$ have the physical interpretation of being nondimensional stresses, and their derivatives with respect to η and ζ are given by

$$\mathring{\varepsilon}_{ij,pq} = \frac{1}{(1+\sigma)}(\varepsilon_{ij,pq} + \delta^j_i B\theta,_{pq}). \tag{17.64}$$

The resultant stress vector in dimensionless form has components of the type

$$N^0_{1k} = N^0_{k1} = \frac{1}{A}\int_{\mathscr{A}(S)} \mathring{\varepsilon}_{ik}\, dA = \mathring{\varepsilon}_{k,pq}\ell^2_{pq} \quad (k = 1, 2, 3),$$

and the components of the couple resultant vector are

$$M^0_{11} = M^0_t = \frac{1}{A}\int_{\mathscr{A}(S)} (\mathring{\varepsilon}_{13}\eta - \mathring{\varepsilon}_{12}\zeta)\, dA = (\mathring{\varepsilon}_{13,(p-1)q} - \mathring{\varepsilon}_{12,p(q-1)})\ell^2_{pq},$$
$$M^0_{22} = M^0_n = \frac{1}{A}\int_{\mathscr{A}(S)} \mathring{\varepsilon}_{11}\zeta dA = \mathring{\varepsilon}_{11,p(q-1)}\ell^2_{pq},$$
$$M^0_{33} = M^0_b = -\frac{1}{A}\int_{\mathscr{A}(S)} \mathring{\varepsilon}_{11}\eta\, dA = -\mathring{\varepsilon}_{11,(p-1)q}\ell^2_{pq},$$

where summation with respect p and q is understood, and we have put

$$\ell_{pq}^2 = \frac{I_{pq}}{AL^2}\,(L=1), \quad I_{pq} = \int\limits_{\mathscr{A}(S)} \eta^p \zeta^q \, dA,$$

$$\ell_{00}^2 = 1, \quad \ell_{10}^2 = \eta_0, \quad \ell_{01}^2 = \zeta_0.$$

In these expressions I_{pq} denotes the moment of inertia of order (p, q), and η_0 and ζ_0 the coordinates of the centroid of the cross-section in the system $(\mathbf{t}, \mathbf{n}, \mathbf{b})$. If the axis connects the centroids of the cross-sections, then $\eta_0 = \zeta_0 = 0$; and if the axes $\mathbf{n}$ and $\mathbf{b}$ are the principal axes of inertia, $\ell_{11} = 0$ and $\ell_{pq} = 0$, provided that at least one of the numbers p or q is odd.

Having defined the stress resultants and the stress-couple resultants, the formal structure of the balance equations may be obtained from the Hamilton principle, which states that the integral

$$\int_{t_0}^{t_1} \left[\int_0^L (T - W)\, dS \right], \tag{17.65}$$

must be stationary for all synchronous motions, that is to say, the candidate functions u_i must have the same values for $t = t_0$ and $t = t_1$. In (17.65) W represents the (dimensionless) strain energy stored in the cross-section:

$$W = (1+\sigma) \int\limits_{\mathscr{A}(S)} \mathring{\varepsilon}_{ij} \varepsilon_{ij} \, dA = (1+\sigma) A \ell_{pq}^2 \mathring{\varepsilon}_{ij,k\ell} \varepsilon_{ij,(p-k)(q-\ell)},$$

and T is the (dimensionless) kinetic energy

$$T = \int\limits_{\mathscr{A}(S)} \dot{u}_i \dot{u}_i \, dA = \ell_{pq}^2 \dot{u}_{i,kh} \dot{u}_{i,(p-k)(q-h)} A,$$

where the dot denotes the derivative with respect a parameter $t = (cT_0/L)$, T_0 is the physical time, and $c^2 = 1/(1+\sigma)$ is the speed of sound corresponding to a material of unit density.

In the functional $T - W$ the variables are

$$r = \frac{\partial \varphi}{\partial t}, \quad p = \frac{\partial \varphi}{\partial S}, \text{ with } \varphi \equiv u_{i,pq},$$

and the Euler equations, which follow from the condition that the functional (17.65) should be stationary, are

$$\frac{\partial}{\partial t}\left(\frac{\partial T}{\partial r}\right) - \frac{\partial}{\partial S}\left(\frac{\partial U}{\partial p}\right) + \frac{\partial U}{\partial \varphi} = 0, \tag{17.66}$$

and they must be supplemented by certain initial and boundary conditions, which depend on the specific problem that we have formulated.

The advantage of this method is that we are in a position to recover all the known theories of curved beams by simply truncating the expansions (17.63) up to a given finite order. For example, let us assume that the displacements have the form

$$u_i = (u_i)_0 + (u_i)_1 \eta + (u_i)_2 \zeta, \tag{17.67}$$

where $(u_i)_0$, $(u_i)_1$, and $(u_i)_2$ are functions of the only variable S. If the beam has small transverse dimensions with respect to the length, then in the formulae for the strains we may neglect the terms quadratic in η and ζ. In particular, it is possible to put

$$\varepsilon_{22} \simeq \frac{\partial u_2}{\partial \eta} = (u_2)_1$$
$$\varepsilon_{33} \simeq \frac{\partial u_3}{\partial \zeta} = (u_3)_2$$
$$\varepsilon_{23} = \varepsilon_{32} \simeq \frac{1}{2}\left(\frac{\partial u_2}{\partial \zeta} + \frac{\partial u_3}{\partial \eta}\right) = \frac{1}{2}[(u_2)_2 + (u_3)_1].$$

For a first theory we assume that the stresses $\mathring{\varepsilon}_{22}$, $\mathring{\varepsilon}_{33}$, and $\mathring{\varepsilon}_{23} = \mathring{\varepsilon}_{32}$ are absent. The assumption that $\mathring{\varepsilon}_{23} = 0$ implies that $\mathring{\varepsilon}_{23} = 0$, and hence the result

$$(u_2)_2 = -(u_3)_1.$$

Furthermore, the assumption that $\mathring{\varepsilon}_{22} = \mathring{\varepsilon}_{33} = 0$ yields the connections

$$\mathring{\varepsilon}_{11} = \varepsilon_{11}, \quad \varepsilon_{22} = \varepsilon_{33} = -\sigma\varepsilon_{11},$$

and, consequently, by disregarding the terms in η and ζ, we have the formulae

$$(u_2)_1 = (u_3)_2 \simeq -\sigma\left[\frac{\partial(u_1)_0}{\partial S} - \kappa_0(u_2)_0\right].$$

In this manner we have found that the displacements (17.67) take the form

$$\begin{aligned} u_1 &= (u_1)_0 + (u_1)_1\eta + (u_1)_2\zeta, \\ u_2 &= (u_2)_0 - \psi\eta - \varphi\zeta, \\ u_3 &= (u_3)_0 + \varphi\eta - \psi\zeta, \end{aligned} \tag{17.68}$$

where φ and ψ have the values

$$\varphi = \frac{1}{2}[(u_3)_1 - (u_2)_2], \quad \psi = \sigma\left[\frac{\partial(u_1)_0}{\partial S} - \kappa_0(u_2)_0\right].$$

In a second theory we put $\varepsilon_{22} = \varepsilon_{33} = \varepsilon_{23} = 0$, so that

$$(u_2)_1 = (u_3)_2 = 0, \quad (u_2)_3 = -(u_3)_2.$$

Then the displacements become

$$\begin{aligned} &u_1 = (u_1)_0 + (u_1)_1\eta + (u_1)_2\zeta \\ &u_2 = (u_2)_0 - \varphi\zeta, \quad u_3 = (u_3)_0 + \varphi\eta. \end{aligned} \tag{17.69}$$

The version (17.68) corresponds to a Saint-Venant's model of curved beams; the version (17.69) describes a beam which is perfectly rigid in each cross-section.

The effect of nonlinear terms can also be taken into account in the problem of the torsion of an initially twisted bar (analyzed in Section 15). Here, however, it is also important to consider the influence of a tensile force acting at the same time as the torsional moment (Rosen 1983), because the strains induced by the two kinds of force cannot be studied separately.

We wish to study an initially twisted bar like the one shown in Figure 17.2, of length ℓ and constant cross-section. Each section is rotated about the longitudinal x

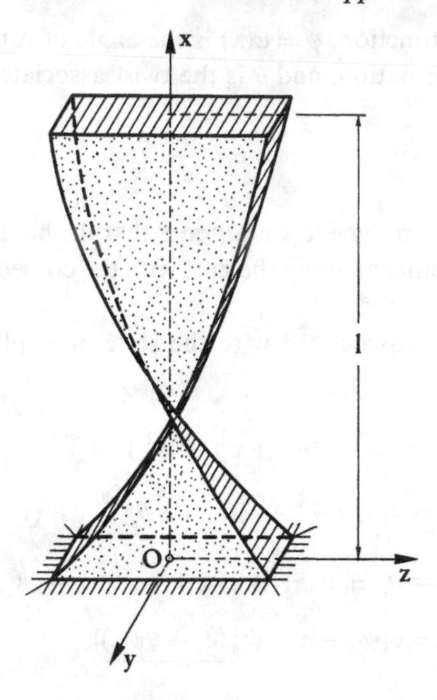

Fig. 17.2

axis of an angle β with respect a fixed section $x = 0$, and the measure of pretwisting is the quantity

$$\kappa = \frac{d\beta}{dx}, \tag{17.70}$$

which is assumed to be a constant. We also limit ourselves to the study of twisted bars the cross-sections of which admit two axes of symmetry, in order to decouple the torsion and extension from the bending.

We choose a system of Cartesian axes, such that the origin O is placed at the centroid of the bottom section, the x axis is drawn vertically upwards, and the y and x axes coincide with the axes of symmetry of the section $x = 0$. A generic cross-section, at a distance x from the basis, is still plane in the undeformed state and its axes of symmetry, denoted by η and ζ, are rigidly rotated with respect the y and z axes. The angle between η and y is set equal to β.

Upon the action of a torque M_t and an axial force T applied at the upper end of the bar, a state of displacement is created at each point, and its vector form is

$$\mathbf{u} = u\mathbf{e}_x + v\mathbf{e}_y + w\mathbf{e}_z, \tag{17.71}$$

where $\mathbf{e}_x$, $\mathbf{e}_y$, and $\mathbf{e}_z$ are unit vectors in the x, y and z directions. Since we now wish to account for large displacements, we assume the following representations for u, v, and w:

$$\begin{aligned} u &= u_1(\mathrm{x}) + \theta(x)\psi(x, y, z), \\ v &= y(\cos\varphi - 1) - z\sin\varphi, \\ w &= y\sin\varphi + z(\cos\varphi - 1), \end{aligned} \tag{17.72}$$

where ψ is a warping function, $\varphi = \varphi(x)$ is the angle of rotation of the cross-section due to the elastic deformation, and θ is the twist associated with φ:

$$\theta = \frac{d\varphi}{dx}.$$

In order to simplify the treatment, we suppose that the bar is free to warp at the ends, so that the torsion is uniform along the bar, with the consequence that θ is a constant and not a function of x.

The nonlinear strains associated with the state of displacement (17.72) have the form

$$\begin{aligned} e_{xx} &= u_{,x} + \tfrac{1}{2}(u_{,x}^2 + v_{,x}^2 + w_{,x}^2) \\ &= \varepsilon_1 + \theta\psi_{,x} + \tfrac{1}{2}[\underline{(\varepsilon_1 + \theta\psi_{,x})^2} + \theta^2(y^2+z^2)], \\ e_{xy} &= u_{,y} + v_{,x} + (u_{,x}u_{,y} + v_{,x}v_{,y} + w_{,x}w_{,y}) \\ &= \theta[\psi_{,y} - z + \underline{\psi_{,y}(\varepsilon_1 + \theta\psi_{,x})}], \\ e_{xz} &= u_{,z} + w_{,x} + (u_{,x}u_{,z} + v_{,x}v_{,z} + w_{,x}w_{,z}) \\ &= \theta[\psi_{,z} + y + \underline{\psi_{,z}(\varepsilon_1 + \theta\psi_{,x})}], \end{aligned} \tag{17.73}$$

where

$$\varepsilon_1 = \frac{\partial u_1}{\partial x}$$

is the elongation, which is assumed to be constant along the bar.

In general, the terms occurring in the definition of the strains are of different orders of magnitude, because pure strains are often much smaller than rotations. This implies that, in a simplified theory, the underlined terms in (17.73) may be neglected. The consequence is that the equilibrium equations are derivable as the Euler equations of the variational problem

$$W - L = \int_0^\ell dx \iint_{\mathscr{A}} \left[\frac{E}{2}e_{xx}^2 + \frac{G}{2}(e_{xy}^2 + e_{xz}^2)\right] dy\,dz - Tu_1(\ell) - M_t\varphi(\ell) = \min, \tag{17.74}$$

in the class of functions u_1 and φ such that $\varphi(0) = u_1(0) = 0$. Then, by substituting the simplified version of (17.73) in (17.74) and recalling that T and M_t are constant, we obtain

$$\begin{aligned} W - L = \int_0^\ell dx \bigg[&\frac{E}{2}(A\varepsilon_1^2 + 2S\varepsilon_1\theta + K\theta^2 + I_p\varepsilon_1\theta^2 + D\theta^3 + \tfrac{1}{4}F\theta^4) \\ &+ \frac{G}{2}J_S\theta^2 - T\varepsilon_1 - M_t\theta\bigg] = \min, \end{aligned} \tag{17.75}$$

where we have put

$$D = \iint_{\mathscr{A}} \psi_{,x}(y^2 + z^2)\,dy\,dz, \quad F = \iint_{\mathscr{A}} (y^2 + z^2)^2\,dy\,dz,$$

$$I_p = \iint_{\mathscr{A}} (y^2 + z^2)\,dy\,dz,$$

$$J_S = \iint_{\mathscr{A}} [(-z + \psi_{,y})^2 + (y + \psi_{,z})^2]\,dy\,dz,$$

$$K = \iint_{\mathscr{A}} \psi_{,x}^2 dy\,dz, \quad S = \iint_{\mathscr{A}} \psi_{,x}\,dy\,dz.$$

The condition for the first variation of $W - L$ to vanish with respect to ε_1 and θ yields the pair of algebraic equations

$$\begin{aligned} &EA\varepsilon_1 + ES\theta + \tfrac{1}{2}EI_p\theta^2 = T, \\ &ES\varepsilon_1 + (GJ_S + EK)\theta + EI_p\varepsilon_1\theta + \tfrac{3}{2}ED\theta^2 + \tfrac{1}{2}EF\theta^3 = M_t. \end{aligned} \tag{17.76}$$

Elimination of ε_1 from these two equations yields

$$\left\{GJ_S + E\left[K - \frac{S^2}{A} + \frac{3}{2}\left(D - \frac{I_pS}{A}\right)\theta + \frac{1}{2}\left(F - \frac{I_p^2}{A}\right)\theta^2\right] + \frac{I_p}{A}\,T\right\}\theta = M_t - \frac{S}{A}\,T. \tag{17.77}$$

The equation so obtained deserves some comment. The most surprising fact is that the presence of a tensile force T increases the torsional rigidity by the amount $(I_p/A)T$, whereas the effective torsional moment is no longer M_t but $[M_t - (S/A)T]$. These two resulting effects are well known by designers of blades propellers and helicopter rotors.

The formulae found so far can be further simplified when the cross-section is very thin, as occurs in a rectangle with sides $2b$ and $2h$ with $h \ll b$. A generic section at a distance x from the origin is defined by the inequalities

$$-b \leqslant \eta \leqslant b, \quad -h \leqslant \zeta \leqslant h,$$

where η and ζ are the local coordinate related to x and y by the formulae

$$\begin{aligned} \eta &= \cos\beta\, y + \sin\beta\, z, \\ \zeta &= -\sin\beta\, y + \cos\beta\, z, \end{aligned}$$

so that, by repeating the argument given in Section 15, we have

$$\frac{\partial\psi}{\partial x} = \frac{\partial\psi}{\partial\eta}\frac{\partial\eta}{\partial x} + \frac{\partial\psi}{\partial\zeta}\frac{\partial\zeta}{\partial x} = \kappa\left(\zeta\,\frac{\partial\psi}{\partial\eta} - \eta\,\frac{\partial\psi}{\partial\zeta}\right) \simeq \kappa(\eta^2 - \zeta^2).$$

It thus turns out that the various constants D and F have the expressions:

$$D = \kappa \iint_{\mathscr{A}} \eta^4 \left(1 - \frac{\zeta^4}{\eta^4}\right) d\eta\, d\zeta, \quad F = \iint_{\mathscr{A}} \eta^4 \left(1 + \frac{\zeta^2}{\eta^2}\right)^2 d\eta\, d\zeta,$$

$$I_p = \iint_{\mathscr{A}} \eta^2 \left(1 + \frac{\zeta^2}{\eta^2}\right) d\eta\, d\zeta, \quad K = \kappa^2 \iint_{\mathscr{A}} \eta^4 \left(1 - \frac{\zeta^2}{\eta^2}\right)^2 d\eta\, d\zeta,$$

$$S = \kappa \iint_{\mathscr{A}} \eta^2 \left(1 - \frac{\zeta^2}{\eta^2}\right) d\eta\, d\zeta,$$

where, sometimes, the terms containing powers of the ratio ζ/η may be disregarded.

18. Justification of the One-dimensional Model

The three-dimensional theory of rods is essentially founded on the idea of representing the position of a point in the deformed configuration of a rod by a formula like (9.1) or, more generally, (8.1), which formally separates the dependence of the unknown functions on the S variable, which denotes arc length along the central line, from the other two variables, which represent the distance of a point in the cross-section from the axis. The second step of the theory is that of choosing a specific form for the function by which the unknowns depend on the variables representing the distance of a point from the axis, with the hope that, provided that the rod is slender, some form of the function of these "transverse" variables might constitute an approximation to the exact law of dependence. In the usual theories, we assume that these functions are polynomials. The third step in the procedure is that of averaging the three-dimensional equations of elasticity over the cross-section in order to obtain a system of ordinary differential equations in only one variable, S.

However, this procedure, purely heuristic, in the sense that the choice of the above-mentioned polynomial functions is justified *a posteriori*, conditioned by the requirement that the classical theories of rods emerge as corollaries of particular choices of the polynomial functions. It is thus understandable that the need has arisen to justify all these attempts in a more rigorous way, and to explain how the results might be interpreted as steps in an asymptotic approach to the study of slender rods.

The method applies to a cylindrical domain $\Omega = \omega \times (0, 1)$, where ω is a bounded regular two-dimensional domain in the plane $x_3 = 0$ of a Cartesian system of coordinates, with the origin at the centroid of ω, and with the x_1 and x_2 axes coinciding with the principal axes of inertia of ω. Without loss of generality, we assume that the area ω is unity, and the length of the cylinder is unitary ($0 < x_3 < 1$).

Now, for $\varepsilon > 0$, we define ω^ε to be a domain of the x_1x_2 plane, such that

$$\omega^\varepsilon \equiv \left\{(x_1, x_2, 0) : \left(\frac{x_1}{\varepsilon}, \frac{x_2}{\varepsilon}, 0\right) \in \omega\right\}$$

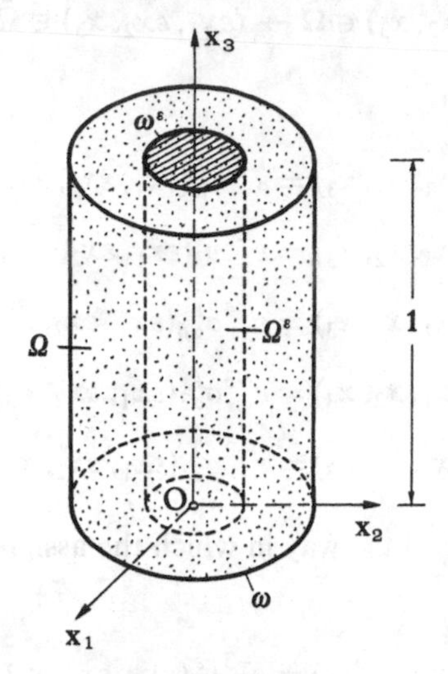

Fig.18.1

and, accordingly, we introduce the cylinder

$$\Omega^\varepsilon \equiv \left\{(x_1, x_2, x_3) : \left(\frac{x_1}{\varepsilon}, \frac{x_2}{\varepsilon}, x_3\right) \in \Omega\right\}.$$

As ε varies the sets Ω^ε describe a family of cylindrical domains the cross-sections ω^ε of which have a diameter of order ε (Figure 18.1). We consider the sets Ω^ε as reference configurations of a family of slender elastic rods the constitutive equations for which are Lamé's equation, with constant moduli λ and μ.

The beam may be loaded and constrained in several ways. For simplicity, however, we suppose that the lower face ω_0^ε at $x_3 = 0$ is rigidly clamped, and only surface tractions h_i^ε are applied on the upper face ω_1^ε at $x_3 = 1$.

It is well known that, if we denote by V^ε the set of (smooth) kinematically admissible displacements (in the present case, the field of displacement v_i^ε vanishing on ω_0^ε), the solution u_i^ε and the associated stresses

$$\sigma_{ij}^\varepsilon(u_k^\varepsilon) = 2\mu\gamma_{ij}^\varepsilon(u_k^\varepsilon) + \lambda\delta_{ij}\gamma_{ss}(u_k^\varepsilon), \tag{18.1}$$

where $\gamma_{ij}^\varepsilon(u_k^\varepsilon) = \frac{1}{2}(u_{i,j}^\varepsilon + u_{j,i}^\varepsilon)$, are the unique solution of the functional equation

$$\int_{\Omega^\varepsilon} \sigma_{ij}^\varepsilon(u_k^\varepsilon)\gamma_{ij}^\varepsilon(v_h^\varepsilon)\, dV = \int_{\omega_1^\varepsilon} h_i^\varepsilon v_i^\varepsilon\, dx_1\, dx_2, \tag{18.2}$$

for all v_i^ε in V^ε.

However, in order to take into account the smallness of ω^ε, we formulate a problem equivalent to problem (18.2), but now posed over a domain that does not depend on ε. Accordingly, we associate each point (x_1, x_2, x_3) of $\bar{\Omega}$ with each point of $\bar{\Omega}^\varepsilon$ through the correspondence:[30]

[30] First used by Ciarlet and Destuynder (1979).

$$(x_1, x_2, x_3) \in \bar{\Omega} \to (\varepsilon x_1, \varepsilon x_2, x_3) \in \bar{\Omega}^\varepsilon.$$

Then, for all v_i^ε in V^ε we set

$$\begin{aligned} v_\alpha(x_1, x_2, x_3) &= \varepsilon^{-1} v_\alpha^\varepsilon(\varepsilon x_1, \varepsilon x_2, x_3), \\ v_3(x_1, x_2, x_3) &= \varepsilon^{-2} v_3^\varepsilon(\varepsilon x_1, \varepsilon x_2, x_3), \\ \sigma_{\alpha\beta}(x_1, x_2, x_3) &= \varepsilon^{-4} \sigma_{\alpha\beta}^\varepsilon(\varepsilon x_1, \varepsilon x_2, x_3), \\ \sigma_{\alpha 3}(x_1, x_2, x_3) &= \varepsilon^{-3} \sigma_{\alpha 3}^\varepsilon(\varepsilon x_1, \varepsilon x_2, x_3), \\ \sigma_{33}(x_1, x_2, x_3) &= \varepsilon^{-2} \sigma_{33}^\varepsilon(\varepsilon x_1, \varepsilon x_2, x_3). \end{aligned} \tag{18.3}$$

In addition, in order to control the way in which the assigned loads depend on ε, we require that

$$\begin{aligned} h_\alpha(x_1, x_2, x_3) &= \varepsilon^{-3} h_\alpha^\varepsilon(\varepsilon x_1, \varepsilon x_2, x_3), \\ h_3(x_1, x_2, x_3) &= \varepsilon^{-2} h_\alpha^\varepsilon(\varepsilon x_1, \varepsilon x_2, x_3), \end{aligned} \tag{18.4}$$

where the functions h_α and h_3 are independent of ε.

An immediate consequence of (18.3) is that the strains have the form

$$\begin{aligned} \gamma_{\alpha\beta}(x_1, x_2, x_3) &= \gamma_{\alpha\varepsilon}^\varepsilon(\varepsilon x_1, \varepsilon x_2, x_3), \\ \gamma_{3\alpha}(x_1, x_2, x_3) &= \varepsilon^{-1} \gamma_{3\alpha}^\varepsilon(\varepsilon x_1, \varepsilon x_2, x_3), \\ \gamma_{33}(x_1, x_2, x_3) &= \varepsilon^{-2} \gamma_{33}^\varepsilon(\varepsilon x_1, \varepsilon x_2, x_3). \end{aligned}$$

It might appear that formulae (18.4) are the result of an unjustified choice, but the explanation lies in the invariance, up to a certain multiplicative power of ε, of the integrals in (18.2). We have, in fact,

$$\int_{\Omega^\varepsilon} \sigma_{ij}^\varepsilon \gamma_{ij}^\varepsilon(v_k)\, dx_1\, dx_2\, dx_3 = \varepsilon^2 \int_\Omega \sigma_{ij} \gamma_{ij}(v_k)\, dx_1\, dx_2\, dx_3,$$

$$\int_{\omega_1^\varepsilon} h_i^\varepsilon v_i^\varepsilon\, dx_1\, dx_2 = \varepsilon^2 \int_{\omega_1} h_i v_i\, dx_1\, dx_2.$$

Replacing the domains Ω^ε and ω_1^ε by Ω and ω_1, the original boundary value problem (18.2) allows the following alternative formulation. We wish to find a state σ_{ij}^ε, u_i^ε such that the equation

$$\int_\Omega \sigma_{ij}^\varepsilon \gamma_{ij}^\varepsilon(v_k)\, dx_1\, dx_2\, dx_3 = \int_{\omega_1} h_i^\varepsilon v_i\, dx_1\, dx_2, \tag{18.5}$$

holds for any v_i belonging to the set V of functions v_i which is kinematically admissible. If we apply (18.3) and (18.4), and we associate the terms of the same order in ε, we arrive at the conclusion that (18.5) is equivalent to the problem

$$\left(\int_{\Omega} \sigma_{33}\gamma_{33}(v_k)\,dx_1\,dx_2\,dx_3 - \int_{\omega_1} h_3 v_3\,dx_1\,dx_2\right)$$
$$+\varepsilon\left(\int_{\Omega} \sigma_{\alpha 3}\gamma_{\alpha 3}(v_k)\,dx_1\,dx_2\,dx_3 - \int_{\omega_1} h_\alpha v_\alpha\,dx_1\,dx_2\right) \tag{18.6}$$
$$+\varepsilon^2 \int_{\Omega} \sigma_{\alpha\beta}\gamma_{\alpha\beta}(v_k)\,dx_1\,dx_2\,dx_3 = 0,$$

for every $v_i \in V$. In order to solve this problem, we make the assumption that u_i and σ_{ij} can be expanded in the formal series

$$u_i = \overset{0}{u_i} + \varepsilon \overset{1}{u_i} + \varepsilon^2 \overset{2}{u_i} + \ldots,$$
$$\sigma_{ij} = \overset{0}{\sigma_{ij}} + \varepsilon \overset{1}{\sigma_{ij}} + \varepsilon^2 \overset{2}{\sigma_{ij}} + \ldots, \tag{18.7}$$

with σ_{ij} related to u_i through the constitutive equations (18.1) and the strain–displacement relation. By substituting (18.7) in (18.6) and equating the terms of the same order in ε to zero, we get a sequence of linear problems in the unknowns $\overset{0}{u_i}, \overset{1}{u_i}, \overset{2}{u_i}, \ldots$, and, of course, in the stress components $\overset{0}{\sigma_{ij}}, \overset{1}{\sigma_{ij}}, \overset{2}{\sigma_{ij}}, \ldots$, which depend on the partial derivatives of the displacements.

The method of asymptotic expansion is remarkable for two reasons. The first is that we can derive all Saint-Venant's solutions in sequence. This means that the terms of zero order in (18.3) permit us to evaluate the influence of the loads h_3 parallel to the longitudinal axes, the terms of order unity take account of loads h_α parallel to the terminal sections, and the terms of higher order are solutions of homogeneous equations. The second argument for which the procedure is important is that it applies also to nonlinear strains and constitutive relations. For instance, it may be extended to the case in which the strains maintain the nonlinear terms and have the form

$$\gamma_{ij} = \tfrac{1}{2}(u_{i,j} + u_{j,i} + u_{r,i}u_{r,j}), \tag{18.8}$$

where γ_{ij} is now the Green–Saint-Venant strain tensor, and the stress, or, more specifically, the second Piola–Kirchhoff stress tensor, has the form

$$T_{\mathrm{AB}} = \gamma_0\delta_{\mathrm{AB}} + \gamma_1\gamma_{\mathrm{AB}} + \gamma_2(\gamma_{\mathrm{AB}})^2, \tag{18.9}$$

where γ_0, γ_1, and γ_2 are functions of the principal invariants of the strain tensor (Cimetière et al. 1988).

The application of the asymptotic expansion of the solutions is now more difficult, but the result is that we can recover, in increasing order of approximation, the nonlinear versions of Saint-Venant's problem in all its six cases. Naturally, the question of the convergence of the formal expansion of solutions, and the question of solving the sequence of nonlinear problems is still unanswered.

Chapter III
Rod Theories: Director Approach

19. Rods as Oriented Bodies

Rods or, more generally, thin bodies are conceptual abstractions of physical bodies having one or two dimensions much smaller than the third. The prototype of a thin body is a cylinder in which the diameter of the cross-section is much smaller than the length of its generators. If we have to determine the stresses and strains in an elastic cylinder under the action of given loads, the solution of equations of three-dimensional elasticity appears as the canonical procedure. Sometimes, however, an alternative method of solution may be suggested by the slenderness of the cylinder. This leads us to attempt a solution which will take the form of a limit of the three-dimensional solution when the diameter of the cylinder approaches zero, while the resultant of loads and the global stiffness of the body tend to finite limits. Obviously, the cylinder must not be conceived as a straight line which can only be lengthened, but account must also be taken of the bending and twisting of this line. This is achieved by regarding the cylinder in the limit as a line endowed with directions that may rotate independently of the deformation of the line to which they are attached. The directions, associated with the axis, are called *directors* and, a material curve, together with a collection of directors assigned to each particle, that is able to deform independently constitutes a *Cosserat rod* (Cosserat and Cosserat 1907).

In general, the present configuration of a Cosserat rod is specified by a vector function of the variable $S\,(S_1 \leqslant S \leqslant S_2)$ and the time t:

$$\mathbf{r} = \mathbf{p}_0(S, t). \tag{19.1}$$

This gives the position vectors of a point of a curve in three-dimensional Euclidean space at S, the arc length of the curve. The directors are a set of vectors $\mathbf{d}_k\,(k = 1, \ldots, K)$, which are also functions of (S, t).

We require that, under the change of frame

$$\mathbf{r} \to \mathbf{c}(t) + \mathbf{Q}(t)\mathbf{r}, \tag{19.2}$$

the vector $\mathbf{p}_0$ transforms according to the rule

$$\mathbf{p}_0(S, t) \to \mathbf{c}(t) + \mathbf{Q}(t)\mathbf{p}_0(S, t), \tag{19.3}$$

and the directors transform according to the rule

$$\mathbf{d}_k \to \mathbf{Q}(t)\mathbf{d}_k. \tag{19.4}$$

This means that, while $\mathbf{p}_0$ behaves like a position vector under a change of frame, the $\mathbf{d}_k$ terms behave like differences between position vectors. In other words, directors are not influenced by pure translations of the frame. In addition, we require that the vectors $\mathbf{p}_{0,S}$, $\mathbf{d}_1$, and $\mathbf{d}_2$ to be independent:

$$\mathbf{p}_{0,S} \cdot (\mathbf{d}_1 \times \mathbf{d}_2) > 0, \tag{19.5}$$

which is called the *continuity condition* (Antman 1972, Sec. 14).

The most important case is that with $K = 2$, because it embodies the simplest director theory of rods that is able to account for bending, torsion, extension, and also for shears and transverse extensions.

The configuration of the rod at time $t = 0$ is conventionally taken as the reference configuration. This is defined by the vector

$$\mathbf{R} = \mathbf{P}_0(S), \tag{19.6}$$

and by a set of vectors $\mathbf{D}_k(S)\,(k = 1, \ldots, K)$, called the directors of the reference state. In general, both the position vector, and the directors of the reference configuration are assumed to be known and, in the most interesting case ($K = 2$), the initial configuration is characterized by the three vectors $\mathbf{R}$, $\mathbf{D}_1$, and $\mathbf{D}_3$, with the continuity constraint

$$\mathbf{P}_{0,S} \cdot (\mathbf{D}_1 \times \mathbf{D}_2) > 0. \tag{19.7}$$

An important preliminary question is that of constructing a differential description of the undeformed rod, starting from a knowledge of the position vector and the two directors. As a first choice, $\mathbf{D}_1$ is placed along the principal normal and $\mathbf{D}_2$ along the binormal, so that the triad $\mathbf{D}_1$, $\mathbf{D}_2$, $\mathbf{P}_{0,S} \equiv \mathbf{D}_3$, is oriented as a right-handed system, coinciding with the principal triad of the curve. The vectors $\mathbf{D}_k$ ($k = 1, 2, 3$) are not necessarily of unit length, but can be taken to be so, without loss of generality. In this case the variation of $\mathbf{D}_k$ with respect to S is defined by the Frenet–Serret formulae

$$\begin{aligned} \mathbf{D}_1' &= -\kappa_0'\mathbf{D}_3 + \tau_0\mathbf{D}_2, \\ \mathbf{D}_2' &= -\tau_0\mathbf{D}_1, \\ \mathbf{D}_3' &= \kappa_0'\mathbf{D}_1, \end{aligned} \tag{19.8}$$

where κ_0' and τ_0 are the principal curvature and the torsion of the curve. We must observe, however, that such a result holds only upon the assumption that the directors remain coincident with the unit normal, binormal, and tangent to the curve as S varies. However, we are also free to choose a set of directors attached to the curve but turning independently of it.

For another special choice of directors, let us consider a single unit vector $\mathbf{D}_1$, normal to the curve, but subtending an angle $\varphi(S)$ with the principal normal $\mathbf{N}$, and an angle $[(\pi/2) - \varphi]$ with the binormal $\mathbf{B}$, so that

$$\mathbf{D}_1 = \mathbf{N}\cos\varphi + \mathbf{B}\sin\varphi. \tag{19.9}$$

The quantity $\varphi' = \varphi_{,S}(S)$ represents the *twist* of $\mathbf{D}_1$ with respect to the axis. Since $\mathbf{D}_1 \cdot \mathbf{N} = \cos\varphi$, we have

$$\mathbf{D}_1' \cdot \mathbf{N} + \mathbf{D}_1 \cdot \mathbf{N}' = -\sin\varphi\,\varphi', \tag{19.10}$$

or, alternatively, by (19.9) and the Frenet–Serret formulae for $\mathbf{N}'$ ($\mathbf{N}' = -\kappa_0'\mathbf{D}_3 + \tau_0\mathbf{B}$),

$$\mathbf{D}_1' \cdot \mathbf{N} + \tau_0 \sin\varphi = -\sin\varphi\,\varphi'. \tag{19.11}$$

Provided that $\varphi \neq 0$, this equation and its variants (Truesdell and Toupin 1960, Sec. 63) permit us to determine the twist in terms of the torsion and the derivative of $\mathbf{D}_1$.

If, instead of one unit vector $\mathbf{D}_1$, we consider two unit vectors $\mathbf{D}_1$ and $\mathbf{D}_2$, there is, in general, no connection between their twists. However, often $\mathbf{D}_1$ and $\mathbf{D}_3$ are normal to one another, so that the twist of the unstrained rod may be defined as the twist of either of these vectors relative to the axis.

Let us now assume that the rod undergoes a deformation. The deformed configuration is thus defined, not only by (19.1), but also by a relation between the new directors $\mathbf{d}_k$ with the original directors $\mathbf{D}_k$:

$$\mathbf{d}_k = \mathbf{d}_k(\mathbf{D}_1, \mathbf{D}_2, \mathbf{D}_3, S, t). \tag{19.12}$$

In general, the directors $\mathbf{d}_k$ are no longer orthogonal or of unit length. However, the common director theories of rods put more restrictions on the law of dependence (19.12). It is usually assumed that $\mathbf{d}_1$ and $\mathbf{d}_2$ remain at right angles and preserve their initial unit length. The plane with the normal

$$\mathbf{d}_3 = \mathbf{d}_1 \times \mathbf{d}_3,$$

is called the *section* of the rod at S. However, unlike the situation in the reference configuration, in which $\mathbf{D}_3$ was tangential to the curve, in the present configuration $\mathbf{d}_3$ is not coaxial with the vector $\mathbf{r}' = \mathbf{p}_{0,S}$. In order to exclude the case in which a section may be sheared so severely so as to contain the tangent to the axis, we require that the continuity condition (19.5) should hold, whatever the magnitude of loads acting on the rod may be. The continuity condition is the counterpart of the condition $J > 0$ of the three-dimensional theory.

We may ask whether it might have been better to choose a triad of directors such that $\mathbf{d}_3^* = \mathbf{r}'/|\mathbf{r}'|$ and with $\mathbf{d}_1^*$ and $\mathbf{d}_2^*$ perpendicular to $\mathbf{d}_3^*$ and perpendicular to one another. The reason for this is that the first triad is more natural both physically and mathematically (Antman 1974). We will explain the advantages of the first basis below, when we define the stress resultants and introduce the constitutive equations.

For the kinematical description of the rod in its deformed state, with the triad $\mathbf{d}_k$ orthonormal, $\mathbf{d}_k$ can be localized with respect to a fixed orthonormal triad $\mathbf{e}_1, \mathbf{e}_2, \mathbf{e}_3$ by means of just three parameters; for example, we can use the Euler angles ψ, θ, and φ employed in Section 17, with the differences that the triad $\mathbf{e}_1, \mathbf{e}_2, \mathbf{e}_3$ is not necessarily a triad coinciding with the direction of the (initially straight) central line and the two principal axes of the cross-section in Figure 17.1, and $\mathbf{d}_1$, $\mathbf{d}_2$, and $\mathbf{d}_3$ are not chosen as the so-called principal torsion–flexure axes of the three-dimensional theory.

In the fixed triad $\mathbf{e}_1, \mathbf{e}_2, \mathbf{e}_3$ we set

$$\begin{aligned} \mathbf{r} &= x_k(S)\mathbf{e}_k, \\ \mathbf{r}'(S) &= x_k'(S)\mathbf{e}_k \equiv y_k(S)\mathbf{d}_k(S) \equiv y_\alpha(S)\mathbf{d}_\alpha(S) + z(S)\mathbf{d}_3(S) \quad (y_3 \equiv z), \end{aligned} \tag{19.13}$$

and substitute (17.15) in (19.13) to obtain

$$
\begin{aligned}
x_1' &= y_1(-\sin\psi\sin\varphi + \cos\psi\cos\varphi\cos\theta) - y_2(\sin\psi\cos\varphi + \cos\psi\sin\varphi\cos\theta) \\
&\quad + z\cos\psi\sin\theta, \\
x_2' &= y_1(\cos\psi\sin\varphi + \sin\psi\cos\varphi\cos\theta) + y_2(\cos\psi\cos\varphi - \sin\psi\sin\varphi\cos\theta) \\
&\quad + z\sin\psi\sin\theta, \qquad (19.14)\\
x_3' &= -y_1\cos\varphi\sin\theta + y_2\sin\varphi\sin\theta + z\cos\theta.
\end{aligned}
$$

From (17.15) we obtain the curvatures by means of the formulae already employed:

$$
\begin{aligned}
\kappa &= \mathbf{d}_2'\cdot\mathbf{d}_3 = \theta'\sin\varphi - \psi'\cos\varphi\sin\theta, \\
\kappa' &= \mathbf{d}_3'\cdot\mathbf{d}_1 = \theta'\cos\varphi + \psi'\sin\varphi\sin\theta, \qquad (19.15)\\
\tau &= \mathbf{d}_1'\cdot\mathbf{d}_2 = \varphi' + \psi'\cos\theta.
\end{aligned}
$$

The functions y_k, κ, κ', and τ constitute the *strains* of the director theory. These quantities are almost universally adopted as appropriate measures of strain in a rod, and they were accepted as far back as the second half of nineteenth century after the publication of Clebsch's (1862) and Love's (1927) treatises. However, it seems that Saint-Venant and Binet preferred the definition of twist as given by (19.11) (Truesdell and Toupin 1960, Sec. 63A).

Another unexpected extension of director theory is that we may also introduce directors to characterize certain privileged directions of heat propagation. These directors are now scalar fields of the type $\lambda_k(S, t)$, and if we take $k = 0$ the heat flow in only one direction is taken into account, which could possibly be along the axis.[31]

The balance equations connecting the mechanical variables must now be added to the system of kinematic variables. However, since Cosserat rods are not classical three-dimensional bodies, they cannot be regarded as trivial applications deducible from three-dimensional theory. The fundamental assumption of the mechanical theory is a *stress principle* for rods, which states that, across any section ideally splitting the rod into two parts, the action of the material on one side upon the material of the other is equipollent to that of a resultant stress vector **n** and a resultant couple **m**.

Let us denote the forces distributed along the axis by **f** and the couples by $\boldsymbol{\ell}$ per unit length. The balance equations at the equilibrium then read

$$[\mathbf{n}]_U^V + \int_U^V \mathbf{f}\,dS = \mathbf{0}, \qquad (19.16)$$

and

$$[\mathbf{m}]_U^V + [\mathbf{r}\times\mathbf{n}]_U^V + \int_U^V (\boldsymbol{\ell} + \mathbf{r}\times\mathbf{f})\,dS = \mathbf{0}, \qquad (19.17)$$

where $[U, V]$ is an arbitrary subinterval of $[S_1, S_2]$. The first equation describes the balance of *linear* momentum, and the second describes the balance of *angular*

[31] This observation is due to Antman (1972, Sec. 14).

momentum. We need not postulate conservation of mass, as we have employed a strictly material description.

In local form, the balance equations become

$$\begin{aligned} &\mathbf{n}' + \mathbf{f} = \mathbf{0}, \\ &\mathbf{m}' + \mathbf{r}' \times \mathbf{n} + \boldsymbol{\ell} + \mathbf{r} \times \mathbf{f} = \mathbf{0}. \end{aligned} \tag{19.18}$$

Traction, displacement and mixed boundary value problems may be formulated by prescribing $\mathbf{n}$, and $\mathbf{m}$, and $\mathbf{r}$, and $\mathbf{r}'$, respectively, or a combination of these at $S = S_1, S_2$, or when the rod is a ring, simply by prescribing the periodic conditions.

To render determinate the rod theories based on the two vector equations (19.18), which are equivalent to six scalar equations, it is necessary to reduce the number of geometric variables to six or less. This reduction may be effected in various ways. The commonest procedure is that of adopting Kirchhoff's hypotheses, which require that planar cross-sections normal to the axis in the reference configuration should remain planar, undeformed, and normal to the same strained axis. Other physically meaningful methods for reducing the kinematical quantities through additional constraints also exist.

The commonly used component version of (19.18) is obtained by setting

$$\mathbf{n}(S) = n_k(S)\mathbf{d}_k(S) \tag{19.19}$$

$$\mathbf{m}(S) = m_k(S)\mathbf{d}_k(S), \tag{19.20}$$

from which, again by using (14.7), we obtain

$$n_1' + \kappa' n_3 - \tau n_2 + f_1 = 0, \tag{19.21a}$$

$$n_2' + \tau n_1 - \kappa n_3 + f_2 = 0, \tag{19.21b}$$

$$n_3' + \kappa n_2 - \kappa' n_1 + f_3 = 0, \tag{19.21c}$$

and

$$m_1' + \kappa' m_3 - \tau m_2 + y_2 n_3 - z n_2 + \ell_1 + (\mathbf{r} \times \mathbf{f}) \cdot \mathbf{d}_1 = 0, \tag{19.22a}$$

$$m_2' + \tau m_1 - \kappa m_3 + z n_1 - y_1 n_3 + \ell_2 + (\mathbf{r} \times \mathbf{f}) \cdot \mathbf{d}_2 = 0, \tag{19.22b}$$

$$m_3' + \kappa m_2 - \kappa' m_1 + y_1 n_2 - y_2 n_1 + \ell_3 + (\mathbf{r} \times \mathbf{f}) \cdot \mathbf{d}_3 = 0. \tag{19.22c}$$

The choice of $\mathbf{n}$ with components along $\mathbf{d}_k$ will illustrate the advantage of having $\mathbf{d}_k$ as a local basis of strain in the deformed rod. In fact, it is the components n_1 and n_2 that are responsible for the shearing, whereas the components along the $\mathbf{d}_1^*$ and $\mathbf{d}_2^*$ directions, perpendicular to $\mathbf{d}_3^* = \mathbf{r}'/|\mathbf{r}'|$, have no direct influence on the shearing. This is illustrated for the two-dimensional case in Figure 19.1 (Antman 1974). We take $n_2 = 0$ and $\mathbf{r}'$, $\mathbf{d}_1$, and $\mathbf{d}_3$ coplanar. It is reasonable to expect that an increase in n_1 will produce a decrease in the angle between $\mathbf{d}_1$ and $\mathbf{r}'$, whereas an increase in n_1^*, the component of $\mathbf{n}$ along $\mathbf{d}_1^*$, will have no such effect.

As for the constitutive relations, the most natural form of these equations, which is valid in the purely elastic case, is

$$\begin{aligned} n_k &= n_k(y_\alpha(S), z(S), \kappa(S), \kappa'(S), \tau(S), S), \\ m_k &= m_k(y_\alpha(S), z(S), \kappa(S), \kappa'(S), \tau(S), S), \end{aligned} \tag{19.23}$$

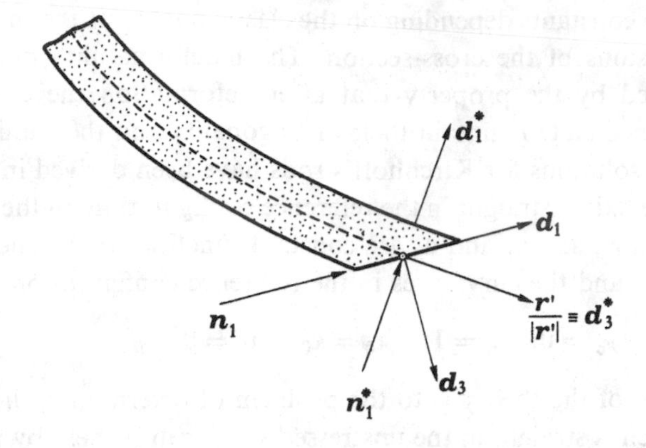

Fig. 19.1

where n_k and m_k are continuous differentiable functions of their arguments. These can assume any real values, except z which is restricted by the continuity condition $\mathbf{r}' \cdot \mathbf{d}_3 > 0$, and S which varies between S_1 and S_2. Equations (19.23) are invariant under rigid motions, and it is possible to show that this condition necessarily implies the form (19.23).[32]

If we substitute the constitutive equations in the balance equations (19.22), we obtain a system of six nonlinear ordinary differential equations in six unknowns. The solution of the system constitutes a difficult problem, which is solvable only in special cases, but at least we have derived a system in which the number of equations equals the number of unknowns.

20. Kirchhoff's Theory

The first complete theory of large displacement but small strains in initially straight elastic rods was proposed by Kirchhoff (1859), who extended an earlier result of Euler (1744). Kirchhoff's theory is an attempt to solve system (19.22), with the following constitutive assumptions: (1) the axis of the rod is inextensible; (2) the stress couple depends linearly on the curvature and the twist; (3) there is no shear between the cross-section and the axis; and (4) the cross-section is undeformable in its plane.

The analytical consequences of Kirchhoff's assumptions are that, whatever the deformation, $\mathbf{d}_3$ must remain coaxial with $\mathbf{r}'$, and, therefore, in (19.13) we must put $y_\alpha = 0$ and $z = |\mathbf{r}'|$. The postulate of inextensibility implies that $|\mathbf{r}'|$ must be unity. Again, Kirchhoff's theory assumes that constitutive relations are not only linear in the corresponding kinematical variables, but also have the special form

$$m_1 = A\kappa, \quad m_2 = B\kappa', \quad m_3 = C\tau, \tag{20.1}$$

[32] This result is due to Green and Laws (1966).

where A, B, and C are constants depending on the elastic quality of the material and the shape and dimensions of the cross-section. The indeformability of the cross-section is characterized by the property that after deformation the directors $\mathbf{D}_1$ and $\mathbf{D}_2$ of the unstrained state maintain their orthogonality and their unit length.

Several remarkable solutions for Kirchhoff's rods have been derived in the particular case in which the axis is straight in the reference configuration; in the deformed state, $\mathbf{d}_3$ coincides with $\mathbf{r}'$, and $\mathbf{d}_1$ and $\mathbf{d}_2$ are constant functions of S. The values of the quantities y_α and z and the curvatures in the reference configuration are then

$$y_\alpha = 0, \quad z = 1, \quad \kappa_0 = \kappa_0' = \tau_0 = 0. \tag{20.2}$$

The first application of the theory is to the problem of determining the forms in which a thin rod, which is straight in the unstressed state, can be held by forces and couples applied at its ends only, when the rod is bent in a principal plane, so that the axis becomes a plane curve and there is no twist. This is the problem of the *elastica*, which was first solved by Euler in 1744.

We take as the plane of bending the plane for which the flexural rigidity is B, so that κ and τ vanish, and the initially straight axis becomes a curve in the plane $\mathbf{d}_1\mathbf{d}_3$ (Figure 20.1) having curvature $\kappa' = \kappa'(S)$.

The load is a single vertical force R, directed downwards, applied at the $S = S_2$ end of the axis. The stress resultants at S, resolved along the directions of $\mathbf{d}_1$ and $\mathbf{d}_3$, are then

$$n_1 = R\sin\theta, \quad n_3 = R\cos\theta,$$

where θ is the angle that $\mathbf{d}_3$ makes with the line of action of the force R. The flexural couple at the section S is then

$$m_2 = B\kappa',$$

where $\kappa' = \mathrm{d}\theta/\mathrm{d}S$. We note that, according to the convention for the signs of moments, m_2 is positive.

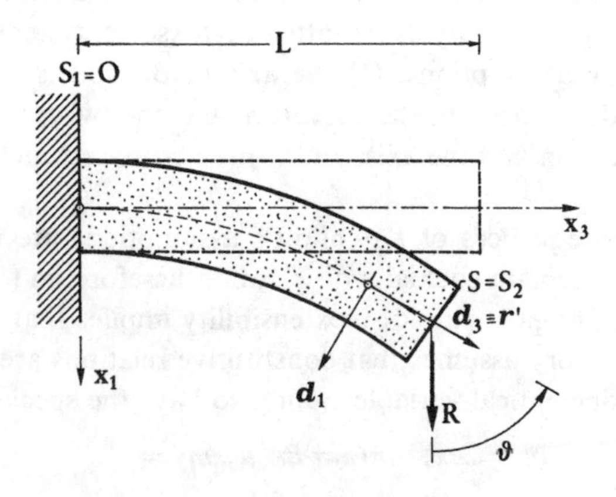

Fig. 20.1

Equation (19.22b), adapted to the present situation, yields

$$m_2' + n_1 = 0,$$

or, in another form,

$$B\frac{d^2\theta}{dS^2} + R\sin\theta = 0, \tag{20.3}$$

with a first integral

$$\frac{1}{2}\left(\frac{d\theta}{dS}\right)^2 + \frac{R}{B}(\cos\theta_0 - \cos\theta) = 0, \tag{20.4}$$

where θ_0 is the value of θ at $S = S_2 = L$. A second integration gives the total length

$$L = \frac{1}{\sqrt{R/B}}\int_0^{\pi/2}\frac{d\phi}{\sqrt{1-k^2\sin^2\phi}} = \frac{K(k)}{\sqrt{R/B}}, \tag{20.5}$$

where $k\sin\phi = \sin(\theta/2)$ and $k = \sin(\theta_0/2)$. Then $K(k)$ is a complete elliptic normal integral of modulus k. Using tables of elliptic integrals, the value of R can then be obtained from a knowledge of k for any value of θ_0.

For a point (x_1, x_2) in the fixed system of axes shown in Figure 20.1, we have

$$\frac{dx_1}{dS} = \cos\theta, \quad \frac{dx_3}{dS} = \sin\theta.$$

Hence, from the above relation, the values of x_1, and x_3 can be expressed in terms of elliptic functions.

The *elastica* takes different forms according to whether or not there are inflexions. At an inflection $d\theta/dS$, and hence the bending moment vanishes, so that a rod can be held in the form of an inflexional *elastica* by a terminal force without couples. The ends are points of inflexion, and all intermediate points of inflexion must lie on the line of action of the terminal force. When there are no inflexions the strained axes are themselves referred to as noninflexional *elastica*. In his original theory, Euler classified all the possible forms of the *elastica* into nine types.

The discussion of all the types of deflection of a rod under the action of terminal forces is particularly important. It is particularly relevant in the study of the stability of an initially straight rod under the action of a compressive force R directed along the undeformed axis (Figure 20.2).

When the straight configuration becomes unstable under an increasing load R, there may exist a neighboring bent state (θ small) of curved form, as shown in Figure 20.2. In this case, $\sin\theta$ is approximated by θ and (20.3) has the approximate solution

$$\theta = C_1\cos\left(S\sqrt{\frac{R}{B}}\right) + C_2\sin\left(S\sqrt{\frac{R}{B}}\right),$$

where C_1 and C_2 are constants, which must be determined by the boundary conditions

$$\theta(0) = 0, \quad \frac{d\theta}{dS}(L) = 0. \tag{20.6}$$

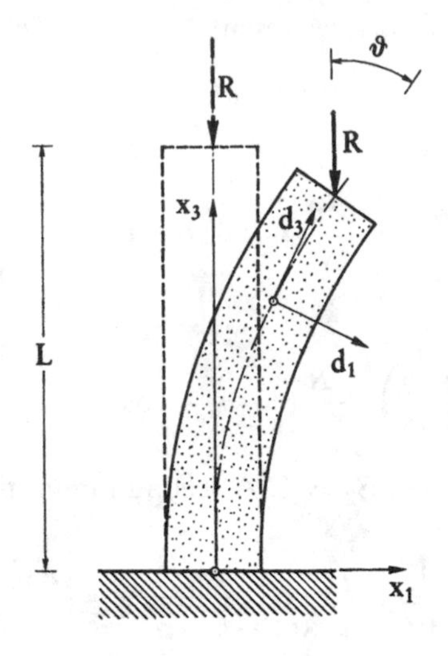

Fig. 20.2

It turns out that nonzero values of the constants C_1 and C_2, compatible with the end conditions, are possible, provided that

$$R \geqslant \frac{1}{4}\frac{\pi^2 B}{L^2}. \tag{20.7}$$

This minimum value of R, for which a possible stable bent configuration arises, is called the *Euler buckling load.*

It should be noted that, in the approximate theory, it is the ratio C_1/C_2 that is determined, and not the individual values of these constants, and the approximate theory can only give the critical load. In order to determine the deflection of the terminal point of the axis in the new configuration under the buckling load, we must return to (20.4). When we remember that $dx_1 = dS \sin\theta$ and that k and ϕ are defined above for (20.5), a simple integration yields

$$x_1(L) = \frac{2k}{\sqrt{R/B}} \int_0^{\pi/2} \sin\varphi \, d\varphi = \frac{2k}{\sqrt{R/B}}. \tag{20.8}$$

An approximate version of this formula can be found on remembering that, for small k, we have the approximation (Jahnke, Emde, and Lösch 1966, p. 62)

$$\frac{2}{\pi}K(k) \simeq 1 + \frac{1}{4}k^2 + \ldots, \tag{20.9}$$

where $K = L\sqrt{R/B}$. Hence, on solving (20.9) with respect to k and substituting in (20.8), we obtain

$$x_1(L) \simeq 4\left[\frac{2L}{\pi} - \left(\frac{B}{R}\right)^{\frac{1}{2}}\right]^{\frac{1}{2}}\left(\frac{B}{R}\right)^{\frac{1}{4}}, \tag{20.10}$$

which is in agreement with a formula registered in Love's (1927, Art. 264) treatise.

A slight generalization of the result is achieved by removing the hypothesis of inextensibility of the axis under longitudinal thrust. In this case the length of the deformed axis is $L[1-(R/EA)]$ and the Euler buckling load of (20.7) must be modified as

$$RL^2\left(1-\frac{R}{EA}\right)^2=\frac{1}{4}\pi^2 B. \tag{20.11}$$

However, the correction is not of importance in practice.

The question of finding the conditions of the buckling of a rod by linearizing (20.3) involves some consideration. We have determined the buckling load by investigating the possible equilibrium configurations close to the rectilinear configuration, but we do not know whether the critical load of the linearized theory is related to the critical load of the nonlinear theory. In addition, because the constants C_1 and C_2 are only partially determined, the impossibility of determining the precise magnitude of the deformation in the buckled configurations is clearly a deficiency of the linearization process.

In order to clarify this situation, it is necessary to analyze the structure of the solutions of (20.3) in more detail. For simplicity, we denote the ratio R/B by λ.

The linearized problem has the solutions

$$\theta(S)=A_n\sin\left(\frac{n\pi}{2L}S\right)\quad n=0,1,3,5,\ldots, \tag{20.12}$$

when

$$\lambda=\lambda_n=\left(\frac{n\pi}{2L}\right)^2,\quad n=0,1,3,5,\ldots. \tag{20.13}$$

For the value $\lambda_0=0$, it has only $\theta=A_0=0$ as a solution, and so λ_0 is not an eigenvalue.

The nonlinear problem has the first integral (20.4), that is

$$\left(\frac{d\theta}{dS}\right)^2=2\lambda(\cos\theta-\cos\theta_0), \tag{20.14}$$

where

$$\theta(L)=\theta_0,\quad 0\leqslant\theta_0\leqslant\pi. \tag{20.15}$$

Since θ is continuous and satisfies (20.15), and the right-hand side of (20.14) must be nonnegative, we must have

$$|\theta|\leqslant\theta_0.$$

Now, from the identity

$$0=-\int_0^L\left(\frac{d^2\theta}{dS^2}+\lambda\sin\theta\right)\theta\,dS=\int_0^L\left[\left(\frac{d\theta}{dS}\right)^2-\lambda\theta\sin\theta\right]dS,$$

and the result

$$\theta \sin\theta \leqslant |\theta|\,|\sin\theta| \leqslant \theta^2,$$

we have the inequality

$$0 \geqslant \int_0^L \left[\left(\frac{d\theta}{dS}\right)^2 - \lambda\theta^2\right] dS. \tag{20.16}$$

This inequality becomes a strict equality only for $\lambda = \lambda_1 = (\pi/2L)^2$, with $\theta = A_1 \sin(\pi S/2L)$, that is, when λ and θ are the first eigenvalue and the first eigenfunction of the linear problem, respectively. This is a consequence of the minimum principle giving a minimum for $\lambda = \lambda_1$, which implies that the integral in (20.16) is positive for $0 < \lambda < \lambda_1$ for all admissible $\theta \neq$ constant. Hence, for $0 < \lambda < \lambda_1$, the only solutions of the nonlinear problem are $\theta \equiv 0$ or $\theta \equiv \pi$.

To study the solutions for $\lambda > \lambda_1$, we return to (20.4) and write it in the form

$$\begin{aligned}\frac{d\theta}{dS} &= 2\sqrt{\lambda}\sqrt{\sin^2\frac{\theta_0}{2} - \sin^2\frac{\theta}{2}} = 2\sqrt{\lambda}\sqrt{k^2 - \sin^2\frac{\theta}{2}}\\ &= 2\sqrt{\lambda}\,k\sqrt{1 - \mathrm{sn}^2(u,k)} = 2k\,\mathrm{cn}(u,k),\end{aligned} \tag{20.17}$$

where $u = S\sqrt{R/B}$, $k = \sin^2(\theta_0/2)$, and $\mathrm{sn}(u,k)$ and $\mathrm{cn}(u,k)$ are Jacobian elliptic functions (Jahnke et al. 1966, p. 73). The function $\mathrm{cn}(u,k)$, regarded as a function of u, vanishes from $u = nK$ $(n = 1, 3, 5, \ldots)$, where $K(k)$ is the elliptic integral as defined before. For $S = L$, we have the boundary condition

$$\mathrm{cn}[u(L), k] = \mathrm{cn}\left(L\sqrt{\frac{R}{B}}, k\right) = 0.$$

This condition is verified, provided that

$$L\lambda = L\lambda_n = L\sqrt{\frac{R}{B}} = nK(k) \quad (n = 1, 3, 5, \ldots), \tag{20.18}$$

or

$$\lambda_n^2 = \frac{R}{B} = \left(\frac{nK}{L}\right)^2 \quad (n = 1, 3, 5, \ldots). \tag{20.19}$$

Thus, for each n we have obtained the response of the rod, that is to say, a relationship between the load parameter R/B and the measure of deformation $k = \sin(\theta_0/2)$. The curves $\lambda_n(k)$ are indicated in Figure 20.3. Since $K(0) = \pi/2$, each curve branches from $k = 0$, $\lambda = \lambda_n$, which are just the eigenvalues of the linearized problem. It thus follows that the linear eigenvalue problem yields the points of the bifurcations for the nonlinear problem.

From (20.18) we find that $d\lambda_n/dk > 0$ for $k > 0$, and hence the curves $\lambda_n = \lambda_n(k)$ are monotonically increasing. In addition, the curves do not intersect and have as a common asymptote the straight line $k = 1$ as λ_n tends to infinity.

The behavior of the rod for $\lambda > \lambda_1$, for instance for $\lambda_n < \lambda < \lambda_{n+2}$ $(n = 1, 3, \ldots)$, must be clarified, because we must decide whether the rod will buckle and, if so,

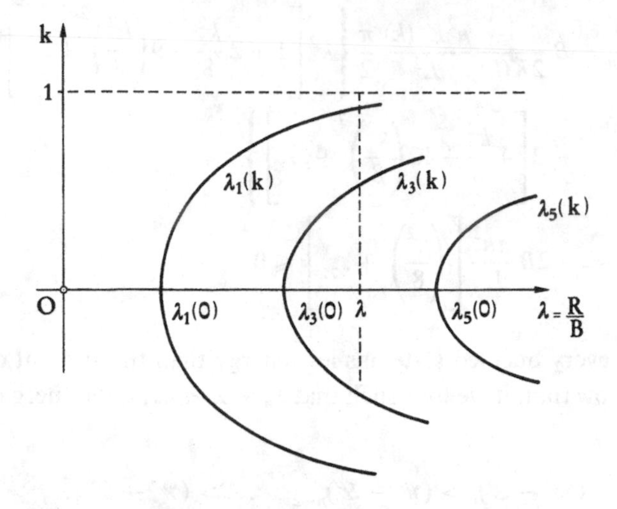

Fig. 20.3

which branch is to be preferred. The answer can be obtained by examining the total potential energy associated with the possible states. The energy of the unbuckled state $\theta = 0$ is clearly zero, provided that we continue to maintain the hypothesis of inextensibility of the axis. The energy of the buckled state is given by the difference

$$\mathscr{W} - \mathscr{L} = \int_0^L \frac{1}{2} B \left(\frac{d\theta}{dS} \right)^2 dS - R \int_0^L (1 - \cos\theta)\, dS, \tag{20.20}$$

where the first term represents the stored strain energy and the second the work done by the thrust R. On using (20.17) and the properties of elliptic functions, we transform (20.20) into the expression

$$\mathscr{W} - \mathscr{L} = \frac{R}{2}\left[k^2 - 2 + 2\frac{E(k)}{K(k)} \right] = \frac{R}{2K(k)}[k^2 K(k) + 2(E(k) - K(k))] \tag{20.21}$$

where $E(k)$ is the elliptic integral of the second kind.

Now the functions $K(k)$ and $E(k)$ have the following expansions, which converge for $k^2 < 1$ (Jahnke et al. 1966, p. 62):

$$K(k) = \frac{\pi}{2}\left[1 + 2\frac{k^2}{8} + 9\left(\frac{k^2}{8}\right)^2 + \ldots \right], \quad E(k) = \frac{\pi}{2}\left[1 - 2\frac{k^2}{8} - 3\left(\frac{k^2}{8}\right)^2 - \ldots \right].$$

On substituting these expressions for $K(k)$ and $E(k)$ into (20.21), and considering the total energy $(\mathscr{W} - \mathscr{L})_n$ corresponding to a buckled state which branches from λ_n, as given by (20.19), we obtain

$$(\mathscr{W}-\mathscr{L})_n = B\frac{1}{2K(k)}\frac{n^2K^2(k)}{L^2}\frac{\pi}{2}\left\{k^2\left[1+2\frac{k^2}{8}+9\left(\frac{k^2}{8}\right)^2+\dots\right]\right.$$
$$\left.-2\left[4\frac{k^2}{8}+12\left(\frac{k^2}{8}\right)^2+\dots\right]\right\}$$
$$= -2B\frac{\pi n^2}{L^2}\left[\left(\frac{k^2}{8}\right)^2+\dots\right] \leqslant 0. \tag{20.22}$$

This means that every buckled state has less energy than the straight configuration.

We can also show that, if we fix λ such that $\lambda_n < \lambda < \lambda_{n+2}$, the energies are ordered in the sequence

$$(\mathscr{W}-\mathscr{L})_n > (\mathscr{W}-\mathscr{L})_{n-2} > \dots > (\mathscr{W}-\mathscr{L})_1. \tag{20.23}$$

After having determined λ (Figure 20.3), let $k_n(\lambda)$ denote the unique root of (20.18):

$$k_n = K^{-1}\left(\frac{L\lambda}{n}\right). \tag{20.24}$$

Since $K(k)$ is increasing monotonically we conclude from (20.24) that, for $m > n$, $k_n < k_m$. We now differentiate (20.21) with respect to k, and obtain

$$\frac{d}{dk}(\mathscr{W}-\mathscr{L}) = -\frac{R}{kK^2}[E-(1-k^2)K]^2 \leqslant 0, \tag{20.25}$$

since, as a consequence of the differentiation formulae for elliptic integrals (Jahnke et al. 1966, p. 64), we have

$$E' = \frac{E-K}{k}, \quad K' = \frac{E}{k(1-k^2)} - \frac{K}{k}.$$

The equality in (20.5) holds only when $k = 0$; otherwise, $[E-(1-k^2)K]$ is strictly positive. It thus follows that the energy $(\mathscr{W}-\mathscr{L})$ decreases strictly monotonically for $k > 0$, and increases strictly monotonically as a function of n, so that (20.23) is established.

Under a given $\lambda\,(> \lambda_1)$, the rod tends to deform in the buckled state of least energy, which is the state branching from λ_1, and will remain in this state for all $\lambda > \lambda_1$.

The existence of a bifurcation in the curve $k = k(\lambda)$ in the $k\lambda$ plane when $\lambda(0)$ exceeds the critical value λ_1 was recognized and discussed by Poincaré, who described such a point as a *point of bifurcation*, at which an exchange of stability takes place because the points on the λ axis with $\lambda > \lambda_1$ are unstable and stability only occurs for points on the curve branching from $\lambda = \lambda_1$. The presence of the other curves branching from the points $\lambda = \lambda_3, \lambda_5, \dots$, is not in conflict with Poincaré's theory, because the curves do not issue from the stable branch but from the straight λ axis, where this has become unstable.

21. Solutions for More General Loads

When the rod is bent and twisted by a more complex system of loads, the integration of Kirchhoff's equations is possible only in some special cases.

If the load consists of a force **R** and a couple **K**, applied at one end, and balanced by another force and couple at the other end, so that the total system of loads is statically equivalent to zero, the stress resultants in a section of the deformed rod at a distance S from the end at which **R** and **K** are applied, can be expressed by means of the Euler angles defined by (17.15). The stress resultants in this section are given by

$$(n_1, n_2, n_3) = (\mathbf{R}\cdot\mathbf{d}_1, \mathbf{R}\cdot\mathbf{d}_2, \mathbf{R}\cdot\mathbf{d}_3). \tag{21.1}$$

The equations of the problem are again (19.22), but they lack the terms representing the influence of the distributed loads and torques **f** and $\boldsymbol{\ell}$. In these equations the couples m_1, m_2, and m_3 are related to the curvatures by constitutive relations like (20.1). However, it is very difficult to obtain their solution, unless we introduce the hypothesis that the strains y_α of the deformed state are negligible and z is approximately equal to unity. Equations (19.22) thus become

$$A\frac{d\kappa}{dS} - (B-C)\kappa'\tau = n_2, \tag{21.2a}$$

$$B\frac{d\kappa'}{dS} - (C-A)\tau\kappa = -n_1, \tag{21.2b}$$

$$C\frac{d\tau}{dS} - (A-B)\kappa\kappa' = 0. \tag{21.2c}$$

On eliminating n_1 and n_2 from (19.21), with the help of (21.2) we find

$$\frac{dn_3}{dS} + A\kappa\frac{d\kappa}{dS} + B\kappa'\frac{d\kappa'}{dS} + (A-B)\tau\kappa\kappa' = 0, \tag{21.3}$$

or, by (21.2c),

$$\frac{d}{dS}[n_3 + \tfrac{1}{2}(A\kappa^2 + B\kappa'^2 + C\tau^2)] = 0.$$

This equation may be integrated directly, and by using (21.1) we obtain a first integral of (21.2):

$$\tfrac{1}{2}(A\kappa^2 + B\kappa'^2 + C\tau^2) + \mathbf{R}\cdot\mathbf{d}_3 = \text{constant}. \tag{21.4}$$

A second integral may be obtained by observing that, if we consider the moment $\mathbf{m} \equiv (m_1, m_2, m_3)$ at any point S on the strained axis, its projection along the fixed direction $\mathbf{e}_3$ must remain constant. Since the components of the unit vector $\mathbf{e}_3$ along the directions $\mathbf{d}_1$, $\mathbf{d}_2$, and $\mathbf{d}_3$ are

$$\mathbf{e}_3\cdot\mathbf{d}_1 = -\sin\theta\cos\varphi, \quad \mathbf{e}_3\cdot\mathbf{d}_2 = \sin\theta\sin\varphi, \quad \mathbf{e}_3\cdot\mathbf{d}_3 = \cos\theta,$$

the above condition yields

$$\mathbf{m}\cdot\mathbf{e}_3 = -A\kappa\sin\theta\cos\varphi + B\kappa'\sin\theta\sin\varphi + C\tau\cos\theta = \text{constant}. \tag{21.5}$$

Having determined two integrals of the equilibrium equations, we might hope to find a third integral of the same equations in order to obtain three finite relations between the Euler angles θ, ψ, and φ. In general, no other integral is known; however, if the flexural rigidities A and B are equal, then from (21.2c) we have at once

$$\tau = \text{constant}, \tag{21.6}$$

and therefore the system of equations (21.2) can be integrated in finite form.

An example of this result is given by a rod which is bent into a helical form,[33] with a constant inclination $\theta = (\pi/2) - \alpha$ to the upward vertical, by a longitudinal force of magnitude R and a twisting couple of magnitude K. Since $d\theta/dS = 0$, (19.15) give

$$\kappa = -\frac{d\psi}{dS}\cos\alpha\cos\varphi, \quad \kappa' = \frac{d\psi}{dS}\cos\alpha\sin\varphi, \quad \tau = \frac{d\varphi}{dS} + \sin\alpha\,\frac{d\psi}{dS},$$

and from (21.4)–(21.6) we obtain

$$\tau = \text{constant}, \quad \kappa^2 + \kappa'^2 = \text{constant}, \quad \frac{d\psi}{dS} = \text{constant}. \tag{21.7}$$

This means that the strained axis is a curve of constant curvature and torsion; that is, a helix traced on a right circular cylinder.

Once we have (21.7), we can determine the constant α, the radius r of the cylinder on which the helix lies, and the constants on the right-hand sides of (21.7) by exploiting the connection between κ, κ', and τ and the stress and moment resultants.

An explicit solution for the problem of the *elastica*, which has some importance in the analysis of mechanical springs, can be obtained for the case of a slender rod of length L, weighted at one end by a vertical load P, while the other end is hinged so that a couple M is applied (Figure 21.1) (Wang 1981a). The form of the *elastica* can assume various configurations as the couple M is changed, and our aim is to find a connection between M and the angle α, which represents the slope of the axis at the origin. If θ denotes the slope at a point S on the abscissa, the equilibrium equation for the *elastica* is

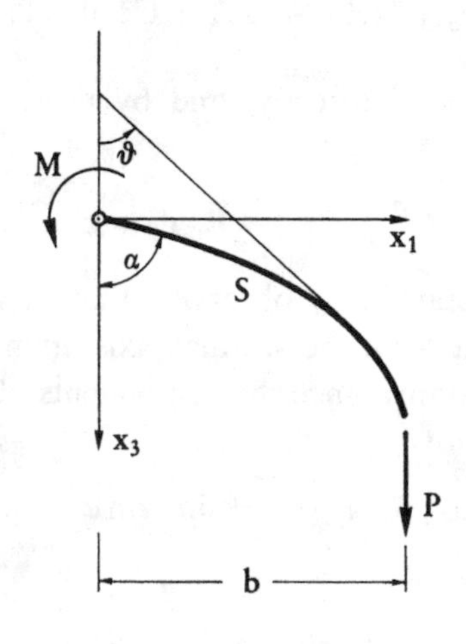

Fig. 21.1

[33] Love (1927, Art. 270), but the solution is due to Kirchhoff.

$$B\frac{d\theta}{dS}=Px_1-M, \tag{21.8}$$

where B is the flexural rigidity, and x_1 the distance of the current section from the x_3 axis. After differentiation with respect to S, (21.8) becomes

$$B\frac{d^2\theta}{dS^2}=P\frac{dx_1}{dS}=P\sin\theta, \tag{21.9}$$

with the boundary conditions

$$\theta(0)=\alpha,\quad \frac{d\theta}{dS}(L)=0. \tag{21.10}$$

The first integral of (21.9) is

$$\frac{1}{2}B\left(\frac{d\theta}{dS}\right)^2=P(\cos\beta-\cos\theta), \tag{21.11}$$

where β is the terminal slope at $S=L$. The formal solution of the above equation is thus

$$\begin{aligned}S&=-\sqrt{\frac{B}{P}}\int_\alpha^\theta\frac{d\theta}{\sqrt{2(\cos\beta-\cos\theta)}}\\&=-\sqrt{2}\sqrt{\frac{B}{P}}\frac{1}{\sqrt{1+\cos\beta}}\times\left[F\left(\sqrt{\frac{2}{1+\cos\beta}},\frac{\theta+\pi}{2}\right)\right.\\&\quad\left.-F\left(\sqrt{\frac{2}{1+\cos\beta}},\frac{\alpha+\pi}{2}\right)\right],\end{aligned} \tag{21.12}$$

where F is an elliptic function of the first kind. The angle β has to be found as one of the roots of the transcendental equation

$$\begin{aligned}L=-\sqrt{2}\sqrt{\frac{B}{P}}\frac{1}{\sqrt{1+\cos\beta}}&\left[F\left(\sqrt{\frac{2}{1+\cos\beta}},\frac{\beta+\pi}{2}\right)\right.\\&\left.-F\left(\sqrt{\frac{2}{1+\cos\beta}},\frac{\alpha+\pi}{2}\right)\right].\end{aligned}$$

The law of dependence of M on α is given by (21.8), written at the origin,

$$M=-B\frac{d\theta}{dS}(0). \tag{21.13}$$

Analysis of the curve $M=M(\alpha)$ for $0\leqslant\alpha\leqslant 2\pi$ shows that, if F/B is small, then M is a single-valued function of α as illustrated by curve I in Figure 21.2; if F/B is large, then the curve behaves like curve II in Figure 21.2, and snapping may occur.

The analysis may be extended to the case in which the weight of the rod itself is taken into account (Bickley 1934; Wang 1981b). The qualitative behavior of the curve $M=M(\alpha)$ is similar to that already described.

The equation describing deformed configurations of heavy rods can be treated mathematically without recourse to an explicit solution (Hsu and Wang 1988). We

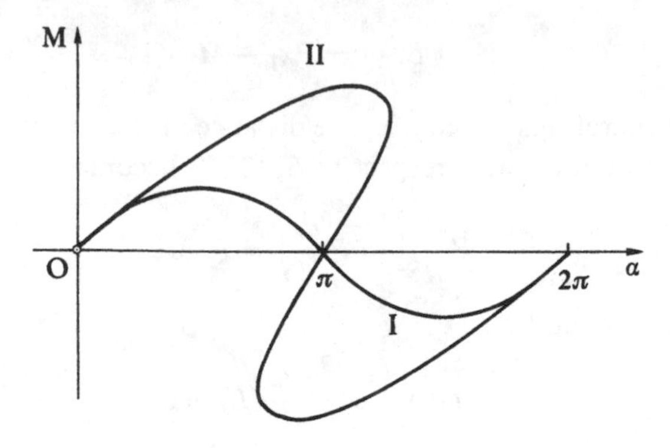

Fig. 21.2

consider again the case shown in Figure 21.1, with no load at the end $S = L$, but under the influence of a vertical body force ρg applied at the points of the strained axis and directed downwards.

Having taken the origin coincident with the hinged end, the equilibrium equation is

$$B\frac{d^2\theta}{dS^2} = \rho g(L - S)\sin\theta, \tag{21.14}$$

with the boundary conditions

$$\theta(0) = \alpha, \quad \frac{d\theta}{dS}(L) = 0. \tag{21.15}$$

On putting $s = S/L$ and $K = [(\rho g L^3)/B]^{1/3}$, (21.14) and (21.15) become

$$\frac{d^2\theta}{ds^2} = K^3(1 - s)\sin\theta, \quad K > 0, \quad 0 < s < 1, \tag{21.16}$$

$$\theta(0) = \alpha, \quad \theta'(1) = 0, \quad -\pi \leqslant \alpha \leqslant \pi, \tag{21.17}$$

and the problem is one of analyzing the solutions of (21.16) under the conditions $K > 0$ and $-\pi \leqslant \alpha \leqslant \pi$.

In order to simplify the notation, let

$$\psi(s) = \theta(1 - s) \quad 0 < s < 1,$$

and reformulate the problem in the form

$$\frac{d^2\psi}{ds^2} = K^3 s \sin\psi, \quad K > 0, \quad 0 < s < 1, \tag{21.18}$$

$$\psi'(0) = 0, \quad \psi(1) = \alpha, \quad -\pi \leqslant \alpha \leqslant \pi. \tag{21.19}$$

In addition, since a solution $\psi(s)$ corresponding to a given α is associated with a solution $-\psi(s)$ when α is replaced by $-\alpha$, we only consider the problem with $0 \leqslant \alpha \leqslant \pi$.

Now (21.18) and (21.19) admit one solution at least, since the right-hand side of (21.18), namely $K^3 s \sin\psi$, is a bounded function for $0 \leqslant s \leqslant 1$. Moreover, the solution is unique, provided that we have $K^3 < \sqrt{45}$. In order to prove this result, let $\psi(s)$ be a solution and write it in the form

$$\psi(s) = \alpha - \int_0^1 K^3 \xi \sin\psi(\xi) G(\xi, s)\, d\xi, \tag{21.20}$$

where

$$G(s, \xi) = 1 - \max(s, \xi).$$

Let $\psi_1(s)$ and $\psi_2(s)$ be two solutions, then we have

$$|\psi_1 - \psi_2| \leqslant K^3 \int_0^1 G(s, \xi)\xi|\psi_1(\xi) - \psi_2(\xi)|\, d\xi$$

$$\leqslant K^3 \left[\int_0^1 G^2(s, \xi)\xi^2\, d\xi\right]^{\frac{1}{2}} \|\psi_1 - \psi_2\|_2,$$

or

$$\|\psi_1 - \psi_2\|_2^2 = \int_0^1 |\psi_1 - \psi_2|^2\, ds \leqslant K^6 \left(\int_0^1\!\!\int_0^1 G^2(s, \xi)\xi^2\, d\xi\, ds\right) \|\psi_1 - \psi_2\|_2^2.$$

As a consequence of the result

$$\int_0^1\!\!\int_0^1 G^2(s, \xi)\xi^2\, d\xi\, ds = \frac{1}{45},$$

it follows that, if $K^6/45 < 1$ or $K^3 < \sqrt{45}$, we must have the identity $\psi_1 = \psi_2$.

In order to understand the bifurcation phenomena that may occur in general, we consider the important case where $\alpha = \pi$, which corresponds to a vertical slope, upwardly directed, at the initial section. Let us make the substitution $v(s) = \psi(s/K) - \pi$, so that (21.18) and (21.19), written with $\alpha = \pi$, take the form

$$\begin{aligned} &v''(s) + s \sin v = 0, \\ &v'(0) = 0, \quad v(K) = 0, \end{aligned} \tag{21.21}$$

and consider the following initial value problem

$$\begin{aligned} &v''(s) + s \sin v = 0, \\ &v'(0) = 0, \quad v(0) = a, \end{aligned} \tag{21.22}$$

where a is a real constant. If we denote the solution of this problem by $v(s, a)$, then from the uniqueness of solutions of ordinary differential equations with Cauchy data we have

$$v(s, 2\pi + a) = v(s, a) + 2\pi,$$
$$v(s, 2\pi - a) = -v(s, a) + 2\pi,$$
$$v(s, a) = -v(s, -a),$$
$$v(s, 0) = 0, \quad v(s, \pi) = \pi,$$

which means that here again we may study $v(s, a)$ only in the interval $0 < a < \pi$.

Now multiply (21.22) by $v'(s)$ and integrate from zero to s, so as to derive

$$\frac{1}{2}[v'(s)]^2 = s \cos v(s) - \int_0^s \cos v(\xi)\, d\xi \geqslant 0. \tag{21.23}$$

Now, if a lies in the interval $0 < a < \pi/2$, then we have $\cos a = \cos v(0) > 0$, and this implies the positive-definiteness of $\cos v(s) > 0$ for all $s \geqslant 0$. If not, then there is a $s_0 > 0$ such that $\cos v(s) > 0$ for all s in the interval $0 \leqslant s < s_0$ with $\cos v(s_0) = 0$. But this contradicts (21.23) when we put $s = s_0$, which proves the results $-\pi/2 \leqslant v(s, a) < \pi/2$ for $0 < a < \pi/2$, $s \geqslant 0$.

If a lies in the interval $\pi/2 \leqslant a < \pi$, then $\cos a = \cos v(0)$ satisfies the inequality $-1 \leqslant \cos a < 0$, and this implies the result $\cos v(s) \neq 1$ for all $s \geqslant 0$. If not, then there is an $s_0 > 0$ such that $\cos v(s_0) = -1$ and $\cos v(s) > -1$ for $0 \leqslant s < s_0$. Again from (21.23), we obtain a contradiction, and hence we must have $-\pi < v(s, a) < \pi$ for $\pi/2 \leqslant a < \pi$ and $s \geqslant 0$.

But we can prove that $v(s, a)$ is oscillatory for any a in the interval $0 < a < \pi$. The function

$$V(s) = [1 - \cos v(s)] + \frac{1}{2}\frac{v'^2}{s}$$

satisfies the inequality

$$V'(s) = -\frac{1}{2}\left(\frac{v'}{s}\right)^2 \leqslant 0.$$

Then we have the results

$$1 - \cos v(s) \leqslant V(s) \leqslant V(0) = 1 - \cos a.$$

As $v(s)$ lies in the interval $-\pi < v(s) < \pi$, then $|v(s)| \leqslant a$ for $s \geqslant 0$. On observing now that (21.22) can be written as

$$v''(s) + s\left[\frac{\sin v(s)}{v(s)}\right]v(s) = 0, \tag{21.24}$$

we put

$$0 < \delta < \min_{0 \leqslant v \leqslant a}\left(\frac{\sin v}{v}\right)$$

and we consider the equation

$$v''(s) + s\delta v(s) = 0. \tag{21.25}$$

This equation is oscillatory for $s \geqslant 0$, and hence, by virtue of Sturm's comparison theorem for ordinary differential equations, the solution of (21.24) is also oscillatory.

In order to characterize the multiplicity of solutions of (21.21), we must consider the family of solutions of (21.22) for $0 \leqslant a \leqslant 2\pi$ and find the value of $v(K)$ corresponding to each a. The values of a for which the equation

$$v(K) = 0 \tag{21.26}$$

is satisfied, represent the possible solutions of our problem. If K is small the solution is unique, but as K becomes larger than $\lambda_1 \simeq 1.98635$, where λ_1 is the first eigenvalue of (21.25) written for $\delta = 1$, then the solution is no longer unique.

Another useful extension of the theory of the *elastica* is necessary when a cantilever beam is subjected to a concentrated load at its tip, the direction of which is not fixed, but can change with the rotation of the terminal section (Figure 21.3). Loads of this type are said to be *nonconservative*. A simple exact solution can be found for a simple beam clamped at its left end, and loaded at the other end by an initially vertical load R, which rotates through an angle $\beta\alpha$ as the tip A rotates through an angle α. This angle α is a consequence of the elastic deformation of the beam. Since β is a constant, we have implicitly assumed that the rotation of the load is proportional to the rotation of the terminal section (Nageswara and Venkateswara 1986).

On taking the origin of the curvilinear abscissa S along the deformed axis at A, the equilibrium equation, a modification of (20.3) becomes

$$B\frac{d^2\theta}{dS^2} + R\cos(\beta\alpha - \theta) = 0, \tag{21.27}$$

where θ is the slope of the strained axis at S, and B is the flexural rigidity. The boundary conditions are

$$\frac{d\theta}{dS}(0) = 0, \quad \theta(L) = 0. \tag{21.28}$$

Once we know the function $\theta = \theta(s)$, the coordinates x_1 and x_3 of the strained section at S are given by

$$x_1 = \int_S^L \sin\theta\, dS, \quad x_3 = \int_S^L \cos\theta\, dS. \tag{21.29}$$

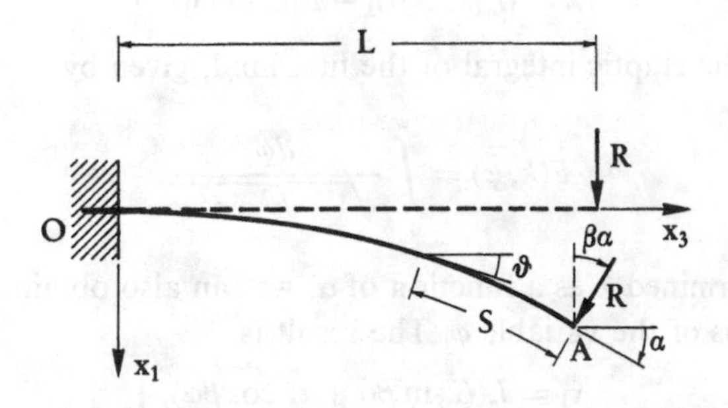

Fig. 21.3

In order to integrate (21.27), we put $\xi = S/L$, so that the field equation and boundary conditions assume the form

$$\theta'' + \lambda \cos(\beta\alpha - \theta) = 0, \tag{21.30a}$$

$$0 < \xi < 1, \tag{21.30b}$$

$$\theta'(0) = 0, \tag{21.30c}$$

$$\theta(1) = 0, \tag{21.30d}$$

$$x_1 = L \int_{\xi}^{1} \sin\theta \, d\xi, \tag{21.30e}$$

$$x_3 = L \int_{\xi}^{1} \cos\theta \, d\xi, \tag{21.30f}$$

where $\lambda = (RL^2)/B$. A first integral of the equation can be obtained by multiplying (21.30a) by θ' and integrating with respect to ξ:

$$\theta'^2 = \{2\lambda[\sin(\beta\alpha - \theta) - \sin(\beta\alpha - \alpha)]\}, \tag{21.31}$$

which satisfies the boundary conditions whenever the slope at the tip is α. From (21.31) we can obtain θ' but, since we expect θ' to be a decreasing function of θ, we take the solution

$$\theta' = -\{2\lambda[\sin(\beta\alpha - \theta) - \sin(\beta - \alpha)]\}^{\frac{1}{2}}. \tag{21.32}$$

On writing $\sin(\beta\alpha - \theta) = 1 - 2k^2 \sin^2\varphi$ and $\sin(\beta\alpha - \alpha) = 1 - 2k^2$, (21.32) is transformed into

$$\varphi' = -[\lambda(1 - k^2 \sin^2\varphi)]^{\frac{1}{2}}, \tag{21.33}$$

and the boundary conditions for φ become

$$\varphi(0) = \frac{\pi}{2}, \tag{21.34a}$$

$$\varphi(1) = \arcsin\left[\frac{1}{k} \sin\frac{(\pi - 2\beta\alpha)}{4}\right]. \tag{21.34b}$$

By integrating this equation together with the condition $\varphi(0) = \pi/2$, we obtain the connection between λ and the tip angle α:

$$\lambda = \{F[k, \varphi(0)] - F[k, \varphi(1)]\}^2, \tag{21.35}$$

where $F(k, \varphi)$ is the elliptic integral of the first kind, given by

$$F(k, \varphi) = \int_0^{\varphi} \frac{d\varphi}{\sqrt{1 - k^2 \sin^2\varphi}}.$$

After having determined λ as a function of α, we can also obtain the expression for x_1 and x_2 in terms of the variable φ. The result is

$$\left.\begin{aligned} x_1 &= L(G \sin\beta\alpha + H \cos\beta\alpha), \\ x_3 &= L(G \cos\beta\alpha - H \sin\beta\alpha), \end{aligned}\right\} \tag{21.36}$$

where $G(\varphi)$ and $H(\varphi)$ have the form

$$G = \frac{2k}{\sqrt{\lambda}}[\cos\varphi(1) - \cos\varphi]$$

$$H = 1 - \frac{2k}{\sqrt{\lambda}}\{E(k,\varphi) - E[k,\varphi(1)]\}$$

and

$$E(k,\varphi) = \int_0^{\varphi} (1 - k^2 \sin^2\varphi)^{\frac{1}{2}}\, d\varphi$$

is the elliptic integral of the second kind, with φ in the interval $\varphi(1) \leqslant \varphi \leqslant \pi/2$. The deflection of the terminal section can be obtained from (21.36) by putting $\varphi = \pi/2$.

If in (21.35) we put $\beta = 0$, we recover the solution for a cantilevered beam loaded at its tip by a load always directed vertically;[34] alternatively, the case $\beta = 1$ corresponds to a load always acting perpendicular to the tangent at the tip. The case $\alpha\beta = \alpha + (\pi/2)$ requires special comment. It corresponds to a beam loaded by a *follower force*, which is horizontal and compressive in the undeformed rectilinear configuration, but acts in the direction of the tangent to the deflection curve at the tip of the bar, as soon as this moves from the straight configuration. For $\alpha\beta = \alpha + (\pi/2)$, the parameter k must necessarily be zero, and the expression for $\varphi(1)$, as given by (21.34b), loses its meaning. Since the solution $\theta \equiv 0$ is an equilibrium state, other bent configurations arise because the fundamental state is not stable. The natural procedure for characterizing the stability of the elastic system is that of studying the equilibrium of configurations that are slightly bent with respect to the rectilinear state. However, as first pointed out by Nikolai (1928, 1930) and later by Beck (1952), the linearized criterion of stability is incapable of determining the critical value of the axial load, so generating a paradox in the classical theory of elastic stability. The explanation of this paradox rests on the fact that, in the presence of nonconservative forces, the static criterion of stability may prove inadequate, and instead we must use a dynamic criterion, which takes account of the circumstance that the passage from the fundamental configuration to a nearby bent configuration is not slow but sudden and, as a consequence, the inertial terms cannot be neglected.

The approximate methods for evaluating the deflection of a Kirchhoff beam subject to a distributed load are commonly based on the device of representing the slope θ by means of trigonometric functions. However, the degree of approximation is much more accurate if we apply an expansion of the solution in terms of Tchebychev polynomials (Schmidt and Da Peppo 1971).

The equation of an initially horizontal beam of length L, of constant flexural rigidity B, loaded by a vertical force P at $S = 0$ and by a uniform load w per unit length is

[34] Already solved by Bisshopp and Drucker (1945).

$$B\frac{d^2\theta}{dS^2}+(P+wS)\cos\theta=0, \quad 0\leqslant\theta\leqslant 2\alpha, \tag{21.37}$$

where $\theta(S)$ is the slope and 2α is the slope at $S=0$. We assume that the end $S=L$ is clamped.

Let us change the asymmetric interval of variation of θ into a symmetric one by putting

$$\theta=\varphi+\alpha,$$

and rewrite (21.37) in the form

$$B\frac{d^2\varphi}{dS^2}+(P+wS)(\cos\alpha\cos\varphi-\sin\alpha\sin\varphi)=0, \quad -\alpha\leqslant\varphi\leqslant\alpha. \tag{21.38}$$

We now use the following expansions (Denman and Schmidt 1968; Schmidt and Da Peppo 1971):

$$\sin\varphi=2J_1(\alpha)T_1\left(\frac{\varphi}{\alpha}\right)-2J_3(\alpha)T_3\left(\frac{\varphi}{\alpha}\right)+\ldots, \tag{21.39}$$

$$\cos\varphi=J_0(\alpha)-2J_2(\alpha)T_2\left(\frac{\varphi}{\alpha}\right)+\ldots, \tag{21.40}$$

where $J_i(\alpha)$ is an ordinary Bessel function and $T_i(\varphi/\alpha)$ is the Tchebychev polynomial of ith order, that is $T_0=1$, $T_1=\varphi/\alpha$, $T_2=2(\varphi/\alpha)^2-1,\ldots$. On substituting (21.39) and (21.40) into (21.38), and retaining only the terms of zeroth and first order, we obtain the linear equation

$$\frac{d^2\varphi}{dS^2}-\frac{2J_1(\alpha)\sin\alpha}{\alpha B}(P+wS)\varphi=-\frac{J_0(\alpha)\cos\alpha}{B}(P+wS). \tag{21.41}$$

On integrating this equation and using the boundary conditions $(d\theta/dS)(0)=0$, $\varphi(0)=\alpha$, and $\varphi(L)=-\alpha$, we determine the approximate solution.

22. Nonclassical Problems for Rods

In the classical theory of the *elastica*, external loads are always regarded as given quantities, as are the boundary conditions, of either geometric or mechanical type. However, in some cases, the quantities customarily assumed as the data of the problem are also unknown and must be determined as part of the solution.

A problem of this kind arises in contact mechanics when we wish to find the equilibrium configuration of a rod initially having the shape of an arc of a ring clamped at one end and impinging on a rigid barrier at the other (Figure 22.1). The initial length of the circular arc (dashed line in Figure 22.1) is L, and the arc AB passes through the sliding clamp A, smoothly touching the rigid barrier at the point B. Let us now suppose that an additional length ΔL is forced through the clamp at A, but this motion is hindered by the reaction exerted by the rigid barrier against the other end B. The barrier, which for simplicity is considered vertical, is not perfectly smooth, but offers a small reaction ΔF inclined at an angle ψ, which has the effect of holding the edge of the ring in place. Whether the ring sticks or slips depends on the value of the angle of friction $\tan^{-1}(\mu)$ between the end B and the

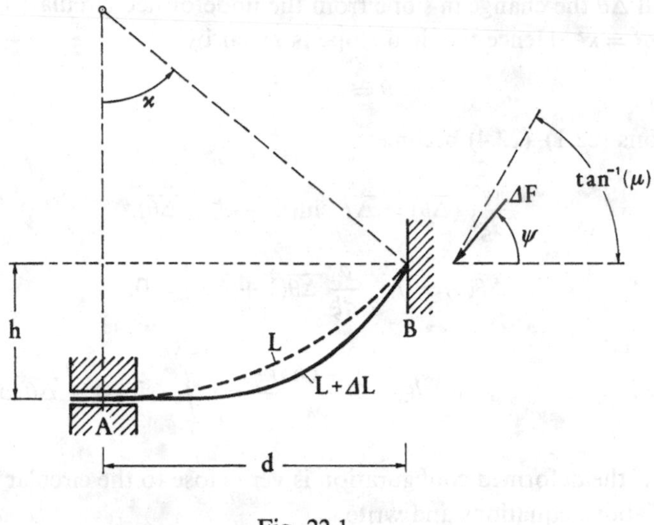

Fig. 22.1

vertical wall. If $\psi > \tan^{-1}(\mu)$, the point B will slide up; if $\psi \leqslant \tan^{-1}(\mu)$, the point will stick. An analysis of the type of contact at B is useful in some practical applications, such as in the making of plain paper copies, in self-threading tapes, and in sewing machines.

A simple solution can be obtained by means of perturbation procedure (Benson 1981). By neglecting gravity and transient effects, the differential equation and the boundary conditions for the deformed beam are

$$B\,\frac{d^2\theta}{dS^2} = \Delta F \sin(\psi - \theta) \tag{22.1}$$

$$\theta(0) = 0, \quad \frac{d\theta}{dS}(L + \Delta L) = \frac{1}{R}, \tag{22.2}$$

where S is the distance along the deformed axis measured from the partial clamp, and $\theta(S)$ is the slope with respect to the horizontal. Before and after the deformation the coordinates of the beam tip remain unchanged:

$$d = R\sin\left(\frac{L}{R}\right) = \int_0^{L+\Delta L} \cos\theta\, dS, \tag{22.3}$$

$$h = R - R\cos\left(\frac{L}{R}\right) = \int_0^{L+\Delta L} \sin\theta\, dS. \tag{22.4}$$

Let us now introduce the dimensionless quantities

$$\xi = \frac{S}{L}, \quad \kappa = \frac{L}{R}, \quad \overline{\Delta L} = \frac{\Delta L}{L}, \quad \overline{\Delta F} = \frac{\Delta F}{(B/L^2)}, \tag{22.5}$$

and let us call $\overline{\Delta\theta}$ the change in slope from the undeformed circular state, which has the value $S/R = \kappa\xi$. Hence the final slope is given by

$$\theta = \kappa\xi + \overline{\Delta\theta}. \tag{22.6}$$

Thus equations (22.1)–(22.4) become:

$$\frac{d^2}{d\xi^2}(\overline{\Delta\theta}) = \overline{\Delta F}\sin(\psi - \kappa\xi - \overline{\Delta\theta}), \tag{22.7}$$

$$\overline{\Delta\theta}(0) = 0, \quad \frac{d}{d\xi}\overline{\Delta\theta}(1 + \overline{\Delta L}) = 0, \tag{22.8}$$

$$\frac{\sin\kappa}{\kappa} = \int_0^{1+\overline{\Delta L}} \cos(\kappa\xi + \overline{\Delta\theta})\,d\xi, \quad \frac{1 - \cos\kappa\xi}{\kappa} = \int_0^{1+\overline{\Delta L}} \sin(\kappa\xi + \overline{\Delta\theta})\,d\xi. \tag{22.9}$$

However, if the deformed configuration is very close to the circular form, we may linearize the above equations and write

$$\frac{d^2}{d\xi^2}(\overline{\Delta\theta}) = \overline{\Delta F}\sin(\psi - \kappa\xi), \tag{22.10}$$

$$\frac{\sin\kappa}{\kappa} = \int_0^{1+\overline{\Delta L}} \cos(\kappa\xi)\,d\xi - \int_0^1 \overline{\Delta L}\sin(\kappa\xi)\,d\xi, \tag{22.11}$$

$$\frac{1 - \cos\kappa\xi}{\kappa} = \int_0^{1+\overline{\Delta L}} \sin(\kappa\xi)\,d\xi + \int_0^1 \overline{\Delta\theta}\cos(\kappa\xi)\,d\xi. \tag{22.12}$$

On integrating the first term on the right-hand side of the last two equations and retaining only the terms of first order in $\overline{\Delta L}$, we obtain

$$\overline{\Delta L}\cos\kappa = \int_0^1 \overline{\Delta\theta}\sin(\kappa\xi)\,d\xi,$$

$$-\overline{\Delta L}\sin\kappa = \int_0^1 \overline{\Delta\theta}\cos(\kappa\xi)\,d\xi.$$

Thus, by eliminating $\overline{\Delta L}$ from these equations we have

$$0 = \int_0^1 \overline{\Delta\theta}\cos(\kappa - \kappa\xi)\,d\xi. \tag{22.13}$$

Next, (22.10) may be integrated, making use of its boundary conditions, to give

$$\overline{\Delta\theta} = \frac{\overline{\Delta F}}{\kappa^2}[\sin\psi - \sin(\psi - \kappa\xi) - \kappa\xi\cos(\psi - \kappa)].$$

It thus follows that, by substituting this value of $\overline{\Delta\theta}$ in (22.13) and integrating the resulting expression with respect to ξ, we arrive at the equation

$$\sin\psi(2\sin\kappa\cos\kappa - \kappa\cos\kappa - \sin\kappa) + \cos\psi(2\cos^2\kappa + \kappa\sin\kappa - 2\cos\kappa) = 0,$$

which can be solved with respect to ψ to give

$$\psi = \tan^{-1}\left[\frac{\kappa \sin \kappa - 2\cos \kappa(1 - \cos \kappa)}{\cos \kappa + \sin \kappa(1 - 2\cos \kappa)}\right]. \tag{22.14}$$

In this manner we have found the reaction angle of the beam in terms of the dimensionless curvature κ. Under the condition $\psi < \tan^{-1}(\mu)$ the beam sticks at B, otherwise it slips under a small advancement $\overline{\Delta L}$ at the left-hand end. We also observe that, for small κ, (22.14) can be replaced by its Taylor expansion

$$\psi \simeq \tfrac{5}{8}\kappa + \ldots, \tag{22.15}$$

which implies that ψ is a linear function of κ, for small κ.

A problem which is, in some respects, the converse of the one just described is represented by the peeling of tough adhesive tape from a plane surface to which the tape was initially glued. This is a typical example of a free boundary problem in mechanics, since the boundary of the tape is not prescribed in advance, but must be found as part of the solution. Problems of this kind are encountered in fracture mechanics, when we must determine the extension of a rectilinear crack, or in seismology, when we must appreciate the boundary of a fault plane over which slipping occurs. The surprising feature is that these different problems, if restricted to a one-dimensional version, require the same mathematical treatment (Sawyers and Rivlin 1974; Burridge and Keller 1978).

We shall consider an *elastica* formed during the process of peeling a tape from a horizontal plane under the effect of a tension T_0 applied at a point of the tape at a very large distance from the point of detachment (Figure 22.2). Let S denote the arc length along the tape measured from a point $S = 0$, and let us assume that we are considering the process of peeling at a given instant t, at which $S = \sigma(t)$ represents the point of separation between the part of the tape already detached and that still glued to the plane surface. If $\mathbf{r}(S, t)$ is the position of the particle S, then $\mathbf{r}_{,S} = \mathbf{d}_3$ is the unit tangent vector to the axis, and $r_{,SS}$ is related to the curvature κ' by the Frenet–Serret formula $\mathbf{r}_{,SS} = \kappa'\mathbf{d}_1$, $\mathbf{d}_1$ being the unit normal directed to the left as one advances in the direction of $\mathbf{d}_3$. The bending moment is of magnitude $B\kappa'$ in the

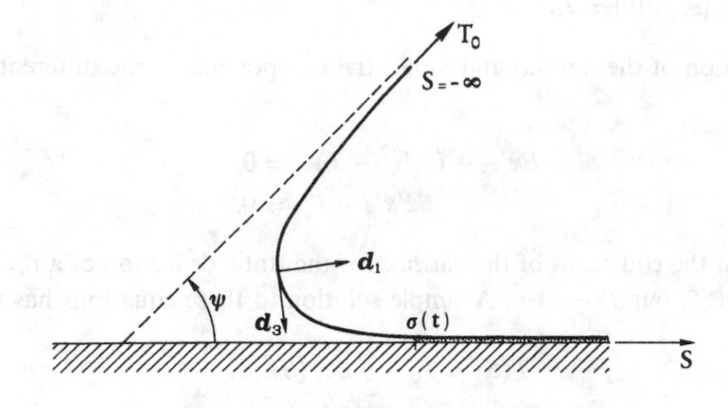

Fig. 22.2

direction $\mathbf{d}_3 \times \mathbf{d}_1$ and the shear force is $-B\kappa_{,S}\mathbf{d}_1$. Since there is a tension $T(S, t)$ along the axis, the total force acting at a section at S is

$$\mathbf{n} = T\mathbf{d}_3 - B\kappa'_{,S}\mathbf{d}_1. \tag{22.16}$$

Let ρ be the density of the material and A the area of the cross-section. The equation of motion of the part $S < \sigma(t)$ of the tape is

$$\rho A\mathbf{r}_{,tt} = (\mathbf{T}\mathbf{d}_3 - B\kappa'_{,S}\mathbf{d}_1)_{,S}. \tag{22.17}$$

At the point of separation $S = \sigma(t)$ we have the conditions

$$\mathbf{r}(\sigma(t), t) = (\sigma(t), 0) \tag{22.18}$$

$$\mathbf{r}_{,S}(\sigma(t), t) = (1, 0). \tag{22.19}$$

In addition, we have at $\sigma(t)$ a condition of rupture, because the detachment of the beam at $\sigma(t)$ occurs just when the bending energy $\frac{1}{2}B\kappa'^2$ per unit length is equal to the energy γ per unit length required to create a local fracture. On putting $K = (2\gamma/B)^{\frac{1}{2}}$ this condition becomes

$$\kappa' = |\mathbf{r}_{,SS}(\sigma(t), t)| = K. \tag{22.20}$$

It is clear that the problem has more boundary conditions than necessary, but in the present case $\sigma(t)$ is also an unknown in addition to $\mathbf{r}$. In order to find $\mathbf{r}(S, t)$ and $\sigma(t)$ we seek a solution of the form

$$\mathbf{r}(S, t) = \mathbf{R}(S') + (vt, 0), \quad S' = S - vt, \quad \sigma(t) = vt, \tag{22.21}$$

which has the property of being stationary if viewed from the point of separation.

We now substitute (22.21) in (22.17) and in the boundary conditions (22.18)–(22.20), noting that $\mathbf{r}_{,tt} = v^2\mathbf{R}_{,SS} = v^2\kappa'\mathbf{d}_1$ and $(d/ds)\mathbf{d}_1 = -\kappa'\mathbf{d}_3$, according to (14.7) with $\tau = 0$. We then drop the primes and write (22.17) in the form

$$\rho Av^2\kappa'\mathbf{d}_1 = (B\kappa'\kappa'_{,S} + T_{,S})\mathbf{d}_3 + (T\kappa' - B\kappa'_{,SS})\mathbf{d}_1, \quad S < 0, \tag{22.22}$$

$$\mathbf{R}(0) = (0, 0), \tag{22.23a}$$

$$\mathbf{R}_{,S}(0) = (1, 0), \tag{22.23b}$$

$$|\mathbf{R}_{,SS}(0)| = K. \tag{22.23c}$$

After separation of the normal and tangential components in the differential equation, we get

$$B\kappa'_{,SS} + (\rho Av^2 - T)\kappa' = 0,$$

$$B\kappa'\kappa'_{,S} + T_{,S} = 0.$$

These are just the equations of the *elastica* for the static deflection of a rod when its tension is not T, but $T - \rho Av^2$. A simple solution to these equations has the form

$$T(S) = T_0 - \frac{1}{2}B\kappa'^2(S), \tag{22.24}$$

$$\kappa'(S) = b\,\mathrm{sech}\,\frac{b(S - S_0)}{2}, \tag{22.25}$$

where T_0 is a constant representing the tension when $\kappa' = 0$, and b is given by

$$b^2 = \frac{4}{B}(T_0 - \rho A v^2). \tag{22.26}$$

To find $\theta(S)$, the angle between $\mathbf{d}_3$ and the horizontal axis, we integrate the equation $(d\theta/dS) = \kappa'$, with $\theta(0) = 0$, and use (22.25) for κ'. This yields

$$\theta(S) = 4\tan^{-1}\left[\tanh\frac{b(S - S_0)}{4}\right] + 4\tan^{-1}\left(\tanh\frac{bS_0}{4}\right). \tag{22.27}$$

In addition, we can determine $\mathbf{R}(S)$ in terms of $\theta(S)$:

$$\mathbf{R}(S) = \int_0^S (\cos\theta(\xi), \sin\theta(\xi))\, d\xi, \quad S < 0, \tag{22.28}$$

which is adapted so that the first two boundary conditions (22.23) are satisfied. By using (22.28) in (22.23c), we get

$$\left|\frac{d\theta}{dS}(0)\right| = |\kappa'| = K.$$

Thus from (22.25) we obtain

$$b\,\mathrm{sech}\,\frac{b(S - S_0)}{2} = K, \tag{22.29}$$

which gives S_0 and completes the solution.

From (22.27) we find that, as S tends to $-\infty$, $\theta(S)$ tends to $\psi - \pi$, where ψ is given by

$$\psi = 4\tan^{-1}\left[\tanh\left(\frac{bS_0}{4}\right)\right].$$

This means that the tape is asymptotic to the direction ψ. To express v in terms of ψ, we note that from the above equation we can derive

$$\tanh\left(\frac{bS_0}{4}\right) = \tan\left(\frac{\psi}{4}\right),$$

then $\mathrm{sech}\,(bS_0/2)$ can be written as follows:

$$\mathrm{sech}\left(\frac{bS_0}{2}\right) = \frac{1 - \tanh^2\left(\frac{bS_0}{2}\right)}{1 + \tanh^2\left(\frac{bS_0}{2}\right)} = \frac{1 - \tan^2\left(\frac{\psi}{4}\right)}{1 + \tan^2\left(\frac{\psi}{4}\right)} = \cos\left(\frac{\psi}{2}\right). \tag{22.30}$$

Thus by squaring (22.29) and using (22.26), (22.30), and the condition of detachment $|\kappa'| = K$, we get

$$\rho\frac{Av^2}{T_0} = 1 - \frac{\gamma}{T_0}(1 + \cos\psi)^{-1}. \tag{22.31}$$

This relation determines the velocity v of the point of separation tip in terms of ψ and the other data. The formula shows that there is a solution only if

$T_0 > \gamma(1+\cos\psi)^{-1}$, and the velocity is greatest when $\psi = 0$, in which case the tape is being pulled forward parallel to the horizontal axis, that is to say, through an angle of 180° from its original direction. Furthermore, v is zero when ψ assumes the value ψ_0 such that

$$\cos\psi_0 = \frac{\gamma}{T_0} - 1.$$

There is no solution with v real if $\pi \geqslant \psi > \psi_0$.

This kind of argument may be applied to study the interesting problem of the longitudinal cleavage of a beam when a longitudinal crack, initially created at one end, is gradually extended along the axis of the beam by two equal and opposite transverse forces F pulling apart the two halves of the terminal section (Figure 22.3). Sometimes, the same effect of the propagation of a crack may be obtained by prescribing the displacement between the two halves of the terminal section. This problem arises when woodmen split a piece of wood longitudinally. The opposing forces F are supplied practically through the insertion of a rigid wedge.

Suppose that an initially rectilinear beam is split from $x = 0$ to $x = \sigma_0$ and its two ends are pulled apart gradually. At first the crack length will remain constant while the two sides of the crack separate, but after the curvature at the crack tip has become large enough the crack starts to elongate. We wish to determine the law of dependence between the prescribed forces F and the displacements U and establish whether a small increase in each of these parameters will result in a small increase in the length of the crack, or result in a catastrophically fast propagation (Burridge and Keller 1978).

We suppose that the end of the beam at $x = 0$ is displaced by an amount U and that the crack length is σ. Since we are interested in predicting the response of specimens made of a slightly deformable material, and in the phase preceding the catastrophic rupture, we apply the static linear beam theory. On denoting the transverse displacement of the upper half-beam by $u(x)$, elementary beam theory permits us to determine $u(x)$ as the solution of the equations

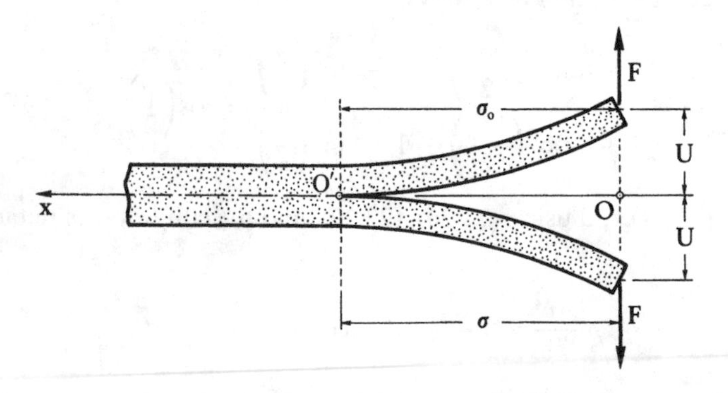

Fig. 22.3

$$u_{,xxxx} = 0, \quad 0 < x < \sigma, \tag{22.32}$$

$$u(0) = U, \quad u_{,xx}(0) = 0, \tag{22.33}$$

$$u(\sigma) = 0, \quad u_{,x}(\sigma) = 0. \tag{22.34}$$

In addition, we require that the curvature at the tip does not exceed a given value K:

$$u_{,xx}(\sigma) \leqslant K. \tag{22.35}$$

Then the solution of (22.32) with its boundary conditions (22.33) and (22.34) is

$$u(x) = \frac{3U}{\sigma^2}\left[\frac{1}{2}(\sigma - x)^2 - \frac{1}{6\sigma}(\sigma - x)^3\right], \quad 0 \leqslant x \leqslant \sigma, \tag{22.36}$$

from which we calculate

$$u_{,xx} = \frac{3U}{\sigma^2}. \tag{22.37}$$

We have thus found the distribution of the transverse displacement $u(x)$ between the end at which a displacement U is prescribed and the tip of separation. In practice, however, the process is slightly different. The tape is initially separated along an interval of extent σ_0, starting from the end $x = 0$, and increasing displacement U is impressed at this end. As long as we have the inequality

$$u_{,xx}(\sigma_0) = \frac{3U}{\sigma_0^2} \leqslant K,$$

the fracture condition at the tip is not violated, and therefore the extent of the separated portion remains equal to σ_0. When U exceeds the value $U_0 = (K\sigma_0^2/3)$, the separation begins to propagate, and the instantaneous value of σ is obtained by setting $3U/\sigma^2 = K$. This yields

$$\sigma = \begin{cases} \sigma_0 & \text{if } U \leqslant \dfrac{K\sigma_0^2}{3}, \\ \left(\dfrac{3U}{K}\right)^{\frac{1}{2}} & \text{if } U > \dfrac{K\sigma_0^2}{3}, \end{cases} \tag{22.38}$$

and the graph of σ as a function of U, shown in Figure 22.4(a), shows that σ is a continuous increasing function of U.

The force required to maintain the displacement U is $F = Bu_{,xxx}(0)$, and can be calculated from (22.36) and (22.38). It is given by

$$F = \begin{cases} \dfrac{3UB}{\sigma_0^3} & \text{if } U \leqslant \dfrac{K\sigma_0^2}{3}, \\ \left(\dfrac{K^3}{3U}\right)^{\frac{1}{2}} B & \text{if } U \geqslant \dfrac{K\sigma_0^2}{3}, \end{cases} \tag{22.39}$$

and the graph of (22.39) is shown in Figure 22.4(b).

We can thus see that the prescription of a displacement U or a force F have radically different effects on the propagation of the crack. If the force F is prescribed at a value greater than $F_0 = KB/\sigma_0$, which represents the maximum of the function

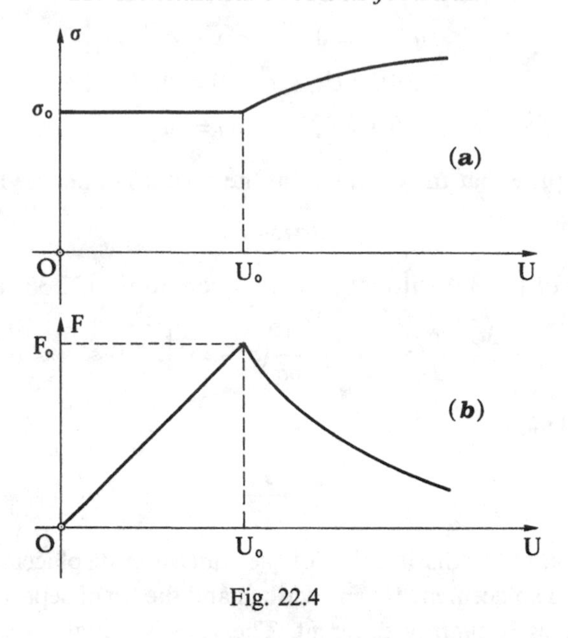

Fig. 22.4

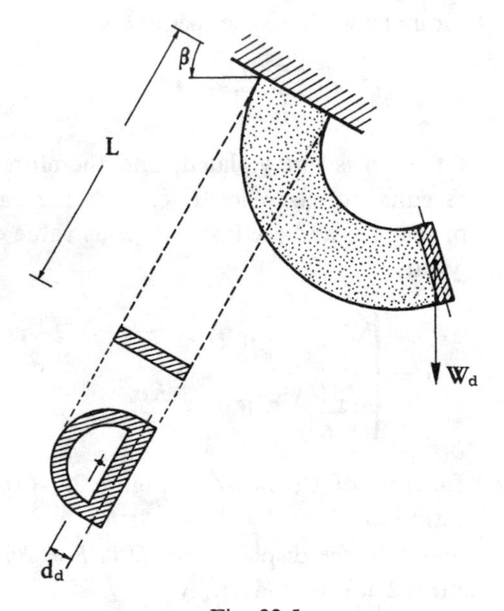

Fig. 22.5

(22.39), there is no static solution, and catastrophic failure will ensue. However, if U is prescribed, there is always a static solution.

The theory of the *elastica* may also be applied to explain the apparently anomalous behavior of a high-flexure manipulator tube element, the cross-section of which is a semi-circular ring composed of a thin corrugated bellows for the curved part, and a flat reinforced strip along the rectilinear side (Figure 22.5). Rigid caps are provided at each end. The tube is pressurized from the inside, and the pressure load is applied at a point of the cross-section situated on the axis of symmetry at a distance d_d from the neutral axis, which is located in the flat strip parallel to the rectilinear edge. When the tube is pressurized, the noncoincidence of the center of the pressure and the centroid of the cross-section causes a deflection of the, initially straight, central line. The resulting pressure rotates to follow the tangent to the deformed axis. By regulating the internal pressure it is possible to lift a payload W_d applied at the movable end. The manipulator arm has many practical applications, especially in medicine.

To study this problem mathematically (Wilson and Snyder 1988), we regard the tube as a slender cantilever beam with an end load composed of P_d, Q_d, and M_d as shown in Figure 22.6; the subscript "d" denotes quantities that have physical dimensions. We fix a system of Cartesian axes x and y with their origin at the clamped end of the beam; the x axis is placed along the unstrained longitudinal axis, and the y axis is perpendicular to it. P_d and Q_d are the components of the end load along the reference axes.

On adopting the customary assumptions of the theory of the *elastica*, the differential equation for the inclination θ is

$$Bk = B\frac{d\theta}{ds_d} = P_d(Y_d - y_d) + Q_d(X_d - x_d) + M_d \tag{22.40}$$

where s_d is the arc length evaluated from the origin, x_d and y_d are the coordinates of the section at s_d, and X_d and Y_d are the coordinates of the terminal section at $s_d = L$. The boundary condition is $\theta(0) = 0$.

In order to solve (22.40), we introduce dimensionless parameters

$$s = \frac{s_d}{L}, \quad x = \frac{x_d}{L}, \quad y = \frac{y_d}{L}, \quad X = \frac{X_d}{L}, \quad Y = \frac{Y_d}{L},$$

$$P = \frac{P_d L^2}{B}, \quad Q = \frac{Q_d L^2}{B}, \quad M = \frac{M_d L}{B},$$

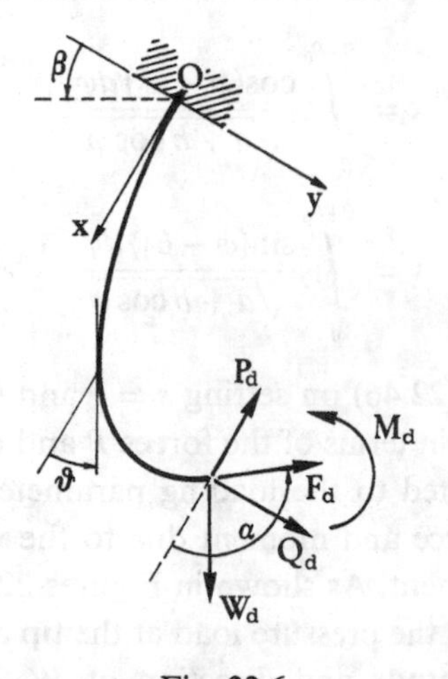

Fig. 22.6

and write the equation in the nondimensional form

$$\frac{d\theta}{ds} = P(Y - y) + Q(X - x) + M, \tag{22.41}$$

with the same boundary condition $\theta(0) = 0$.

On differentiating (22.41) with respect to s and using the relations

$$\frac{dx}{ds} = \cos\theta, \quad \frac{dy}{ds} = \sin\theta,$$

(22.41) becomes

$$\frac{d^2\theta}{ds^2} = -P\sin\theta - Q\cos\theta. \tag{22.42}$$

We must now add the boundary condition

$$\frac{d\theta}{ds}(1) = M, \tag{22.43}$$

because we know the bending moment at $s = 1$.

The first integral of (22.42) is

$$\frac{1}{2}\left(\frac{d\theta}{ds}\right)^2 = \frac{1}{2}M^2 + P(\cos\theta - \cos\alpha) + Q(\sin\alpha - \sin\theta), \tag{22.44}$$

where α is the slope of the elastic curve at $s = 1$. If we now make the substitutions

$$a = M^2 - b\cos(\alpha + \theta_0), \quad b = 2\sqrt{P^2 + Q^2}, \tag{22.45}$$

with $\theta_0 \arctan(Q/P)$ for $-\pi < \theta_0 \leqslant \pi$, we can take the positive square root of the right-hand side of (22.44) and integrate. This yields

$$s = \int_{\theta_0}^{\theta+\theta_0} \frac{d\phi}{\sqrt{a + b\cos\phi}} \tag{22.46}$$

where $\phi = \theta + \theta_0$. The nondimensional x, y coordinates at s are

$$\begin{aligned} x &= \int_{\theta_0}^{\theta+\theta_0} \frac{\cos(\varphi - \theta_0)\,d\varphi}{\sqrt{a + b\cos\varphi}}, \\ y &= \int_{\theta_0}^{\theta+\theta_0} \frac{\sin(\varphi - \theta_0)\,d\varphi}{\sqrt{a + b\cos\varphi}}, \end{aligned} \tag{22.47}$$

and the angle α is given by (22.46) on setting $s = 1$ and $\theta = \alpha$.

We thus have the solution in terms of the forces P and Q and the couple M. These quantities must now be related to the loading parameters of our problem. Let us define W_d and N_d as the force and moment due to the weight, respectively, of the payload at the tip of the element. As shown in Figures 22.5 and 22.6, gravity acts at an angle β to the x axis. F_d is the pressure load at the tip acting at a distance d_d from the neutral axis. The magnitude and direction of W_d remain unchanged as the element rotates, while F_d follows the slope at the tip. The couple N_d varies with both α and the configuration of the payload. On introducing the nondimensional parameters

$$F = \frac{F_d L^2}{B}, \quad W = \frac{W_d L^2}{B}, \quad N = \frac{N_d L}{B}, \quad d = \frac{d_d}{L}, \tag{22.48}$$

the relations between these nondimensional loads and the nondimensional loads given by equations (22.45) are

$$P = -F\cos\alpha - W\cos\beta, \quad Q = F\sin\alpha + W\sin\beta,$$
$$M = N + Fd. \tag{22.49}$$

The solution can be computed by fixing W and β and letting F vary.[35] The slope α of the tip increases gradually as F increases, as does the curvature of the element. However, shortly after the tip rotates past 90°, the vertical component of F opposes W so that a relatively small increment of pressure load causes a "flip-over," or a large change in the shape of the *elastica*.

Another, apparently abnormal, phenomenon that can be explained using the theory of the *elastica* is the kinking of submarine cables when the initial tension is temporarily reduced (Yabuta, Yoshizawa, and Kojima 1982). It has frequently been noted that loops occur in oceanic cables, which are perfectly straight under a given tensile force, as soon as the tension is relaxed, and when the cable is retensioned the loops decrease in diameter, kink, and may cause damage. The explanation of this kinking is relatively simple if we assume that a state of initial twist accompanied by tension preexists in the cable. The twist is due to the fact that the wires composing the cable are helically stranded so that torsional strains are induced under tension. Then, if the tensile load is decreased, perhaps as a result of wave motion, a reversed torsional load is induced, which may also cause loops. This results from a transfer of torsional strain energy to bending strain energy. However, when the cable is retensioned, the loop gradually decreases in diameter, and eventually reopens, transferring back the bending strain energy to torsional strain energy. The initial straight twisted stated is shown in Figure 22.7(a), the cable with a loop is shown in Figure 22.7(b), and in Figure 22.7(c) the loop has become very small under increasing load and may eventually disappear.

To study the problem, we neglect the cable weight and the elongation of the cable, and assume that the cable is linearly elastic, with a bending stiffness B and a torsional rigidity C. The cable loop is regarded as a closed curve branching from the line of action of the tensile force, as represented in Figure 22.7(b). The points of the central line of the loop are identified by Cartesian coordinates x and y with respect to a system having its origin at the center and the x axis parallel to the line of action of P. Alternatively, polar coordinates r and θ can be specified, so that the loop crossing point is at $x = 0$, $\theta = -(\pi/2)$. If the loop is perfectly planar, the strain energy is purely flexural. However, if the loop leaves the xy plane and becomes displaced so as to form a cylindrical helix with the z axis perpendicular to the xy plane, part of the

[35] Wilson and Snyder (1988) described the deformation of the *elastica* numerically, but it would be interesting to treat the problem in terms of bifurcation theory.

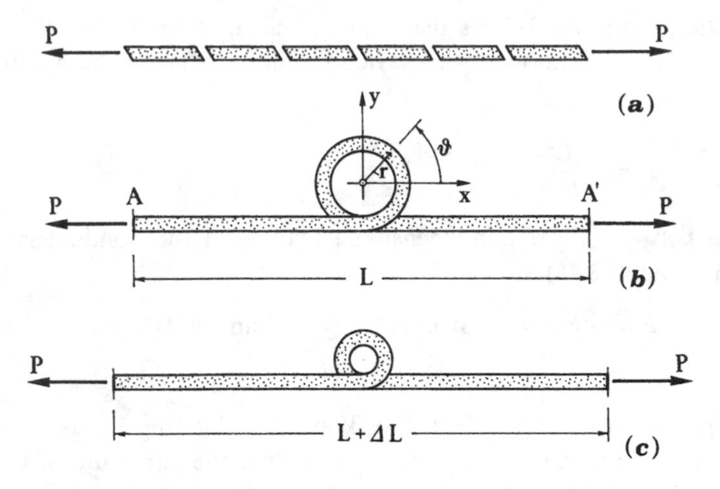

Fig. 22.7

strain energy is transferred to torsional strain energy. In this configuration, the coordinates of the central line of the loop have the expressions

$$x = r\cos\theta, \quad y = r\sin\theta, \quad z = z. \tag{22.50}$$

The curvature κ and torsion φ can be calculated from (22.50) according to the formula for the differential geometry (Eisenhart 1947)

$$\begin{aligned}\kappa^2 = {}& [(r^2 + r'^2 + z'^2)(r''^2 + 4r'^2 + r^2 - 2rr'' + z''^2) - (r'r'' + rr' + z'z'')^2] \\ & \times [r^2 + r'^2 + z'^2]^{-3},\end{aligned} \tag{22.51}$$

$$\begin{aligned}\varphi = {}& [z'(r^2 - 4rr'' + 3r''^2 - 2r'r'' + 6r'^2) + z''(4rr' + rr''' - 3rr'') \\ & + z'''(r^2 + rr'' + 2r'^2)] \times [r^4 + 4r^2r'^2 - 2r^3r'' + r^2r'^2 + r^2(z''^2 + z'^2)]^{-1},\end{aligned} \tag{22.52}$$

where the primes denote differentiation with respect to θ. The total twist of the cable is given by

$$\tau = \tau_0 + \varphi, \tag{22.53}$$

where τ_0 indicates the residual twist of the straight configuration which cannot be converted to bending energy of the loop.

In the intermediate helical configuration we thus have the bending strain energy W_B and the torsional energy W_T:

$$W_B = \frac{1}{2}\int_0^L B\kappa^2\, ds, \quad W_T = \frac{1}{2}\int_0^L C\tau^2\, ds, \tag{22.54}$$

where L is the length of cable under consideration. The mutual displacement ΔL between the ends A and A$'$ of the cable when it passes from the configuration shown in Figure 22.7(a) to the one shown in Figure 22.7(b) is the perimeter of the loop. As a consequence, the work done by the external forces has the value

$$\mathscr{L} = P\,\Delta L, \tag{22.55}$$

provided that we assume P to remain constant during the formation of the loop. We observe that this work is negative when the cable passes from state (a) to state (b), and positive when it passes from state (b) to state (c) in Figure 22.7.

The total potential energy of the piece of cable of length L is thus given by

$$E = W_{\mathrm{B}} + W_{\mathrm{T}} - \mathscr{L}, \tag{22.56}$$

and the stability of a loop is determined by the critical points of this energy.

The analysis of stability is complicated by the fact that $r = r(\theta)$, the equation of the loop, is not known, but needs to be determined. However, progress could be made if the form of the loop were known, and so we tentatively assume that it takes the shape of an arc of a circular helix having the parametric equations

$$r = R = \text{constant} \begin{cases} z = R\theta\delta, & 0 \leqslant \theta \leqslant \pi, \\ z = 0, & -\dfrac{\pi}{2} \leqslant \theta \leqslant 0, \\ z = R\pi\delta, & \pi \leqslant \theta \leqslant \frac{3}{2}\pi, \end{cases} \tag{22.57}$$

where δ, the displacement in the z direction, is assumed to be very small. Outside the loop, the shape of the cable is regarded as perfectly straight. The curvature κ is then easily derived from (22.51) and (22.57) as

$$\begin{aligned} &\kappa^2 = \frac{1}{R^2}\frac{1}{(1+\delta^2)^2}, \quad 0 < \theta < \pi, \\ &\kappa^2 = \frac{1}{R^2}, \quad -\frac{\pi}{2} \leqslant \theta < 0, \quad \pi < \theta \leqslant \frac{3}{2}\pi. \end{aligned} \tag{22.58}$$

The torsion φ, which is induced by a partial reopening of the loop, is also given by (22.51) and (22.57):

$$\begin{aligned} &\varphi = \frac{1}{R}\frac{\delta}{1+\delta^2}, \quad 0 \leqslant \theta \leqslant \pi, \\ &\varphi = 0, \quad \theta < 0, \quad \pi < \theta. \end{aligned} \tag{22.59}$$

The arc element has the expression

$$ds = d\theta(R^2 + z'^2)^{\frac{1}{2}} = \begin{cases} R\,d\theta(1+\delta^2)^{\frac{1}{2}}, & 0 \leqslant \theta \leqslant \pi, \\ R\,d\theta, & \theta < 0, \quad \pi < \theta. \end{cases}$$

The bending strain energy and the torsional strain energy are evaluated by putting the curvature equal to zero in the straight part of the cable and assuming the torsion to be τ_0 everywhere except along the arc $0 \leqslant \theta \leqslant \pi$ of the circular loop:

$$W_{\mathrm{B}} = \frac{1}{2}\frac{\pi B}{R}(1+\delta^2)^{-\frac{3}{2}} + \frac{1}{2}\pi\frac{B}{R}, \tag{22.60}$$

$$W_{\mathrm{T}} = \frac{1}{2}\pi C R\left[\frac{1}{R}\frac{\delta}{(1+\delta^2)} + \tau_0\right]^2 (1+\delta^2)^{\frac{1}{2}} + \frac{1}{2}C\tau_0^2 L_1, \tag{22.61}$$

where

$$L_1 = L - \int_0^{\pi} \sqrt{1+\delta^2}\, R\, d\theta.$$

The work $\mathscr{L}$ is considered to be divided into two parts $\mathscr{L}_1$ and $\mathscr{L}_2$: the first represents the work done by the forces P applied at the ends of a portion of cable of length L during the formation of a loop from the state shown in Figure 22.7(a) to the state shown in Figure 22.7(b); the second part is the work done in a small reopening of the cable during the passage from the state shown in Figure 22.7(b) to that in Figure 22.7(c). The work $\mathscr{L}_1$ is clearly

$$\mathscr{L}_1 = -P(2\pi R); \tag{22.62}$$

the work $\mathscr{L}_2$ may be expressed in terms of the angle of the helix

$$\beta = \frac{1}{r} z', \tag{22.63}$$

and is given by

$$\mathscr{L}_2 = P \int_0^{\pi} (1 - \cos\beta) \sin\theta \, ds,$$

or, more simply, on putting $1 - \cos\beta \simeq \frac{1}{2}(z'/r)^2$ by

$$\mathscr{L}_2 = P \int_0^{\pi} \frac{1}{2}\left(\frac{z'}{r}\right)^2 \sin\theta \, ds = PR\delta^2(1+\delta^2)^{\frac{1}{2}}. \tag{22.64}$$

The total potential energy E is then given by the sum

$$E = W_B + W_T - \mathscr{L}_1 - \mathscr{L}_2.$$

If we exploit the condition that δ is much smaller than unity, the energy assumes the following simpler form

$$E = \frac{\pi B}{R}\left(1 - \frac{3}{4}\delta^2\right) + \pi \frac{C}{2R}\delta^2 + \frac{1}{2}\pi C\tau_0\delta(1-\delta^2) + \frac{1}{2}C\tau_0^2 L + 2\pi PR - PR\delta^2. \tag{22.65}$$

The equilibrium states of the cable are defined by the pairs δ, R, such that

$$\frac{\partial E}{\partial \delta} = -\frac{3\pi B}{2R}\delta + \frac{\pi C}{R}\delta + \frac{\pi C\tau_0}{2}(1 - 3\delta^2) - 2PR\delta = 0,$$

$$\frac{\partial E}{\partial R} = -\frac{\pi B}{R^2}\left(1 - \frac{3}{4}\delta^2\right) - \frac{\pi C}{2R^2}\delta^2 + 2\pi P - P\delta^2 = 0,$$

and the stability of the equilibrium is determined by the nature of the Hessian form associated with (22.65).[36]

[36] The stability criterion used by Yabuta et al. (1982) is wrong.

In order to have a very simple criterion of stability, let us consider the case in which $\tau_0 = 0$, and the cable has no residual twist. Then the equilibrium is determined by the solution

$$\delta = 0, \quad R = \left(\frac{B}{2P}\right)^{\frac{1}{2}}. \tag{22.66}$$

After calculating the second derivatives of E in the equilibrium state, the Hessian form becomes

$$H = \begin{Vmatrix} \dfrac{\pi C}{R}\left[1 - \left(\dfrac{3}{2} + \dfrac{1}{\pi}\right)\dfrac{B}{C}\right] & 0 \\ 0 & \dfrac{2\pi B}{R^3} \end{Vmatrix}.$$

The result is that the Hessian form is negative definite whenever the first term of the matrix H is negative, and a loop is unstable. Kinking is thus difficult when bending stiffness is large and torsional rigidity is small. This confirms the common experience that wires that are stiff in bending do not generate loops.

The *elastica* theory has several practical applications, even in circumstances of everyday experience which seem to have escaped any attempt at a mathematical approach. One of these unusual problems is the "spaghetti problem," named and studied by Carrier (1949), which endeavors to determine the motion of a dangling inextensible string when it is drawn upwards through a hole in a rigid wall. If the string initially has a transverse motion, the kinetic energy of the string does not change while it is being drawn through the hole and being squeezed into an ever-shrinking length. As a result, on reaching the hole, the tail of the string must have infinite transverse velocity; this explains why sucking spaghetti improperly leads to a spot of sauce on one's blouse or tie (Mansfield and Simmonds 1987)!

The spaghetti problem has an equally interesting reverse problem, which consists of finding out what happens when one spits out a piece of spaghetti which, if under-cooked, is not totally limp. The problem arises not only for spaghetti but also in the study of the motion of a sheet of paper issuing from a rigid horizontal guide into a uniform gravitational field. The same dynamic effect as in the spaghetti problem explains why the ends of a sheet of paper extruded by a copier are often curved.

To study this last problem, we model the paper as forming an *elastica* of length L, bending stiffness B, and mass per unit area m. The paper is expelled by a guide and its free part is subject to a weight mg per unit area, g being the acceleration due to gravity (Figure 22.8). Let us call V the constant velocity at which the paper issues from the guide.

With reference to the fixed Cartesian axes X and Y (Figure 22.8), the sheet initially occupies the interval $0 \leqslant X \leqslant L$ of the horizontal X axis. At an instant T after the onset of motion, the coordinates of the points of the deformed paper elastica, denoted by $\hat{X}$ and $\hat{Y}$, satisfy the conditions

$$\hat{X} = X + VT, \quad \hat{Y} = 0 \quad 0 < X < L - VT, \tag{22.67}$$

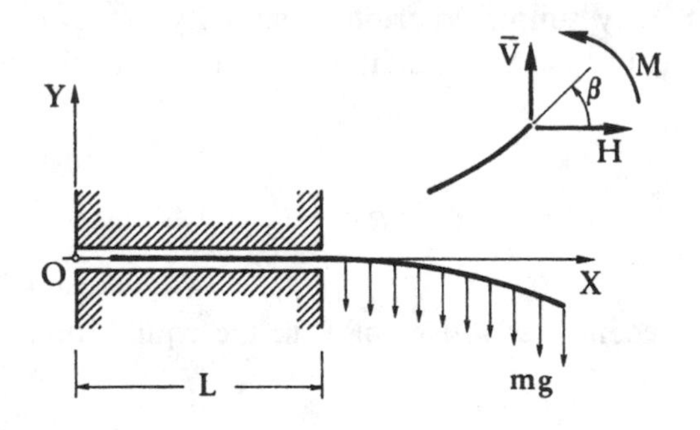

Fig. 22.8

inside the guide, and the balance equations

$$\frac{dH}{dX} = m\frac{d^2\hat{X}}{dT^2}, \quad \frac{d\bar{V}}{dX} - mg = m\frac{d^2\hat{Y}}{dT^2}, \tag{22.68}$$

$$\frac{dM}{dX} + \bar{V}\cos\beta - H\sin\beta = 0, \tag{22.69}$$

for $L - VT < X < L$, outside the guide. As indicated in Figure 22.8, H, $\bar{V}$, and M denote the resultants generated by the material to the right of a particle initially at $(X, 0)$ in the material to the left; the angle β represents the inclination of the deformed axis with respect to the X axis.

In addition, we have the constitutive relation

$$M = B\frac{d\beta}{dX}, \tag{22.70}$$

and the conditions of inextensibility

$$\frac{d\hat{X}}{dX} = \cos\beta, \quad \frac{d\hat{Y}}{dX} = \sin\beta. \tag{22.71}$$

In order to integrate equations (22.68) and (22.69), we introduce the functions

$$P(X, T) = \int_L^X \hat{X}(\xi, T)\, d\xi, \quad Q(X, T) = \int_L^X \hat{Y}(\xi, T)\, d\xi. \tag{22.72}$$

Because the end $X = L$ is free, (22.68) can be integrated once with respect to X to yield

$$H = m\frac{d^2P}{dT^2}, \quad \bar{V} = m\left[g(X - L) + \frac{d^2Q}{dT^2}\right],$$

and, on substituting these relations along with (22.70) in (22.69), we get

$$B\frac{d^2\beta}{dX^2} - mg(L - X)\cos\beta = m\left(\frac{d^2P}{dT^2}\sin\beta - \frac{d^2Q}{dT^2}\cos\beta\right). \tag{22.73}$$

But, after combining (22.71) and (22.72), we also have the equations

$$\frac{d^2P}{dx^2} = \cos\beta, \quad \frac{d^2Q}{dX^2} = \sin\beta, \tag{22.74}$$

which, together with (22.73), constitute the field equations outside the guide. The associated boundary conditions are that, at the wall, the velocity is purely horizontal and β is zero, while at the free end the bending moment is zero. In terms of P, Q, and β these conditions assume the form

$$\frac{dQ}{dX} = 0, \quad \beta = 0 \text{ at } X = L - VT, \tag{22.75}$$

$$\frac{d\beta}{dX} = 0, \quad P = Q = 0 \text{ at } X = L. \tag{22.76}$$

We still have to state the appropriate initial conditions. However, these are not completely obvious because all the paper is initially contained within the guide. The initial conditions therefore apply only at the point $(L, 0)$, and may be obtained by requiring that the conditions at the end of the guide ($X = L - VT$) match those at the free end ($X = L$) as T tends to zero.

Let us introduce dimensionless quantities by putting

$$P = L^2 p, \quad Q = L^2 q,$$

$$X = Lx = L(1 - s), \quad T = \frac{L}{V}\,t.$$

On substituting these expressions in (22.73) and (22.74) and setting

$$\mu = \frac{mgL^3}{B}, \quad \nu = VL\sqrt{\frac{m}{B}},$$

we obtain the dimensionless field equations

$$\frac{d^2\beta}{ds^2} - \mu s\cos\beta = \nu^2\left(\frac{d^2p}{dt^2}\sin\beta - \frac{d^2q}{dt^2}\cos\beta\right), \tag{22.77}$$

$$\frac{d^2p}{ds^2} = \cos\beta, \quad \frac{d^2q}{ds^2} = \sin\beta, \tag{22.78}$$

which must be solved in the triangular domain $0 < s < t$ and $0 < t < 1$, subject to the boundary conditions

$$\frac{d\beta}{ds}(0, t) = p(0, t) = q(0, t) = 0,$$

$$\beta(t, t) = \frac{dq}{ds}(t, t) = 0,$$

$$\frac{dp}{ds}(t, t) = -1.$$

As a result of direct computation, the solution can be found as a formal expansion in powers of s and t, of the form

$$\beta = \frac{1}{6}\mu(s^3 - t^3) + \frac{\mu v^2}{24}(s^4 t - 6s^3 t^2 + 5t^5) + \ldots, \tag{22.79}$$

$$p = -s + \frac{s^2}{2} - st + \ldots, \tag{22.80}$$

$$q = \frac{\mu}{120}(s^5 - 10s^2 t^3 + 15st^4) + \ldots. \tag{22.81}$$

The dimensionless coordinates at the end of the elastica, relative to the end of the guide, defined as

$$\hat{x}_e = [\hat{X}(L, T) - L]\frac{1}{L} = -\frac{dp}{ds}(0, t) - 1,$$

$$\hat{y}_e = \hat{Y}(L, T)\frac{1}{L} = -\frac{dq}{ds}(0, t),$$

become

$$\hat{x}_e = t + \ldots,$$
$$\hat{y}_e = -\frac{\mu t^4}{8} + \ldots. \tag{22.82}$$

Of course, this solution is crude, but might give a rough idea of the form of the bent sheet for short times t and at short distances from the end of the guide.

23. Initially Curved Rods

When a rod is naturally bent and twisted, the general equations are still as given in (19.22), but now the constitutive equations, although linear, must be modified. In its unstressed state the center line of the rod is a tortuous curve, the shape of which may be defined by considering an orthogonal triad of directors $\mathbf{D}_i$ chosen, for instance, in such a way that $\mathbf{D}_3$ is tangential to the axis of the curve and the other two directors coincide with the principal axes of the cross-sections at their centroids. As the origin of this triad moves along the curve with unit velocity, the components of the angular velocity of the moving triad, referred to the instantaneous position of the axes, are denoted by κ_0, κ_0', and τ_0, where τ_0, the component along the tangent, is the initial twist, and κ_0 and κ_0', the components along the other axes, are the initial curvatures.

When the rod is bent and twisted further, we can construct a second triad $\mathbf{d}_i$, such that $\mathbf{d}_3$ is directed along the strained central line, the plane $\mathbf{d}_1\mathbf{d}_3$ contains the unit vector $\mathbf{D}_1$, and $\mathbf{d}_2$ is perpendicular to the $\mathbf{d}_1\mathbf{d}_3$ plane and directed in such a way that the new triad is right-handed. As before, the components of the angular velocity of this triad, as its origin moves along the strained axis, define the twist τ_1 and the curvatures κ_1 and κ_1'. If we introduce the bending stiffness with respect the $\mathbf{d}_1$ and $\mathbf{d}_2$ axes, and denote them by A and B, and we call C the torsional rigidity, the relations between stress couples and the changes in curvature assume the form

$$m_1 = A(\kappa_1 - \kappa_0), \quad m_2 = B(\kappa_1' - \kappa_0'), \quad m_3 = C(\tau_1 - \tau_0). \tag{23.1}$$

These formulae are known as Clebsch's *constitutive relations* (Love 1927, Art. 259).

Now, after these preliminaries, let us return to Kirchhoff's equations for the case in which the sections of the rod have equal moments of inertia, so that $A = B$, and the rod is a circular spiral traced in its unstressed configuration. This is a spiral on a circular cylinder of radius r having a constant inclination α with respect to the planes perpendicular to the generators of the cylinder. As a consequence of the assumption $A = B$, all pairs of orthogonal axes in the plane of the cross-section issuing from its centroid are principal axes. We may then choose the triad $\mathbf{D}_i$, such that $\mathbf{D}_3$ is the unit vector tangential to the center line, $\mathbf{D}_1$ coincides with the principal normal to this curve pointing towards the center of curvature, and $\mathbf{D}_2$ coincides with the binormal and is oriented so that the triad $\mathbf{D}_1, \mathbf{D}_2, \mathbf{D}_3$ is right-handed. In this particular reference system the curvatures and the twist have the expressions

$$\kappa_0 = 0, \quad \kappa_0' = \frac{\cos^2\alpha}{r}, \quad \tau_0 = \frac{\sin\alpha\cos\alpha}{r}. \tag{23.2}$$

These quantities define the initial unstressed state. Let us now apply terminal forces and couples to the rod. The new state of strain is defined by the formulae

$$\kappa_1 = 0, \quad \kappa_1' = \frac{\cos^2\alpha_1}{r_1}, \quad \tau_1 = \frac{\sin\alpha_1\cos\alpha_1}{r_1}, \tag{23.3}$$

where r_1 and α_1 are the radius and the inclination of the new helix. It then follows that the stress couples are given by the equations

$$m_1 = 0, \quad m_2 = B\left(\frac{\cos^2\alpha_1}{r_1} - \frac{\cos^2\alpha}{r}\right), \quad m_3 = C\left(\frac{\sin\alpha_1\cos\alpha_1}{r_1} - \frac{\sin\alpha\cos\alpha}{r}\right),$$

and the stress-resultants have the form

$$n_1 = 0, \quad n_3 = n_2\tan\alpha_1,$$

$$n_2 = C\frac{\cos^2\alpha_1}{r_1}\left(\frac{\sin\alpha_1\cos\alpha_1}{r_1} - \frac{\sin\alpha\cos\alpha}{r}\right) - B\frac{\sin\alpha_1\cos\alpha_1}{r_1}\left(\frac{\cos^2\alpha_1}{r_1} - \frac{\cos^2\alpha}{r}\right).$$

These are the basic formulae governing the theory of spiral springs. From them we can calculate the resulting force R parallel to the axis of the cylinder of magnitude

$$R = n_3\sin\alpha_1 + n_2\cos\alpha_1, \tag{23.4}$$

and the couple K parallel to the same axis

$$K = m_3\sin\alpha_1 + m_2\cos\alpha_1. \tag{23.5}$$

If R and K are given, we can determine r_1 and α_1; alternatively, if we know r_1 and α_1, the calculation of R and K is immediate. Very often, the deformation is small so that we can write $r + \delta r$, $\alpha + \delta\alpha$ for r_1 and α_1, and linearize the equations.

The equations are simpler in the case of a planar deformation, that is to say, when a rod, the axis of which was initially contained in a plane, is loaded by forces and couples that maintain the strained configuration in the same plane. In this case it is even possible to find exact solutions when forces are applied along the length of the beam.

An example of a case of this type is the equilibrium of a heavy wire suspended from two fixed points at the same level.[37] Let us call θ the inclination of the tangent to the axis with respect to the horizonal, and let γ be the weight of the wire per unit length. As the strained axis is contained in a vertical plane passing through the ends of the wire, the curvatures are

$$\kappa = 0, \quad \kappa' = \frac{d\theta}{ds}, \tag{23.6}$$

and the twist τ is zero. The self-weight of the wire acts vertically, and its components along the tangent and the normal to the strained axis are

$$f_3 = -\gamma \sin\theta, \quad f_1 = -\gamma \cos\theta. \tag{23.7}$$

The equilibrium equations become

$$\begin{aligned} m_2' + n_1 &= 0, \\ n_1' + n_3\kappa' - \gamma\cos\theta &= 0, \\ n_3' - n_1\kappa' - \gamma\sin\theta &= 0, \end{aligned}$$

where $m_2 = B\kappa' = B(d\theta/ds)$. Elimination of n_1 and n_3 from the last equations yields

$$B\left[\frac{d}{ds}\left(\frac{d^3\theta/ds^3}{d\theta/ds}\right) + \frac{d\theta}{ds}\frac{d^2\theta}{ds^2}\right] - \gamma\left[2\sin\theta + \frac{(d^2\theta/ds^2)\cos\theta}{(d\theta/ds)^2}\right] = 0. \tag{23.8}$$

This equation admits a first integral after multiplying both sides by $\cos\theta$:

$$B\left(\frac{(d^3\theta/ds^3)\cos\theta}{d\theta/ds} + \frac{d^2\theta}{d\theta^2}\sin\theta\right) + \gamma\frac{\cos^2\theta}{(d\theta/ds)} = T_0, \tag{23.9}$$

where the constant of integration T_0 is the tension n_3 at a point $\theta = 0$. A subsequent integration of (23.9) yields

$$B\frac{d^2\theta}{ds^2}\sec\theta + \gamma s = T_0 \tan\theta, \tag{23.10}$$

where s is measured from the point $\theta = 0$.

For the case in which the wire is stretched taut under high tension so that any additional strains are small everywhere, (23.10) can be simplified by putting $\sec\theta \simeq 1$ and $\tan\theta \simeq \theta$. The resulting differential equation is linear, with constant coefficients.

The solution just found may be extended further to take account of extensibility in the axis. The theory for fixed end supports is particularly important in the case of an initially rectilinear bar subjected to transverse loads. Rods with constraints at the ends are called tie rods and their equilibrium configurations under transverse loads may be determined by applying Kirchhoff's theory, but removing the hypothesis of inextensibility of the central line. The problem is well illustrated by the case of a tie rod pinned at its unyielding supports and subjected to a single vertical load P (Figure 23.1) (Huddleston and Dowd 1979).

[37] According to Love (1927, Art. 273A), this solution is due to A. E. Young (1915).

Let us consider a tie rod AB, which is straight in its unstrained state, AB being a segment of length L, loaded by a single transverse force at C, at a distance a from A and b from B. After the deformation the rod becomes an arc connecting the fixed points A and B, and C, the point of application of the force P, will shift to a new point C_1. On taking a system of axes x and y, as shown in Figure 23.1, the position of a point D_1, which initially had coordinates $(x, 0)$, is now $(x + u, v)$, where u and v are the displacement components, while the position of C_1 is determined by the coordinates $[a + u(a), v(a)]$.

As before, let us call B the flexural rigidity of the rod in the xy plane, and let EA be its extensional rigidity. At a generic point D_1 on the deformed axis, the slope of the curve is denoted by $\theta(s)$, where s is the arc length measured from A. The bending moment at D_1 is then

$$M = B\kappa' = B\frac{d\theta}{ds}, \tag{23.11}$$

where M is expressed by

$$M = \begin{cases} Rv - P\left[\dfrac{b - u(a)}{L}\right][x + u(x)] & 0 \leqslant x \leqslant a, \\ Rv - P\left[\dfrac{a + u(a)}{L}\right]\{L - [x + u(x)]\} & a \leqslant x \leqslant L. \end{cases} \tag{23.12}$$

The axial force at D_1 is denoted by N, where N has the form

$$N = \begin{cases} R\cos\theta + \left[\dfrac{b - u(a)}{L}\right]P\sin\theta & 0 \leqslant x \leqslant a, \\ R\cos\theta - \left[\dfrac{a + u(a)}{L}\right]P\sin\theta & a \leqslant x \leqslant L. \end{cases} \tag{23.13}$$

The ratio

$$\frac{N}{EA} = \varepsilon = \frac{ds}{dx} - 1$$

represents the elongation of the strained axis, and the coordinates of P_1 are related to the slope by the geometric conditions

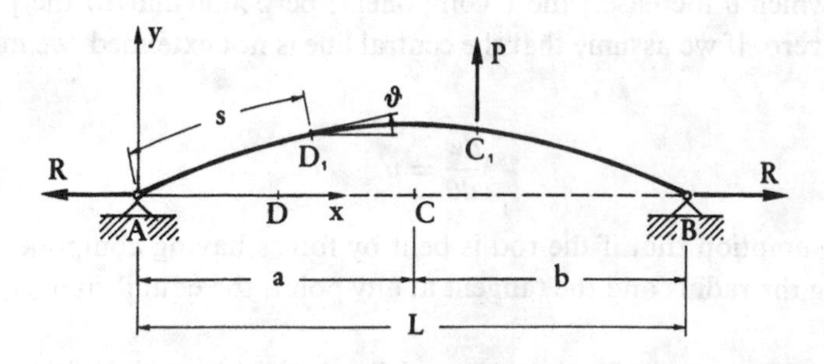

Fig. 23.1

$$\frac{d[x+u(x)]}{ds} = (1+\varepsilon)\cos\theta,$$
$$\frac{dv(x)}{ds} = (1+\varepsilon)\sin\theta. \tag{23.14}$$

As the ends A and B are fixed, we must have the boundary conditions

$$u(0) = v(0) = 0, \quad u(L) = v(L) = 0; \tag{23.15}$$

in addition, the bending moment must also vanish at the ends, so that we have

$$\frac{d\theta}{ds}(0) = \frac{d\theta}{ds}(L) = 0. \tag{23.16}$$

The unknowns of the problem are the displacements u and v, the rotation θ, the stress resultant N, the stress couple M, and the shear force Q. The stress resultants are related by the equilibrium equations

$$\frac{dN}{ds} + Q\frac{d\theta}{ds} = 0,$$
$$\frac{dQ}{ds} - N\frac{d\theta}{ds} = 0,$$
$$\frac{dM}{ds} - Q = 0.$$

The full system can be integrated by a shooting method in which the unknown quantities $\theta(0)$, R, and $u(a)$ are given tentative values. This converts the problem into an initial-value type one, which gives us $u(L)$ and $v(L)$ and the recomputed $u(a)$, so that we can adjust the initial data to improve the values of $u(L)$, $v(L)$, and $u(a)$.[38]

The specific influence of extensibility in a naturally curved rod can be well explained in terms of the deformation of a circular ring of radius a, when it is bent and stretched in its plane. In this case it is convenient to introduce the angle θ between the radius drawn from the center of the circle to any point, and a chosen radius. The initial curvature κ_0' is then given by

$$\frac{d\theta}{ds} = \kappa_0' = \frac{1}{a}. \tag{23.17}$$

We also note that, if the plane of the circle is a principal plane of the rod at any point, then κ_0 and τ_0 vanish. Let us call u the displacement component directed along the radius inwards and w the component directed along the tangent to the circle in the sense in which θ increases; the v component, perpendicular to the plane of the circle, is now zero. If we assume that the central line is not extended, we must add the condition

$$\frac{dw}{d\theta} = u. \tag{23.18}$$

Under the assumption that if the rod is bent by forces having components f_r and f_θ directed along the radius and the tangent at any point, the equilibrium equations are

[38] This method has been applied by Huddleston and Dowd (1979), but with a slight modification to (23.11), where they put $M = B(d\theta/dx)$.

$$\frac{dn_1}{d\theta} + n_3 + f_r a_1 = 0, \tag{23.19a}$$

$$\frac{dn_3}{d\theta} - n_1 + f_\theta a_1 = 0, \tag{23.19b}$$

$$\frac{dm_2}{d\theta} + n_1 a_1 = 0, \tag{23.19c}$$

where $a_1 = a_1(\theta)$ is the radius of the central line at any point after the deformation. In (23.19c) the stress couple m_2 is related to the change in curvature by the equation

$$m_2 = B(\kappa_1' - \kappa_0') = B\left(\frac{1}{a_1(\theta)} - \frac{1}{a}\right), \tag{23.20}$$

where B is the flexural rigidity for bending in the plane of the rod. The above equations have been written without making any assumption about the magnitude of strains, but if we suppose these to be small, we can put

$$\kappa_1' \simeq \frac{1}{a} + \frac{1}{a}\frac{d}{d\theta}\left(\frac{1}{a}\frac{du}{d\theta} + \frac{w}{a}\right),$$

where a is the radius of the undeformed ring. From this equation, using (23.18), we derive

$$m_2 = \frac{B}{a^2}\left(\frac{d^3w}{d\theta^3} + \frac{dw}{d\theta}\right). \tag{23.21}$$

Hence we find that n_1 and n_3 are expressible in terms of w through the equations

$$n_1 = -\frac{B}{a^3}\left(\frac{d^4w}{d\theta^4} + \frac{d^2w}{d\theta^2}\right), \quad n_3 = -f_\theta a + \frac{B}{a^3}\left(\frac{d^5w}{d\theta^5} + \frac{d^3w}{d\theta^3}\right), \tag{23.22}$$

and that w satisfies the equation[39]

$$\frac{B}{a^3}\left(\frac{d^6w}{d\theta^6} + 2\frac{d^4w}{d\theta^4} + \frac{d^2w}{d\theta^2}\right) = a\left(\frac{df_r}{d\theta} - f_\theta\right). \tag{23.23}$$

Let us now remove the assumption (23.18) and try to reformulate the problem, still maintaining the hypothesis that strains are small. Then the extension of the central line is

$$\varepsilon = \frac{1}{a}\left(\frac{dw}{d\theta} - u\right), \tag{23.24}$$

and the tension n_3 is

$$n_3 = EA\varepsilon. \tag{23.25}$$

The expression for the couple m_2 and the shearing force n_1 depends on the flexural rigidity B, while the expression for n_3 depends on the extensional rigidity EA. However, the contribution of the terms containing B may be omitted, to a first

[39] According to Love (1927, Art. 292), the result is due to Lamb.

approximation, and the equations to be satisfied by u and w are (23.19a) and (23.19b), that is

$$\frac{EA}{a}\left(\frac{dw}{d\theta}-u\right)+f_r a=0, \quad \frac{EA}{a}\left(\frac{d^2w}{d\theta^2}-\frac{du}{d\theta}\right)+f_\theta a=0. \tag{23.26}$$

The result is that the assumption of inextensibility in a curved rod is one that introduces considerable sensitivity, because it leads to a mathematical description and results that may be far from those furnished by the set of equations involving extensibility.

The effect of the self-weight in a circular elastic ring hung at its highest point cannot be evaluated by direct integration of the equation of the *elastica*, but an approximate solution can be found by using a perturbation method (Wang and Watson 1980). Let us consider a thin circular ring suspended at the top A and subjected to the action of its own weight (Figure 23.2), defined by a constant value ρg per unit length. Let L be the perimeter of half the ring, EI be the flexural rigidity, and F the lateral force exerted across a section at O. On choosing a system of axes x' and y' with their origin at O and directed as shown in Figure 23.2, the equation of the heavy *elastica* can be written as

$$EI\,\frac{d^2\theta}{ds'^2}=-F\sin\theta-g\rho s'\cos\theta, \tag{23.27}$$

$$\frac{dx'}{ds'}=\cos\theta, \quad \frac{dy'}{ds'}=\sin\theta, \tag{23.28}$$

where θ is the local angle of inclination and s' is the arc length measured from O. The boundary conditions are

$$\theta(0)=x'(0)=y'(0)=0, \tag{23.29}$$

$$\theta(L)=\pi, \quad x'(L)=0. \tag{23.30}$$

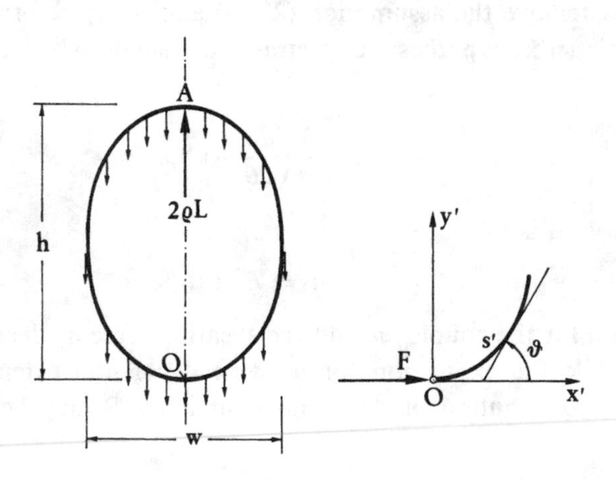

Fig. 23.2

In order to treat these equations, we normalize all lengths by putting

$$s = \frac{s'}{L}, \quad x = \frac{x'}{L}, \quad y = \frac{y'}{L},$$

and the governing equations become

$$\frac{d^2\theta}{ds^2} = -A\sin\theta - Bs\cos\theta \tag{23.31}$$

$$\frac{dx}{ds} = \cos\theta, \quad \frac{dy}{ds} = \sin\theta, \tag{23.32}$$

where A and B are nondimensional quantities defined by

$$A = \frac{FL^2}{EI}, \quad B = \frac{\rho g L^3}{EI}.$$

The boundary conditions are

$$\theta(0) = x(0) = y(0) = 0,$$
$$\theta(1) = \pi, \quad x(1) = 0. \tag{23.33}$$

No solution of (23.31) and (23.32) in closed form is available, because A is also unknown. However, an approximate solution can be found for small values of the constant B, which measures the relative importance of density and length to rigidity. A small value of B means relatively stiff rings with almost circular shape. Since the effect of gravity is small, we expect A also to be small. Now, let

$$B = \varepsilon \ll 1, \quad A = \alpha\varepsilon,$$

where α is of the order of unity, and consider the formal expansions

$$\theta = \theta_0(s) + \varepsilon\theta_1(s) + \dots, \tag{23.34a}$$
$$x = x_0(s) + \varepsilon x_1(s) + \dots, \tag{23.34b}$$
$$y = y_0(s) + \varepsilon y_1(s) + \dots, \tag{23.34c}$$
$$\alpha = \alpha_0 + \varepsilon\alpha_1 + \dots. \tag{23.34d}$$

When these expansions are substituted in the field equations and boundary conditions, we find that the zero-order terms must satisfy the equations

$$\frac{d^2\theta_0}{ds^2} = 0, \quad \frac{dx_0}{ds} = \cos\theta_0, \quad \frac{dy_0}{ds} = \sin\theta_0, \tag{23.35}$$
$$\theta_0(0) = x_0(0) = y_0(0) = 0, \quad \theta_0(1) = \pi, \quad x_0(1) = 0, \tag{23.36}$$

and thus the solution is

$$\theta_0 = \pi s, \tag{23.37a}$$
$$x_0 = \frac{\sin\pi s}{\pi}, \tag{23.37b}$$
$$y_0 = \frac{1 - \cos\pi s}{\pi}. \tag{23.37c}$$

The equations of first order are

$$\frac{d^2\theta_1}{ds^2} = \alpha_0 \sin\theta_0 - s\cos\theta_0, \quad \frac{dx_1}{ds} = -\theta_1 \sin\theta_0,$$

$$\frac{dy_1}{ds} = \theta_1 \cos\theta_0, \tag{23.38}$$

$$\theta_1(0) = x_1(0) = y_1(0) = 0, \quad \theta_1(1) = x_1(1) = 0. \tag{23.39}$$

The corresponding solution is

$$\theta_1 = \frac{s}{\pi^2} - \frac{3\sin\pi s}{2\pi^3} + \frac{s\cos\pi s}{\pi^2}, \tag{23.40}$$

$$x_1 = \frac{3s}{4\pi^3} - \frac{\sin 2\pi s}{2\pi^4} + \frac{s\cos\pi s}{\pi^3} - \frac{\sin\pi s}{\pi^4}, \tag{23.41}$$

$$y_1 = \frac{s^2}{4\pi^2} - \frac{3}{2\pi^4} + \frac{s\sin 2\pi s}{4\pi^3} + \frac{\cos 2\pi s}{2\pi^4} + \frac{s\sin\pi s}{\pi^3} + \frac{\cos\pi s}{\pi^4}, \tag{23.42}$$

$$\alpha_0 = -\frac{1}{2\pi}. \tag{23.43}$$

An idea of the final configuration of the ring is given by its maximum extended vertical length

$$y(1) = \frac{2}{\pi} + \varepsilon\left(\frac{1}{4\pi^2} - \frac{2}{\pi^4}\right) + 0(\varepsilon^2), \tag{23.44}$$

and by its maximum width at $\theta = \pi/2$. To calculate the latter, we use (23.34a), (23.37a), and (23.40) to obtain the corresponding arc length

$$s^* = \frac{1}{2} + \varepsilon\left(\frac{3}{2\pi^4} - \frac{1}{2\pi^3}\right) + 0(\varepsilon^2).$$

On substituing this value of s in (23.37b) and (23.41) we obtain the maximum width

$$2x(s^*) = \frac{2}{\pi} - \varepsilon\left(\frac{2}{\pi^4} - \frac{1}{2\pi^3}\right) + 0(\varepsilon^2). \tag{23.45}$$

One of the most important applications of the theory of naturally curved rods is in the analysis of wire ropes. Over the years, many types of wire rope have been designed and tested for various applications, but the theoretical analysis of the problem is very complex. It is thus not surprising that, until 1950, the only reliable criteria for designing wire ropes were based on extensive programs involving testing under both static and fatigue loading conditions. According to Phillips and Costello (1985), the first assessment of useful design criteria taken from existing experimental data was done by Drucker and Tachau (1945), and the first attempts at theoretical models predicting the stresses in the individual wires of a strand were made by Hall (1951) and Hruska (1951).

Wire ropes are complex structures formed by helical wires of circular cross-section wound round a central rectilinear wire called the *core* (Figure 23.3). The wires and core form a *strand*, and complex cables are formed by the union of strands which are also formed into helices. Sometimes the strands have no core and the single wires are wound directly around one another. The cross-section of a strand with a core is

Fig. 23.3

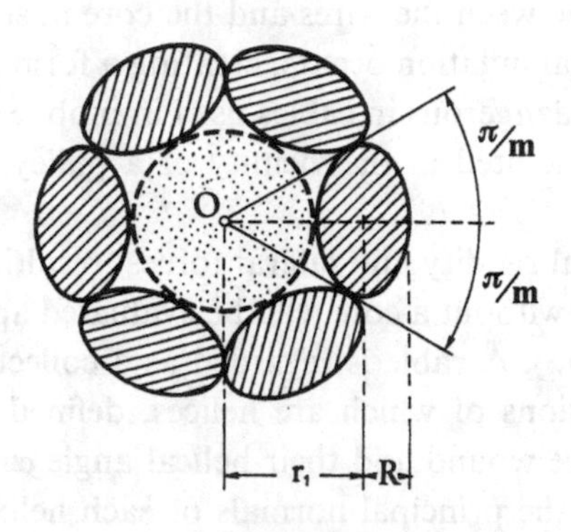

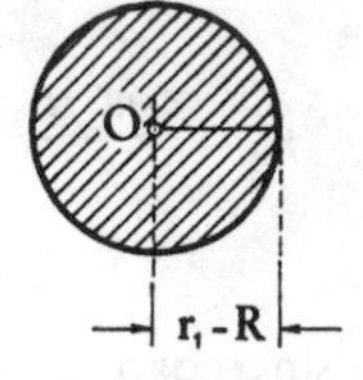

Fig. 23.4

sketched in Figure 23.4 where, for clarity, the section of the core has also been drawn separately. Since the wires are aligned along the helices, their horizontal cross-sections are ellipses; the horizontal section of the core is a circle. Wire ropes have been employed since the beginning of the nineteenth century, but calculations of their behavior have been extremely crude. As ropes are essentially destined to bear axial tensile forces, the immediate initial assumption is that both the wires and the core simply carry an axial tensile force, parallel to the central line of each element. The sum of the tensile force in the core and of the vertical components of the tensile forces in the wires must balance the total axial force in the strand. All the formulae proposed in manuals of mechanical engineering rest on this assumption and the difference between one formula and another depends only on the criterion used to assign the fraction of load absorbed by the wires and that absorbed by the core.

Traditional methods of calculating the stresses in wire ropes have been resolved by a large sequence of experiments accompanying the technological progress in structural design. However, these tests have in fact confirmed that there are some effects in the deformation and the generation of stresses in a wire rope that are not negligible. First, the flexural rigidity of each wire is often very high, especially in wires made of high tensile steel. The flexural rigidity of the wires is also important in very flexible ropes, such as those employed by rock climbers, in connection with knots, when these are under severe tension. Secondly, the wires and the core, being in contact with one other, transmit transverse forces between them. This has the result of altering, sometimes radically, the state of the stress calculated under the assumption that each of these components is independent. The situation is even more complicated if we take into account the fact that the transverse forces are purely compressive, and may disappear as soon as loads are removed and contact is lost. In the limit, the separation between the wires and the core in some parts of the cord is so high that a strange local inflation occurs, creating a form like a hollow bird cage. This effect is particularly dangerous in cables used in mobile situations, because, if a bird-cage condition is generated in the vicinity of a pulley, the cable is at risk of immediate damage.

The influence of flexural rigidity and lateral forces, resulting from the packing of wires or cords in a strand without a core, can be evaluated approximately as follows (Phillips and Costello 1973). A cable is regarded as a collection of m smooth wires the unstrained configurations of which are helices, defined by the radius r of the cylinder on which they are wound and their helical angle α. If we take a system of directors coinciding with the principal normals of each helix, the curvatures of the unstrained helix are (see (23.2))

$$\kappa_0 = 0, \quad \kappa_0' = \frac{\cos^2 \alpha}{r},$$

and the twist is

$$\tau_0 = \frac{\sin \alpha \cos \alpha}{r}.$$

When the cable is loaded, the new configuration of each wire, provided that it is still a helix coaxial with the initial helix, is characterized by two curvatures and a twist, typically

$$\kappa_1 = 0, \quad \kappa_1' = \frac{\cos^2 \alpha_1}{r_1}, \quad \tau_1 = \frac{\sin \alpha_1 \cos \alpha_1}{r_1},$$

where r_1 and α_1 are the radius and angle of the new helix. Let us call N the shear force along the normal to the helix, N' the shear force along the binormal, and T the tension along the tangent. These forces have previously been denoted by n_1, n_2, and n_3, respectively. The stress couples are represented by a couple G' round the binormal, a couple G round the normal, and a twisting couple H round the tangent. In the notation used for the general theory these couples were denoted by m_1, m_2, and m_3, respectively. Here, N and G vanish, since κ_0 and κ_1 vanish. Let us now assume that the exterior body forces and body couples per unit length are zero. We now wish to describe the effect of the contact force per unit length on each wire as a constant body force X directed along the principal normal. In these circumstances we suppose that stress resultants and stress couples are independent of the variable s_1, the arc length along the central line of each strained wire. It thus follows that the equilibrium equations, relative to the strained configuration, reduce to

$$-N'\tau_1 + T\kappa_1' + X = 0, \tag{23.46}$$

$$-G'\tau_1 + H\kappa_1' - N' = 0. \tag{23.47}$$

As, by virtue of the constitutive equations, we know the values G' and H,

$$G' = B(\kappa_1' - \kappa_0'), \quad H = C(\tau_1 - \tau_0),$$

and by eliminating N' from (23.46) and (23.47) we can express X in the form

$$X = [C(\tau_1 - \tau_0)\kappa_1' - B(\kappa_1' - \kappa_0')\tau_1]\tau_1 - T\kappa_1'. \tag{23.48}$$

Hence the contact force is determined by the tensile force T and the curvatures and twists. The total axial force and the total axial twisting moment on the cable are given by

$$F = m(T\sin\alpha_1 + N'\cos\alpha_1), \tag{23.49}$$

$$M = m(H\sin\alpha_1 + G'\cos\alpha_1 + Tr_1\cos\alpha_1 - N'r_1\sin\alpha_1). \tag{23.50}$$

The force F and the moment M form a wrench the axis of which coincides with the axis of the helix.

An improvement in the theory may be achieved by specifying the magnitude of the contact forces mutually exerted between the single wires and the positions of these forces (Costello and Phillips 1973, 1974). Let us consider the cross-section of a wire, traced perpendicularly to its helicoidal longitudinal axis in the strained configuration. This cross-section is a circle of radius R, the center O_1 of which is at a distance r_1 from the axis of the helix and describes a curve, which, in a Cartesian x, y, z system having the z axis coinciding with the axis of the helix, has the parametric equations

$$x = r_1\cos\varphi, \quad y = r_1\sin\varphi, \quad z = k\varphi,$$

where k is a constant related to the pitch h of the helix by the equation $h = 2\pi k$. The points of the circle of radius R where it touches the contiguous wires are placed at an angle β symmetrically with respect to the binormal $\mathbf{b}$, along which the force per unit length X is directed (Figure 23.5). On denoting the magnitude of the effective contact forces per unit length at the contact points by Q, the relation

$$Q = -\frac{X}{2\cos\beta}, \tag{23.51}$$

must hold, and hence once we know X and β we can find Q. Since X is given by (23.48), it remains for us to determine β. For this purpose, let us consider the section

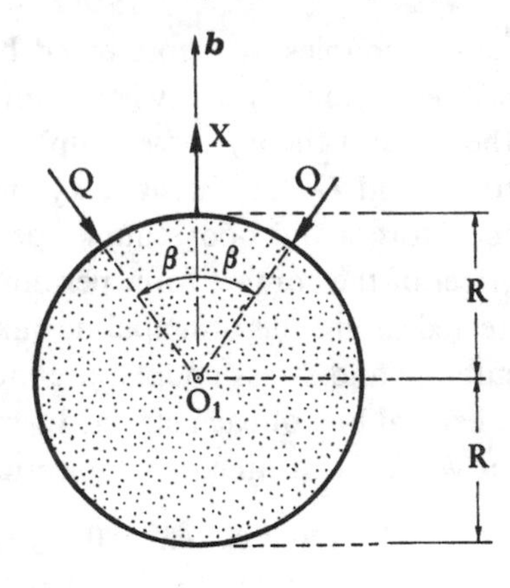

Fig. 23.5

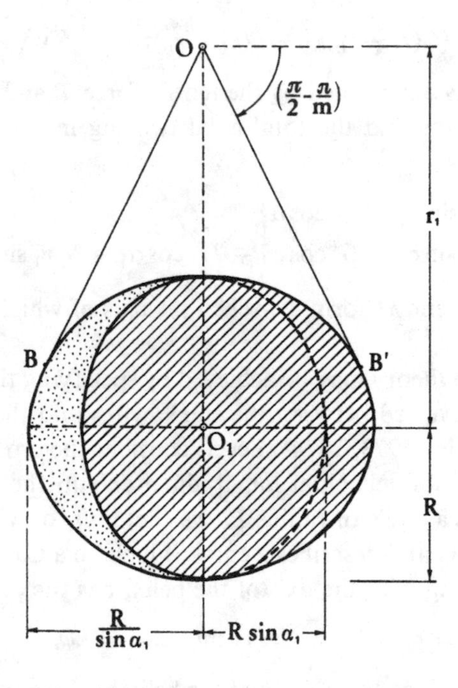

Fig. 23.6

of the helical tubular wire with a plane perpendicular to the axis of the helix. This section is approximately an ellipse with half-axes $(R/\sin\alpha_1)$ and R, where α_1 is the helical angle (Figure 23.6). As there are m wires, the angle made by the tangents to the ellipse drawn from O is $2\pi/m$. The distance of O_1 from each of the points of contact of the tangents B and B$'$ is

$$\overline{O_1B} = \overline{O_1B'} = R\left[\cos^2\left(\frac{\pi}{2} - \frac{\pi}{m}\right) + \frac{\sin^2\left(\frac{\pi}{2} - \frac{\pi}{m}\right)}{\sin^2 \alpha_1}\right]^{\frac{1}{2}},$$

whence we are able to express r_1 as a function of α_1:

$$r_1 = \frac{\overline{OB'}}{\cos\left(\frac{\pi}{2} - \frac{\pi}{m}\right)} = R\left[1 + \frac{\tan^2\left(\frac{\pi}{2} - \frac{\pi}{m}\right)}{\sin^2 \alpha_1}\right]^{\frac{1}{2}}.$$

Having now obtained r_1, the angle β in Figure 23.5 can be evaluated approximately by assuming that the point of application of each force Q is not far from the point of contact of each of the tangents drawn from O to the circle of radius R and the center at O_1. It thus follows that we have

$$\cos\beta = \frac{R}{r_1} = \left[1 + \frac{\tan^2\left(\frac{\pi}{2} - \frac{\pi}{m}\right)}{\sin^2 \alpha_1}\right]^{-\frac{1}{2}}. \tag{23.52}$$

This formula is simple, but not exact. It can be improved further, but the result is much more complicated (Costello and Phillips 1974; Kunoh and Leech 1985).

So far we have not discussed possible loss of contact due to a local separation of the wires. When the wires are separated in the deformed configuration, the inequality

$$\frac{r_1}{R} > \left[1 + \frac{\tan^2\left(\frac{\pi}{2} - \frac{\pi}{m}\right)}{\sin^2 \alpha_1}\right]^{\frac{1}{2}}$$

must hold and X must vanish. We use (23.48), and recall that in a circular section of radius R, like that of wires, the bending stiffness B and the torsional rigidity C are given by

$$B = \frac{\pi E R^4}{4}, \quad C = \frac{\pi E R^4}{4(1+\sigma)},$$

where E is Young's modulus and σ is Poisson's ratio. The resultant stress T is either prescribed, or, provided that the wires are extensible, related to the elongation ε by an equation of the form

$$T = \pi E R^2 \varepsilon.$$

In the second case, on substituting the values of B, C, and T in (23.48) and using the relations

$$\kappa_1 = 0, \quad \kappa_1' = \frac{\cos^2\alpha_1}{r_1}, \quad \tau_1 = \frac{\sin\alpha_1 \cos\alpha_1}{r_1},$$

we find that r_1 satisfies a quadratic equation for which the only physically meaningful root is (Phillips and Costello 1979)

$$r_1 = [-b - (b^2 - 4ac)^{\frac{1}{2}}]\frac{1}{2a},$$

where

$$a = 4\varepsilon, \quad b = -\left[\kappa_0' \sin\alpha_1 - \frac{\tau_0 \cos\alpha_1}{(1+\sigma)}\right] \sin\alpha_1,$$

$$c = \frac{\sigma}{(1+\sigma)} \sin^2\alpha_1 \cos^2\alpha_1.$$

In these expressions, ε and α_1 are still unknown because they are part of the solution. But, if ε and α_1 are given, we can express all the other relevant quantities (r_1, N', X, etc.) in terms of ε and α_1. As, in general, the data for the problem are F and M, as expressed by equations (23.49) and (23.50), respectively, the unknowns ε and α_1 must be determined, at least in principle, by solving these two equations with respect to ε and α_1. Owing to the nonlinear nature of the governing equations, they are often solved numerically (Phillips and Costello 1979).

None of the above results takes into consideration the presence of an elastic core, which could absorb a certain part of the applied axial force F and twisting moment M. In order to treat this situation as well, it is necessary to divide each of the quantities into two parts (Huang 1979):

$$\bar{F} = F_w + F_c,$$
$$\bar{M} = M_w + M_c, \tag{23.53}$$

where F_w and M_w are carried by the helical wires, and F_c and M_c are carried by the central core. Therefore, the core is under the action of F_c and M_c and lateral forces X along the line of contact with the wires. In order to evaluate the deformation of the core, we replace the line forces X on the mantle of the core by a statically equivalent uniformly distributed lateral normal pressure of intensity

$$q = \frac{mX}{2\pi(r_1 - R)\sin\alpha_1},$$

where $(r_1 - R)$ is the radius of the core (see Figure 23.4) and α_1 is the slope of the central axis of the helix of each wire. The axial strain of the core is therefore

$$\varepsilon_c = \frac{1}{E\pi(r_1 - R)}\left(\frac{F_c}{(r_1 - R)} + \sigma \frac{mX}{\sin\alpha_1}\right), \tag{23.54}$$

and the transverse strain in the core is

$$\varepsilon_c' = -\frac{1}{E\pi(r_1 - R)}\left(\frac{mX}{2\sin\alpha_1}(1-\sigma) + \sigma \frac{F_c}{(r_1 - R)}\right). \tag{23.55}$$

Solving these equations we obtain the results

$$F_c = \frac{2E\pi(r_1 - R)^2}{(1+\sigma)}\left[\frac{1}{2}(1-\sigma)\varepsilon_c - \sigma\varepsilon_c'\right], \tag{23.56}$$

$$X = -\frac{2E\pi(r_1 - R)\sin\alpha_1}{m(1+\sigma)(1-2\sigma)}(\varepsilon_c' - \sigma\varepsilon_c). \tag{23.57}$$

It should be noted that the transverse strain ε_c' is related to r_1 and R by

$$\varepsilon_c' = \frac{r_1 - R}{r - R_0} - 1, \tag{23.58}$$

where r is the radius of the axes of the helices of the wires in the unstrained state, and R_0 is the radius of the cross-section of each undeformed wire. The angle of twist per unit length of the core is related to the twisting moment M_c by the formula

$$\Omega = \frac{M_c}{C} = \frac{4(1+\sigma)}{E\pi(r_1 - R)^4} M_c. \tag{23.59}$$

The axial force due to the wires is given by (23.49), which now reads

$$F_w = m(T \sin\alpha_1 + N' \cos\alpha_1),$$

and the associated twisting moment, according to (23.50), is

$$M_w = m(H \sin\alpha_1 + G' \cos\alpha_1 + Tr_1 \cos\alpha_1 - N' r_1 \sin\alpha_1).$$

Let us consider a wire of M_0 turns in the undeformed state, and assume that, after the deformation, it has M turns. Hence the undeformed and deformed lengths of the wire measured in the direction of the axis of the strand are, respectively,

$$L_0 = 2\pi M_0 k_0, \quad L = 2\pi M k,$$

where k_0 and k are two constants related to the pitches of the two helices by

$$h_0 = 2\pi k_0, \quad h = 2\pi k.$$

The undeformed and deformed lengths of the wire are then

$$\ell_0 = 2\pi M_0 \rho, \quad \ell = 2\pi M \rho_1,$$

where

$$\rho = \sqrt{r^2 + R_0^2}, \quad \rho_1 = \sqrt{r_1^2 + R^2},$$

R_0 and R being the radii of the cross-sections of a wire before and after the deformation. The axial strain of the strand is therefore

$$\varepsilon = \frac{L}{L_0} - 1 = \frac{Mk}{M_0 k_0} - 1, \tag{23.60}$$

and the axial strain of the wire is

$$\varepsilon_w = \frac{\ell}{\ell_0} - 1 = \frac{M\rho_1}{M_0 \rho} - 1. \tag{23.61}$$

The lateral contraction is then

$$\frac{R}{R_0} - 1 = -\sigma\varepsilon_w. \tag{23.62}$$

After elimination of ε_w and M/M_0 from the last three equations, we have

$$\varepsilon = \frac{k\rho}{k_0\rho_1}\left[1 + \frac{1}{\sigma}\left(1 - \frac{R}{R_0}\right)\right] - 1. \tag{23.63}$$

In the case where there is no rotational slip between the central core and the wires, the angle of twist of the strand per unit length of the strand is

$$\Omega = \frac{2\pi}{L}(M - M_0) = \frac{1}{k_0}\left(\frac{M}{M_0} - 1\right). \tag{23.64}$$

After elimination of ε_w and M/M_0 from (23.61), (23.62), and (23.64), we obtain

$$\Omega = \frac{1}{k_0}\left\{\frac{\rho}{\rho_1}\left[1 + \frac{1}{\sigma}\left(1 - \frac{R}{R_0}\right)\right] - 1\right\}. \tag{23.65}$$

Thus, if the ends of the strand are unconstrained, elongation of the strand will be coupled with torsion. On the other hand, if the ends of the strand are fixed, Ω vanishes and we have the following condition of constraint

$$\rho_1 = \rho\left[1 + \frac{1}{\sigma}\left(1 - \frac{R}{R_0}\right)\right]. \tag{23.66}$$

In both cases, the procedure for solving the problem is as follows. For given values of the inclination of the unstrained helix α, the number of wires m, Poisson's ratio σ, the total twisting moment $\bar{M}$, and the ratio of the radius of the cross-section of each wire R/R_0, we assume a tentative value of k. We then find a solution and, eventually, we start the procedure again, using a correct value of k.

The method also allows us to ascertain if there is any separation between wires. It is found that, when the central core and surrounding wires are made of the same material, the extension of the strain always gives rise to a separation between helical wires.

24. The Stability of Rods

One of the merits of Kirchhoff's theory is that it allows an investigation of the stability of a rod under combined states of stress. It uses a systematic method, in which all approximations that are applicable can be put into practice.

The simplest of the cases concerned with a loss of stability under combined loads occurs when a rod, assumed straight in the unstrained state, is subjected to a thrust R and a couple K. Let us consider the case for which the cross-section is doubly symmetric so that the principal moments of inertia A and B are equal, and assume a system of fixed unit vectors such that $\mathbf{e}_3$ coincides with the axis of the rod and $\mathbf{e}_1$ and $\mathbf{e}_2$ are placed in the perpendicular plane containing the lower end of the rod (Figure 24.1). In a strained state, a generic cross-section, initially placed at a distance S from the origin, is moved to a new position defined by an arc of length S_1 and by a triad of directors $\mathbf{d}_1, \mathbf{d}_2, \mathbf{d}_3$ coinciding with the principal torsion–flexure axes of the rod at the section. The directions of these axes are defined by Euler's angles θ, ψ, and φ, of the new triad with respect the fixed axes. The direction cosines of the axes x, y, and z are expressible in terms of Euler's angles, and their detailed form is obtainable from equations (17.15) by observing that we have

$$\ell_1 = \mathbf{d}_1 \cdot \mathbf{e}_1, \quad m_1 = \mathbf{d}_1 \cdot \mathbf{e}_2, \ldots\ldots, n_3 = \mathbf{d}_3 \cdot \mathbf{e}_3. \tag{24.1}$$

By using these direction cosines in the equations we can express the resultant couples acting on the section in terms of the force R and the couple K:

$$\begin{aligned} A\kappa\ell_1 + B\kappa'\ell_2 + C\tau\ell_3 &= -Ry, \\ A\kappa m_1 + B\kappa' m_2 + C\tau m_3 &= Rx, \\ A\kappa n_1 + B\kappa' n_2 + C\tau n_3 &= K, \end{aligned} \tag{24.2}$$

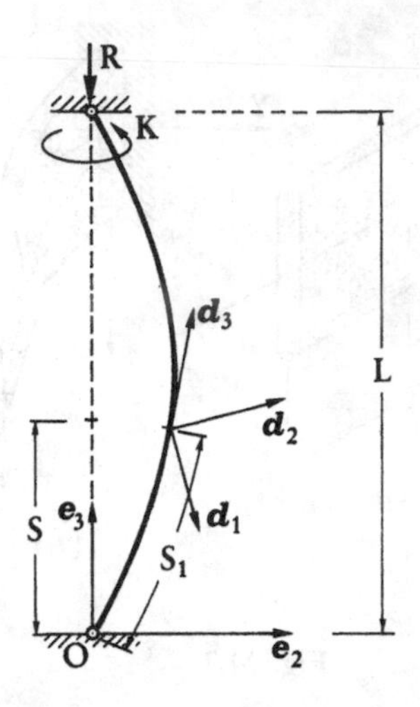

Fig. 24.1

where x and y are the coordinates of any point on the strained axis. When we replace κ, κ', and τ using (19.15) and $\ell_1, \ldots, n_3$ using (24.1), we obtain a system of three ordinary differential equations in θ, ψ, φ, and x, y. A relationship between x, y and Euler's angles can be found using the formulae of the differential geometry, but their precise expressions are not necessary.

Since the treatment of system (24.2) is very complicated, we approximate the equations by considering small deviations from the unstrained rectilinear state, and study the stability of the solutions by applying the linearized criterion for stability. In an approximate analysis, the coefficients $\ell_1, m_1, \ldots,$ are given by the equations (Love 1927, Art. 258)

$$\begin{aligned} &\ell_1 = 1, \quad && m_1 = \beta, \quad && n_1 = -\frac{dx}{dS}, \\ &\ell_2 = -\beta, \quad && m_2 = 1, \quad && n_2 = -\frac{dy}{dS}, \\ &\ell_3 = \frac{dx}{dS}, \quad && m_3 = \frac{dy}{dS}, \quad && n_3 = 1, \end{aligned} \tag{24.3}$$

where the quantity β is, to first order, equal to $(\psi + \varphi)$, or the angle through which a cross-section is rotated about the central line. The curvatures and the twist are given by (19.15), and, to the same order of approximation, we have

$$\kappa = -\frac{d^2y}{dS^2}, \quad \kappa' = \frac{d^2x}{dS^2}, \quad \tau = \frac{d\beta}{dS}. \tag{24.4}$$

On substituting (24.3) and (24.4) in (24.2), and neglecting the terms of higher order, we obtain the simpler system

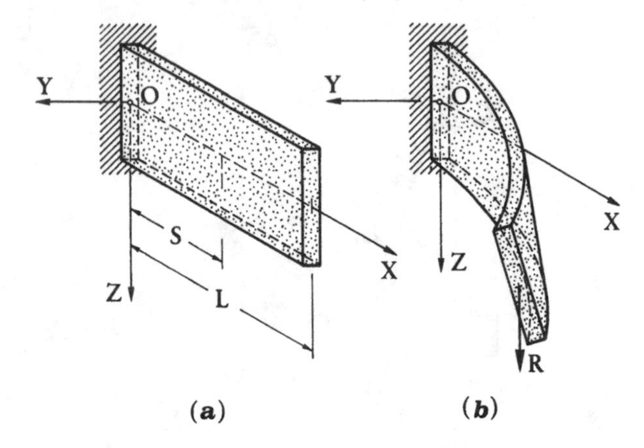

Fig. 24.2

$$
\begin{aligned}
-B\frac{d^2y}{dS^2} + C\tau\frac{dx}{dS} &= +Ry,\\
B\frac{d^2x}{dS^2} + C\tau\frac{dy}{dS} &= -Rx, \qquad (24.5)\\
C\tau &= K,
\end{aligned}
$$

or, alternatively, the pair of equations

$$
\begin{aligned}
-B\frac{d^2y}{dS^2} + K\frac{dx}{dS} - Ry &= 0,\\
B\frac{d^2x}{dS^2} + K\frac{dy}{dS} + Rx &= 0.
\end{aligned} \qquad (24.6)
$$

If the beam is hinged at its end, the boundary conditions are

$$x(0) = x(L) = y(0) = y(L). \qquad (24.7)$$

The combinations of K and R for which (24.6) and (24.7) allow nontrivial solutions determine the conditions of instability of the structure. In practice, as we have undertaken a linearized analysis, only the combinations giving the lowest eigenvalue of the problem are of interest regarding the onset of instability.

The situation is well illustrated by considering two limiting cases. If K is zero, then system (24.6) splits into two separate equations, from which we obtain the result that the critical value of R is exactly Euler's load $(\pi^2 B)/L^2$. Alternatively, if R is zero, then the first eigenvalue of the problem is $K = (2\pi B)/L$, and a behavior well known to designers of wires having a moderate flexibility occurs in which, when the torsional couple exceeds a certain critical value, the wire suddenly becomes a spiral.

Let us now consider the case of a cantilevered rod the cross-section of which has a flexural rigidity B with respect to one principal axis, which is much larger than the flexural rigidity with respect to the other principal axis, and much larger than the torsional rigidity. This would be the case for a rectangular cross-section in which one pair of sides is much longer than the other pair (Figure 24.2(a)). The rod is built in at one end and bent by a vertical transverse load R applied at the centroid of the other end, and placed in the plane of highest flexural rigidity. Let us choose a system of Cartesian axes such that the origin O coincides with the centroid of the clamped section, the X axis is the central line in the unstrained state, and the Y and Z axes are the principal axes of inertia of the initial section. If the length L of the beam is not too great and the load R is sufficiently small, the rod will simply bend in the YZ plane. However, if the magnitude of the load exceeds a certain limit, the plane bent state will become unstable, and the rod will be distorted and twisted out of its plane. This kind of buckling is known as Michell–Prandtl *instability*[40] (Figure 24.2(b)).

The equations that govern the problem are again Kirchhoff's equations (24.2), where the right-hand sides must now be modified. Let us call S the distance of a point on the central axis from the origin in the unstressed state; after the deformation, the initially rectilinear central axis is displaced into a curved arc, and a point of the deformed axis will have coordinates x, y, z with respect to the fixed axes, and its arc length measured from the origin will be S_1. Let us call x_1, y_1, and z_1 the coordinates of the centroid of the end of the rod where the load R is applied.

The equilibrium equations, written for an intermediate deformed section, become

$$\begin{aligned} A\kappa\ell_1 + B\kappa'\ell_2 + C\tau\ell_3 &= R(y_1 - y), \\ A\kappa m_1 + B\kappa' m_2 + C\tau m_3 &= -R(x_1 - x), \\ A\kappa n_1 + B\kappa' n_2 + C\tau n_3 &= 0, \end{aligned} \tag{24.8}$$

which again can be transformed into three differential equations in Euler's angles θ, ψ, and φ and in the displacements x, y, and z. Note that, in this case, x_1 and y_1 are also unknown.

In our linearized analysis of stability, let us use the direction cosines defined by the scheme (19.14), put θ nearly equal to $\frac{1}{2}\pi$, and assume that ψ and φ are small. At this level of approximation, we may take x_1 to be equal to L and x to be equal to S. The approximate version of (19.15) becomes

$$\kappa = -\frac{d\psi}{dS}, \quad \kappa' = \frac{d\theta}{dS}, \quad \tau = \frac{d\varphi}{dS}. \tag{24.9}$$

On recalling that A and C are small compared to B, (24.8) can be replaced by

$$\begin{aligned} -B\psi\frac{d\theta}{dS} + C\frac{d\varphi}{dS} &= R(y_1 - y), \\ B\frac{d\theta}{dS} &= -R(L - S), \\ A\frac{d\psi}{dS} + B\varphi\frac{d\theta}{dS} &= 0. \end{aligned} \tag{24.10}$$

As $dy/dS = m_3 = \sin\theta\sin\psi = \psi$, nearly, we derive from the first two equations the new equation

$$C\frac{d^2\varphi}{dS^2} + \frac{d}{dS}[R(L-S)\psi] = -R\psi, \tag{24.11}$$

[40] Love (1927, Art. 272). Note that, in contrast to the customary notation, the central line of the unstrained rod is denoted by X.

and from the last two equations we deduce the result

$$A\frac{d\psi}{dS} = R(L-S)\varphi, \tag{24.12}$$

and, eliminating ψ from (24.11) and (24.12), we obtain

$$C\frac{d^2\varphi}{dS^2} + \frac{R^2}{A}(L-S)^2\varphi = 0. \tag{24.13}$$

This is an ordinary linear differential equation in φ, which is reducible to a Bessel equation. In addition, we know that φ vanishes for $S = 0$, where the cross-section is clamped, and that $d\varphi/dS$ vanishes at the free end $S = L$.

The combination of the coefficients in (24.13), which corresponds to the first eigenvalue of the problem, is

$$\gamma^4 = \frac{R^2L^4}{AC},$$

where γ is a number very nearly equal to 2. It then follows that for a load R such that

$$R > \gamma^2\frac{\sqrt{AC}}{L^2},$$

the rod, bent in the plane of greatest flexural rigidity, becomes unstable. Note that the critical value of R does not depend on B.

Both Michell's and Prandtl's theories use as unknown functions Euler's angles of the strained axis and, consequently, are unable to embrace all the types of boundary condition that may occur in beam theory when these conditions also involve the displacements of the central line. To overcome this lack of completeness it is necessary to find the connection between Euler's angles and the displacements u, v, and w along the fixed axes X, Y, and Z.[41] These relations are again obtainable from the scheme (19.24), by observing that, if S_1 is the arc length of the deformed axis, we have

$$\frac{dx}{dS_1} = \ell_3, \quad \frac{dy}{dS_1} = m_3, \quad \frac{dz}{dS_1} = n_3, \tag{24.14}$$

that is to say, since $x = X + u$, $y = Y + v$, and $z = Z + w = S + w$, it follows that

$$\frac{du}{dS_1} = \sin\theta\cos\psi, \quad \frac{dv}{dS_1} = \sin\theta\sin\psi, \quad \frac{dS}{dS_1} + \frac{dw}{dS_1} = \cos\theta. \tag{24.15}$$

Since the elongation ε is small, differentiation with respect to S_1 may be replaced by differentiation with respect to S. Thus the second derivatives of u and v with respect to S become

$$u'' = \frac{d^2u}{dS^2} = \cos\theta\,\theta'\cos\psi - \sin\theta\sin\psi\,\psi',$$

$$v'' = \frac{d^2v}{dS^2} = \cos\theta\,\theta'\sin\psi + \sin\theta\cos\psi\,\psi',$$

[41] Celigoj (1979), with a slight, but unessential, change in sign.

where the primes denote differentiation with respect to S. From these equations we immediately have the relation

$$-u'v'' + u''v' = -\psi' \sin^2 \theta.$$

The following equalities are also easily verified

$$u'' \cos(\varphi + \psi) + v'' \sin(\varphi + \psi) = \theta' \cos\theta \cos\varphi + \psi' \sin\theta \sin\varphi,$$
$$u'' \sin(\varphi + \psi) - v'' \sin(\varphi + \psi) = \theta' \cos\theta \sin\varphi - \psi' \sin\theta \cos\varphi.$$

The result is that (19.15) can be written in the form

$$\begin{aligned}\kappa &= \theta' \sin\varphi - \psi' \sin\theta \cos\varphi = u'' \sin(\varphi + \psi) - v'' \cos(\varphi + \psi) + (1 - \cos\theta) \sin\varphi,\\ \kappa' &= \theta' \cos\varphi + \psi' \sin\theta \sin\varphi = u'' \cos(\varphi + \psi) + v'' \sin(\varphi + \psi) - (1 - \cos\theta) \cos\varphi,\\ \tau &= \varphi' + \psi' \cos\theta,\end{aligned} \tag{24.16}$$

or

$$\tau \cos\theta = \psi' + \varphi' \cos\theta - \psi' \sin^2\theta = \psi' + \varphi' \cos\theta - u'v'' + u''v'.$$

If we use the approximations of the linearized theory (24.4), and put

$$\beta = \varphi + \psi, \quad \kappa^{(1)} = -v'', \quad \kappa'^{(1)} = u'', \quad \tau^{(1)} = \beta',$$

we obtain an explicit formula for the first- and second-order terms in the curvatures and the twist:

$$\begin{aligned}\kappa^{(2)} &= u''\beta - v'' = u''\beta + \kappa^{(1)},\\ \kappa'^{(2)} &= u'' + v''\beta = v''\beta + \kappa'^{(1)},\\ \tau^{(2)} &= \beta' - u'v'' + u''v = -u'v'' + u''v' + \tau^{(1)}.\end{aligned} \tag{24.17}$$

If we accept the ordinary linear theory, according to which the stress couples are connected (as before, but with a change of superscripts) with the curvatures and the twist by equations of the form

$$G^{(2)} = A\kappa^{(2)}, \quad G'^{(2)} = B\kappa'^{(2)}, \quad H^{(2)} = C\tau^{(2)}, \tag{24.18}$$

we can write (24.18) in the more explicit form

$$\begin{aligned}G^{(2)} &= -Av'' + \frac{A}{B} G'^{(1)}\beta,\\ G'^{(2)} &= Bu'' - \frac{B}{A} G^{(1)}\beta,\\ H^{(2)} &= C\beta' + \frac{C}{A} G^{(1)}u' + \frac{C}{B} G'^{(1)}v',\end{aligned} \tag{24.19}$$

which show that, in contrast to the first-order theory, each stress couple in the second-order theory is connected with all the kinematical functions u', v', and β and their derivatives. In other words, in the second-order theory, a twist β about the Z axis generates stress couples round the other axes, and so on.

An illustrative application of the theory is the combined application of a longitudinal compressive force $T = -P$ along the Z axis, two transverse distributions of uniform loads along the longitudinal axis, and possible bending moments at the ends. If the transverse loading is represented by f_x and f_y per unit length along the

X and Y axes, respectively, the field equations, derived from Kirchhoff's theory with the application of (24.19) become

$$
\begin{aligned}
Av^{iv} + Pv'' + \left(1 - \frac{A}{B}\right)(G'^{(1)}\beta)'' &= f_y, \\
Bu^{iv} + Pu'' + \left(1 - \frac{B}{A}\right)(G^{(1)}\beta)'' &= f_x, \\
C\beta'' - \left(1 - \frac{C}{A}\right)G^{(1)}u'' - \left(1 - \frac{C}{B}\right)G'^{(1)}v'' &= 0.
\end{aligned}
\tag{24.20}
$$

It is easy to see that these equations are not independent but coupled, and so their solution, although simple, may be cumbersome. But, to illustrate the consequences of (24.20), let us study the stability of the solutions under the action of a pure bending couple M_0 about the X axis ($f_y = f_x = G'^{(1)} = P = 0$ and $G^{(1)} = M_0$), and assume that A is much larger than B or C. If the lateral displacements vanish at the ends, in addition to the stress couples (except M_0), a simple computation yields the condition of critical equilibrium

$$
\left(\frac{\pi}{L}\right)^2 = \left(1 - \frac{B}{A}\right)\left(1 - \frac{C}{A}\right)\frac{M_0^2}{BC}, \tag{24.21}
$$

a formula which was first found by Chwalla (1939, 1943).

If we work only with the linear terms, we obtain (Vielsack 1975)

$$
\left(\frac{\pi}{L}\right)^2 = \left(1 - \frac{B}{A}\right)\frac{M_0^2}{BC}, \tag{24.22}
$$

which, for $A \gg B$, C, can be approximated by

$$
\left(\frac{\pi}{L}\right)^2 = \frac{M_0^2}{BC}, \tag{24.23}
$$

which is the classical formula obtained by Prandtl (1899).

The variational method is the most powerful one for treating the linear stability of elastic structures. The method is based on the fact that the critical compressive thrust P on a simple rod clamped at its ends minimizes the ratio

$$
P = \frac{\int_0^L Bv''^2\, dz}{\int_0^L v'^2\, dz}, \tag{24.24}
$$

where L is the length and B is the flexural rigidity. The minimum is contained in the class of functions $v(z)$, which have piecewise continuous second derivatives, such that

$$
v(0) = v'(0) = v(L) = v'(L) = 0.
$$

This ratio is known as *Rayleigh's quotient*.

The variational formulation of problems of buckling is not restricted to beams, but can also be extended to cases of structural stability involving frames, plates, or shells, provided that the Rayleigh quotient (24.24) is defined as the ratio

$$\lambda = \frac{W}{L} \tag{24.25}$$

where W is the strain energy arising from the buckling and L is the change in the potential energy of the loads due to buckling. If the loads acting on the structure are proportionally increased by a factor λ, the minimum of the ratio (24.25) defines the value of the multiplicative parameter of the loads at which the elastic equilibrium becomes unstable.

Once the condition defining the onset of structural instability has been formulated in the form of a minimum principle, it is possible to apply the general theory of minima to determine the dependence of the solutions on the data. For instance, on examining (24.24), it is clear that an increase in the flexural rigidity B will not decrease the numerator of the right-hand side of (24.24), and so P does not decrease. Alternatively, a relaxation of the boundary conditions, such as, for example, the suppression of one of the end constraints on $v(z)$, will enlarge the class of admissible functions while having the effect of not increasing the minimum of P. All these results have been codified by Courant (Courant and Hilbert 1953, Vol. 1, Chap. VI) in the form of a list of theorems on the variational characterization of eigenvalues of self-adjoint problems, such as the case of elastic stability under conservative loads. The consequence has been that there is now a widely held conjecture that reinforcing a structure, either by increasing its stiffness or by providing additional constraints, increases (not decreases) the critical load. However, although accepted almost as dogma, the principle has been proved false, at least when the restrictions are not chosen carefully (Tarnai 1980).

In order to elucidate how the additional constraints modify the critical load, it is necessary to distinguish two kinds of constraint: those that impede the displacement components arising during buckling and acting in the buckling directions; and those that prevent the other displacement components, and not occurring in the equations of buckling. The former are constraints on solutions of the eigenvalue problem and called *proper constraints*; the latter are constraints on displacement, not found in the eigenvalue problem, and are called *improper constraints*. For example, in the case of in-plane buckling of a compressed straight bar, the lateral components of displacement of points on the axis will be eigen-displacements; but the axial component of the displacement will not.

Provided that the structure is of a linearly elastic material admitting a stored energy function and the external loads are conservative, only bifurcation problems need to be considered, so that the critical load factor means the load factor associated with a point of bifurcation. The fundamental result is then that an additional constraint on any component of the displacement that is found in the buckling eigenvalue problem cannot decrease the critical load factor, but any additional constraint on any component of the displacement that is not found in the eigenvalue problem can decrease the critical load factor. It then follows that the corollaries of the variational theory of eigenvalues apply only to constraints on eigen-displacements.

In order to show the necessity of narrowing the class of admissible displacements so that the variational theory can be applied, let us call e the class of eigen-displacements and f the class of displacements orthogonal to e. Instead of the Rayleigh quotient (24.25), let us consider its reciprocal

$$\frac{1}{\lambda} = \frac{L}{W}.$$

We do this because, in certain problems, the infimum of Rayleigh's quotient may be equal to $-\infty$, while the supremum of its reciprocal remains bounded. Suppose now that we add a constraint such that the strain energy becomes W_1 and the work done by the external forces, with respect to a unit value of the load factor λ, becomes L_1. The Rayleigh quotients for the original and modified structures, regarded as a function of e and f, are then

$$\frac{1}{\lambda(e,f)} = \frac{L(e,f)}{W(e,f)}, \quad \frac{1}{\lambda_1(e,f)} = \frac{L(e,f)}{W_1(e,f)}. \tag{24.26}$$

If the modified structure is obtained by increasing the stiffness of a proper constraint, then we can keep the displacements of the type f fixed, and regard e as admissible functions. In this case the strain energy W_1 has the form $W_1 = W + \Delta W$, with $\Delta W \geqslant 0$, but the work of external forces remains unchanged ($L_1 = L$). Thus, a comparison of the two values in (24.26) yields

$$\frac{1}{\lambda(e,f)} = \frac{L(e,f)}{W(e,f)} \geqslant \frac{L(e,f)}{W(e,f) + \Delta W(e,f)} = \frac{L_1(e,f)}{W_1(e,f)} = \frac{1}{\lambda_1(e,f)}, \tag{24.27}$$

and this inequality holds in the limit for the extremum of (24.27) with respect to e. On the other hand, if we increase the stiffness of an improper constraint, then the admissible displacements are of both class e and of class f. In this case, provided that W_1 is still larger than W, L_1 is not necessarily equal to L, and the inequality

$$\frac{L(e,f)}{W(e,f)} \geqslant \frac{L_1(e,f)}{W_1(e,f)}, \tag{24.28}$$

is no longer valid.

As an example, let us consider a single elastic bar supported at both ends by two linearly elastic springs with rigidities c_A and c_B (Figure 24.3). Obviously, the spring forces are $P_A = c_A w_A$ and $P_B = c_B w_B$, where w_A and w_B are the displacements at A and B. The load P is in equilibrium, so that $P = P_A + P_B$. Furthermore, up to the onset of instability, the shape of the bar is straight, and so the displacements of the end points are equal ($w_A = w_B$). At the point of bifurcation the spring B sustains the Euler load $P_E = (\pi^2 EJ)/L^2$. Thus the critical load for the system is

$$P_{cr} = \left(1 + \frac{c_A}{c_B}\right) P_E. \tag{24.29}$$

This expression shows that an increase in c_B, the stiffness of the spring B, decreases the critical load.

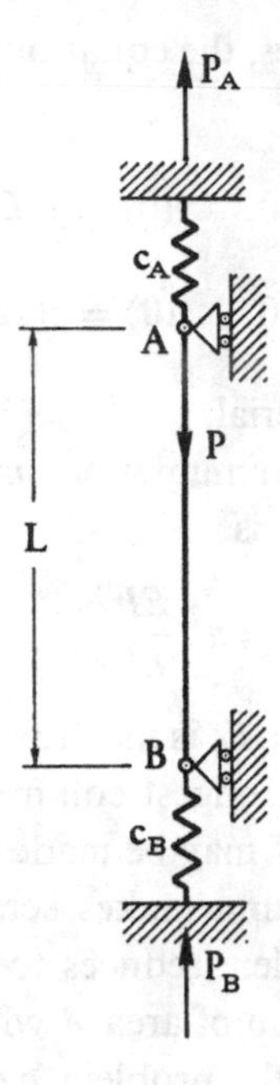

Fig. 24.3

The linearized version of the problem of the stability of a compressed column can sometimes be formulated in an inverse form. In the classical formulation, the shape of the cross-section, the length, and the boundary conditions are prescribed, and the critical load is the unknown. However, the problem is very often formulated differently. The designer knows the length of the column, the end conditions, and the total volume of material to be used in the construction of the profile. The objective then is to determine the shape of the column that has the largest critical buckling load, provided that the cross-section, although variable, remains convex. This problem, which is of considerable importance in technical applications, has been studied at different levels of generality, depending on the nature of the restrictive control imposed on the shape of the cross-section and on the way in which it varies along the longitudinal axis.

In a first approach, we begin by considering cylindrical columns of arbitrary cross-section (Keller 1960). Let L and V denote the length and the volume of the column and $A = V/L$ be the area of its cross-section. Let the x axis pass through the centroids of the cross-sections of the column in its straight, unbuckled state. Let y and z be a system of orthogonal axes, placed in the plane perpendicular to the x axis and containing the end $x = 0$ of the column. Axes y and z are the principal axes of the inertia of the cross-section and may be chosen so that I', the moment of inertia about the y axis, is the largest, and I, the moment of inertia about the z axis, is the smallest.

Let a compressive load T_0 be applied along the longitudinal axis when the column is pinned at its ends. Under buckling conditions, with deflections of the central line

$y(x)$ and $z(x)$ parallel to the y and z axes, the equations governing the deformed state are

$$y_{xx} + \frac{T_0}{EI}\, y = 0, \quad y(0) = y(L) = 0, \tag{24.30}$$

$$z_{xx} + \frac{T_0}{EI'}\, z = 0, \quad z(0) = z(L) = 0, \tag{24.31}$$

E being Young's modulus of the material.

As we have supposed that I is the minimum moment of inertia, the critical value of T_0, at which Euler's instability occurs, is

$$T_0 = \pi^2 \frac{EI}{L^2}. \tag{24.32}$$

The strongest column is that for which T_0 is as large as possible. Since E, L, and V and therefore A, are prescribed, the strongest column is one for which I is largest. However, if the section is not convex, I may be made arbitrarily large; therefore, the problem of finding the strongest column makes sense only if the cross-section is required to be convex. Thus the problem reduces to a purely geometrical problem of determining the convex plane domain of area A which has the largest I. It is easy to show that the domain that solves the problem has the same moment of inertia about every axis passing through the centroid and lying in the plane of the cross-section. However, the domain is not a circle but an equilateral triangle, the moment of inertia and critical load T_0 of which, corresponding to a given area A, are given by

$$I = \frac{A^2}{6\sqrt{3}} = 0.096225\, A^2, \tag{24.33a}$$

$$T_0 = \frac{\pi^2 E A^2}{6\sqrt{3} L^2} = \frac{EA^2}{L^2}\, 0.949703. \tag{24.33b}$$

If the cross-section were circular, the numerical factor in (24.33b) would be $\pi/4 = 0.785$, that is about 20.9% less.

So far we have assumed that the cross-section is constant. Let us now remove this restriction and assume that all cross-sections are similar and that the principal axes of inertia are parallel, but the area $A(x)$ may vary subject to the condition

$$\int_0^L A(x)\, dx = V. \tag{24.34}$$

The moments of inertia can be related to A by the equations

$$I(x) = \alpha A^2(x), \quad I'(x) = \alpha' A^2(x),$$

where α and α' are two constants depending on the shape of the cross-section. Then the equation for the buckled state becomes

$$y_{xx} + \frac{T_0}{E\alpha A^2(x)}\, y = 0, \quad y(0) = y(L) = 0,$$

or, after the introduction of the new variable $\xi = x/L$ and the parameter $\lambda = (T_0 L^2)/(\alpha E)$,

$$y'' + \frac{\lambda}{A^2(\xi)}\, y = 0, \quad y(0) = y(1) = 0, \tag{24.35}$$

$$\int_0^1 A(\xi)\, d\xi = \frac{V}{L}. \tag{24.36}$$

We want to find the distribution of $A(\xi)$ that maximizes the lowest eigenvalue of (24.35). For this purpose we introduce a family of functions $A(\xi, \varepsilon)$, which are smoothly dependent on the parameter ε, satisfy (24.36), and are such that $A(\xi, 0) = A(\xi)$. Then, for each ε (24.35) will have a solution $y(\xi, \varepsilon)$ with the corresponding eigenvalue $\lambda(\varepsilon)$, both of which are differentiable functions of ε. Next, we differentiate (24.35) and (24.36) with respect to ε, obtaining

$$y_\varepsilon'' + \frac{\lambda}{A^2}\, y_\varepsilon = 2\lambda A_\varepsilon \frac{1}{A^3}\, y - \lambda_\varepsilon \frac{1}{A^2}\, y, \quad y_\varepsilon(0) = y_\varepsilon(1) = 0, \tag{24.37}$$

$$\int_0^1 A_\varepsilon\, d\xi = 0. \tag{24.38}$$

The derivative y_ε satisfies an inhomogeneous linear differential equation and y satisfies the corresponding homogeneous problem. If we multiply both sides of (24.37) by y and integrate with respect to ξ from 0 to 1, we find

$$\int_0^1 \left(2\lambda A_\varepsilon \frac{1}{A^3}\, y - \lambda_\varepsilon \frac{1}{A^2}\, y \right) y\, d\xi = 0,$$

but, at $\varepsilon = 0$, λ_ε is a maximum, so $\lambda_\varepsilon = 0$. Thus the previous equation becomes

$$\int_0^1 A_\varepsilon \frac{1}{A^3}\, y^2\, d\xi = 0. \tag{24.39}$$

Since this condition must hold for every $A_\varepsilon(\xi)$ satisfying (24.38), it follows that $(1/A^3)y^2$ is a constant, which we denote by $c^{\frac{3}{2}}$. Thus, for the strongest column we have

$$y^2 = c^{\frac{3}{2}} A^3. \tag{24.40}$$

To find $y(\xi)$, we eliminate A from (24.35) by means of (24.40), and obtain

$$y'' + \lambda c y^{-\frac{1}{3}} = 0, \quad y(0) = y(1) = 0.$$

The solution of this equation is

$$\left| \xi - \frac{1}{2} \right| = \frac{1}{2} + \frac{1}{\pi} \left(\frac{y}{y_0} \right)^{\frac{1}{3}} \left[1 - \left(\frac{y}{y_0} \right)^{\frac{2}{3}} \right]^{\frac{1}{2}} - \frac{1}{\pi} \sin^{-1} \left(\frac{y}{y_0} \right)^{\frac{1}{3}}, \tag{24.41}$$

where

$$y_0^{\frac{2}{3}} = \frac{2}{3\pi} (\lambda c)^{\frac{1}{2}}.$$

On replacing $y(\xi)$ in (24.41) by A, with the aid of (24.40) we obtain the following expression for $A(\xi)$:

$$\left|\xi - \frac{1}{2}\right| = \frac{1}{2} + \frac{1}{\pi}\left[\frac{A}{A_0} - \left(\frac{A}{A_0}\right)^2\right]^{\frac{1}{2}} - \frac{1}{\pi}\sin^{-1}\left(\frac{A}{A_0}\right)^{\frac{1}{2}}, \qquad (24.42)$$

where the constant $A_0 = A(\frac{1}{2})$ can be determined as a function of V/L by inserting (24.42) in (24.36). This yields

$$\frac{V}{L} = 2\int_0^{\frac{1}{2}} A(\xi)\,d\xi = 2\int_0^{A_0} \xi\,dA = \frac{3}{4}A_0. \qquad (24.43)$$

In order to find λ, let us use (24.40) to eliminate y from (24.35), which then becomes

$$AA'' + \frac{1}{2}(A')^2 + \frac{2}{3}\lambda = 0. \qquad (24.44)$$

From (24.42) we find the values $A'(\frac{1}{2}) = 0$ and $A''(\frac{1}{2}) = (\pi^2 A_0)/2$. Therefore, upon inserting these values in (24.44) and using (24.43) for A_0, we obtain

$$\lambda = \frac{4}{3}\pi^2\left(\frac{V}{L}\right)^2,$$

and the corresponding value of the critical load is

$$T_0 = \frac{4}{3}\pi^2\alpha\frac{V^2E}{L^4}. \qquad (24.45)$$

This is $\frac{4}{3}$ times as large as the value for a cylindrical column of the same volume, length, and cross-section.

If we combine this result and the one relative to a constant equilateral triangular cross-section, we have to put $\alpha = 1/(6\sqrt{3})$, and we then have

$$T_0\frac{2\pi^2}{9\sqrt{3}}\frac{V^2E}{L^4} = \frac{V^2E}{L^4}\,1.26627. \qquad (24.46)$$

The critical buckling load is 1.612 times as large as that for a circular cylindrical column. We may ask if a further increase in critical load can be achieved by removing the restriction that the cross-sections remain similar, that is to say, by allowing a twist τ_0 in the unstressed state. However, the surprising answer is that (24.46) also holds for columns in the pre-twisted state (Keller 1960).

The results just found may appear very restrictive, because only the particular boundary condition of a pinned beam has been considered. The problem thus arises of trying to find out whether the method still works for more complicated end constraints, and, if so, of finding the necessary modifications to the procedure.

In order to solve these problems (Tadjabakhsh and Keller 1962), let us remember that the equations governing the buckling of a column in a plane, say the xz plane, under the assumptions of the linearized theory, are an equilibrium equation

$$-\frac{d^2G'}{dx^2} + T_0 y_{xx} = 0, \tag{24.47}$$

where G' is the bending moment about the y axis, and a moment–curvature equation of the form

$$G' = EI\kappa' = -EIy_{xx}. \tag{24.48}$$

Assuming that the end $x = 0$ is clamped, the boundary conditions are $y(0) = y_x(0) = 0$. At the other end $x = L$, we consider two possibilities: either that it is clamped $[y(L) = y_x(L) = 0]$ or that it is free

$$G'(1) = 0, \quad \frac{d}{dx} G'(L) - T_0 y_x(L) = 0. \tag{24.49}$$

When we combine (24.47) and (24.48), we obtain a fourth-order differential equation for $y(x)$, associated with two types of boundary conditions. In order to solve these problems we introduce the new independent variable $\xi = X/L$, and the auxiliary function

$$\varphi(\xi) = L^2 A^2(\xi L) y_{xx}(\xi L), \tag{24.50}$$

and we define λ as before:

$$\lambda = \frac{T_0 L^2}{E\alpha}. \tag{24.51}$$

Then, $\varphi(\xi)$ must satisfy the following second-order equation:

$$\varphi_{\xi\xi} + \frac{\lambda}{A^2}\varphi = 0. \tag{24.52}$$

In order to express the boundary conditions in terms of φ, we integrate the preceding equation once with respect to ξ, from $\xi = 0$, after having replaced φ/A^2 by y_{xx} according to (24.50). On using the boundary conditions $y(0) = y_x(0) = 0$, we obtain

$$y_x(\xi L) = \frac{1}{\lambda L}[\varphi_\xi(0) - \varphi_\xi(\xi)], \tag{24.53}$$

and a subsequent integration with respect to ξ from $\xi = 0$ to ξ yields

$$y(\xi L) = \lambda[\xi\varphi_\xi(0) - \varphi(\xi) + \varphi(0)]. \tag{24.54}$$

This means that, in terms of $\varphi(\xi)$, the boundary conditions of a doubly clamped beam at $\xi = 1$ become

$$\begin{aligned} \varphi_\xi(0) - \varphi_\xi(1) &= 0, \\ \varphi_\xi(0) - \varphi(1) + \varphi(0) &= 0, \end{aligned} \tag{24.55}$$

and the analogous conditions for a beam free at $\xi = 1$ are

$$\begin{aligned} \varphi(1) &= 0, \\ \varphi_\xi(0) &= 0. \end{aligned} \tag{24.56}$$

We have already seen that, as a consequence of the condition of optimality, the cube of the maximizing function $A(\xi)$ is proportional to the square of the corresponding solution $\varphi(\xi)$. Since $\varphi(\xi)$ is the solution of a homogeneous problem, we may multiply it by a constant so that the proportionality factor is unity, and then

$$\varphi^2 = A^3. \tag{24.57}$$

In order to determine $\varphi(\xi)$, we eliminate A from (24.52) by means of the above equality, and obtain

$$\varphi_{\xi\xi} + \lambda\varphi^{-\frac{1}{3}} = 0. \tag{24.58}$$

The general solution of this equation and the corresponding $A(\xi)$, determined with the aid of (24.57), can be expressed in terms of a function $\theta(\xi)$ with the conditions

$$\varphi(\xi) = A_0^{\frac{3}{2}} \sin^3 \theta(\xi), \tag{24.59}$$

$$A(\xi) = A_0 \sin^2 \theta(\xi), \tag{24.60}$$

$$\theta - \frac{1}{2} \sin 2\theta + a = 2\left(\frac{\lambda}{3}\right)^{\frac{1}{2}} \frac{1}{A_0}\, \xi, \tag{24.61}$$

where A_0 and a are constants to be determined in each case so that $\varphi(\xi)$ satisfies the appropriate pair of boundary conditions. To this end, we evaluate, from (24.59) and (24.61),

$$\varphi_\xi(\xi) = (3\lambda_0 A_0)^{\frac{1}{2}} \cos \theta(\xi). \tag{24.62}$$

Let us first consider the clamped–clamped case, in which (24.55) hold, and use the preceding formulae to write

$$\cos \theta(0) - \cos \theta(1) = 0, \tag{24.63}$$

$$3\lambda \frac{1}{A_0} \cos \theta(0) + \sin^3 \theta(0) - \sin^3 \theta(1) = 0. \tag{24.64}$$

These equations are satisfied if we set $\theta(0) = -(\pi/2)$ and $\theta(1) = \frac{3}{2}\pi$. Any other solution differs from this by the addition of an integer multiple of π to both $\theta(0)$ and $\theta(1)$, or by the addition of an integer multiple of 2π to $\theta(1)$, or by both additions. As A and φ, given by (24.59) and (24.60), are periodic functions of θ (ignoring the sign of φ, which is arbitrary), the addition of multiples of π to $\theta(1)$ does not yield a different solution. However, the addition of an integer multiple of 2π to $\theta(1)$ does yield a different solution, because φ has additional nodes. As we are seeking the lowest eigenvalue, we must consider the eigenfunction with the fewest nodes, which is the one given by the foregoing choice. By using these values of $\theta(0)$ and $\theta(1)$ in (24.61), we obtain

$$a = \frac{\pi}{2}, \quad A_0 = \frac{1}{\pi}\left(\frac{\lambda}{3}\right)^{\frac{1}{2}},$$

and then (24.61) becomes

$$\theta - \frac{1}{2} \sin 2\theta + \frac{\pi}{2} = 2\pi\xi. \tag{24.65}$$

From (24.53) and (24.62) we see that the slope y_x vanishes only at the ends and at the midpoint of the column.

In order to determine λ, we use (24.60) and (24.65) in (24.36), and obtain

$$\frac{V}{L} = \frac{A_0}{\pi} \int_{-\frac{\pi}{2}}^{\frac{3}{2}\pi} \sin^4 \theta \, d\theta = \frac{3}{4} A_0, \tag{24.66}$$

which gives the results

$$A_0 = \frac{4}{3} \frac{V}{L},$$
$$\lambda = 3\pi^2 A_0^2 = \frac{16}{3} \pi^2 \left(\frac{L}{V}\right)^2. \tag{24.67}$$

The area of the cross-section, given by (24.60), can now be written as

$$A(\xi) = \frac{4}{3} \frac{V}{L} \sin^2 \theta(\xi), \tag{24.68}$$

with θ given by (24.65). From (24.68) we find the results

$$A(0) = A\left(\frac{1}{2}\right) = A(1) + \frac{4}{3} \frac{V}{L},$$
$$A\left(\frac{1}{4}\right) = A\left(\frac{3}{4}\right) = 0.$$

This means that the column is shaped in such a way that the cross-section is equal to A_0 at the ends and at the midpoint, and is equal to zero at the points on the abscissae $\xi = \frac{1}{4}$ and $\xi = \frac{3}{4}$ (Figure 24.4).

The solution just found also provides the solution for the clamped–free case. In fact, the profile of the strongest column clamped at one end and free at the other can be obtained from the diagram shown in Figure 24.4, if one stretches the scale of ξ by a factor of 4. The solution for the clamped–free case is

$$A(\xi) = \frac{4}{3} \frac{V}{L} \sin^2 \theta(\xi), \quad -\frac{\pi}{2} \leqslant \theta \leqslant 0, \tag{24.69}$$

$$\theta - \frac{1}{2} \sin 2\theta + \frac{\pi}{2} = \frac{\pi}{2} \xi, \quad 0 \leqslant \xi \leqslant 1, \tag{24.70}$$

$$\lambda = \frac{1}{3} \pi^2 \left(\frac{V}{L}\right)^2.$$

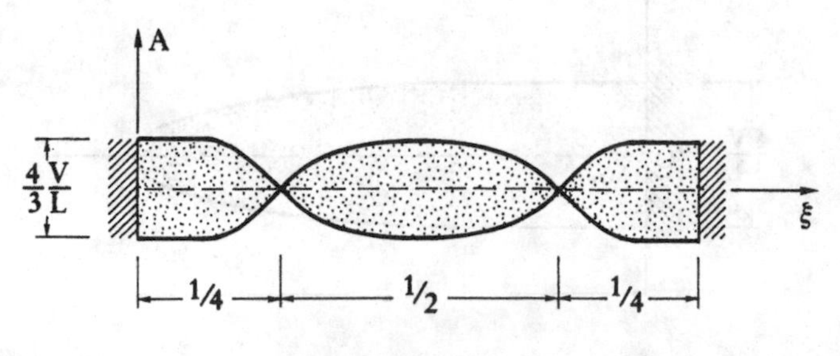

Fig. 24.4

Here again, the buckling load is $\frac{4}{3}$ that of a cylindrical column, and we can verify that this solution satisfies all the boundary conditions of the problem. The contour of the profile is shown in Figure 24.5.

Notwithstanding its simplicity, the above result has been queried in connection with some points (Olhoff and Rasmussen 1977). The clamped–clamped case is more complex than has been hitherto assumed, because it is necessary to take into account the possibility that the optimum buckling load is a double eigenvalue, so that the structure of the associated eigenfunctions corresponding to a double eigenvalue is larger than the one corresponding to a single eigenvalue. We may find shapes of other columns that have equal volume and length but possess higher values of the fundamental buckling load.

In order to check the nonoptimality of the column shown in Figure 24.4, we introduce a function $\alpha_1(\xi) = A(\xi)(L/V)$, so that the condition (24.36) can be written in the form

$$\int_0^1 \alpha_1(\xi)\, d\xi = 1, \tag{24.71}$$

and the differential equation (24.52), expressed in terms of $y(\xi)$, becomes

$$(A^2 y'')'' + \lambda y'' = 0,$$

or, alternatively,

$$(\alpha_1^2 y'')'' = -\bar{\lambda} y'', \quad \bar{\lambda} = \frac{T_0 L^4}{E\alpha V^2}, \tag{24.72}$$

with the boundary conditions

$$y(0) = y'(0) = y(1) = y'(1). \tag{24.73}$$

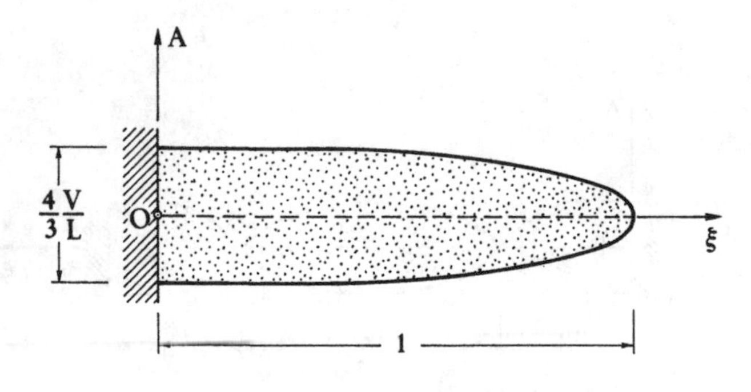

Fig. 24.5

On introducing Rayleigh's quotient

$$\bar{\lambda} = \frac{\int_0^1 \alpha_1^2 y''^2 \, d\xi}{\int_0^1 y'^2 \, d\xi}, \tag{24.74}$$

the optimization problem consists of determining the function $\alpha_1(\xi)$ that maximizes (24.74) in the class of continuous functions with piecewise continuous second derivatives, these functions satisfying the boundary conditions (24.73) and the constraint (24.71). On forming a functional based on (24.74) together with (24.71), a Lagrangian multiplier 2β can be introduced, and the necessary condition for the functional to be stationary with respect to arbitrary admissible variations of $\alpha_1(\xi)$ is the requirement

$$\alpha_1(\xi) = \frac{\beta}{y''^2}. \tag{24.75}$$

Deriving the condition for the functional to be stationary with respect to variations in $y(\xi)$, recovers the differential equation (24.72) along with the continuity conditions of the functions $(\alpha_1^2 y'')' + \bar{\lambda} y'$ and $\alpha_1^2 y''$. At an interior point, where the bending moment $\alpha_1^2 y''$ vanishes, a discontinuity of the slope y' is possible because α_1 vanishes when y'' tends to infinity (see (24.75)), and such behavior of y' would then imply a discontinuity of the shear force $-(\alpha_1^2 y'')'$.

The four coupled differential and integral equations (24.71), (24.72), (24.74), and (24.75) form a system from which we must derive, at least in principle, the unknowns $\alpha_1(\xi)$, $y(\xi)$, $\bar{\lambda}$, and β.

This is the customary, or single-mode, formulation. Let us now assume that the problem admits two eigenfunctions, say $y_1(\xi)$ and $y_2(\xi)$, corresponding to the same eigenvalue $\bar{\lambda}$, which is expressed as

$$\bar{\lambda} = \int_0^1 \alpha_1^2 y_i''^2 \, d\xi, \quad (i = 1, 2). \tag{24.76}$$

When $\bar{\lambda}$ is double, y_1 and y_2 need not to be mutually orthogonal, but they can be normalized to give the result

$$\int_0^1 y_i'^2 \, d\xi = 1, \quad (i = 1, 2). \tag{24.77}$$

Now, for greater generality, let us introduce the additional constraint $\alpha_1(\xi) \geqslant \bar{\alpha}_1 \geqslant 0$ on this design variable. This minimum constraint can be expressed by means of an auxiliary nonnegative function $g(\xi)$, such that

$$g^2(\xi) = \alpha_1(\xi) - \bar{\alpha}_1. \tag{24.78}$$

The function $g(\xi)$ is also said to be a *slack variable.*

We then define the functional

$$\lambda^* = \int_0^1 \alpha_1^2 y''^2\, d\xi - \gamma\left(\int_0^1 \alpha_1^2 y_1''^2\, d\xi - \int_0^1 \alpha_1 y_2''^2\, d\xi\right) - \sum_{i=1}^{2} \eta_i \left(\int_0^1 y_i'^2\, d\xi - 1\right)$$
$$- 2\beta\left(\int_0^1 \alpha_1\, d\xi - 1\right) + 2\int_0^1 \mu(\xi)[g^2 - \alpha_1(\xi) + \bar{\alpha}_1]\, d\xi, \tag{24.79}$$

which is introduced to replace the numerator of the Rayleigh quotient in (24.74). This ensures that, with λ^* from (24.79), a double normalized eigenvalue is obtained in (24.76), equations (24.77) and (24.71) on $\alpha_1(\xi)$ having been taken into account. The quantities γ, η_i, 2β, and $2\mu(\xi)$ are the Lagrangean multipliers associated with additional constraints on the first term in (24.79), and are effected by the functions $\alpha_1(\xi)$, $g(\xi)$, $y_1(\xi)$, and $y_2(\xi)$, respectively. The Euler equations ensuring the stationary conditions of the functional λ^* with respect to variations in these functions are as follows. Variations in $\alpha_1(\xi)$ and $g(\xi)$ lead to

$$\alpha_1[(1-\gamma)y_1''^2 + \gamma y_2''^2] - \beta + \mu(\xi) = 0, \tag{24.80}$$

$$\mu(\xi)g(\xi) = 0. \tag{24.81}$$

In order to solve (24.80) it is necessary to remove the explicit appearance of the function $\mu(\xi)$. Let us call I_u the subinterval of [0, 1] in which $g(\xi) \neq 0$, and call I_c the complementary subinterval where $g(\xi) \equiv 0$. For $\xi \in I_u$ we then have $\alpha_1(\xi) > \bar{\alpha}_1$, that is to say the area of the cross-section is unconstrained, which gives $\mu(\xi) \equiv 0$. For $\xi \in I_c$ we have $\alpha_1(\xi) \equiv \bar{\alpha}_1$, and the cross-section is constrained. It then follows that (24.80) and (24.81) can be replaced by

$$\alpha_1(\xi) = \begin{cases} \dfrac{\beta}{(1-\gamma)y_1''^2 + \gamma y_2''^2}, & \text{for } \xi \in I_u, \\ \bar{\alpha}_1, & \text{for } \xi \in I_c. \end{cases} \tag{24.82}$$

The conditions for λ^* to be stationary under arbitrary admissible variations in $y_i(\xi)$ yield the Euler differential equations for $y_i(\xi)$, and the Lagrangian multipliers η_i are determined from (24.77). The differential equations which must be satisfied by $y_i(\xi)$ are

$$(\alpha_1^2 y_i'')'' = -\bar{\lambda} y_i'',$$
$$y_i(0) = y_i'(0) = y_i(1) = y_i'(1) = 0, \quad (i = 1, 2) \tag{24.83}$$

with y_i subject to the normalization (24.77).

In order to express the Lagrangean multipliers β and γ in terms of the other quantities, we substitute (24.82) in the volume constraint (24.71) and obtain

$$\beta = \frac{1 - \bar{\alpha}_1 \displaystyle\int_{I_c} d\xi}{\displaystyle\int_{I_u} \frac{d\xi}{(1-\gamma)y_1''^2 + \gamma y_2''^2}}. \tag{24.84}$$

Then, on subtracting the two equations comprising (24.76), substituting $\alpha_1(\xi)$ from (24.82), and using (24.84), we obtain the following implicit equation for γ:

$$\int_{I_u} \frac{y_1''^2 - y_2''^2}{(1-\gamma)y_1''^2 + \gamma y_2''^2}\, d\xi + \bar{\alpha}_1^2 \left[\frac{\displaystyle\int_{I_u} \frac{d\xi}{(1-\gamma)y_1''^2 + \gamma y_2''^2}}{1 - \bar{\alpha}_1 \displaystyle\int_{I_c} d\xi} \right]^2 \int_{I_c} (y_1''^2 - y_2''^2)\, d\xi = 0. \quad (24.85)$$

Finally, substitution of (24.82) and (24.84) in (24.76) yields the explicit form of $\bar{\lambda}$:

$$\bar{\lambda} = \left[\frac{1 - \bar{\alpha} \displaystyle\int_{I_c} d\xi}{\displaystyle\int_{I_u} \frac{d\xi}{(1-\gamma)y_1''^2 + \gamma y_2''^2}} \right]^2 \int_{I_u} \frac{y_1''^2\, d\xi}{[(1-\gamma)y_1''^2 + y_2''^2]^2} + \bar{\alpha}_1^2 \int_{I_c} y_1''^2\, d\xi. \quad (24.86)$$

Equations (24.82)–(24.86) constitute a strongly coupled nonlinear eigenvalue problem in the unknowns $\bar{\lambda}$, $\alpha_1(\xi)$ (and, thereby, the subintervals I_u and I_c), $y_i(\xi)$, β, and γ. If we drop the constraint (24.78), we must put $\mu(\xi) = 0$ in (24.79), and we obtain an unconstrained optimization problem as a special case of the foregoing problem by putting $\bar{\alpha}_1 = 0$, which implies that I_c vanishes.

On assuming $\gamma = 0$ and $\eta_2 = 0$ in the functional λ^*, the bimodal optimization formulation reduces to a single-mode problem. This leads to a restriction of the class in which the optimum is sought, and consequently to a higher value of the fundamental buckling load.

The result just obtained can also be improved upon (Barnes 1988). The solutions found for the case of a column clamped at each end show that the extreme shapes of the unconstrained problem have two points at which the cross-section vanishes. This means that the coefficient of y'' becomes singular at some points in the intervals $0 \leqslant \xi \leqslant 1$, with the consequence that, if we do not restrict the class of admissible functions, as has been done before, there is no solution in the class of shapes allowed by the mathematical eigenvalue problem, and the supremum of the buckling load is not attained for any reasonable shape.

To explain this anomaly, let us consider the case of a column pinned at each end, so that the buckling load is the first eigenvalue λ_1 of the Sturm–Liouville problem:

$$y'' + \lambda \frac{1}{A^2} y = 0, \quad y(0) = y(1) = 0. \quad (24.87)$$

This eigenvalue problem is a good mathematical model of the buckling problem if it is well posed. To be precise about this concept, let $\mathscr{C}$ be the class of all shape functions $A(\xi)$ that satisfy the following four conditions:

(1) $\lambda_1(A)$ exists.
(2) $A(\xi) \geqslant 0$. (24.88)
(3) $\int_0^1 A(\xi) d\xi = V$.
(4) The functional $\lambda_1(A)$ is continuous in A. This means that, if B is any function that satisfies conditions (1)–(3), then for each $\varepsilon > 0$ there is a value $\delta_\varepsilon > 0$ such that if,

$$\sup_{0 \leqslant \xi \leqslant 1} |A(\xi) - B(\xi)| < \delta_\varepsilon,$$

then

$$| \lambda_1(A) - \lambda_1(B) | < \varepsilon.$$

If we seek the solution in the class $\mathscr{C}$, then the problem is well posed. However, if we remove the assumption of continuity, it may occur that a very small change in a solution like (24.69) yields a buckling load very far from $\lambda = \frac{1}{3}\pi^2(V/L)^2$ and much smaller than this. To prove this result, let us introduce two small positive numbers δ and η, and define the function $A_\delta(\xi)$ as

$$A_\delta(\xi) = \begin{cases} A(\delta - \xi) & \text{if } 0 \leqslant \xi \leqslant \delta, \\ A(\xi) & \text{if } \delta < \xi \leqslant 1, \end{cases} \tag{24.89}$$

where, according to (24.69), $A(\xi)$ has the form

$$A(\xi) = \frac{4}{3}\frac{V}{L}\sin^2\theta(\xi).$$

We then consider a function $B(\xi)$, defined as

$$B(\xi) = \eta + (1 - \eta)A_\delta(\xi). \tag{24.90}$$

$B(\xi)$ is piecewise continuous and bounded from below by the positive constant η. It follows that $B^{-2}(\xi)$ is bounded from above, and so it belongs to the class $\mathscr{C}$. Given $\varepsilon_1 > 0$, we first select δ and η so small that, for all ξ, we have $|A(\xi) - B(\xi)| < \varepsilon_1$.

Next, we show how to make $\lambda_1(B) < \varepsilon_2$. The necessary condition of extremum (24.40) requires that $A(\xi) = (1/\sqrt{c})y^{\frac{2}{3}}$. Therefore the function $A^{-2}(\xi)$ behaves like $\xi^{-\frac{4}{3}}$ as $\xi \to 0$, and $(A_\delta)^{-2}$ behaves like $(\delta - \xi)^{-\frac{4}{3}}$ as $\xi \to \delta$. This is a nonintegrable singularity so that, if $\delta > 0$ is fixed, then

$$\lim_{\eta \to 0} \int_0^1 B^{-2}(\xi)\, d\xi = \int_0^1 (A_\delta)^{-2}\, d\xi = \infty.$$

Now, if we substitute the trial function $u = \xi(1 - \xi)$ into the variational characterization of the first eigenvalue, we obtain

$$\lambda_1(B) = \min_u \frac{\int_0^1 u'^2\, d\xi}{\int_0^1 u^2 B^{-2}(\xi)\, d\xi} \leqslant \frac{1}{6\int_0^1 \xi^2(1 - \xi)^2 B^{-2}(\xi)\, d\xi}.$$

The singularity of $B^{-2}(\xi)$ occurs at $\xi = \delta > 0$, while the zero value for the trial function u is at $\xi = 0$ and $\xi = 1$. It follows that, if we hold δ fixed and let $\eta \to 0$, then $\int_0^1 \xi^2(1 - \xi)^2 B^{-2}(\xi)\, d\xi \to \infty$, and so $\lambda_1(B)$ can be made less than and given number ε_2, while $\lambda_1(A)$ is fixed.

The unfortunate behavior of this eigenvalue problem is not related to the kinds of boundary condition or the multiplicity of the eigenvalues involved, but solely to the singularity of the coefficient in the differential equation. It is thus not surprising that the natural way of resetting the problem is to exclude the singularity in advance by imposing a lower positive bound, and sometimes an upper bound, on the function $A(\xi)$:

$$h \leqslant A(\xi) \leqslant H. \tag{24.91}$$

These are called *technological constraints*, because, in practice, a beam whose cross-section is disrupted at some points or is excessively large in others, is useless. On putting $\rho(\xi) = A^{-2}(\xi)$, $b = h^{-2}$, and $a = H^{-2}$, (24.87) becomes

$$y'' + \lambda\rho(\xi)y = 0, \quad y(0) = y(1) = 0, \tag{24.92}$$

with the function $\rho(\xi)$ restricted to the class of functions defined by the conditions

$$\int_0^1 \rho(\xi)\, d\xi = M, \quad a \leqslant \rho \leqslant b.$$

In this class we introduce the metric

$$d_0(\rho_1, \rho_2) = \max_{0 \leqslant \xi \leqslant 1} \left| \int_0^\xi [\rho_1(t) - \rho_2(t)]\, dt \right|. \tag{24.93}$$

Now, by a theorem of Krein (1955), the nth eigenvalue of (24.92) is a continuous function on the compact metric space endowed with the norm (24.93), and therefore the maximizing function exists.

An additional problem is to see whether it is better to introduce a different metric from (24.93), such as, for example,

$$d_1(\rho_1, \rho_2) = \max_{0 \leqslant \xi \leqslant 1} |\rho_1(\xi) - \rho_2(\xi)|,$$

or

$$d_2(\rho_1, \rho_2) = \max_{0 \leqslant \xi \leqslant 1} |\rho_1(\xi) - \rho_2(\xi)| + \max_{0 \leqslant \xi \leqslant 1} |\rho_1'(\xi) - \rho_2'(\xi)|.$$

The choice of metric is quite arbitrary, but very often it is suggested immediately by the physics of the problem. In our treatment we have used the metric $d_1(\rho_1, \rho_2)$, because no directions or surface areas are involved. The metric $d_2(\rho_1, \rho_2)$ is too strong, the metric $d_0(\rho_1, \rho_2)$ is too weak, and because $\rho(\xi)$ is bounded away from zero, the metric $d_1(\rho_1, \rho_2)$ is sufficiently strong to give a precise meaning to the continuity of $\lambda_1(\rho)$ with respect to the function ρ.

Once the problem is well posed, the question of constructing a solution satisfying the equality and inequality constraints arises. In general, the construction is not simple, but some interesting results can be achieved by applying Pontryagin's *maximum principle*, which is well known in optimal control theory (Barnes 1988).

The analysis of the stability of rods beyond the critical load cannot be treated by linearized equations, but requires the application of bifurcation theory in the full nonlinear description. An example of how the method works is exemplified in the study of the equilibrium configurations of an inextensible elastic ring subjected to a (dimensionless) hydrostatic pressure p. We assume that the ring is a circle of unit radius in its undeformed state, and that after deformation it is a closed curve in the xz plane (Figure 24.6).

In order to write down the equilibrium equation of the problem (Tadjbakhsh 1969), let us introduce the coordinate s of the arc length in the deformed state, so that $x(s)$ and $z(s)$ are the Cartesian coordinates of any point on the deformed axis, $\theta(s)$ is the angle between the tangent and the x axis, and $\kappa' = d\theta/ds$ is the curvature. The direction cosines of the tangent can clearly be expressed in terms of $\theta(s)$ by the derivatives

$$\frac{dx}{ds} = \sin\theta(s), \quad \frac{dz}{ds} = \cos\theta(s), \tag{24.94}$$

and the components of the exterior forces per unit of length of the central line have the form

$$f_1 = p\cos\theta, \quad f_3 = -p\sin\theta. \tag{24.95}$$

The equilibrium equations are again those used in Section 23 to derive (23.8), with the difference that the body forces are now given by (24.95). Then, on repeating the argument used in those conditions, and for simplicity putting $B = 1$, we derive the equation

$$\frac{d}{ds}\left(\frac{\frac{d^3\theta}{ds^3}}{\frac{d\theta}{ds}}\right) + \frac{d\theta}{ds}\frac{d^2\theta}{ds^2} - p\frac{d}{ds}\left(\frac{1}{\frac{d\theta}{ds}}\right) = 0,$$

which can be integrated to yield

$$\frac{d^3\theta}{ds^3} + \frac{1}{2}\left(\frac{d\theta}{ds}\right)^3 - c\left(\frac{d\theta}{ds}\right) - p = 0, \tag{24.96}$$

where c is an arbitrary constant of integration.

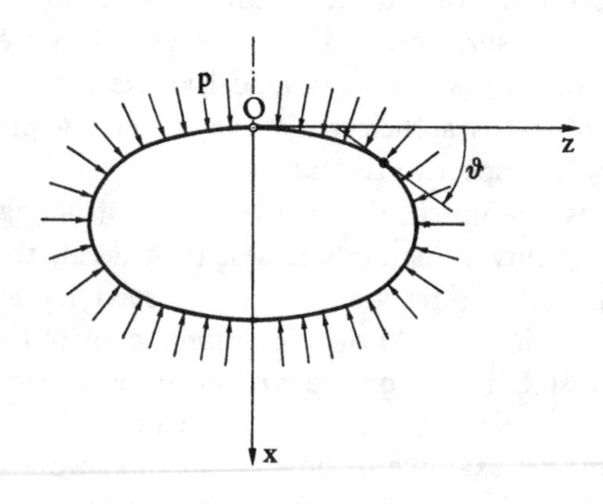

Fig. 24.6

In order to study the behavior of solutions of (24.96), we introduce the new variables

$$v(s) = \frac{d\theta}{ds} - 1, \quad \mu = \frac{3}{2} - c, \quad \beta = c + p - \frac{1}{2} = p + 1 - \mu,$$

so that (24.96) can be transformed into the equation

$$v_{ss} + \mu v = \beta - \frac{1}{2}v^2(v+3), \tag{24.97}$$

where we have put $d^2v/ds^2 = v_{ss}$, and the pressure p is related to the coefficients by the equation

$$p = \mu - \beta + 1.$$

It is obvious that (24.97) admits the trivial solution $v =$ constant for suitable values of μ and β, but we wish to see if there are also nontrivial solutions and to discover their structure. In particular, there are good physical reasons to expect that a nontrivial solution must satisfy the following conditions

$$\theta(s + 2\pi) = \theta(s) + 2\pi,$$

$$x(s + 2\pi) = x(s), \quad z(s + 2\pi) = z(s).$$

It then follows that a generic eigenfunction, which we denote by $v(s)$ for brevity, must be periodic with period $2\pi/n\,(n = 1, 2, \ldots)$. In addition, if we consider only smooth, simple, closed curves, any such curve has at least four points at which the curvature has an extremum,[42] and hence we may assume that we have $n \geqslant 2$.

Since (24.97) is invariant under the change of independent variable $s^* = \pm s +$ constant, it is sufficient to study its solutions in the half-period $(0, \pi/n)$ with the boundary conditions

$$v_s(0) = v_s\left(\frac{\pi}{n}\right) = 0,$$

while the periodicity of v implies the condition

$$\int_0^{\frac{\pi}{n}} v(s)\, ds = 0.$$

Let us now set $a = v(0)$ and introduce the new dependent variable w given by

$$v = aw,$$

and consider, for the moment, the following initial-value problem for w:

$$w_{ss} + \mu w = \gamma - \tfrac{3}{2}aw^2 - \tfrac{1}{2}a^2w^3, \quad s > 0$$
$$w_s(0) = 0, \quad w(0) = 1, \tag{24.98}$$

[42] This property is called the four vertex theorem (Antman 1969b).

where $\gamma = \beta/a$. For $a = 0$, the above equation has the solution

$$w = \varphi(s, \mu, \gamma) = \left(1 - \frac{\gamma}{\mu}\right) \cos \sqrt{\mu} s + \frac{\gamma}{\mu}.$$

From the theory of nonlinear ordinary differential equations, (24.98) possesses a unique solution $w(s, \mu, \gamma, a)$, provided that a is sufficiently small. Using this knowledge, we wish to find a pair of values of μ and γ such that the solution satisfies the conditions

$$\begin{aligned} w_s\left(\frac{\pi}{n}, \mu, \gamma, a\right) &\equiv b_1\left(\frac{\pi}{n}, \mu, \gamma, a\right) = 0, \\ \int_0^{\frac{\pi}{n}} w(s, \mu, \gamma, a)\, ds &\equiv b_2\left(\frac{\pi}{n}, \mu, \gamma, a\right) = 0. \end{aligned} \tag{24.99}$$

For $a = 0$, these conditions are verified for the values

$$\mu = n^2, \quad \gamma = 0.$$

For $a \neq 0$, by the implicit function theorem, (24.99) have the solutions $\mu = \mu(a)$ and $\gamma = \gamma(a)$ in the neighborhood of $a = 0$, provided that the Jacobian satisfies the condition

$$J = \frac{\partial(b_1, b_2)}{\partial(\mu, \gamma)} \neq 0, \quad \text{for } a = 0.$$

We can also directly verify the equality

$$J = \left[\varphi_{s\mu}\left(\frac{\pi}{n}, \mu, \gamma\right) \int_0^{\frac{\pi}{n}} \varphi_\gamma(s, \mu, \gamma)\, ds - \varphi_{s\gamma}\left(\frac{\pi}{n}, \mu, \gamma\right) \int_0^{\frac{\pi}{n}} \varphi_\mu(s, \mu, \gamma)\, ds \right]_{\substack{\gamma = 0 \\ \mu = n^2}}$$

$$= \frac{\pi^2}{2n^2} \neq 0.$$

Thus bifurcation occurs for the value

$$p = n^2 - 1,$$

and by a regular perturbation we find the results

$$v(s) = a \cos ns + \frac{a^2}{4n^2} (\cos 2ns - \cos ns) + O(a^2),$$

$$p = (n^2 - 1) + \frac{3a^2}{8}\left(1 - \frac{1}{n^2}\right) + O(a^2).$$

The diagram of p as a function of a (Figure 24.7) shows that the branches that bifurcate from $p = n^2 - 1$ behave like parabolae and have vertices located on the $a = 0$ axis.

The theory of bifurcation can be extended further to treat more complex load conditions, such as those due to the simultaneous action of an axial force and a couple applied at the ends of an initially staight rod clamped at the ends, but allowed

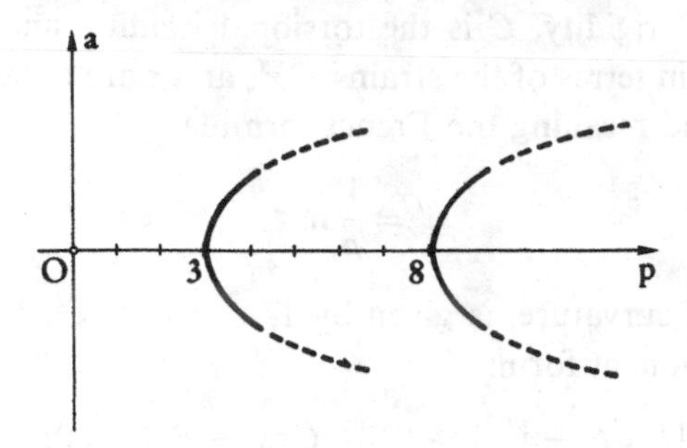

Fig. 24.7

to twist (Zachmann 1979). Let us consider a slender, nonextensible rod with a circular cross-section, and choose the units so that the length of the rod is unity. In its undeformed state the coordinates of the central line of the rod are

$$(X, Y, Z) = (0, 0, s), \quad 0 \leq s \leq 1.$$

Under the action of an axial force $P\mathbf{k}$ and a twisting couple $M\mathbf{k}$, where $\mathbf{k}$ is the unit vector along the Z axis, the parametric equations of the deformed central line have the form

$$[x(s), y(s), s + \zeta(s)], \quad 0 \leq s \leq 1.$$

If the ends of the rod are clamped, but the upper end is capable of transmitting the axial force and the couple, the following boundary conditions must be imposed:

$$\begin{aligned} &x(0) = y(0) = x'(0) = y'(0) = 0, \\ &x(1) = y(1) = x'(1) = y'(1) = 0, \\ &\zeta(0) = 0. \end{aligned} \tag{24.100}$$

These conditions ensure that the loads are conservative.

In order to write down the equilibrium equations in the strained configuration, we introduce the unit vector tangent to the strained central line

$$\mathbf{t} = [x'(s), y'(s), 1 + \zeta'(s)], \tag{24.101}$$

and observe that, in the absence of a body force, the equations are

$$\mathbf{n}' = \mathbf{0}, \quad \mathbf{m}' + \mathbf{t} \times \mathbf{n} = \mathbf{0}.$$

This implies the condition

$$\mathbf{n} = \text{constant} = -P\mathbf{k},$$

provided that the axial force is compressive, and the second equation of equilibrium yields

$$\mathbf{m}'' + P\mathbf{k} \times \mathbf{t} = \mathbf{0}, \tag{24.102}$$

and $\mathbf{m}$ can be expressed vectorially as

$$\mathbf{m} = B(\mathbf{t} \times \mathbf{t}') + C\tau\mathbf{t}, \tag{24.103}$$

where B is the flexural rigidity, C is the torsional rigidity, and τ is the twist. The equilibrium equations in terms of the strains κ, κ', and τ are obtained by substituting (24.103) in (24.102) and recalling the Frenet formula

$$\mathbf{t}' = \frac{1}{\rho}\,\mathbf{n},$$

where ρ, the radius of curvature, is given by $1/\rho^2 = \kappa^2 + \kappa'^2$. In particular, we can write (24.103) in component form:

$$\{B[y'\zeta'' - y''(1+\zeta')] + C\tau x' - Py\}'' = 0, \tag{24.104a}$$

$$\{B[x''(1+\zeta') - x'\zeta''] + C\tau y' + Px\}'' = 0, \tag{24.104b}$$

$$\{B(x'y'' - x''y') + C\tau(1+\zeta')\}'' = 0. \tag{24.104c}$$

The boundary conditions on x and y require both of them to vanish at the ends with their first derivatives, and this implies that $\zeta'(0) = \zeta'(1) = 0$. By integrating (24.104c) twice, and using the boundary conditions, we obtain

$$B(x'y'' - x''y') + C\tau(1+\zeta') = M,$$

since at the ends we have $C\tau = M$.

From the value $|\mathbf{t}| = 1$, we have the relation

$$1 + \zeta' = (1 - x'^2 - y'^2)^{\frac{1}{2}},$$

and hence we can derive the equation

$$\zeta'' = -(x'x'' + y'y'')(1+\zeta')^{-1}.$$

On using the last two equations we can eliminate ζ' and ζ'' from (24.104a) as follows:

$$\begin{aligned}[y'\zeta'' - y''(1+\zeta')] &= [-(y'x'x'' + y'^2y'') - y''(1+\zeta')^2](1+\zeta')^{-1}\\ &= [x'(x'y'' - x''y') - y''](1+\zeta')^{-1}\\ &= \left(x'\frac{M}{B} - y''\right)(1 - x'^2 - y'^2)^{-\frac{1}{2}} - \frac{C\tau}{B}\,x'.\end{aligned} \tag{24.105}$$

On putting $2\lambda = M/B$ and $\mu = P/B$, and substituting (24.105) in (24.104a), we obtain

$$[(y'' - 2\lambda x')(1 - x'^2 - y'^2)^{-\frac{1}{2}} + \mu y]'' = 0. \tag{24.106}$$

By operating in exactly the same way with (24.104b), we obtain the equation

$$[(x'' + 2\lambda y')(1 - x'^2 - y'^2)^{-\frac{1}{2}} + \mu x]'' = 0. \tag{24.107}$$

The last two equations describe the problem completely. Once these equations have been solved, $\zeta(s)$ can be recovered from

$$\zeta'(s) = -1 + (1 - x'^2 - y'^2)^{\frac{1}{2}}, \quad \zeta(0) = 0,$$

and the torsion τ can be evaluated by means of Frenet's formulae.

If $x = y = 0$, the rod is in an unbuckled state; any nontrivial solution of (24.106) and (24.107) represents a buckled state. In order to investigate the values of the parameters λ and μ for which buckled states are possible, let us assume as a new unknown the complex-valued function $z(s) = x(s) + iy(s)$, $0 \leq s \leq 1$, satisfying the conditions $z(0) = z'(0) = z(1) = z'(1) = 0$, and introduce the operator

$$G(z, \lambda, \mu) = [(z'' - 2i\lambda z')(1 - |z'|^2)^{-\frac{1}{2}} + \mu z]''.$$

Then the system (24.106), (24.107) is equivalent to the single differential equation

$$G(z, \lambda, \mu) = 0. \tag{24.108}$$

If we neglect the nonlinear terms in (24.108) we obtain the linearized equation

$$L(z, \lambda, \mu) = (z'' - 2i\lambda z' + \mu z)'' = 0, \tag{24.109}$$

the general solution of which, under the condition $\mu \neq 0$, is

$$z(s) = a + bs + c\exp[i(\lambda + \kappa)s] + d\exp[i(-\lambda + \kappa)s],$$

where $\kappa = (\lambda^2 + \mu^2)^{\frac{1}{2}}$ and a, b, c, and d are complex constants. The constants must be determined by imposing the boundary conditions, and some computation shows that, if $\mu \neq 0$, (24.109) has a nontrivial solution if, and only if,

$$\mu \sin\kappa + 2\kappa(\cos\kappa - \cos\lambda) = 0. \tag{24.110}$$

This equation has been derived by other authors, using different procedures (Trosch 1952; Beck 1955).

If there is no axial thrust on the rod, then $\mu = 0$ and the general solution of (24.109) is

$$z(s) = a + bs + cs^2 + d\exp(2i\lambda s).$$

In this case, (24.109) admits nontrivial solutions if, and only if,

$$\lambda = \tan\lambda. \tag{24.111}$$

The computations leading to (24.110) and (24.111) show that the matrices of the coefficients of the homogeneous systems in a, b, c, and d have a rank of at least three, in both cases. This means that the eigenpoints are geometrically simple in the sense that if we have

$$L(z^*, \lambda^*, \mu^*) = 0$$

with $z^* \neq 0$, then z^* is unique up to a multiplicative constant.

To study the nonlinear problem, we reformulate equation (24.108) as

$$G(z, \lambda, \mu) = L(z, \lambda, \mu) + N(z, \lambda) = 0, \tag{24.112}$$

where $L(z, \lambda, \mu)$ is defined by (24.109) and $N(z, \lambda, \mu)$ is given by

$$N(z, \lambda) = \{[z'' - 2i\lambda z' - (z'' - 2i\lambda z')(1 - |z'|^2)^{\frac{1}{2}}](1 - |z'|^2)^{-\frac{1}{2}}\}''. \tag{24.113}$$

We already know the loci in the $\lambda\mu$ plane that define the values of the $\lambda\mu$ parameters for which the linearized problem admits nontrivial solutions. We now ask if it is possible to say something about the relationships between these curves and those representing the points of bifurcation of the nonlinear equation (24.112). A partial answer to this question is given by a bifurcation theorem for nonlinear operators. This states that the bifurcation points of an equation such as (24.112) can occur only at points on the spectrum of the linearized equation (24.109).[43] This result, however, leaves unanswered the question of the actual existence of bifurcation points. The answer to this question cannot be derived solely from an examination of the linearized problem.

The methods of nonlinear analysis can also be applied to treat more complicated cases of buckling, such as the loss of stability of a column under a load λ, a random initial displacement, and a nonlinear restoring force (Fraser and Budiansky 1969; Day 1980). To be more specific, we study a column hinged at its ends, of unit flexural rigidity, in its undeformed state occupying the interval $0 \leq x \leq n\pi$, under an axial load 2α having a restoring force responding to a cubic law, and an initial transverse random displacement $w_0(x)$. If we denote the lateral displacement by $w(x)$, the nondimensional form of balance equation is

$$w^{iv} + 2\alpha(w + w_0)'' + w - \varepsilon w^3 = 0, \tag{24.114a}$$

$$w(0) = w(n\pi) = w''(0) = w''(n\pi) = 0. \tag{24.114b}$$

The linearized, homogeneous problem associated with (24.113) is

$$\begin{gathered} w^{iv} + 2\alpha w'' + w = 0, \\ w(0) = w(n\pi) = w''(0) = w''(n\pi) = 0, \end{gathered} \tag{24.115}$$

which possesses the solutions

$$w_m(x) = \sin\left(\frac{m}{n}x\right), \quad m = 1, 2, \ldots, \tag{24.116}$$

and the corresponding eigenvalues

$$\alpha_m = \frac{1}{2}\left[\left(\frac{m}{n}\right)^2 + \left(\frac{n}{m}\right)^2\right], \quad m = 1, 2, \ldots.$$

Thus the linearized equation has a discrete spectrum, the minimum of which occurs for $m = n$ when $\alpha_m = 1$. If n is large, then the values of α_m for m near to n are very close to one, and an initial deflection proportional to any one of these modes greatly reduces the buckling load of the column.

In order to characterize the onset of buckling in the nonlinear problem, we write w and w_0 as series expansions of the classical buckling modes (24.116)

$$w(x) = \sum_{k=1}^{\infty} a_k \sin\left(\frac{k}{n}x\right), \quad w_0(x) = \sum_{k=1}^{\infty} \bar{a}_k \sin\left(\frac{k}{n}x\right),$$

and substitute these series in (24.114a). Then, on multiplying the result by $\sin[(m/n)x]$ and integrating over the length of the column, we obtain an infinite set of coupled nonlinear algebraic equations for the coefficients a_m:

$$\alpha_m a_m - \alpha(a_m + \bar{a}_m) - \frac{\varepsilon}{8}\left(\frac{n}{m}\right)^2 I_m = 0, \tag{24.117}$$

[43] For the proof, which is very simple, see Berger (1969).

where

$$I_m = \frac{8}{n\pi}\int_0^{n\pi}\left[\sum_{k=1}^{\infty} a_k \sin\left(\frac{k}{n}x\right)\right]^2 \sin\left(\frac{m}{n}x\right) dx$$
$$= \sum_p\sum_q\sum_r a_p a_q a_r(\delta_{rm}\delta_{pq} + \delta_{(p-q)(r-m)} + \delta_{(p+q)(r+m)} - \delta_{(p+q)(r-m)} - \delta_{(p-q)(r+m)}),$$

where δ_{ij} are the Kronecker symbol, and the coefficients $\bar{a}_m$ of the initial deflection are assumed known. Equation (24.117) constitutes the solution of (24.114) through Galerkin's procedure. Since the most significant contribution to the solution is expected to come from the nth term and its neighbors, only a few terms on either side of the nth term are retained. For instance, we may retain N terms on either side of the nth one so that $2N+1$ equations of (24.117) are reduced to a system in $2N+1$ unknowns a_m. It is also convenient to rearrange (24.117) and write the truncated set of equations as

$$g_m = (a_m + \bar{a}_m)^{-1}\left[\alpha_m a_m - \frac{\varepsilon}{8}\left(\frac{n}{m}\right)^2 I_m\right] = \alpha, \quad m = n-N, \ldots, n, \ldots, n+N. \tag{24.118}$$

Instead of using Galerkin's method, (24.114) may be solved by a method called *equivalent linearization*, which consists of replacing the nonlinear restoring force $w - \varepsilon w^3$ by a linear force $\bar{\varepsilon}w$ provided that we choose the parameter $\bar{\varepsilon}$ such that the integral

$$\Phi = \int_0^{n\pi} (\bar{\varepsilon}w - w + \varepsilon w^3)^2\, dx,$$

is a minimum. This minimum occurs for the value

$$\bar{\varepsilon} = 1 - \varepsilon \frac{\int_0^{n\pi} w^4\, dx}{\int_0^{n\pi} w^2\, dx}.$$

The linearized equation becomes

$$w^{iv} + 2\alpha(w'' + w_0'') + \bar{\varepsilon}w = 0, \quad w(0) = w(n\pi) = w''(0) = w''(n\pi) = 0, \tag{24.119}$$

and, if we substitute the series expansion

$$w(x) = \sum_{k=1}^{\infty} a_k \sin\left(\frac{k}{n}x\right)$$

in (24.117), we obtain an infinite set of algebraic equations for the coefficients a_m:

$$a_m = \frac{2\alpha \bar{a}_m \left(\frac{m}{n}\right)^2}{\left[\left(\frac{m}{n}\right)^4 - 2\alpha\left(\frac{m}{n}\right)^2 + \bar{\varepsilon}\right]}. \tag{24.120}$$

As $\bar{\varepsilon}$ depends on $w(x)$, (24.120) are again a nonlinear algebraic system of infinite equations, which can be solved by truncation, as in Galerkin's method.

In order to get an idea of the efficiency of the two methods, we can terminate the solution of (24.118) and (24.120) after only one term, by putting $N = 0$ and $m = n$. Then both procedures yield the single equation

$$\frac{a_n}{a_n + \bar{a}_n}\left(1 - \frac{3}{8}\varepsilon a_n^2\right) = \alpha.$$

This solution is very rough and is acceptable only for short columns ($n = 1$), but may give an idea of the influence of the nonlinear terms in the description of the buckled configurations of a beam.

25. Dynamical Problems for Rods

When we want to describe the motion of a rod, the influence of inertial forces must be taken into account. These forces are generally regarded as body forces, and are simply added to the equilibrium equation. However, in doing this, we make the tacit assumption that the mode of distribution of these forces does not seriously invalidate the hypotheses and approximations implicit in the deduction of the purely statical equilibrium equations. In other words, this assumption implies that, when the rod is in motion, the internal strain in the portion between two neighboring cross-sections is the same as it would be if that portion were in equilibrium under tractions at its ends, which produce in it instantaneous extension, curvature, and twist. Although it may appear surprising, no complete justification of this assumption has yet been given (Love 1927, Art. 277), not even for a specific case. However, it seems legitimate to state that the assumption gives a better approximation in the case of relatively slow motions than in the case of relatively fast motions, where the expressions "slow" and "fast" have meanings in the context of lower or higher frequencies of vibration of the rod.

Provided that we accept the assumption that the constitutive equations are unaffected by the motion, the dynamical equations of elastic rods can be derived in a fairly simple way (Eliseyev 1988). For each particle of the axis in its instantaneous configuration, we may define a radius vector

$$\mathbf{r} = \mathbf{r}(S, t), \tag{25.1}$$

and a rotation tensor

$$\mathbf{P} = \mathbf{P}(S, t), \tag{25.2}$$

which allow us to represent the orthonormal triad $\mathbf{d}_k$ relative to a fixed orthonormal triad $\mathbf{D}_k$ by means of the relation

$$\mathbf{d}_k = \mathbf{P}\mathbf{D}_k. \tag{25.3}$$

If the triad $\mathbf{D}_k$ coincides with the orthonormal basis $\mathbf{e}_k$, then an explicit form for the components of $\mathbf{P}$ in terms of Euler angles is as given in (17.15).

The derivatives of $\mathbf{d}_k$ with respect to S can be represented as

$$\mathbf{d}'_k = \boldsymbol{\Omega} \times \mathbf{d}_k, \tag{25.4}$$

where $\boldsymbol{\Omega} = (\kappa, \kappa', \tau)$ is the curvature vector. If we call $\mathbf{R}$ and $\boldsymbol{\Omega}_0$ the values of $\mathbf{r}$ and $\boldsymbol{\Omega}$ in the reference configuration, the quantities

$$\boldsymbol{\Gamma} = \mathbf{r}' - \mathbf{P}\mathbf{R}', \quad \boldsymbol{\kappa} = \boldsymbol{\Omega} - \mathbf{P}\boldsymbol{\Omega}_0, \tag{25.5}$$

are measures of the deformation.[44] In Kirchhoff's model, the strain $\boldsymbol{\Gamma}$ is assumed to be zero.

The velocity vector $\mathbf{v}$ and the angular velocity $\boldsymbol{\omega}$ are defined by

$$\dot{\mathbf{r}} = \mathbf{v}, \quad \dot{\mathbf{d}}_k = \boldsymbol{\omega} \times \mathbf{d}_k,$$

so that, after using the equalities $(\dot{\mathbf{r}})' = (\mathbf{r}')^{\cdot}$, $(\dot{\mathbf{d}}_k)' = (\mathbf{d}'_k)^{\cdot}$ it follows that we have

$$\mathbf{v}' - \boldsymbol{\omega} \times \mathbf{r}' = \dot{\boldsymbol{\Gamma}} - \boldsymbol{\omega} \times \boldsymbol{\Gamma}, \quad \boldsymbol{\omega}' = \dot{\boldsymbol{\kappa}} - \boldsymbol{\omega} \times \boldsymbol{\kappa}.$$

These quantities characterize the kinematic properties of the motion. The inertial properties are defined by the *eccentricity vector*

$$\boldsymbol{\varepsilon} = \frac{1}{A} \iint_A \mathbf{x}\, dA \tag{25.6}$$

where $\mathbf{x}$ is the radius vector relating any point of the cross-section to its centroid, so that a generic point of the rod is located by $\mathbf{r}_1 = \mathbf{r} + \mathbf{x}$. The *inertia tensor* with respect to the same point is given by

$$\mathbf{J} = \frac{1}{A} \iint_A (|\mathbf{x}|^2 \mathbf{I} - \mathbf{x} \otimes \mathbf{x})\, dA, \tag{25.7}$$

where $\mathbf{I}$ denotes the unit tensor and $\otimes$ is the tensor product.

The inertial properties of the rod are then characterized by the functions $\rho(S)$, $\varepsilon(S, t)$, and $\mathbf{J}(S, t)$, and by the time derivatives

$$\dot{\boldsymbol{\varepsilon}} = \boldsymbol{\omega} \times \boldsymbol{\varepsilon}, \quad \overline{(\mathbf{J}\boldsymbol{\omega})}^{\cdot} = \boldsymbol{\omega} \times (\mathbf{J}\boldsymbol{\omega}). \tag{25.8}$$

In order to find the contribution of the inertial forces to the equations of mechanical balance, we observe that the velocity and acceleration of any point on the cross-section have the form

$$\dot{\mathbf{r}}_1 = \dot{\mathbf{r}} + \boldsymbol{\omega} \times \mathbf{x}, \tag{25.9a}$$

$$\ddot{\mathbf{r}}_1 = \ddot{\mathbf{r}} + \dot{\boldsymbol{\omega}} \times \mathbf{x} + \boldsymbol{\omega} \times (\boldsymbol{\omega} \times \mathbf{x}). \tag{25.9b}$$

[44] Note that the generalized strains defined by (25.5) are not the same as those in Clebsch's theory described at the beginning of Section 23.

Let us now consider a variation in the position $\mathbf{r}_1$ of the type

$$\delta\mathbf{r}_1 = \delta\mathbf{r} + \delta\mathbf{o} \times \mathbf{x}, \tag{25.10}$$

where $\delta\mathbf{r}$ represents a small-translation vector and $\delta\mathbf{o}$ is a small-rotation vector, and evaluate the work done by inertial forces applied to the portion of the rod contained between two cross-sections of the abscissae $S = U$ and $S = V$:

$$\delta\mathscr{L} = -\int_U^V \rho\, dS \frac{1}{A} \iint_A (\ddot{\mathbf{r}}_1 \cdot \delta\mathbf{r}_1)\, dA,$$

where ρ is the density per unit length of the rod. By substituting (25.9b) and (25.10) in the above expression we obtain

$$\delta\mathscr{L} = -\int_U^V \rho\, dS[(\overline{\mathbf{v} + \boldsymbol{\omega} \times \boldsymbol{\varepsilon}})^{\boldsymbol{\cdot}} \cdot \mathbf{r} + (\boldsymbol{\varepsilon} + \dot{\mathbf{v}} + \dot{(\overline{\mathbf{J}\boldsymbol{\omega}})}) \cdot d\mathbf{o}].$$

If we postulate that the total work of the forces applied to that portion of the rod is zero, we can write the variational equation as

$$[\mathbf{m} \cdot \delta\mathbf{o} + \mathbf{n} \cdot \delta\mathbf{r}]_U^V + \int_U^V \{[f - \rho(\dot{\mathbf{v}} + \ddot{\boldsymbol{\varepsilon}})] \cdot d\mathbf{r} + [\mathbf{m} - \rho(\boldsymbol{\varepsilon} \times \dot{\mathbf{v}} + \dot{(\overline{\mathbf{J}\boldsymbol{\omega}})})] \cdot d\mathbf{o}\}\, dS = 0, \tag{25.11}$$

where, as usual, $\mathbf{n}$ and $\mathbf{m}$ denote the force and couple resultants, and $\mathbf{f}$ and $\boldsymbol{\ell}$ are the body forces and body couples per unit length of the central line. Since (25.11) is true for an arbitrary interval, we have

$$\begin{aligned} \mathbf{n}' + \mathbf{f} &= \rho(\dot{\mathbf{v}} + \ddot{\boldsymbol{\varepsilon}}), \\ \mathbf{m}' + \mathbf{r}' \times \mathbf{n} + \boldsymbol{\ell} &= \rho(\boldsymbol{\varepsilon} \times \dot{\mathbf{v}} + \dot{(\overline{\mathbf{J}\boldsymbol{\omega}})}), \end{aligned} \tag{25.12}$$

which expresses the balance of momentum and angular momentum.

The balance equations are not sufficient for finding the unknown quantities $\mathbf{n}$, $\mathbf{m}$, $\mathbf{v}$ and $\boldsymbol{\omega}$, and so additional constitutive equations are needed. For small strains, these equations can be written as

$$\begin{aligned} \mathbf{m} &= \mathbf{P}\mathbf{m}_0 + \mathbf{a}\boldsymbol{\kappa} + \mathbf{c}\boldsymbol{\Gamma}, \\ \mathbf{n} &= \mathbf{P}\mathbf{n}_0 + \mathbf{b}\boldsymbol{\Gamma} + \mathbf{c}\boldsymbol{\kappa}, \end{aligned} \tag{25.13}$$

where $\mathbf{n}_0$ and $\mathbf{m}_0$ are the force and moment vectors in the reference configuration, and $\mathbf{a}$, $\mathbf{b}$, and $\mathbf{c}$ are elastic tensors. The tensors $\mathbf{a}$ and $\mathbf{b}$ are symmetric; $\mathbf{c}$ is not. On setting $\boldsymbol{\Gamma} = 0$ and getting rid of the equation for $\mathbf{n}$, Kirchhoff's model is recovered from (25.13). In writing (25.13) in the framework of director theory, the constitutive equations are either postulated or derived by means of the analysis of the three-dimensional model.

It may be interesting to note that, if we accept a general linear constitutive equation such as (25.13), the behavior of a rod may differ appreciably from that predicted by applying simpler models. An example of this is the pure static deformation of a simple beam compressed at its top (Eliseyev 1988). In its unstrained configuration the beam occupies the interval $0 \leqslant S = z \leqslant \ell$ of the z axis of a Cartesian system of axes (Figure 25.1). The beam is clamped at the end $S = 0$ and compressed by a dead force Q at the end $S = \ell$. We assume that the constitutive equations of the material do not have the general form (25.13), but have $\mathbf{m}_0 = \mathbf{n}_0 = 0$, and the elastic tensors **a**, **b**, and **c** are given by

$$\mathbf{a} = \begin{Vmatrix} a_1 & 0 & 0 \\ 0 & a_2 & 0 \\ 0 & 0 & a_3 \end{Vmatrix}, \quad \mathbf{b} = \begin{Vmatrix} b_1 & 0 & 0 \\ 0 & b_2 & 0 \\ 0 & 0 & b_3 \end{Vmatrix}, \quad \mathbf{c} = 0,$$

where the indices 1, 2, and 3 denote the components along the directions **i**, **j**, and **k** in the Cartesian x, y, z system.

On assuming that the strained configuration is still planar and located in the xz plane, the slope of the central line is defined by only one angle θ, and the directors of the strained rods are expressed by

$$\mathbf{d}_1 = \mathbf{i}\cos\theta - \mathbf{k}\sin\theta, \quad \mathbf{d}_2 = \mathbf{j}, \quad \mathbf{d}_3 = \mathbf{i}\sin\theta + \mathbf{k}\cos\theta.$$

Furthermore, the strain–displacement relations and the elasticity equations yield

$$\boldsymbol{\kappa} = \theta'\mathbf{j} = \mathbf{a}^{-1}\mathbf{m}, \quad \boldsymbol{\Gamma} = \mathbf{r}' - \mathbf{d}_3 = \mathbf{b}^{-1}\mathbf{n}.$$

Thus, on inserting these equations into the equilibrium equations,

$$\mathbf{n} = \text{constant} = -Q\boldsymbol{\kappa},$$
$$\mathbf{m}' + \mathbf{r}' \times \mathbf{n} = \mathbf{0},$$

we arrive at the single equation

$$\mathbf{a}_2\theta'' + Q\sin\theta + \tfrac{1}{2}Q^2(b_1^{-1} - b_3^{-1})\sin 2\theta = 0, \tag{25.14}$$

with the boundary conditions $\theta(0) = 0$ and $\theta'(\ell) = 0$.

If we linearize (25.14), we find that buckling occurs when

$$Q + \frac{Q^2}{F} = \frac{\pi^2 a_2}{4\ell^2} = P_{\text{cr}}, \tag{25.15}$$

where $F = (b_1 b_3)/(b_3 - b_1)$ and P_{cr} is Euler's critical load. For $F > 0$, that is, if the tensile stiffness b_3 is greater than the shear stiffness b_1, then (25.15) admits a negative

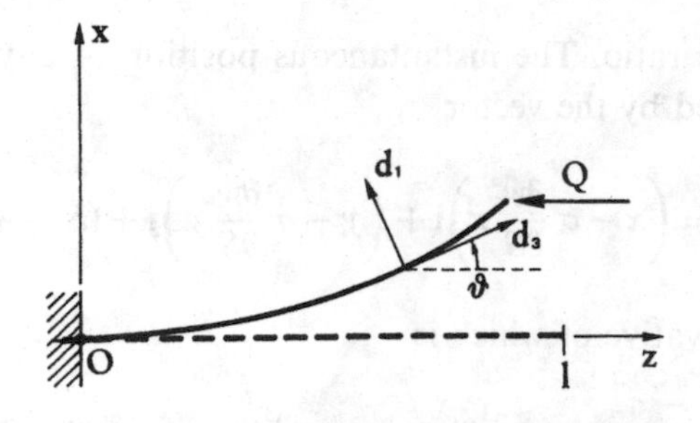

Fig. 25.1

root as well as the expected positive root. This means that, at least in theory, buckling under tension is not excluded.

In the treatment of the full dynamical case it is often convenient to make a similar hypothesis about the form of the constitutive relations. For instance, if we put $\boldsymbol{\Gamma}_\alpha = 0\ (\alpha = 1, 2)$, we characterize a rod with tension but without shear. However, constitutive restrictions may also be made on the inertial terms. For example, taking the values

$$\boldsymbol{\varepsilon} = \mathbf{0}, \quad \mathbf{J} = J\,\mathbf{d}_3 \otimes \mathbf{d}_3, \tag{25.16}$$

means neglecting any kind of motion of the particles of a cross-section relative to the centroid other than rotation about the axial direction $\mathbf{d}_3$.

Assumptions like (25.16) make it possible to justify many of the classical attempts to describe the motion of a beam, taking into account the lateral displacements of the particles. The method can also be slightly generalized by removing the hypothesis of rigidity of the cross-section.

An example of how the procedure works is given by examining the purely extensional vibrations of a straight beam when the inertia of the lateral motion is taken into account (Love 1927, Art. 278). Let us consider the motion of a cylindrical beam of length L and cross-sectional area A, made of a linear elastic material of Young's modulus E. Let w be the displacement of the centroid of the cross-section parallel to the central line, the distance of which from a given origin placed on the axis in the equilibrium state is S. If we neglect the lateral motion, the equation of motion is simply

$$n_3' = \rho A \frac{\partial^2 w}{\partial t^2}, \tag{25.17}$$

where ρ is the density of the material and n_3 is the axial force. As $\partial w/\partial S$ represents the elongation, the force n_3 is equal to $EA(\partial w/\partial S)$.

If we want to take account of the lateral motion due to the transverse elastic deformation of the bar, we must observe that the lateral displacements of a point on the cross-section, of coordinates x, y with respect to axes drawn through the centroid, are

$$u = -\sigma x \frac{\partial w}{\partial S}, \quad v = -\sigma y \frac{\partial w}{\partial S}, \tag{25.18}$$

where σ is Poisson's ratio. The instantaneous position of any point on the cross-section is thus defined by the vector

$$\mathbf{r}_1 = \left(x - \sigma \frac{\partial w}{\partial S} x\right)\mathbf{i} + \left(y - \sigma \frac{\partial w}{\partial S} y\right)\mathbf{j} + (S + w)\mathbf{k},$$

the second time derivative of which is

$$\ddot{\mathbf{r}}_1 = -\sigma \frac{\partial^3 w}{\partial S\, \partial t^2} x\mathbf{i} - \sigma \frac{\partial^3 w}{\partial S\, \partial t^2} y\mathbf{j} + \frac{\partial^2 w}{\partial t^2}\mathbf{k}.$$

The work done by inertial forces over a section of the beam $U \leq S \leq V$ is

$$\delta \mathscr{L} = -\int_U^V \rho\, dS \iint_A \ddot{\mathbf{r}}_1 \cdot \delta \mathbf{r}_1\, dA, \tag{25.19}$$

where

$$\delta \mathbf{r}_1 = -\sigma \frac{\partial \delta w}{\partial S} x\mathbf{i} - \sigma \frac{\partial \delta w}{\partial S} y\mathbf{j} + \delta w\mathbf{k},$$

δw being a small increment of the displacement field conforming to the conditions $\delta w(U) = \delta w(V) = 0$. Thus, on inserting the expression for $\ddot{\mathbf{r}}_1$ and $\delta \mathbf{r}_1$, in (25.19) and integrating by parts, with K the radius of gyration of the cross-section about its centroid, we obtain

$$\delta \mathscr{L} = -\int_U^V \rho\, dS\, A\left(\frac{\partial^2 w}{\partial t^2} - \sigma^2 K^2 \frac{\partial^4 w}{\partial S^2\, \partial t^2}\right) \delta w.$$

Hence, by postulating that the total work of the forces applied to the portion of rod contained between the two sections $S = U$ and $S = V$ is zero, we obtain precisely equation (6.25), derived here by means of Hamilton's principle.

The same kind of argument may be employed to study the torsional vibration of a rod, taking into account the inertial motion during which the cross-sections do not remain planar but warp (Love 1927, Art. 279). If ψ denotes the relative rotation of two cross-sections, regarded as rigid around their centroids, the derivative $\partial\psi/\partial S$ is the twist of the rod. The displacement components of a point of a cross-section parallel to the axes x and y are

$$u = -\psi y, \qquad v = \psi x.$$

If φ is the torsion function, the longitudinal displacement is

$$w = \varphi \frac{\partial \psi}{\partial S}.$$

The position vector of any point on the rod in its strained configuration is then

$$\mathbf{r}_1 = (x - \psi y)\mathbf{i} + (y + \psi x)\mathbf{j} + \left(S + \varphi \frac{\partial \psi}{\partial S}\right)\mathbf{k},$$

from which the derivative and increment follow:

$$\ddot{\mathbf{r}}_1 = -\frac{\partial^2 \psi}{\partial t^2} y\mathbf{i} + \frac{\partial^2 \psi}{\partial t^2} x\mathbf{j} + \varphi \frac{\partial^3 \psi}{\partial S\, \partial t^2},$$

$$\delta \mathbf{r}_1 = -\delta\psi y\,\mathbf{i} + \delta\psi x\,\mathbf{j} + \varphi \frac{\partial \delta\psi}{\partial S}\,\mathbf{k}.$$

From these two equations, by again applying (25.19), we can find the work done by the inertial forces. On the other hand, the work done by the elastic forces corresponding to a rotation $\delta\psi$ is

$$\delta W = [m_3 \delta\psi]_U^V = \int_U^V dS\, C \frac{\partial\psi}{dS}\frac{\partial\delta\psi}{\partial S},$$

where C is the torsional rigidity. By imposing the condition that the total work corresponding to a variation $\delta\psi$ vanishes at U and V, we obtain the differential equation

$$\rho A K^2 \frac{\partial^2\psi}{\partial t^2} - \rho\left(\iint_A \varphi^2\, dA\right)\frac{\partial^4\psi}{\partial S^2\,\partial t^2} = C\frac{\partial^2\psi}{\partial S^2}. \tag{25.20}$$

The method also allows us to derive the equations of the planar transverse vibrations of a beam when we take into account the longitudinal motion (Love 1927, Art. 280). Let us assume that the rod vibrates in a principal plane, which we take as the xz plane, with the z axis coinciding with the straight undeformed axis and the x axis perpendicular to the axis of the rod directed downwards. If u is the displacement component along the x axis, $\partial^2 u/\partial S^2$ represents the curvature of the strained central line, and consequently the flexural couple is $m_2 = B(\partial^2 u/\partial S^2)$, where B is the flexural rigidity with respect to the y axis, orthogonal to the xz plane. The flexural rigidity B can be also written as $B = EAK'^2$, K' being the radius of gyration of the cross-section about the y axis. The transverse displacement u is accompanied by a longitudinal displacement $w = -x(\partial u/\partial S)$, where the minus sign indicates that a positive couple m_2, that is, a couple directed clockwise in the xz plane, generates a contraction in the lower fibers of the beam. The position vector of a point on the cross-section in an instantaneously strained configuration thus has the form

$$\mathbf{r}_1 = u\mathbf{i} + \left(S - x\frac{\partial u}{\partial S}\right)\mathbf{k}$$

from which we calculate the results

$$\ddot{\mathbf{r}}_1 = \frac{\partial^2 u}{\partial t^2}\mathbf{i} - x\frac{\partial^3 u}{\partial S\,\partial t^2}\mathbf{k},$$

$$\delta\mathbf{r}_1 = \delta u\,\mathbf{i} - x\frac{\partial\delta u}{\partial S}\mathbf{k}.$$

On repeating the variational argument, we write down the equation

$$\int_U^V dS\left[B\frac{\partial^2\delta u}{\partial S^2} - \rho\left(\frac{\partial^2 u}{\partial t^2}\delta u + x^2\frac{\partial^3 u}{\partial S\,\partial t^2}\frac{\partial\delta u}{\partial S}\right)\right] = 0, \tag{25.21}$$

and, provided we choose δu such that δu and $\partial\delta u/\partial S$ vanish at U and V, the pointwise version of (25.21) becomes

$$\rho\frac{\partial^2 u}{\partial t^2} = -EK'^2\frac{\partial^4 u}{\partial S^4}. \tag{25.22}$$

The equations we have derived have the advantage of considering the effect of motions at right angles to the privileged direction along which dynamical balance is required. However, they are restricted by the hypothesis that strains and displacements are small, so that all approximations of linearized theory apply. Nevertheless, the equations are not satisfactory when displacements and strains are large, because the influence of nonlinear terms is no longer negligible. Modification is also necessary when the motion of a slender, originally straight, elastic rotating rod is studied. This could be rotating either longitudinally about its axis (Figure 25.2(a)) or about a line perpendicular to its axis at the midpoint (Figure 25.2(a′)) (Wang 1982). If the rod is free and its angular velocity Ω is low, the rod remains straight during these two types of rotation, but when Ω is sufficiently large the rod may assume a curved configuration as well, as shown in Figure 25.2(b,b′).

To derive the equations of motion, we put the origin of a Cartesian x', y' system at the midpoint of the undeformed rod with the x' axis coinciding with the axis of rotation. Let us call θ the local angle of inclination of the rod to the axis of rotation, and s the arc length measured from the origin. The length of the rod is 2ℓ, C is its flexural rigidity to inflexions in the $x'y'$ plane, and ρ is the mass per unit length. After introducing the nondimensional quantities

$$x = \frac{x'}{\ell}, \quad y = \frac{y'}{\ell}, \quad s = \frac{s'}{\ell}, \tag{25.23}$$

$$u \equiv \int_1^s y\, ds, \quad J^4 \equiv \frac{\rho \Omega^2 \ell^4}{C},$$

the local balance of momentum gives

$$\frac{d^2\theta}{ds^2} = J^4 u \cos\theta, \quad \frac{d^2 u}{ds^2} = \sin\theta, \tag{25.24}$$

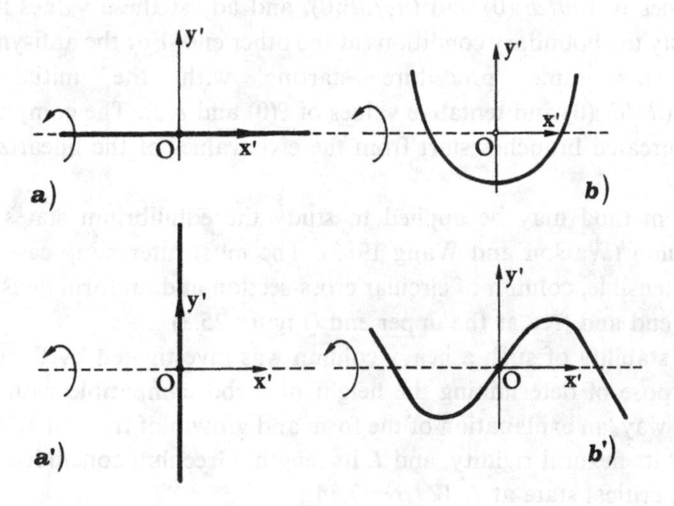

Fig. 25.2

with the associated boundary conditions, which may be of two types:

For symmetric configurations: $\theta(0) = u(0) = u(1) = \frac{d\theta}{ds}(1) = 0.$

For antisymmetric configurations: $\frac{d\theta}{ds}(0) = \frac{du(0)}{ds} = u(1) = \frac{d\theta}{ds}(1) = 0.$

These boundary conditions hold when we limit ourselves to studying motions that are symmetric or antisymmetric with respect to the y' axis, but, of course, other modes are possible.

The stability of the straight configurations may be investigated easily by assuming that both θ and u are small, so that (25.24) may be replaced by the single linear equation

$$\frac{d^4\theta}{ds^4} = J^4\theta. \tag{25.25}$$

The general solution is

$$\theta = C_1 \sinh Js + C_2 \cosh Js + C_3 \sin Js + C_4 \cos Js,$$

and the corresponding eigenvalues are $J = 2.365020, 5.497804, \ldots,$ for the symmetric case, and $J = 3.926602, 7.068583, \ldots,$ for the antisymmetric case. These values of J also represent points of bifurcation in the nonlinear problem (25.24).

When θ is not small, the solution can be obtained numerically in the following way. We put

$$t \equiv Js, \quad v \equiv J^2 s,$$

and transform (25.24) into

$$\frac{d^2\theta}{dt^2} = v\cos\theta, \quad \frac{d^2v}{dt^2} = \sin\theta.$$

Then, for the symmetric case, we start with the initial conditions $\theta(0) = v(0) = 0$ and tentative values of $(d\theta/dt)(0)$ and $(dv/dt)(0)$, and adjust these values iteratively in order to satisfy the boundary conditions at the other end. For the antisymmetric case we repeat the same procedure starting with the initial conditions $(d\theta/dt)(0) = (dv/dt)(0)$ and tentative values of $\theta(0)$ and $v(0)$. The computations confirm that bifurcated branches start from the eigenvalues of the linearized equation (25.25).

The same method may be applied to study the equilibrium states of a heavy rotating column (Watson and Wang 1983). The most interesting case is that of a vertical, inextensible, column of circular cross-section and uniform density, clamped at the lower end and free at the upper end (Figure 25.3).

The static stability of such a heavy column was investigated by Greenhill (1881) with the purpose of determining the height of a rod compatible with its stability, giving, in his way, an explanation of the form and growth of trees. If W is the weight of the rod, B its flexural rigidity, and L its length, Greenhill concluded that the rod will reach its critical state at $L^2W/B = 7.84$.

In order to study the dynamical case, let us choose coordinate axes r' and z' with the origin placed at the fixed end, and the z' axis coinciding with the axis of

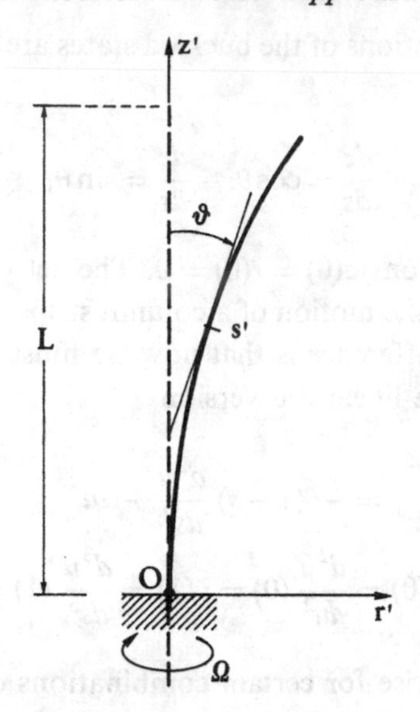

Fig. 25.3

the unstrained rod. Let us call s' the arc length from the origin, and θ the local inclination. The local equation of momentum balance gives

$$m' + \rho g(L - s')\sin\theta + F\cos\theta = 0, \tag{25.26}$$

where m, the bending moment, is proportional to the curvature

$$m = B\,\frac{d\theta}{ds'},$$

and F is the centrifugal force acting from s' to L

$$F = \int_{s'}^{L} \rho r'^2 \Omega^2\, ds',$$

where ρ is the density and Ω is the angular velocity. On normalizing all lengths by L and dropping the primes, (25.26) becomes

$$\frac{d^2\theta}{ds^2} = -\beta(1 - s)\sin\theta + \alpha u \cos\theta, \tag{25.27}$$

where we have put

$$\beta = \frac{\rho g L^3}{B}, \quad \alpha = \frac{\rho L^4 \Omega^2}{B}, \quad \frac{du}{ds} = r, \quad u = \int_1^s r\, ds.$$

The boundary conditions are

$$\theta(0) = \frac{d\theta}{ds}(1) = 0.$$

Once we have $\theta(s)$, the equations of the buckled states are obtained by integrating the equations

$$\frac{dz}{ds} = \cos\theta, \quad \frac{dr}{ds} = \sin\theta,$$

with the boundary conditions $z(0) = r(0) = 0$. The set of equations just derived is also capable of describing the motion of a column suspended at the top end and free at the bottom. The only difference is that now we must regard β as negative.

Equation (25.27) has the linearized version

$$\begin{gathered} \frac{d^4u}{ds^4} = -\beta(1-s)\frac{d^2u}{ds^2} + \alpha u, \\ \frac{du}{ds}(0) = \frac{d^2u}{du^2}(0) = u(1) = \frac{d^3u}{ds^3}(1) = 0, \end{gathered} \tag{25.28}$$

and nontrivial solutions arise for certain combinations of the parameters α and β. These combinations determine a family of curves in the $\alpha\beta$ plane, corresponding to which there are particular eigensolution branches. The behavior of the curves is roughly represented in Figure 25.4 in terms of the variables $K = \beta^{\frac{1}{3}}$ and $J = \alpha^{\frac{1}{4}}$. It is interesting to observe that J must always be greater than or equal to zero, whereas K could also be negative, which would correspond to the case in which the beam is suspended rather than supported from below. If we take $J = 0$ and $K \geq 0$, the column does not rotate about its axis, and is simply compressed by its own weight.

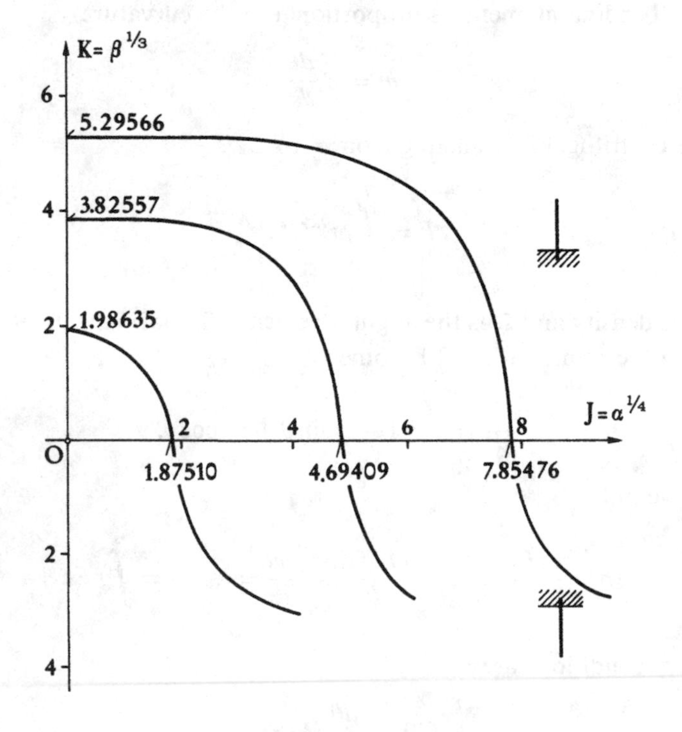

Fig. 25.4

In this case the values $K = 1.98635, 3.82557$, and 5.29566 represent the critical compressive loads found by Greenhill.

The linearized criterion of stability can be applied to determine the onset of buckling in the motion of a column subjected to a follower force applied at its top in the presence of damping. It is universally believed that "damping" has a dissipative effect and, as such, it must improve the stability of an elastic body in motion. However, in 1952, Ziegler proposed a counterexample which seems to disprove the conjecture about the stabilizing effect of damping. Ziegler considered a simple model, namely an inverted double pendulum subject to a tangential load at its unsupported end, and applied the linearized criterion of stability in order to evaluate the critical load both for the undamped system (denoted by P_{el}) and for the same system with a slight viscous damping acting at the joints (denoted by P_{dam}). The result was $P_{\text{dam}} < P_{\text{el}}$, and the conclusion followed that the addition of slight damping not only lowers the critical load but also produces an abrupt fall in its value as soon as an infinitely small amount is present. Ziegler's conclusion was questioned by Plaut (1971). First of all, the significance of the value P_{el} as a "critical value" is incorrect. The undamped system is unstable if the applied load P is greater than P_{el}; however, for $P < P_{\text{el}}$, the critical load cannot be determined via a linearized analysis. Secondly, for the damped system, $P > P_{\text{dam}}$ produces instability, while $P < P_{\text{dam}}$ implies asymptotic stability. Therefore, P_{el} and P_{dam} have different meanings, and the value of making any comparison between them is open to question.

The extension of Ziegler's problem to a cantilever beam subject to a follower force at its free end, with internal rotatory damping, leads to analogous results (Lottati and Elishakoff 1987) but a careful discussion of the solution may clarify the apparent paradox.

The governing equation of the motion of a cantilevered column subjected to a follower force and endowed with rotatory damping, proportional to the rotation of the cross-section, assuming small displacements, is

$$B\frac{\partial^4 w}{\partial x^4} + P\frac{\partial^2 w}{\partial x^2} - C\frac{\partial^3 w}{\partial x^2\,\partial t} + \rho A\frac{\partial^2 w}{\partial t^2} = 0, \tag{25.29}$$

where B is the flexural rigidity, ρ is the mass density, P is the follower force, C is the rotatory damping coefficient, x is the axial coordinate, and $w(x, t)$ is the transverse displacement of the column. The boundary conditions are

$$w(0) = \frac{\partial w}{\partial x}(0) = 0, \quad \frac{\partial^2 w}{\partial x^2}(\ell) = \left(B\frac{\partial^3 w}{\partial x^3} - C\frac{\partial^2 w}{\partial x\,\partial t}\right)_{x=\ell} = 0. \tag{25.30}$$

By introducing the notation

$$\xi = \frac{x}{\ell}, \quad t = \sqrt{\frac{\rho\ell^4 A}{B}}\,\tau, \quad W = \frac{w}{\ell}, \quad p = \frac{P\ell^2}{B}, \quad d = \frac{C}{\sqrt{b\rho A}},$$

and seeking solutions of the form

$$W(\xi, \tau) = A_1 e^{\sigma\tau} y(\xi),$$

the equation of our original problem becomes

$$\frac{d^4y}{d\xi^4}+p\,\frac{d^2y}{d\xi^2}-d\sigma\,\frac{d^2y}{d\xi^2}+\sigma^2y=0, \tag{25.31}$$

with the boundary conditions

$$y(0)=\frac{dy}{d\xi}(0)=0,\quad \frac{d^2y}{d\xi^2}(1)=\left(\frac{d^3y}{d\xi^3}-d\sigma\,\frac{dy}{d\xi}\right)_{\xi=1}=0. \tag{25.32}$$

In order to find the eigensolutions of (25.31) and (25.32), we put

$$W(\xi,t)=A_1\,e^{\sigma\tau}\,e^{r\xi},$$

where A_1 and r may take complex values, and σ may also be a complex number of the type $\sigma=\alpha+i\omega$. The characteristic equation is

$$r^4+(p-d\sigma)r^2+\sigma^2=0.$$

This equation has four roots r_j, so the solution is

$$W=\sum_{j=1}^{4}a_j\exp(\sigma\tau+r_j\xi),$$

and nontrivial solutions exist if the characteristic determinant vanishes. Hence we have

$$\det\begin{Vmatrix} \cdot & 1 & \cdot & \cdot \\ \cdot & r_j & \cdot & \cdot \\ \cdot & r_j^2\,e^{r_j} & \cdot & \cdot \\ \cdot & r_j(r_j^2-d\sigma)e^{r_j} & \cdot & \cdot \end{Vmatrix}=0, \tag{25.33}$$

where we have written only the jth column explicitly ($j=1$–4). The solution of the transcendental equation in r, derived from the expansion of the determinant in (25.33), can be found numerically only for small values of the damping factor d, which happens to be the most important case. For each value of d from zero to one, the corresponding values of p_{cr} and ω_{cr} are represented in Figure 25.5, where p is the nondimensional form of P and ω is the imaginary part of σ. The form of these curves, which exhibit a sharp decrease on leaving the origin, explains the so-called Ziegler paradox. It is true that damping causes destabilization because the curves $p_{\mathrm{cr}}(d)$ and $\omega_{\mathrm{cr}}(d)$ are strongly decreasing for small values of d; but there is no jump in p_{cr} or ω_{cr} when rotatory damping occurs. This means that Ziegler's jump is not an effect of physics, but only of the inadequacy of the numerical approximation of the eigenvalues in (25.31) and (25.32) using Galerkin's method and employing certain polynomials known as Duncan's functions (Duncan 1937).

The cases analyzed so far are not entirely satisfactory, because in attempting to establish the onset of instability they use the linearized criterion of stability, and they require recourse to numerical procedures in an attempt to say something about the behavior of solutions in the postcritical range. However, because it considers only motions of small amplitude, linear theory can only be valid for small times. The

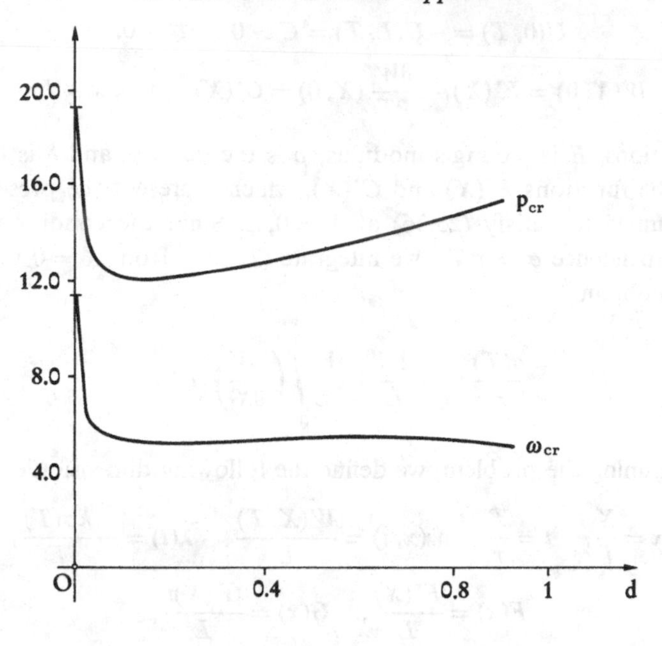

Fig. 25.5

analysis of the stability of a column for long times thus requires the use of a nonlinear dynamical theory (Reiss and Matkowsky 1971).

Let us study the motion of a column whose centroidal axis, in the unstrained configuration, coincides with the segment $[0, L]$ of the X axis, and the cross-section of which lies in the YZ plane. Let A denote the area of the cross-section and B the flexural rigidity of the rod with respect to the Y axis. The rod is simply supported at its ends and a compressive axial displacement C, which is independent of time T, is applied at each end. At $T = 0$ the column is subjected to a given initial velocity and an initial displacement in the XZ plane, and we wish to find the subsequent motion (*i.e.* whether it diverges or decays). If we denote the axial displacement by $U(X, T)$, the transverse displacement by $W(X, T)$, and the axial stress and strain by $\sigma(X, T)$ and $e(X, T)$, respectively, the equation of the motion and the associated boundary and initial conditions are as follows:

$$\frac{B}{A}\frac{\partial^4 W}{\partial X^4} - \frac{\partial}{\partial X}\left(\sigma\frac{\partial W}{\partial X}\right) + \rho\frac{\partial^2 W}{\partial T^2} + \delta\frac{\partial W}{\partial T} = 0, \quad \text{for } 0 < X < L, \quad T > 0, \tag{25.34}$$

$$\frac{\partial \sigma}{\partial X} = 0, \tag{25.35a}$$

$$\frac{\sigma}{E} = e = \frac{\partial U}{\partial X} + \frac{1}{2}\left(\frac{\partial W}{\partial X}\right)^2, \tag{25.35b}$$

$$W = \frac{\partial^2 W}{\partial X^2} = 0 \quad \text{for } X = 0, L, \quad T \geq 0, \tag{25.36}$$

$$U(0, T) = -U(L, T) = C > 0, \quad T > 0, \tag{25.37}$$

$$W(X, 0) = F^*(X), \quad \frac{\partial W}{\partial T}(X, 0) = G^*(X), \quad 0 < x < L. \tag{25.38}$$

In these equations, E is Young's modulus, ρ is the density, and δ is the damping coefficient. The functions $F^*(X)$ and $G^*(X)$, which represent the prescribed initial data, are assumed to satisfy (25.36) at $X = 0, L$. Since the condition $\partial\sigma/\partial X = 0$ implies the dependence $\sigma = \sigma(T)$, we integrate (25.35b) from $X = 0$ to $X = L$ and use (25.37) to obtain

$$\frac{\sigma(T)}{E} = -\frac{2C}{L} + \frac{1}{2L}\int_0^L \left(\frac{\partial W}{\partial X}\right)^2 dX. \tag{25.39}$$

Before examining the problem, we define the following dimensionless variables:

$$x = \frac{X}{L}, \quad t = \frac{T}{r}, \quad w(x, t) = \frac{W(X, T)}{L}, \quad \lambda(t) = -\frac{k\sigma(T)}{E},$$

$$F(x) = \frac{F^*(X)}{L}, \quad G(x) = \frac{G^*(X)r}{L}, \tag{25.40}$$

and the dimensionless parameters

$$k = \frac{L^2 EA}{B}, \quad r^2 = \frac{\rho L^2 k}{E}, \quad \Gamma = \frac{\delta L^2 k}{2Er}, \quad c = \frac{C}{L} > 0, \tag{25.41}$$

so that (25.34)–(25.39) can be rewritten in the nondimensional form:

$$\frac{\partial^4 w}{\partial x^4} + \lambda \frac{\partial^2 w}{\partial x^2} + \frac{\partial^2 w}{\partial t^2} + 2\Gamma \frac{\partial w}{\partial t} = 0, \quad 0 < x < 1, \quad t > 0, \tag{25.42}$$

$$\lambda = 2ck - \frac{k}{2}\int_0^1 \left(\frac{\partial w}{\partial x}\right)^2 dx, \quad t > 0, \tag{25.43}$$

$$w = \frac{\partial^2 w}{\partial x^2} = 0, \quad x = 0, 1, \quad t > 0, \tag{25.44}$$

$$w(x, 0) = F(x), \quad \frac{\partial w}{\partial t}(x, 0) = G(x), \quad 0 \le x \le 1. \tag{25.45}$$

A first insight into the nature of the solutions of the problem may be obtained by considering the case where $w(x, t)$ is time independent, and is a function of the only variable x. In this case the unbuckled state

$$w \equiv 0, \quad \lambda = 2ck, \tag{25.46}$$

is a solution of the static problem for all values of c, and is the only solution for the conditions

$$\lambda \neq \lambda_n = (n\pi)^2, \quad n = 1, 2, \ldots.$$

If we have $\lambda = \lambda_n$, the problem has the eigensolutions

$$w(x) = A_n \varphi_n(x), \quad \varphi_n(x) = \sqrt{2}\sin(\sqrt{\lambda_n}\, x), \tag{25.47}$$

where, for $n = 1, 2, \ldots$, the amplitudes A_n are related to c by

$$c = c_n\left(1 + k\frac{A_n^2}{2}\right), \quad c_n = \frac{\lambda_n}{2k}, \tag{25.48}$$

and the eigenfunctions $\varphi_n(x)$ are normalized by the condition

$$\int_0^1 \varphi_n^2(x)\,dx = 1.$$

The coefficients c_n can be interpreted as critical end shortenings, because when the bar is shortened by an amount c_n the straight configuration is no longer stable. A qualitative picture of the structure of the static solution is given in Figure 25.6. For each n, (25.48) has real buckled solutions $A_n(c)$ if, and only if, the condition $c \geqslant c_n$ is satisfied. These solutions branch in pairs from the unbuckled state $A_n = 0$ at $c = c_n$, and they exist for all $c > c_n$ as increasing functions of c. For $c < c_1$ the only solution is the trivial one; for $c_n < c < c_{n+1}$, there are exactly $2n + 1$ solutions, comprising the $2n$ buckled states and the unbuckled state. For fixed c in the interval $c_n < c < c_{n+1}$ we may evaluate the potential energies $V_1, V_2, \ldots$, associated with the $2n + 1$ solutions (five for $c_2 < c < c_3$), and these energies may be ordered according to the relations $V_1 < V_2 < \ldots < V_n < V_0$, where V_0 is the energy of the unbuckled state. Since the static criterion for buckling asserts that the system chooses the deformed state with least energy, after the first critical value c_1 the column, if progressively compressed, deforms into one of the buckled states branching from c_1 and remains in that state for $c > c_1$.

If we wish to study the dynamical problem without invoking the nonlinear theory, we may assume, in the first instance, that the initial disturbances are sufficiently small for us to assume that the subsequent motions have small amplitudes. Under these hypotheses we write

$$F(x) = \varepsilon f(x), \quad G(x) = \varepsilon g(x), \quad \text{for } 0 \leq x \leq 1 \tag{25.49}$$

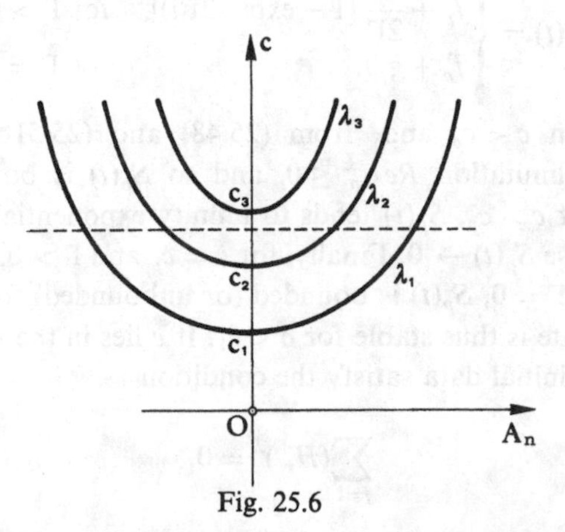

Fig. 25.6

where ε is a small parameter, and we represent the solution in the form

$$w(x, t) = \varepsilon v(x, t). \tag{25.50}$$

On substituting $\varepsilon v(x, t)$ in (25.42)–(25.45) and linearizing the resulting equations with respect to ε, we obtain

$$\frac{\partial^4 v}{\partial x^4} + \lambda \frac{\partial^2 v}{\partial x^2} + \frac{\partial^2 v}{\partial t^2} + 2\Gamma \frac{\partial v}{\partial t} = 0, \text{ for } 0 < v < 1, \quad t > 0, \quad \lambda = 2ck, \tag{25.51a}$$

$$v = \frac{\partial^2 v}{\partial x^2} = 0, \text{ for } x = 0, 1 \quad t > 0, \tag{25.51b}$$

$$v(x, 0) = f(x), \quad \frac{\partial v}{\partial t}(x, 0) = g(x), \text{ for } 0 < x < 1. \tag{25.51c}$$

Provided that $f(x)$ and $g(x)$ are sufficiently smooth, we consider their Fourier expansions

$$f(x) = \sum_{n=1}^{\infty} f_n \varphi_n(x), \quad g(x) = \sum_{n=1}^{\infty} g_n \varphi_n(x), \tag{25.52}$$

where f_n and g_n are the Fourier coefficients of f and g, and we take for $v(x, t)$, the formal solution

$$v(x, t) = \sum_{p=1}^{\infty} S_p(t)\varphi_p(x), \tag{25.53}$$

where $S_p(t)$ have the following alternative expressions. For $\lambda \neq \lambda_p$ and $p = 1, 2, \ldots,$ the expressions are

$$S_p(t) = H_p^+ \exp(\sigma_p^+ t) + H_p^- \exp(\sigma_p^- t),$$
$$H_p^+ = -\left[\frac{\sigma_p^- f_p + g_p}{\sigma_p^+ - \sigma_p^-}\right], \quad H_p^- = +\left[\frac{\sigma_p^+ f_p - g_p}{\sigma_p^+ - \sigma_p^-}\right], \tag{25.54}$$

$$\sigma_p^{\pm} = -\Gamma \pm \left[\Gamma^2 - \lambda_p(\lambda_p - \lambda)\right]^{\frac{1}{2}}, \tag{25.55}$$

and for $\lambda = \lambda_p$ and $p = 1, 2, \ldots,$ they are

$$S_p(t) = \begin{cases} f_p + \dfrac{g_p}{2\Gamma}[1 - \exp(-2\Gamma t)], & \text{for } \Gamma > 0, \\ f_p + g_p t, & \text{for } \Gamma = 0. \end{cases} \tag{25.56}$$

Under the condition $c < c_p$ and, from (25.48) and (25.51b) and with $\lambda < \lambda_p$, (25.55) implies the limitation $Re\, \sigma_p^{\pm} \leq 0$, and so $S_p(t)$ is bounded for all $t > 0$. Correspondingly, for $c > c_p$, $S_p(t)$ tends to infinity exponentially as $t \to \infty$, unless $H_p^+ = 0$, in which case $S_p(t) \to 0$. Finally, for $c = c_p$ and $\Gamma > 0$, $S_p(t)$ is bounded for all t; for $c = c_p$ and $\Gamma = 0$, $S_p(t)$ is bounded (or unbounded) for $g_p = 0$ (or $g_p \neq 0$).

The unbuckled state is thus stable for $c < c_1$. If c lies in the interval $c_r < c < c_{r+1}$ for some $n \geqslant 1$ and initial data satisfy the condition

$$\sum_{j=1}^{n} (H_j^+)^2 = 0,$$

then the unbuckled state is linearly stable with respect to the data.

But the linear dynamical theory is, presumably, valid only for motions of small amplitude, so that linear instability need not imply that the solutions of the nonlinear problem are unbounded. To gain a more precise knowledge of the behavior of solutions when the initial data are not small, we must study the nonlinear equation. We shall, however, assume that the initial data are *monochromatic*, that is, F and G satisfy

$$F(x) = F_n \varphi_n(x), \quad H(x) = G_n \varphi_n(x).$$

In this case the solution can be written in the form

$$w(x, t) = B(t)\varphi_n(x), \tag{25.57}$$

where $B(t)$ is a solution of the initial value problem

$$B'' + 2\Gamma B' + K_n(\beta_n B + B^3) = 0, \quad \beta_n \equiv \frac{4}{\lambda_n}(c_n - c), \quad K_n \equiv \frac{k\lambda_n^2}{2}, \tag{25.58}$$

$$B(0) = F_n, \quad B'(0) = G_n.$$

For simplicity, we will omit the subscripts from K_n, β_n, F_n, and G_n, and consider first the undamped column ($\Gamma = 0$). An exact solution to (25.58) can be obtained in terms of elliptic functions, but it is more convenient to examine the solutions in the phase plane (Stoker 1950). Thus, in the usual way, we multiply (25.58) by B' and integrate from 0 to t to obtain

$$V^2 + K\left(\beta B^2 + \frac{B^4}{2}\right) = H = G^2 + K\left(\beta F^2 + \frac{F^4}{2}\right), \tag{25.59}$$

where

$$V = B'. \tag{25.60}$$

It follows from (25.59) and (25.60) that the initial-value problem (25.58) is equivalent to the initial-value problem

$$\frac{dV}{dB} = -\frac{K(\beta B + B^3)}{V}, \quad V(F) = G, \tag{25.61}$$

in the phase plane.

If $c \leqslant c_n$, then $B \geqslant 0$, and the unbuckled state $B = V = 0$ is the only singular point of (25.61) and this point is a center. As the integral curves (25.59) of (25.61) are closed in the phase plane, $B(t)$ is a periodic function, and thus for $c \leq c_n$ the column oscillates periodically about the unbuckled state. If, on the other hand, we have $c > c_n$ and $\beta < 0$, then there are three singular points of (25.61):

$$B = V = 0, \tag{25.62}$$

$$B = \pm B_0, \quad V = 0, \quad \text{where } B_0 = \sqrt{(-\beta)}. \tag{25.63}$$

The origin (25.62) corresponds to the unbuckled state and the points (25.63) correspond to the two buckled states that branch from $c = c_n$. A sketch of (25.59) is given in Figure 25.7 for fixed K and β and some values of H. The curve $H = 0$, which is usually called *separatrix*, passes through the origin; for $H < 0$, (25.59) has two branches of closed curves, each containing one of the singular points (25.63). Thus, for $H < 0$, $B(t)$ is a periodic function of one sign and the column oscillates periodically about one of the static buckled states that branch from $c = c_n$. For

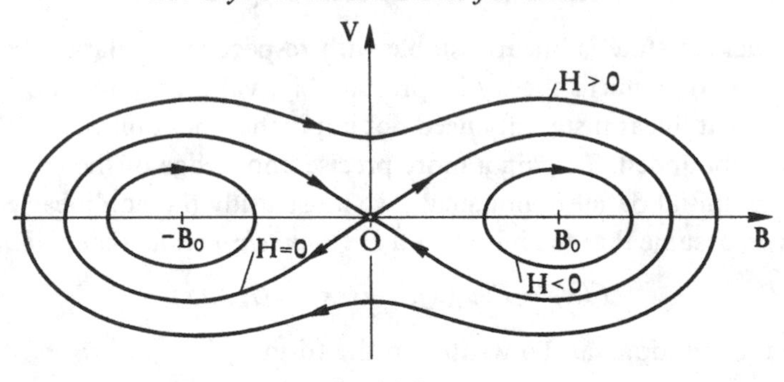

Fig. 25.7

$c > c_n$ and for each $H > 0$, (25.59) gives a closed curve containing the three singular points (25.62) and (25.63). Hence $B(t)$ is a periodic function that changes sign twice in each period. Thus the column sways between a neighborhood of each of the static buckled states that branch from c_n and passes twice in each period through the unbuckled state. We refer to this motion of the column as the "swaying mode." For $H = 0$, then we have $\lim_{t \to \infty} w(x, t) = 0$ and the unbuckled state is said to have a strong nonlinear stability with respect to the data.

If the column is damped and we have $\Gamma > 0$, the first-order initial value problem equivalent to (25.61) is

$$\frac{dV}{dB} = -\frac{2\Gamma V + K(\beta B + B^3)}{V}, \quad V(F) = G. \tag{25.64}$$

When $c \leq c_n \, (\beta > 0)$ the origin $B = V = 0$ is the only singular point. It is a stable spiral (stable node) for $\Gamma^2 - 4K\beta < 0 \, (\geqslant 0)$, and the unbuckled state is stable with respect to the data.

With $c > c_n$, (25.64) has three singular points. The origin is a saddle point and the singularities (25.63) are stable spirals (stable nodes) for $\Gamma^2 + 8K\beta < 0 \, (\geqslant 0)$. For simplicity, we will consider only stable spirals. Figure 25.8 shows a sketch of the separatrix, that is, the solutions of (25.64) that branch from the origin. For any initial data not on the separatrix, the solution of (25.64), or equivalently of

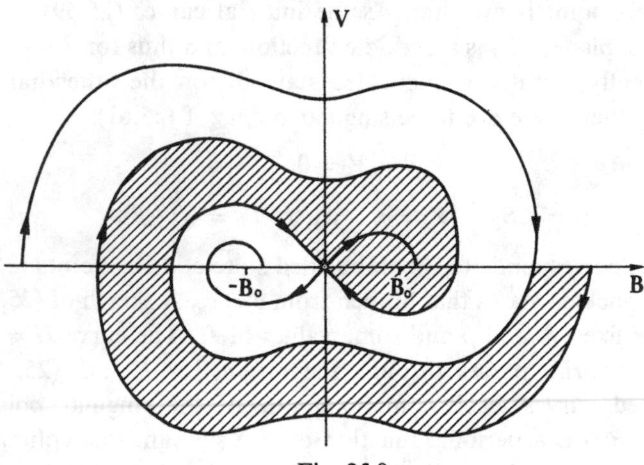

Fig. 25.8

(25.58), will be captured by one of the singular points (25.63) for $t \to \infty$. For sufficiently small $|F|$ and $|G|$, the motion is polarized about one of the static buckled states and then captured by this state as $t \to \infty$. For sufficiently large $|F|$ and $|G|$, the motion occurs in the swaying mode for a finite time and then becomes polarized and damped about one of the static buckled states. The final state need not be the one nearest to the initial point, because, if the initial point is in an unshaded region, then the final state will be the buckled state at $-B_0$ (or at B_0 for a shaded region).

26. Kirchhoff's Problem for Nonlinearly Elastic Rods

Kirchhoff's classical theory is founded on the assumption that the relative displacements of the parts of a long thin rod may be by no means small, and yet the strains that occur in any part of the rod may be small enough to satisfy the assumptions of the linear elastic theory. A consequence of this fact is that a semilinear theory may be formulated resting on the constitutive assumptions that the stress couples depend linearly on the curvatures and the twist, the axis of the rod is inextensible, the cross-sections remain perpendicular to the strained axis, and no deformations occur within the cross-section. These assumptions constitute Kirchhoff's hypotheses.

However, in some circumstances, not all Kirchhoff's hypotheses are unconditionally valid, and thus the question arises of formulating a more general theory of rods that removes Kirchhoff's constraints but without compromising the mathematical structure of the resulting equations, and that is still capable of finding the most relevant properties of solutions (Antman 1974, 1995). We assume merely that the constitutive equations give the stress resultant and couples as arbitrary nonlinear functions of appropriate strain variables, where these functions need not be derivable from a strain energy function. In this way we remove the requirement that the material is hyperelastic.[45] In the semilinear theory it is possible to show that an initially straight prismatic rod with equal principal flexural rigidities admits helical solutions solely under the action of terminal loads. The same result holds in the nonlinear theory, but the variety of solutions is much wider than is obtained in Kirchhoff's theory.

The constitutive equations that we introduce are those for the relationships between the stress and couple resultants $\mathbf{n}$ and $\mathbf{m}$ and the kinematic quantities $\mathbf{r}'$, $\mathbf{d}_p$, and $\mathbf{d}_p'$ defined in Section 19. If the constitutive equations are to be unaffected by rigid motions, it can be shown (Green and Laws 1966) that they must have the form of (19.23), that is

$$\begin{aligned} n_a(S) &= n_a[y_\alpha(S), z(S), \kappa(S), \kappa'(S), \tau(S), S], \\ m_a(S) &= m_a[y_\alpha(S), z(S), \kappa(S), \kappa'(S), \tau(S), S], \end{aligned} \tag{26.1}$$

where $a = 1$, 2, or 3, and $\alpha = 1$ or 2. It is customary to take n_a and m_a to be independent of S, which means that the material is homogeneous and the reference configuration prismatic. In addition, we assume that n_a and m_a are continuously differentiable functions of their arguments in their domains, with z restricted by the constraint $z = \mathbf{r}' \cdot \mathbf{d}_3 > 0$.

[45] The hyperelastic case has been studied by Ericksen (1970).

To be useful, the constitutive equation (26.1) must be restricted by other constraints, suggested by physical arguments. These restrictions are of two types. The first type of restriction, said to be of *strict monotonicity*, requires that the symmetric part of the matrix

$$\left\| \begin{array}{cc} \dfrac{\partial n_a}{\partial y_b} & \dfrac{\partial n_a}{\partial u_b} \\ \dfrac{\partial m_a}{\partial y_b} & \dfrac{\partial m_a}{\partial u_b} \end{array} \right\| \tag{26.2}$$

is positive definite, where we have put $y_b \equiv (y_1, y_2, z)$ and $u_b \equiv (\kappa, \kappa', \tau)$. This implies the reasonable consequence that n_1, n_2, n_3, m_1, m_2, and m_3 are monotonically increasing functions of y_1, y_2, y_3, u_1, u_2, and u_3, respectively, for fixed values of the remaining arguments. The second restriction, referred to as that of *coercivity*, requires that

$$\frac{n_a(y_b, u_b)y_a + m_a(y_b, u_b)u_a}{(y_a y_a + u_a u_a)^{\frac{1}{2}}} \tag{26.3}$$

tends to infinity as $(y_a y_a + u_a u_a)$ tends to infinity and approaches $-\infty$ as $y_3 \equiv z \to 0$. This condition essentially ensures that the resultant n_a and m_a become large as the corresponding strains become large, and that an infinitely large force is needed to violate the condition $\mathbf{r}' \cdot \mathbf{d}_3 > 0$.

Strict monotonicity and coercivity guarantee that the algebraic equations

$$n_a = n_a(y_b, u_b), \quad m_a = m_a(y_b, u_b), \tag{26.4}$$

can be uniquely resolved with respect to y_b and u_b. Moreover, the equations

$$n_a(y_b, u_\rho, u_3) = n_a, \quad m_3 \equiv q(y_b, u_\rho, u_3) \tag{26.5}$$

can be solved uniquely for y_b, $v = u_3$ in terms of n_a, u_ρ, and q:

$$y_a = y_a^*(n_b, u_\rho, q), \quad v = v^*(n_b, u_\rho, q). \tag{26.6}$$

We also define

$$m_\rho^*(n_b, u_\rho, q) \equiv m_\rho[y_b^*(n_c, u_\sigma, q), u_\rho, v^*(n_c, u_\sigma, q)] \tag{26.7}$$

so that (26.6) and

$$m_\rho = m_\rho^*(n_a, u_\sigma, q) \tag{26.8}$$

represent a set of constitutive equations equivalent to (26.1). These equations differ from the first set because they assume n_a, u_σ, and q as new set of independent constitutive variables.

Kirchhoff's theory is also founded on the hypotheses that the directors $\mathbf{d}_1$ and $\mathbf{d}_2$ are principal axes of inertia of the cross-section, that the moments of inertia with respect to these axes are equal, and the constitutive equations have the forms

$$m_1 = EIu_1 = A\kappa, \quad m_2 = EIu_2 = B\kappa', \quad (A = B).$$

Since these relations represent the class of rods the mechanical response of which is indistinguishable from that of rods having a circular cross-section, the natural extension of this property to the nonlinear constitutive equation is that y_ρ^* and m_ρ^* are isotropic vector functions of the vectors n_ρ and u_ρ and of the scalars p and q, and that z^* and v^* are isotropic scalar functions of the same arguments. Isotropy also implies that n_1, n_2, m_1, and m_2 vanish in the reference state.

In order to get a picture of the possible strained configurations of a nonlinear rod loaded only at its ends, we start from the equilibrium equation $\mathbf{n}' = \mathbf{0}$, from which we obtain

$$\mathbf{n} = \text{constant} = \mathbf{n}(L). \tag{26.9}$$

If we choose a Cartesian base of vectors $\mathbf{e}_k$ such that we have

$$\mathbf{n}(L) = N\mathbf{e}_3,$$

the components of $\mathbf{n}$ with respect to $\mathbf{d}_a$ are

$$n_1 = -N \sin\theta \cos\varphi, \quad n_2 = N \sin\theta \sin\varphi, \quad n_3 \equiv p = N\cos\theta, \tag{26.10}$$

where θ and φ are Euler's angles. As a consequence of (26.9), the equilibrium equation

$$\mathbf{m}' + \mathbf{r}' \times \mathbf{n} = \mathbf{0}, \tag{26.11}$$

can be integrated;

$$\mathbf{m} + \mathbf{r} \times \mathbf{n}(L) = \text{constant} = \mathbf{m}(L) + N\mathbf{r}(L) \times \mathbf{e}_3,$$

giving the following results:

$$\mathbf{m} \cdot \mathbf{e}_3 = -m_1 \sin\theta \cos\varphi + m_2 \sin\theta \sin\varphi + q\cos\theta = \mathbf{m}(L) \cdot \mathbf{e}_3. \tag{26.12}$$

We seek solutions for which θ is constant, $0 \leqslant \theta \leqslant \pi$, so that (19.15) and (26.10) reduce to

$$u_1 = -u\cos\varphi, \tag{26.13a}$$

$$u_2 = u\sin\varphi, \tag{26.13b}$$

$$v = \varphi' + \psi' \cos\theta, \tag{26.13c}$$

$$n_1 = -n\cos\varphi, \quad n_2 = n\sin\varphi, \quad p = N\cos\theta, \tag{26.14}$$

with

$$u \equiv \psi' \sin\theta, \quad n = N\sin\theta.$$

A consequence of isotropy is that, if we define

$$y = -\cos\varphi\, y_1 + \sin\varphi\, y_2, \quad m = -\cos\varphi\, m_1 + \sin\varphi\, m_2, \tag{26.15}$$

then y and m satisfy the constitutive relations

$$y = y(n, p, u, q), \quad m = m(n, p, u, q), \tag{26.16}$$

and y_ρ, m_ρ, z, and v can be derived from (26.16) in the reduced forms

$$y_1 = -\cos\varphi\, y(n, p, u, q), \tag{26.17a}$$

$$y_2 = \sin\varphi\, y(n, p, u, q), \tag{26.17b}$$

$$z = z(n, p, u, q), \tag{26.17c}$$

$$m_1 = -\cos\varphi\, m(n, p, u, q), \tag{26.17d}$$

$$m_2 = \sin\varphi\, m(n, p, u, q), \tag{26.17e}$$

$$v = v(n, p, u, q). \tag{26.17f}$$

Now, the component of (26.11) along the $\mathbf{d}_3$ direction is given by (19.22c) where $\ell_3 = f_1 = f_2 = 0$:

$$m_3' + y_1 n_2 - y_2 n_1 + u_1 m_2 - u_2 m_1 = 0. \tag{26.18}$$

Furthermore, since we have made the substitution $m_3 \equiv q$, we can write

$$q' = u_2 m_1 - u_1 m_2 + y_2 n_1 - y_1 n_2.$$

Hence, by using (26.17) we obtain

$$q' = (-\sin\varphi\cos\varphi + \sin\varphi\cos\varphi)(um + yn) = 0,$$

and we may conclude that

$$q = \text{constant} = \mathbf{m}(L) \cdot \mathbf{d}_3(L). \tag{26.19}$$

We now show that u must also be a constant. On substituting (26.17b) and (26.19) in (26.12) we obtain

$$\sin\theta\, m(n, p, u, q) = \mathbf{m}(L) \cdot [\mathbf{e}_3 - \mathbf{d}_3(L)\cos\theta]. \tag{26.20}$$

A consequence of coercivity is the condition $\partial m/\partial u > 0$ and that $m(n, p, u, q)$ tends to $\pm\infty$ as u tends to $\pm\infty$, for fixed n, p, and q. Thus, if $\sin\theta \neq 0$, we can solve (26.20) uniquely for u in terms of the other (constant) variables. For $\sin\theta = 0$, the substitution $u \equiv \psi' \sin\theta$ implies the result $u = 0$. In either case, on specifying the terminal load and θ, a real constant value for u is uniquely defined.

In order to obtain a picture of the possible configurations corresponding to the solutions $\theta, u =$ constant, we consider the elongation

$$\frac{ds}{dS} = [\mathbf{r}'(S) \cdot \mathbf{r}'(S)]^{\frac{1}{2}} = (y^2 + z^2)^{\frac{1}{2}},$$

where s is the arc length parameter of the deformed axis. If $\sin\theta \neq 0$, then the constancy of u and θ implies that ψ' is a uniquely determined constant; moreover, (26.13c) and (26.17f) imply that

$$\varphi' = v(n, p, u, q) - \psi' \cos\theta$$

is also a uniquely determined constant. For $\sin\theta = 0$, all we can conclude from (26.13c) and (26.17f) is that $\varphi' \pm \psi' = v(o, p, o, q)$ is a constant. The ambiguity of sign depends on the use of spherical coordinates, and leads to no ambiguity of the solution. In both cases we have the result

$$\frac{ds}{dS} = (y^2 + z^2)^{\frac{1}{2}} \equiv \lambda = \text{constant}.$$

It thus follows that, on substituting this equation and the constitutive representations for y_a in (19.14) we obtain

$$\frac{dx_1}{ds} = a\cos\psi, \quad \frac{dx_2}{ds} = a\sin\psi, \quad \frac{dx_3}{ds} = b, \tag{26.21}$$

where

$$\psi = \lambda^{-1}s\psi' + \text{constant}, \tag{26.22a}$$
$$\lambda a = -y\cos\theta + z\sin\theta, \tag{26.22b}$$
$$\lambda b = y\sin\theta + z\cos\theta. \tag{26.22c}$$

For $\sin\theta \neq 0$ we have $\psi' =$ constant and (26.21) give the equations of a circular helix (perhaps degenerate) of radius $|\lambda a/\psi'|$ and pitch b. For $\sin\theta = 0$, then (26.16) and (26.17), associated with the isotropy equation

$$y(-n, p, -u, q) = -y(n, p, u, q),$$

allow us to conclude that we have the zero values

$$n = 0, \quad u = 0, \quad y = 0. \tag{26.23}$$

For this case, also $a = 0$, so that (26.21) represent a straight line along the x_3 axis. Note that a and b cannot vanish simultaneously because of the condition $z = \mathbf{r}' \cdot \mathbf{d}_3 > 0$. We also note the relations

$$\frac{1}{\lambda}\mathbf{n} \times \mathbf{r}' = Na(\sin\varphi\mathbf{d}_1 + \cos\varphi\mathbf{d}_2), \quad \frac{1}{\lambda}\mathbf{r}' \cdot \mathbf{n} = Nb, \tag{26.24}$$

so that Na is the magnitude of the component of the resultant force in the plane perpendicular to the axis and Nb is the component of this force along the axis.

In order to be sure that solutions of this kind actually exist, we must show that all equilibrium equations are satisfied. It is clear that equations beginning with n_1', n_2', p', or q' are trivially satisfied by virtue of (26.9) and (26.19). If we now examine the remaining two equilibrium equations, taking special note of the constancy of y and m, we find that these equations are satisfied if, and only if, we have the condition

$$\psi'(m\cos\theta - q\sin\theta) = N(y\cos\theta - z\sin\theta). \tag{26.25}$$

This equation offers a criterion of classification for the possible solutions. Equations (26.21) imply that the strained axis will be straight when $a\psi' = 0$. Now $a = 0$ if we have the relation

$$y\cos\theta = z\sin\theta,$$

in which case (26.25) reduces to

$$\psi'(m\cos\theta - q\sin\theta) = 0, \tag{26.26}$$

and these two conditions, associated with $z > 0$, restrict the six variables θ, y, z, ψ', m, and q. Under these circumstances the axis is straight and parallel to the direction $\mathbf{e}_3$ of the terminal force. With $\psi' = 0$, (26.25) implies the null value

$$Na = 0.$$

Equations (26.21) imply that the axis will become circular if, and only if, the conditions $b = 0$ and $\psi' \neq 0$ apply. From (26.22c), the result $b = 0$ follows from

$$y\sin\theta + z\cos\theta = 0. \tag{26.27}$$

Because a and b cannot vanish simultaneously, the coefficient of N in (26.25) cannot vanish. Moreover, $\sin\theta$ cannot vanish, because if it did so it would cause (26.27) to violate the condition $z > 0$. We thus have the two equations (26.25) and (26.27) and the three inequalities $z > 0$, $\psi' \neq 0$, and $\sin\theta \neq 0$ restricting the seven variables θ, y, z, ψ', N, m, and q. The last case for analysis is that represented by deformations with a helical axis. As we require the helix to be nondegenerate, the condition $ab\psi' \neq 0$ must be satisfied. Then, since here again equation (26.25) must be verified by all possible solutions, it follows that there will be nondegenerate helical solutions if there are numbers θ, ψ', N, and q satisfying the conditions

$$\begin{aligned}\mathscr{H}(\theta, \psi', N, q) \equiv \psi'[&m(N\sin\theta, N\cos\theta, \psi'\sin\theta, q)\cos\theta - q\sin\theta]\\ &- N[y(N\sin\theta, N\cos\theta, \psi'\sin\theta, q)\cos\theta\\ &- z(N\sin\theta, N\cos\theta, \psi'\sin\theta, q)\sin\theta]\\ &= 0, \quad \text{for } ab\psi' \neq 0. \end{aligned} \tag{26.28}$$

We note that $\sin\theta \geqslant 0$, which is a result of the condition that $0 \leq \theta \leq \pi$, and the same argument as before proves $a = 0$ for $\sin\theta = 0$.

So far we have considered elastic rods with a general constitutive equation of the type (26.1). It is therefore natural to ask whether restricting our analysis to hyperelastic rods and planar deformations might lead to more specific results. The answer is affirmative (Antman 1968, 1969a) and, for a certain simple form of constitutive equation, the equilibrium equations of planar deformation can be integrated by quadratures, even in the case of naturally curved rods.

If we let S be the arc length of the undeformed rod, $\Phi(S)$ be the angle made by the tangent to the undeformed axis with a given direction, and $\varphi(S)$ be the tangential angle to the deformed rod, then the constitutive equations have the form

$$m = \frac{\partial W}{\partial \mu}, \tag{26.29a}$$

$$n = \frac{\partial W}{\partial \delta}, \tag{26.29b}$$

$$W = W(\mu, \delta, S), \tag{26.29c}$$

where $m \equiv m_2$ is the bending moment in the plane of the rod, $n \equiv n_3$ is the axial force, and μ and δ are strains defined by

$$\delta = \frac{ds}{dS} - 1, \quad \mu = \frac{ds}{dS}\kappa' - \kappa'_0, \tag{26.30}$$

where κ' and κ'_0 are the curvatures of the deformed and undeformed states, respectively. We observe that, for a uniform extension of a circular arc into another circular arc of different radius, the quantity $\kappa' - \kappa'_0$ changes, but μ remains zero. Hence the strain measure μ isolates the effect of bending.

To ensure that (26.29) represent physically reasonable constitutive equations, we require that m be a monotonically increasing function of μ for fixed δ, n be a monotonically increasing function of δ for fixed μ, and (26.29a) and (26.29b) be invertible so as to give μ and δ as functions of m and n. All these properties are guaranteed by requiring the Hessian matrix

$$\begin{Vmatrix} W_{\mu\mu} & W_{\mu\delta} \\ W_{\delta\mu} & W_{\delta\delta} \end{Vmatrix}$$

for the strain energy function W to be positive definite.

The case we consider is that of a rod subjected to a constant hydrostatic pressure p perpendicular to the axis of the rod in its deformed state. The equilibrium equations (19.21a), (19.21c), and (19.22b) then read

$$q' + \mu n + p = 0, \tag{26.31a}$$
$$n' - \mu q = 0, \tag{26.31b}$$
$$m' - y_1 n + zq + (\mathbf{r} \times \mathbf{f}) \cdot \mathbf{d}_2 = 0, \tag{26.31c}$$

where we have put $q \equiv n_1$, to denote the shear force in the $\mathbf{d}_1$ direction (Figure 26.1). Now, if x_1, 0, and x_3 are the coordinates of a point P on the axis, we have the geometric relations

$$\frac{ds}{dS}\kappa' = \frac{ds}{dS}\frac{d\varphi}{ds} = \mu + \kappa_0', \tag{26.32a}$$
$$\frac{dx_3}{ds} = \cos\varphi, \tag{26.32b}$$
$$\frac{dx_1}{ds} = \sin\theta, \tag{26.32c}$$

which are equivalent to the following:

$$\left.\begin{aligned} \frac{d\varphi}{dS} = \mu + \kappa_0', \quad \frac{dx_3}{dS} &= (1+\delta)\cos\varphi, \\ \frac{dx_1}{dS} &= (1+\delta)\sin\varphi. \end{aligned}\right\} \tag{26.33}$$

In order to write the equilibrium equations in a more tractable form, we observe that the vector $\mathbf{f}$, which represents the hydrostatic load, has the components $(p\mathbf{d}_1, 0, 0)$. Moreover, the vector position $\mathbf{r}$ is given by

$$\mathbf{r} = x_1\mathbf{e}_1 + x_3\mathbf{e}_3,$$

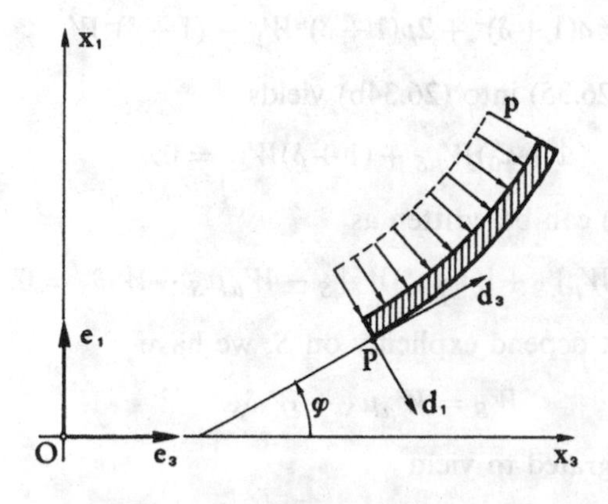

Fig. 26.1

where $\mathbf{e}_1$ and $\mathbf{e}_3$ are unit vectors along the coordinate axes, and the vector $\mathbf{r}'$ is given by

$$\mathbf{r}' = \frac{dx_1}{ds}\,\mathbf{e}_1 + \frac{dx_3}{ds}\,\mathbf{e}_3 \equiv y_1\mathbf{d}_1 + z\mathbf{d}_3,$$

from which we derive

$$y_1 = \frac{dx_1}{ds}\,\mathbf{e}_1 \cdot \mathbf{d}_1 + \frac{dx_3}{ds}\,\mathbf{e}_3 \cdot \mathbf{d}_1 = 0,$$
$$z = \frac{dx_1}{ds}\,\mathbf{e}_1 \cdot \mathbf{d}_3 + \frac{dx_3}{ds}\,\mathbf{e}_3 \cdot \mathbf{d}_3 = 1.$$

Since the last term in (26.31c) vanishes, the equilibrium equations reduce to

$$q' + \mu n + p = 0,$$
$$n' - \mu q = 0,$$
$$m' + q = 0.$$

Thus, if we eliminate q from these equations and write the remaining two equations in terms of the undeformed arc length S, we obtain

$$\frac{d}{dS}\left(\frac{\dfrac{dm}{dS}}{1+\delta}\right) - (\mu + \kappa_0')n - p(1+\delta) = 0, \qquad (26.34a)$$

$$(\mu + \kappa_0')\frac{dm}{dS} + (1+\delta)\frac{dn}{dS} = 0. \qquad (26.34b)$$

The constitutive equations are (26.29), which can be written as

$$m = W_\mu, \quad n = W_\delta. \qquad (26.35)$$

In order to solve (26.34) in the case in which κ_0' and p are constants, we multiply (26.34a) by $(dm/dS)(1+\delta)^{-1}$ and integrate the resulting expression to obtain

$$m'^2 + (1+\delta)^2 n^2 - 2pm(1+\delta)^2 = a(1+\delta)^2, \qquad (26.36)$$

where a is an arbitrary constant of integration. This equation is valid for any constitutive assumption. Substitution of (26.35) into (26.36) yields

$$W_{\mu S}^2 = a(1+\delta)^2 + 2p(1+\delta)^2 W_\mu - (1+\delta)^2 W_\delta^2, \qquad (26.37)$$

and the substitution of (26.35) into (26.34b) yields

$$(\mu + \kappa_0')W_{\mu S} + (1+\delta)W_{\delta S} = 0. \qquad (26.38)$$

As κ_0' is constant, (26.38) can be written as

$$[(\mu + \kappa_0')W_\mu]_{,S} + [(1+\delta)W_\delta]_{,S} - W_\mu \mu_S - W_\delta \delta_S = 0, \qquad (26.39)$$

and, because W does not depend explicitly on S, we have

$$W_S = W_\mu \mu_S + W_\delta \delta_S.$$

Thus (26.39) can be integrated to yield

$$(\mu + \kappa_0')W_\mu + (1+\delta)W_\delta - W = b, \qquad (26.40)$$

where b is another arbitrary constant of integration. Equation (26.40) is just an algebraic relation between μ and δ. To obtain μ and δ explicitly, we write (26.38) in the form

$$(\mu + \kappa_0')(W_{\mu\mu}\mu_S + W_{\mu\delta}\delta_S) + (1+\delta)(W_{\delta\mu}\mu_S + W_{\delta\delta}\delta_S) = 0. \tag{26.41}$$

Let us assume the restriction

$$(\mu + \kappa_0')W_{\mu\delta} + (1+\delta)W_{\delta\delta} \neq 0, \tag{26.42}$$

which involves both the strains and the strain energy W. Then, from (26.41), we obtain the value

$$-\delta_S = \frac{(\mu + \kappa_0')W_{\mu\mu} + (1+\delta)W_{\mu\delta}}{(\mu + \kappa_0')W_{\mu\delta} + (1+\delta)W_{\delta\delta}}\mu_S \tag{26.43}$$

and therefore

$$W_{\mu S} = W_{\mu\mu}\mu_S + W_{\mu\delta}\delta_S = (1+\delta)\frac{(W_{\mu\mu}W_{\delta\delta} - W_{\mu\delta}^2)}{(\mu + \kappa_0')W_{\mu\delta} + (1+\delta)W_{\delta\delta}}\mu_S, \tag{26.44}$$

so that (26.37) becomes

$$\mu_S^2 = (a + 2pW_\mu - W_S^2)\left[\frac{(\mu + \kappa_0')W_{\mu\delta} + (1+\delta)W_{\delta\delta}}{W_{\mu\mu}W_{\delta\delta} - W_{\mu\delta}^2}\right]^2, \tag{25.45}$$

where the denominator does not vanish by virtue of the positive definiteness of the Hessian matrix of W.

If we solve (26.40) for δ as a function of μ, which is possible if (26.42) holds, then the right-hand side of (26.45) may be regarded as a function of μ only. Once we have μ, we can derive δ from (26.40), and therefore $\varphi(S)$, $x_1(S)$, and $x_3(S)$ can be obtained by further integration. The method fails in the case in which (26.42) is violated, but then we may assume that the numerator in (26.43) does not vanish. It is now possible to carry out operations analogous to those used when considering μ, and we arrive at the equation

$$\delta_S^2 = (a + 2pW_\mu - W_\delta^2)\left[\frac{(\mu + \kappa_0')W_{\mu\mu} + (1+\delta)W_{\mu\delta}}{W_{\mu\mu}W_{\delta\delta} - W_{\mu\delta}^2}\right]^2, \tag{26.46}$$

where the right-hand side is now regarded as a function of δ.

As an illustration, we take

$$W = \tfrac{1}{2}B\mu^2 + \tfrac{1}{2}F\delta^2, \tag{26.47}$$

where B and F are given constants. Then the constitutive equations are linear in μ and δ: $m = B\mu$ and $n = F\delta$. Since, by the continuity condition, $s' = 1 + \delta > 0$, (26.40) can be written as

$$1 + \delta = \left[c - (\mu + \kappa_0)\frac{B}{F}\right]^{\frac{1}{2}}, \tag{26.48}$$

where c is an arbitrary constant. Then (26.45) becomes

$$(B\mu_S)^2 = \left[c - (\mu + \kappa_0')^2 \frac{B}{F}\right][a + 2Bp\mu + F^2]\left\{\left[c - (\mu + \kappa_0')^2 \frac{B}{F}\right]^{\frac{1}{2}} - 1\right\}^2. \qquad (26.49)$$

With the change of variables $\mu + \kappa_0' = [c(F/B)]^{\frac{1}{2}} \sin\psi$ and $w = \tan(\psi/2)$ we convert (26.49) into an equation that can be integrated in terms of elliptic functions. However, very often it is more convenient to study (26.49) using a qualitative technique (Antman 1969).

Conditions like the positive definiteness of the Hessian matrix made with the second derivatives of the strain energy function W with respect to its arguments μ and δ are not only suggested by physics, but also by the need of a well position in the mathematical problem describing the equilibrium state.

Let us assume, for simplicity, that the rod is nonextensible and has the constitutive equation

$$m = W_\mu, \quad W = W(\mu, S), \qquad (26.50)$$

and the material constraint of inextensibility $ds/dS = 1$, or $\delta = 0$. In this case we impose the following physically reasonable restrictions on W:

$$W_{\mu\mu} > 0, \qquad (26.51)$$

$$W(\mu, S) \geqslant K|\mu|^\alpha + \gamma(S), \quad K > 0, \quad \alpha > 1, \qquad (26.52)$$

where K and α are constants and $\gamma(S)$ is an integrable function.

We denote the length of the rod by ℓ, so that $0 \leqslant S \leqslant \ell$, and consider either the case in which the ends are hinged ($d\varphi/dS$ prescribed) or the case in which they are built-in (φ prescribed). Moreover, we assume that the end points are fixed so that, recalling (26.32b), there are two equations

$$\int_0^\ell \cos\varphi(S)\, dS = \Delta, \quad \int_0^\ell \sin\varphi(S) = 0, \qquad (26.53)$$

where $\Delta < \ell$ is a given number.

We set $\psi = (\phi - \Phi)$, so that $\mu = \psi_S$. Then the total potential energy for a rod subjected to a hydrostatic pressure p has the form

$$V(\psi) = V_1 + pV_2, \qquad (26.54)$$

with

$$V_1 = \int_0^\ell W(\psi_S, S)\, dS, \qquad (26.55)$$

$$V_2 = \frac{1}{2}\int_0^\ell (xy_S - yx_S)\, dS = \frac{1}{2}\int_0^\ell dS \int_0^S \sin[\varphi(S) - \varphi(\xi)]\, d\xi. \qquad (26.56)$$

Since the governing equations for the equilibrium state are exactly the Euler equation characterizing the minimum of (26.54) subject to (26.53), we can prove the existence of a solution to our problem once we have shown that the functional $V(\psi)$ has a minimum

For this purpose we introduce the Sobolev space $W^{1,\alpha}(0,\ell)$ $(\alpha > 1)$ equipped with the norm

$$\|\psi\| = \left[\int_0^\ell (|\psi|^\alpha + |\psi_S|^\alpha)\, dS \right]^{\frac{1}{\alpha}},$$

and see whether all conditions ensuring the existence of a minimum of V among the functions $\psi \in W^{1,\alpha}(0,\ell)$, satisfying the boundary condition either of hinged or of clamped ends, and the constraints (26.53), are satisfied. We can now check that the following properties are verified:

(a) $V_1 \to \infty$ as $\|\psi\| \to \infty$, as an immediate consequence of (26.52).
(b) V_1 is a weakly lower semicontinuous functional, as a consequence of (26.51).
(c) The functionals V_2, $\int_0^\ell \cos\varphi(S)\, dS$, and $\int_0^\ell \sin\varphi(S)\, dS$, are weakly continuous, as can be proved by applying the definition of weak continuity.
(d) V_2 is a number bounded from above and below by two finite numbers M and m.
(e) The set of admissible functions such that $m \leqslant V_2 \leqslant M$ is not empty.

Conditions (a) and (d) imply the conditions $V \to \infty$ as $\|\psi\| \to \infty$, and hence we need only seek a minimum for V in some bounded subset of $W^{1,\alpha}$. Conditions (b) and (c) state that V is weakly semicontinuous, and conditions (d) and (e) ensure that the subset of $W^{1,\alpha}$, on which solutions are sought, is weakly closed and non-empty. Thus we can conclude that there is an element ψ_0 for which V attains its absolute minimum.

Once we have proved the existence of a minimum for V, we may ask how regular the minimum is. The answer to this question is provided by a theorem by De Giorgi (1957), stating that, if $W(\mu, S)$ satisfies the inequality

$$|W_\mu| < C(|\mu|^\alpha + 1), \quad C = \text{constant}, \tag{26.57}$$

then the minimizing function ψ_0 is twice continuously differentiable on $[0,\ell]$ and, therefore, is a classical solution of Euler's equation.

The theory of constitutively nonlinear rods may be extended to the treatment of problems of buckling. When a rod is linearly elastic, but the displacements are not small, the bifurcation theory gives a precise description of how solutions behave in the neighborhood of critical points. The question thus arises of whether an analogous behavior can be confirmed for rods having a more general constitutive equation, or, more importantly, whether the material nonlinearity offers new branches of buckled states due to the interaction of the constitutive and geometrical nonlinear effects. The answer is affirmative: nonlinear rods exhibit a new type of bifurcation, which is called *subcritical* (Antman and Marlow 1992).

As the phenomenon also occurs when the deformation of a column remains in a plane, we study the problem for the planar equilibrium states of nonlinearly elastic rods subjected only to terminal loads.

Let $(\mathbf{i}, \mathbf{j}, \mathbf{k})$ be a fixed, right-handed, orthonormal basis for the Euclidean three-dimensional space, and let us consider an initially straight rod, which, in its deformed configurations, remains in the **ij** plane. The strained configuration of the rod is described by a vector function $\mathbf{r}(s)$, which represents the position of any point of the axis as a function of the arc length s, and a vector function

$$\mathbf{b}[\theta(s)] \equiv \cos\theta(s)\mathbf{i} + \sin\theta(s)\mathbf{j}, \tag{26.58}$$

describing the orientation of the cross-section of the rod at s. Setting

$$\mathbf{a}(\theta) \equiv -\mathbf{k} \times \mathbf{b}(\theta), \tag{26.59}$$

$$\mathbf{r}' = \nu\mathbf{a} + \eta\mathbf{b}, \tag{26.60}$$

we assume that the strains characterizing the deformation of the rod are $(\nu, \eta, \theta' \equiv \mu)$. In the initial, straight, configuration, the strains are

$$\nu = 1, \quad \eta = 0, \quad \theta(s) = 0. \tag{26.61}$$

Let us introduce the identity $\mathbf{n}(s) \equiv N(s)\,\mathbf{a}(s) + H(s)\,\mathbf{b}(s)$, and let $\mathbf{n}(s)$ and $M(s)\mathbf{k}$ be the resultant contact force and couple across a section in a deformed configuration. Since the loads are applied only at the ends of the rod, the equilibrium equations are

$$\mathbf{n}' = \mathbf{0}, \quad M' + \mathbf{k}\cdot(\mathbf{r}' \times \mathbf{n}) = 0, \tag{26.62}$$

which have the components

$$N' = H\theta', \tag{26.63a}$$

$$H' = -N\theta', \tag{26.63b}$$

$$M' = N\eta - H\nu. \tag{26.63c}$$

The material of the rod is taken to be elastic, and the stress forces and the stress couples are connected with the strains by equations of the form

$$N(s) = N[\nu(s), \eta(s), \mu(s), h, \tau, s] \tag{26.64a}$$

$$M(s) = M(\ldots), \tag{26.64b}$$

$$H(s) = H(\ldots), \tag{26.64c}$$

where h represents a positive parameter that accounts for the possible variability in the thickness of the cross-section, and τ is another parameter representing the possible asymmetry of the cross-section.

The constitutive equations must, of course, be restricted by the customary conditions of physical admissibility. These conditions require that the matrix

$$\frac{\partial(N, H, M)}{\partial(\nu, \eta, \mu)} \tag{26.65}$$

be positive definite, and with $N \to \infty$ for $\nu \to \infty$, and $H \to \infty$ as $\eta \to \infty$, $M \to \pm\infty$ as μ approaches its positive and negative extremes in its range of variability.

In order to derive the governing equations, we scale s so that we have $0 \leqslant s \leqslant 1$ and arrange the $(\mathbf{i}, \mathbf{j}, \mathbf{k})$ vectors so that we have

$$\mathbf{r}(s) = s\mathbf{i}, \quad \mathbf{r}(0) = \mathbf{0}. \tag{26.66}$$

If the column is subjected to a compressive thrust of magnitude λ at the end $s = 1$, so as to have $\mathbf{n}(1) \cdot \mathbf{i} = -\lambda$, the stress resultants at s become

$$N = -\lambda \cos\theta, \quad H = \lambda \sin\theta,$$

and the equilibrium equation (26.63c) yields

$$M' + \lambda(\nu \sin\theta + \eta \cos\theta) = 0. \tag{26.67}$$

Since the conditions of positivity (26.65) and of the growth of the constitutive functions N, H, and M as separate functions of ν, η, and μ, respectively, ensure that the mapping (26.64) has the inverse

$$\nu = \nu(N, H, M, h, \tau, s), \quad \eta = \eta(\ldots), \quad \mu = \mu(\ldots), \tag{26.68}$$

we obtain, from the definition $\mu \equiv \theta'$ and from (26.67), the values

$$\theta' = \mu(-\lambda \cos\theta, \lambda \sin\theta, M, h, \tau, s), \tag{26.69}$$

$$\begin{aligned} M' = &-\lambda[\nu(-\lambda \cos\theta, \lambda \sin\theta, M, h, \tau, s) \sin\theta \\ &+ \eta(-\lambda \cos\theta, \lambda \sin\theta, M, h, \tau, s) \cos\theta]. \end{aligned} \tag{26.70}$$

Thus the boundary value problem is reduced to solving (26.69) and (26.70) subject to certain boundary conditions, depending on the constraints imposed at the ends. For instance, if the ends are clamped, the conditions are

$$\theta(0) = \theta(1) = 0, \quad \mathbf{n}(1) \cdot \mathbf{j} = 0; \tag{26.71}$$

if the ends are hinged, the boundary conditions are

$$M(0) = M(1) = 0, \quad \mathbf{r}(1) \cdot \mathbf{j} = 0. \tag{26.72}$$

If the column is uniform, the constitutive functions are independent of s. In this case we put

$$\nu^0 = \nu(-\lambda, 0, 0, h, \tau), \quad \eta^0 = \eta(-\lambda, 0, 0, h, \tau), \quad \mu^0 = \mu(-\lambda, 0, 0, h, \tau), \tag{26.73}$$

and let us denote by subscripts the partial derivatives of constitutive functions. Then the linearization of (26.69) and (26.70) about the trivial solution may be written in the form

$$\theta' = \mu_M^0 M, \quad M' = -\lambda(\nu^0 + \lambda\eta_H^0)\theta,$$

and these two equations are equivalent to the single equation

$$\theta'' = -q(\lambda, h, \tau)\theta, \tag{26.74}$$

where we have set

$$q(\lambda, h, \tau) \equiv \lambda\mu_M^0(\nu^0 + \lambda\eta_H^0). \tag{26.75}$$

The nontrivial solutions of (26.74) are

$$(\theta, M) = \left(\frac{\mu_M^0}{n\pi} \sin n\pi s, \cos n\pi s\right), \quad \text{for } \theta(0) = \theta(1) = 0,$$

and

$$(\theta, M) = \left(\cos n\pi s, -\frac{n\pi}{\mu_0}\sin n\pi s\right), \quad \text{for } M(0) = M(1) = 0,$$

and the corresponding eigenvalues are

$$q(\lambda, h, \tau) = n^2\pi^2 \tag{26.76}$$

in both cases.

To analyze the nonlinear problem, we assume that q has the form shown in Figure 26.2: $q(0, h, \tau) = 0$; $q(\lambda, h, \tau)$, regarded as a function of λ, increases to a maximum λ^{**} at which $q_{\lambda\lambda} < 0$, and then decreases asymptotically to zero for $\lambda \to \infty$. Moreover, we require that, for each $\lambda > 0$, $q(\lambda, h, \tau)$ decreases to zero for $h \to \infty$. These constitutive properties are not artificial, but are suggested by the effective response of many materials. The most significant property is that $q(\lambda, h, \tau) \to 0$ as $\lambda \to \infty$, a property that is markedly different from the one assumed in the linearized case, where (26.75) is linear in λ.

Our purpose is to study the behavior of solutions of (26.69) and (26.70) near a simple eigenvalue λ^* of the linearized problem (26.74), for which (26.76) holds and $q_\lambda(\lambda^*, 0,) \neq 0$. The nonlinear problem, subject to one of the two types of boundary conditions assumed before, is treated by a method proposed by Poincaré, called the *shooting method*, which consists of replacing the boundary conditions (26.71) and (26.72) with the initial conditions

$$\theta(0) = M(0) = m,$$

and adjusting the parameter m in order to satisfy the boundary condition at $s = 1$. The initial-value problem admits the trivial solution for $m = 0$ and, by the Cauchy–Kowalewski theorem, it admits a unique solution defined for $s \in [0, 1]$, provided that $|m|$ is sufficiently small, and this solution depends smoothly on λ, η, τ, and m. As the solutions of the boundary value problem must also satisfy the boundary condition at $s = 1$, the eigensolutions of (26.69) and (26.70) are given by the values of m for which

$$\theta(1, \lambda, h, \tau, m) = 0 \tag{26.77}$$

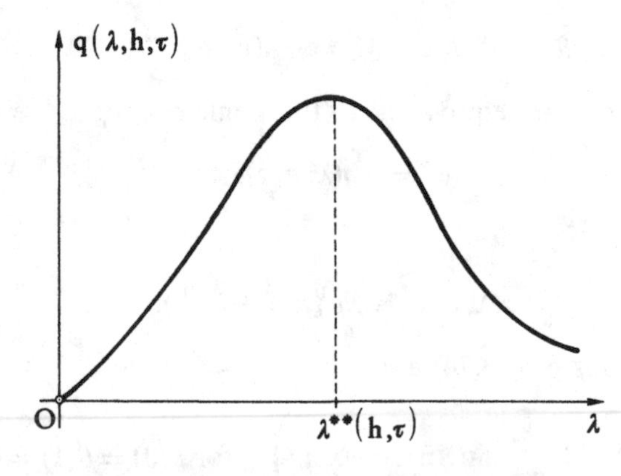

Fig. 26.2

in the case of clamped ends, and

$$M(1, \lambda, h, \tau, m) = 0 \tag{26.78}$$

for hinged ends.

Let us consider, for brevity, (26.77). We do not know the solution explicitly, but with a standard perturbation method we readily reach the result

$$\theta_m(s, \lambda, h, \tau, 0) = \frac{\overset{0}{\mu}_M(\lambda, h, \tau)}{\sqrt{q(\lambda, h, \tau)}} \sin \sqrt{q(\lambda, h, \tau)}\, s. \tag{26.79}$$

When (26.79) holds, we find that (26.79) agrees with the linearized solution

$$\theta = \frac{\overset{0}{\mu}_M}{n\pi} \sin n\pi s,$$

together with the results

$$\theta_m(1, \lambda, h, \tau, 0) = 0, \text{ when } q(\lambda, h, \tau) = n^2\pi^2, \tag{26.80}$$

$$\theta_{m\lambda}(1, \lambda, h, \tau, 0) = (-1)^n \frac{\overset{0}{\mu}_M(\lambda, h, \tau) q_\lambda(\lambda, h, \tau)}{2n^2\pi^2}, \text{ when } q(\lambda, h, \tau) = n^2\pi^2, \tag{26.81}$$

$$\theta_{m\lambda\lambda}(1, \lambda, h, \tau, 0) = (-1)^n \frac{\overset{0}{\mu}_M(\lambda, h, \tau) q_{\lambda\lambda}(\lambda, h, \tau)}{2n^2\pi^2}, \quad \text{when } q(\lambda, h, \tau) = n^2\pi^2, \quad q_\lambda(\lambda, h, \tau) = 0. \tag{26.82}$$

Equation (26.81) states that an eigenvalue λ satisfying (26.76) is simple when $q_\lambda(\lambda, h, \tau) \neq 0$. In addition, it is possible to prove the two results

$$\theta_{mm}(1, \lambda, h, \tau, 0) = 0, \text{ when } q(\lambda, h, \tau) = n^2\pi^2, \tag{26.83}$$

$$\theta_{mm\lambda}(1, \lambda, h, \tau, 0) = 0, \text{ when } q(\lambda, h, \tau) = n^2\pi^2, \quad q_\lambda(\lambda, h, \tau) = 0. \tag{26.84}$$

In general, however, we find that the condition $\theta_{mmm}(1, \lambda, h, \tau, 0) \neq 0$ is satisfied for $q(\lambda, h, \tau) = n^2\pi^2$, whether or not we have $q_\lambda = 0$. Finally, we note that all derivatives of $\theta(1, \lambda, h, \tau, 0)$ with respect to λ are zero.

We now study the behavior of solutions of (26.69) and (26.70) near a simple eigenvalue λ^* of the linearized problem; that is, when (26.76) holds with $q_\lambda \neq 0$. However, one result of the *singularity theory* (Golubitsky and Schaeffer 1985) states that the bifurcation diagram for (26.77) is strongly equivalent to that for the equation

$$m[\varepsilon m^2 + \delta(\lambda - \lambda^*)] = 0, \tag{26.85}$$

in the $[\gamma, (\lambda - \lambda^*)]$ plane, where

$$\varepsilon = \operatorname{sign} \theta_{mmm}(1, \lambda, h, \tau, 0), \quad \delta = \operatorname{sign} \theta_{m\lambda}(1, \lambda, h, \tau, 0). \tag{26.86}$$

When the product $\varepsilon\delta$ is unity, that is when $\varepsilon\delta = 1$, the bifurcating branches do not have the same behavior as shown in Figure 20.3, but the one sketched in Figure 26.3. This kind of behavior is called *subcritical bifurcation*, and is a consequence of the failure of $q(\lambda, h, \tau)$ to be increasing.

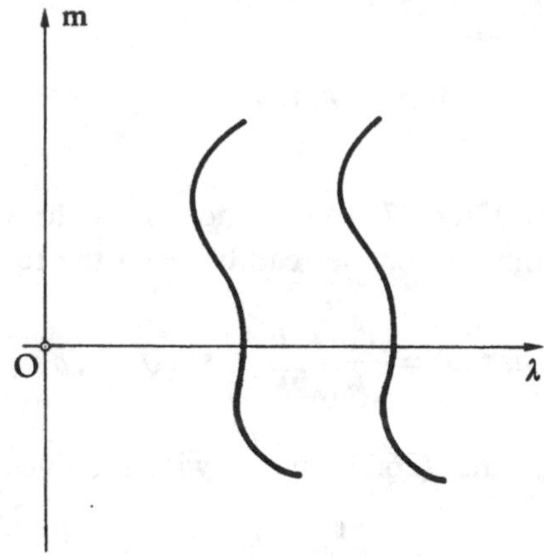

Fig. 26.3

Under similar circumstances, the diagrams shown in Figure 26.3 are not even symmetric with respect to the λ axis, and the bifurcation is then said to be *transcritical.*

A simple example of a reasonable constitutive equation for which the theory applies is the specific family of unshearable columns in pure compression (Antman and Pierce 1990). The relations (26.64a) and (26.64b) can be chosen with the forms

$$N = 2hK\left[1 - \frac{\nu}{(\nu^2 - h^2\mu^2)^2}\right], \quad M = \frac{2h^3 K\mu}{(\nu^2 - h^2\mu^2)^2}, \tag{26.87}$$

where $K > 0$ is a given number. These relations have the inverse

$$\nu = \left(1 - \frac{N}{2hK}\right)\delta^4(N, M), \quad \mu = \frac{M}{2h^3 K}\,\delta^4(N, M), \tag{26.88}$$

with

$$\delta^2(N, M) = \nu^2 - h^2\mu^2 = \left[\left(1 - \frac{N}{2hk}\right)^2 + \left(\frac{M}{2h^2 K}\right)^2\right]^{-\frac{1}{3}}. \tag{26.89}$$

The linearized version of (26.88) for $N = -\lambda$ and $M = 0$ is

$$\nu^0 = \left(1 + \frac{\lambda}{2hk}\right)^{-\frac{1}{3}}, \quad \mu_M^0 = \frac{\lambda}{2h^3 K}\left(1 + \frac{\lambda}{2hk}\right)^{-\frac{4}{3}},$$

and hence, applying definition (26.75) with $\eta \equiv 0$, we obtain

$$q(\lambda, h, \tau) = \lambda\mu_M^0\nu^0 = \frac{\lambda}{2h^3 K}\left(1 + \frac{\lambda}{2hk}\right)^{-\frac{5}{3}}. \tag{26.90}$$

The graph of this function has the behavior shown in Figure 26.2.

For constitutively nonlinear rods a central problem is that of determining the detailed form of the bifurcation diagrams. However, we can also deal with the inverse problem, namely that of determining the material properties for which the bifurcating branch emanating from the lowest buckling load has a prescribed global form. This question arises in several problems of mechanical engineering, such as in the design of switches or sensors, the nontrivial bifurcated configurations of which must follow certain prescribed patterns (Antman and Adler 1987).

In order to formulate the problem, let s [$\in (0, 1)$] be the scaled arc length of an initially straight vertical, inextensible and unshearable, uniform column clamped at the end $s = 0$ and subjected to a compressive axial force λ at the end $s = 1$. The equilibrium equation describing planar deformations of the column is

$$M' + \lambda \sin \theta(s) = 0, \tag{26.91}$$

where θ is measured between the tangent to the strained axis and the vertical and $M = M[\theta'(s)]$, is the bending moment. The boundary conditions corresponding to the geometrical and mechanical constraints are

$$\theta(0) = 0, \quad M[\theta'(1)] = 0. \tag{26.92}$$

We require the constitutive function M to satisfy the physically reasonable assumptions

$$M(0) = 0, \quad M_\mu(\mu) > 0, \quad M(\mu) \to \infty \text{ for } \mu \to \infty, \tag{26.93}$$

where $\mu = \theta'(s)$ denotes the curvature. These properties ensure that the function

$$m = M(\mu),$$

can be inverted and written in the form

$$\mu = \mu(m), \tag{26.94}$$

with

$$\mu(0) = 0, \quad \mu_m(m) > 0, \quad \mu(m) \to \infty \text{ for } m \to \infty. \tag{26.95}$$

Under these hypotheses it is possible to define the stored energy of the column as

$$W(\mu) \equiv \int_0^\mu M(\xi)\, d\xi,$$

giving $M = W_\mu$. Then, on introducing the Legendre transformation of W, that is, the function

$$V(m) \equiv m\mu(m) - W, \tag{26.96}$$

we can replace (26.91) by the equivalent Hamiltonian system:

$$\theta'(s) = V_m[m(s)], \tag{26.97}$$

$$m'(s) = -\lambda \sin \theta(s), \tag{26.98}$$

$$\theta(0) = 0, \quad m(1) = 0, \tag{26.99}$$

which admits the energy integral

$$V[m(s)] = \lambda[\cos \theta(s) - 1] - V[m(0)]. \tag{26.100}$$

If we replace $\cos \theta$ by $\pm[1 - (m'/\lambda)^2]^{\frac{1}{2}}$, obtained from (26.98), and rearrange the terms, (26.100) becomes

$$[m'(s)]^2 = \{V[m(0)] - V[m(s)]\}(2\lambda - \{V[m(0)] - V[m(s)]\}). \tag{26.101}$$

The first buckling mode of (26.97), (26.98), and (26.99) is characterized by the conditions

$$0 \leqslant \theta(s) < \pi, \quad m'(s) < 0 \text{ for } s > 0, \quad m(s) > 0 \text{ for } s < 1. \tag{26.102}$$

The fact that $\theta(s)$ cannot attain the limiting value π is a consequence of (26.97)–(26.99); the fact that m' is negative and m is positive follows from (26.98). Equation (26.101) then yields

$$1 = \int_{m(0)}^{1} \frac{ds}{dm}\,dm = \int_{0}^{m(0)} \{V[m(0)] - V(m)\}^{-\frac{1}{2}}(2\lambda - \{V[m(0)] - V(m)\})^{-\frac{1}{2}}\,dm. \tag{26.103}$$

Now conditions (26.95) and the constitutive properties of the strain energy W ensure the following conditions:

$$V(0) = 0, \quad V_m(0) = 0, \quad V_{mm}(m) > 0, \quad V(m) \to \infty \text{ for } m \to \infty. \tag{26.104}$$

It then follows that $V(m)$ has an inverse, Q, for $m \geqslant 0$, such that $m = Q(v)$ is equivalent to $v = V(m)$ for $m \geqslant 0$. Moreover, conditions (26.104) imply that $Q(v)$ behaves like a constant multiplied by $\sqrt{v}$ for small v. Accordingly, we set

$$Q_v(v) = F(v)v^{-\frac{1}{2}}, \qquad t = V[m(0)], \tag{26.105}$$

and reduce (26.103) to the form

$$1 = \int_0^t \frac{F(v)\,dv}{[v(t-v)(2\lambda - t + v)]^{\frac{1}{2}}}. \tag{26.106}$$

We now prescribe $\lambda = \Lambda(t)$, substitute it in (26.106), and seek a value for F that satisfies the resulting equation for all $t \geqslant 0$. Equation (26.106) is a linear integral equation of Volterra type with a singular kernel. In order that the integrand in (26.106) allows interpretation, we require the condition that

$$\Lambda(t) = \lambda > \frac{t}{2}. \tag{26.107}$$

Provided that we find F from (26.106), we can reconstruct M using (26.105) to obtain

$$Q(v) = \int_0^v F(z)z^{-\frac{1}{2}}\,dz. \tag{26.108}$$

From the equation $m = Q(v)$, we obtain the result $V_m(m) = \{Q_v[V(m)]\}^{-1}$, and hence (26.97) gives

$$\mu(m) = \theta'(s) = \{Q_v[V(m)]\}^{-1},$$

and (26.105) allows us to write

$$\mu[Q(v)] = \frac{\sqrt{v}}{F(v)},$$

$$Q(v) = M\left[\frac{\sqrt{v}}{F(v)}\right]. \tag{26.109}$$

The integral equation (26.106) is linear, but it has no known solution in closed form. It is then necessary to resort to numerical methods for its solution (Antman and Adler 1987).

Thus far we have assumed that the constitutive relations between the resultant stresses and couples and the strains have certain properties of growth suggested by a long sequence of observations made on physical materials. These properties are known as the conditions of *physical admissibility* of an elastic material. There are, however, a great number of phenomena involving bodies whose associated strain energy expressions violate the customary assumptions of continuum mechanics. The strain energy for these phenomena is a nonconvex function of the kinematic variables, a condition necessary for explaining the onset of strongly localized deformations in elastic bodies, such as necks in bars, dimples in shells, or dislocation.

The appearance of these localized effects can be illustrated by means of the simple model of the stretching or compression of a rod which is straight in its undeformed state (Berdichevskii and Truskinovskii 1985). We assume that the rod deforms in a plane, so that the strained position of its axis is defined by two components of the displacement vector: the longitudinal displacement $u(x)$ and the transverse displacement $w(x)$, where $x(0 \leqslant x \leqslant \ell)$ denotes the distance of a point of the undeformed axis from the left-hand end.

Provided that the strains are small, the elongation ε and the curvature κ are given by the

$$\varepsilon_x = \frac{du}{dx} + \frac{1}{2}\left(\frac{dw}{dx}\right)^2, \quad \kappa = \frac{d^2w}{dx^2}. \tag{26.110}$$

The strain energy density Φ can be expressed as the sum of the extensional and flexural energy:

$$\Phi = \tfrac{1}{2} S\,\varepsilon_x^2 + \tfrac{1}{2} B\kappa^2, \tag{26.111}$$

where $S = EA$ is the extensional rigidity of the cross-section and B is the flexural rigidity. On introducing the parameter $\varepsilon^2 = B/S$, (26.111) can be rewritten in the form

$$\frac{2\Phi}{S} = \left[\frac{du}{dx} + \frac{1}{2}\left(\frac{dw}{dx}\right)^2\right]^2 + \varepsilon^2\left(\frac{d^2w}{dx^2}\right)^2, \tag{26.112}$$

or, alternatively, on putting $\theta = dw/dx$, in the new form

$$\frac{2\Phi}{S} = \left(\frac{du}{dx} + \frac{1}{2}\theta^2\right)^2 + \varepsilon^2\left(\frac{d\theta}{dx}\right)^2. \tag{26.113}$$

Let us now assume that, under an axial compressive load, the rod has a fixed linear longitudinal displacement $u = -ax = -\frac{1}{2}\theta_0^2 x$, and a rotation $\theta(x)$. The strain-energy density has the expression

$$\frac{2\Phi}{S} = F(\theta) + \varepsilon^2 \left(\frac{d\theta}{dx}\right)^2, \quad F(\theta) = \frac{1}{4}(\theta^2 - \theta_0^2)^2, \tag{26.114}$$

and we see that it contains two terms: a nonconvex function $F(\theta)$ possessing two local minima with equal values of $F(\theta)$, and the square of $(d\theta/dx)$. The equilibrium configurations of the rod are the stationary points of the functional

$$E = W - \mathscr{L} = \int_0^\ell \Phi \, dx - M_1\theta(\ell) + M_0\theta(0), \tag{26.115}$$

where M_0 and M_1 are the moments applied at the ends of the rod. From (26.114) and (26.115) we find that the vanishing of the first variation

$$\delta E = \int_0^\ell \delta\Phi \, dx - M_1 \delta\theta(\ell) + M_0 \delta\theta(0) = 0 \tag{26.116}$$

yields the Euler differential equation

$$\begin{gathered} \varepsilon^2 \frac{d^2\theta}{dx^2} - \frac{1}{2}(\theta^2 - \theta_0^2)\theta = 0, \quad 0 < x < \ell, \\ M_0 = \varepsilon^2 S \frac{d\theta}{dx}(0), \quad M_1 = \varepsilon^2 S \frac{d\theta}{dx}(\ell). \end{gathered} \tag{26.117}$$

If the moments at the ends are zero, the solution can be written explicitly, because a first integral is

$$\varepsilon^2 \left(\frac{d\theta}{dx}\right)^2 - \frac{1}{4}(\theta^2 - \theta_0^2)^2 = \text{constant} = -\frac{1}{4}h_0^4$$

and the complete solution assumes the form

$$x(\theta) = 2\varepsilon \int_{-\sqrt{\theta_0^2 - h_0^2}}^{\theta} [(\theta^2 - \theta_0^2)^2 - h_0^4]^{-\frac{1}{2}} \, d\theta, \tag{26.118}$$

where the constant h_0 can be obtained by the condition $x(\theta_0^2 - h_0^2) = 1$. If the rod is thin enough, which implies $\varepsilon/b \ll 1$, the integration limits in (26.118) can be referred to infinity, and the solution takes the form

$$\theta_\varepsilon = \theta_0 \tanh \frac{\theta_0 x}{2\varepsilon}. \tag{26.119}$$

This equation describes the creation of a fin. When $\varepsilon \to 0$, the continuous solution θ_ε converges to the discontinuous solution

$$\theta = \begin{cases} -\theta_0, & x < 0 \\ \theta_0, & x > 0, \end{cases}$$

with a jump $2\theta_0$ at the origin. All the energy is concentrated in the fin because the strain energy is zero where the curvature is also zero. Thus the energy is purely extensional and its value is $\frac{2}{3}S\varepsilon\theta_0^2$.

The one-dimensional equilibrium theory of a bar also provides a simple model for studying certain types of material instabilities, which are usually attributed to non-elastic effects; for example, the occurrence of a flat interval in the uniaxial stress–strain curve followed by a monotonic dependence when the tension is relaxed. These effects are generally considered as being typical of the plastic behavior of a material.

The phenomenon has been interpreted by Ericksen (1975), in a remarkable paper devoted to the treatment of a simple bar the initial configuration of which is the interval $0 \leqslant X \leqslant L$ of the X axis. The deformed configurations are described by functions of the type $x = x(X)$, which are assumed to be of class $\mathscr{C}^2$ in their common interval of definition $[0, L]$, with $x' = dx/dX > 0$, in order to guarantee conservation of matter. For simplicity, we consider only a homogeneous bar and assume that its strain energy has the form

$$W = \int_0^L w(x')\,dX, \tag{26.120}$$

where $w(x')$ is a function of class $\mathscr{C}^3$ defined in a finite interval, say $0 < a \leqslant x' \leqslant b < \infty$.

In order to obtain the total mechanical energy we must distinguish between a *soft device*, in which the bar is firmly fixed at the end $x(0)$ and subject to a dead load σ at the end $x(L)$, and a *hard device*, in which the bar is again fixed at $x(0)$, but the end $x(L)$ occupies a prescribed position $x(L) = \bar{x}$. In the first case, the total energy is

$$E = \int_0^L w(x')\,dX - \sigma\bar{x}, \tag{26.121}$$

with the restriction $x(0) = 0$. In the second case the corresponding energy is

$$E = \int_0^L w(x')\,dX, \quad x(0) = 0, \quad x(L) = \bar{x}. \tag{26.122}$$

According to the energy criterion, equilibrium configurations must satisfy the Euler–Lagrange equations:

$$\frac{d}{dX}\left(\frac{dw}{dx'}\right) = 0,$$

or

$$\frac{dw}{dx'} = \text{constant}.$$

In addition, the equilibrium is *infinitesimally stable* or *neutrally infinitesimally stable* at y' if we have

$$\frac{d^2w}{dx'^2}(y') \geqslant 0, \quad a \leqslant y' \leqslant b. \tag{26.123}$$

Let us now assume that the law of dependence of dw/dx' on x' has the form indicated in Figure 26.4. An equilibrium configuration violating (26.123) is unobservable because, under the least disturbance, the bar tends to jump to another, more stable, configuration. If starting with σ near σ_1 we monotonically increase σ, taking care to avoid disturbances, we would expect x' to increase smoothly until it nears σ_3, and then suddenly to jump, at the same load, to the stable value $\bar{x}'_1$, thereafter increasing smoothly until it approaches σ_4, where it goes outside the range of the theory. Similarly, if we decrease σ, x' should decrease smoothly until it nears σ_2, where it should jump to the stable value $\bar{x}'_2$. As a consequence, the curve will describe a hysteresis loop. In the presence of stronger disturbances, the hysteresis loop should decrease, shrinking to zero if sufficiently higher disturbances are applied to the system. The limiting case is represented by van der Waal's fluid, for which the extrema x'_1 and x'_2 are never reached because the system jumps in both loading and unloading along the same horizontal interval AB in Figure 26.4, characterized by the geometrical property that the hatched areas are equivalent. The segment AB is known as *Maxwell's line*.

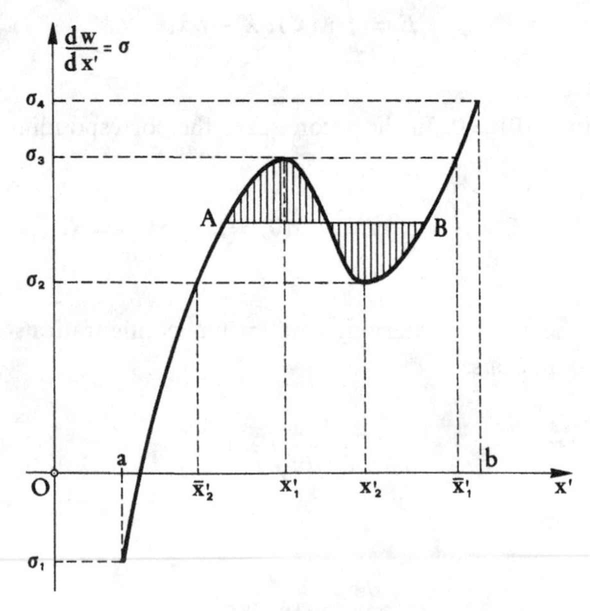

Fig. 26.4

27. Dynamical Problems in Nonlinear Rods

If not much is known about the statics of constitutive nonlinear rods, even less can be said about their dynamics. In static problems all the obtainable results rest heavily on the existence of a strain energy function which possesses certain properties of convexity and growth at infinity and is able to ensure the existence of minimizers for the total energy. When inertial terms are present, the kinetic energy is no longer negligible, and the variational techniques fail because the solutions are not minimizers of a physically meaningful functional.

A full nonlinear treatment is lacking, and only partial results are known. A surprising aspect of the theory is that the dynamic equations of the planar motion of a simple rod become nonlinear, even in classical elasticity and linearized curvature, provided that we take account of the influence of transverse deflection of the axial displacement. This effect was found by Kirchhoff (1859), but the mathematical treatment of Kirchhoff's equations is more recent.

We consider a prismatic rod of cross-section area A, the longitudinal axis of which, in its undeformed state, coincides with the x axis of a Cartesian system x, y, z. The rod occupies the interval $0 \leqslant x \leqslant \ell$ of the x axis, and the y and z axes are parallel to the principal axes of inertia of each cross-section. We assume that the axis deforms only in the xz plane, so that the displacement of a point on the axis, having an abscissa x in the undeformed rectilinear state, has only two components: $u(x, t)$ in the x direction, and $w(x, t)$ in the z direction, where t denotes the time.

The longitudinal strain of the central line, expressed in terms of the displacement, has the form

$$\varepsilon_{x0} = \sqrt{1+\lambda_{xx}} - 1, \tag{27.1}$$

with

$$\lambda_{xx} = 2\frac{\partial u}{\partial x} + \left(\frac{\partial u}{\partial x}\right)^2 + \left(\frac{\partial w}{\partial x}\right)^2. \tag{27.2}$$

If we wish to consider the influence of the transverse displacement $w(x, t)$ on the axial elongation, we cannot neglect the term $(\partial w/\partial x)^2$ with respect to $\partial u/\partial x$, while neglecting $(\partial u/\partial x)^2$ in (27.2) is allowed. Expanding the square root in (27.1) in a binomial series and retaining only the terms of first order, we obtain

$$\varepsilon_{x0} = \frac{1}{2}\lambda_{xx} = \frac{\partial u}{\partial x} + \frac{1}{2}\left(\frac{\partial w}{\partial x}\right)^2. \tag{27.3}$$

If the z axis is directed upwardly, the curvature generated by a positive couple around the y axis is positive and can be put equal to $\partial^2 w/\partial x^2$, approximatively. The elongation of the fibers at a distance z from the xy plane is, therefore,

$$\varepsilon_x = \varepsilon_{x0} - z\frac{\partial^2 w}{\partial x^2} = \frac{\partial u}{\partial x} + \frac{1}{2}\left(\frac{\partial w}{\partial x}\right)^2 - z\frac{\partial^2 w}{\partial x^2}. \tag{27.4}$$

Assuming that the material is linearly elastic, with an extensional rigidity S and a flexural rigidity B, the strain energy per unit length is given by

$$\varphi = \frac{1}{2} S \left[\frac{\partial u}{\partial x} + \frac{1}{2} \left(\frac{\partial w}{\partial x} \right)^2 \right]^2 + \frac{1}{2} B \left(\frac{\partial^2 w}{\partial x^2} \right)^2,$$

and the total strain energy stored in the beam is

$$W = \frac{1}{2} S \int_0^{\ell} \left[\frac{\partial u}{\partial x} + \frac{1}{2} \left(\frac{\partial w}{\partial x} \right)^2 \right]^2 dx + \frac{1}{2} B \int_0^{\ell} \left(\frac{\partial^2 w}{\partial x^2} \right)^2 dx. \tag{27.5}$$

In order to obtain the total energy, we must express the total kinetic energy as

$$T = \frac{1}{2} \rho A \int_0^{\ell} \left[\left(\frac{\partial u}{\partial t} \right)^2 + \left(\frac{\partial w}{\partial t} \right)^2 \right] dx, \tag{27.6}$$

where ρ denotes the density of the material.

The differential equations of the motion can be obtained by applying Hamilton's principle to the functional $W - T$:

$$\delta \int_{t_0}^{t_1} dt(W - T) = 0, \tag{27.7}$$

allowing all variations δu and δw, which vanish at the times t_1 and t_2 and at the points where the kinematic-type boundary conditions are imposed. The Euler equations associated with the variational equation (27.7) can be derived by the usual technique of integration by parts and equating to zero the coefficients of δu and δw under the sign of double integration. The result is the following pair of differential equations:

$$S \frac{\partial}{\partial x} \left[\frac{\partial u}{\partial x} + \frac{1}{2} \left(\frac{\partial w}{\partial x} \right)^2 \right] = \rho A \frac{\partial^2 u}{\partial t^2}, \tag{27.8a}$$

$$B \frac{\partial^4 w}{\partial x^4} - S \frac{\partial}{\partial x} \left\{ \frac{\partial w}{\partial x} \left[\frac{\partial u}{\partial x} + \frac{1}{2} \left(\frac{\partial w}{\partial x} \right)^2 \right] \right\} = \rho A \frac{\partial^2 w}{\partial t^2}, \tag{27.8b}$$

which were first derived by Kirchhoff (1876). The boundary and initial conditions are not given here because they depend on the particular type of constraint at the ends and on the initial data.

System (27.8) may be reduced to a single differential equation for the function $w(x, t)$, provided that we make the following assumption: the x component $\partial u/\partial t$ of the velocity vector, evaluated at points on the central line of the rod, is negligible with respect to the transverse velocity $\partial w/\partial t$. The consequence of this is that we may omit the right-hand side of (27.8a) and integrate the resulting equation with respect to x, to obtain

$$\frac{\partial u}{\partial x} + \frac{1}{2} \left(\frac{\partial w}{\partial t} \right)^2 = \varepsilon_{x0}(t), \tag{27.9}$$

which shows that the elongation of the central line does not depend on x, but only on the time. Integration of (27.9) with respect to x from $x = 0$ to $x = \ell$, and subsequent division by ℓ yields

$$\varepsilon_{x0}(t) = \frac{1}{\ell}\int_0^\ell \left[\frac{\partial u}{\partial x} + \frac{1}{2}\left(\frac{\partial w}{\partial x}\right)^2\right] dx = \frac{1}{\ell}\left[u(\ell, t) - u(0, t) + \frac{1}{2}\int_0^\ell \left(\frac{\partial w}{\partial x}\right)^2 dx\right].$$

Using this relation we obtain the following integrodifferential equation for $w(x, t)$:

$$\rho A\,\frac{\partial^2 w}{\partial t^2} + B\,\frac{\partial^4 w}{\partial t^4} - \frac{S}{\ell}\left[u(\ell, t) - u(0, t) + \frac{1}{2}\int_0^\ell \left(\frac{\partial w}{\partial x}\right)^2 dx\right]\frac{\partial^2 w}{\partial x^2} = 0. \tag{27.10}$$

To integrate this equation we may, as a first attempt, apply the method of separation of variables (Kauderer 1958, Ziff. 88; Narashima 1968). Let us assume, for definiteness, that the ends of the beam are hinged, so that w and $\partial^2 w/\partial x^2$ are zero at $x = 0$ and $x = \ell$ for all times, and let us seek a solution of the form

$$w(x, t) \simeq \sum_{m=1}^{\infty} q_m(t)\sin\frac{m\pi x}{\ell}, \tag{27.11}$$

where the functions $q_m(t)$ are to be determined. By substituting (27.11) in (27.10) and performing the integration from 0 to ℓ for the term $(\partial w/\partial x)^2$, we obtain the equation

$$\sum_{m=1}^{\infty}\left\{\rho A\,\frac{d^2 q_m}{dt^2} + B\,\frac{\pi^4}{\ell^4}\,m^4 q_m + \frac{S}{\ell}\left[u(\ell, t) - u(0, t) + \frac{\pi^2}{4\ell}\sum_{\nu=1}^{\infty}\nu^2 q_\nu^2\right]\frac{\pi^2 m^2}{\ell^2}\right\} \times \sin\frac{m\pi}{\ell} = 0. \tag{27.12}$$

As this equation must be satisfied identically for any x, the functions $q_m(t)$ must be solutions of the system of infinite differential equations

$$\rho A\,\frac{d^2 q_m}{dt^2} + \frac{\pi^2}{\ell^2}\left\{B\,\frac{\pi^2}{\ell^2}\,m^2 + \frac{S}{\ell}\left[u(\ell, t) - u(0, t) + \frac{\pi^2}{4\ell}\sum_{\nu=1}^{\infty}\nu^2 q_\nu^2\right]\right\}q_m = 0, \quad (m = 1, 2, \ldots). \tag{27.13}$$

In order to solve this system we must know the functions $u(\ell, t)$ and $u(0, t)$, which represent the longitudinal displacements of the ends during the specified time. In addition, we must know the initial conditions for $w(x, t)$, that is to say we must know the two functions

$$f(x) = w(x, 0), \quad g(x) = \frac{\partial w}{\partial x}(x, 0),$$

which represent the transverse displacement and velocity at time $t = 0$. These functions must be expanded in two Fourier series:

$$f(x) = \sum_{m=1}^{\infty} f_m \sin\frac{m\pi}{\ell}\,x, \quad g(x) = \sum_{m=1}^{\infty} g_m \sin\frac{m}{\ell}\,x, \tag{27.14}$$

where f_m and g_m are constants. Each equation in system (27.13) is then accompanied by two initial conditions f_m and g_m $(m = 1, 2, \ldots)$.

However, despite the unavoidable approximation connected with the method of separation of variables, the solution of system (27.13) is still very difficult, and its validity is questionable because, in order to ensure the legitimacy of the method, we must prove *a posteriori* that the formal expansion (27.11) is uniformly convergent (Weinberger 1965, Sec. 6). Another objection is that Kirchhoff's hypothesis of omitting the longitudinal velocity in the kinetic energy is not justified under close examination. Both these criticisms can be partially overcome. However, before we discuss them, we will attempt to derive qualitative information about the solutions for a particular case.

Let us look for forced motions generated by a transverse load of the type

$$p(x, t) = \sum_{m=1}^{\infty} p_m(t) \sin \frac{m\pi}{\ell} x, \tag{27.15}$$

where $p_m(t)$ are functions of time, and assume that the ends are prevented from shifting longitudinally, so that the difference $u(\ell, t) - u(0, t)$ is zero. Thus, instead of (27.13), the system of equations of motion becomes

$$\rho A \frac{d^2 q_m}{dt^2} + \frac{\pi^2}{\ell^2} m^2 \left(B \frac{\pi^2}{\ell^2} m^2 + \frac{S}{\ell} \frac{\pi^2}{4\ell} \sum_{\nu=1}^{\infty} \nu^2 q_\nu^2 \right) q_m = p_m(t), \quad m = 1, 2, \ldots. \tag{27.16}$$

To a first approximation, we assume that only the first term in (27.11) is appreciable and the rest are small. In this case we can replace system (27.13) by a single differential equation for $q_1(t)$:

$$\rho A \frac{d^2 q_1}{dt^2} + \frac{\pi^2}{\ell^2} \left(B \frac{\pi^2}{\ell^2} + \frac{S}{\ell} \frac{\pi^2}{4\ell} q_1^2 \right) q_1 = p_1. \tag{27.17}$$

If the forcing term is harmonic, we can write $p_1 = \cos \omega t$ without loss of generality, and put

$$q_1 = Q_1 \cos \omega t, \tag{27.18}$$

where Q_1 is a constant. By substituting this expression for q_1 in (27.17) using the identity $\cos^3 \omega t = \frac{1}{4}(3 \cos \omega t + \cos 3\omega t)$, and disregarding the term $\cos 3\omega t$, we obtain

$$\left(\frac{\pi^4}{\ell^4} B - \rho A \omega^2 \right) Q_1 + \frac{3\pi^4}{16\ell^4} S Q_1^3 = 1, \tag{27.19}$$

which is an algebraic equation for evaluating Q_1.

Here we have a resonance phenomenon, and it is of interest to note the behavior of the solutions in the neighborhood of the *singularity* at $S = 0$, when ω has a value making $(\pi^4/\ell^4)B - \rho A\omega^2$ very small. The second term in (27.19) is then not negligible, and finding an approximate solution in this neighborhood requires the introduction of new variables Ω and η, which are defined by the transformation

$$\frac{\pi^4}{\ell^4} B - \rho A \omega^2 = -S^{\frac{1}{3}} \Omega, \quad Q_1 = S^{-\frac{1}{3}} \eta_1, \tag{27.20}$$

in which the "stretching" factors $S^{\frac{1}{3}}$ and $S^{-\frac{1}{3}}$ are incorporated. Equation (27.19) then becomes

$$-\Omega\eta_1 + \frac{3\pi^4}{16\ell^4}\eta_1^3 = 1. \tag{27.21}$$

Terms then balance in the limit: for $\Omega \to 0$, Q_1 approaches a finite limit; for $|\Omega| \to \infty$, the second term in (27.21) becomes negligible and Q_1 approximates to $[(\pi^4/\ell^4)B - \rho A\omega^2]^{-1}$.

The rigorous treatment of system (27.13), under the hypothesis that the ends remain at a fixed distance $u(\ell, t) - u(0, t) = H$, can be performed with full generality (Dickey 1970b). By putting

$$\alpha_0 = \frac{B\pi^4}{\rho A\ell^4}, \quad \alpha_1 = \frac{SH\pi^2}{\rho A\ell^3}, \quad \alpha_1 = \frac{S\pi^4}{4\rho A\ell^4}, \tag{27.22}$$

system (27.13) can be written as

$$q_j'' + j^2\left(\alpha_0 j^2 + \alpha_1 + \alpha_2 \sum_{\nu=1}^{\infty} \nu^2 q_\nu^2\right) q_j = 0, \quad j = 1, 2, \ldots. \tag{27.23}$$

In addition, the fact that $w(x, t)$ must satisfy the initial conditions implies the conditions

$$q_j(0) = f_j, \quad q_j'(0) = g_j, \quad j = 1, 2, \ldots, \tag{27.24}$$

where f_j and g_j are the coefficients of expansions (27.14).

Before treating the general case, let us assume, for the moment, that the initial data are representable as sums of the type

$$f(x) = \sum_{j=1}^{N} f_j \sin j\frac{\pi x}{\ell}, \quad g(x) = \sum_{j=1}^{N} g_j \sin j\frac{\pi x}{\ell}, \tag{27.25}$$

so that it is reasonable to look for solutions of the form

$$w(x, t) = \sum_{j=1}^{N} q_j(t) \sin j\frac{\pi x}{\ell}, \tag{27.26}$$

where the functions q_j satisfy the system of equations

$$q_j'' + j^2\left(\alpha_0 j^2 + \alpha_1 + \alpha_2 \sum_{\nu=1}^{\infty} \nu^2 q_\nu^2\right) q_j = 0, \quad j = 1, \ldots, N. \tag{27.27}$$

System (27.27) can be solved by using the method of successive approximations, giving a solution for all $t \geqslant 0$. However, in practice, to prove this result it is only necessary to show that $|q_j|$ and $|q_j'|$ are bounded for all $t \geqslant 0$, and this can be verified easily by multiplying (27.27) by q_j', summing over j, and integrating the resulting expression. The result is

$$V_N^2 + \alpha_0 Y_N^2 + \alpha_1 Z_N^2 + \frac{\alpha_2}{2} Z_N^4 = h_N, \tag{27.28}$$

where we have put

$$V_N^2 = \sum_{j=1}^{N} q_j'^2, \quad Y_N^2 = \sum_{j=1}^{N} j^4 q_j^2, \quad Z_N^2 = \sum_{j=1}^{N} j^2 q_j^2, \tag{27.29}$$

and h_N is a number depending only on the initial conditions. For $\alpha_1 \geqslant 0$, and hence $h_N > 0$, it follows immediately from (27.28) that $|q_j|$ and $|q_j'|$ are bounded. For $\alpha_1 < 0$ the number h_N may be positive, negative, or zero, depending on the initial conditions. Even so, (27.28) shows that V_N^2, Y_N^2, and Z_N^2 are bounded, as are $|q_j|$ and $|q_j'|$.

Let us now consider general initial conditions and define the functions $q_{j,N}$ to be solutions of (27.27) satisfying the initial conditions (27.24) for $j \leqslant N$, and $q_{j,N} = q_{j,N}' = 0$ for $j > N$. These functions $q_{j,N}$ are also solutions of (27.23) and satisfy the initial conditions (27.24) for $j \leqslant N$ and $q_{j,N}(0) = q_{j,N}'(0) = 0$ for $j > N$. The $q_{j,N}$ are then solutions of the system

$$q_{j,N}'' + j^2 A_{j,N} q_{j,N} = 0, \quad j = 1, 2, \ldots, \infty, \tag{27.30}$$

where

$$A_{j,N} = \alpha_0 j^2 + \alpha_1 + \alpha_2 \sum_{\nu=1}^{\infty} \nu^2 q_{\nu,N}^2. \tag{27.31}$$

The series in (27.31) has, of course, a finite sum, since we have $q_{j,N} = 0$ for $j > N$. Then, on multiplying (27.30) by $q_{j,N}'$, summing the equations over j, and integrating the resulting expression, we obtain

$$V_N^2 = \alpha_0 Y_N^2 + \alpha_1 Z_N^2 + \frac{\alpha_2}{2} Z_N^4 = h_N, \tag{27.32}$$

where

$$h_N = \sum_{j=1}^{N} g_j^2 + \alpha_0 \sum_{j=1}^{N} j^4 f_j^2 + \alpha_1 \sum_{j=1}^{N} j^2 F_j^2 + \frac{\alpha_2}{2} \left(\sum_{j=1}^{N} j^2 f_j^2 \right)^2. \tag{27.33}$$

Now, if we assume that the initial data satisfy the conditions

$$\sum_{j=1}^{N} g_j^2 < \infty, \quad \sum_{j=1}^{N} j^4 f_j^2 < \infty, \tag{27.34}$$

then, from (27.33) giving the expression for h_N, we finally have

$$\lim_{N \to \infty} h_N = h < \infty. \tag{27.35}$$

This means that the quantities V_N^2, Y_N^2, and Z_N^2 in (27.28) are bounded independently of N and regardless of the sign of α_1, so there are constants M_1, M_2, and M_3 satisfying the inequalities

$$V_N^2 \leqslant M_1, \tag{27.36a}$$

$$Y_N^2 \leqslant M_2, \tag{27.36b}$$

$$Z_N^2 \leqslant M_3, \tag{27.36c}$$

and, consequently, the following inequality is obtained:

$$|A_{j,N}| \leqslant \alpha_0 j^2 + |\alpha_1| + \alpha_2 M_3. \tag{27.37}$$

On the other hand, we can call on the Schwarz inequality to bound the time derivative of $A_{j,N}$,

$$A'_{j,N} = 2\alpha_2 \sum_{\nu=1}^{\infty} \nu^2 q_{j,N} q'_{j,N}, \tag{27.38}$$

that is,

$$|A'_{j,N}| \leqslant 2\alpha_2 \left(\sum_{\nu=1}^{\infty} \nu^4 q_{j,N}^2 \right)^{\frac{1}{2}} \left(\sum_{\nu=1}^{\infty} q'^2_{j,N} \right)^{\frac{1}{2}} \leqslant 2\alpha_2 (Y_N^2 V_N^2)^{\frac{1}{2}} \leqslant 2\alpha_2 (M_2 M_1)^{\frac{1}{2}}. \tag{27.39}$$

Since (27.37) and (27.39) show that $A_{j,N}$ and $A'_{j,N}$ are uniformly bounded independently of N and for $0 \leqslant t \leqslant t^* < \infty$, the sequence $A_{j,N}$ therefore forms a bounded equicontinuous set of functions in any closed subinterval of $[0, t^*]$, and, by the Ascoli–Arzelá theorem, it is possible to choose a subsequence A_{j,N_i} that converges uniformly to a continuous function A_j in the interval $[0, t^*]$. In addition, let q_j be the solution of the linear equation

$$q''_j + j^2 A_j q_j = 0, \quad j = 1, 2, \ldots, \infty, \tag{27.40}$$

$$q_j(0) = f_j, \quad q'_j(0) = g_j, \tag{27.41}$$

then we can show that there is a subsequence q_{j,N_i} converging to q_j on the interval $[0, t^*]$.

We must now prove that the solutions of (27.40) and (27.41) are also solutions of (27.23). For this purpose it is only necessary to show that $A_j(t)$ can be obtained as a limit:

$$A_j(t) = \lim_{N_i \to \infty} \left(\alpha_0 j^2 + \alpha_1 + \alpha_2 \sum_{\nu=1}^{\infty} \nu^2 q_{\nu,N_i}^2 \right). \tag{27.42}$$

In fact, we have $q_{j,N_i} \to q_j$ for $N_i \to \infty$, and, furthermore, by (27.36b) we have

$$q_{j,N_i}^2 \leqslant \frac{M^2}{j^4}.$$

This implies the inequality

$$\left| A_j - \alpha_0 j^2 - \alpha_1 - \alpha_2 \sum_{\nu=1}^{\infty} \nu^2 q_j^2 \right| \leqslant |A_j - A_{j,N_i}| + \alpha_1 \sum_{\nu=1}^{K} \nu^2 |q_{\nu,N_i}^2 - q_\nu^2| + 2\alpha_1 M_2 \sum_{\nu=K+1}^{\infty} \frac{1}{\nu^4}, \tag{27.43}$$

where, by first choosing K and then N_i, the right-hand side of (27.43) can be made arbitrarily small.

The proof of the existence of a solution does not ensure that the solution is unique. However, we do expect uniqueness in order to be sure that we have a good model, and we know that uniqueness is also useful in constructing the solution because the function A_j can be obtained as the limit of a whole sequence and not as the limit of subsequences. To prove uniqueness, we assume the existence of two sets of solutions, say p_j and q_j, and that

$$B_j = \alpha_0 j^2 + \alpha_1 + \alpha_2 \sum_{\nu=1}^{\infty} \nu^2 p_\nu^2. \tag{27.44}$$

We define a function u_j,

$$u_j = q_j - p_j, \tag{27.45}$$

such that u_j must be a solution of the differential equation

$$u_j'' + j^2 A_j u_j = j^2(B_j - A_j)p_j, \tag{27.46}$$

$$u_j(0) = u_j'(0) = 0. \tag{27.47}$$

Equivalently, (27.46) may be rewritten as

$$u_j'' + j^2(A_j - \alpha_1)u_j = j^2(B_j - A_j)p_j - j^2\alpha_1 u_j, \tag{27.48}$$

where

$$A_j - \alpha_1 = \alpha_0 j^2 + \alpha_2 \sum_{\nu=1}^{\infty} \nu^2 q_\nu^2 \geqslant \alpha_0 j^2 \geqslant \alpha_0 > 0. \tag{27.49}$$

Our purpose is to prove that the only solution of (27.48) satisfying (27.47) is the trivial solution. On defining

$$v_j = \frac{(u_j')^2}{j^2(A_j - \alpha_1)} + u_j^2 \geqslant 0, \tag{27.50}$$

after differentiating (27.50) and using (27.48), we find the value

$$v_j' = -\frac{A_j'}{A_j - \alpha_1}\frac{(u_j')^2}{j^2(A_j - \alpha_1)} + 2\frac{B_j - A_j}{A_j - \alpha_1} p_j u_j' - 2\alpha_1 \frac{u_j u_j'}{A_j - a_1}, \tag{27.51}$$

or

$$v_j' \leqslant \frac{|A_j'|}{A_j - \alpha_1} v_j + 2\frac{|B_j - A_j|}{A_j - \alpha_1}|p_j||u_j'| + 2\alpha_1 \frac{|u_j||u_j'|}{A_j - \alpha_1}. \tag{27.52}$$

This inequality can be rewritten as

$$\left(\frac{d}{dt} v_j\right) \exp\left(-\int_0^t \frac{|A_j'|}{A_j - \alpha_1} d\tau\right) \leqslant 2\left(\frac{|B_j - A_j|}{A_j - \alpha_1}|p_j||u_j'| + |\alpha_1| \frac{|u_j||u_j'|}{A_j - \alpha_1}\right),$$

whence we have

$$v_j \leqslant 2\exp\left(\int_0^t \frac{|A_j'|}{A_j - \alpha_1} d\tau\right) \cdot \int_0^t \left(\frac{|B_j - A_j|}{A_j - \alpha_1}|p_j||u_j'| + |\alpha_1| \frac{|u_j||u_j'|}{A_j - \alpha_1}\right) d\tau. \tag{27.53}$$

The quantities $|A_j'|$ and $|B_j - A_j|$ are independent of j; therefore, (27.53) can be multiplied by j^2 and summed over j to obtain (see (27.49))

$$\sum_{j=1}^{\infty} j^2 v_j \leqslant 2\exp\left(\frac{1}{\alpha_0}\int_0^t |A_j'|\,d\tau\right)$$
$$\cdot \int_0^t \left(|B_j - A_j| \sum_{j=1}^{\infty} \frac{j^2|p_j||u_j'|}{A_j - \alpha_1} + |\alpha_1| \sum_{j=1}^{\infty} \frac{j^2|u_j||u_j'|}{A_j - \alpha_1}\right) d\tau. \tag{27.54}$$

However, from the Schwarz inequality, we have

$$|B_j - A_j| \leqslant \alpha_2 \sum_{\nu=1}^{\infty} \nu^2 |p_\nu + q_\nu||u_\nu| \leqslant \alpha_2 \left[\sum_{\nu=1}^{\infty} \nu^2 (p_\nu + q_\nu)^2 \cdot \sum_{\nu=1}^{\infty} \nu^2 u_\nu^2\right]^{\frac{1}{2}}, \tag{27.55}$$

$$\sum_{j=1}^{\infty} \frac{j^2|p_j||u_j'|}{A_j - \alpha_1} \leqslant \left[\sum_{j=1}^{\infty} \frac{j^4 p_j^2}{A_j - \alpha_1} \cdot \sum_{j=1}^{\infty} \frac{(u_j')^2}{A_j - \alpha_1}\right]^{\frac{1}{2}} \leqslant \frac{1}{\alpha_0^{\frac{1}{2}}} \left[\sum_{j=1}^{\infty} j^2 p_j^2 \cdot \sum_{j=1}^{\infty} \frac{(u_j')^2}{A_j - \alpha_1}\right]^{\frac{1}{2}}, \tag{27.56}$$

$$\sum_{j=1}^{\infty} j^2 \frac{|u_j||u_j'|}{A_j - \alpha_1} \leqslant \left[\sum_{j=1}^{\infty} \frac{(u_j')^2}{A_j - \alpha_1} \sum_{j=1}^{\infty} \frac{j^4 u_j^2}{A_j - \alpha_1}\right]^{\frac{1}{2}} \leqslant \frac{1}{\alpha_0^{\frac{1}{2}}} \left[\sum_{j=1}^{\infty} \frac{(u_j')^2}{A_j - \alpha_1} \cdot \sum_{j=1}^{\infty} j^2 u_j^2\right]^{\frac{1}{2}}. \tag{27.57}$$

Combining all these inequalities, we derive

$$\sum_{j=1}^{\infty} j^2 v_j \leqslant 2\exp\left(\frac{1}{\alpha_0}\int_0^t |A_j'|\,d\tau\right) \cdot \int_0^t G(t) \left[\sum_{j=1}^{\infty} \frac{(u_j')^2}{A_j - \alpha_1} \cdot \sum_{j=1}^{\infty} j^2 u_j^2\right]^{\frac{1}{2}} d\tau, \tag{27.58}$$

where

$$G(t) = \frac{1}{\alpha_0^{\frac{1}{2}}} \left\{\alpha_2 \left[\sum_{j=1}^{\infty} j^2 (p_j + q_j)^2 \cdot \sum_{j=1}^{\infty} j^2 p_j^2\right]^{\frac{1}{2}} + |\alpha_1|\right\}. \tag{27.59}$$

The function $G(t)$ and the exponential in (27.58) are bounded on any finite interval. Therefore, there is a value $t_c > 0$ such that

$$t\,G(t)\exp\left(\frac{1}{\alpha_0}\int_0^t |A_j'|\,d\tau\right) < 1, \quad \text{for } 0 \leqslant t \leqslant t_c. \tag{27.60}$$

At this point, uniqueness in $[0, t_c]$ follows by contradiction. Assume that the integrand in (27.58) does not vanish identically for $0 \leqslant t \leqslant t_c$. Then it will attain a maximum at $t^* \neq 0$. The inequality (27.58), written for $t = t^*$, yields

$$\sum_{j=1}^{\infty} j^2 v_j(t^*) \leqslant 2t^* \exp\left(\frac{1}{\alpha_0}\int_0^{t^*} |A_j'|\,d\tau\right) \cdot G(t^*) \left\{\sum_{j=1}^{\infty} \frac{(u_j')^2}{A_j - \alpha_1} \cdot \sum_{j=1}^{\infty} j^2 u_j^2\right\}_{t=t^*}^{\frac{1}{2}}.$$

As we have $0 \leqslant t^* \leqslant t_c$, then from (27.60) we obtain the inequality

$$\sum_{j=1}^{\infty} j^2 v_j(t^*) = \left[\sum_{j=1}^{\infty} \frac{(u_j')^2}{A_j - \alpha_1} + \sum_{j=1}^{\infty} j^2 u_j^2\right]_{t=t^*} < 2\left[\sum_{j=1}^{\infty} \frac{(u_j')^2}{A_j - \alpha_1} \cdot \sum_{j=1}^{\infty} j^2 u_j^2\right]_{t=t^*}^{\frac{1}{2}},$$

or

$$\left\{\left(\sum_{j=1}^{\infty} \frac{(u_j')^2}{A_j - \alpha_1}\right)^{\frac{1}{2}} - \left(\sum_{j=1}^{\infty} j^2 u_j^2\right)^{\frac{1}{2}}\right\}_{t=t^*}^{2} < 0. \qquad (27.61)$$

This contradiction proves the theorem of uniqueness.

However, we can show that this result may be extended to the whole interval $0 \leqslant t < \infty$. Let us assume that there are points with $q_j \neq p_j$, that is, $u_j \neq 0$. Let $t_0 \geqslant t_c > 0$ be the greatest lower bound of these points. At t_0, we must have $u_j(t_0) = u_j'(t_0) = 0$, and thus it is possible to repeat the above argument to give the identity $u_j \equiv 0$ in some interval on the right of t_0. Therefore, t_0 cannot be the greatest lower bound of points for which $u_j \neq 0$.

The same techniques may also be suitable for treating the case in which a transverse motion of the beam takes place in a medium offering a certain viscous resistance proportional to the transverse velocity. Supposing, for simplicity, that the ends are fixed, so as to give $\alpha_1 = 0$, then system (27.23) must be replaced by the set of infinite equations

$$q_j'' + K^2 q_j' + j^2\left(\alpha_0 + \alpha_2 \sum_{j=1}^{\infty} \nu^2 q_\nu^2\right) q_j = 0, \quad j = 1, 2, \ldots, \infty, \qquad (27.62)$$

$$q_j(0) = f_j, \quad q_j'(0) = g_j, \qquad (27.63)$$

where the constants K^2, α_0, and α_2 are assumed positive.

The development described previously is completely applicable to the new system (Dickey 1970a). However, the results regarding the existence and uniqueness of a solution are even simpler because, using the constant K^2, which represents the dissipation, renders it easier to obtain estimates. The main result is the inequality

$$K^2 > \frac{\alpha_2 M_1}{\alpha_0}\left(\sum_{j=1}^{\infty} j^4 f_j^2 + \frac{1}{\alpha_0} \sum_{j=1}^{\infty} j^2 g_j^2\right)^{\frac{1}{2}}, \qquad (27.64)$$

with M_1 given by (27.36a). The system then has exactly one solution for $0 \leqslant t < \infty$.

A nonlinear equation similar to (27.10) also describes the dynamical snapping of an elastic arch (Reiss 1980). We consider a shallow elastic arch of length L, cross-sectional area A, and cross-sectional moment of inertia I, subjected to a lateral force per unit length $P(X)$, where X is the horizontal coordinate of the arch ($0 \leqslant X \leqslant L$). The vertical form of the undeformed arch is given by $Y = W^0(X)$, where W^0 is a specified function. If the deformed vertical curve of the arch is denoted by $y(X, T)$, where T is the time, then the function

$$W(X, T) = y(X, T) - W^0(X), \qquad (27.65)$$

denotes the vertical displacement of the arch.

We now introduce the following dimensionless variables and parameters

$$x = \frac{X}{L}, \quad t = \frac{\pi^2}{2L}\left(\frac{E}{\rho}\right)^{\frac{1}{2}} T, \quad w(x,t) = \frac{1}{L}\,W(X,t),$$
$$w^0(x) = \frac{W^0(X)}{L}, \quad p(x) = \frac{4}{\pi^4}\frac{L}{EA}\,P(X), \quad Z = \frac{L^2 A}{2J}, \tag{27.66}$$

where ρ is the mass density per unit length and E is the Young's modulus of the material. With positions (27.66), the original equation of motion

$$\rho A W_{TT} + EJW_{XXXX} - \frac{EA}{2L}\left[\int_0^L (W_X - W_X^0)^2 dX\right](W_{XX} - W_{XX}^0) - P(X) = 0,$$

after the term $(W_X^0)^2$ has been neglected, can be converted into

$$w_{xxxx} - \left[Z\int_0^1 (w_x^2 - 2w_x^0 w_x)\,dx\right] w_{xx} + \frac{\pi^4}{2} Z w_{tt}$$
$$= -\left[Z\int_0^1 (w_x^2 - 2w_x^0 w_x)\,dx\right] w_{xx}^0 + \frac{\pi^4}{2} Zp. \tag{27.67}$$

If the arch is simply supported, the boundary conditions are

$$w(0,t) = w(1,t) = w_{xx}(0,t) = w_{xx}(1,t) = 0, \tag{27.68}$$

and the initial conditions are

$$w(x,0) = f(x), \quad w_t(x,0) = g(x), \tag{27.69}$$

where $f(x)$ and $g(x)$ are two prescribed dimensionless functions.

We analyze (27.67) for the special case of initially sinusoidal arches with sinusoidal initial data and pressures; that is, we assume that

$$w_0 = h\sin\pi x, \quad p = q\sin\pi x,$$
$$f = f_0\sin\pi x, \quad g = g_0\sin\pi x, \tag{27.70}$$

and seek solutions of the form

$$w(x,t) = A(t)\sin\pi x, \tag{27.71}$$

where $A(t)$ is an unknown function. Then, by inserting (27.70) and (27.71) into (27.67), we find that $A(t)$ must satisfy the initial-value problem

$$A'' + \alpha A - 3hA^2 + A^3 - q = 0,$$
$$A(0) = f_0, \quad A'(0) = g_0, \tag{27.72}$$

where the parameter α is defined as

$$\alpha = 2h^2 + \frac{8}{Z}.$$

In order to illustrate the structure of the solutions, we first examine the static case in which A does not depend on the time and the equilibrium positions satisfy the cubic equation

$$q = \alpha A - 3hA^2 + A^3. \tag{27.73}$$

The number of real roots of (27.73) depends on the sign of the discriminant

$$D \equiv h^2 - \frac{\alpha}{3}.$$

For $D < 0$, (27.73) has a unique real solution that is a monotonically increasing function of q. For $D = 0$, there is a unique solution, but the graph of $A(q)$ has a point with a vertical tangent. For $D > 0$, the response curve has the graph shown in Figure 27.1. There is a unique static solution for $q < q_\ell$ and $q > q_u$. For $q_\ell < q < q_u$ there are three solutions. As q increases from zero, the equilibrium states of the arch follow the lower branch; but, if q exceeds q_u, the arch jumps dynamically towards the upper branch. Similarly, if the arch is in the upper branch, and q is decreased below q_ℓ, then the arch will suddenly jump towards the lower branch. These dynamical transitions are called *snap buckling*.

The jumps occurring at the points q_u and q_ℓ cannot be described by means of a purely static analysis because the inertia is no longer negligible during the snapping. In order to analyze the dynamical case, we assume the values $q = q_u$ and $A = A_u$, where the arch is in equilibrium at a limit point. A small initial disturbance ε is then given to this equilibrium state:

$$A(0) = A_u + \varepsilon, \quad A'(0) = 0. \tag{27.74}$$

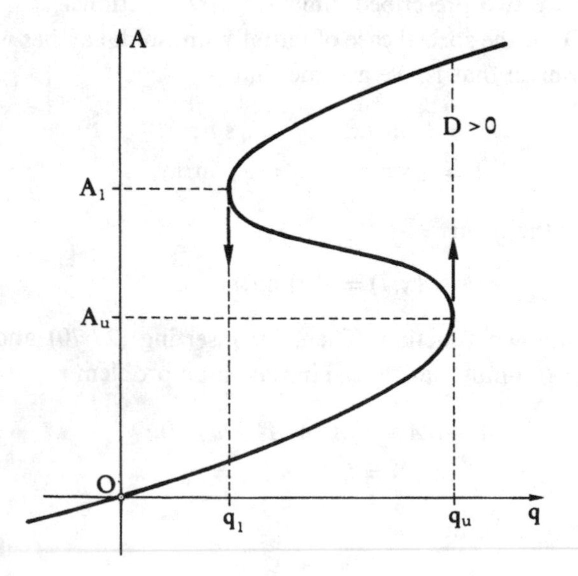

Fig. 27.1

In order to be specific we take the case $\varepsilon < 0$, although a similar analysis applies for $\varepsilon > 0$. It is convenient to transform the dependent variable in (27.72) by putting

$$A(t) = A_u + v(t) \tag{27.75}$$

where $v(t)$ represents the displacement from A_u. By inserting (27.74), (27.75), and $q = q_u$ into (27.72), the initial-value problem is transformed to

$$v'' - (\tfrac{3}{4}\delta)v^2 + v^3 = 0, \quad v(0) = \varepsilon, \quad v'(0) = 0, \tag{27.76}$$

where $\delta = 4D^{\frac{1}{2}}$.

A first integral of (27.76) is

$$(v')^2 = \tfrac{1}{2}G^2(v) \equiv \tfrac{1}{2}(3\mathscr{E} + \delta v^3 - v^4), \tag{27.77}$$

where $\mathscr{E}$, which represents the energy, is expressed in terms of the initial data by

$$4\mathscr{E} = -\delta\varepsilon^3 + \varepsilon^4.$$

We denote the maximum amplitude of the solution of (27.76) by $a(\varepsilon)$, and the velocity v' is zero when the amplitude is either ε or $a(\varepsilon)$. It follows that both these values are two real roots of $G(v) = 0$, while the remaining two roots are complex conjugates, with values of, say, $c(\varepsilon)$ and $\bar{c}(\varepsilon)$. Then, from the tables of elliptic integrals (Byrd and Friedman 1971, p. 133), we can write the solution of (27.76) as

$$v(t) = \frac{(aB + \varepsilon H) - (aB - \varepsilon H)\varphi(\tau)}{(B + H) + (H - B)\varphi(\tau)}, \tag{27.78}$$

where $\varphi(\tau)$ is the Jacobian elliptic function

$$\varphi(\tau) = cn\tau, \tag{27.79}$$

and τ is a "slowed" function of the time, defined by

$$\tau \equiv \left(\frac{BH}{2}\right)^{\frac{1}{2}}. \tag{27.80}$$

The constants B and H in (27.78) are functions of ε, and have the expressions

$$B^2 \equiv (\varepsilon - b)^2 + e^2, \quad H^2 \equiv (a - b)^2 + e^2, \tag{27.81}$$

with b and e given by

$$b \equiv \tfrac{1}{2}(c + \bar{c}), \quad e^2 \equiv \tfrac{1}{4}(c - \bar{c})^2. \tag{27.82}$$

The function $\varphi(\tau)$ is periodic, of period $4K$, where $K(k)$ is the complete elliptic integral of the first kind, with the modulus k given by

$$k^2 = \frac{(a - \varepsilon)^2 - (H - B)^2}{4BH}. \tag{27.83}$$

The fact that the function $\varphi(\tau)$ is periodic implies that, according to the dynamic analysis, the solution is oscillatory after the first jump at q_u. The jump occurs smoothly because the initial velocity is zero, but the system returns to the initial state over a relatively short interval τ.

It remains to discuss the legitimacy of the neglect by Kirchhoff of the longitudinal kinetic energy and of some nonlinear terms in the strain energy. In order to deal with this question, it is necessary to write the equations of motion considering the longitudinal strain, not in the semilinearized form (27.4), but in the complete form

$$\varepsilon_x = \sqrt{1 + \lambda_{xx}} - 1 + z\kappa', \tag{27.84}$$

where λ_{xx} is given by (27.2) and κ' is the curvature, which can be written as

$$\kappa' = -\frac{\dfrac{\partial^2 w}{\partial x^2}}{\left[1 + \left(\dfrac{\partial w}{\partial x}\right)^2\right]^{\frac{3}{2}}}. \tag{27.85}$$

Kirchhoff's equations can be justified as the first step in a systematic procedure of approximation.

Chapter IV
Theories of Cables

28. Equilibrium Equations

Strings are slender bodies like rods, but are characterized by the properties that they cannot withstand compression or bending and can adapt their shape to any form of loading. They are important because they have wide technical applications, such as those required for the construction of suspension bridges, wire meshes, musical instruments, and nets of textile material. Strings are the vehicle through which some of the methods of mathematical physics have found their simplest applications.

Strings were used by Stevin in 1586 to given an experimental demonstration of the law of the triangle of forces. It seems likely[46] that, in 1615, Beeckman solved the problem of finding the shape of a cable modeled as a string under a load uniformly distributed in a plane, and found that the string hangs in a parabolic arc. The problem of finding the configuration of a chain hanging under its own weight was considered by Galilei (1638), in his *Discorsi*, who concluded, erroneously, that the chain assumes a parabolic form. Subsequently, Leibniz, Huygens and James Bernoulli, apparently independently, discovered the solution now known as the catenary, taking its name from the Latin for a suspended chain. In investigating the catenary, different approaches were employed: Huygens relied on geometrical considerations, while Leibniz and James Bernoulli relied on the calculus. However, at the same time, Hooke (1675) found that a moment-free arch supports its own weight with a curve which is an inverted catenary. Another long-debated problem was that of *velaria*, that is, the curve assumed by an inextensible weightless string subjected to a normal force of constant magnitude. This is the form of a cylindrical sail under the action of a uniform wind, the name having its origin in this example. Huygens incorrectly stated that the curve is a parabola, but James Bernoulli aptly proved that the velaria is a circle.

The vibration of taut strings was extensively studied by d'Alembert, Daniel Bernoulli and Lagrange in the early part of the eighteenth century. Daniel Bernoulli (1733) found a solution for the natural frequencies of a chain hanging from one end. D'Alembert (1797) derived the general solution for the equation governing the small transverse displacements of a taut string stretched between

[46] According to Truesdell (1960).

two fixed pegs. To illustrate his equations of motion, Lagrange (1759) used a discrete sequence of beads as a model of a taut string. The importance of these achievements lies not only in terms their practical applications, but in their exceptional theoretical results, using which new methods for the solution of partial differential equations were established.

From the point of view of technical applications, cable theory had a firm grounding in Europe by the nineteenth century. However, outside Europe, the early civilizations of the Far East and Central America mastered the construction of suspension bridges without any knowledge of theory.

The deduction of the equilibrium equations for an elastic string with fixed ends subjected to a distributed vertical load is done in the usual way. Let $\{\mathbf{i}, \mathbf{j}, \mathbf{k}\}$ be a fixed orthonormal basis, with the $\mathbf{j}$ vector pointing upwards. Let $s \in [0, 1]$ be the arc length identifying a material particle of the string in its unstretched state. The position of the same particle in a deformed configuration is defined by a vector $\mathbf{r}(s)$, and its derivative with respect to s, or, more precisely, the modulus,

$$\nu = |\mathbf{r}'|, \tag{28.1}$$

represents the elongation of the string. The unit vector

$$\mathbf{e}(s) = \frac{\mathbf{r}'}{\nu} \tag{28.2}$$

is tangential to the curve, and, without loss of generality, we may assume that the boundary conditions for $\mathbf{r}$ are

$$\mathbf{r}(0) = \mathbf{0}, \quad \mathbf{r}(1) = a\mathbf{i} + b\mathbf{j}, \quad a \geqslant 0, b \geqslant 0. \tag{28.3}$$

A perfectly flexible string is characterized by the property that the resultant contact force $\mathbf{n}(s)$ exerted across any intermediate section $0 < s < 1$ is directed along the tangent $\mathbf{e}(s)$ without any contact couple. If $\mathbf{f}$ is the external force per unit length applied to the string, the equilibrium equations have the form

$$\mathbf{n}' + \mathbf{f} = \mathbf{0}. \tag{28.4}$$

We say that a pair of functions $(\mathbf{r}, \mathbf{n})$ is a classical solution of (28.4) if we have $\mathbf{r} \in \mathscr{C}^2[0, 1]$ and $\mathbf{n} \in \mathscr{C}^1[0, 1]$, with $\nu > 0$. A solution is called "planar" if $\mathbf{r}$ and $\mathbf{n}$ lie in the $\mathbf{ij}$ plane.

An immediate consequence of (28.4) is that, if $\mathbf{f}$ has the form

$$\mathbf{f}(s) = -F'(s)\mathbf{j}, \quad F'(s) > 0, \quad F(0) = 0, \tag{28.5}$$

then any classical solution of (28.4) is planar (Antman 1979b). To prove this result, let us integrate (28.4) from 0 to s and use (28.5), together with the property that $\mathbf{n}(s)$, being tangential to the curve, may also be written as $\mathbf{n} = \mathbf{e}(s)n(s)$. The result is

$$n\mathbf{e} - F\mathbf{j} = \mathbf{n}(0), \tag{28.6a}$$

$$n\mathbf{e} \cdot \mathbf{k} = \mathbf{n}(0) \cdot \mathbf{k}. \tag{28.6b}$$

If n should vanish at a point s_0 in $[0, 1]$, then (28.6b) would imply $\mathbf{n}(0) \cdot \mathbf{k} = 0$ and, therefore, that n lies in the $\mathbf{ij}$ plane. Suppose now that we have $n(s) \neq 0$ for each $s \in [0, 1]$. Then (28.2), (28.3), and (28.6b) imply that we have the result

$$0 = (a\mathbf{i} + b\mathbf{j}) \cdot \mathbf{k} = \int_0^1 \nu(s)\mathbf{e}(s) \cdot \mathbf{k}\, ds = \mathbf{n}(0) \cdot \mathbf{k} \int_0^1 \frac{\nu(s)}{n(s)}\, ds,$$

whence the condition $\mathbf{n}(0) \cdot \mathbf{k} = 0$ follows. Equation (28.6b) then implies that $\mathbf{n}$ lies in the $\mathbf{ij}$ plane. The result confirms our intuition that a thread, held at its ends, subject to its own weight, remains in a plane. The nonplanar equilibrium states of a real thread are due to the presence of small flexural and torsional stiffness that can dominate the small self-weight of the thread.

The properties of the equilibrium configurations of an ideal string subjected to a load $\mathbf{f}(s)$ satisfying conditions (28.5) are independent of the choice of constitutive representation for n. To obtain a more specific description of the possible deformed states of the string, we must introduce suitable constitutive functions. Elastic strings are characterized by a constitutive equation of the form

$$n(s) = N[\nu(s), s], \tag{28.7}$$

where N is a continuously differentiable function of $[\nu(s), s]$ with

$$N_\nu(\nu, s) > 0, \quad N(1, s) = 0,$$
$$N(\nu, s) \to \begin{Bmatrix} +\infty \\ -\infty \end{Bmatrix} \text{ as } \nu \to \begin{Bmatrix} +\infty \\ 0 \end{Bmatrix}. \tag{28.8}$$

Figure 28.1 shows, qualitatively, the graph of N as a function of ν. The assumption (28.8) ensures that (28.7) can be inverted and written in the equivalent form

$$\nu(s) = \hat{\nu}[n(s), s]. \tag{28.9}$$

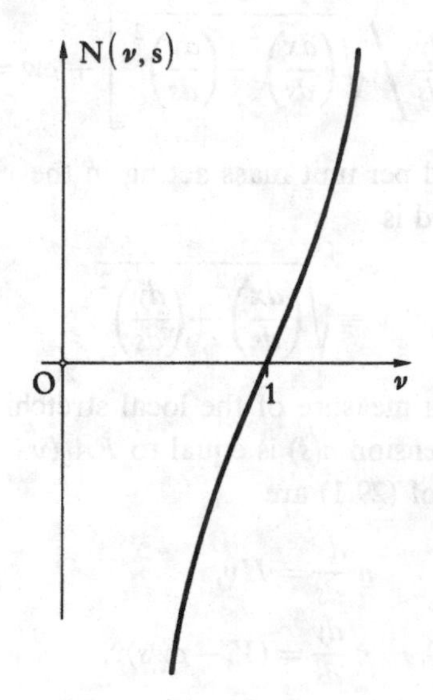

Fig. 28.1

Equation (28.7) is sometimes replaced by a linear constitutive relation that is a natural extension of Hooke's law. The common form of the linearized constitutive equation is

$$n(s) = EA_0[\nu(s) - 1], \tag{28.10}$$

where E is Young's modulus and A_0 is the (uniform) cross-sectional area in the unstrained profile. Equation (28.9) does not satisfy the constitutive requirement (28.8), because $n(s)$ tends to a finite limit as ν tends to zero, but it can be adopted as an approximation to (28.7) in the neighborhood of $\nu = 1$.

29. Classical Solutions for Linearly Elastic Strings

Even though the differential equations of equilibrium of a stretched cord were derived by James Bernoulli, it seems that the solution of the symmetrically suspended elastic catenary, that is, of a cord deformed only by its own weight, was due to Routh (1891, Chap. 10).

The cable under consideration is suspended between two fixed points A and B that have Cartesian coordinates $(0, 0)$ and (ℓ, h), respectively. The unstrained length of the cable is $L_0 = 1$, where L_0 is necessarily not greater than $(\ell^2 + h^2)^{\frac{1}{2}}$, although it obviously cannot be much less if we wish to apply Hooke's law. If s denotes the arc-length parameter in the unstretched state measured from the point $(0, 0)$, the equilibrium equations (28.4), written in terms of components along the x and y axes, are

$$\frac{d}{ds}\left[n\frac{dx}{ds}\Bigg/\sqrt{\left(\frac{dx}{ds}\right)^2 + \left(\frac{dy}{ds}\right)^2}\,\right] = 0, \tag{29.1a}$$

$$\frac{d}{ds}\left[n\frac{dy}{ds}\Bigg/\sqrt{\left(\frac{dx}{ds}\right)^2 + \left(\frac{dy}{ds}\right)^2}\,\right] + \rho w = 0, \tag{29.1b}$$

where ρw is the vertical load per unit mass acting in the y direction.

The elongation of the cord is

$$\nu \equiv \sqrt{\left(\frac{dx}{ds}\right)^2 + \left(\frac{dy}{ds}\right)^2}, \tag{29.2}$$

and the quantity $(\nu - 1)$ is a measure of the local stretching per unit length. If the cord is linearly elastic, the tension $n(s)$ is equal to $EA_0(\nu - 1)$. Provided that ρw is a constant, the first integrals of (29.1) are

$$n\frac{dx}{ds} = H\nu, \tag{29.3a}$$

$$n\frac{dy}{ds} = (V - \rho w s)\nu, \tag{29.3b}$$

where V is the vertical reaction at the support $s = 0$, and H is the horizontal component of the cable tension.

In order to find the solutions $x(s)$, $y(s)$, and $n(s)$, we square and add (29.3) and obtain

$$n(s) = [H^2 + (V - \rho ws)^2]^{\frac{1}{2}}. \tag{29.4}$$

Once we have $n(s)$, we also know $\nu(s)$ from the constitutive relation. Therefore (29.3a) becomes

$$\frac{dx}{ds} = \frac{H}{EA_0} + \frac{H}{[H^2 + (V - \rho ws)^2]^{\frac{1}{2}}},$$

which, on using the end condition $x(0) = 0$, has the solution

$$x(s) = \frac{Hs}{EA_0} + \frac{H}{\rho w}\left[\sinh^{-1}\left(\frac{V}{H}\right) - \sinh^{-1}\left(\frac{V - \rho ws}{H}\right)\right]. \tag{29.5}$$

By following the same procedure as that employed for determining $x(s)$, we can solve (29.3b) with the boundary condition $y(0) = 0$. The final result is

$$y(s) = \frac{\rho ws}{EA_0}\left(\frac{V}{\rho w} - \frac{s}{2}\right) + \frac{H}{\rho w}\left\{\left[1 + \left(\frac{V}{H}\right)^2\right]^{\frac{1}{2}} - \left[1 + \left(\frac{V - \rho ws}{H}\right)^2\right]^{\frac{1}{2}}\right\}. \tag{29.6}$$

In (29.5) and (29.6), the reactions V and H are unknown, but we still have to satisfy the boundary conditions $x(1) = \ell$ and $y(1) = h$. This means that V and H must be solutions of the following system of transcendental equations:

$$\ell = \frac{H}{EA_0} + \frac{H}{\rho w}\left[\sinh^{-1}\left(\frac{V}{H}\right) - \sinh^{-1}\left(\frac{V - \rho w}{H}\right)\right], \tag{29.7}$$

$$h = \frac{\rho w}{EA_0}\left(\frac{V}{\rho w} - \frac{1}{2}\right) + \frac{H}{\rho w}\left\{\left[1 + \left(\frac{V}{H}\right)^2\right]^{\frac{1}{2}} - \left[1 + \left(\frac{V - \rho w}{H}\right)^2\right]^{\frac{1}{2}}\right\}. \tag{29.8}$$

The solution of these two simultaneous equations for H and V can only be found numerically, although, in some limiting cases, it is possible to explore the properties of the solutions without needing to resort to a numerical solution (Irvine 1981, Chap. 1).

If the supports are at the same level, then $h = 0$ and (29.8) supplies the expected result $V = \frac{1}{2}\rho w$. Equation (29.7), written in terms of the single dependent variable H, becomes

$$\frac{\rho w}{2H} = \sinh\left(\frac{\rho w\ell}{2H} - \frac{\rho w}{2EA_0}\right). \tag{29.9}$$

If the cable is inextensible, (29.9) reduces to

$$\frac{\rho w}{2H} = \sinh\left(\frac{\rho w\ell}{2H}\right), \tag{29.10}$$

and a solution exists only for $\ell \leqslant 1$. When ℓ is little different from unity and the cable is relatively inextensible, a situation common in practice, (29.5), (29.6), and (29.9) can be shown to give the approximate solution

$$y(s) = \frac{1}{2}\left(1 + \frac{H}{EA_0}\right)\frac{\rho w}{H}\, s(1-s), \tag{29.11}$$

where H is solution of the cubic equation

$$H^3 + EA_0(1-\ell)H^2 - EA_0\,\frac{(\rho w)^2}{24} = 0, \tag{29.12}$$

which, by Descartes' rule of signs, admits just one positive root regardless of the sign of $(1-\ell)$. If EA_0 is large and ℓ only slightly less than one, an approximate solution of (29.12) is

$$H = \frac{\rho w}{[24(1-\ell)]^{\frac{1}{2}}}. \tag{29.13}$$

When $h = 0$ and the string is initially pulled with a force T_0, we can simplify the differential equations (29.1) by expressing s as a function of x, provided that $dx/ds \neq 0$, that is, the string is never vertical (Weinberger 1965, Sec. 1). We write

$$y = v(x),$$

where

$$v[x(s)] = y(s). \tag{29.14}$$

By the chain rule we have

$$\frac{dy}{ds} = \frac{dv}{dx}\frac{dx}{ds},$$

and thus (29.1b) becomes

$$\frac{d}{ds}\left[n\,\frac{\dfrac{dv}{dx}\cdot\dfrac{dx}{ds}}{\sqrt{\left(\dfrac{dx}{ds}\right)^2 + \left(\dfrac{dy}{ds}\right)^2}}\right] + \rho w = 0. \tag{29.15}$$

On expanding (29.15), and using (29.1a) and (29.2), we obtain

$$\frac{n}{\nu}\left(\frac{dx}{ds}\right)^2\frac{d^2v}{dx^2} + \rho w = 0. \tag{29.16}$$

This is a single differential equation for the unknown function v. However, the coefficient of d^2v/dx^2 depends on the unknown function $x(s)$. In order to eliminate this dependence, we assume that the slope dx/ds is small, with $\nu \simeq 1$. In this case the tension $n(s)$ is practically equal to T_0, and the equilibrium equation becomes

$$T_0\,\frac{\partial^2 v}{\partial x^2} + \rho w = 0, \tag{29.17}$$

with the boundary conditions $v(0) = v(1) = 0$. For $\rho w =$ constant, the solution of (29.17) is simply

$$T_0 v(x) = \frac{\rho w}{2}\, x(1-x). \tag{29.18}$$

If, in addition to the cable self-weight, a concentrated vertical load F_1 is hung from the cable at a point s_1, the equilibrium configuration is again obtained by solving (29.1a) (29.1b), but with the difference that the integration must be split into the two separate intervals $0 \leqslant s < s_1$ and $s_1 < s \leqslant 1$. The first integral of (29.1a) remains unaltered, while that of (29.1b) reads

$$n\frac{dy}{ds} = (V - \rho ws)\nu, \quad 0 \leqslant s < s_1, \tag{29.19a}$$

$$n\frac{dy}{ds} = (V - F_1 - \rho ws)\nu, \quad s_1 < s \leqslant 1. \tag{29.19b}$$

The constitutive relation $n(s) = EA_0(\nu - 1)$ again holds everywhere except at the point of loading. The boundary conditions are also the same but, in addition, we must have the requirement that at s_1 the following conditions of continuity hold:

$$\begin{gathered} x^+(s_1) = x^-(s_1), \quad y^+(s_1) = y^-(s_1), \\ \left[n(s_1)\frac{dy}{ds}(s_1)\right]^+ - \left[n(s_1)\frac{dy}{ds}(s_1)\right]^- + F_1 = 0, \end{gathered} \tag{29.20}$$

where the symbols $(\ )^+$ and $(\ )^-$ denote the limit of a quantity as s approaches s_1 from the right and from the left, respectively. From (29.3a) and (29.19a) we get

$$n(s) = [H^2 + (V - ws)^2]^{\frac{1}{2}}, \quad 0 \leqslant s < s_1, \tag{29.21}$$

and from (29.3a) and (29.19b) we have

$$n(s) = [H^2 + (V - F_1 - ws)^2]^{\frac{1}{2}}, \quad s_1 < s \leqslant 1. \tag{29.22}$$

Once we know $n(s)$, we can obtain $x(s)$ and $y(s)$ in both intervals ($0 \leqslant s < s_1$ and $s_1 < s \leqslant 1$). The final expressions for $n(s)$, $x(s)$, and $y(s)$ contain the quantities H and V, which remain to be determined. The conditions by which H and V may be found are the boundary conditions

$$x(s) = \ell, \quad y(s) = h. \tag{29.23}$$

Since (29.23) are two transcendental equations, their solutions must be obtained numerically.

A particular property of the response of a cable under successive equal increments of the load is that the corresponding increments in the deflection of some characteristic points are not equal, each one being smaller than the preceding one. The nonlinear stiffening was probably known to Rankine (Irvine 1981, Chap. II), but the exact analysis of the problem was somewhat restricted because the methods of solution were cumbersome. A simplification can be made when the profile of the cable is sufficiently flat; for example, cables with a relatively low sag, which are employed in many technical applications.

Consider, for example, a point load P acting at a distance s_1 from the left-hand support of a cable hanging under its own weight between two supports at the same level. If the profile is flat, the equation governing the vertical equilibrium of the cable before the application of the point load is (29.18). Provided that the additional vertical movements of the cable produced by P are small, so that the profile remains relatively shallow, (29.19) can be replaced by the equations

$$(T_0 + \delta T_0)\frac{d}{ds}(y + \delta y) = P(1 - x_1) + \frac{\rho w}{2}(1 - 2x), \quad 0 \leqslant x < x_1,$$
$$(T_0 + \delta T_0)\frac{d}{ds}(y + \delta y) = -Px_1 + \frac{\rho w}{2}(1 - 2x), \quad x_1 < x \leqslant 1,$$

where δT_0 is the increment in the horizontal component of the tension due to the point load, and δy is the additional deflection induced by P. After expanding these equations and removing the self-weight terms, which cancel identically, we obtain

$$(T_0 + \delta T_0)\frac{d}{ds}(\delta y) = P(1 - x_1) - \delta T_0 \frac{dy}{dx}, \quad 0 \leqslant x < x_1,$$
$$(T_0 + \delta T_0)\frac{d}{ds}(\delta y) = -Px_1 - \delta T_0 \frac{dy}{dx}, \quad x_1 < x \leqslant 1.$$

A subsequent integration, after the boundary conditions have been satisfied, yields

$$\begin{aligned} \delta y &= \frac{1}{(1 + \delta T_0/T_0)}\left[x(1 - x_1) - \frac{\rho w}{2P}\frac{\delta T_0}{T_0}x(1 - x)\right], \quad 0 \leqslant x < x_1, \\ \delta y &= \frac{1}{(1 + \delta T_0/T_0)}\left[x_1(1 - x) - \frac{\rho w}{2P}\frac{\delta T_0}{T_0}x(1 - x)\right], \quad x_1 < x \leqslant 1. \end{aligned} \tag{29.24}$$

With the solution represented in this form, the ratio $\delta T_0/T_0$ is still unknown. In order to determine $\delta T_0/T_0$, we observe that the length of an arc element of a cable under the action of its self-weight is $dS = \sqrt{dx^2 + dy^2}$. The length of the same element after the application of a load is $dS_1 = [(dx + d\delta x)^2 + (dy + d\delta y)^2]^{\frac{1}{2}}$. The additional elongation of the cable, corrected to second-order quantities, is then

$$\delta e = \frac{dS_1 - dS}{dS} \simeq \frac{d\delta x}{dS}\frac{dx}{ds} + \frac{d\delta y}{dS}\frac{dy}{ds} + \frac{1}{2}\left(\frac{d\delta y}{dS}\right)^2. \tag{29.25}$$

Since the cable is linearly elastic, we know, by Hooke's law, that δn is represented by

$$\delta n = EA\,\delta e, \tag{29.26}$$

where δn is the increment in tension and EA is the extensional rigidity before the application of P. However, to the second order, we may approximate δn by $\delta n \simeq \delta T_0(dS/dx)$, so that (29.25) can be written as

$$\frac{\delta T_0}{EA}\left(\frac{dS}{dx}\right)^3 = \frac{d\delta x}{dx} + \frac{d\delta y}{dx}\frac{dy}{dx} + \frac{1}{2}\left(\frac{d\delta y}{dx}\right)^2. \tag{29.27}$$

If we integrate (29.27) from 0 to 1 with respect to x, and put

$$L_e = \int_0^1 \left(\frac{dS}{dx}\right)^3 dx,$$

which is known because we know the configuration of the cable under its self-weight, we obtain

$$\frac{\delta T_0}{EA}L_e = \delta x(1) - \delta x(0) + \int_0^1 \frac{d\delta y}{dx}\frac{dy}{dx}\,dx + \frac{1}{2}\int_0^1 \left(\frac{d\delta y}{dx}\right)^2 dx.$$

If, as often occurs, δx and δy vanish at the supports, this equation integrates to

$$\frac{\delta T_0}{EA} L_e = \frac{\rho w}{T_0} \int_0^1 \delta y \, dx + \frac{1}{2} \int_0^1 \left(\frac{d\delta y}{dx} \right)^2 dx. \tag{29.28}$$

Under the point of application of P the additional slope $d\delta y/dx$ is discontinuous, and the last integral in (29.28), when integrated by parts, gives

$$\frac{1}{2} \int_0^1 \left(\frac{d\delta y}{dx} \right)^2 dx = -\frac{1}{2} \left(\frac{d\delta y}{dx} \bigg|_{x_1^-}^{x_1^+} + \frac{d^2\delta y}{dx^2} \int_0^{x_1} \delta y \, dx + \frac{d^2 \delta y}{dx^2} \int_{x_1}^{1} \delta y \, dx \right).$$

On substituting this result in (29.28), and performing the integration using (29.24), we obtain the following equation for the ratio $\delta T_0 / T_0$:

$$\left(\frac{\delta T_0}{T_0} \right)^3 + \left(2 + \frac{\lambda^2}{24} \right) \left(\frac{\delta T_0}{T_0} \right)^2 + \left(1 + \frac{\lambda^2}{12} \right) \left(\frac{\delta T_0}{T_0} \right) - \lambda^2 x_1 \left(1 - x_1 \right) \frac{P}{2\rho w} \left(1 + \frac{P}{\rho w} \right) = 0, \tag{29.29}$$

where $\lambda^2 = (\rho w / T_0)^2 / (T_0 L_e / EA)$. The parameter λ^2 accounts for geometric and elastic effects. For light metal cables, λ^2 is small and becomes smaller as the material becomes more extensible. On the other hand, when $\rho w / T_0$ has a more appreciable effect, as in a suspension bridge cable, λ^2 may reach values of the order of 10^3.

In a cable loaded by an additional point force P superimposed on the self-weight ρw, the total energy is given as the sum of three contributory factors.[47] Suppose that, in a free-hanging uniform cable suspended between two supports at the same level, a small point force P is gradually applied at some point x_1. The work associated with the load is

$$\mathcal{L} = P \, \delta y_1, \tag{29.30}$$

where δy_1 is the vertical additional deflection at x_1. The strain energy can be evaluated by considering the following linearized version of (29.25):

$$\delta e \simeq \frac{dy}{dx} \frac{d\delta y}{dx} + \frac{1}{2} \left(\frac{d\delta y}{dx} \right)^2 \tag{29.31}$$

to second order. The strain energy is given by the sum $W_e + W_g$, where W_e is the elastic strain energy of extension given by

$$W_e = \frac{1}{2} \int_0^1 \delta T_0 \, \delta e \, dx \simeq \frac{1}{2} \delta T_0 \int_0^1 \frac{dy}{dx} \frac{d\delta y}{dx} \, dx, \tag{29.32}$$

and W_g is the gravitational energy arising as a second-order difference of two larger terms, namely

[47] In Irvine's (1980) definition of "energy," there is no clear distinction between strain energy and exterior work.

$$W_g = \int_0^1 T_0\,\delta e\,dx - \int_0^1 \rho w\,\delta y\,\frac{ds}{dx}\,dx,$$

where dS/dx is the secant of the angle subtended between the tangent to the cable and the horizontal. The last equation may be rewritten as

$$W_g = \frac{1}{2}T_0 \int_0^1 \left(\frac{d\delta y}{dx}\right)^2 dx + \int_0^1 \left(T_0\,\frac{dy}{dx}\,\frac{d\delta y}{dx} - \rho w\,\delta y\,\frac{dS}{dx}\right) dx.$$

As y and $\partial y/\partial x$ are continuous, and so is δy, integration by parts yields

$$W_g = \frac{1}{2}T_0 \int_0^1 \left(\frac{d\delta y}{dx}\right)^2 dx + T_0\,\frac{dy}{dx}\,\delta y\bigg|_0^1 - \int_0^1 \left(T_0\,\frac{d^2y}{dx^2} + \rho w\,\frac{dS}{dx}\right)\delta y\,dx. \tag{29.33}$$

The second term on the right-hand side vanishes, because $\delta y(0) = \delta y(1) = 0$, as so does the third term, because the equilibrium equation before the application of P is just

$$T_0\,\frac{d^2y}{dx^2} + \rho w\,\frac{dS}{dx} = 0.$$

On using this equation, the strain energy of extension becomes

$$W_e \simeq \frac{1}{2}\delta T_0\,\frac{\rho w}{T_0}\int_0^1 \delta y\,dx. \tag{29.34}$$

Thus the total additional energy is given by

$$\mathscr{E} = W_e + W_g - \mathscr{L}. \tag{29.35}$$

This expression is useful when applying direct variational methods.

The technical applications of the theory of strings are infinite in number, but in many cases the solutions are not explicit. Examples of such cases are the problem of an inextensible cable immersed in a uniform steady liquid flow, the problem of the construction of a flexible barrage, and fishing nets (Simpson and Tabarrok 1976). Consider an initially slack cable of length $2S$ that connects two fixed pegs a distance $2L$ apart ($L < S$). We introduce rectangular coordinates x, y in such a way that the origin O is placed at the vertex of the deformed profile, the x axis being horizontal to and the y axis coinciding with the vertical axis of symmetry (Figure 29.1). Let us denote the arc length of any point of the deformed chain by s, where $-S \leqslant s \leqslant S$. If we call T the tension along the chain and f_x and f_y are the forces per unit length exerted by the fluid, the equilibrium equations are simply

$$\begin{aligned}(Tx')' + f_x &= 0,\\ (Ty')' + f_y &= 0,\end{aligned} \tag{29.36}$$

where the prime denotes differentiation with respect to s. If we consider the symmetry and the fixed condition at the end points, the boundary conditions are

$$x(0) = y(0) = y'(0) = 0, \quad x(s) = L. \tag{29.37}$$

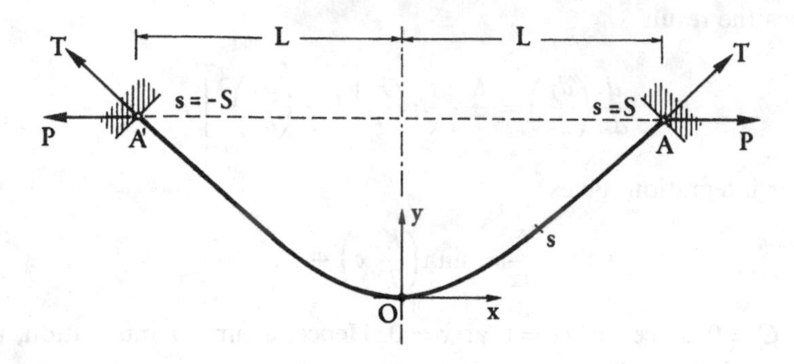

Fig. 29.1

If the fluid density is ρ and the hydrodynamic reference diameter of the chain is c, with c constant, then the forces due to the action of a steady stream, moving with a velocity V in the negative direction of the y axis, are

$$\begin{aligned} f_x &= \tfrac{1}{2}\rho V^2 c C_D \cos^2\psi \sin\psi, \\ f_y &= -\tfrac{1}{2}\rho V^2 c C_D \cos^3\psi, \end{aligned} \tag{29.38}$$

where C_D is the normal-flow drag factor of the chain section, and ψ is the local yaw angle at s. The forces (29.38) represent the case in which the fluid is inviscid and exerts a purely normal force $f_n = \frac{1}{2}\rho V^2 c C_D \cos^2\psi$, the components of which are f_x and f_y along the coordinate axes.

On writing $K = \frac{1}{2}\rho V^2 c C_D$, and observing the relations $\cos\psi = x'$ and $\sin\psi = y'$, (29.36) become

$$(Tx')' + K(x')^2 y' = 0, \tag{29.39a}$$

$$(Ty')' - K(x')^3 = 0. \tag{29.39b}$$

On multiplying (29.39a) by x' and (29.39b) by y', and adding, we arrive at the result

$$T' + T(x'x'' + y'y'') = 0.$$

From $x'^2 + y'^2 = 1$, the coefficient of T in the equation above vanishes, and hence we derive

$$T = \text{constant.} \tag{29.40}$$

Again, multiplication of (29.39a) by y' and (29.39b) by x' followed by subtraction yields

$$T(x''y' - y''x') + K(x')^2 = 0,$$

which gives

$$T\left(\frac{y'}{x'}\right)' = K \quad \text{or} \quad \left(\frac{dy}{dx}\right)' = \frac{K}{T} = \text{constant.} \tag{29.41}$$

This implies the result

$$\frac{d}{dx}\left(\frac{dy}{dx}\right)=\frac{K}{T}\frac{ds}{dx}=\frac{K}{T}\left[1+\left(\frac{dy}{dx}\right)^2\right]^{\frac{1}{2}},$$

which, after integration, gives

$$\frac{dy}{dx}=\sinh\left(\frac{K}{T}x\right)+C,$$

leading to $C=0$, since $dy/dx=0$ at $x=0$. Hence, a further integration, using the condition $y=0$ at $x=0$, gives

$$y=\frac{T}{K}\left[\cosh\left(\frac{K}{T}x\right)-1\right]. \tag{29.42}$$

We thus obtain the unexpected result that the equilibrium shape is a catenary.

The tension T may be obtained from the condition

$$S=\int_0^L\sqrt{1+\left(\frac{dy}{dx}\right)^2}\,dx=\frac{T}{K}\sinh\left(\frac{K}{T}L\right). \tag{29.43}$$

In addition, from Figure 29.1, we find that the total "towing" force is

$$F=2T\tanh\left(\frac{KL}{T}\right), \tag{29.44}$$

and the "separating force" is

$$P=T\,\mathrm{sech}\left(\frac{KL}{T}\right). \tag{29.45}$$

If we wish to consider the influence of viscosity, a natural assumption is that the viscous force per unit length is purely tangential and proportional to the tangential component of the velocity, which is $-V\sin\psi$. If f_t is this force, we can write

$$f_t=-\bar{\kappa}Vy'=-\kappa y', \tag{29.46}$$

where $\bar{\kappa}$ and κ are constants. Equations (29.40) and (29.41) are now

$$T'-\kappa y'=0, \tag{29.47a}$$

$$T\left(\frac{y'}{x'}\right)'=K \quad \text{(as before)}. \tag{29.47b}$$

From (29.47a) we derive

$$T=H+\kappa y, \tag{29.48}$$

where H is the tension at the point (0, 0). Then, on substituting (29.48) in (29.47b) we have the equation

$$(H+\kappa y)\frac{d^2y}{dx^2}-K\left[1+\left(\frac{dy}{dx}\right)^2\right]^{\frac{1}{2}}=0. \tag{29.49}$$

This equation does not appear to lend itself to a solution in a known form, but, in view of the fact that we often have $\kappa \ll K$, an approximate solution can be achieved by using a perturbation method (Simpson and Tabarrok 1976).

Another problem that can be solved by means of manipulating the equilibrium equation, is the determination of the shape of a boom of logs in tow, when it is dragged with constant velocity in relatively still water (Irvine 1981, Chap. I). Under the action of a concentrated tugging force and the friction arising from the turbulent flow of water around the logs, the contour of the boom tends to assume a profile typical of the meridional section of a liquid droplet under gravity just before detaching itself from the point of suspension (Figure 29.2). Let us assume rectangular coordinates x, y such that the y axis coincides with the axis of symmetry of the towed boom, and x marks the thickness of the boom.

The simplest way of describing the profile of the enveloping cable is by assuming that the logs in the raft experience a state of hydrostatic compressive stress that increases in magnitude directly the rear of the boom is approached. The shaping force is thus everywhere perpendicular to the cable, and the equation of equilibrium is

$$T_0 \frac{d\psi}{ds} = -\tau y, \tag{29.50}$$

where T_0 is the constant tension in the cable, ψ is the angle between the tangent to the curve and the x axis, s is an arc length measured from the origin, and τ is a coefficient that accounts for the turbulent passage of water around the logs. Recalling that $dy/ds = \sin\psi$, on integration, (29.50) yields

$$\tau y^2 = 2T_0 \cos\psi + C, \tag{29.51}$$

where C is a constant determined from the condition $\psi = (\pi/2) + \alpha$ at $y = 0$ and $x = 0$, 2α being the angle between the line of logs in the boom at their junction with the tug's hawser. Therefore

$$\tau y^2 = 2T_0(\cos\psi + \sin\alpha). \tag{29.52}$$

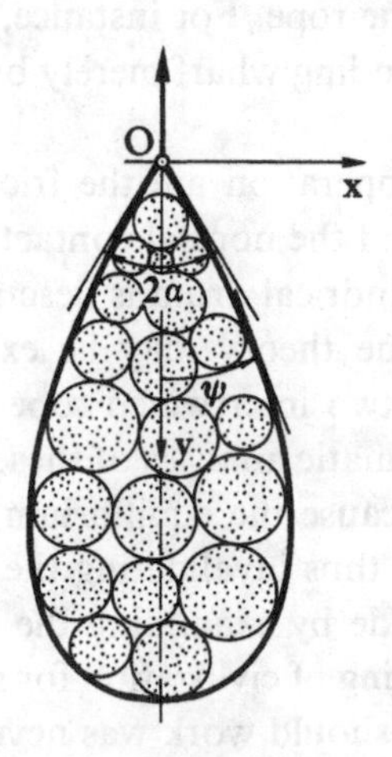

Fig. 29.2

Substitution of this equation in (29.50) provides

$$s = \left(\frac{T_0}{2\tau}\right)^{\frac{1}{2}} \int_{\psi}^{\frac{\pi}{2}+\alpha} \frac{d\psi}{(\cos\psi + \sin\alpha)^{\frac{1}{2}}}.$$

To reduce the last integral to a more familiar form, let $\psi = (\pi/2) + \theta$, and let us write

$$\sin\varphi = \left(\frac{1+\sin\theta}{1+\sin\alpha}\right)^{\frac{1}{2}}.$$

Since $-\pi/2 \leqslant \theta \leqslant \alpha$, it follows that we have $0 \leqslant \sin\varphi \leqslant 1$, and θ is the angle between the tangent to the profile and the y axis. After some manipulation we obtain the standard form for an elliptic integral of the first kind:

$$s(\varphi) = \left(\frac{T_0}{\tau}\right)^{\frac{1}{2}} \int_{\varphi}^{\frac{\pi}{2}} \frac{d\varphi}{(1 - m\sin^2\varphi)^{\frac{1}{2}}}, \tag{29.53}$$

with $m = (1 + \sin\alpha)/2$. Once we have s, we can find $x(s)$ and $y(s)$ by integrating the equations $x' = \cos\psi$ and $y' = \sin\psi$. In addition, we can calculate the total area A enclosed by the cable, observing that for overall equilibrium in the y direction we must have

$$2T_0\cos\alpha = A\tau. \tag{29.54}$$

In addition, the power required to tow the boom at various velocities can also be estimated.

30. Influence of Friction

One of the most important applications of ropes is based on the possibility of exploiting their high degree of flexibility, so that one end of the rope can be wound round a body, such as a post or another rope, in order to fix a further body tied to the other end of the rope. For instance, everybody knows how quickly sailors can secure a ship to a landing wharf merely by winding a hawser many times round a capstan or post.

The forces involved in this operation are the friction between the rope and the post, the tension in the rope, and the normal contact forces. The planar equilibrium of a rope wrapped round a cylindrical drum is described and discussed in all books on technical mechanics, but the theory may be extended to a rope lying on an arbitrary curved surface or to two intertwined ropes. Surprisingly, in view of their historical importance, the systematic analysis of these problems is recent (Maddocks and Keller 1987), probably because the equilibrium configurations are a long way from the initial configurations, thus invalidating the use of linearized analysis.

A first approach may be made by examining the behavior of knots. Knots have been employed since the beginning of civilization for a multitude of purposes, but the specific reasons as to why they should work was never fully understood. In order to determine whether or not a knot in a rope will hold, we consider a rope in which the tension $T(s)$ varies with the position s, from zero at the loose end to some positive value at the loaded end. The physical mechanism that causes T to vary with s is the friction between different points of the rope or between one rope and another. If $T(s)$ were constant no knot could hold, so we may ask how large the friction factor μ must be in order for a given knot to remain firm.

Let us begin by analyzing the symmetric square of the reef knot shown in the Figure 30.1, where the tension is T_1 at 1, T_2 at 2, T_3 at 3, and zero at 0. We assume that the tension jumps from zero at the loose end to a value T_1 just to the left of the crossover at point A. If N is the normal force squeezing the two ropes together at A, then the law of friction yields

$$T_1 \leqslant 2\mu N, \tag{30.1}$$

the factor 2 arising because the piece of rope under consideration is in contact with two other pieces of rope at A. The normal force N is essentially equal to the tension in the vertical rope at A. To find N we assume that the tension in this rope increases from its value T_1 at B by the factor $e^{\mu\pi}$, because the rope is practically wrapped around a cylinder along an arc of π radians giving $N = T_1 e^{\mu\pi}$. On combining this relation with (30.1) and eliminating T_1, which is assumed to be positive, we get

$$1 \leqslant 2\mu e^{\mu\pi}. \tag{30.2}$$

This is the condition for the tight knot to hold. Exactly the same condition could have been obtained by considering the jump in tension $T_3 - T_2$.

We now consider a symmetrical sheepshank (a device for shortening a single rope) like the one shown in Figure 30.2. The tension is zero in the two loops 3, and we assume that it also zero in the segment 4. By symmetry $T_1 = T_2$. Then the force balance requires the equality

$$T_2 = \tfrac{1}{2} T_0. \tag{30.3}$$

The tension jumps from T_0 to $T_0 - \delta$ and from T_2 to $T_2 + \delta$ in the two parts of the rope that cross at A. The normal force between these parts is approximately $T_0/2$, so that the condition for equilibrium is

$$T_0 - (T_0 - \delta) = \delta \leqslant \mu \frac{T_0}{2}. \tag{30.4}$$

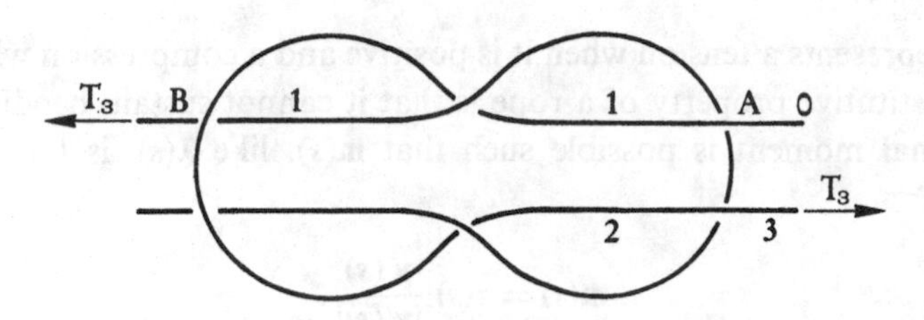

Fig. 30.1

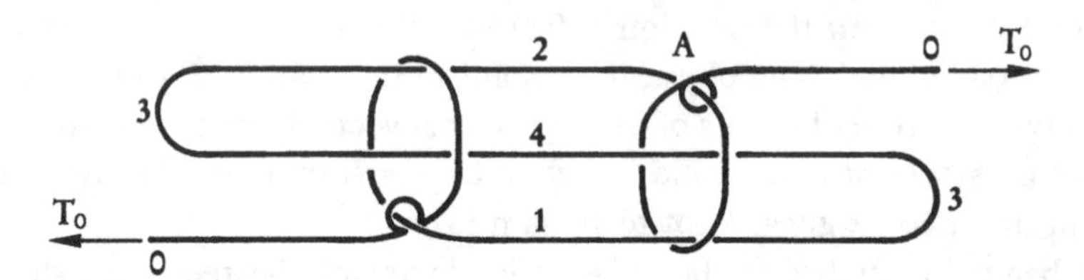

Fig. 30.2

The tension in the rope wrapped round segments 1 and 4 does not decrease by more than $e^{-2\pi\mu}$, which we obtain by considering the limiting case when the wrapping is 2π. Thus we have

$$T_2 + \delta \geqslant e^{-2\pi\mu}(T_0 - \delta). \tag{30.5}$$

On using (30.3) and (30.4) in (30.5) we get

$$\frac{1}{2}(1+\mu) \geqslant e^{-2\pi\mu}\left(1 - \frac{\mu}{2}\right). \tag{30.6}$$

This is the condition for the sheepshank to hold, and the same condition results when we examine the jump in tension from T_1 in segment 1 to zero in segment 3.

In order to solve more complicated problems concerning the contact of a rope with a given surface, it is necessary to find the tension and the frictional forces as functions of the arc length s. The rope, in its reference configuration, is a long, thin cylinder, the central line of which is parameterized through the variable s. Any deformed configuration of the central line is described by a vector function $\mathbf{x}(s)$. The stress resultant and the couple across each orthogonal cross-section are a force $\boldsymbol{\lambda}(s)$ and a moment $\mathbf{m}(s)$, both acting at the point $\mathbf{x}(s)$. The equilibrium equations are, as usual,

$$\boldsymbol{\lambda}'(s) + \mathbf{f} = \mathbf{0}, \quad \mathbf{m}'(s) + \mathbf{x}'(s) \times \boldsymbol{\lambda}(s) + \boldsymbol{\ell} = \mathbf{0}, \tag{30.7}$$

where the prime denotes a derivative with respect to s, and $\mathbf{f}$ and $\boldsymbol{\ell}$ are the external force and moment densities, respectively.

A characteristic property of a rope is that it is unable to sustain shear forces, so that $\boldsymbol{\lambda}(s)$ must always be parallel to the tangent to the central line:

$$\boldsymbol{\lambda}(s) = T(s)\,\frac{\mathbf{x}'(s)}{|\mathbf{x}'(s)|}, \tag{30.8}$$

where $T(s)$ represents a tension when it is positive and a compression when negative. Another constitutive property of a rope is that it cannot sustain bending moments, but a torsional moment is possible such that $\mathbf{m}(s)$, like $\boldsymbol{\lambda}(s)$, is tangential to the central line:

$$\mathbf{m}(s) = \tau(s)\,\frac{\mathbf{x}'(s)}{|\mathbf{x}'(s)|}, \tag{30.9}$$

where $\tau(s)$ represents the twist.

The external load densities on the rope arise because of its contact with the surface S′ of another body. The contact condition is usually described by requiring the line $\mathbf{x}(s)$ to lie at a fixed distance a above the surface S′. In addition, if $N(s)$ is the contact force normal to S′, there is a tangential contact force due to friction and its magnitude is bounded by $\mu N(s)$. The contact also determines the moment density $\ell(s)$; but, in order to evaluate this effect it is necessary to construct a detailed model of the contact region. This difficulty can be overcome by the expedient of assuming that the twist and contact moments will adjust to one another so as to satisfy the moment balance equations.

Under these circumstances the central line $\mathbf{x}(s)$ must lie in a surface S, parallel to the prescribed surface S′ and at a distance a from it. The balance of forces provides the equation

$$-(T\mathbf{e})' = N\mathbf{n} + F_1\mathbf{e} + F_2\mathbf{n}\times\mathbf{e} + \mathbf{P}, \tag{30.10}$$

where $\mathbf{e}(s) = \mathbf{x}'/|\mathbf{x}|$ denotes the unit tangent to the central line, $T(s) \geqslant 0$ is the tension in the rope, $\mathbf{n}$ is the unit outward normal to the surface S, $N(s)$ is the normal contact force exerted by the surface, $F_1(s)$ and $F_2(s)$ are the tangential and transverse components of the frictional force, and $\mathbf{P} \equiv (P_1, P_2, P_3)$ is the external force per unit length. The components of (30.10) in the directions $\mathbf{e}$, $\mathbf{n}\times\mathbf{e}$, and $\mathbf{n}$ are

$$-T' = F_1 + P_1, \tag{30.11}$$

$$-\kappa_t T = F_2 + P_2, \tag{30.12}$$

$$-\kappa_n T = N + P_3, \tag{30.13}$$

where $\kappa_t = \mathbf{e}'\cdot\mathbf{n}\times\mathbf{e}$ is the geodesic curvature of the curve $\mathbf{x}(s)$ on the surface S, and $\kappa_n = \mathbf{e}'\cdot\mathbf{n}$ is the normal curvature of the surface in the direction $\mathbf{e}$. These equations must be supplemented by the condition that the magnitude of the friction force density cannot exceed the normal force density N multiplied by μ:

$$F_1^2 + F_2^2 \leqslant (\mu N)^2. \tag{30.14}$$

Now, on solving the equilibrium equations for F_1, F_2, and N, and substituting the resulting expression in (30.14), we obtain

$$(T' + P_1)^2 \leqslant \mu^2(P_3 + \kappa_n T)^2 - (P_2 + \kappa_t T)^2. \tag{30.15}$$

In the special case $\mathbf{P} = \mathbf{0}$, (30.15) simplifies to

$$T'^2 \leqslant (\mu^2\kappa_n^2 - \kappa_t^2)T^2, \tag{30.16}$$

which shows that a necessary condition for $\mathbf{x}(s)$ to be an equilibrium configuration with $\mathbf{P} = \mathbf{0}$ and $T(s) \neq 0$ is

$$\mu^2\kappa_n^2 - \kappa_t^2 \geqslant 0. \tag{30.17}$$

Provided that (30.17) holds and that $T(s)$ is a function satisfying (30.16), then (30.11)–(30.13) are satisfied by $\mathbf{P} = \mathbf{0}$, $F_1 = -T'$, $F_2 = -\kappa_t T$, and $N = -\kappa_n T$. This choice will also satisfy (30.14), because $T(s)$ satisfies (30.16). Thus (30.17) is both necessary and sufficient for $\mathbf{x}(s)$ to be an equilibrium curve on S with $\mathbf{P} = \mathbf{0}$ and $T \neq 0$. When (30.17) holds we may integrate (30.16) to give

$$T(s_0)\exp\left[-\int_{s_0}^{s}(\mu^2\kappa_n^2-\kappa_t^2)^{\frac{1}{2}}\,d\sigma\right]\leqslant T(s)\leqslant T(s_0)\exp\left[+\int_{s_0}^{s}(\mu^2\kappa_n^2-\kappa_t^2)^{\frac{1}{2}}\,d\sigma\right].\quad(30.18)$$

For $\kappa_t(s)=0$, the curve $\mathbf{x}(s)$ is a geodesic on S and the integrands in (30.18) are, respectively, as small as possible and as great as possible, so that the geodesics are the curves of maximal growth or decay of the tension. Moreover, in the absence of friction, they are the only curves of equilibrium, as (30.18) shows.

There are several important corollaries of (30.17), which are obtained simply by specifying the nature of the surface S. If a rope with nonzero tension lies along a circle of radius $r\leqslant R$ on a sphere of radius R, then $\kappa_n=R^{-1}$ and $\kappa_t^2=r^{-2}-R^{-2}$, and (30.17) yields

$$r\geqslant(1+\mu^2)^{-\frac{1}{2}}R.\quad(30.19)$$

Thus the smallest possible radius of the circle is $(1+\mu^2)^{-\frac{1}{2}}R$, with $r=R$ when $\mu=0$. If the rope lies in a plane, $\kappa_n=0$, then (30.17) shows that the only equilibrium configuration with $T\neq 0$ is a straight line.

Consider a rope lying along a geodesic on a cylindrical surface having the closed curve C as cross-section, and assume that C is a convex curve. A geodesic on a cylinder is a generalized helix, that is, a curve making some constant angle $[(\pi/2)-\alpha]$ with the generators of the cylinder. The normal curvature along a geodesic is $\kappa_n(s)=\kappa(s)\cos^2\alpha$, where $\kappa(s)$ is the curvature of C. The curvature $\kappa(s)$ of C is given by $d\theta(t)/dt$, where t is the arc length and $\theta(t)$ is the angle between the tangent to C at t and a fixed direction in the plane containing C. Furthermore, for a geodesic $t=s\cos\alpha$ and $\kappa_t=0$, so the integrand in (30.18) becomes $\mu\kappa_n(s)=\mu\kappa(s)\cos^2\alpha=\mu\cos^2\alpha\,(d\theta/dt)$. Since we have $dt/ds=\cos\alpha$, (30.18) yields

$$T(s_0)\exp(-\mu\cos\alpha\,\Delta\theta)\leqslant T(s)\leqslant T(s_0)\exp(+\mu\cos\alpha\,\Delta\theta),\quad(30.20)$$

with

$$\Delta\theta=\theta(s\cos\alpha)-\theta(s_0\cos\alpha).\quad(30.21)$$

The inequality (30.20) determines the maximum factor by which the tension can increase and the minimum factor by which it can decrease along a rope lying on a geodesic on a cylinder. The case $\alpha=0$ is the most famous formula of the statics of ropes, and is called the *capstan condition*.[48]

Let us now consider a closed loop of rope encircling a cylinder and pulled by a point load P, as depicted in Figure 30.3. The rope lies in a plane normal to the generators, so that we have $\alpha=\pi/2$ and $\Delta\theta=(\pi/2)+\delta_1+\delta_2$, where δ_1 and δ_2 are the angles made by the line of action of P with the two branches of rope leaving the cylinder. The tensions T_1 and T_2 in the two straight parts of the rope satisfy inequality (30.20), which assumes the form

$$\exp[-\mu(\pi+\delta_1+\delta_2)]\leqslant\frac{T_1}{T_2}\leqslant\exp[+\mu(\pi+\delta_1+\delta_2)].\quad(30.22)$$

[48] The formula was given by Euler (1775).

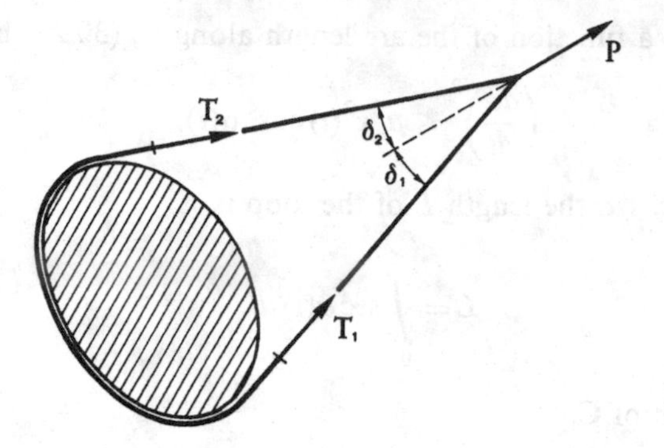

Fig. 30.3

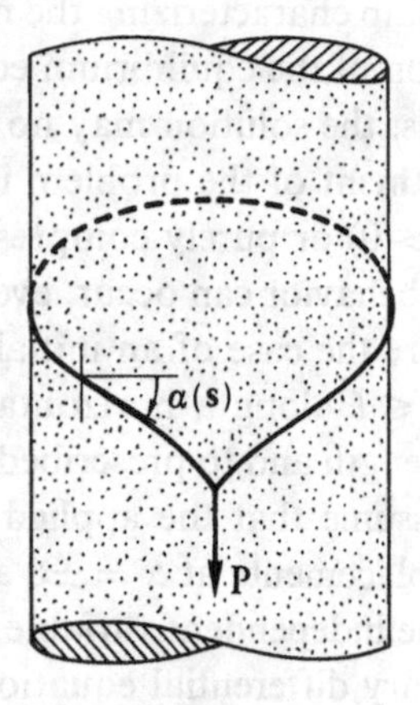

Fig. 30.4

However, balance of the forces requires the conditions

$$T_1 \sin \delta_1 = T_2 \sin \delta_2, \quad T_1 \cos \delta_1 + T_2 \cos \delta_2 = P.$$

If we now substitute the first of these conditions in (30.22) we have

$$\exp[-\mu(\pi + \delta_1 + \delta_2)] \leqslant \frac{\sin \delta_1}{\sin \delta_2} \leqslant \exp[+\mu(\pi + \delta_1 + \delta_2)],$$

which determines all the possible directions of the applied force.

Finally, let us consider a loop completely in contact with a cylinder and with the load directed along a generator, as illustrated in Figure 30.4. Let $\alpha(s)$ denote the angle between the central line and the cross-sectional curve C. Then we have $\kappa_n(s) = \kappa(s) \cos^2 \alpha$ and $\kappa_t = d\alpha/ds$, so that (30.17) becomes

$$\left(\frac{d\alpha}{ds}\right)^2 \leqslant \mu^2 \kappa^2(s) \cos^4 \alpha(s). \tag{30.23}$$

If κ is considered as a function of the arc length along C, (30.23) becomes

$$\left(\frac{d\alpha}{dt}\right)^2 \leqslant \mu^2\kappa^2(t)\cos^4\alpha(t),$$

and, since ds is $dt/\cos\alpha$, the length L of the loop is

$$L = \int_0^\ell \sec\alpha(t)\,dt, \tag{30.24}$$

where ℓ is the length of C.

31. Equilibrium States for Nonlinear Elastic Strings

When the stress–strain relationship characterizing the material constituting the string is nonlinearly elastic, the solution of the equilibrium equations becomes increasingly difficult because, for given loads, the solution may no longer be unique, or may not exist at all. Moreover, the treatment of the problem is substantially different if the solutions are purely tensile ($T > 0$) or purely compressive ($T < 0$) (Dickey 1969).

As all these possible types of behavior can occur, even in the planar equilibrium of a string, we simply consider here the case of an initially straight string of length 2ℓ occupying the interval $-\ell \leqslant x \leqslant \ell$ along a horizontal x axis, subject to a vertical load $p(x)$ per unit undeformed length and to prescribed longitudinal displacements at the ends. For simplicity, we assume that the applied force is symmetrical, that is, $p(x) = p(-x)$, and that the displacements at $x = \pm\ell$ are equal in magnitude but of opposite sign. On taking x as the independent variable, the equilibrium equations for the string are the pair of ordinary differential equations (Carrier 1945)

$$\frac{d}{dx}(T\cos\theta) = 0, \quad \frac{d}{dx}(T\sin\theta) = -p(x), \tag{31.1}$$

where T is the stress resultant and θ is the slope of the deformed axis with respect the horizontal x axis. The slope θ is related to the horizontal and vertical displacements u and v by

$$\cos\theta = \frac{1+u_x}{1+e}, \quad \sin\theta = \frac{v_x}{1+e}, \tag{31.2}$$

where e is the strain, defined as

$$e = [(1+u_x)^2 + v_x^2]^{\frac{1}{2}} - 1. \tag{31.3}$$

As we will study only symmetrical deformations, we may restrict the analysis to the half-interval $0 \leqslant x \leqslant \ell$, and write the boundary conditions as follows

$$v_x(0) = v(\ell) = u(0) = 0, \quad u(\ell) = \lambda, \tag{31.4}$$

where λ is known.

Finally, assuming that the string is perfectly elastic, the tension T is related to the strain e by a constitutive equation of the form

$$T = f(e). \tag{31.5}$$

Before proceeding to a discussion of the tensile solution, we introduce the new variables

$$\xi = \frac{x}{\ell}, \quad \mu = \frac{u}{\ell}, \quad \omega = \frac{v}{\ell}, \quad P = \frac{p\ell}{A}, \tag{31.6}$$

where A is the cross-sectional area of the string. In terms of these quantities, (31.1)–(31.3) become

$$\frac{d}{d\xi}(T\cos\theta) = 0, \quad \frac{d}{d\xi}(T\sin\theta) = -P(\xi), \tag{31.7}$$

$$\cos\theta = \frac{1+\mu_\xi}{1+e}, \quad \sin\theta = \frac{\omega_\xi}{1+e}, \tag{31.8}$$

$$e = [(1+\mu_\xi)^2 + \omega_\xi^2]^{\frac{1}{2}} - 1, \tag{31.9}$$

and the boundary conditions are

$$\omega_\xi(0) = \omega(1) = \mu(0) = 0, \quad \mu(1) = \frac{\lambda}{\ell} = \nu. \tag{31.10}$$

On integrating (31.7) we obtain

$$T^2 = B^2 + \left[\int_0^\xi P(\tau)\,d\tau\right]^2 = B^2 + \tilde{P}(\xi)^2, \tag{31.11}$$

where B is a constant of integration to be determined from (31.10). On combining this result with (31.8) we find

$$\mu(\xi) + \xi = B\int_0^\xi \frac{1+e}{T}\,d\tau, \tag{31.12}$$

$$\omega(\xi) = -\int_\xi^1 \frac{\tilde{P}(\tau)(1+e)}{T}\,d\tau. \tag{31.13}$$

These relations yield an exact solution, providing that the constant B can be chosen such that, according to (31.10) and (31.12):

$$\mu(1) + 1 = \nu + 1 = B\int_0^1 \frac{1+e}{T}\,d\tau, \tag{31.14}$$

with both T and e also dependent on B.

In order to discuss the solutions of (31.14), we simplify the notation by putting

$$I(B,\tilde{P}) = B\int_0^1 \frac{1+e}{T}\,d\tau, \tag{31.15}$$

which, for a given $\tilde{P}$, defines a curve in the BI plane. The intersections of this curve with the straight line $\nu + 1 = I$ determine those values of B that satisfy (31.14). If T is the positive root of (31.11), the solution is tensile, and there are no intersections for

$B < 0$. However, in order to conclude that intersection points exist when $B > 0$, it is necessary for the constitutive function $f(e)$ to possess the following, physically reasonable, properties:

(a) $f(e)$ is defined for $-1 \leqslant e_- \leqslant e < e_+ \leqslant \infty$, where $e_- < 0$ and $e_+ > 0$;
(b) $f(0) = 0$;
(c) $0 \leqslant df/de < \infty$ for $e_- < e < e_+$;
(d) there is a unique function g with $e = g(T)$ for $e_- < e < e_+$.

Corresponding to e_+ and e_-, we can define T_+ and T_- as

$$T_+ = \lim_{e \to e_+} f(e), \quad T_- = \lim_{e \to e_-} f(e). \tag{31.16}$$

As a consequence of its definition, $g(T)$ possesses the following properties:

(a′) $g(T)$ is defined for $T_- < T < T_+$;
(b′) $g(0) = 0$;
(c′) $0 < dg/dT \leqslant \infty$ for $T_- < T < T_+$;
(d′) $\operatorname{sign} g(T) = \operatorname{sign} T$ for $T_- < T < T_+$;
(e′) $g(T) > -1$ for $T_- < T < T_+$.

In the case of tensile solutions, (31.15) may be written in the form

$$I(B, \tilde{P}) = B \int_0^1 \frac{1 + g(T)}{T} \, d\tau = B \int_0^1 \frac{1 + g\{[B^2 + \tilde{P}(\tau)^2]^{\frac{1}{2}}\}}{[B^2 + \tilde{P}(\tau)^2]^{\frac{1}{2}}} \, d\tau. \tag{31.17}$$

On differentiating the function $I(B, \tilde{P})$ with respect to B we find

$$\frac{\partial I(B, \tilde{P})}{\partial B} = \int_0^1 \frac{TB^2 g'(T) + \tilde{P}(\tau)^2[1 + g(T)]}{T^3} \, d\tau, \tag{31.18}$$

where the prime indicates differentiation with respect to T. In the case of tension, the integrand in (31.18) is clearly positive, and hence, for a given $\tilde{P}$, $I(B, \tilde{P})$ is a monotonically increasing function of B. This implies that the solution is unique.

In order to discuss the existence of solutions to (31.14), we assume for simplicity the case when $T_+ = \infty$. This case corresponds to the situation in which there is a well-defined strain regardless of the magnitude of the tension. For instance, Hooke's materials and almost all materials of nonlinear elasticity fall into this category. In addition, we assume that $1/|\tilde{P}(\xi)|$ is integrable. Now the monotonicity of $I(B, \tilde{P})$ implies the limiting conditions

$$\lim_{B \to 0} I(B, \tilde{P}) = \inf_{B > 0} I(B, \tilde{P}),$$

and

$$\lim_{B \to \infty} I(B, \tilde{P}) = \sup_{B > 0} I(B, \tilde{P}).$$

In particular, these two limits can be evaluated explicitly. For the infimum we have

$$0 \leqslant \lim_{B>0} I(B, \tilde{P}) \leqslant B\{1 + g([B^2 + \|\tilde{P}\|^2]^{\frac{1}{2}})\} \int_0^1 \frac{d\tau}{(B^2 + \tilde{P}(\tau)^2)^{\frac{1}{2}}}$$

$$\leqslant \lim_{B \to 0} B\{1 + g([B^2 + \|\tilde{P}\|^2]^{\frac{1}{2}})\} \int_0^1 \frac{d\tau}{|\tilde{P}(\tau)|} = 0,$$

where $\|\tilde{P}\| = \max_{0 \leqslant \xi \leqslant 1} |\tilde{P}(\xi)|$, with the result

$$\inf_{B>0} I(B, \tilde{P}) = 0.$$

Similarly, we can find the limiting value

$$\lim_{B \to \infty} I(B, \tilde{P}) = \int_0^1 \lim_{B \to \infty} \frac{1 + g(B[1 + \tilde{P}(\tau)^2/B^2]^{\frac{1}{2}})}{(1 + \tilde{P}(\tau)^2/B^2)^{\frac{1}{2}}} \, d\tau = 1 + e_+,$$

giving

$$\sup_{B>0} I(B, \tilde{P}) = 1 + e_+.$$

It is clear that, if we have $0 < \nu + 1 < 1 + e_+$, the straight line $I = \nu + 1$ will intersect the curve $I = I(B, \tilde{P})$, proving in this way that a tensile solution exists.

If we seek compressive solutions, we must, as before, solve (31.14), but T must be chosen as the negative root of (31.11). Therefore the compressive solutions are determined by the intersections of the straight line $I = \nu + 1$ with the curve

$$I(B, \tilde{P}) = B \int_0^1 \frac{1 + g(\tau)}{T} \, d\tau = -B \int_0^1 \frac{1 + g(-[B^2 + \tilde{P}(\tau)^2]^{\frac{1}{2}})}{[B^2 + \tilde{P}(\tau)^2]^{\frac{1}{2}}} \, d\tau. \tag{31.19}$$

This curve has no intersection with the line $I = \nu + 1$ for $B > 0$; thus it is only necessary to consider the case $B < 0$. In the tensile case we have seen that there is a solution for both positive and negative displacements, but this is not the case for compressive solutions. More precisely, there are no compressive solutions for $\nu \geqslant 0$. In fact, let us assume, on the contrary, that (31.14) has a compressive solution for $\nu \geqslant 0$. The integral in (31.14) may be evaluated by using the mean-value theorem; hence (31.14) becomes

$$\nu + 1 = B \, \frac{1 + g(T^*)}{T^*}, \tag{31.20}$$

where $T^* = T(\xi^*)$, ξ^* being a value in the interval $0 < \xi < 1$. On combining this equation with (31.11), we find that, at $\xi = \xi^*$, the compression T^* must satisfy the equation

$$T^{*2} = \frac{(\nu + 1)^2 T^{*2}}{[1 + g(T^*)]^2} + \tilde{P}(\xi^*)^2,$$

or, equivalently,

$$[(1 + g(T^*))^2 - (\nu + 1)^2]T^{*2} = \tilde{P}(\xi^*)^2. \tag{31.21}$$

However, for $T^* < 0$ and $\nu \geqslant 0$, the left-hand side of (31.21) is negative, as a consequence of the monotonicity of $g(T)$, while the right-hand side is positive. This is the desired contradiction.

As an illustration, consider a string that satisfies Hooke's law

$$T = Ee, \tag{31.22}$$

where E is the Young's modulus. In this case we may take $e_- = -1$ and $T_- = -E$. Let us assume further that the force p per unit undeformed length is a constant, so that we have

$$\tilde{P}(\xi) = \frac{p\ell}{a}\xi = P\xi. \tag{31.23}$$

We can then calculate

$$\begin{aligned} I(B, \tilde{P}) &= -B \int_0^1 \frac{1 - [B^2 + P^2\tau^2]^{\frac{1}{2}}/E}{[B^2 + P^2\tau^2]^{\frac{1}{2}}}\, d\tau \\ &= -\frac{B}{|P|} \ln([|P| + (B^2 + P^2)^{\frac{1}{2}}]/|B|) + \frac{B}{E}, \end{aligned} \tag{31.24}$$

and the domain of $I(B, \tilde{P})$ is $B_- = -(E^2 - P^2)^{\frac{1}{2}} < B < 0$. When $|P|$ is small there are two compressive solutions for $\nu + 1$ in the interval (see (31.15))

$$I(B_-, \tilde{P}) < \nu + 1 < \sup_{B_- < B < 0} I(B, \tilde{P}). \tag{31.25}$$

For $\nu + 1 < I(B_-, \tilde{P})$ there is one compressive solution; if $\nu + 1$ exceeds the right-hand side of (31.25), there are no compressive solutions.

There are some important practical cases in which (31.7) can be integrated explicitly, even if the stress–strain law is left in the generic form (31.5) (Antman 1979, 1995). One example is the equation of the catenary, in which the vertical load on the string is its own weight, that is (see (28.5)),

$$\mathbf{f}(x) = -(\rho A)(x)g\mathbf{j}.$$

$$F(x) = \int_0^x (\rho A)(\sigma)g\, d\sigma, \tag{31.26}$$

where $(\rho A)(x)$ is the mass density per unit unstretched length and g is the acceleration due to gravity. The coordinates x^*, y^* of any point of the deformed state are related to the displacements u and v by the relations

$$x^* = x + u, \quad y^* = v.$$

A first integral of (31.7) is

$$T\cos\theta = \alpha, \tag{31.27a}$$

$$T\sin\theta = \beta + F(x), \tag{31.27b}$$

where α and β are constants, to be determined later. Then, from (31.27), we derive

$$\tan\theta = \frac{dy^*}{dx^*}(x) = \frac{\beta + F(x)}{\alpha}. \tag{31.28}$$

Moreover, if we introduce the elongation $\nu = e + 1$ (see (28.1)), and we write the constitutive equation in the equivalent form

$$T = f(e) = N[\nu(x), x], \tag{31.29}$$

the properties of $f(e)$, postulated above, ensure that we can invert (31.29) and write

$$\nu(x) = \hat{\nu}[T(x), x],$$

as we have done in deriving (28.9). Now the elongation $\nu(x)$ is related to x^*, y^* by the geometrical condition

$$\frac{dx}{dx^*} = \frac{1}{\nu(x)}\sqrt{1 + \left(\frac{dy^*}{dx^*}\right)^2}, \tag{31.30}$$

and hence, on recalling (31.27a), we can write (31.30) in the alternative form

$$\nu(x) = \hat{\nu}\left[\alpha\sqrt{1 + \left(\frac{dy^*}{dx^*}\right)^2}, x\right]. \tag{31.31}$$

As $F(x)$ is invertible, (31.28) is equivalent to

$$x = F^{-1}\left(\alpha\frac{dy^*}{dx^*} - \beta\right). \tag{31.32}$$

By differentiating (31.28) with respect to x^* and substituting (31.31) and (31.32) in the resulting equation, we get

$$\alpha\frac{d^2y^*}{dx^{*2}} = \frac{F'\left[F^{-1}\left(\alpha\frac{dy^*}{dx^*} - \beta\right)\right]\sqrt{1 + \left(\frac{dy^*}{dx^*}\right)^2}}{\hat{\nu}\left[\alpha\sqrt{1 + \left(\frac{dy^*}{dx^*}\right)^2}, F^{-1}\left(\alpha\frac{dy^*}{dx^*} - \beta\right)\right]}. \tag{31.33}$$

This is the differential equation of the elastic catenary. Since the right-hand side of (31.33) is strictly positive, it follows that $\alpha(d^2y^*/dx^{*2}) > 0$, which confirms the expected result that a vertically loaded string assumes an equilibrium configuration strictly concave in the positive direction of the load.

Equation (31.33) can, in principle, be integrated by quadratures. The two integration constants and the two unknown parameters α and β are found by recourse to (31.30) and (31.33) and taking

$$y^*(0) = 0, \quad y^*(a) = b, \quad x(0) = 0, \quad x(a) = 1 \tag{31.34}$$

as the boundary conditions of the enlarged system. The use of the auxiliary equations (31.28)–(31.31) allows us to reduce the problem to the single equation (31.33) with the conditions

$$y^*(0) = 0, \quad y^*(a) = b, \quad \frac{dy^*}{dx^*}(0) = \frac{\beta}{\alpha},$$

$$\int_0^a \frac{\sqrt{1 + \left(\frac{dy^*}{dx^*}\right)^2}}{\hat{v}\left[\alpha\sqrt{1 + \left(\frac{dy^*}{dx^*}\right)^2}, \; F^{-1}\left(\alpha\frac{dy^*}{dx^*} - \beta\right)\right]} dx^* = 1. \tag{31.35}$$

If the string is uniform so that $F' = \gamma$, and if the material is inextensible so that $\hat{v} \equiv 1$, then (31.33) reduces to the classical equation of the catenary

$$\alpha \frac{d^2y^*}{dx^{*2}} = \gamma\sqrt{1 + \left(\frac{dy^*}{dx^*}\right)^2}. \tag{31.36}$$

Another significant application of the theory is the integration of the equation for a suspension bridge, in which the force per unit of horizontal length in the deformed configuration is prescribed. In this case we can take

$$\mathbf{p}(x) = -G'[x^*(x)]\frac{dx^*}{dx}\mathbf{j}, \quad G(0) = 0, \tag{31.37}$$

with $G'(x^*) > 0$. Then the function F in (28.5) is given by

$$F(x) = G[x^*(x)]. \tag{31.38}$$

If we have $x = x(x^*)$ as the arc length in the reference state, the abscissa of which in the deformed configuration is x^*, then the equivalent of (31.28) is

$$\tan\theta = \frac{dy^*}{dx^*} = \frac{\beta + G(x^*)}{\alpha}. \tag{31.39}$$

Then, for $\alpha \neq 0$, we obtain y^* by integrating (31.39). In particular, if the load is uniform so that we have $G(x^*) = \mu x^*$, we find that y^* describes a parabola, no matter what the form of the constitutive function $\hat{v}$.

32. Inverse Problems in Strings

In studying the equilibrium of a string it is tacitly assumed that the position of the ends and the loads are prescribed, whereas the tension and the strained configuration are unknown. This is not, however, the only way in which the problem can be formulated, nor the most interesting for applications. In fact, very often, we have a curve connecting two fixed points, say (0, 0) and (0, 1) in the xz plane, which represents the central line of a beam loaded by given loads acting tangentially and perpendicularly to the curve. If the profile of the curve is fixed, and we know the bending stiffness and the extensional stiffness of the beam, we should be able, at least in principle, to determine the bending moment and the axial force at any cross-section of the bar, and also the elastic displacement of any point of the central line. Suppose that the shape of the curve is slightly modified, and that, as a consequence, the resulting stress and couple change, it is well known that these changes are, in general not necessarily continuous. In particular, we may ask whether it is possible to adjust the curve in such a way that, while leaving the magnitude and the direction of the loads unchanged, but not their points of application, the bending moment and the shear force can be made to vanish at each cross-section, and the normal force to be made purely compressive. A curve satisfying these properties is called a *funicular of compression*, and determining it was of vital importance for building structures such as masonry arches, tunnels, and domes, the basic material of which (stone) is unable to sustain tension. The problem of finding the compressive funicular curve of a one-dimensional structure is exactly that of finding the form of a string supporting a prescribed system of loads with a pure compressive axial force. Of course, we must not anticipate that a solution always exists, or that it is unique.

In order to give an idea of how the problem may be treated, let us consider a plane arch, of variable cross-section A, the central line of which, in the deformed configuration, is determined by a vector $\mathbf{r} = \mathbf{r}(s)$, where in this state s is an arc length parameter ($s \in [0, 1]$). The unit tangent to the curve $\mathbf{r}$ at s is the vector

$$\mathbf{e} = \frac{\mathbf{r}'(s)}{|\mathbf{r}(s)|}. \tag{32.1}$$

A funicular arch has the defining property that the resultant force $\mathbf{n}(s)$ is purely tangential to $\mathbf{r}$ at s:

$$\mathbf{n}(s) = n(s)\mathbf{e}(s). \tag{32.2}$$

Without loss of generality we can take the boundary conditions to be

$$\begin{aligned} \mathbf{r}(0) &= \mathbf{0}, \\ \mathbf{r}(1) &= a\mathbf{i} + b\mathbf{j}, \quad a \geqslant 0, b \geqslant 0, \end{aligned} \tag{32.3}$$

where $\mathbf{i}$ and $\mathbf{j}$ are the first two unit vectors of a fixed orthonormal basis $\{\mathbf{i}, \mathbf{j}, \mathbf{k}\}$. We confine our attention to planar solutions, characterized by the property that $\mathbf{r}$ and $\mathbf{n}$ remain in the $\mathbf{ij}$ plane.

In a funicular arch the couple resultant $\mathbf{m}(s)$ is identically zero, and therefore the equilibrium equation reduces to (see (19.28))

$$\mathbf{n}' + \mathbf{f} = \mathbf{0},$$

where $\mathbf{f}(s)$ denotes the exterior force per unit length. In the planar case, $\mathbf{f}$ is a vector lying in the $\mathbf{ij}$ plane. The equilibrium equation can be written by taking its components with respect to a local basis $\mathbf{d}_1, \mathbf{d}_3$, such that $\mathbf{d}_3 \equiv \mathbf{e}$ and $\mathbf{d}_1$ is perpendicular to $\mathbf{d}_3$ and oriented so that the pair $(\mathbf{d}_1, \mathbf{d}_3)$ is right-handed. With respect to the local basis, the vector $\mathbf{n}$ has the components $(0, n(s))$, while (f_1, f_3) are the components of $\mathbf{f}$. Denoting the curvature by κ', we can write

$$\kappa' n(s) = -f_1, \tag{32.4}$$

$$n'(s) = -f_3. \tag{32.5}$$

Once $\mathbf{f} \equiv (f_1, f_3)$ is given, then (32.4) and (32.5) yield the profile of the arch and the axial load $n(s)$. By integrating (32.4) and (32.5) we obtain (Farshad 1976)

$$n(s) = C \exp\left(\int \kappa' \frac{f_3}{f_1}\, ds\right), \tag{32.6}$$

for $f_1 \neq 0$, and

$$\kappa' \equiv 0, \quad n(s) = -C\left(\int f_3\, ds\right), \tag{32.7}$$

for $f_1 = 0$.

As for the state of deformation, the position vector $\mathbf{r}(s)$ may be represented as the vector sum

$$\mathbf{r}(s) = \mathbf{R}(s) + \mathbf{u}(s), \tag{32.8}$$

where $\mathbf{R}(s)$ is the position vector of the central line in its undeformed state and $\mathbf{u}$ is the displacement vector. We put $u_1 = \mathbf{u} \cdot \mathbf{d}_1$, $u_2 = \mathbf{u} \cdot \mathbf{d}_3$ to denote the displacement components along the normal and the tangent to the curve. On assuming that the deformations are small, we define the quantities

$$\varepsilon_3 = u_3' + \kappa' u_1, \tag{32.9}$$

$$\gamma_1 = u_1' - \kappa' u_3, \tag{32.10}$$

as the "strains" of our problem. In particular, ε_3 is the extension and γ_1 the shear strain.

If we assume that the material is linearly elastic, the axial force $n(s)$ is related to ε_3 by the equation $n = EA\varepsilon_3$. On the other hand, we have excluded the presence of shear forces, and therefore, if we wish the theory to apply to rods of finite shear rigidity, we must impose the constraint

$$\gamma_1 = u_1' - \kappa' u_3 = 0. \tag{32.11}$$

In order to derive the consequences of (32.6), (32.7), and (32.11), we define the following nondimensional quantities

$$w = \frac{u_1}{R_0}, \quad v = \frac{u_3}{R_0}, \quad ds = \rho\, d\phi, \quad Q = \frac{n}{EA}, \quad a = \frac{A}{A_0},$$

where A_0 is a reference cross-sectional area, R_0 is a reference radius of curvature, and ρ is the local radius of curvature ($\kappa' = 1/\rho$). In this case the geometrical constraint (32.11) becomes

$$v = w_\varphi,$$

and, consequently, (32.9) becomes $\varepsilon_3 = (1/\rho)\,(w_{\varphi\varphi} + w)$, and the dimensionless axial force is given by

$$Q = \frac{a}{\rho}(w_{\varphi\varphi} + w). \tag{32.12}$$

In the nondimensional variables, equation (32.6) becomes

$$Q = C\exp\left(\int \frac{f_3}{f_1}\, d\varphi\right), \tag{32.13}$$

while substitution of (32.6) in (32.4) yields

$$\rho = -\frac{C}{f_1}\exp\left(\frac{f_3}{f_1}\, d\varphi\right). \tag{32.14}$$

On differentiating (32.12) with respect to φ and replacing w_φ by v we obtain

$$v_{\varphi\varphi} + v = \left(\frac{\rho}{a} Q\right)_\varphi. \tag{32.15}$$

For the purpose of illustrating the consequences of the equations just derived, let us assume that the load is hydrostatic with constant magnitude p_0, so that we have $f_3 = 0$ and $f_1 = p_0$. Relations (32.13) and (32.14) readily show that ρ is a constant which can be put equal to R_0, and Q is also constant and equal to $-(p_0 R_0/EA_0)$. Moreover, (32.12) and (32.15) become

$$w_{\varphi\varphi} + w = -\frac{p_0 R_0^2}{EA_0 a} = -\frac{\beta}{a}, \quad v_{\varphi\varphi} + v = -\beta \frac{a_\varphi}{a^2}, \tag{32.16}$$

after the substitution

$$\beta = \frac{p_0 R_0^2}{EA_0}.$$

Let us suppose that the ends of the arch are clamped at the positions $\pm\varphi$. At these ends the boundary conditions are $w = v = 0$. In order to find a solution, we assume, provisionally, that the cross-section is a constant, so that $a = 1$. As a solution to (32.16) we obtain

$$w(\varphi) = -\beta\left(1 - \frac{\cos\varphi}{\cos\varphi_0}\right), \quad v(\varphi) = \beta \frac{\sin\varphi}{\cos\varphi_0}.$$

However, the expression for $v(\varphi)$ does not satisfy the boundary condition $v(\pm\varphi_0) = 0$. Hence a funicular arch with clamped ends must have a variable section. In order to take this into account we introduce $a(\phi)$ as an even function of ϕ, and a_φ as an odd function of φ, and we expand the ratio a_φ/a^2 as a Fourier series of the form

$$\frac{a_\varphi}{a^2} = \sum_{n=1}^{\infty} a_n \sin n\varphi, \tag{32.17}$$

the coefficients a_n being unknown, for the moment. By integrating (32.17) we obtain

$$a(\varphi) = \frac{1}{1 - \sum_{n=1}^{\infty} \frac{2}{n} a_n \sin^2 \frac{n\varphi}{2}}. \tag{32.18}$$

Once we have $a(\varphi)$, we can obtain $v(\varphi)$ and $w(\varphi)$ by integration of (32.16), utilizing the boundary conditions. It remains to determine the coefficients a_n, but we still have the condition $v = w_\varphi$, which yields a recursive relation for the a_n terms.

The problem of the funicular arch is a particular case of the inverse problem for strings. More frequently, the problem is one of optimization; that is to say, one of finding the law of variation in the cross-section that minimizes or maximizes a given functional of the solution.

A very simple case, but nevertheless of interest for applications, is that of the suspended rope with minimum elongation (Verma and Keller 1984). A heavy rope hangs vertically from a fixed support and carries a weight W at its lower end. The rope is stretched elastically by the load and its own weight, and we wish to minimize the elongation of the lower end as a result of suitable shaping of the cross-section.

Let us assume that the unstretched length L, the total volume V, the mass density ρ, and the elastic modulus E are given, so that the only variable is the cross-sectional area $A(x)$, subject to the volume condition

$$\int_0^L A(x)\,dx = V, \tag{32.19}$$

where x is the distance from the upper end in the unstretched state. We seek the distribution of area $A(x)$, satisfying (32.19), which minimizes the total elongation.

To solve this problem, we introduce the downward displacement $y(x)$ and assume that the associated strain dy/dx is small, so that Hooke's law applies. The stress at x is just the total downward force at x divided by the area $A(x)$, so that Hooke's law yields

$$\frac{dy}{dx} = \left[W + \int_0^x \rho g\, A(x')\,dx'\right](EA(x))^{-1}. \tag{32.20}$$

As the top of the rope is fixed, the displacement there is zero, that is, $y(0) = 0$, and therefore the solution of (32.20) satisfying this condition is

$$y(x) = \int_0^x \left[W + \rho g \int_{x'}^L A(x'')\,dx''\right](EA(x'))^{-1}\,dx', \tag{32.21}$$

and we wish to find $A(x)$ that minimizes $y(L)$ subject to (32.19).

For this purpose, we introduce $B(x)$ defined by

$$B(x) = \frac{W}{\rho g} + \int_x^L A(x')\,dx', \tag{32.22}$$

and consider the functional

$$\begin{aligned} y(L) - \lambda\left[\int_0^L A(x)\,dx - V\right] &= \int_0^L \left\{\frac{\rho g}{EA}\left[\frac{W}{\rho g} + \int_x^L A(x')\,dx'\right] - \lambda A\right\} dx + \lambda V \\ &= \int_0^L \left\{-\frac{\rho g B}{EB'} + \lambda B'\right\} dx + \lambda V. \end{aligned} \tag{32.23}$$

The functional is stationary provided that the Euler equation

$$\left(\frac{d}{dx}\frac{\partial}{\partial B'} - \frac{\partial}{\partial B}\right)\left(-\frac{\rho g B}{EB'} + \lambda B'\right) = 0 \tag{32.24}$$

holds. On carrying out the differentiations we obtain

$$BB'' = (B')^2. \tag{32.25}$$

By writing $B''/B' = B'/B$ and integrating, we get $B' = -KB$, where K is a constant. The solution of this equation is

$$B(x) = B(L)\, e^{K(L-x)} \tag{32.26}$$

and on setting $x = L$ in (32.26) and in (32.22), and equating the results, we get $B(L) = W/\rho g$. Then (32.22) gives $A(x) = -B'(x)$, so (32.26) yields

$$A(x) = \frac{KW}{\rho g}\, e^{K(L-x)}. \tag{32.27}$$

In order to determine K, we substitute (32.27) in (32.19) and find the value

$$K = L^{-1} \ln\left(1 + \frac{\rho g V}{W}\right), \tag{32.28}$$

so that (32.27) then becomes

$$A(x) = \frac{W}{\rho g L} \ln\left(1 + \frac{\rho g V}{W}\right)\left(1 + \frac{\rho g V}{W}\right)^{(1-x/L)}. \tag{32.29}$$

This is the only function that satisfies (32.19) and makes $y(L)$ stationary. Then, if $y(L)$ has a minimum, this is the function that yields that minimum. By using (32.29) in (32.21) we find

$$y(L) = \frac{\rho g L^2}{E \ln(1 + \rho g V/W)}. \tag{32.30}$$

In a uniform rope $A = V/L =$ constant, and the displacement $y_u(L)$ is given by

$$y_u(L) = \frac{\rho g L^2}{E}\left(\frac{1}{2} + \frac{W}{\rho g V}\right). \tag{32.31}$$

The ratio

$$\frac{y(L)}{y_u(L)} = \frac{1}{\left(\frac{1}{2} + W/\rho g V\right) \ln\left(1 + \frac{\rho g V}{W}\right)}$$

decreases monotonically as the ratio $\rho g V/W$ increases. It is unity at $\rho g V/W = 0$ and falls to zero for $\rho g V/W = \infty$.

The strain dy/dx is given by

$$\frac{dy}{dx}(x) = -\frac{\rho g B}{E B'} = \frac{\rho g}{E K}, \tag{32.32}$$

and is constant for the optimal rope; therefore, as a consequence, the stress is also constant.

This last property may seem an incidental consequence of the problem just considered, but it in fact reflects a rather general feature of the distributions of optimal stresses in homogeneous media. The stresses corresponding to the optimal solutions tend to become the most uniform that is allowably possible. This result reflects the intuitive expectation that, in structures in which the overall rigidity is minimized while constant weight is maintained, the stress tends to adopt a sort of "cooperativeness."

The variational method can be extended to other optimization problems by writing the Euler equations for each case, although integrating them is often difficult. An example of an application is that of a cable of span L anchored between two supports A and B at different levels, under the action of a system of vertical loads (Figure 32.1) (Thevendran and Wang 1985). The straight line joining the supports makes an angle θ with the horizontal, measured clockwise, and a vertical load $w(x)$, per unit horizontal length, is to be supported by the cable. We assume that the specific weight of the material is prescribed, so that, if we know the total weight of the cable, we also know its volume. In addition, a constitutive constraint must be applied, which requires the stress everywhere in the cable to be equal to a maximum permissible value σ_0. At the same time, elastic strains are neglected and the material is regarded as rigid. The problem we pose is then that of finding the minimum volume of material that must be employed in order to carry the load $w(x)$, everywhere at constant stress σ_0.

Once the function $w(x)$ is given, we know that its resultant W, acting at a horizontal distance aL from A, is

$$aL = \frac{1}{W}\int_0^L x\,w(x)\,dx, \quad a \leqslant 1, \tag{32.33}$$

where

$$W = \int_0^L w(x)\,dx.$$

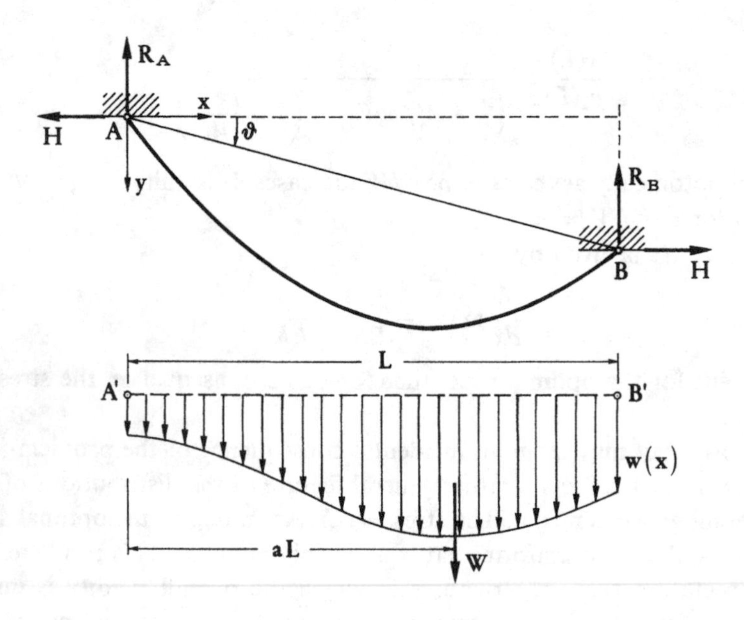

Fig. 32.1

If H is the (still unknown) horizontal reaction exerted by the cable on the supports, the vertical reactions at the supports, R_A and R_B, are

$$R_A = W(1-a) + H\tan\theta$$
$$R_B = Wa - H\tan\theta. \qquad (32.34)$$

The equation $y = y(x)$ of the cable is determined by the condition that its profile is the funicular curve of the loads:

$$y(x) = \frac{M(x)}{H} + x\tan\theta = h(x) + x\tan\theta,$$

where $M(x) = R_A x - \int_0^x w(\xi)(x-\xi)\,d\xi$, has the static interpretation of bending moment in a simply supported horizontal beam A′B′ loaded by the vertical distribution of forces $w(x)$ (Figure 32.1).

The total volume Φ of the cable is given by the relation

$$\Phi\sigma_0 = \int_0^L H(1+S^2)^{\frac{1}{2}}[1+y'^2(x)]^{\frac{1}{2}}\,dx, \qquad (32.35)$$

where S is the maximum slope of the cable. This maximum slope is located at either of the supports A or B, and is not known in advance. It can be written as

$$S = \max\{S_A = R_A/H,\ S_B = R_B/H\}, \qquad (32.36)$$

or, alternatively, by (32.34),

$$S = \frac{W}{2H} + K\left[\frac{W(1-2a)}{2H} + \tan\theta\right] \qquad (32.37)$$

where the parameter K is defined as follows

$$K = \begin{cases} +1, & \text{for } R_A > R_B \quad \text{and } S = S_A \\ -1, & \text{for } R_A < R_B \quad \text{and } S = S_B \\ \ \ 0, & \text{for } R_A = R_B \quad \text{and } S = \dfrac{W}{2H}. \end{cases} \qquad (32.38)$$

Let us consider the case in which θ is fixed. For minimum cable weight, we find the stationary value of (32.35) with respect to the variation of H. This condition reads

$$\sigma_0\frac{d\Phi}{dH} = \int_0^L \left\{\left[(1+S^2)^{\frac{1}{2}}(1+y'^2)^{\frac{1}{2}} + \frac{HS(1+y'^2)}{(1+S^2)^{\frac{1}{2}}}\right]\frac{dS}{dH}\right.$$
$$\left. + H(1+S^2)^{\frac{1}{2}}\frac{y'}{(1+y')^{\frac{1}{2}}}\frac{dy'}{dH}\right\}dx = 0. \qquad (32.39)$$

Now, by differentiation with respect to x, we have

$$y'(x) = h'(x) + \tan\theta, \qquad (32.40)$$

and therefore from this equation and (32.37) we can calculate

$$\frac{dy'}{dH} = \frac{dh'}{dH} = \frac{-(y' - \tan\theta)}{H}, \tag{32.41}$$

$$\frac{dS}{dH} = -\frac{W}{2H^2}[1 + K(1 - 2a)] = \frac{-(S - K\tan\theta)}{H}. \tag{32.42}$$

Substitution of these equations in (32.38) and some simplifications yields

$$\int_0^L \{1 - (y'S)^2 + \tan\theta[KS(1 + y'^2) + y'(1 + S^2)]\}\frac{dx}{(1 + y'^2)^{\frac{1}{2}}} = 0, \tag{32.43}$$

which furnishes the necessary optimality condition. It is worth noting that, for $\theta = 0$, (32.40) and (32.43) reduce to

$$y' = h' \quad \text{and} \quad \int_0^L \{1 - (h'S)^2\}\frac{dx}{(1 + h'^2)^{\frac{1}{2}}} = 0. \tag{32.44}$$

Unfortunately, the solution of (32.43) is only possible in some particular cases.

33. Unilateral Problems for Strings

The prototype of a unilateral problem for a string is the determination of the equilibrium configuration of a string stretched between two fixed pegs a distance ℓ apart, but constrained to pass over a given area A in the plane of the string called the *obstacle*[49] (Figure 33.1).

We introduce rectangular coordinates (x, y) such that one peg is at the origin $(0, 0)$ and the other is on the x axis at $(\ell, 0)$. In this system of coordinates the upper part of the boundary of the obstacle A is a curve described by the equation $y = \psi(x)$.

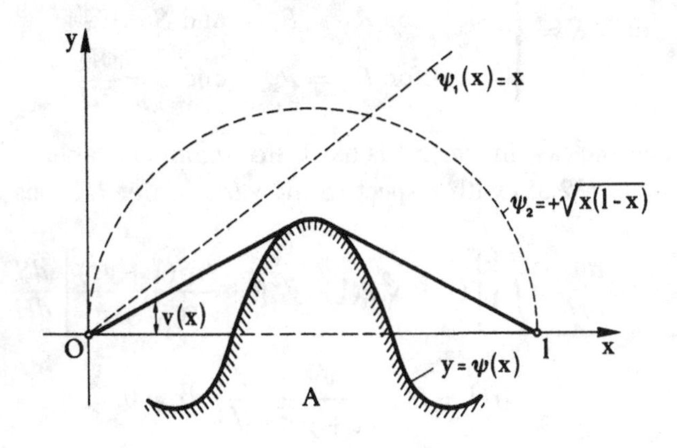

Fig. 33.1

[49] The first formulation of this problem was due to Kneser (1900, p. 178).

We assume that the weight of the string is negligible, and that its transverse displacement $v(x)$ and the slope $v'(x)$ are sufficiently small that we can assume the approximation

$$v'(x) = \tan\alpha \approx \sin\alpha, \tag{33.1}$$

α being the slope at x, and disregard the variation of the tension T resulting from longitudinal elongation. This allows us to take $T \approx T_0 =$ constant.

With these assumptions, the most natural way to formulate the equilibrium problem is to look for the configuration $v(x)$ that minimizes the strain energy

$$\mathscr{W} = \frac{T_0}{2}\int_0^{\ell} v'^2(x)\,dx, \tag{33.2}$$

among all the configurations that conform to the geometrical constraints $v(0) = v(\ell) = 0$ and $v \geqslant \psi$.

However, the problem is not completely defined, because we also require that $v'(x)$ is square integrable on $(0, \ell)$ in order for the integral (33.2) to have meaning; it is thus natural to put $v \in H_0^1(0, \ell)$. In addition, on assuming that $\psi(x)$ is continuous in $(0, \ell)$, we require the conditions $\psi(0^+) < 0$ and $\psi(0^-) < 0$. These hypotheses ensure that the set $K \equiv \{v \in H_0^1(0, \ell);\ v(x) \geqslant \psi(x)\}$ is a nonempty convex subset of $H_0^1(0, \ell)$. For instance, K would be empty if $\psi(x)$ were one of the two functions ψ_1 or ψ_2 represented in Figure 33.1, because of the conditions that $\psi_1(1) = 1 > 0$ and $\psi_2 \notin H_0^1(0, \ell)$.

Under these assumptions, the variational problem can be reduced to the following: to find a $u \in K$ such that $\mathscr{W}(u) \leqslant \mathscr{W}(v)$, $\forall\, v \in K$, with $\mathscr{W}(u)$ given by (33.2) and K defined as above. As $\mathscr{W}$ is convex and continuous, and K is a convex and closed subset of $H_0^1(0, \ell)$, the problem then admits solutions. The solution is unique because $\mathscr{W}(u)$ is strictly convex (see Baiocchi and Capelo 1978, Chap. 1).

Having proved the existence of a unique solution, a second important consideration is the regularity of the solution. In particular, we might expect that, by increasing the regularity of ψ, u will continue to become smoother. However, this regularity cannot be pushed beyond a certain limit. Figure 33.2 shows an example in which $\psi \in C^\infty(0, \ell)$ and $u \notin C^2(0, \ell)$, because u'' is only piecewise continuous. This bounded regularity is typical of all problems involving obstacles.

The dynamic case, that is, the motion of a vibrating string in contact with an obstacle, is more complicated because we must account for the impulsive forces applied at the points of the string impinging against the obstacle. In order to simplify the analysis of the problem, we assume (Amerio and Prouse 1975) that, in its equilibrium configuration, the string is stretched between the points $(0, 0)$ and $(0, \ell)$ of the x axis, and that the instantaneous position of a generic particle is described by the transverse displacement $y = y(x, t)$. A further simplification can be made by assuming that no exterior forces act on the string; that the obstacle is constituted by a plane wall $y = K > 0$, which confines the string to the half-plane $y \leqslant K$; and, finally, that the impact is perfectly elastic, so that, if (x_0, t_0) is a contact point, where $y = K$, we have

$$y_t(x_0, t_0^+) = -y_t(x_0, t_0^-) < 0. \tag{33.3}$$

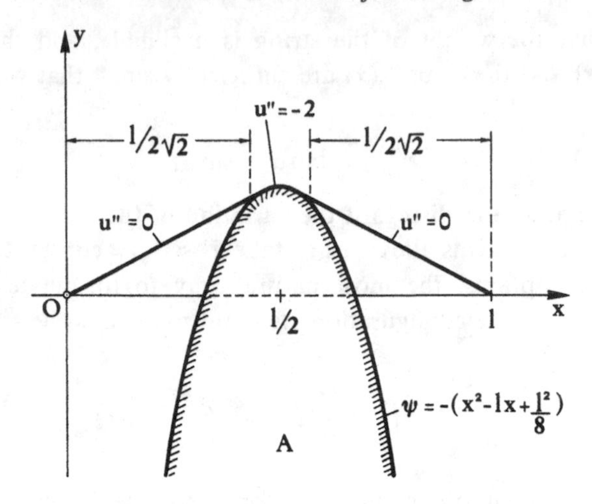

Fig. 33.2

After making a suitable choice of the units for length and time, the equation governing the free motion of the string is

$$y_{tt} - y_{xx} = 0 \quad \text{for} \quad 0 < x < \ell, \ t > 0, \tag{33.4a}$$

with the boundary condition

$$y(0, t) = y(\ell, t) = 0, \tag{33.4b}$$

and the initial conditions

$$y(x, 0) = \varphi(x), \quad y_t(x, 0) = \psi(x), \tag{33.5}$$

where φ and ψ represent the initial displacement and the initial velocity, respectively. It is assumed that φ is continuously differentiable and ψ is continuous, with the exception of a finite number of discontinuities of the first kind. Moreover, we require the condition that $\varphi(x) < K$.

Let us assume, for the moment, that there is no obstacle, so that the motion is that of a free string. Then the solution to (33.4) and (33.5) is just d'Alembert's solution, given by

$$y(x, t) = \frac{1}{2}[\varphi(x + t) + \varphi(x - t)] + \frac{1}{2}\int_{x-t}^{x+t} \psi(\bar{x})\, d\bar{x}. \tag{33.6}$$

The functions $\varphi(x)$ and $\psi(x)$ are given initially for $0 \leqslant x \leqslant \ell$. However, they can be extended as odd functions about $x = 0$ and $x = \ell$, and regarded as periodic functions of period 2ℓ defined for $-\infty < x < \infty$. Equation (33.6) then gives the solution for all times. For instance, the solution at (x_0, t_0) is defined by the initial data contained in the interval M_0N_0 (Figure 33.3). This is obtained by intersecting the x axis with the characteristic straight lines issuing from (x_0, t_0).

Provided that t_0 is sufficiently small, the condition $\varphi(x) < K$ ensures the result $y(x, t) < K$ in the triangle with a base M_0N_0 and a vertex at (x_0, t_0). This triangle is also called the *domain of dependence* of the point (x_0, t_0).

We now denote by $t(x_0)$ the maximum value of t for which the condition $y(x, t) \leqslant K$ applies over the whole triangle. This means that, taking the value $t' > t(x_0)$, and considering the triangle with vertex (x_0, t'), there is a point in this triangle at which $y > K$. Hence $t(x_0)$ represents the maximum time for which the motion of the point x_0 is not influenced by the action of the wall. If we define a value of t in this way for each point in the range $0 \leqslant x \leqslant \ell$, we can construct a line λ having the equation $t = t(x)$, and characterized by the property that below λ the motion is free. For this reason, λ is called the *line of influence of the wall*. It is possible to show that λ is a Lipschitz continuous function ($|t(x'') - t(x')| \leqslant |x' - x''|$) and its graph consists of a finite number of arcs with $|t'(x)| < 1$ and of characteristic segments. In addition, we know both y and y_t on λ. The points of λ at which $y(x_0, t_0) = K$ and $y_t(x_0, t_0^-) > 0$ are called *points of impact*, and are characterized by the condition

$$y_t(x_0, t_0^+) = -y_t(x_0, t_0^-) < 0. \tag{33.7}$$

For times above λ we must again solve (33.4), but now with a view to the initial data prescribed on λ. As λ can be extended by reflection to the whole interval $-\infty < x < \infty$ as a periodic function of period 2ℓ, we can express $y(x, t)$ for $t > \lambda$ by a formula due to Riemann, which generalizes (33.6). Above λ, for small times, the motion is again free, and a second line of influence can then be defined and the procedure repeated. As the distance between two successive lines of influence is greater or equal to a quantity $\delta > 0$, the process of successive extensions can be iterated for all times.

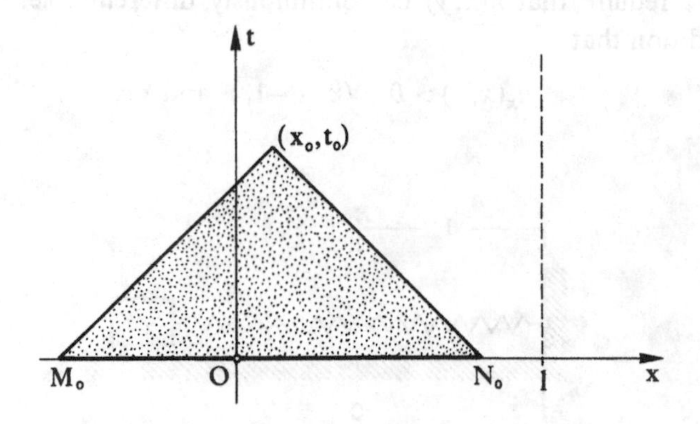

Fig. 33.3

34. Dynamical Problems for Strings

The analysis of the motion of a nonlinear elastic string is complicated by the fact that, even for smooth data, the solutions may exhibit discontinuities in the tension, slope, and velocity of the string, which are generically defined by *shocks*. Shocks might arise when the initial data are discontinuous, but, if the equations are nonlinear, very smooth data might evolve over time to generate wild solutions.

Shocks are even present in the simple problem of the motion of a mass point attached to a spring (Antman 1988a). A spring is confined to a horizontal groove along the x axis (Figure 34.1). The left-hand end of the spring is fixed to a wall at $x = -1$; the right-hand end is tied to a unit-mass point sliding along the horizontal x axis. We assume, for simplicity, that the natural, unstretched, length of the spring is unity, and that $x(t)$ denotes the position of the mass point with respect the point 0, at unit distance from the wall. The mass point is subjected to a restoring force which depends on the position $x(t)$ and the velocity $\dot{x}(t)$ of the mass point. We denote this force by $-n(x(t), \dot{x}(t))$, where n is a given function. The function n has the form

$$n(x, y) = \varphi'(x) + \nu g(x, y), \tag{34.1}$$

where

$$\varphi'(x) = n(x, 0), \quad \text{with} \quad \varphi'(0) = \varphi(0) = 0, \tag{34.2}$$

and

$$g(x, 0) = 0, \quad \nu \geqslant 0. \tag{34.3}$$

The term $\varphi'(x)$ represents the force exerted on the mass point when it occupies the position x and is not moving; g accounts for the forces that arise from internal friction between the spring and the groove when the mass point moves. Then the equation of motion for the mass is

$$\ddot{x}(t) + \varphi'(x(t)) + \nu g(x(t), \dot{x}(t)) = 0, \tag{34.4}$$

with some initial conditions on $x(t)$ and $\dot{x}(t)$.

The constitutive functions $\varphi'(x)$ and $g(x, \dot{x})$ cannot, however, be completely arbitrary, but must satisfy certain reasonable conditions of regularity and growth. More precisely, we require that $n(x, y)$ be continuously differentiable, and satisfies the further condition that

$$n_x(x, y) > 0 \quad \forall x > -1, \quad \text{and } \forall y, \tag{34.5}$$

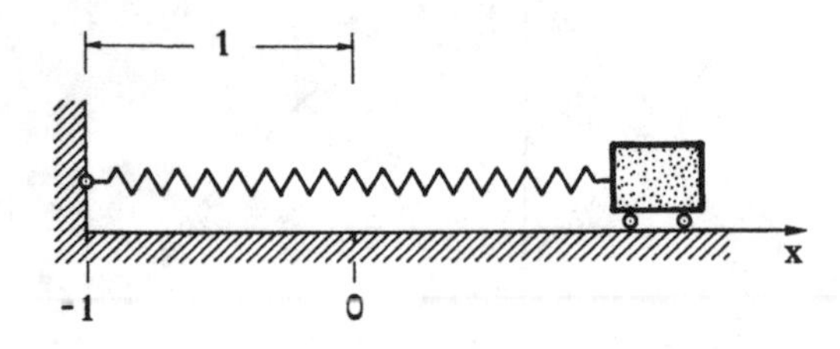

Fig. 34.1

so that an increase in the length of the spring is accompanied by an increase in the tensile force. In addition, we assume that, for each fixed y, we have

$$n(x, y) \to \begin{Bmatrix} +\infty \\ -\infty \end{Bmatrix} \text{ for } x \to \begin{Bmatrix} +\infty \\ -1 \end{Bmatrix}. \tag{34.6}$$

Conditions (34.5) and (34.6) yield corresponding conditions on φ and φ'. The function φ, which has the physical meaning of the stored energy of the spring, may have a graph like the one denoted by φ_1 or φ_2 in Figure 34.2. That $\varphi(x)$ need not approach $+\infty$ for $x \to -1$ explains the onset of shocks. As for the dissipative function $g(x, y)$, a natural requirement is

$$g_y(x, y) > 0 \quad \forall x, \tag{34.7}$$

so that, for $\nu > 0$, an increase in the end velocity of the spring is accompanied by an increase in the tensile force. We adopt a strong version of (34.7), which is required to hold when the spring is under compression ($x < 0$) and is being compressed ($\dot{x} < 0$). The condition is that there is a positive-valued function $\psi'(x)$ satisfying the condition

$$g(x, y) \leqslant \psi'(x)y \quad \text{for} -1 < x \leqslant 0, \ y \leqslant 0. \tag{34.8}$$

We note that, if (34.7) is an equality, then (34.5) holds for $-1 < x \leqslant 0$ and $y \leqslant 0$ if, and only if, we have $\psi''(x) \leqslant 0$. For definiteness, we also take $\psi(0) = 0$.

The qualititative properties of the solutions of (34.4) may be described by means of a representation in the phase plane. If the motion is undamped, then $\nu = 0$ and, after multiplying (34.4) by $\dot{x}(t)$ and integrating the resulting equation, we obtain the energy equality

$$\tfrac{1}{2}\dot{x}(t)^2 + \varphi(x(t)) = \tfrac{1}{2}\dot{x}(0)^2 + \varphi(x(0)), \tag{34.9}$$

which holds over time intervals for which $x(t)$ is twice differentiable and with $x(t) > -1$. On putting $y(t) = \dot{x}(t)$, the phase portrait of (34.4) consists of the family of curves

$$\mathscr{C}(E)\colon \ \tfrac{1}{2}y^2 + \varphi(x) = E = \text{constant}. \tag{34.10}$$

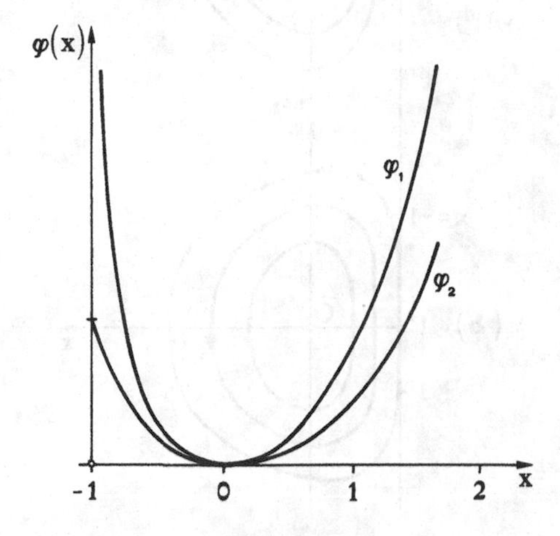

Fig. 34.2

If we have $\varphi(x) \to \infty$ for $x \to -1$, the phase portrait is represented by the family of curves shown in Figure 34.3(a), in which no trajectory can reach the wall $x = -1$. However, if φ has a finite limit, then the phase portrait has a completely different character, as illustrated in Figure 34.3(b). For large enough E, the trajectories actually touch the wall $x = -1$ tangentially, and we say that the spring suffers a *shock*. The problem of describing the mathematical properties of this shock then arises.

In order to do this we must reconsider the complete equation (34.4). Suppose that φ has a finite limit for $x \to -1$ (or, equivalently, that φ' is integrable for $-1 < x \leqslant 0$). Then g is called a *shock absorber* if, for each $\nu > 0$, there is no trajectory starting to the right of the line $x = -1$ that reaches this line. To determine the conditions characterizing a shock absorber, we integrate (34.4) from 0 to t to obtain

$$0 = \dot{x}(t) - \dot{x}(0) + \nu \int_0^t g(x(\tau), \dot{x}(\tau))\, d\tau + \int_0^t \varphi'(x(\tau))\, d\tau. \tag{34.11}$$

We study this equation over a time interval $[0, t]$ in which $(x, \dot{x})$ lies entirely in the quadrant $\{Q: -1 < x \leqslant 0, y < 0\}$, starting with the initial conditions $x(0) < 0$ and $\dot{x}(0) < 0$. Condition (34.8) implies the result

$$0 \leqslant \dot{x}(t) - \dot{x}(0) + \nu \int_0^t \psi'(x(\tau))\dot{x}(\tau)\, d\tau + \int_0^t \varphi'(x(\tau))\, d\tau. \tag{34.12}$$

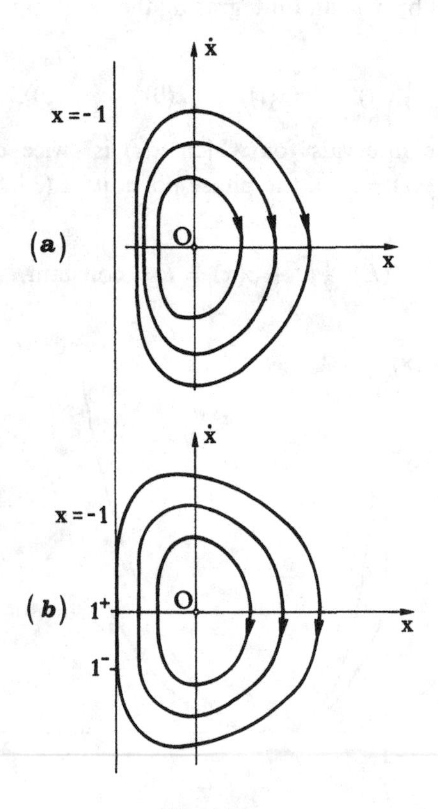

Fig. 34.3

As $\varphi'(0) = 0$, condition (34.5) also implies the result that $\varphi'(x) < 0$ for $x < 0$. Then inequality (34.12) yields

$$\nu\psi(x(t)) \geqslant \dot{x}(0) - \dot{x}(t) + \nu\psi(x(0)) - \int_0^t \varphi'(x(\tau))\,d\tau \geqslant \dot{x}(0) + \nu\psi(x(0)). \qquad (34.13)$$

Now, in Q the inequality $\dot{x}(0) < 0$ holds. If we have $\psi(x) \to -\infty$ for $x \to -1$, then we also have

$$x(t) \geqslant \psi^{-1}[\psi(x(0)) + \dot{x}(0)/\nu] > -1 \text{ for } \nu > 0, \qquad (34.14)$$

so that g is a shock absorber. If $\psi(x)$ has a finite limit α for $x \to -1$, then there are initial data, including those with $\dot{x}(0) < \nu\alpha$, for which shock must occur. The requirement that $\psi(x) \to -\infty$ for $x \to -1$ means that as the spring is pushed against the wall $x = -1$ the internal friction increases without bound.

Let us now suppose that the elastic energy $\varphi(x)$ has a finite limit for $x \to -1$, in which case the phase portrait of the undamped problem is as shown in Figure 34.3(b), and let g be a shock absorber. Our purpose is to characterize the solutions of the undamped problem as limits of the solution of the damped problem with a shock absorber as ν tends to zero. Let us assume that ν is small. Then, away from the wall $x = -1$, where the solution ceases to be regular, the trajectories of (34.4) are close to those of the undamped equation. However, while the trajectories corresponding to the dissipative case never touch the wall and slowly wind down to the equilibrium state $(x, y) = (0, 0)$, the behavior of the phase portrait in the limit as $\nu \to 0$ is entirely different. The trajectory starting from a point $(x(0), \dot{x}(0))$ sufficiently far from the origin will necessarily touch the axis $x = -1$ along a segment having the points 1^- and 1^+ as ends (Figure 34.3(b)). After 1^+ the velocity $\dot{x}(t)$ becomes positive and, therefore, the point mass will leave the wall. To determine what is happening on the segment $1^-, 1^+$, we revert to the case in which ν is small, and study the phase portrait of (34.4) in the strip $\{(x, y)\colon -1 < x < -1 + \varepsilon, y < 0\}$. For this purpose, let $y(t)$, defined over (t_1, t_2), correspond to a trajectory lying in this strip, and let $\hat{t}(y)$ denote its inverse, defined over (y_1, y_2), where $y_1 = y(t_1)$ and $y_2 = y_2(t_2)$. The time lapse in transversing this portion of trajectory is

$$t_2 - t_1 = \int_{y_1}^{y_2} \hat{t}'(y)\,dy = \int_{y_1}^{y_2} \frac{dy}{\dot{y}(\hat{t}(y))} = -\int_{y_1}^{y_2} \frac{dy}{\varphi'[x(\hat{t}(y))] + \nu g[x(\hat{t}(y)), y]}.$$

As we have $x < -1 + \varepsilon$, we find the inequality

$$t_2 - t_1 \leqslant \frac{y_2 - y_1}{\varphi'(-1 + \varepsilon)}. \qquad (34.15)$$

For $\varepsilon \to 0$, we also have $(t_2 - t_1) \to 0$. This means that the jump from 1^- to 1^+ occurs instantaneously, so that $\dot{x}(t)$ is discontinuous; a discontinuity in the velocity field is a defining property of shock.

The problem studied so far is also called the *zero-dimensional shock* (Antman 1988a), because it exhibits the typical features of shock phenomena in a system with only one degree of freedom. The treatment of continuous bodies is more difficult, but the general principles are similar (Shearer 1985). We consider an elastic string the configuration of which at time t is specified by a position vector $\mathbf{r}(x, t)$, with $0 \leqslant x \leqslant 1$. Let $n(x, y)$ denote the tension in the string, which acts tangentially to the string. In the absence of external forces, the equations of motion for the string are

$$\left[n(x, t)\,\frac{\mathbf{r}_x(x, t)}{|\mathbf{r}_x(x, t)|}\right]_x = \mathbf{r}_{tt}, \quad 0 < x < 1, \quad t > 0, \tag{34.16}$$

where we have assumed that the density per unit length is uniform and equal to unity. The tension $n(x, t)$ depends explicitly only on the scalar $|\mathbf{r}_x|$, which represents the instantaneous elongation of the string:

$$n(x, t) = T(|\mathbf{r}_x|). \tag{34.17}$$

The function $T(\xi)$ is defined for $0 < \xi < +\infty$, and satisfies the constitutive requirement

$$T'(\xi) > 0, \text{ for all } \xi, \tag{34.18}$$

illustrating that the tension increases with increasing local elongation. However, (34.18) is insufficient to render the problem well posed, and physically complete. It is convenient to let the reference configuration correspond to an unstretched equilibrium state, so that we have

$$T(1) = 0. \tag{34.19}$$

In addition, we require two other material properties. The first is that we have either (a) $T''(\xi) < 0$ for all ξ, or (b) there is $\xi_I > 1$ such that

$$(\xi - \xi_I)T''(\xi) > 0, \text{ for all } \xi > 0. \tag{34.20}$$

Case (a) would simplify the analysis, and may be included in (b) by simply taking ξ_I very large. We observe that a piecewise linear stress–strain law violates (34.20). The second material property is the inequality

$$T'(\xi) > \frac{T(\xi)}{\xi}, \text{ for all } \xi. \tag{34.21}$$

A typical graph of a function $T(\xi)$, satisfying all the above requirements, is illustrated in Figure 34.4.

With these assumptions, we formulate the problem of the planar motion of the string in the following form. We put $\mathbf{r}_x = (p, q)$, $\xi = |\mathbf{r}_x|$, and $\mathbf{r}_t = (u, v)$, and write (34.16) as the first-order system

$$\frac{\partial}{\partial t}\begin{bmatrix} p \\ q \\ u \\ v \end{bmatrix} = \frac{\partial}{\partial x}\mathbf{F}(p, q, u, v), \tag{34.22}$$

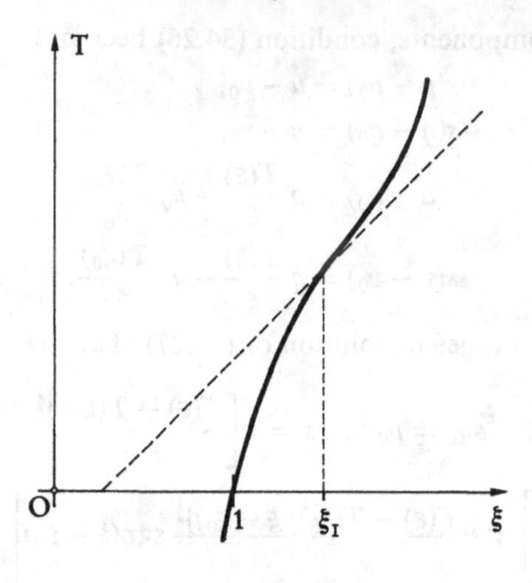

Fig. 34.4

where

$$\mathbf{F}(p, q, u, v) = \begin{bmatrix} u \\ v \\ pT(\xi)/\xi \\ qT(\xi)/\xi \end{bmatrix}. \tag{34.23}$$

The characteristic values for the system (34.21) are the eigenvalues of the Jacobian matrix $\mathbf{F}'(\mathbf{U})$, where $\mathbf{U} = (p, q, u, v)$, namely,

$$\lambda_{\pm}(\mathbf{U}) = \lambda_{\pm}(\xi) = \pm[T'(\xi)]^{\frac{1}{2}}, \quad \mu_{\pm}(\mathbf{U}) = \mu_{\pm}(\xi) = \pm\left[\frac{T(\xi)}{\xi}\right]^{\frac{1}{2}}. \tag{34.24}$$

In order to study the properties of the discontinuities of the solutions, we consider system (34.22) for the interval $-\infty < x < +\infty$ with $t > 0$, that is, the motion of an infinitely long string under given initial conditions. Let $\mathbf{U} = (p_0, q_0, u_0, v_0)$ be a given state of the string and $\mathbf{U}(p, q, u, v)$ be another state separated from the first by a line $x = st$, across which the solutions may be discontinuous. On this line the solutions must satisfy the Rankine–Hugoniot conditions

$$[\mathbf{U}]\, dx = [\mathbf{F}]\, dt \tag{34.25}$$

where $[\mathbf{U}]$ and $[\mathbf{F}]$ denote the jumps $(\mathbf{U} - \mathbf{U}_0)$ and $(\mathbf{F} - \mathbf{F}_0)$ and $\mathbf{U}$ and $\mathbf{F}$, respectively. An alternative form of (34.25) is

$$-s(\mathbf{U} - \mathbf{U}_0) = \mathbf{F}(\mathbf{U}) - \mathbf{F}(\mathbf{U}_0). \tag{34.26}$$

Written in terms of components, condition (34.26) becomes

$$\begin{aligned}
-s(p-p_0) &= u-u_0,\\
-s(q-q_0) &= v-v_0,\\
-s(u-u_0) &= p\,\frac{T(\xi)}{\xi}-p_0\,\frac{T(\xi_0)}{\xi_0},\\
-s(v-v_0) &= q\,\frac{T(\xi)}{\xi}-q_0\,\frac{T(\xi_0)}{\xi_0}.
\end{aligned} \tag{34.27}$$

We can distinguish two types of solution of (34.27). The first is

$$\begin{gathered}
\xi \neq \xi_0, \quad pq_0 = p_0q, \quad s = \pm\left[\frac{T(\xi)-T(\xi_0)}{\xi-\xi_0}\right]^{\frac{1}{2}},\\
\begin{bmatrix} u \\ v \end{bmatrix} = \begin{bmatrix} u_0 \\ v_0 \end{bmatrix} \mp \frac{\{[T(\xi)-T(\xi_0)](\xi-\xi_0)\}^{\frac{1}{2}}}{\xi_0}\operatorname{sgn}(\xi-\xi_0)\begin{bmatrix} p_0 \\ q_0 \end{bmatrix}.
\end{gathered} \tag{34.28}$$

The second of these equations shows that the slope p/q of the string is unchanged across the line $x = st$, while the tension $T(\xi)$ undergoes a jump. Moreover, only the tangential component of the velocity is discontinuous. These solutions are called *genuinely nonlinear shocks*.

Another solution of (34.27) is

$$\begin{gathered}
\xi = \xi_0, \quad s = \pm\left[\frac{T(\xi)}{\xi}\right]^{\frac{1}{2}} \text{ for } (p,q) = \xi(\cos\theta, \sin\theta),\\
\begin{bmatrix} u \\ v \end{bmatrix} = \begin{bmatrix} u_0 \\ v_0 \end{bmatrix} \pm \frac{[\xi T(\xi)]^{\frac{1}{2}}}{\xi_0}\begin{bmatrix} p_0 \\ q_0 \end{bmatrix} \mp [\xi T(\xi)]^{\frac{1}{2}}\begin{bmatrix} \cos\theta \\ \sin\theta \end{bmatrix}.
\end{gathered} \tag{34.29}$$

Across the line $x = st$, the slope of the string $tg\,\theta$ jumps, but the tension is continuous. These solutions are called *contact discontinuities*.

Equations (34.23) admit another type of solution, called *rarefaction waves*. These are solutions of (34.23) of the form $\mathbf{U} = \mathbf{U}(x/t)$, with $x/t = \lambda_{\pm}[\mathbf{U}(x/t)]$. For such solutions we must have

$$\mathbf{U}'(\eta) = \mathbf{w}_{\pm}(\mathbf{U}(\eta)), \tag{34.30}$$

where $\mathbf{w}_{\pm}(\mathbf{U})$ is the right eigenvector of $-\mathbf{F}'(\mathbf{U})$ corresponding to the eigenvalue $\lambda_{\pm}(\mathbf{U})$, normalized by the condition $\mathbf{w}_{\pm} \cdot \operatorname{grad}\lambda_{\pm} = 1$. Let us suppose that a rarefaction wave separates constant states $\mathbf{U}_0$ on the left from $\mathbf{U}$ on the right. Then (34.30) leads to

$$\begin{gathered}
\xi \neq \xi_0, \quad qp_0 = pq_0,\\
\begin{bmatrix} u \\ v \end{bmatrix} = \begin{bmatrix} u_0 \\ v_0 \end{bmatrix} - \frac{1}{\xi_0}\int_{\xi_0}^{\xi} \lambda_{\pm}(\nu)\,d\nu \begin{bmatrix} p_0 \\ q_0 \end{bmatrix}
\end{gathered} \tag{34.31}$$

with $\lambda_{\pm}(\xi) = \eta$.

Genuinely nonlinear shock and rarefaction waves are collectively referred to as *longitudinal waves*. These, however, must satisfy an additional "admissibility" condition, which is imposed in order to guarantee the uniqueness of the solution for the initial-value problem

$$\mathbf{U}_t = \mathbf{F}(\mathbf{U})_x, \quad -\infty < x < \infty, \quad t > 0,$$
$$\mathbf{U}(x, 0) = \begin{cases} \mathbf{U}_L, & \text{for } x < 0, \\ \mathbf{U}_R, & \text{for } x > 0, \end{cases} \tag{34.32}$$

where $\mathbf{U}_L \neq \mathbf{U}_R$ are the initial data. Problem (34.32) is known as the *Riemann problem* associated with system (34.22). The condition of admissibility requires that, if the shock $x = st$ separates into states $\mathbf{U}_0$ on the left and $\mathbf{U}_1$ on the right, then

$$s\left[\frac{T(\xi) - T(\xi_0)}{\xi - \xi_0} - \frac{T(\xi_1) - T(\xi_0)}{\xi_1 - \xi_0}\right] \geqslant 0 \tag{34.33}$$

for all ξ between ξ_0 and ξ_1.

Another condition that must be satisfied by the solutions of (34.16) is the following. Suppose that we put

$$\eta = \frac{1}{2}(\mathbf{r}_t \cdot \mathbf{r}_t) + \int_1^{|\mathbf{r}_x|} T(\nu)\, d\nu, \quad Q = -\mathbf{r}_t \cdot \frac{\mathbf{r}_x}{|\mathbf{r}_x|}, \tag{34.34}$$

which represent energy and energy flux, respectively. Then any smooth solution of (34.16) satisfies the conservation law

$$\eta_t + Q_x = 0, \tag{34.35}$$

while admissible genuinely nonlinear shocks satisfy the inequality

$$\eta_t + Q_x \leqslant 0. \tag{34.36}$$

More precisely, since these derivatives may not exist across a discontinuity, inequality (34.36) holds in the sense of distributions. Shock wave solutions of (34.32) satisfying (34.36) are not necessarily admissible in the sense of (34.32). However, it may be proven that they do satisfy the following *viscosity criterion*. Let us replace $T(\xi)$ by $\hat{T}_\varepsilon(\xi, \xi_t) = T(\xi) + \varepsilon\xi_t$, where the terms $\varepsilon\xi_t$ represent a small perturbation of $T(\xi)$ due to dissipation. Then, every admissible shock wave solution of (34.23) is the pointwise limit, for $\xi \to 0_+$, of a traveling wave solution of the modified system

$$\left[\hat{T}_\varepsilon \frac{\mathbf{r}_x}{|\mathbf{r}_x|}\right]_x = \mathbf{r}_{tt}. \tag{34.37}$$

All the results are remarkable in that they describe the behavior of solutions in the neighborhood of a line of discontinuity, but do not offer a systematic procedure for constructing them. The only case that may be treated in a fairly general way is the Riemann problem, but here also we have the connection between the initial data on the line $-\infty < x < \infty$, $t = 0$, and the solution is involved and not easily interpretable. Furthermore, the assumptions made so far are not always unconditionally acceptable. The solution of the Riemann problem rests heavily on assumption (34.21), which, in particular, establishes a fixed order for the characteristic speeds. However, (34.21) is not satisfied for all materials. A second questionable assumption is that of having considered $T(\xi)$ to be defined for all $\xi > 0$; namely, that the string may undergo arbitrarily large stretches without breaking. If we assume instead the interval $0 < \xi \leqslant \xi_{max}$, where ξ_{max} ($\xi_{max} < \xi_I$) corresponds to the tension $T(\xi_{max}) < \infty$

at which the string will break, then the Riemann problem can be solved for $\mathbf{U}_{\mathrm{L}}$ and $\mathbf{U}_{\mathrm{R}}$, but only in a restricted region of the space.

The general theory of stability of motion can be extended further to investigate the motion of a closed loop of a homogeneous string in the absence of body forces (Healey 1990). This problem can be analyzed without putting any particular restriction on the constitutive matrix of the material. However, the problem has been neglected for a long time, probably because observing the associated phenomena is difficult due to the effects of gravity.

Consider a closed loop of string of total length 2π in some stress-free homogeneous, reference configuration. We identify a material point in his natural state by its arc length s ($s \in [0, 2\pi]$). Let $\mathbf{r}(s, t)$ denote the position of a particle s measured from a fixed origin, and let $\mathbf{r}_t$ represent the velocity $\mathbf{v}$ of the particle. Then the equation for the linear momentum balance is

$$\mathbf{v}_t = (T\mathbf{r}_s)_s, \tag{34.38}$$

where T is the tension in the string. If the string is inextensible, we have the constraint $|\mathbf{r}_s|^2 \equiv 1$. On the other hand, if the string is hyperelastic, its constitutive equation is

$$T = W'(|\mathbf{r}_s|), \tag{34.39}$$

where W denotes the strain energy density. We shall assume that $W(\xi)$ is defined for $0 < \xi < \infty$, with $W'(1) = 0$ and $W''(\xi) > 0$ over $(0, \infty)$.

If the loop rotates about a fixed axis with an angular velocity $\boldsymbol{\omega}$, we can introduce an orthonormal basis $\{\mathbf{i}, \mathbf{j}, \mathbf{k}\}$ such that $\boldsymbol{\omega}$ is parallel to $\mathbf{k}$. During the motion, the total energy of the string is stationary under the constraint of constant angular momentum. For an inextensible string, the solution is a critical point of the functional

$$\mathscr{H}_1(\mathbf{v}, \mathbf{r}, \boldsymbol{\omega}, T) = \int_0^{2\pi} [\tfrac{1}{2}|\mathbf{v}|^2 + \tfrac{1}{2}T(|\mathbf{r}_s|^2 - 1)]\, ds - \boldsymbol{\omega} \cdot \left[\int_0^{2\pi} \mathbf{r} \times \mathbf{v}\, ds - 2\pi\mu\mathbf{k}\right], \tag{34.40}$$

where T and $\boldsymbol{\omega}$ are Lagrange multipliers, and $\mu \neq 0$ is the prescribed magnitude of the angular momentum. For a hyperelastic string we must search for the extrema of

$$\mathscr{H}_2(\mathbf{v}, \mathbf{r}, \boldsymbol{\omega}) = \int_0^{2\pi} [\tfrac{1}{2}|\mathbf{v}|^2 + W(|\mathbf{r}_s|)]\, ds - \boldsymbol{\omega} \cdot \left[\int_0^{2\pi} \mathbf{r} \times \mathbf{v}\, ds - 2\pi\mu\mathbf{k}\right]. \tag{34.41}$$

The first variation of $\mathscr{H}_1$ is given by

$$\begin{aligned} \delta\mathscr{H}_1 &= \frac{d}{d\alpha}\mathscr{H}_1(\mathbf{v} + \alpha\boldsymbol{\zeta},\ \mathbf{r} + \alpha\boldsymbol{\eta},\ \boldsymbol{\omega}, T)\Big|_{\alpha=0} \\ &= \int_0^{2\pi} [\mathbf{v} \cdot \boldsymbol{\zeta} + T(\mathbf{r}_s \cdot \boldsymbol{\eta}_s) - \boldsymbol{\omega} \cdot (\boldsymbol{\eta} \times \mathbf{v} + \mathbf{r} \times \boldsymbol{\zeta})]\, ds. \end{aligned} \tag{34.42}$$

Integration by parts yields the Euler equations

$$\mathbf{v} = \boldsymbol{\omega} \times \mathbf{r}, \tag{34.43a}$$

$$(T\mathbf{r}_s)_s = \boldsymbol{\omega} \times \mathbf{v}, \tag{34.43b}$$

subject to the constraints $|\mathbf{r}_s|^2 \equiv 1$ and

$$\int_0^{2\pi} \mathbf{r} \times \mathbf{v}\, ds = 2\pi\mu\mathbf{k}. \tag{34.44}$$

Equation (34.43a) implies that the extremizers are rigid rotations, with angular velocity $\boldsymbol{\omega}$. The tension T secures the inextensibility constraint, and condition (34.44) prescribes the total angular momentum. In the absence of external forces the total linear momentum is conserved; consequently, the center of mass of the loop translates at a constant velocity, which can be taken to be zero. We therefore append to the condition of inextensibility and (34.44) the side condition

$$\int_0^{2\pi} \mathbf{r}\, ds = \mathbf{0}, \tag{34.45}$$

which fixes the center of mass at the origin.

A simple solution of (34.43) has the form

$$\boldsymbol{\omega} = \omega\mathbf{k}, \quad \mathbf{r} = \mathbf{e}(s + \omega t) \tag{34.46}$$

where

$$\mathbf{e}(\nu) \equiv \cos(\nu)\mathbf{i} + \sin(\nu)\mathbf{j}. \tag{34.47}$$

Solution (34.46) corresponds to a steady circular rotation. Substitution of (34.47) in (34.43)–(34.45) yields

$$\begin{aligned} \boldsymbol{\omega}_e &= \mu\mathbf{k}, \quad \mathbf{r}_e = \mathbf{e}(s + \mu t), \\ T_e &= \mu^2, \quad \mathbf{v}_e = \mu\mathbf{e}'(s + \mu t), \end{aligned} \tag{34.48}$$

which is both a solution of (34.38) and a critical point of $\mathscr{H}_1$.

Similarly, the Euler equations for (34.41) are

$$\mathbf{v} = \boldsymbol{\omega} \times \mathbf{r}, \tag{34.49a}$$

$$\left[W'(\mathbf{r}_s)\frac{\mathbf{r}_s}{|\mathbf{r}_s|}\right]_s = \boldsymbol{\omega} \times \mathbf{v}, \tag{34.49b}$$

subject to (34.44) and (34.45). As, in this case, the string is extensible, we seek a solution of the form

$$\boldsymbol{\omega} = \omega\mathbf{k}, \quad \mathbf{r} = \lambda\mathbf{e}(s + \omega t) \quad (\lambda > 0), \tag{34.50}$$

and substitution of this in (34.44) and (34.49b) gives

$$\omega = \frac{\mu}{\lambda^2}, \quad \lambda^3 W'(\lambda) = \mu^2. \tag{34.51}$$

The properties of W ensure that function (34.51) is a monotonically increasing map from $(1, \infty)$ onto $(0, \infty)$. Hence, we can invert (34.51) to obtain λ as a unique function of μ^2, denoted by $\lambda = \tilde{\lambda}(\mu^2)$, where $\tilde{\lambda}$ is $\mathscr{C}^1$ and monotonically increasing. Therefore we can compute

$$\boldsymbol{\omega}_e = \tilde{\omega}(\mu)\mathbf{k},$$
$$\tilde{\mathbf{r}}_e = \tilde{\lambda}(\mu^2)\,\mathbf{e}[s + \tilde{\omega}(\mu)t],$$

and

$$\tilde{\mathbf{v}}_e \equiv \tilde{\omega}(\mu)\,\tilde{\lambda}(\mu^2)\,\mathbf{e}'[s + \tilde{\omega}(\mu)t], \quad \text{with } \tilde{\omega}(\mu) \equiv \frac{\mu}{[\tilde{\lambda}(\mu^2)]^2}, \tag{34.52}$$

which represent both a critical point of $\mathscr{H}_2$ and a solution of (34.38) and (34.39).

In order to analyze the stability of the circular configuration of inextensible strings we need to examine the second variation of $\mathscr{H}_1$ evaluated at the critical point (34.48):

$$\begin{aligned}\delta^2\mathscr{H}_1^\circ(\boldsymbol{\zeta}, \boldsymbol{\eta}, \mu) &\equiv \frac{d^2}{d\alpha^2}\mathscr{H}_1(\mathbf{v}_e + \alpha\boldsymbol{\zeta}, \mathbf{r}_e + \alpha\boldsymbol{\eta}, \mu\mathbf{k}, \mu^2)|_{\alpha=0} \\ &= \int_0^{2\pi} [|\boldsymbol{\zeta}|^2 + \mu^2|\boldsymbol{\eta}_s|^2 - 2\mu\mathbf{k}\cdot(\boldsymbol{\eta}\times\boldsymbol{\zeta})]\,ds.\end{aligned} \tag{34.53}$$

We say that a pair of functions $(\boldsymbol{\zeta}, \boldsymbol{\eta})$ is an *admissible variation* if, for all t, $\boldsymbol{\zeta}(.,t)$ is absolutely continuous (that is, continuous with a classical derivative almost everywhere) and $\boldsymbol{\eta}(.,t)$ is periodic with period 2π, of class $\mathscr{C}^1$, with $\boldsymbol{\eta}_s(.,t)$ absolutely continuous, and the following linearized constraints hold:

$$\mathbf{e}'(s + \mu t)\cdot\boldsymbol{\eta}_s \equiv 0, \tag{34.54a}$$

$$\int_0^{2\pi} [\mu\boldsymbol{\eta}\times\mathbf{e}'(s + \mu t) + \mathbf{e}(s + \mu t)\times\boldsymbol{\zeta}]\,ds = \mathbf{0}, \tag{34.54b}$$

$$\int_0^{2\pi} \boldsymbol{\eta}\,ds = \mathbf{0}. \tag{34.54c}$$

Now, by virtue of the Poincaré inequality, we have

$$\int_0^{2\pi} |\boldsymbol{\eta}_s|^2\,ds \geqslant \sigma^2 \int_0^{2\pi} |\boldsymbol{\eta}|^2\,ds,$$

where σ^2 is the minimum eigenvalue of

$$\boldsymbol{\eta}_{ss} + \sigma^2\boldsymbol{\eta} = \mathbf{0},$$

subject to (34.54) and the condition of periodicity in s. In particular, a simple calculation using (34.54c) yields the value $\sigma^2 = 1$. Thus we can derive from (34.53) the inequality

$$\delta^2 \mathscr{H}_1^e(\boldsymbol{\zeta}, \boldsymbol{\eta}, \mu) \geq \int_0^{2\pi} [|\boldsymbol{\zeta}|^2 + \mu^2|\boldsymbol{\eta}|^2 - 2\mu\mathbf{k}\cdot(\boldsymbol{\eta}\times\boldsymbol{\zeta})]\, ds$$

$$= \int_0^{2\pi} [(\zeta_1 + \mu\eta_2)^2 + (\zeta_2 - \mu\eta_1)^2 + \mu^2(\eta_3)^2 + (\zeta_3)^2]\, ds \geqslant 0, \tag{34.55}$$

where $\boldsymbol{\eta} = \eta_1\mathbf{i} + \eta_2\mathbf{j} + \eta_3\mathbf{k}$, etc. We have thus proved that the second variation of $\mathscr{H}_1$, evaluated at the initial point, is positive semidefinite.

This result can be simplified by eliminating the constraint (34.54b) and replacing it by the new condition

$$\int_0^{2\pi} [\mathbf{e}'(s+\mu t)\cdot\boldsymbol{\eta} - \mu\mathbf{e}(s+\mu t)\cdot\boldsymbol{\zeta}]\, ds = 0. \tag{34.56}$$

Within this modified class of admissible functions we can prove that the second variation of $\mathscr{H}_1$ is strictly positive definite. This implies that solutions (36.48) are stable.

The behavior of extensible solutions is, however, different. The second variation of $\mathscr{H}_2$ evaluated using (34.52) is

$$\delta^2 \mathscr{H}_2^e(\boldsymbol{\zeta}, \boldsymbol{\eta}, \mu) \equiv \frac{d^2}{d\alpha^2} \mathscr{H}_2(\tilde{\mathbf{v}}_e + \alpha\boldsymbol{\zeta}, \tilde{\mathbf{r}}_e + \alpha\boldsymbol{\eta}, \tilde{\omega}_e)|_{\alpha=0}$$

$$= \int_0^{2\pi} \left(|\boldsymbol{\zeta}|^2 + W'[\tilde{\lambda}(\mu^2)]\frac{1}{\tilde{\lambda}(\mu^2)}|\boldsymbol{\eta}_s|^2 + \left\{ W''[\tilde{\lambda}(\mu^2)] - \frac{W'[\tilde{\lambda}(\mu^2)]}{\tilde{\lambda}(\mu^2)} \right\}(\mathbf{e}'\cdot\boldsymbol{\eta}_s)^2 - 2\tilde{\omega}(\mu)\mathbf{k}\cdot(\boldsymbol{\eta}\times\boldsymbol{\zeta}) \right) ds. \tag{34.57}$$

We place the same smoothness requirements as before on the admissible variation $(\boldsymbol{\zeta}, \boldsymbol{\eta})$, but the linearized constraints are as follows:

$$\int_0^{2\pi} \{\tilde{\omega}(\mu)\boldsymbol{\eta} \times \mathbf{e}'[s + \tilde{\omega}(\mu)t] + \mathbf{e}[s + \tilde{\omega}(\mu)t] \times \boldsymbol{\zeta}\}\, ds = \mathbf{0}, \tag{34.58a}$$

$$\int_0^{2\pi} \boldsymbol{\eta}\, ds = \mathbf{0}. \tag{34.58b}$$

In order to discuss the sign of the second variation of $\mathscr{H}_2$, it is convenient to introduce the distinction between *stiff strings*, for which

$$\frac{d}{d\nu}\left[W'(\nu)\frac{1}{\nu} \right] > 0 \quad \forall \nu \in (1, \infty), \tag{34.59}$$

and *soft strings*, for which (34.59) does not hold. However, even for soft strings, there is a number $\lambda_* > 1$ such that we have

$$\frac{d}{d\nu}\left[\frac{W'(\nu)}{\nu}\right] > 0, \text{ for } 1 \leqslant \nu < \lambda_* \text{ and } \frac{d}{d\nu}\left[\frac{W'(\nu)}{\nu}\right]_{\nu=\lambda_*} = 0. \tag{34.60}$$

In fact, our restrictions on W imply that the function $g(\nu) \equiv \{[W''(\nu)/\nu] - [W'(\nu)/\nu^2]\}$ is continuous over $(0, \infty)$, with $g(1) = W''(1) > 0$. Thus we have $g > 0$ in some open interval containing $\nu = 1$. By the definition of a soft string, (34.60) requires that g has at least one zero value over $(0, \infty)$, the smallest of which is denoted by λ_*. Two typical constitutive laws for stiff and soft strings are illustrated in Figure 34.5. Note that our previous restrictions on W require even a soft string to have $W''(\nu) > 0$ for $\nu \in (0, \infty)$, so that the graphs of $W'(\nu) > 0$ for stiff and soft materials have the forms shown in Figures 34.5(a) and 34.5(b), respectively. Condition (34.60) shows that all strings are stiff initially.

The second variation (34.57) is not positive-definite in the class $(\boldsymbol{\zeta}, \boldsymbol{\eta})$ satisfying constraints (34.58). It is thus necessary to remove (34.58a) and replace it by two orthogonality conditions, analogous to (34.56), of the form

$$\begin{aligned} &\int_0^{2\pi} \{\mathbf{e}'(s + \tilde{\omega}(\mu)t] \cdot \boldsymbol{\eta} - \tilde{\omega}(\mu)\mathbf{e}[s + \tilde{\omega}(\mu)t] \cdot \boldsymbol{\zeta}\}\, ds = 0, \\ &\int_0^{2\pi} \{\mathbf{e}[s + \tilde{\omega}(\mu)t] \cdot \boldsymbol{\eta} + \tilde{\omega}(\mu)\mathbf{e}'[s + \tilde{\omega}(\mu)t] \cdot \boldsymbol{\zeta}\}\, ds = 0. \end{aligned} \tag{34.61}$$

Within this modified class $(\boldsymbol{\eta}, \boldsymbol{\zeta})$, the second variation $\delta^2 \mathscr{H}_2^e$ is strictly positive, and consequently stiff strings are stable for all $\mu \neq 0$, and soft strings are stable for all μ under the condition $0 < \mu^2 < \mu_*^2 \equiv \lambda_*^3 W'(\lambda_*)$. The result agrees with the intuitive expectation that the motion can become unstable only in a deformable string, provided that the material is soft according to definition (34.59).

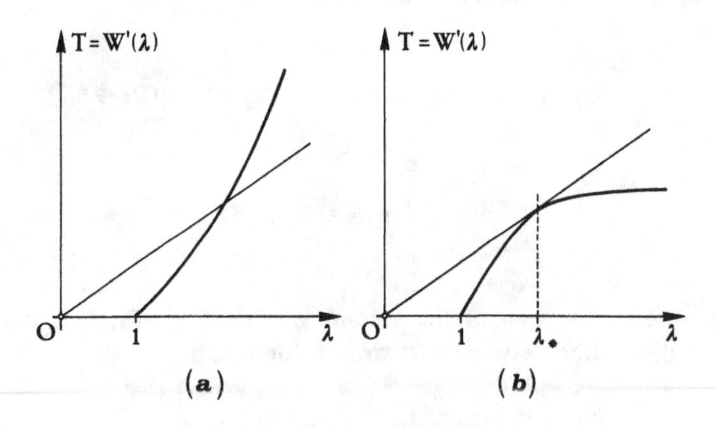

Fig. 34.5

The small motions of a string subject to a constant tension T_0 are governed by an equation like (6.8) where a solution can be constructed explicitly by means of d'Alembert's method. Unfortunately, d'Alembert's solution is valid only for linear hyperbolic second-order equations with constant coefficients, and is not extendible to the simple case of a string in which the tension possibly changes with time. In such a case the problem can be treated by exploiting the notion of the *adiabatic invariant*, introduced by Lorentz in 1911. Lorentz had in mind a mechanical clock, affected by seasonal variations in temperature. The motion is described by the differential equation

$$\ddot{x} + \omega(t)^2 x = 0, \tag{34.62}$$

where $\omega/2\pi$ is the instantaneous frequency of vibration. Lorentz conjectured that the ratio

$$I = \frac{1}{\omega}(\omega^2 x^2 + \dot{x}^2),$$

is approximately constant, so that the total energy is approximately proportional to the frequency. This explains the definition of the adiabatic invariant given to the quantity I.

The same argument applies to an equation of the form (Day 1987)

$$\varphi(t)u_{xx} = b(x)u_{tt}, \quad 0 < x < 1, \quad t \geqslant 0, \tag{34.63}$$

with the boundary conditions $u(0, t) = u(1, t) = 0$, and the initial conditions

$$u(x, 0) = u_0(x), \quad u_t(x, 0) = v_0(x).$$

In (34.63), $b(x)$ is taken to be positive and independent of t; the coefficient $\varphi(t)$ is required to be positive, to start at the initial value $\varphi(0) = 1$, and to tend slowly to a limiting value $\varphi(\infty)$, for $t \to \infty$, which need *not* be close to the initial value. In precise terms, φ will be said to be *slowly* varying if we have $\varphi(0) = 1$, and $\varphi \to \varphi(\infty)$, $\dot{\varphi} \to 0$, $\ddot{\varphi} \to 0$ for $t \to \infty$; in addition, there are positive constants ε and c_k $(k = 1, \ldots, 7)$ such that

$$\left.\begin{aligned} &c_1 \leqslant \varphi \leqslant c_2, \quad |\dot{\varphi}| \leqslant c_3\varepsilon, \quad |\ddot{\varphi}| \leqslant c_4\varepsilon^2, \\ &\int_0^\infty |\dot{\varphi}|\, dt \leqslant c_5, \quad \int_0^\infty |\ddot{\varphi}|\, dt \leqslant c_6\varepsilon, \quad \int_0^\infty |\dddot{\varphi}|\, dt \leqslant c_7\varepsilon^2. \end{aligned}\right\} \tag{34.64}$$

These conditions are satisfied, in particular, whenever $\varphi(t) = \Phi(\varepsilon t)$, where $\phi(0) = 1$, $\phi \to \Phi(\infty)$, $\dot{\Phi} \to 0$, $\ddot{\Phi} \to 0$ as $t \to \infty$, and

$$\begin{aligned} &c_1 \leqslant \Phi \leqslant c_2, \quad |\dot{\Phi}| \leqslant c_3, \quad |\ddot{\Phi}| \leqslant c_4, \\ &\int_0^\infty |\dot{\Phi}|\, dt \leqslant c_5, \quad \int_0^\infty |\ddot{\Phi}|\, dt \leqslant c_6, \quad \int_0^\infty |\dddot{\Phi}|\, dt \leqslant c_7. \end{aligned}$$

It is easily checked that, when φ is slowly varying, the same is true of $\psi = \varphi^p$, where p is any real exponent, the cases $p = -\frac{1}{2}$ and $p = -\frac{3}{2}$ being the ones of interest.

The treatment of the problem follows a procedure suggested by Littlewood (1963). Let us introduce the integrals

$$E = J + K, \quad F = \int_0^1 bu^2\,dx,$$
$$J = \varphi \int_0^1 u_x^2\,dx, \quad K = \int_0^1 bu_t^2\,dx, \tag{34.65}$$

where, to within irrelevant factors, J is the potential energy, K is the kinetic energy, and E is the total energy.

It is easy to recognize that the time derivative of E can be calculated using the formula

$$\dot{E} = \left(\frac{\dot{\varphi}}{\varphi}\right) J. \tag{34.66}$$

Next, we observe the results

$$\dot{F} = 2\int_0^1 bu\,u_t\,dx,$$
$$\ddot{F} = 2\int_0^1 bu_t^2\,dx + 2\int_0^1 bu\,u_{tt}\,dx,$$

so that, on integrating by parts and using the boundary conditions, we obtain the relation

$$\ddot{F} = 2K - 2J = 2E - 4J.$$

Thus, in (34.66) we can replace J by $\frac{1}{2}E - \frac{1}{4}\ddot{F}$, and obtain

$$\dot{E} = \left(\frac{\dot{\varphi}}{\varphi}\right)\left(\frac{1}{2}E - \frac{1}{4}\ddot{F}\right). \tag{34.67}$$

If (34.67) is regarded as a first-order differential equation for E, we introduce the integration factor $\varphi^{-\frac{1}{2}}$ and put $\psi = \varphi^{-\frac{1}{2}}$ and $I = \psi E$, so that (34.67) reduces to the equation

$$\dot{I} = \tfrac{1}{2}\dot{\psi}\ddot{F}, \tag{34.68}$$

while (34.66) can be written in the form

$$\dot{E} = -\left(\frac{\dot{\psi}}{2\psi}\right) J. \tag{34.69}$$

By virtue of the Poincaré inequality, we can find a positive constant μ allowing the inequality

$$F = \int_0^1 bu^2\,dx \leqslant \mu \int_0^1 u_x^2\,dx = \mu\psi^2 J. \tag{34.70}$$

The function I is an adiabatic invariant, if the limit $I(\infty)$ exists, and

$$I = I(\infty) + O(\varepsilon). \tag{34.71}$$

In order to prove that I tends to a limit, it will be enough to show that E tends to a limit $E(\infty)$, because for $t \to \infty$, $I \to I(\infty) = \psi(\infty)E(\infty)$. Since we have $E = J + K \geqslant J$, (34.69) yields

$$|\dot{E}| = \frac{|\dot{\psi}|}{2\psi} J \leqslant \frac{|\dot{\psi}|}{2\psi} E \leqslant \frac{1}{2} d_1^{-1} |\dot{\psi}| E, \tag{34.72}$$

or, alternatively,

$$|\dot{E}| \leqslant \tfrac{1}{2} d_1^{-1} d_3 \varepsilon E, \tag{34.73}$$

where d_1 and d_3 are suitable constants. The existence of these constants is a consequence of (34.65) and of the relation $\psi = \varphi^p$ with $p = -\frac{1}{2}$. This means that the function ψ also satisfies a set of inequalities of the form

$$d_1 \leqslant \psi \leqslant d_2, \tag{34.74a}$$

$$|\dot{\psi}| \leqslant d_3 \varepsilon, \tag{34.74b}$$

$$|\ddot{\psi}| \leqslant d_4 \varepsilon^2, \tag{34.74c}$$

$$\int_0^\infty |\dot{\psi}| \, dt \leqslant d_5, \tag{34.74d}$$

$$\int_0^\infty |\ddot{\psi}| \, dt \leqslant d_6 \varepsilon, \tag{34.74e}$$

$$\int_0^\infty |\dddot{\psi}| \, dt \leqslant d_7 \varepsilon^2. \tag{34.74f}$$

Inequality (34.73) ensures that E is either strictly positive on $0 \leqslant t < \infty$ or vanishes identically. In fact, suppose we have $E(t_0) = 0$ at some t_0. Let t_1 be any time greater than t_0, and let M be the maximum value attained by E on the interval $0 \leqslant t_0 \leqslant t_1$. Over this interval the inequality

$$E(t) = \int_{t_0}^{t} \dot{E}(t) \, dt \leqslant \frac{1}{2} d_1^{-1} d_3 \varepsilon \int_{t_0}^{t} E \, dt,$$

holds and an inductive argument allows us to establish the estimates

$$E(t) \leqslant \frac{M}{k!} \left[\frac{1}{2} d_1^{-1} d_3 \varepsilon (t - t_0) \right]^k,$$

for every $k = 1, 2, 3, \ldots$. On taking the limit for $k \to \infty$, we conclude that E vanishes identically for every interval $0 \leqslant t \leqslant t_1$, with $t_1 > t_0$; that is, E vanishes identically for $0 \leqslant t < \infty$. The same is true of I, and in this case (34.71) holds. Thus, E may be supposed to be strictly positive. We then derive from (34.72) the inequality

$$|\dot{\overline{\ln E}}| \leqslant \frac{1}{2} d_1^{-1} |\dot{\psi}|.$$

Thus, the integral $\int_0^\infty |\dot{\overline{\ln E}}|\, dt$ converges and is bounded by

$$\tfrac{1}{2} d_1^{-1} \int_0^\infty |\dot{\psi}|\, dt \leqslant \tfrac{1}{2} d_1^{-1} d_5. \tag{34.75}$$

On combining the last two inequalities, we can write

$$\left| \ln \frac{E(t)}{E(0)} \right| \leqslant \int_0^t |\dot{\overline{\ln E}}|\, dt \leqslant \tfrac{1}{2} d_1^{-1} d_5,$$

and we therefore obtain

$$E(t) \leqslant E(0) \exp(\tfrac{1}{2} d_1^{-1} d_5), \tag{34.76}$$

the right-hand side of which is independent of ε.

Returning now to (34.68), an integration between t and some large time T yields

$$I(T) - I(t) = \frac{1}{2} [\dot{\psi} \dot{F}]_t^T - \frac{1}{2} \int_t^T \ddot{\psi} \dot{F}\, dt. \tag{34.77}$$

By combining (34.70) and (34.74a), we obtain the inequalities

$$F \leqslant \mu \psi^2 J \leqslant \mu \psi^2 E \leqslant \mu d_2^2 E,$$

while the Schwarz inequality gives

$$|\dot{F}| = 2 \left| \int_0^1 bu \frac{\partial u}{\partial t}\, dx \right| \leqslant 2(FK)^{\frac{1}{2}} \leqslant 2(FE)^{\frac{1}{2}}.$$

This implies that $|\dot{F}|$ is bounded, because so are F and E. On the other hand, (34.74e) gives $|\dot{\psi}| \to 0$ at infinity, and therefore $\dot{\psi}(T)\dot{F}(T)$ in (34.77) tends to zero as $T \to \infty$. Thus from (34.77) we derive

$$I(t) = I(\infty) + \tfrac{1}{2} \dot{\psi}(t) \dot{F}(t) + \tfrac{1}{2} \int_t^\infty \ddot{\psi} \dot{F}\, dt,$$

and hence

$$|I(t) - I(\infty)| \leqslant C \left[\frac{1}{2} |\dot{\psi}(t)| + \frac{1}{2} \int_t^\infty |\ddot{\psi}|\, dt \right],$$

where C is a constant. Finally, on using (34.74) again, we obtain

$$|I(t) - I(\infty)| \leqslant C \tfrac{1}{2} (d_3 + d_6) \varepsilon,$$

which is exactly (34.71).

35. Peeling and Slipping

So far we have considered the motion of a string regarded as an isolated mechanical system, in which the possible interactions with other bodies have been included among the extraneous forces and the boundary conditions. By a suitable adjustment of both of the latter, it is possible to simulate, at least in principle, any kind of influence that the surrounding bodies might have on the motion of the string. However, situations occur in which this conceptual separation is no longer possible, and a more detailed description of the exchange of forces between the string and its environment is necessary. There are certain problems of contact or detachment in which the mutual forces also determine the length of the string or its constitutive properties, which are usually considered known.

An illustrative example of this kind of problem is the peeling of an adhesive tape initially adhered to a plane surface (Burridge and Keller 1978). The tape is glued to the surface along a straight line, which is taken as the x axis of an (x, y) Cartesian system of coordinates (Figure 35.1). A concentrated force of magnitude $2f_0$ is suddenly applied to the tape in the direction of the positive y axis. If the force is applied at the origin at $t = 0$, we wish to determine the extent of the detached position as a function of time, and also the motion of this detached part. For this purpose, we denote the end points of the detached part by $\sigma_1(t)$ and $\sigma_2(t)$, and the displacement of the tape in the y direction by $u(x, t)$.

We assume that the tape is under a constant tension T_0 and that the displacement $u(x, t)$ is small, so that the equation of motion is

$$\rho u_{tt} - Tu_{xx} = 2f_0\,\delta(x), \quad t > 0, \quad u(x, 0) = u_t(x, 0) = 0, \tag{35.1}$$

where $\delta(x)$ denotes the delta function. The tape will begin to peel at $x = 0$, $t = 0$, and at the instant t a triangular part $\sigma_2(t) \leqslant x \leqslant \sigma_1(t)$ will be completely detached from the plane. The symmetry of the problem shows that $u(x, t) = u(-x, t)$ and $\sigma_2(t) = -\sigma_1(t)$. Furthermore, for constant load, a natural assumption is that $\sigma_1(t)$ is proportional to t:

$$\sigma_1(t) = vt, \quad \sigma_2(t) = -vt, \quad t \geqslant 0. \tag{35.2}$$

In addition, at σ_1 and σ_2 we have

$$u[\sigma_1(t), t] = u[\sigma_2(t), t] = 0. \tag{35.3}$$

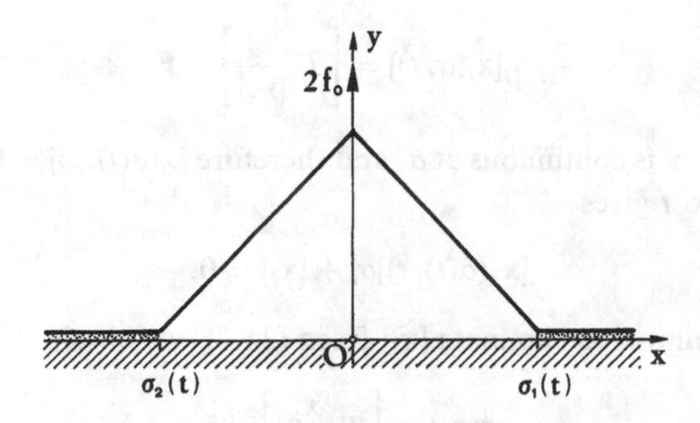

Fig. 35.1

The solution of (35.1) satisfying the boundary conditions (35.3) is

$$u(x, t) = (vt - |x|)\frac{f_0}{T}, \quad |x| \leqslant vt. \tag{35.4}$$

In this formula, $\sigma_1(t) = vt$ is still unknown and must be determined using another mechanical equation called the *breaking condition*. It is the derivation of this condition that constitutes the novelty of the problem, because it requires an accurate analysis of the one-dimensional continuum in the neighborhood of the points $|x| = vt$. To do this, we consider an interval, say $a \leqslant x \leqslant b$, containing in its interior one of the points of instantaneous detachment, generally denoted by $\sigma(x, t)$. Then the balance of linear momentum for the piece of tape $a \leqslant x \leqslant b$ reads:

$$\overline{\left(\int_a^b \rho \mathbf{x}_t \, dx\right)}^{\bullet} = \varphi(b, t) - \varphi(a, t) + \int_a^b \mathbf{f} \, dx - \mathbf{F}, \tag{35.5}$$

where $\mathbf{x}(x, t)$ is the position at time t of the particle at a distance x from the origin in the straight reference configuration, $\mathbf{f}(x, t)$ is the external force density, $-\mathbf{F}$ is the concentrated cohesive force at the moving boundary $x = \sigma(t)$, and φ is the force exerted on the part with $x' > x$ upon the part of the continuum with $x' < x$. The left-hand side of (35.5) can be rewritten in the form

$$\begin{aligned}\overline{\left(\int_a^{\sigma(t)} \rho \mathbf{x}_t \, dx + \int_{\sigma(t)}^b \rho \mathbf{x}_t \, dx\right)}^{\bullet} &= \sigma_t \rho \mathbf{x}_t(\sigma^-, t) - \sigma_t \rho \mathbf{x}_t(\sigma^+, t) + \int_a^b (\rho \mathbf{x}_t)_t \, dx \\ &= -\sigma_t[\rho \mathbf{x}_t(\sigma, t)] + \int_a^b (\rho \mathbf{x}_t)_t \, dx,\end{aligned} \tag{35.6}$$

where, for any function $g(x, t)$, $g(\sigma^\pm, t)$ is the limit of $g(x, t)$ as x approaches σ from above or below, and $[g(\sigma, t)] = g(\sigma^+, t) - g(\sigma^-, t)$. On letting a tend to σ from below and b tend to σ from above, (35.5) becomes

$$-\sigma_t[\rho \mathbf{x}_t(\sigma, t)] = [\varphi(\sigma, t)] - \mathbf{F}. \tag{35.7}$$

If we suppose that ρ is continuous and $\varphi = T(\mathbf{x}_x/|\mathbf{x}_x|)$, then (35.7) yields

$$-\rho\sigma_t[\mathbf{x}_t(\sigma, t)] = \left[T\frac{\mathbf{x}_x}{|\mathbf{x}_x|}\right] - \mathbf{F}. \tag{35.8}$$

On the other hand, $\mathbf{x}$ is continuous at σ, and therefore $[\mathbf{x}(\sigma(t), t)] = 0$. Differentiating this with respect to t gives

$$[\mathbf{x}_x(\sigma(t), t)]\sigma_t + [\mathbf{x}_t] = 0. \tag{35.9}$$

If we use this condition to eliminate $[\mathbf{x}_t]$ from (35.7), we obtain

$$\rho\sigma_t^2[\mathbf{x}_x] = \left[T\frac{\mathbf{x}_x}{|\mathbf{x}_x|}\right] - \mathbf{F}. \tag{35.10}$$

If $T/[\mathbf{x}_x]$ is continuous, (35.10) can be rewritten as

$$\left(\frac{T}{|\mathbf{x}_x|} - \rho\sigma_t^2\right)[\mathbf{x}_x] = \mathbf{F}. \tag{35.11}$$

This condition has been derived without making any assumption about the form of $\mathbf{x}$, but in linear theory it has the form $\mathbf{x} = (x, u(x, t))$ with u small. Then we have $\mathbf{x}_x \simeq (1, u_x)$, $|\mathbf{x}_x| \simeq 1$, and $[\mathbf{x}_x] \simeq (0, [u_x])$. At the ends $u_x = 0$ for $x > \sigma_1$, and $u_x = 0$ for $x < \sigma_2$. Thus (35.11) yields

$$(T - \rho\sigma_t^2)u_x[\sigma(t), t] = \begin{cases} -F_y \text{ at } \sigma = \sigma_1, \\ \;\;\, F_y \text{ at } \sigma = \sigma_2, \end{cases} \tag{35.12}$$

where F_y is the vertical component of the force, directed downwards, exerted by the adhesive on the tape at σ_1 and $-\sigma_2$. Equation (35.12) holds at the onset of detachment of the tape at $\sigma = \sigma_1 = \sigma_2$, but in this case F_y must have a given constant value, still denoted by F_y, in order for breaking to occur.

Now, from (35.4) and (35.12) we obtain an equation for v. From the first equation we derive

$$u_x = \begin{cases} -f_0/T \text{ for } 0 < x \leqslant vt, \\ +f_0/T \text{ for } -vt \leqslant x < 0, \end{cases} \tag{35.13}$$

and therefore (35.12) gives

$$v = \left(\frac{T}{\rho}\right)^{\frac{1}{2}}\left(1 - \frac{F}{f_0}\right)^{\frac{1}{2}}. \tag{35.14}$$

This formula shows that v increases with f_0, tending to the velocity of sound $(T/\rho)^{\frac{1}{2}}$ as f_0 becomes infinite and becoming zero at $f_0 = F$.

In order to take friction into account, we consider a string stretched along the x axis on a rough plane, which is taken as the x, y plane. The string, initially undisplaced and at rest, is subject to an increasing concentrated force $2tf_0$ applied at $x = 0$, $y = 0$, starting at $t = 0$, and directed along the positive y axis. The string is also held against the plane by a force per unit length P, which is assumed constant for simplicity, so that there is a frictional force $\mu_0 P$ which resists the motion. Here μ_0 represents the coefficient of friction between the string and the plane. If the frictional force is large enough, the string may not move at all, or part of it may move and part may not. As before, we denote the end points of the moving part by $\sigma_1(t)$ and $\sigma_2(t)$, and the displacement of the string in the y direction by $u(x, t)$. The equation of motion is

$$\rho u_{tt} - T u_{xx} = 2tf_0\delta(x) - \mu_0 P, \quad t > 0, \quad u(x, 0) = u_t(x, 0) = 0, \tag{35.15}$$

which must hold for $\sigma_2 \leqslant x \leqslant \sigma_1$, with $\sigma_2(t) = -\sigma_1(t)$ for symmetry. The boundary conditions (35.3) still hold, but instead of (35.12) we have

$$[u_x(\sigma_1(t), t)] = [u_x(\sigma_2(t), t)] = 0. \tag{35.16}$$

This condition follows from (35.12), with $F_y = 0$ because the contact of the string with the x axis is due only to friction. A solution of (35.15) satisfying (35.16) and with v arbitrary is

$$u(x, t) = \frac{1}{T}\left(\frac{\rho v f_0}{T} + P\mu_0\right)\left(\frac{x^2}{2} - |x|vt\right) + \frac{vf_0t^2}{2T}, \quad |x| \leqslant vt. \tag{35.17}$$

Then (35.3) gives the velocity v in (36.17) as

$$\frac{v}{c} = \left[1 + \left(\frac{\mu_0 Pc}{2f_0}\right)^2\right]^{\frac{1}{2}} - \frac{\mu_0 Pc}{2f_0}, \tag{35.18}$$

where $c = (T/\rho)^{\frac{1}{2}}$ is the speed of propagation of the waves in the string. The last equation shows that v/c increases from zero for $2f_0/\mu_0 Pc = 0$, to unity for $2f_0/\mu_0 Pc = \infty$.

The theory can be extended without difficulty to the peeling of an adhesive tape from a plane surface, permitting the displacements to be large (Burridge and Keller 1978). Let $\mathbf{x}(x, t)$ be the position of a particle at time t labeled by the parameter x, which we shall take as the arc length of the tape at $t = 0$, when it coincides with the x axis. If the tape were initially attached to the wall and at rest, the initial condition would be

$$\mathbf{x}(x, 0) = (x, 0), \quad \mathbf{x}_x(x, 0) = (0, 0). \tag{35.19}$$

Let us assume that the tape is pulled by a steady concentrated force of magnitude $2f_0$ in the y direction applied at the origin for $t > 0$. This force then has the form $\mathbf{f}(x, t) = (0, 2f_0\delta(x))$. The tape will begin to peel at $x = 0$, $t = 0$. If $\sigma_1(t)$ and $\sigma_2(t)$ are the boundaries of the detached part, the symmetry of the problem requires that $\mathbf{x}(x, t) = \mathbf{x}(-x, t)$ and $\sigma_2(t) = -\sigma_1(t)$. The equation of motion of the detached part is

$$\rho\mathbf{x}_{tt} = \left(\frac{T\mathbf{x}_x}{|\mathbf{x}_x|}\right) + \mathbf{f}, \quad \sigma_2(t) \leqslant x \leqslant \sigma_1(t). \tag{35.20}$$

The tension T is related to the elongation $|\mathbf{x}_x|$ by a stress–strain law of the form

$$T = T(|\mathbf{x}_x|, x), \tag{35.21}$$

which describes the elastic properties of the string at x. If the tape is inextensible, (35.21) is replaced by $|\mathbf{x}_x| = 1$ and $T = T_0$, which is a given value of the tension.

Furthermore, in this case the problem must be integrated by the breaking conditions at $\sigma_1(t)$ and $\sigma_2(t)$. We assume that two concentrated forces $\mathbf{F}_j = (0, F_j)$ directed along the positive y axis must be exerted by the tape at the points of detachment σ_j $(j = 1, 2)$ in order to break the adhesive bond. If the detachment is symmetrical we can assume the conditions $\mathbf{F}_1 = \mathbf{F}_2 = \mathbf{F}$. Then the balance of momentum of the tape in the neighborhoods of σ_1 and σ_2 yields

$$-\rho\sigma_t[\mathbf{x}_t] = \left(\frac{T\mathbf{x}_x}{|\mathbf{x}_x|}\right) - \mathbf{F}, \tag{35.22}$$

where $\sigma = \sigma_1 = -\sigma_2$, and $-\mathbf{F}$ is the force exerted on the tape by the wall. The preceding equations are sufficient to determine the unknowns of the problem, namely $\mathbf{x}(x, t)$, T, and $\sigma(t)$.

We seek a solution of the form

$$\mathbf{x}(x, t) = (x, 0) + (vt - |x|)(1, k), \quad |x| \leqslant vt, \tag{35.23}$$

where v and k are constants to be determined. This solution satisfies the boundary conditions $\mathbf{x}(x, t) = 0$ at $x = vt = \sigma_1(t)$, and at $x = -vt = \sigma_2(t)$. Outside the interval $|x| \leqslant \sigma$, the tape remains attached to the x axis. Because of the condition $|\mathbf{x}_x| = k$, (35.21) gives $T = T(k)$. Equation (35.20) is satisfied for $x \neq 0$, while at $x = 0$ it yields

$$T \frac{\mathbf{x}_x}{|\mathbf{x}_x|} = (0, -f_0), \tag{35.24}$$

or, alternatively, on recalling (35.23),

$$T(k) = f_0, \tag{35.25}$$

from which we can determine k. If the tape is inextensible, $k = 1$ and (35.25) is used to determine T. Next, by using (35.23) in (35.22), and taking account of (35.25),

$$\rho v^2(1, k) = (0, -F) + T_0(1, 0) + f_0(0, 1), \tag{35.26}$$

where $T_0 = T(1)$ represents the tension for $|x| > \sigma(t)$. Solving the y component of (35.26) with respect to v, we get

$$v = \left(\frac{f_0}{k\rho}\right)^{\frac{1}{2}}\left(1 - \frac{F}{f_0}\right)^{\frac{1}{2}}. \tag{35.27}$$

Furthermore, the equation for x, after elimination of v^2, yields

$$\frac{1}{k} = \frac{T_0}{f_0 - F}. \tag{35.28}$$

The solution of the problem for large displacements maintains the two special features of the linear problem, namely the triangular form of the detached configuration and the existence of the condition $f_0 \geqslant F$ in order for the velocity v to be real.

An application of the theory of the slipping of a string on a rough surface arises in the lateral motion of a tape on a cylindrical guide surface when, for some reason, the ingoing and outgoing branches of the tape lie in different planes and not the plane perpendicular to the axis of the cylinder (Figure 35.2) (Ono 1979). The problem of controlling this unfavorable lateral motion arises in many fields of engineering, such as textile technology and magnetic digital recording.

The physical model we consider is simplified by several assumptions. The tangential forces due to friction are proportional to the normal forces, but the coefficient of friction varies with the direction in which the string slides. More precisely, we introduce a coefficient μ_φ between the string and the cylinder in the circumferential direction, and a coefficient μ_z in the axial direction. We assume, in addition, that the string is perfectly flexible and inextensible, whatever the tension. Finally, we assume that the lateral motion is sufficiently small and slow such that only first-order terms of the lateral displacement and its derivatives with respect to the space and time variables need to be retained.

The string is partially wound round the lateral surface of a horizontal cylinder having a circular cross-section of radius a. The cylinder is referred to a system of cylindrical coordinates (r, φ, z), with the z axis placed along the longitudinal axis of symmetry. We take a point O on the z axis as the origin of a Cartesian system (x, y, z), and denote by $\{\mathbf{i}, \mathbf{j}, \mathbf{k}\}$ the fixed orthonormal basis associated with the

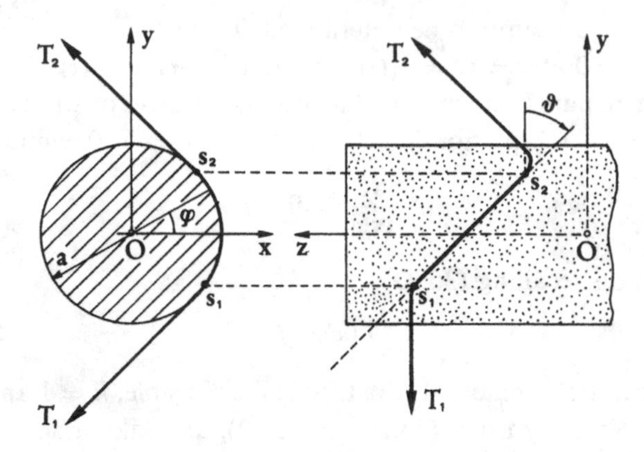

Fig. 35.2

axes x, y, and z. The unit vectors $\{\mathbf{e}_r, \mathbf{e}_\varphi, \mathbf{e}_z\}$, defined at each point parallel to the (r, φ, z) coordinate lines, constitute a local base of reference. Between the two triads we have the following obvious relationships:

$$\mathbf{e}_r = \mathbf{i}\cos\varphi + \mathbf{j}\sin\varphi, \quad \mathbf{e}_\varphi = -\mathbf{i}\sin\varphi + \mathbf{j}\cos\varphi, \quad \mathbf{e}_z = \mathbf{k}. \tag{35.29}$$

The position of the string at time t is defined by the vector

$$\mathbf{r} = \mathbf{r}(s, t), \tag{35.30}$$

where s is the arc length measured along the string from a prescribed section. Note that, because s is measured in an instantaneous configuration, s itself depends on t. The vector $\mathbf{r}$ can also be represented as the sum

$$\mathbf{r}(s, t) = a\mathbf{e}_r + z\mathbf{k}, \tag{35.31}$$

for all points of the string placed on the surface of the cylinder. From (35.29) we calculate

$$\begin{aligned} \frac{\partial \mathbf{e}_r}{\partial s} &= (-\mathbf{i}\sin\varphi + \mathbf{j}\cos\varphi)\frac{\partial\varphi}{\partial s} = \mathbf{e}_\varphi \frac{\partial\varphi}{\partial s}, \\ \frac{\partial \mathbf{e}_\varphi}{\partial s} &= -(\mathbf{i}\cos\phi + \mathbf{j}\sin\phi)\frac{\partial\varphi}{\partial s} = -\mathbf{e}_r \frac{\partial\varphi}{\partial s}. \end{aligned} \tag{35.32}$$

From (33.31) and (35.32) we find

$$\begin{aligned} \frac{\partial \mathbf{r}}{\partial s} &= a\frac{\partial\varphi}{\partial s}\mathbf{e}_\varphi + \frac{\partial z}{\partial s}\mathbf{k}, \\ \frac{\partial^2 \mathbf{r}}{\partial s^2} &= -a\left(\frac{\partial\varphi}{\partial s}\right)^2 \mathbf{e}_r + a\frac{\partial^2\varphi}{\partial s^2}\mathbf{e}_\varphi + \frac{\partial^2 z}{\partial s^2}\mathbf{k}, \\ \frac{\partial \mathbf{r}}{\partial t} &= a\frac{\partial\varphi}{\partial t}\mathbf{e}_\varphi + \frac{\partial z}{\partial t}\mathbf{k}, \\ \frac{\partial^2 \mathbf{r}}{\partial t^2} &= -a\left(\frac{\partial\varphi}{\partial t}\right)^2 \mathbf{e}_r + a\frac{\partial^2\varphi}{\partial t^2}\mathbf{e}_\varphi + \frac{\partial^2 z}{\partial t^2}\mathbf{k}. \end{aligned} \tag{35.33}$$

The velocity of the string is

$$\frac{d\mathbf{r}}{dt} = \frac{\partial \mathbf{r}}{\partial t} + \frac{\partial \mathbf{r}}{\partial s} v, \tag{35.34}$$

where $v = ds/dt$ is the *transport velocity* of the string. The acceleration can then be obtained in the form

$$\begin{aligned} \frac{d^2\mathbf{r}}{dt^2} &= \frac{\partial}{\partial t}\left(\frac{d\mathbf{r}}{dt}\right) + v\frac{\partial}{\partial s}\left(\frac{d\mathbf{r}}{ds}\right) = \left(a\frac{\partial^2 \varphi}{\partial t^2} + 2av\frac{\partial^2 \varphi}{\partial s \partial t} + a\frac{\partial \varphi}{\partial s}\frac{\partial \varphi}{\partial t} + av^2\frac{\partial^2 \varphi}{\partial s^2}\right)\mathbf{e}_\varphi \\ &- \left[a\left(\frac{\partial \varphi}{\partial t}\right)^2 + 2av\frac{\partial \varphi}{\partial s}\frac{\partial \varphi}{\partial t} + av^2\left(\frac{\partial \varphi}{\partial s}\right)^2\right]\mathbf{e}_r + \left(\frac{\partial^2 z}{\partial t^2} + 2v\frac{\partial^2 z}{\partial s \partial t} + v^2\frac{\partial^2 z}{\partial s^2} + \frac{\partial z}{\partial s}\frac{\partial v}{\partial t}\right)\mathbf{k}. \end{aligned} \tag{35.35}$$

We now consider the forces acting on a small element ds, that is, the tension $\mathbf{T}$, the normal reaction $\mathbf{N}$, and the frictional force $\mathbf{F}$. The sum of tensions may be written as

$$d\mathbf{T} = T(s+ds)\mathbf{t}(s+ds) - T(s)\mathbf{t}(s), \tag{35.36}$$

where $T = |\mathbf{T}|$ and $\mathbf{t}$ is the unit vector to the string at s. On expanding (35.36) in a Taylor series and observing the results $\mathbf{t} = \partial\mathbf{r}/\partial s$, $\partial\mathbf{t}/\partial s = \partial^2\mathbf{r}/\partial s^2$, we get

$$d\mathbf{T} = T(s)\frac{\partial^2 \mathbf{r}}{\partial s^2}ds + \frac{\partial \mathbf{r}}{\partial s}\frac{\partial T}{\partial s}ds, \tag{35.37}$$

after having neglected the terms of higher order in ds. By substituting (35.33) in (35.37) we obtain

$$d\mathbf{T} = -Ta\left(\frac{\partial \varphi}{\partial s}\right)^2 ds\,\mathbf{e}_r + \left(\frac{\partial T}{\partial s}a\frac{\partial \varphi}{\partial s} + Ta\frac{\partial^2 \varphi}{\partial s^2}\right)ds\,\mathbf{e}_\varphi + \left(\frac{\partial T}{\partial s}\frac{\partial z}{\partial s} + T\frac{\partial^2 z}{\partial s^2}\right)ds\,\mathbf{k}. \tag{35.38}$$

The frictional forces can be written, according our first assumption, as

$$\mathbf{F} = -\begin{bmatrix}\mu_\varphi & 0 \\ 0 & \mu_z\end{bmatrix}|\mathbf{N}|\frac{\dfrac{d\mathbf{r}}{dt}}{\left|\dfrac{d\mathbf{r}}{dt}\right|}, \tag{35.39}$$

where $\mathbf{N} = N\mathbf{e}_r$ is the normal reacting force and we have introduced the matrix

$$\begin{bmatrix}\mu_\varphi & 0 \\ 0 & \mu_z\end{bmatrix}$$

in order to account for the different coefficients of friction in the φ and z directions. By using (35.31), (35.33), and (35.34), we have

$$\begin{aligned} \left|\frac{d\mathbf{r}}{dt}\right| &= \left[\left(a\frac{\partial \varphi}{\partial t} + av\frac{\partial \varphi}{\partial s}\right)^2 + \left(\frac{\partial z}{\partial t} + v\frac{\partial z}{\partial s}\right)^2\right]^{\frac{1}{2}} \\ &= v\left\{1 + \frac{2}{v}\left(a^2\frac{\partial \varphi}{\partial t}\frac{\partial \varphi}{\partial s} + \frac{\partial z}{\partial t}\frac{\partial z}{\partial s}\right) + \frac{1}{v^2}\left[a^2\left(\frac{\partial \varphi}{\partial t}\right)^2 + \left(\frac{\partial z}{\partial t}\right)^2\right]\right\}^{\frac{1}{2}}. \end{aligned} \tag{35.40}$$

The dynamic equation governing the motion of an element ds is

$$\rho\, ds\, \frac{d^2\mathbf{r}}{dt^2} = d\mathbf{T} + \mathbf{F} + \mathbf{N}, \tag{35.41}$$

which can be expressed in terms of components by using the formulae derived above. As these equations are highly nonlinear, a first possible procedure for simplifying them is to exploit the assumption that the lateral displacement and its derivatives with respect to s and t are small. This implies that, if θ is the slope of the string with respect the xy plane, we may take

$$\begin{gathered} \frac{a}{v}\frac{\partial\varphi}{\partial t} = 0, \quad \frac{1}{v}\frac{\partial z}{\partial t} \ll 1, \quad \frac{\partial z}{\partial s} = \sin\theta \ll 1, \\ a\,\frac{\partial\varphi}{\partial s} = \cos\theta \simeq 1. \end{gathered} \tag{35.42}$$

On using these relations and neglecting the small terms in (35.40), we get

$$\left|\frac{d\mathbf{r}}{dt}\right| \simeq \frac{1}{v},$$

so that (35.39) becomes

$$\mathbf{F} = -\frac{N}{v}\begin{bmatrix} \mu_\varphi & 0 \\ 0 & \mu_z \end{bmatrix}\frac{d\mathbf{r}}{dt}. \tag{35.43}$$

By further substituting (35.35), (35.38), and (35.43) in (35.41), we can derive the three scalar equations

$$N = \left[a(T - \rho v^2)\left(\frac{\partial\varphi}{\partial s}\right)^2 + \rho a\left(\frac{\partial\varphi}{\partial t}\right)^2 - 2a\rho v\,\frac{\partial\varphi}{\partial s}\frac{\partial\varphi}{\partial t} \right] ds, \tag{35.44}$$

$$\begin{aligned} &\left(\rho a\,\frac{\partial^2\varphi}{\partial t^2} + 2a\rho v\,\frac{\partial^2\varphi}{\partial s\,\partial t} + a\rho\,\frac{\partial\varphi}{\partial s}\frac{\partial v}{\partial t} + a\rho v^2\,\frac{\partial^2\varphi}{\partial s^2} - a\,\frac{\partial\varphi}{\partial s}\frac{\partial T}{\partial s} - Ta\,\frac{\partial^2\varphi}{\partial s^2}\right) ds \\ &+ \frac{\mu_\varphi N}{v}\left(a\,\frac{\partial\varphi}{\partial t} + av\,\frac{\partial\varphi}{\partial s}\right) = 0, \end{aligned} \tag{35.45}$$

$$\left(\rho\,\frac{\partial^2 z}{\partial t^2} + 2\rho v\,\frac{\partial^2 z}{\partial s\,\partial t} + \rho v^2\,\frac{\partial^2 z}{\partial s^2} + \rho\,\frac{\partial z}{\partial s}\frac{\partial v}{\partial s} - \frac{\partial T}{\partial s}\frac{\partial z}{\partial s} - T\,\frac{\partial^2 z}{\partial s^2}\right) ds + \frac{\mu_z N}{v}\left(\frac{\partial z}{\partial t} + v\,\frac{\partial z}{\partial s}\right) = 0. \tag{35.46}$$

Also, substitution of (35.44) in (35.45) and (35.46) yields the system

$$\begin{aligned} &\rho\left(a\,\frac{\partial^2\varphi}{\partial t^2} + 2av\,\frac{\partial^2\varphi}{\partial s\,\partial t} + av^2\,\frac{\partial^2\varphi}{\partial s^2} + a\,\frac{\partial\varphi}{\partial s}\frac{\partial v}{\partial t}\right) \\ &= a\,\frac{\partial}{\partial s}\left(T\,\frac{\partial\varphi}{\partial s}\right) - \mu_\varphi\left[a(T - \rho v^2)\left(\frac{\partial\varphi}{\partial s}\right)^2 - \rho a\left(\frac{\partial\varphi}{\partial t}\right)^2 - 2a\rho v\,\frac{\partial\varphi}{\partial s}\frac{\partial\varphi}{\partial t}\right]\left(\frac{a}{v}\frac{\partial\varphi}{\partial t} + a\,\frac{\partial\varphi}{\partial s}\right), \end{aligned} \tag{35.47}$$

$$\rho\left(a\frac{\partial^2 z}{\partial t^2}+2v\frac{\partial^2 z}{\partial s\,\partial t}+v^2\frac{\partial^2 z}{\partial s^2}+\frac{\partial z}{\partial s}\frac{\partial z}{\partial t}\right)$$
$$=\frac{\partial}{\partial s}\left(T\frac{\partial z}{\partial s}\right)-\mu_z\left[a(T-\rho v^2)\left(\frac{\partial\varphi}{\partial s}\right)^2-\rho a\left(\frac{\partial\varphi}{\partial t}\right)^2-2a\rho v\frac{\partial\varphi}{\partial t}\frac{\partial\varphi}{\partial s}\right]\left(\frac{1}{v}\frac{\partial z}{\partial t}+a\frac{\partial z}{\partial s}\right). \tag{35.48}$$

The foregoing equations can be simplified further if we observe that the inertial forces are negligibly small with respect to the stretching force, normal reacting force, and frictional forces. This allows us to assume the following results:

$$\frac{\partial\varphi}{\partial t}=\frac{\partial^2\varphi}{\partial t^2}=0,\quad a\frac{\partial^2\varphi}{\partial s^2}=-\sin\theta\frac{\partial\theta}{\partial s}\ll 1,\quad \frac{\partial z}{\partial s}=\sin\theta\ll 1,\quad a\frac{\partial\varphi}{\partial s}=\cos\theta\simeq 1.$$

Accordingly, (35.47) and (35.48) are reduced to

$$\frac{\partial T}{\partial s}=\frac{\mu_\varphi}{a}T, \tag{35.49a}$$

$$\frac{\partial}{\partial s}\left(T\frac{\partial z}{\partial s}\right)=\frac{\mu_z}{av}T\left(\frac{\partial z}{\partial t}+v\frac{\partial z}{\partial s}\right). \tag{35.49b}$$

On integrating the first of these equations under the condition that $T=T_1$ at $s=s_1$, we obtain the known equation

$$T=T_1\exp\left[\frac{\mu_\varphi}{a}(s-s_1)\right].$$

Also, on putting this equation into (35.49b) we have

$$\frac{\partial^2 z}{\partial s^2}-\frac{(\mu_z-\mu_\varphi)}{a}\frac{\partial z}{\partial s}-\frac{\mu_z}{av}\frac{\partial z}{\partial t}=0. \tag{35.50}$$

This is the basic equation governing the lateral motion of an axially moving string on a cylindrical guide surface. The interesting property of this equation is that, on neglecting the inertial terms, we obtain a single partial differential equation of parabolic type. In the particular case where $\mu_z=\mu_\varphi$, (35.50) is exactly the same as the one-dimensional heat equation.

Chapter V
Theories of Membranes

36. Constitutive Properties of Membranes

A *curved membrane* can be described geometrically by means of its middle surface, its contour, and its thickness. We shall take, for the moment, the thickness to be constant and equal to $2h$, so that the upper and lower faces of the membrane are two surfaces each at a distance h from the middle surface and situated on opposite sides of it. We draw a closed curve s on the middle surface and consider the surface described by the normals to the middle surface drawn from the points of s and bounded by the two outer surfaces of the membrane with spacing $2h$. We call this surface $\mathscr{C}$. The edge of the membrane is a surface like $\mathscr{C}$. If s is smooth, we can define the outer normal $\mathbf{v}$ at any point on s, so that $\mathbf{v}$ lies in the tangent plane to the middle surface, while $\mathbf{s}$ is the tangent to the line s, and $\mathbf{n}$ the normal to the middle surface, so that $\mathbf{v}$, $\mathbf{s}$, and $\mathbf{n}$ form a right-handed triad.

Let δs be a short arc of the curve s and take two generating lines of $\mathscr{C}$ drawn through the extremities of δs so as to include an area δA of $\mathscr{C}$. The tractions on the area δA are statically equivalent to a force at the centroid of δA together with a couple. The components of the forces in the $\mathbf{v}$, $\mathbf{s}$, and $\mathbf{n}$ directions are denoted by δT, δS, and δN, and those of the couple are denoted by δH, δG, and δK. So far we have introduced no specific hypothesis that distinguishes membranes from other continua having the same geometrical definition, called *shells*. The distinction arises from the constitutive behavior of membranes of being unable to transmit couples or forces perpendicular to the middle surface. In mathematical terms, this property is equivalent to the fact that, when δs is diminished indefinitely, the limits $\delta H/\delta s$, $\delta G/\delta s$, and $\delta K/\delta s$ are zero, as is the limit of $\delta N/\delta s$. Only the limits $\delta T/\delta s$ and $\delta S/\delta s$ can be finite. Expressed in another way, a membrane is the two-dimensional analog of a string, that is to say, it is a two-dimensional continuum such that the only forces interacting between its parts are tangential to it. We denote the limit of $\delta T/\delta s$ by T, and the limit of $\delta S/\delta s$ by S. T is the *tension* in the membrane and S is the *shearing force* tangential to the middle surface.

The forces T and S are called the characteristics of the stress in the membrane and a nontrivial problem is to find a rigorous relationship between them and the individual stress components defined in the three-dimensional continuum composed of the solid layer of thickness $2h$, which defines the membrane. The difficulty arises because stresses are automatically defined in a strained configuration, whereas in general we only know the unstrained configuration of the body. This difficulty can be overcome by assuming that the strains are so small that we can identify the strained with unstrained configurations by defining stresses, as we do when we accept the approximations of the linear theory. However, membranes are bodies characterized by a high degree of deformability, and membrane theory is required to apply to clothes, inflatable shells, liquid drops, biological tissues, and other cases in which linear theory is clearly inadequate. Furthermore, as membranes are bodies capable of transmitting only tensile forces, we have the complication that, if the external loads and boundary conditions are such that they generate compressions in some parts of the membrane, these parts automatically go out of service. However, their shape and extent are not known in advance and must be determined as a part of the solution.

In contrast to the rod and plate theories, membrane theory has not been considered as a fundamental subject in structural mechanics, probably because the traditional concept of structure is related to an essentially stiff object and not to a soft body. It is therefore not surprising that a systematic theory of membranes has only been developed for specialized purposes, such as in acoustics (where a membrane is a fundamental element for transmitting and receiving sound waves (Rayleigh 1894)), in heat conduction, and in fluid mechanics (where some two-dimensional problems are governed by the same equation as that describing the equilibrium of a plane membrane).

The revival of interest in membrane theory occurred only in the second half of the present century, when materials such as rubber, paper, and gauze became frequently employed for many purposes.

37. Plane Membranes

If the middle surface of a membrane is initially planar, we can represent the region so occupied as a domain D of the xy plane bounded by a closed continuously differentiable curve C, to which the membrane is rigidly attached. However, as the physical membrane is a three-dimensional body of thickness $2h$ and cross-section D, instead of local stresses we consider their averages T and S. Furthermore, we assume that the motion is described with sufficient approximation by the single function $z = u(x, y, t)$, representing the transverse displacement of the middle surface. In this case the problem is purely two-dimensional, since T, S, and u depend only on the variables (x, y, t). We also make assumptions similar to those made in deriving the equation for a vibrating string; that is, we assume that displacement gradients in the x and y directions are small, as are the velocities in these directions. In addition, we assume that the membrane, in its planar equilibrium position, is subjected to a constant tensile force T_0 per unit length along the normal to C in the xy plane, so that the resultant stresses in the interior are $S = 0$ and $T = T_0 =$ constant. During the motion, the membrane is stretched, and therefore the resulting stresses change, but we neglect these variations and assume that S is very small and $T \simeq T_0$. Then, if $\rho(x, y)$ is a continuous function representing the mass per unit area and $F(x, y, t)$ is the force per unit mass in the z direction, the motion is approximately described by the equation

$$\frac{\partial^2 u}{\partial t^2} - c^2(x, y)(u_{xx} + u_{yy}) = F(x, y, t), \tag{37.1}$$

where $c^2 = T_0/\rho$, and with suitable initial and boundary conditions.

Equation (37.1) describes the small motions of a planar membrane. If we wish to consider the equilibrium shape of the membrane, we must know the force $F(x, y)$, independent of t, and prescribe the boundary conditions. For example, we might fix the vertical displacement $f(x, y)$ of the boundary. Then, for convenience taking $c = 1$, the tranverse displacement is a function $u(x, y)$, independent of t, such that

$$\nabla^2 u = u_{xx} + u_{yy} = -F(x, y) \text{ in } D, \tag{37.2a}$$

$$u(x, y) = f(x, y) \text{ on } C. \tag{37.2b}$$

Problem (37.2) is called a *Dirichlet problem* for the Laplace operator. The Laplace operator is elliptic and a natural expectation is that the problem admits a solution and that this solution is unique. This is the case if D is bounded, and the data are like those prescribed in (37.2), with both F and f sufficiently smooth. However, we may also think of formulating the problem in a different way (Weinberger 1965, Sec. 10). For instance, we might consider the following boundary-value problem

$$\begin{aligned} &\nabla^2 u = 0, && \text{for } \ 0 < x < 1, \ y > 0; \\ &u(x, 0) = f(x), && \text{for } \ 0 \leqslant x \leqslant 1; \\ &\frac{\partial u}{\partial y}(x, 0) = 0, && \text{for } \ 0 \leqslant x \leqslant 1; \\ &u(0, y) = u(1, y) = 0. \end{aligned} \tag{37.3}$$

This problem admits a unique solution, but the stability with respect to the data breaks down, as shown by a well-known example due to Hadamard. The function

$$u_n(x, y) = e^{-\sqrt{n}} \cosh(4n + 1)\pi y \sin(4n + 1)\pi x, \tag{37.4}$$

where n, any positive integer, is the solution of (37.3) with $f = e^{-\sqrt{n}} \sin(4n + 1)\pi x$. By choosing n sufficiently large we can render f and its derivatives of any order arbitrarily small. On the other hand,

$$u_n(\tfrac{1}{2}, y) = e^{-\sqrt{n}} \cosh(4n + 1)\pi y, \tag{37.5}$$

can be made arbitrarily large for any fixed $y > 0$ by choosing n sufficiently large. This shows that the solution is not continuous with respect to the initial data. On recalling that $u(x, y)$ represents the deflection of a membrane, this lack of continuity implies that it is impossible to compute the deflection at an interior point from measurements subject to an error in the datum prescribed on the segment $0 \leqslant x \leqslant 1$, $y = 0$. Any measurement, no matter how accurate, will fail to distinguish between two initial distributions differing by u_n if n is sufficiently large.

Another remarkable property of elliptic second-order equations is that their solutions obey a *maximum principle*. Let $u(x, y)$ be a solution of the equation

$$\nabla^2 u = -F(x, y) \text{ in } D, \tag{37.6}$$

for $F < 0$, and let u be continuous in $D + C$. Then it attains its maximum M somewhere in $D + C$. If the point of maximum (x_0, y_0) is in D, and the function $u(x, y)$ is continuous in (x_0, y_0) with its first and second partial derivatives, then we must have

$$\frac{\partial u}{\partial x} = \frac{\partial u}{\partial y} = 0, \quad \frac{\partial^2 u}{\partial x^2} \leqslant 0, \quad \frac{\partial^2 u}{\partial y^2} \leqslant 0 \quad \text{at } (x_0, y_0).$$

This means that we have $\nabla^2 u \leqslant 0$ in (x_0, y_0), which contradicts the hypothesis that $F < 0$. Therefore, the maximum of $u(x, y)$ must occur on C. Viewed as the deflection of a membrane, the maximum principle states that a strictly downwards directed force $F(x, y)$ at any point on the membrane cannot produce an upward bulge. The consequences of the maximum principle are particularly useful in approximating solutions and describing them qualitatively (Protter and Weinberger 1967).

A formidable amount of effort has been spent on devising methods for solving (37.2) in explicit form. The most powerful of these methods is based on a suitable change of independent variables that preserves the equation $\nabla^2 u = 0$. More precisely, let $u(x, y)$ be a harmonic function, and let us introduce new coordinates ξ, η via the transformation

$$\begin{aligned} x &= x(\xi, \eta), \\ y &= y(\xi, \eta), \end{aligned} \tag{37.7}$$

such that the transformed function

$$U(\xi, \eta) = u[x(\xi, \eta), y(\xi, \eta)] \tag{37.8}$$

is a harmonic function of ξ and η. By the chain rule we can write

$$\begin{aligned} \nabla^2 U = U_{\xi\xi} + U_{\eta\eta} = {} & u_{xx}(x_\xi^2 + x_\eta^2) + 2u_{xy}(x_\xi y_\xi + x_\eta y_\eta) \\ & + u_{yy}(y_\xi^2 + y_\eta^2) + u_x(x_{\xi\xi} + x_{\eta\eta}) + u_y(y_{\xi\xi} + y_{\eta\eta}). \end{aligned}$$

Now U is harmonic, provided that u is also harmonic and the following conditions are satisfied

$$x_\xi^2 + x_\eta^2 = y_\xi^2 + y_\eta^2, \tag{37.9a}$$

$$x_\xi y_\xi + x_\eta y_\eta = 0, \tag{37.9b}$$

$$x_{\xi\xi} + x_{\eta\eta} = 0, \tag{37.9c}$$

$$y_{\xi\xi} + y_{\eta\eta} = 0. \tag{37.9d}$$

On solving the first two equations for x_ξ and x_η, we find

$$x_\xi = \pm y_\eta, \quad x_\eta = \mp y_\xi. \tag{37.10}$$

If we choose the upper signs in these two equations, we have the Cauchy–Riemann equations, which state that $z = x + iy$ is analytic in $\zeta = \xi + i\eta$. The result is the same if we reverse the signs, but with y replaced by $-y$. In either case, (37.9c) and (37.9d) are a consequence of the Cauchy–Reimann equations. This means that, if we make the transformation

$$z = f(\zeta), \tag{37.11}$$

where $f(\zeta)$ is an analytic function, harmonic functions of x and y are transformed into harmonic functions in ξ and η. Note that the change of y into $-y$ is irrelevant, because it is equivalent to the change of $f(\zeta)$ into $\overline{f(\zeta)}$. For the function $f(\zeta)$ we also assume the conditions $f'(\zeta) \neq 0$ and $f(\zeta_1) \neq f(\zeta_2)$ whenever $\zeta_1 \neq \zeta_2$. This ensures that $f(\zeta)$ has an analytic inverse function.

The function $f(\zeta)$, satisfying the above properties, takes a domain D^* in the ζ plane into a domain D in the z plane, and the boundary C^* of D^* into the boundary C and D. Furthermore, the mapping is *conformal* because it preserves the angles between curves. The knowledge of the mapping $f(\zeta)$ allows the Dirichlet problem to be solved for a domain D whenever we can solve the problem in D^*. If the solution in D^* is explicit, so is the solution in D. As an example (Weinberger 1965, Sec. 52), let us consider an initially planar membrane occupying a bounded, simply connected domain D in the $z = x + iy$ plane, with a boundary C. The membrane is clamped along C and subject to the action of a unit point load applied at a point $z_1 = x_1 + iy_1$ contained within D. The explicit solution of this problem is, in general, difficult, but it is almost immediate if we know the analytical function $\zeta = g(z)$, which is the one-to-one conformal mapping from D in the z plane into the circle $|\zeta| < 1$. Let us call $\zeta_1 = g(z_1)$ the image of z_1 in the ζ plane. The deflection of a circular membrane of unit radius, fixed along the boundary and loaded by a unit transverse force at ζ_1, is classical and reads

$$-\frac{1}{2\pi}\ln|\zeta - \zeta_1| + \frac{1}{2\pi}\ln|1 - \zeta\bar{\zeta}_1|. \tag{37.12}$$

Formula (37.12) is known as Green's function for the circle $|\zeta| < 1$. Preserving the harmonic property in conformal mapping ensures that the function

$$-\frac{1}{2\pi}\ln|g(z) - g(z_1)| + \frac{1}{2\pi}\ln|1 - g(z)\overline{g(z_1)}| \tag{37.13}$$

is Green's function for D. Note that $g(z)$ is the inverse of an analytic function $f(\zeta)$ with $f'(\zeta) \neq 0$ for $|\zeta| < 1$, but not necessarily for $|\zeta| = 1$. It must be continuous for $|\zeta| \leqslant 1$.

If the membrane, instead of being fixed along C, is subject along C to vertical forces proportional to a given function $h(x, y)$, the field equation of the problem is still (37.2a), but the boundary condition becomes

$$\frac{\partial u}{\partial n} = h(x, y) \text{ on } C. \tag{37.14}$$

The new problem, which is called a Neumann problem, admits a solution provided that the following compatibility equation holds:

$$\iint_D F(x, y)\, dS + \oint_C h(x, y)\, ds = 0. \tag{37.15}$$

Condition (37.15) expresses the natural requirement that, if vertical forces are prescribed in D and on the boundary C, the resultant of these forces must be zero.

The Neumann problem can also be treated by conformal mapping. Let us suppose, for simplicity, that $F(x, y) \equiv 0$. Then the transformation $z = f(\zeta)$ maps harmonic functions in the z plane into harmonic functions in the ζ plane. However, the boundary conditions $\partial u/\partial n = h$ can also be transferred by a conformal mapping (Weinberger 1965, Sec. 52). To prove this property, let $z = z(t)$ be the equation of the curve C in the z plane. Then we have the identity

$$\frac{\partial u}{\partial n} = \frac{1}{\left|\frac{dz}{dt}\right|}\left(u_x \frac{dy}{dt} - u_y \frac{dx}{dt}\right) = \frac{1}{\left|\frac{dz}{dt}\right|} \operatorname{Im}\left[(u_x - iu_y)\frac{dz}{dt}\right].$$

Now, under the mapping $\zeta = g(z)$, the curve C goes into C^* of the equation $\zeta = g[z(t)]$. By the chain rule, if U is given by (37.8), we find

$$\frac{\partial U}{\partial n} = \frac{1}{\left|\frac{d\zeta}{dt}\right|} \operatorname{Im}\left[(U_\xi - iU_\eta)\frac{d\zeta}{dt}\right] = \frac{1}{\left|\frac{d\zeta}{dt}\right|} \operatorname{Im}\left\{[u_x(x_\xi - ix_\eta) + u_y(y_\xi - iy_\eta)]\frac{d\zeta}{dt}\right\}$$

$$= \frac{1}{\left|\frac{d\zeta}{dt}\right|}\left[(u_x - iu_y)\frac{dz}{d\zeta}\frac{d\zeta}{dt}\right] = \frac{1}{\left|\frac{d\zeta}{dt}\right|}\left|\frac{dz}{dt}\right|\frac{\partial u}{\partial n} = \frac{1}{|g'|}\frac{\partial u}{\partial n}. \tag{37.16}$$

The normal derivative of U on C^* is thus given in terms of the normal derivative of u at the corresponding points on C.

The method of conformal mapping fails when u and its gradients are large. When this occurs the equilibrium equation is nonlinear and its solution is possible only in some particular cases. In 1907, Föppl proposed a semilinear theory for studying moderately large deformations of membranes. This theory proved to be of particular interest in the study of radially symmetric deformations of an initially planar, circular membrane, under normal loads. The equations derived by Föppl are the result of three sets of assumptions. The first is that the deformed state of the membrane is described both by the normal displacement and by the displacement components parallel to the middle surface; a second hypothesis is that stresses and strains are related by Lamé's equations; and the third assumption is that large strains are taken into account. For an axisymmetric deformation the strains take the form

$$\varepsilon_r = \frac{du}{dr} + \frac{1}{2}\left(\frac{dw}{dr}\right)^2, \quad \varepsilon_\theta = \frac{u}{r}, \tag{37.17}$$

where u and w are the displacements in the radial and normal directions with w positive downwards and u positive if away from the center. These strains are related to the stresses σ_r and σ_θ in the radial and circumferential directions by Hooke's law

$$E\varepsilon_r = \sigma_r - \sigma\sigma_\theta, \quad E\varepsilon_\theta = \sigma_\theta - \sigma\sigma_r, \tag{37.18}$$

where E is Young's modulus and σ is the Poisson ratio. Let us call R the radius of the membrane, h its thickness and p the pressure. For equilibrium we must have

$$(r\sigma_r)_r - \sigma_\theta = 0, \quad \sigma_r \frac{dw}{dr} = -\frac{pr}{2h}. \tag{37.19}$$

At the boundary $r = R$, either the stress is prescribed

$$\sigma_r(R) = S, \tag{37.20}$$

or the radial displacement is prevented

$$u(R) = 0. \tag{37.21}$$

In both cases, the condition $w(R) = 0$ is assumed and all quantities are taken to remain finite at $r = 0$.

Solutions to this problem were found by Hencky (1915) through expansion of σ_r as a power series in r and evaluation of the coefficients from the differential equation and the boundary conditions. However, it is not clear whether the formal power series converges, or even whether the problem admits a solution. Existence and uniqueness theorems for the problem, with either boundary condition (37.20) or (37.21), can be obtained by transforming the differential problem into an integral equation and solving this by iteration (Dickey 1967). A different method, which is less direct but still constructive, is the "shooting method," but its application requires some care because the differential equation is singular (Callegari and Reiss 1970).

To illustrate the iteration procedure, we introduce the following notation:

$$\rho \equiv \frac{r}{R}, \quad 0 \leqslant \rho \leqslant 1, \quad \kappa^3 = \frac{Ep^2 R^2}{64h^2}, \tag{37.22}$$

so that (37.19) can be reduced to a single second-order differential equation for the radial stress

$$\sigma_r^2 \left(\frac{d^2\sigma_r}{d\rho^2} + \frac{3}{\rho} \frac{d\sigma_r}{d\rho} \right) + 8\kappa^3 = 0. \tag{37.23}$$

This is the equation derived by Hencky and treated by the method of power expansion in ρ. In order to derive an integral equation equivalent to (37.23), it is convenient to begin by considering the boundary condition (37.20). If we have $p \neq 0$, it is possible to divide (37.23) by $4^{\frac{1}{3}}\kappa$ and to rewrite the equation in the form

$$\frac{d^2\alpha}{d\rho^2} + \frac{3}{\rho} \frac{d\alpha}{d\rho} + \frac{2}{\alpha^2} = 0 \tag{37.24}$$

with $\alpha = \sigma_r / 4^{\frac{1}{3}}\kappa$. In this case, the boundary condition (37.20) becomes

$$\alpha(1) = \frac{S}{4^{\frac{1}{3}}\kappa} = \lambda; \tag{37.25}$$

moreover, $\alpha(r)$ must be regular at the origin.

Let us now multiply (37.24) by ρ^3 and integrate the result from 1 to ρ:

$$\rho^3 \alpha'(\rho) - \alpha'(1) = -2 \int_1^{\rho} \frac{\tau^3 \, d\tau}{\alpha^2}, \tag{37.26}$$

where the prime denotes differentiation with respect to ρ. A second integration yields

$$\alpha(\rho) = \alpha(1) - \frac{\alpha'(1)}{2\tau^2}\bigg|_1^\rho - 2\int_1^\rho \frac{1}{\omega^3}\left(\int_1^\omega \frac{\tau^3\,d\tau}{\alpha^2}\right)d\omega,$$

or, on using (37.25),

$$\alpha(\rho) = \lambda + \frac{\alpha'(1)}{2}\left(1 - \frac{1}{\rho^2}\right) - 2\int_1^\rho \frac{1}{\omega^3}\left(\int_1^\omega \frac{\tau^3\,d\tau}{\alpha^2}\right)d\omega. \tag{37.27}$$

The double integral in (37.27) may be simplified by means of an integration by parts:

$$\alpha(\rho) = \lambda + \frac{\alpha'(1)}{2}\left(1 - \frac{1}{\rho^2}\right) + \frac{1}{\rho^2}\int_1^\rho \frac{\tau^3\,d\tau}{\alpha^2} - \int_1^\rho \frac{\tau\,d\tau}{\alpha^2}. \tag{37.28}$$

In order to eliminate the singularity of the type $1/\rho^2$, we multiply (37.28) by ρ^2 so that

$$\rho^2\alpha(\rho) = \lambda\rho^2 + \frac{\alpha'(1)}{2}(\rho^2 - 1) + \int_1^\rho \frac{\tau^3\,d\tau}{\alpha^2} - \rho^2\int_1^\rho \frac{\tau\,d\tau}{\alpha^2}. \tag{37.29}$$

For $\alpha(\rho) \neq 0$ with $0 \leqslant \rho \leqslant 1$, we found, by letting $\rho \to 0$, that

$$\alpha'(1) = -2\int_0^1 \frac{\tau^3\,d\tau}{\alpha^2},$$

and hence, on substituting this expression for $\alpha'(1)$ in (37.27), we obtain

$$\alpha(\rho) = \lambda - \int_0^1 \frac{\tau^3\,d\tau}{\alpha^2} + \frac{1}{\rho^2}\int_0^\rho \frac{\tau^3\,d\tau}{\alpha^2} + \int_\rho^1 \frac{\tau\,d\tau}{\alpha^2}. \tag{37.30}$$

This equation has no singularity at $\rho = 0$, $\alpha(0)$ being

$$\alpha(0) = \lambda + \int_0^1 \frac{\tau - \tau^3}{\alpha^2}\,d\tau. \tag{37.31}$$

In order to solve (37.30) we rewrite (37.30) in the form

$$\alpha(\rho) = \lambda + \int_0^\rho \frac{\tau^3/\rho^2 - \tau^3}{\alpha^2}\,d\tau + \int_\rho^1 \frac{\tau - \tau^3}{\alpha^2}\,d\tau. \tag{37.32}$$

As $\rho \leqslant 1$, the numerator is positive, $\tau^3/\rho^2 - \tau^3 \geqslant 0$; similarly, the condition $0 \leqslant \tau \leqslant 1$ implies that $\tau - \tau^3 \geqslant 0$. We thus have the lower bound

$$\alpha(\rho) \geqslant \lambda. \tag{37.33}$$

Having obtained a lower bound, it is possible to obtain an upper bound by putting λ in place of α in (37.32):

$$\alpha(\rho) \leqslant \lambda + \frac{1}{\lambda^2}\left(\frac{1}{\rho^2}\int_0^\rho \tau^3\,d\tau - \int_0^\rho \tau^3\,d\tau + \int_\rho^1 \tau\,d\tau - \int_\rho^1 \tau^3\,d\tau\right) = \lambda + \frac{1}{4\lambda^2}(1-\rho^2). \tag{37.34}$$

In order to create an iterative sequence from (37.32), we define the $(n+1)$th term α_{n+1} as follows:

$$\alpha_{n+1} = \lambda + \int_0^\rho \frac{\tau^3/\rho^2 - \tau^3}{\alpha_n^2}\,d\tau + \int_\rho^1 \frac{\tau - \tau^3}{\alpha_n^2}\,d\tau, \tag{37.35}$$

with $\alpha_0 = \lambda$. We can now show that the terms form a nested sequence of functions, that is to say

$$\alpha_0 \leqslant \alpha_2 \leqslant \ldots \leqslant \alpha_{2n} \leqslant \alpha_{2n+1} \leqslant \ldots \leqslant \alpha_3 \leqslant \alpha_1. \tag{37.36}$$

The proof is achieved by induction. Defining $I(\gamma)$ by

$$I(\gamma) = \int_0^\rho \left(\frac{\tau^3}{\rho^2} - \tau^3\right)\gamma\,d\tau + \int_\rho^1 (\tau - \tau^3)\gamma\,d\tau,$$

and noting that I is positive if its argument γ is positive, in this notation the iteration scheme (37.35) becomes

$$\alpha_{n+1} = \lambda + I\left(\frac{1}{\alpha_n^2}\right). \tag{37.37}$$

For $n = 0$, we write $\alpha_1 = \lambda + I(1/\alpha_0^2) \geqslant \lambda = \alpha_0$, giving $\alpha_1 \geqslant \alpha_0$. For $n = 1$, we have

$$\alpha_2 - \alpha_0 = \lambda + I\left(\frac{1}{\alpha_1^2}\right) - \lambda = I\left(\frac{1}{\alpha_1^2}\right) \geqslant 0,$$

$$\alpha_1 - \alpha_2 = I\left(\frac{1}{\alpha_0^2}\right) - I\left(\frac{1}{\alpha_1^2}\right) = I\left[\frac{\alpha_1 + \alpha_0}{\alpha_1^2\alpha_2^2}(\alpha_1 - \alpha_0)\right] \geqslant 0,$$

since we have $\alpha_1 \geqslant \alpha_0$, and the inequalities $\alpha_1 \geqslant \alpha_2 \geqslant \alpha_0$ follow. In order to complete the proof for $n = 1$, we observe the results

$$\alpha_3 - \alpha_2 = I\left(\frac{1}{\alpha_2^2}\right) - I\left(\frac{1}{\alpha_1^2}\right) = I\left[\frac{\alpha_1 + \alpha_2}{\alpha_1^2\alpha_2^2}(\alpha_1 - \alpha_2)\right] \geqslant 0,$$

$$\alpha_1 - \alpha_3 = I\left(\frac{1}{\alpha_0^2}\right) - I\left(\frac{1}{\alpha_2^2}\right) = I\left[\frac{\alpha_2 + \alpha_0}{\alpha_2^2\alpha_0^2}(\alpha_2 - \alpha_0)\right] \geqslant 0,$$

from which the results $\alpha_0 \leqslant \alpha_2 \leqslant \alpha_3 \leqslant \alpha_1$ are obtained. This proves that the inequalities (37.36) hold for $n = 1$. The argument can be repeated for any n.

The iteration procedure defines a bounded monotonically increasing sequence $\{\alpha_{2n}\}$ and a bounded monotonically decreasing sequence $\{\alpha_{2n+1}\}$. Both sequences are composed of functions which are not only continuous, but also equicontinuous, because all functions have a uniform bound for all n. In fact, differentiation of (37.35) yields

$$\frac{d\alpha_{n+1}}{d\rho} = -\frac{2}{\rho^3}\int_0^\rho \frac{\tau^3}{\alpha_n^2}\,d\tau,$$

or

$$\left|\frac{d\alpha_{n+1}}{d\rho}\right| \leqslant \frac{\rho}{2\lambda^2} \leqslant \frac{1}{2\lambda^2}.$$

Hence the two sequences $\{\alpha_{2n}\}$ and $\{\alpha_{2n+1}\}$ converge to two limit functions α_- and α_+, respectively, such that we have

$$\lim_{n\to\infty} \alpha_{2n} = \alpha_- \leqslant \alpha_+ = \lim_{n\to\infty} \alpha_{2n+1},$$

where α_- and α_+ are continuous.

Thus a solution of (37.30) exists if $\alpha_- = \alpha_+$. To decide under which conditions this is the case, we introduce a new independent variable $\gamma = \alpha_+ - \alpha_-$. The variable γ satisfies the linear integral equation

$$\gamma = I\left(\frac{\alpha_+ + \alpha_-}{\alpha_+^2\alpha_-^2}\,\gamma\right), \tag{37.38}$$

or, equivalently, the linear ordinary differential equation

$$\frac{d^2\gamma}{d\rho^2} + \frac{3}{\rho}\frac{d\gamma}{d\rho} + 2\,\frac{\alpha_+ + \alpha_-}{\alpha_+^2\alpha_-^2}\,\gamma = 0, \tag{37.39}$$

$$\gamma'(0) = \gamma(1) = 0. \tag{37.40}$$

Then the integral equation (37.38) will have a solution if (37.39), together with the boundary conditions (37.40), has only the trivial solution. A sufficient condition to ensure this fact may be found by observing that, from (37.33), we have

$$\frac{\alpha_+ + \alpha_-}{\alpha_+^2\alpha_-^2} = \frac{1 + (\alpha_-/\alpha_+)}{\alpha_+\alpha_-^2} \leqslant \frac{2}{\lambda^3},$$

and the related equation in γ

$$\frac{d^2\gamma}{d\rho^2} + \frac{3}{\rho}\frac{d\gamma}{d\rho} + \frac{4}{\lambda^3}\,\gamma = 0. \tag{37.41}$$

The general solution of (37.41) under the boundary condition $\gamma'(0) = 0$ is

$$\gamma = A\,\frac{1}{\rho}\,J_1(2\lambda^{-\frac{3}{2}}\rho), \tag{37.42}$$

where J_1 is the first-order Bessel function. Thus (37.41) has only the trivial solution for $2\lambda^{-\frac{3}{2}} < j_{11}$, where j_{11} ($\simeq 3.84$) is the smallest root of J_1. Now, on returning to (37.39), Sturm's comparison theorem shows that this equation has a unique solution provided that we have $\lambda^3 > 4/j_{11}^2$.

This result proves the existence of a unique solution for the membrane problem under boundary conditions of the type (37.20). We must now discuss the case of the boundary conditions (37.21). The main idea is very simple. As we are able to solve the problem with prescribed tension λ at the boundary, we vary λ and try to adjust it so that condition (37.21) is satisfied. In order to write these conditions explicitly, it is necessary to express the displacements u and w in terms of α. Let us define a new variable β as

$$\beta = \frac{\sigma_\theta}{4^{\frac{1}{3}}\kappa}, \tag{37.43}$$

so that the strain–displacement equations may be written in the form

$$\varepsilon_r = \frac{1}{R}\frac{du}{d\rho} + \frac{1}{2R^2}\left(\frac{dw}{d\rho}\right)^2, \tag{37.44a}$$

$$\varepsilon_\theta = \frac{u}{R\rho}. \tag{37.44b}$$

The stress–strain relations then become

$$E\varepsilon_r = 4^{\frac{1}{3}}\kappa(\alpha - \sigma\beta), \quad E\varepsilon_\theta = 4^{\frac{1}{3}}\kappa(\beta - \sigma\alpha), \tag{37.45}$$

and the equilibrium equations reduce to

$$\frac{d}{d\rho}(\rho\alpha) = \beta, \tag{37.46a}$$

$$\alpha\,\frac{dw}{d\rho} = -\frac{4^{\frac{2}{3}}\kappa^{\frac{1}{2}}}{E^{\frac{1}{2}}}\,R\rho. \tag{37.46b}$$

Using (37.46a), we can find β in terms of α:

$$\beta = \alpha + \rho\,\frac{d\alpha}{d\rho},$$

where we already have α from (37.30). By differentiating (37.30) with respect to ρ we obtain

$$\frac{d\alpha}{d\rho} = -\frac{2}{\rho^3}\int_0^\rho \frac{\tau^3\,d\tau}{\alpha^2}, \tag{37.47}$$

and so we can calculate β as

$$\beta = \lambda - \int_0^1 \frac{\tau^3\,d\tau}{\alpha^2} - \frac{1}{\rho^2}\int_0^\rho \frac{\tau^3\,d\tau}{\alpha^2} + \int_\rho^1 \frac{\tau\,d\tau}{\alpha^2}. \tag{37.48}$$

The strains ε_r and ε_θ can be determined with the aid of (37.30), (37.48), and (37.45):

$$\frac{E\varepsilon_\theta}{4^{\frac{1}{3}}\kappa} = (1-\sigma)\lambda - (1-\sigma)\int_0^1 \frac{\tau^3\,d\tau}{\alpha^2} - \frac{1+\sigma}{\rho^2}\int_0^\rho \frac{\tau^3\,d\tau}{\alpha^2} + (1-\sigma)\int_\rho^1 \frac{\tau\,d\tau}{\alpha^2}, \tag{37.49}$$

$$\frac{E\varepsilon_r}{4^{\frac{1}{3}}\kappa} = (1-\sigma)\lambda - (1-\sigma)\int_0^1 \frac{\tau^3\,d\tau}{\alpha^2} + \frac{1+\sigma}{\rho^2}\int_0^\rho \frac{\tau^3\,d\tau}{\alpha^2} + (1-\sigma)\int_\rho^1 \frac{\tau\,d\tau}{\alpha^2}. \tag{37.50}$$

The radial displacement u can be computed from (37.44b) and (37.49):

$$u = \frac{4^{\frac{1}{3}}\kappa R}{E}\left[(1-\sigma)\lambda\rho - (1-\sigma)\rho\int_0^1 \frac{\tau^3\,d\tau}{\alpha^2} - \frac{1+\sigma}{\rho}\int_0^\rho \frac{\tau^3\,d\tau}{\alpha^2} + (1-\sigma)\int_\rho^1 \frac{\tau\,d\tau}{\alpha^2}\right]. \quad (37.51)$$

From (37.46b) we derive

$$w = -\frac{4^{\frac{2}{3}}\kappa^{\frac{1}{2}}R}{E^{\frac{1}{2}}}\int_\rho^1 \frac{\tau\,d\tau}{\alpha}. \quad (37.52)$$

On introducing the dimensionless quantities

$$U = \frac{Eu}{4^{\frac{1}{3}}\kappa R}, \quad W = \frac{E^{\frac{1}{2}}w}{4^{\frac{2}{3}}\kappa^{\frac{1}{2}}R},$$

the equations for U and W are

$$U = (1-\sigma)\lambda\rho - (1-\sigma)\rho\int_0^1 \frac{\tau^3\,d\tau}{\alpha^2} - \frac{(1+\sigma)}{\rho}\int_0^\rho \frac{\tau^3 d\tau}{\alpha^2} + (1-\sigma)\rho\int_\rho^1 \frac{\tau\,d\tau}{\alpha^2}, \quad (37.53)$$

$$W = -\int_\rho^1 \frac{\tau\,d\tau}{\alpha}. \quad (37.54)$$

In the case in which λ is large, so that $1/\lambda^2$ is small, inequalities (37.33) and (37.34) require the approximation $\alpha \simeq \lambda$. On placing this approximate value of α in (37.48), (37.53), and (37.54), we obtain the results

$$\beta \simeq \lambda, \quad U \simeq (1-\sigma)\lambda\rho, \quad W \simeq -\frac{1}{\lambda}(1-\rho^2).$$

After having expressed the displacements in terms of λ, the existence of a solution to Hencky's problem will be proved by showing that it is always possible to find a value for $\lambda > 4/j_{11}^2$ such that the corresponding displacement has $U(1) = 0$. The proof involves three requirements: (a) $U(1;\lambda)$ is a continuous function of λ; (b) there is a value of λ with $\lambda > 4/j_{11}^2$ and $U(1;\lambda) > 0$; and (c) there is a value of $\lambda > 4/j_{11}^2$ with $U(1,\lambda) < 0$.

To show the continuity of $U(1;\lambda)$, we first discuss the qualitative behavior of α as a function of λ. Let us consider two solutions α_2 and α_1 corresponding to two values of λ: λ_2 and λ_1, with $\lambda_2 > \lambda_1$. Thus (37.30) implies the values

$$\alpha_2(0) = \lambda_2 + \int_0^1 (\tau - \tau^3)\frac{1}{\alpha_2^2}\,d\tau, \quad \alpha_1(0) = \lambda_1 + \int_0^1 (\tau - \tau^3)\frac{1}{\alpha_1^2}\,d\tau,$$

with the consequence that we can rewrite (37.30) in the form

$$\alpha_2 = \alpha_2(0) + \int_0^\rho \left(\frac{\tau^3}{\rho^2} - \tau\right) \frac{1}{\alpha_2^2}\, d\tau,$$

$$\alpha_1 = \alpha_1(0) + \int_0^\rho \left(\frac{\tau^3}{\rho^2} - \tau\right) \frac{1}{\alpha_1^2}\, d\tau.$$

On subtracting these two expressions we obtain

$$\alpha_2 - \alpha_1 = \alpha_2(0) - \alpha_1(0) + \int_0^\rho \left(\tau - \frac{\tau^3}{\rho^2}\right) \frac{\alpha_2 + \alpha_1}{\alpha_2^2 \alpha_1^2} (\alpha_2 - \alpha_1)\, d\tau, \tag{37.55}$$

with $(\tau - \tau^3/\rho^2) > 0$, since we have $\tau < \rho$. Let us assume the condition $\alpha_2 - \alpha_1 < 0$ for some value of $\rho = \rho^*$. Then there are three conceivable cases (see Figure 37.1). Note that the point ρ'' always exists because of the conditions $\alpha_2(1) = \lambda_2 > \lambda_1 = \alpha_1(1)$. Case (a) is not possible because $\alpha_2(0) - \alpha_1(0)$, together with (37.55), written with $\rho = \rho'$, implies the inequality $\alpha_2(\rho') - \alpha_1(\rho') > 0$. Similarly, cases (b) and (c) are not possible from the result $\alpha_2(0) - \alpha_1(0) \leqslant 0$ and the fact that the integral from 0 to ρ'' in (37.55) is negative, resulting in $\alpha_2(\rho'') - \alpha_1(\rho'') < 0$. We thus obtain, by contradiction, the result that $\alpha_2 - \alpha_1 \geqslant 0$. Moreover, $\alpha_2 - \alpha_1$ is a continuous function of ρ, and therefore, when ρ is sufficiently close to 1, we have the inequality $\alpha_2 > \alpha_1$.

This result yields a slight extension of the condition $\lambda^3 > 4/j_{11}^2$, which ensures the existence of a solution of (37.24) with the boundary condition (37.25). The condition can, in fact, be replaced by $\lambda^3 \geqslant 4/j_{11}^2$. If we take a sequence of functions $\{\lambda_n\}$ with $\lambda_n > \lambda_{n+1}$ and $\lambda_n \to 4/j_{11}^2$, this defines a sequence of functions $\{\alpha_n\}$ with the property $\alpha_n \geqslant \alpha_{n+1}$. Thus $\{\alpha_n\}$ is a bounded monotonically decreasing sequence of functions with $\alpha_n \to \alpha$, which is again a solution of the problem with $\lambda^3 = 4/j_{11}^2$.

In (37.55) the integrand is positive. Thus $\alpha_2 - \alpha_1$ is a monotonically increasing function of ρ and its maximum will occur at $\rho = 1$. We can then write

$$\max_{0 \leqslant \rho \leqslant 1} (\alpha_2 - \alpha_1) = \alpha_2(1) - \alpha_1(1) = \lambda_2 - \lambda_1, \tag{37.56}$$

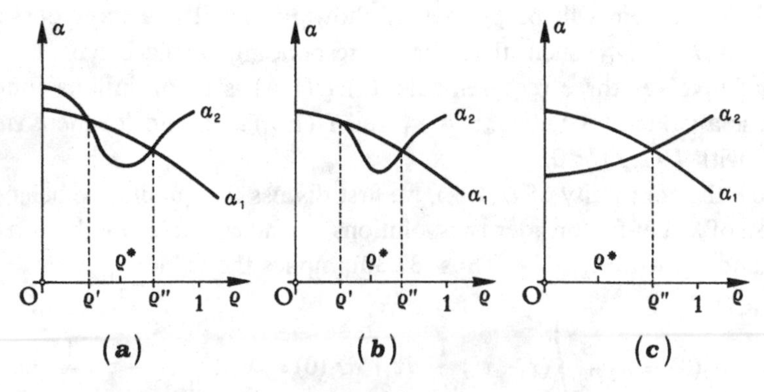

Fig. 37.1

and this equation implies the inequality $|\alpha_2(\rho;\lambda_2) - \alpha_1(\rho;\lambda_1)| \leqslant |\lambda_2 - \lambda_1|$; that is, the continuity of $\alpha(\rho;\lambda)$ as a function of λ. However, β, U, and W can be expressed in terms of quadratures of α, and, since we have $\alpha \geqslant \lambda$, it follows that α, β, U, and W are continuous functions of λ. This completes the proof of property (a).

The proof of property (b) gives no difficulty, because if λ is very large the solution actually approaches the solution obtained in classical linear membrane theory: $U(1) = (1-\sigma)\lambda > 0$. It then remains to prove that there is a value $\lambda^3 > 4/j_{11}^2$ such that we have $U(1) < 0$. Equation (37.53) implies the value

$$U(1) = (1-\sigma)\lambda - 2\int_0^1 \frac{\tau^3\,d\tau}{\alpha^2},$$

and, from inequality (37.34), it follows that

$$U(1) \leqslant (1-\sigma)\lambda - 2\int_0^1 \frac{\tau^3\,d\tau}{\left(\lambda + \dfrac{1-\tau^2}{4\lambda^2}\right)^2},$$

which, by direct integration, becomes

$$U(1) \leqslant \lambda\left[-3-\sigma-16\lambda^3\left(\log\frac{4\lambda^3}{4\lambda^3+1}\right)\right]. \tag{37.57}$$

On putting $\lambda^3 = 4/j_{11}^2$, which is allowed for the extensions of the existence theorem made previously, we obtain the result $U(1) < 0$ for any $-1 < \sigma < \frac{1}{2}$.[50]

The theory of the equilibrium of circular membranes can be extended in one of two ways: by formulating a set of equations derived from Föppl's approximations, or by introducing the stress–strain relations of nonlinear elasticity. The advantage of the latter approach is that, by analyzing the fully nonlinear problem, we can find a formal mathematical justification of Föppl's equation as the first term of an asymptotic expansion of the exact theory, with the possibility of obtaining higher order approximations, should we wish to do so (Dickey 1983).

In order to derive the exact equation of equilibrium, we consider a point with coordinates $(x, y, 0)$ that assumes a new position (ξ, η, ζ) after the deformation. In polar coordinates we can write

$$(r\cos\theta, r\sin\theta, 0) \to [(r+u)\cos\theta, (r+v)\sin\theta, w],$$

where u is the radial displacement and w is the vertical displacement. The assumption of radial symmetry requires us to take $u = u(r)$ and $v = v(r)$, both being independent of θ. An element of length dS before the deformation will be transformed into an element of length dS^* after the deformation, where

[50] Dickey (1967) suggested taking $\lambda^3 = \frac{1}{3}$, but this choice does not satisfy the inequality $U(1) < 0$ for $\sigma < 0$.

$$(dS)^2 = (dr)^2 + r(d\theta)^2, \tag{37.58}$$

$$(dS^*)^2 = [(1+u')^2 + w'^2](dr)^2 + \left(1+\frac{u}{r}\right)^2 r^2(d\theta)^2. \tag{37.59}$$

Accordingly, we can define the radial and circumferential nonlinear strains as

$$\mathscr{E}_r = [(1+u')^2 + w'^2]^{\frac{1}{2}} - 1, \tag{37.60a}$$

$$\mathscr{E}_\theta = \left(1+\frac{u}{r}\right) - 1 = \frac{u}{r}. \tag{37.60b}$$

In order to formulate the equilibrium equations of the circular membrane under a vertical pressure p, which for simplicity is assumed constant, we take an element of the deformed membrane and indicate the forces on the positive faces (Figure 37.2). The unit normal to the face $r =$ constant is

$$\frac{(1+u')\mathbf{r} + w'\boldsymbol{\kappa}}{[(1+u')^2 + w'^2]^{\frac{1}{2}}},$$

and the unit normal to the face $\theta =$ constant is $\boldsymbol{\theta}$, where

$$\mathbf{r} = \mathbf{i}\cos\theta + \mathbf{j}\sin\theta, \tag{37.61}$$

$$\boldsymbol{\theta} = -\mathbf{i}\sin\theta + \mathbf{j}\cos\theta, \tag{37.62}$$

where **i**, **j**, and **k** are unit vectors in the x, y, and z directions. The equilibrium equations become

$$\frac{\sigma_r h(r+u)[(1+u')\mathbf{r} + w'\boldsymbol{\kappa}]}{[(1+u')^2 + w'^2]}\Delta\theta\bigg|_r^{r+\Delta r} + \sigma_\theta h[(1+u')^2 + w'^2]^{\frac{1}{2}}\boldsymbol{\theta}\,\Delta r\bigg|_\theta^{\theta+\Delta\theta} + pr\boldsymbol{\kappa}\,\Delta r\,\Delta\theta = \mathbf{0}. \tag{37.63}$$

On dividing (37.63) by $h\,\Delta r\,\Delta\theta$ and taking the limit for $\Delta r \to 0$ and $\Delta\theta \to 0$, we find that

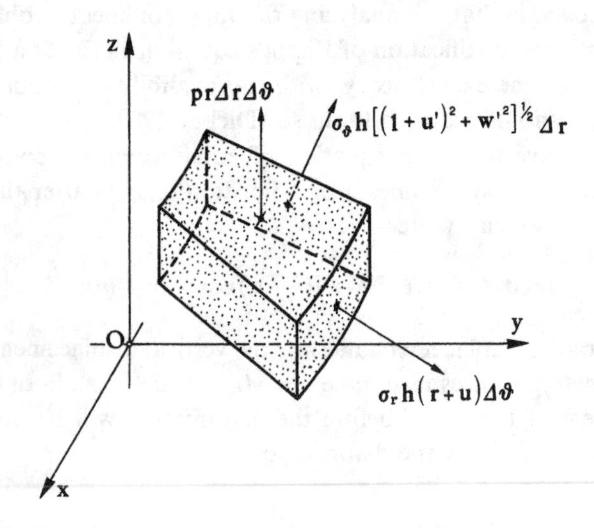

Fig. 37.2

$$\frac{d}{dr}\left\{\frac{\sigma_r(r+u)[(1+u')\mathbf{r}+w'\boldsymbol{\kappa}]}{[(1+u')^2+w'^2]^{\frac{1}{2}}}\right\}-\sigma_\theta[(1+u')^2+w'^2]^{\frac{1}{2}}\mathbf{r}+\frac{pr\boldsymbol{\kappa}}{h}=\mathbf{0},$$

or, in terms of the components,

$$\frac{d}{dr}\left\{\frac{\sigma_r(r+u)(1+u')}{[(1+u')^2+w'^2]^{\frac{1}{2}}}\right\}-\sigma_\theta[(1+u')^2+w'^2]^{\frac{1}{2}}=0, \tag{37.64}$$

$$\frac{d}{dr}\left\{\frac{\sigma_r(r+u)w'}{[(1+u')^2+w'^2]^{\frac{1}{2}}}\right\}+\frac{pr}{h}=0. \tag{37.65}$$

For the boundary conditions we prescribe either the radial displacement at $r=R$

$$u(R)=\mu, \tag{37.66}$$

or the radial stress

$$\sigma_2(R)=\sigma, \tag{37.67}$$

where R is the radius of the undeformed membrane. We also assume that $w(R)=0$, and that all quantities remain finite at $r=0$.

Equations (37.64) and (37.65) can now be rewritten as

$$\frac{d}{dr}\left[\frac{r\Sigma_r(1+u')}{1+\mathscr{E}_r}\right]-\Sigma_\theta=0, \tag{37.68}$$

$$\frac{d}{dr}\left(\frac{r\Sigma_r w'}{1+\mathscr{E}_r}\right)+\frac{pr}{h}=0, \tag{37.69}$$

where we have introduced the Lagrange stresses

$$\Sigma_r=(1+\mathscr{E}_\theta)\sigma_r,\quad \Sigma_\theta=(1+\mathscr{E}_r)\sigma_\theta. \tag{37.70}$$

If the material is hyperelastic, there is a function $W(\mathscr{E}_r, \mathscr{E}_\theta)$ for which

$$\sigma_r=\frac{\partial W}{\partial \mathscr{E}_r},\quad \sigma_\theta=\frac{\partial W}{\partial \mathscr{E}_\theta}, \tag{37.71}$$

leading to

$$\Sigma_r=(1+\mathscr{E}_\theta)\frac{\partial W}{\partial \mathscr{E}_r},\quad \Sigma_\theta=(1+\mathscr{E}_r)\frac{\partial W}{\partial \mathscr{E}_\theta}. \tag{37.72}$$

It will be assumed that this system can be solved for $\mathscr{E}_r$ and $\mathscr{E}_\theta$ as functions of Σ_r and Σ_θ. A sufficient condition is that the Jacobian satisfies the condition

$$\frac{\partial(\Sigma_r, \Sigma_\theta)}{\partial(\mathscr{E}_r, \mathscr{E}_\theta)}\neq 0. \tag{37.73}$$

It is easily verified that, if we have

$$\frac{\partial^2 W}{\partial \mathscr{E}_r^2}\frac{\partial^2 W}{\partial \mathscr{E}_\theta^2}-\left(\frac{\partial^2 W}{\partial \mathscr{E}_r\,\partial \mathscr{E}_\theta}\right)^2\neq 0 \tag{37.74}$$

for $\mathscr{E}_r=\mathscr{E}_\theta=0$, (37.73) will be satisfied when the strains are sufficiently small. In any case we assume that we can find the inverse functions

$$\mathscr{E}_r = e_r(\Sigma_r, \Sigma_\theta), \quad \mathscr{E}_\theta = e_\theta(\Sigma_r, \Sigma_\theta). \tag{37.75}$$

Equations (37.68) and (37.69) may be integrated to give

$$\frac{r\Sigma_r(1+u')}{1+\mathscr{E}_r} = \int_0^r \Sigma_\theta \, d\tau, \quad \frac{r\Sigma_r w'}{1+\mathscr{E}_r} = \frac{pr^2}{2h}. \tag{37.76}$$

By combining these equations we obtain

$$r\Sigma_r = \left[\frac{p^2r^4}{4h^2} + \left(\int_0^r \Sigma_\theta \, d\tau\right)^2\right]^{\frac{1}{2}}, \tag{37.77}$$

$$w' = \frac{-pr^2(1+u')}{2h\left(\int_0^r \Sigma_\theta \, d\tau\right)}. \tag{37.78}$$

In (37.77) we have chosen the positive root in order to describe the expected fact that the membrane is under tension. Combining (37.77) and (37.78) with (37.60a) gives

$$\mathscr{E}_r + 1 = e_r(\Sigma_r, \Sigma_\theta) + 1 = \frac{(1+u')r\Sigma_r}{\left(\int_0^r \Sigma_\theta \, d\tau\right)}, \tag{37.79}$$

where we have assumed the condition that $u' > -1$. Equation (37.79) is equivalent to

$$1 + u' = \frac{\left(\int_0^r \Sigma_\theta \, d\tau\right)}{r\Sigma_r} + \frac{e_r(\Sigma_r, \Sigma_\theta)\left(\int_0^r \Sigma_\theta \, d\tau\right)}{r\Sigma_r}. \tag{37.80}$$

On the other hand, from (37.60b) we have

$$u' = \frac{d}{dr}[re_\theta(\Sigma_r, \Sigma_\theta)],$$

so that

$$1 + \frac{d}{dr}[re_\theta(\Sigma_r, \Sigma_\theta)] = \frac{S}{r} + \frac{e_r(\Sigma_r, \Sigma_\theta)S}{\Sigma_r}, \tag{37.81}$$

where we have put

$$rS = \int_0^r \Sigma_\theta \, d\tau = \frac{r\Sigma_r(1+u')}{1+\mathscr{E}_r}.$$

In terms of S we have

$$\Sigma_r = \left(\frac{p^2r^2}{4h^2} + S^2\right)^{\frac{1}{2}}, \quad \Sigma_\theta = (rS)'. \tag{37.82}$$

On substituting (37.82) in (37.81) we obtain a second-order differential equation for determining S. The boundary conditions are either

$$e_\theta\left[\left(\frac{p^2r^2}{4h^2}+S^2\right)^{\frac{1}{2}}, (rS)'\right]\Bigg|_{r=R} = \frac{u(R)}{R} = \frac{\mu}{R}, \tag{37.83}$$

or

$$\frac{[p^2r^2/4h^2+S^2]^{\frac{1}{2}}}{1+e_\theta[(p^2r^2/4h^2+S^2)^{\frac{1}{2}}, rS']}\Bigg|_{r=R} = \sigma. \tag{37.84}$$

In order to explain the role of Föppl's theory, let us assume that the constitutive relations are derivable from a strain-energy density of the form

$$W = \frac{E}{2(1-\sigma^2)}(\mathscr{E}_2^2+\mathscr{E}_\theta^2+2\sigma\mathscr{E}_2\mathscr{E}_\theta), \tag{37.85}$$

so that the Lagrange stresses (see (37.72)) are

$$\begin{aligned}\Sigma_r &= \frac{E}{1-\sigma^2}(1+\mathscr{E}_\theta)(\mathscr{E}_r+\sigma\mathscr{E}_\theta),\\ \Sigma_\theta &= \frac{E}{1-\sigma^2}(1+\mathscr{E}_r)(\mathscr{E}_\theta+\sigma\mathscr{E}_r).\end{aligned} \tag{37.86}$$

Equation (37.86) can be solved for $\mathscr{E}_r$ and $\mathscr{E}_\theta$ to give

$$\mathscr{E}_r = e_r(\Sigma_r, \Sigma_\theta) = \frac{1}{E}(\Sigma_r-\sigma\Sigma_\theta)+\frac{1}{E^2}[\sigma\Sigma_r^2-(1-\sigma)\Sigma_r\Sigma_\theta-\sigma^2\Sigma_\theta^2]+O[(\Sigma_r^2+\Sigma_\theta^2)^{\frac{3}{2}}], \tag{37.87}$$

$$\mathscr{E}_\theta = e_\theta(\Sigma_r, \Sigma_\theta) = \frac{1}{E}(\Sigma_\theta-\sigma\Sigma_r)+\frac{1}{E^2}[\sigma\Sigma_\theta^2-(1-\sigma)\Sigma_r\Sigma_\theta-\sigma^2\Sigma_r^2]+O[(\Sigma_r^2+\Sigma_\theta^2)^{\frac{3}{2}}]. \tag{37.88}$$

After introducing the new variables

$$\rho = \frac{r}{R}, \quad \kappa^3 = \frac{p^2R^2}{4h^2}, \quad S = \kappa T, \tag{37.89}$$

(37.81) becomes

$$1+\frac{d}{d\rho}\left\{\rho e_\theta\left[\kappa(\kappa\rho^2+T^2)^{\frac{1}{2}}, \kappa\frac{d}{d\rho}(\rho T)\right]\right\} = \frac{T}{(\kappa\rho^2+T^2)^{\frac{1}{2}}} + \frac{Te_r\left[\kappa(\kappa\rho^2+T^2)^{\frac{1}{2}}, \kappa\frac{d}{d\rho}(\rho T)\right]}{(\kappa\rho^2+T^2)^{\frac{1}{2}}}. \tag{37.90}$$

The boundary conditions corresponding to (37.83) and (37.84) are

$$e_\theta\left[\kappa(\kappa\rho^2+T^2)^{\frac{1}{2}},\kappa\frac{d}{d\rho}(\rho T)\right]\Bigg|_{\rho=1}=\frac{\mu}{R}=\frac{\lambda}{R}\kappa, \tag{37.91}$$

$$\left.\frac{\kappa[\kappa\rho^2+T^2]^{\frac{1}{2}}}{1+e_\theta\left[\kappa(\kappa\rho^2+T^2)^{\frac{1}{2}},\kappa\frac{d}{d\rho}(\rho T)\right]}\right|_{\rho=1}=\sigma=s\kappa, \tag{37.92}$$

where, for convenience, we have introduced the quantities $\lambda=\mu/\kappa$ and $s=\sigma/\kappa$. At $\rho=0$ we prescribe T to be finite.

The functions $e_r(\Sigma_r,\Sigma_\theta)$ and $e_\theta(\Sigma_r,\Sigma_\theta)$ (see (37.87) and (37.88)) can be written in terms of T:

$$\begin{aligned}e_r\left[\kappa(\kappa\rho^2+T^2)^{\frac{1}{2}},\kappa\frac{d}{d\rho}(\rho T)\right]&=\frac{\kappa}{E}\left[\frac{d}{d\rho}(\rho T)-\sigma(\kappa\rho^2+T^2)^{\frac{1}{2}}\right]\\&+\frac{\kappa^2}{E^2}\left\{\sigma(\kappa\rho^2+T^2)^{\frac{1}{2}}-(1-\sigma)(\kappa\rho^2+T^2)^{\frac{1}{2}}\frac{d}{d\rho}(\rho T)-\sigma^2\left(\frac{d}{d\rho}(\rho T)\right)^2\right\}+O(\kappa^3),\end{aligned} \tag{37.93}$$

$$\begin{aligned}e_\theta\left[\kappa(\kappa\rho^2+T^2)^{\frac{1}{2}},\kappa\frac{d}{d\rho}(\rho T)\right]&=\frac{\kappa}{E}\left[\frac{d}{d\rho}(\rho T)-\sigma(\kappa\rho^2+T^2)^{\frac{1}{2}}\right]\\&+\frac{\kappa^2}{E^2}\left\{\sigma\left[\frac{d}{d\rho}(\rho T)\right]^2-(1-\sigma)(\kappa\sigma^2+T^2)^{\frac{1}{2}}\frac{d}{d\rho}(\rho T)-\sigma^2(\kappa\rho^2+T^2)\right\}+O(\kappa^3).\end{aligned} \tag{37.94}$$

A consequence of (37.93) and (37.94) is that

$$\frac{1}{\kappa}e_r\left[\kappa(\kappa\rho^2+T^2)^{\frac{1}{2}},\kappa\frac{d}{d\rho}(\rho T)\right]=\frac{1}{E}\left[T-\sigma\frac{d}{d\rho}(\rho T)\right]+O(\kappa), \tag{37.95}$$

$$\frac{1}{\kappa}e_\theta\left[\kappa(\kappa\rho^2+T^2)^{\frac{1}{2}},\kappa\frac{d}{d\rho}(\rho T)\right]=\frac{1}{E}\left[\frac{d}{d\rho}(\rho T)-\sigma T\right]+O(\kappa). \tag{37.96}$$

In addition, we have

$$\frac{1}{\kappa}[1-T(\kappa\rho^2+T^2)^{-\frac{1}{2}}]=\frac{\rho^2}{2T^2}+O(\kappa). \tag{37.97}$$

On using (37.95)–(37.97), (37.90) can be written as

$$\frac{d}{d\rho}\left\{\frac{\rho}{E}\left(\frac{d}{d\rho}(\rho T)-\sigma T\right)\right\}+\frac{\rho^2}{2T^2}-\frac{1}{E}\left(T-\sigma\frac{d}{d\rho}(\rho T)\right)+O(\kappa)=0.$$

This equation can be rearranged to show that, neglecting terms of order $O(\kappa)$, it is equivalent to

$$\frac{d^2T}{d\rho^2}+\frac{3}{\rho}\frac{dT}{d\rho}+\frac{E}{2T^2}=0,$$

which is just Föppl's equation. The associated boundary conditions, accurate to $O(\kappa)$, are either

$$\frac{1}{E}\left[\frac{d}{d\rho}(\rho T)-\sigma T\right]\bigg|_{\rho=1}=\frac{\lambda}{R},$$

or

$$T(1)=s,$$

which are again Föppl's boundary conditions. In principle, we could derive more accurate theories by retaining terms of order $O(\kappa^2)$, and so on.

38. Theories of Networks

Networks are two-dimensional bodies formed by two systems of parallel threads that cross each other at a constant angle. The commonest example of network is a piece of cloth constituted by two orders of fiber, warps and woofs, which, when the cloth is undeformed, are interwoven at right angles. Cloth is deformed mainly by changing the angle between the threads of the warp and the woof; the stretching of the fibers is usually negligible in comparison with the finite distortion due to the angular distortion. If a network is regarded as a combination of one-dimensional threads, the natural way to study the equilibrium and deformation is to apply the theory of perfectly flexible and inextensible strings to each component of the mesh. However, in 1878, Tchebychev (1878, 1907) suggested a continuum model for cloth in which the fibers are treated as continuously distributed, so that it is possible to consider the network as a particular type of membrane that is able to transmit normal stresses but not shear forces. In addition, the normal stresses can only be tensile.

If we take only purely kinematic considerations into account, networks offer a nontrivial problem known as *clothing a surface* (Servant 1902, 1903). Let us suppose we have an, initially planar, orthogonal network of inextensible fibers, and we wish to lay the network onto a given curved surface without stretching the fibers but simply allowing the angle between them to alter. In order to formulate this problem properly, we consider a fabric or net composed of fibers that initially lie in the xy plane of a Cartesian reference system (Pipkin 1984). The network is regarded as a continuum so that every line $x=$ constant or $y=$ constant is a fiber. The net is deformed, and a particle initially at x, y moves to the position $\mathbf{r}(x, y)$ in three-dimensional space.

In a Tchebychev net, the fibers are inextensible, so that a segment of length dx of a fiber with $y=$ constant is mapped by the deformation onto the arc $d\mathbf{r}=\mathbf{r}_{,x}\,dx$, with length $ds=|\mathbf{r}_{,x}|dx$. For the length of the element to be unchanged, $\mathbf{a}=\mathbf{r}_{,x}$ must be a unit vector. Similarly, $\mathbf{b}=\mathbf{r}_{,y}$ is a unit vector. The vectors $\mathbf{a}$ and $\mathbf{b}$ are tangential to the deformed fibers. In terms of $\mathbf{a}$ and $\mathbf{b}$, we have

$$d\mathbf{r}=\mathbf{a}\,dx+\mathbf{b}\,dy, \tag{38.1}$$

with $\mathbf{a} \cdot \mathbf{a} = \mathbf{b} \cdot \mathbf{b} = 1$. In addition, $\mathbf{a}$ and $\mathbf{b}$ must satisfy the integrability condition for (38.1):

$$\mathbf{a}_{,y} = \mathbf{b}_{,x}. \tag{38.2}$$

Let us now assume that we are given a surface $\mathbf{r} = \mathbf{r}(u, v)$, where u and v are parameters. Then the problem of clothing the surface is the problem of expressing u and v in terms of the coordinates x, y of a Tchebychev net. For the given surface $\mathbf{r}(u, v)$, the first fundamental form is

$$ds^2 = E\,du^2 + 2F\,du\,dv + G\,dv^2, \tag{38.3}$$

where

$$E = \mathbf{r}_{,u} \cdot \mathbf{r}_{,u}, \quad F = \mathbf{r}_{,u} \cdot \mathbf{r}_{,v}, \quad G = \mathbf{r}_{,v} \cdot \mathbf{r}_{,v}. \tag{38.4}$$

If the clothing is perfectly adherent, the vectors $\mathbf{a}$ and $\mathbf{b}$ can be expressed in terms of u and v:

$$\mathbf{a} = \mathbf{r}_{,u} u_x + \mathbf{r}_{,v} v_x, \quad \mathbf{b} = \mathbf{r}_{,u} u_y + \mathbf{r}_{,v} v_y. \tag{38.5}$$

Then, the condition of inextensibility gives

$$\mathbf{a} \cdot \mathbf{a} = Eu_x^2 + 2Fu_x v_x + Gv_x^2 = 1, \tag{38.6a}$$
$$\mathbf{b} \cdot \mathbf{b} = Eu_y^2 + 2Fu_y v_y + Gv_y^2 = 1. \tag{38.6b}$$

By differentiating (38.6a) with respect to y and (38.6b) with respect to x we convert the first-order nonlinear system into a second-order quasilinear system in u and v. This system is linear in u_{xy} and v_{xy}, and so can be solved algebraically to give *Servant's equations*:

$$\begin{aligned} 2D^2 u_{xy} &= A_1 u_x u_y + B_1(u_x v_y + u_y v_x) + C_1 v_x v_y, \\ 2D^2 v_{xy} &= A_2 u_x u_y + B_2(u_x v_y + u_y v_x) + C_2 v_x u_y, \end{aligned} \tag{38.7}$$

with

$$D^2 = EG - F^2. \tag{38.8}$$

D^2 is positive because (38.3) is positive definite. The remaining coefficients are defined by

$$\begin{aligned} A_1 &= 2FF_u - GE_u - FE_v, \\ B_1 &= FG_u - GE_u, \\ C_1 &= GG_u + FG_v - 2GF_v, \\ A_2 &= EE_v + FE_u - 2EF_u, \\ B_2 &= FE_v - EG_u, \\ C_2 &= 2FF_v - EG_v - FG_u. \end{aligned} \tag{38.9}$$

In system (38.7) the coefficients are known functions of u and v. The system is hyperbolic with the fibers as characteristics. The simplest problem is that in which the data are prescribed along two interesting characteristics; that is, when the positions of two fibers, say $x = 0$ and $y = 0$, are prescribed. This means that we are given the functions $u(x, 0)$, $v(x, 0)$, $u(0, y)$, and $v(0, y)$. These data should satisfy (38.6), but this requirement can be ignored at first. The solutions can be constructed by iteration, and their existence rests on the proof that the iteration converges.[51] We begin with an arbitrary guess for the right-hand sides of (38.7), say zero. Then an integration with respect to y gives

$$u_x(x,y) = u_x(x,0), \quad v_x(x,y) = v_x(x,0), \tag{38.10}$$

and a second integration with respect to x yields

$$\begin{aligned} u &= u(x,0) + u(0,y) - u(0,0), \\ v &= v(x,0) + v(0,y) - v(0,0). \end{aligned} \tag{38.11}$$

This approximation is now used to evaluate the right-hand sides of (38.7) and the process of integration is repeated to obtain a new approximation.

Assuming that a solution has been obtained, we can evaluate the quantities

$$\begin{aligned} Eu_x^2 + 2Fu_xv_x + Gv_x^2 &= [f(x)]^2, \\ Eu_y^2 + 2Fu_yv_y = Gv_y^2 &= [g(y)]^2, \end{aligned} \tag{38.12}$$

where the right-hand sides are positive because (38.3) is positive definite. Then the net x, y can be converted into a Tchebychev net x', y' satisfying (38.6) by changing the variables: $dx' = f(x)\,dx$, $dy' = g(y)\,dy$. If the initial values $u(x,0)$ and $v(x,0)$ are chosen to satisfy (38.6), that is $ds = dx$ along $y = 0$, then $f = 1$ and the transformation to x' is unnecessary. Similarly, the solution of (38.7) automatically satisfies (38.6b), if it is satisfied by the initial data $u(0,y)$ and $v(0,y)$, so that $g = 1$.

The solution $u(x,y)$ and $v(x,y)$ gives the position on the surface that will be occupied by the particle initially at x, y. However, the mapping need not be uniquely invertible, in the sense that more than one particle may occupy the same position on the surface. Furthermore, the existence of a solution does not in itself imply that the whole surface is actually covered, because some points u, v on the surface may fail to appear in the solution. This happens when **a** and **b** become parallel and the net folds back on itself. In order to characterize this phenomenon more precisely, we introduce the angle of shear γ of the deformed surface, defined by the relation

$$\sin\gamma = \mathbf{a}\cdot\mathbf{b}.$$

A unit vector **n** perpendicular to the deformed sheet is given by

$$\mathbf{a}\times\mathbf{b} = J\mathbf{n}, \quad J = \cos\gamma. \tag{38.13}$$

If the surface normal **n** is specified in advance by the shape of the body being covered, then $J(=\cos\gamma)$ can be either positive or negative, depending on which side of the sheet is in contact with the body, and in practice there is the possibility that J might change sign on the surface. A locus defined by the curve $J = 0$ is an envelope of fibers along which **a** and **b** are parallel to one another and tangential to the locus. The sheet turns back on itself at an envelope, and so fails to cover that part of the surface beyond the envelope.

Servant's theory is purely geometric. If stresses are taken into account, they must be related by equilibrium equations. The equations for the case of an originally

[51] The proof was given by Bieberbach (1926).

planar sheet with initially straight, orthogonal, fibers, and purely normal transverse loading, were derived by Rivlin (1955, 1959). However, the result may easily be generalized by also taking shear resistance into account (Pipkin 1980).

Let $\mathbf{t}_a\,dy$ be the force exerted across an arc $\mathbf{b}\,dy$ by that part of material initially to the right of the arc on the material initially to its left. Let $\mathbf{t}_b\,dx$ be the force exerted across $\mathbf{a}\,dx$ by the material initially above. Then the force exerted from right to left across the vector arc $\mathbf{a}\,dx$ is $-\mathbf{t}_b\,dx$. The force across an arbitrary arc $d\mathbf{r} = \mathbf{a}\,dx + \mathbf{b}\,dy$ is then $\mathbf{t}_a\,dy - \mathbf{t}_b\,dx$. Furthermore, we assume that a purely normal load $-p\mathbf{n}\,dA$ is applied to an element of area dA, where $dA = |J|\,dx\,dy$.

The equilibrium of an arbitrary part D of the sheet with a boundary C requires the conditions

$$\oint_C (\mathbf{t}_a\,dy - \mathbf{t}_b\,dx) - \iint_D \mathbf{n}p|J|\,dx\,dy = \mathbf{0}, \tag{38.14}$$

$$\oint_C \mathbf{r} \times (\mathbf{t}_a\,dy - \mathbf{t}_b\,dx) - \iint_D \mathbf{r} \times \mathbf{n}p|J|\,dx\,dy = \mathbf{0}. \tag{38.15}$$

For smooth fields $\mathbf{t}_a$ and $\mathbf{t}_b$, the curvilinear integrals can be converted into area integrals by using Green's theorem:

$$\iint_D [\mathbf{t}_{a,x} + \mathbf{t}_{b,y} - \mathbf{n}p|J|]\,dx\,dy = \mathbf{0}, \tag{38.16}$$

$$\iint_D [(\mathbf{r} \times \mathbf{t}_a)_x + (\mathbf{r} \times \mathbf{t}_b)_y - \mathbf{r} \times \mathbf{n}|J|]\,dx\,dy = \mathbf{0}. \tag{38.17}$$

These integrals must vanish for arbitrary choices of D, and then the integrands must vanish, so that we obtain the local version of the equilibrium equations:

$$\mathbf{t}_{a,x} + \mathbf{t}_{b,y} = \mathbf{n}p|J|, \tag{38.18}$$

$$\mathbf{a} \times \mathbf{t}_a + \mathbf{b} \times \mathbf{t}_b = \mathbf{0}. \tag{38.19}$$

For constitutive assumptions, we suppose that the stresses vectors $\mathbf{t}_a$ and $\mathbf{t}_b$ lie in the local tangential plane, so that they can be resolved into two components along the directions $\mathbf{a}$ and $\mathbf{b}$:

$$\mathbf{t}_a = T_a\mathbf{a} + S\mathbf{b}, \quad \mathbf{t}_b = T_b\mathbf{b} + S'\mathbf{a}, \tag{38.20}$$

where the shear components S and S' must be equal in order to satisfy (38.19). The components T_a and T_b are the reactions resulting from the constraint of inextensibility, and are to be determined from equilibrium considerations rather than regarded as functions of the strains. They are called *fiber tensions*. The shear stress is a specified function of γ of the form

$$S = S' = S(\gamma). \tag{38.21}$$

In many cases the shear stresses are negligible, and we can use the condition $S(\gamma) \equiv 0$.

The components of the equilibrium equations along the directions **a**, **b**, and **n**, can be obtained by substituting (38.20) and (38.21) into (38.18), and observing that the quantities

$$\kappa_a = \mathbf{n} \cdot \mathbf{a}_{,x}, \quad \kappa_b = \mathbf{n} \cdot \mathbf{b}_{,y} \tag{38.22}$$

represent the normal curvatures of the lines of **a** and **b**; in addition, the quantities

$$\gamma_x = \mathbf{n} \times \mathbf{a} \cdot \mathbf{a}_{,x}, \quad \gamma_y = \mathbf{b} \times \mathbf{n} \cdot \mathbf{b}_{,y} \tag{38.23}$$

(note the different sign convention) represent the geodesic curvatures of the lines of **a** and **b**, respectively. From (38.22) and (38.23) we can derive

$$\mathbf{a}_{,x} = \gamma_x \mathbf{n} \times \mathbf{a} + \kappa_a \mathbf{n}, \quad \mathbf{b}_{,y} = \gamma_y \mathbf{b} \times \mathbf{n} + \kappa_b \mathbf{n}, \tag{38.24}$$

which are known as the *Gauss equations* for the surface. In a Tchebychev net the vectors **a** and **b** are of unit length. Since any derivative of a unit vector is perpendicular to it, the two equal members of (38.2) are orthogonal to **a** and **b**, and thus perpendicular to the surface:

$$\mathbf{a}_{,y} = \mathbf{b}_{,x} = \tau \mathbf{n}. \tag{38.25}$$

By using these expressions for the derivatives of **a** and **b** in (38.18), and then taking the inner products with **n**, **n** × **a**, and **n** × **b** in succession, we obtain the three component equations. The normal component is

$$T_a \kappa_a + T_b \kappa_b + 2S\tau = p|J|, \tag{38.26}$$

and, after some rearrangement, the tangential components become

$$(JT_a)_x + T_b \gamma_y + JS_{,y} = 0, \tag{38.27}$$

$$(JT_b)_y + T_a \gamma_x + JS_{,x} = 0. \tag{38.28}$$

Note that the curvatures κ_a, κ_b, and τ do not appear in the tangential equilibrium equations. Consequently, T_a, T_b, and S are *bending invariants*, in the sense that a stress field that is in equilibrium on one surface is also in equilibrium on any surface obtained from the first one by an isometric mapping. In particular, stress fields that are in equilibrium on developable surfaces are also in equilibrium in the plane.

In the typical problem of clothing the shape of the surface to be covered is already known; then, in (38.27) and (38.28), γ, $J (= \cos\gamma)$, and $S(\gamma)$ are known functions of x and y. The tangential equations of equilibrium are a first-order hyperbolic system for T_a and T_b, with the fibers as characteristics. If the traction boundary data prescribe T_a at one point on each **a** line and T_b at one point on each **b** line, the tension can be found. Then the normal force p required to support the deformation is given directly by (38.26).

There is, however, an additional complication, because the unknowns T_a, T_b, and p are subject to certain, physically necessary, inequalities. First, if the normal **n** to the surface points toward the body being covered, the value of p should be positive if the surface is to support the sheet passively. Secondly, if the forces tangential to the surface are to be purely tensile, we must have the inequalities

$$T_a \geqslant 0, \quad T_b \geqslant 0 \quad T_a T_b \geqslant S^2. \tag{38.29}$$

This is an example of a problem with *unilateral constraints* in its solutions.

The equilibrium equations (38.26)–(38.28) are valid for those parts of the network in which the pressure is non-negative and the stresses satisfy the inequalities (38.29). The parts in which the fibers are shortened are called *slack regions*; their shape and extent is not known in advance, and must be determined as part of the solution. In slack regions the deformations are highly arbitrary, and consequently we cannot hope to have a unique solution, at least, in the classical sense. In addition, it is not said that solutions always exist.

In order to illustrate how the problem can be treated, we first set out the properties of possible solutions (Pipkin 1982, 1986). We study, in particular, the case of an initially planar network composed of inextensible fibers that are initially parallel to the x and y directions. On deformation, the particle with the coordinates x, y moves to the position $\mathbf{r} = \mathbf{r}(x, y)$. The fiber element dx parallel to the x axis is mapped into $\mathbf{r}_{,x}\, dx = \mathbf{a}\, dx$, and the element dy becomes $\mathbf{r}_y\, dy = \mathbf{b}\, dy$. For the one-sided constraint of inextensibility, we postulate that $|\mathbf{a}|$ and $|\mathbf{b}|$ cannot be greater than one. In this respect the theory of this network is more general than that for Tchebychev nets, which are characterized by the condition that $|\mathbf{a}|$ and $|\mathbf{b}|$ must be strictly equal to one. For $|\mathbf{a}| = 1$ at some point, we say that the $\mathbf{a}$ fiber is *fully extended* at that point. For $|\mathbf{a}| < 1$, we say that the fiber is a *slack* there.

The fiber tensions T_a and T_b can be expressed in terms of a single vector function $\mathbf{F}(x, y)$ such that[52]

$$\mathbf{F}_{,y} = T_a \mathbf{a}, \quad -\mathbf{F}_{,y} = T_b \mathbf{b}. \tag{38.30}$$

The absence of compressive tensions requires that T_a and T_b must be non-negative:

$$T_a \geqslant 0, \quad T_b \geqslant 0. \tag{38.31}$$

If a fiber is contracted, the tension vanishes; that is, the condition $|\mathbf{a}| < 1$ implies the condition $T_a = 0$, and $|\mathbf{b}| < 1$ implies $T_b = 0$. In addition, we have the kinematic constraints

$$|\mathbf{a}| \leqslant 1, \quad |\mathbf{b}| \leqslant 1. \tag{38.32}$$

The tensions T_a and T_b are not specified by constitutive relations, but take whatever values the equilibrium requires; in particular, they can be infinite.

A boundary-value problem for a sheet initially occupying a planar domain D with a boundary C is commonly formulated by prescribing values $\mathbf{x}_0(x, y)$ on some portion C_p of C and prescribing the function $\mathbf{F}(x, y) = \mathbf{F}_0(x, y)$ on the remainder C_t of the boundary. Prescribing $\mathbf{F}$ corresponds to assigning boundary tractions as dead loads. A deformation is kinematically admissible if it satisfies the boundary conditions on C_p and the constraints (38.32). A stress function $\mathbf{F}$ is statically admissible if satisfies the boundary conditions on C_t and the constraints (38.31).

As D is a domain, two points (x_1, y_1) and (x_2, y_2) in D can be connected by a path composed of segments $x =$ constant and $y =$ constant, the total length of which is

[52] Introduced by Rivlin (1959).

$|x_1 - x_2| + |y_1 - y_2|$. The constraint of inextensibility requires that these two points never be further apart than that length:

$$|\mathbf{r}(x_1, y_1) - \mathbf{r}(x_2, y_2)| \leqslant |x_1 - x_2| + |y_1 - y_2|. \tag{38.33}$$

This constraint is a Lipschitz condition, which guarantees that $\mathbf{r}$ is absolutely continuous. As a consequence, the derivatives $\mathbf{a} = \mathbf{r}_{,x}$ and $\mathbf{b} = \mathbf{r}_{,y}$ exist almost everywhere and satisfy (38.1).

By letting (x_1, y_1) be a current point and on choosing (x_2, y_2) to be a point on C_p, where $\mathbf{r}$ is prescribed, we conclude from (38.33) that $\mathbf{r}$ is bounded by, say, R:

$$|\mathbf{r}(x, y)| \leqslant R. \tag{38.34}$$

From (38.33) it also follows that the deformations form an equicontinuous family. Then, from every infinite sequence of deformations we can extract a subsequence which converges to an admissible deformation (by the Ascoli–Arzelá theorem).

The set of admissible functions is convex, because, if $\mathbf{r}_1(x, y)$ and $\mathbf{r}_2(x, y)$ have the same value $\mathbf{x}_0(x, y)$ on C_p and satisfy (38.33), then so does the function

$$\alpha\mathbf{r}_1(x, y) + (1 - \alpha)\mathbf{r}_2(x, y), \quad 0 \leqslant \alpha \leqslant 1. \tag{38.35}$$

In order to prove the existence of solutions, we consider the total energy. The strain energy is zero because the fibers are inextensible. The work done by the loads is a linear functional of $\mathbf{r}(x, y)$ defined by

$$\mathscr{E} = -\int_{C_t} \mathbf{r} \cdot (T_a \mathbf{a}\, dy - T_b \mathbf{b}\, dx) - \iint_D \mathbf{r} \cdot \mathbf{f}\, dx\, dy, \tag{38.36}$$

where $\mathbf{f}$ is a prescribed body force. On writing

$$T_a \mathbf{a}\, dy - T_b \mathbf{b}\, dx = \mathbf{T}\, ds,$$

where $\mathbf{T}$ is a prescribed function on C_t, the energy becomes

$$\mathscr{E} = -\int_{C_t} \mathbf{r} \cdot \mathbf{T}\, ds - \iint_D \mathbf{r} \cdot \mathbf{f}\, dx\, dy. \tag{38.37}$$

With $|\mathbf{r}_1| \leqslant R$, the energy is bounded:

$$|\mathscr{E}| \leqslant R \int_{C_t} |\mathbf{T}|\, ds + R \iint_D |\mathbf{f}|\, dx\, dy.$$

Let $\mathscr{E}_0$ be the greatest lower bound, and let $\{\mathbf{r}_n\}$ be a minimizing sequence. Then $\{\mathbf{r}_n\}$ has a convergent subsequence approaching $\mathbf{r}_0$, say, and $\mathscr{E}(\mathbf{r}_0) = \mathscr{E}_0$. Thus there is an admissible deformation that minimizes the energy.

In passing, we note that, because the energy is a linear functional defined on a convex set, energy minimization by direct variational methods would lead to a problem of *convex programming*.

39. Membranes Stretched against an Obstacle

The classical problem describing the equilibrium configuration of a planar membrane subject to a given normal load and a given displacement at the boundary is mathematically formulated by the boundary problem (37.2), provided that the displacement and the displacement gradients are small. We have seen that, not only is the problem solvable, but also the solution can be written explicitly whenever we know the conformal mapping that takes the given domain into the unit circle.

In other cases the state of stress and strain in an initially planar membrane is not generated by surface forces or boundary displacements, but by pushing a rigid indenting object or prescribed shape into the planar domain D initially occupied by the membrane, while the boundary C is maintained fixed (Figure 39.1). Under the action of the indentor, part of the surface of the deformed membrane adheres to the indentor and part of the surface is free, although it is in contact everywhere above it. If the contact is smooth, the force exerted by the indentor on the membrane can only be directed normally to the contact region in the sense of the outward normal to the indentor. We are given the total vertical displacement of the indentor. However, the shape of the contact region and the forces of mutual contact are unknown and must be determined as part of the solution. The problem is known as that of the "membrane with *obstacle*" and constitutes an example of a free boundary problem.

In the presence of symmetries the problem can be solved directly (Yang and Hsu 1971), even allowing for large strains. This is the case for the indentation of a circular membrane by a smooth sphere. The unstrained membrane is a circle of radius a, when a smooth sphere of radius R is brought into contact with the membrane at its center (Figure 39.2). Through further indentation by the sphere, the membrane is deformed into an axisymmetric surface composed partly of a piece of the sphere and partly of a surface of revolution the shape of which must be determined. We also need to find the reaction of the membrane in resisting the penetration of the sphere.

We shall use plane polar coordinates to describe the position of a point P on the undeformed membrane. The corresponding point P′ in the deformed membrane is identified by the distance ρ from the axis of symmetry and the meridian arc length ξ

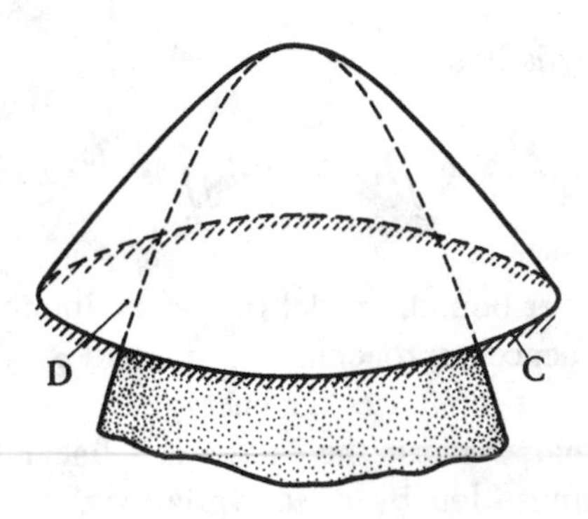

Fig. 39.1

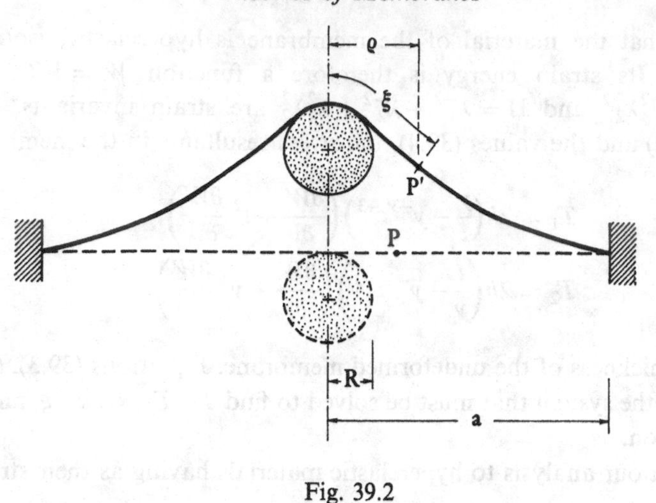

Fig. 39.2

measured from the vertex of the deformed membrane and the point P′. Both ρ and ξ are functions of the single independent variable r:

$$\xi = \xi(r), \quad \rho = \rho(r). \tag{39.1}$$

The stretches in the meridian and circumferential directions are

$$\lambda_1 = \frac{d\xi}{dr}, \quad \lambda_2 = \frac{\rho}{r}. \tag{39.2}$$

As the noncontact region is free, the equilibrium equations in the meridian and circumferential directions are

$$\begin{aligned} &\frac{dT_1}{d\rho} + \frac{1}{\rho}(T_1 - T_2) = 0, \\ &\kappa_1 T_1 + \kappa_2 T_2 = 0, \end{aligned} \tag{39.3}$$

where T_1 and T_2 are the stress resultants in the meridian and circumferential directions, and κ_1 and κ_2 are the principal curvatures in the same directions. On introducing the variables

$$x = \frac{d\rho}{dr}, \tag{39.4a}$$

$$y = \frac{d\xi}{dr}, \tag{39.4b}$$

$$t = \frac{\rho}{r}, \tag{39.4c}$$

the principal curvatures can be expressed in terms of these variables by the formulae

$$\begin{aligned} \kappa_1 &= -\frac{yx' - xy'}{y^2\sqrt{y^2 - x^2}}, \\ \kappa_2 &= \frac{\sqrt{y^2 - x^2}}{\rho y}, \end{aligned} \tag{39.5}$$

where the prime denotes differentiation with respect to r.

We assume that the material of the membrane is hyperelastic, isotropic, and incompressible. Its strain energy is therefore a function $W = W(\mathrm{I}, \mathrm{II})$, where $\mathrm{I} = \lambda_1^2 + \lambda_2^2 + \lambda_1^{-2}\lambda_2^{-2}$ and $\mathrm{II} = \lambda_1^{-2} + \lambda_2^{-2} + \lambda_1^2\lambda_2^2$ are strain invariants. Using the definitions (39.2) and the values (39.4), the stress resultants in the membrane are

$$T_1 = 2h\left(\frac{y}{t} - y^{-3}t^{-3}\right)\left(\frac{\partial W}{\partial \mathrm{I}} + t^2\frac{\partial W}{\partial \mathrm{II}}\right),$$
$$T_2 = 2h\left(\frac{t}{y} - y^{-3}t^{-3}\right)\left(\frac{\partial W}{\partial \mathrm{I}} + y^2\frac{\partial W}{\partial \mathrm{II}}\right), \tag{39.6}$$

where h is the thickness of the undeformed membrane. Equations (39.3), (39.5), and (39.6) comprise the system that must be solved to find T_1, T_2, κ_1, κ_2, ξ, and ρ, in the noncontact region.

Let us restrict our analysis to hyperelastic materials having as their strain-energy function

$$W = C_1(\mathrm{I} - 3) + C_2(\mathrm{II} - 3) = C_1[\mathrm{I} - 3 + \alpha(\mathrm{II} - 3)], \tag{39.7}$$

where C_1 and C_2 are material constants and $\alpha = C_2/C_1$. Equation (39.7) characterizes *Mooney's material.* The governing differential equations for the variables x, y, and t in the noncontact region become

$$\frac{dy}{d\bar{r}} = F(x, y, t, \bar{r}), \tag{39.8a}$$

$$\frac{dx}{d\bar{r}} = \frac{x}{y}F(x, y, t, \bar{r}) + G(x, y, t, \bar{r}), \tag{39.8b}$$

$$\frac{dt}{d\bar{r}} = \frac{1}{\bar{r}}(x - t), \tag{39.8c}$$

where $\bar{r} = r/R$, and we have

$$F = \frac{1}{\bar{r}}\frac{y}{t}\{x(y^2t^4 - 3) - t(y^4t^2 - 3) + \alpha[t^3(y^4t^2 + 1) - xy^2(y^2t^4 + 1)]\}$$
$$\times (y^4t^2 + 3)^{-1}(1 + \alpha t^2)^{-1}, \tag{39.9}$$

$$G = \frac{1}{\bar{r}t}(y^2 - x^2)(y^2t^4 - 1)(1 + \alpha y^2)(y^4t^2 - 1)^{-1}(1 + \alpha t^2)^{-1}. \tag{39.10}$$

In the contact region the membrane follows the surface of the indentor, and hence ξ and ρ must be related by

$$\bar{\rho} = \sin\bar{\xi}, \tag{39.11}$$

where $\bar{\rho} = \rho/R$ and $\bar{\xi} = \xi/R$. From (39.11) we also derive

$$x = y\cos\bar{\xi}, \quad t = \sin\left(\frac{\bar{\xi}}{\bar{r}}\right). \tag{39.12}$$

Assuming that there is no friction between the sphere and the indentor, the equilibrium equation in the meridian direction is homogeneous. On substituting (39.9) into (39.8a) we obtain

$$\frac{dy}{d\bar{r}} = \frac{1}{\bar{r}}\frac{y}{t}\{x(y^2t^4 - 3) - t(y^4t^2 - 3) + \alpha[t^3(y^2t^2 + 1) - xy^2(y^2t^4 + 1)]\}$$
$$\times (y^4t^2 + 3)(1 + \alpha t^2), \tag{39.13}$$

and from equation (39.4b) we have

$$\frac{d\bar{\xi}}{d\bar{r}} = y.$$

The second equation of equilibrium gives the pressure distribution between the sphere and the membrane:

$$p_n(\bar{r}) = \kappa_1 T_1 + \kappa_2 T_2 = \frac{2h}{R}\left[\frac{y}{t} + \frac{t}{y} - \frac{2}{y^3t^3} + \alpha\left(2yt - \frac{1}{y^3t} - \frac{1}{t^3y}\right)\right]. \tag{39.14}$$

The differential equations must be completed by the boundary conditions. At the pole $\bar{r} = 0$, we have the condition of symmetry, that is

$$\bar{\xi}(0) = 0, \tag{39.15}$$

and we know the vertical displacement δ of the membrane. This displacement can be related to x and y by

$$\frac{\delta}{R} = \int_0^{\bar{r}_c} [y^2(\bar{r}) - x^2(\bar{r})]^{\frac{1}{2}}\, d\bar{r} + \int_{\bar{r}_c}^{\bar{a}} [y^2(\bar{r}) - x^2(\bar{r})]^{\frac{1}{2}}\, d\bar{r}, \tag{39.16}$$

where $\bar{a} = a/R$ and $\bar{r}_c = r_c/R$, r_c being the radius at which contact is lost. At $\bar{r} = \bar{r}_c$ we require that $\rho(\bar{r})$ and $T_1(\bar{r})$ are continuous. The continuity of ρ implies that t is also continuous at $\bar{r}_c$ (see (39.4c)). At $\bar{r}_c$, y is also continuous because it is a continuous function of T_1 and t. The only quantity that is not continuous at $\bar{r}_c$ is the pressure p_n between the sphere and the membrane.

The solution of the system can be obtained numerically. Once this has been obtained, we can calculate other quantities of interest, such as the slope θ of the membrane, given by

$$\cos\theta = \frac{d\bar{\rho}}{d\bar{\xi}} = \frac{x(\bar{r})}{y(\bar{r})}, \tag{39.17}$$

and the total vertical load P, obtained from $T_1(\bar{a})$ as

$$P = 2\pi R\bar{\rho}(\bar{a})T_1(\bar{a})\left[1 - \frac{x^2(\bar{a})}{y^2(\bar{a})}\right]^{\frac{1}{2}}. \tag{39.18}$$

As a further control on the validity of the solution, there are the requirements that $T_1 \geqslant 0$ and $T_2 \geqslant 0$, since the membrane cannot carry compressive stresses.

The results of the numerical calculations may find application in the design of tyres, when it may be necessary to determine the effects of the indentation of a tyre by an object of small dimensions.

Problems of the unilateral contact of membranes with rigid obstacles can also be solved easily for cylindrical membranes when they enter into contact with two rigid surfaces (Callegari and Keller 1974). The membrane is a long cylinder of circular cross-section, progressively inflated by a gas enclosed inside it. As the membrane is deformable, the radius of the cross-section increases progressively during the inflation until it interferes with the rigid cylindrical surfaces (Figure 39.3). Our purpose is to determine the cross-section C of the membrane, the tension T, the pressure p of the gas, and the force F on the rigid surfaces. We assume that the stress–strain law of the membrane is a function relating L, the length of C, with T; that is, $L = L(T)$. We also assume that the equation of state of the gas determines the pressure p as a function of the area A enclosed by C; $p = p(A)$. Furthermore, we assume that there is no friction between the membrane and the rigid surfaces.

We treat both the case in which C is closed and the case in which C is open. In the first case, C consists of two circular arcs separated by two arcs of contact with the rigid surfaces; in the second case, the ends of C are fixed on one of the two surfaces. We wish to find the common radius R of the two lateral circular arcs and the end point of the two contact arcs.

We first consider the case in which C is closed. The tension T in the membrane is constant because no tangential forces act upon it. Therefore, whenever the membrane is free, it assumes the form of a circle of radius R, which is related to T by the equilibrium equation

$$\frac{T}{R} = p(A) - p_0, \tag{39.19}$$

p_0 being the pressure outside the membrane. Now, T determines L, which can be expressed in terms of the angles φ_1 and φ_{-1} of the two circular arcs and the lengths of the two arcs of contact with the rigid surfaces. This yields the relation

$$L(T) = R(\varphi_1 + \varphi_{-1}) + \int_{x_{1,-1}}^{x_{11}} [1 + b'^2(x)]^{\frac{1}{2}}\, dx + \int_{x_{0,-1}}^{x_{01}} [1 + t'^2(x)]^{\frac{1}{2}}\, dx, \tag{39.20}$$

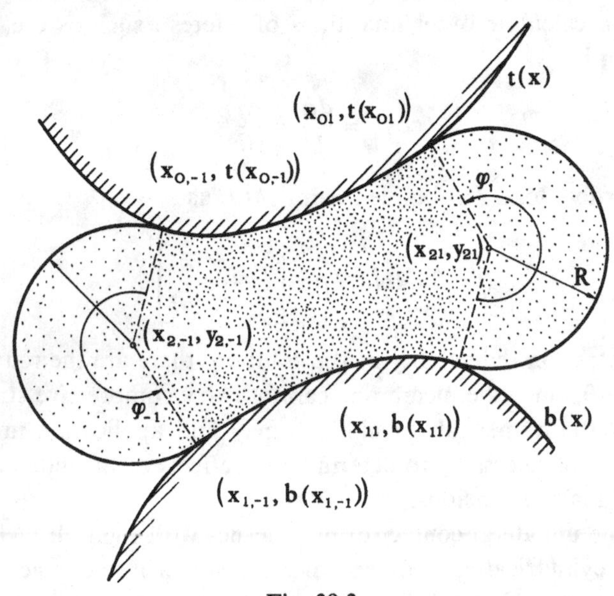

Fig. 39.3

where $y = b(x)$ and $y = t(x)$ are the equations for the bottom and top surfaces, respectively, and the limits of integration are then end points of the two intervals of contact.

Let (x_{2j}, y_{2j}), $j = \pm 1$, denote the coordinates of the centers of the two circular arcs. R is the distance of the center of each circle to its point of tangency with the two rigid surfaces. This fact and the tangency condition yield

$$x_{2j} = x_{1j} - Rb'(x_{1j})[1 + b'^2(x_{1j})]^{-\frac{1}{2}} = x_{0j} + Rt'(x_{0j})[1 + t'^2(x_{0j})]^{-\frac{1}{2}}, \quad j = \pm 1, \tag{39.21}$$

$$y_{2j} = b(x_{1j}) + R[1 + b'^2(x_{1j})]^{-\frac{1}{2}} = t(x_{0j}) - R[1 + t'^2(x_{0j})]^{-\frac{1}{2}}, \quad j = \pm 1. \tag{39.22}$$

The angles φ_1 and φ_{-1} are found to be

$$\varphi_j = +j \tan^{-1} t'(x_{0j}) - j \tan b'(x_{1j}), \quad j = \pm 1. \tag{39.23}$$

Finally, the area A enclosed by C is

$$A = \int_{x_{0,-1}}^{x_{01}} t(x)\, dx - \int_{x_{1,-1}}^{x_{11}} b(x)\, dx + \sum_{j=-1}^{+1} \Bigg\{ \frac{R^2}{2} \varphi_j + j y_{2j}(x_{1j} - x_{0j}) - \frac{1}{2} j(x_{0j} - x_{2j})[R^2 - (x_{0j} - x_{2j})^2]^{\frac{1}{2}} - \frac{1}{2} j(x_{1j} - x_{2j})[R^2 - (x_{1j} - x_{2j})^2]^{\frac{1}{2}} \Bigg\}. \tag{39.24}$$

This expression for A may be substituted in (39.19) about the resulting expression for T used in (39.20). Then, (39.20) and the second equalities (39.21) and (39.22) with $j = \pm 1$ yield five equations for the five unknowns: R, x_{01}, $x_{0,-1}$, x_{11}, and $x_{1,-1}$.

When the rigid surfaces are symmetrical about $x = 0$, we can seek a symmetrical solution by setting $x_{0,-1} = -x_{01}$ and $x_{1,-1} = -x_{11}$, which reduces the problem to three equations for R, x_{01}, and $x_{0,-1}$. If the rigid surfaces are symmetrical about both axes, we can set $x_{01} = -x_{0,-1} = -x_{1,-1} = x_{11}$, which yields two equations for R and x_{11}. This last case applies to a membrane between two parallel planes $y = \pm b$. In this case the solution is not unique because the membrane can be translated parallel to the planes, but the shape of the deformed membrane is unique.

A further simplification results for inextensible membranes, which have $L(T) =$ constant. For these, (39.19) need not be used in (39.20) and the shape of C is determined by geometrical quantities alone. The tension T is then determined by the pressure difference $p(A) - p_0$.

Once the solution has been found, we must verify that the normal force exerted by the rigid surface upon the membrane is directed towards the membrane. If ρ is the radius of curvature of the rigid surface, the condition for contact is

$$\frac{1}{\rho} < \frac{1}{R}, \tag{39.25}$$

where ρ is positive if the center of curvature is on the same side of the surface as the membrane.

As an example, we solve the problem of the symmetrical membrane between the planes $y = \pm b$. Then (39.23) yields $\varphi_1 = \varphi_{-1} = \pi$, and the second equality in (39.22) yields $R = b$. Next, (39.24) yields $A = \pi R^2 + 4bx_{11}$, and (39.20) becomes

$$L = [bp(\pi b^2 + 4bx_{11}) - bp_0] = 2\pi b + 4x_{11}, \tag{39.26}$$

which is an equation in x_{11}. In the inextensible case L is constant and (39.26) yields $x_{11} = \frac{1}{4}(L - 2\pi b)$. Thus x_{11} varies from $x_{11} = 0$ at $b = L/2\pi$, to $x_{11} = L/4$ at $b = 0$.

When the membrane is extensible the simplest constitutive equation is the linear stress–strain law $L = L_0[1 + (T/T_0)]$, with L_0 and T_0 constants; the constitutive equation for the gas is the polytropic equation of state $p = KA^{-\gamma}$, where $K > 0$ and $\gamma > 1$ are constants. Then (39.26) becomes

$$L_0 + \frac{L_0 bK}{T_0}(\pi b^2 + 4bx_{11})^{-\gamma} - \frac{L_0 bp_0}{T_0} = 2\pi b + 4x_{11}. \tag{39.27}$$

As b tends to zero, we find from (39.27) that x_{11} tends to infinity and is approximated by

$$x_{11} \simeq \frac{1}{4}\left(\frac{L_0 K}{T_0}\right)^{\frac{1}{(\gamma+1)}} b^{\frac{(1-\gamma)}{(1+\gamma)}}. \tag{39.28}$$

Let us now suppose that C is open and that the ends of C are fixed at two points x_{01} and $x_{0,-1}$ of the top surface which are given. Then (39.19) still hold, as does (39.20), provided that the last integral is omitted. The first equalities in (39.21) and (39.22) hold, but not the second because the circles are not tangent to the upper surface at x_{0j}. The distance from the centre of each circle to the end point is still R, so we have, instead of the last two equalities in (39.21) and (39.22), the single relation

$$R^2 = [x_{0j} - x_{2j}]^2 + [t(x_{0j}) - y_{2j}]^2, \quad j = \pm 1. \tag{39.29}$$

The angle φ_j is now given by

$$\varphi_j = j\tan^{-1}\frac{t(x_{0j}) - y_{2j}}{x_{0j} - x_{2j}} + j\tan^{-1}\frac{1}{b'(x_{1j})}, \quad j = \pm 1. \tag{39.30}$$

The area A enclosed by C and the top surface is given, as in the previous case, by (37.24).

Again, using (39.24) for A in (39.19) and (39.19) for T in (39.20), we obtain an equation from which A and T are eliminated. We can also use the first equalities in (39.21) and (39.22) to eliminate x_{2j} and y_{2j} from (39.29) and (39.30). We can then substitute the expression for φ_j given by (39.30) in (39.20). In this way, (39.20) and (39.29) (the latter written for $j = \pm 1$) comprise a system of three equations for R, x_{11}, and $x_{1,-1}$.

When $b(x)$ and $t(x)$ are symmetrical about the axis $x = 0$, then $x_{0,-1} = -x_{01}$, and we can seek a symmetric solution by putting $x_{1,-1} = -x_{11}$, so that (39.20) and (39.29) become two equations for R and x_{11}. We now seek a solution for a membrane constrained to line between the plane $y = -b$ and the plane $y = b$, and such that we have $x_{0,-1} \leqslant x \leqslant x_{01}$ with $x_{0,-1} = -x_{01}$. As a result (39.20), (39.29), (39.30), and (39.24) become

$$L[Rp(A) - Rp_0] = 2R\varphi_1 + 2x_{11}, \tag{39.31}$$

$$R = b + \frac{1}{4b}(x_{01} - x_{11})^2, \tag{39.32}$$

$$\varphi_1 = \sin^{-1}\left(\frac{x_{01} - x_{11}}{R}\right), \tag{39.33}$$

$$A = R^2\varphi_1 + 2b(x_{01} + x_{11}) - R(x_{01} - x_{11}). \tag{39.34}$$

On substituting (39.32)–(39.34) in (39.31) we obtain a single equation for x_{11}. However, instead we can eliminate x_{11} and get the following equations for φ_1:

$$\frac{L - 2x_{01}}{4b} = \frac{\varphi_1 - \sin\varphi_1}{1 - \cos\varphi_1}.$$

The numerical solution to the obstacle problem for a membrane can be tackled by means of several different techniques, each of which has its own specific field of application. However, in spite of these special applications, the method proposed by Ritz has the overall advantage of giving a sequence of approximate values to the solutions to increasing degrees of precision, so that, if we terminate the procedure in the first steps, useful approximate solutions are easily calculable. When unilateral constraints are involved, in general the solution requires a sequence of problems in convex programming, but this difficulty can be avoided by enlarging the number of unknown parameters (Feng and Yang 1973, Feng, Tielking, and Huang 1974, Tielking and Feng 1974). The method can be applied to nonsymmetric membranes characterized by a nonlinear stress–strain relation. A detailed illustration of the procedure can be carried out for a rectangular membrane of Mooney-type material. We assume that the, initially flat, membrane is inflated against a plane wall positioned at a fixed distance above it.

Before the deformation, the membrane is a parallelepiped of thickness h and sides of length $2a$ and $2b$. We choose a system of rectangular axes so that the membrane occupies the three-dimensional domain:

$$-a \leqslant x_1 \leqslant a, \tag{39.35a}$$

$$-b \leqslant x_2 \leqslant b, \tag{39.35b}$$

$$-\frac{h}{2} \leqslant x_3 \leqslant \frac{h}{2}. \tag{39.35c}$$

After the deformation, the middle surface of the membrane, which was a rectangle in the x_1x_2 plane, becomes a surface, with the parametric equations

$$y_i = y_i(x_1, x_2), \quad i = 1, 2, 3, \tag{39.36}$$

where y_i are assumed to be three continuously differentiable functions of their arguments. We also assume that lines normal to the undeformed middle surface remain normal during the deformation. The thickness, however, may change and has the form $h\lambda_3(x_1, x_2)$, where $\lambda_3(x_1, x_2)$ is the extension along filaments originally perpendicular to the middle surface. The assumption that the extension λ_3 does not depend on x_3 allows the deformed position Y_i of any point of the membrane to be expressed in the form

$$Y_i(x_1, x_2, x_3) = y_i(x_1, x_2) + \lambda_3 x_3 n_i(x_1, x_2), \tag{39.37}$$

where n_i are the components of the unit vector normal to the deformed middle surface. The components n_i are determined by observing that the base vectors $\mathbf{g}_1$ and $\mathbf{g}_2$ are tangential to the deformed middle surface, and defined as

$$\mathbf{g}_1 = \frac{\partial \mathbf{r}}{\partial x_1}, \quad \mathbf{g}_2 = \frac{\partial \mathbf{r}}{\partial x_2}, \tag{39.38}$$

where $\mathbf{r}$ is the position vector (y_1, y_2, y_3). Then $\mathbf{n} \equiv (n_1, n_2, n_3)$ is given by

$$\mathbf{n} = \frac{\mathbf{g}_1 \times \mathbf{g}_2}{|\mathbf{g}_1 \times \mathbf{g}_2|}. \tag{39.39}$$

The metric tensor in the undeformed state is simply

$$G_{ij} = G^{ij} = \delta_{ij} = \begin{bmatrix} 1 & 0 & 0 \\ 0 & 1 & 0 \\ 0 & 0 & 1 \end{bmatrix}. \tag{39.40}$$

After the deformation, the covariant components of the metric tensor are computed from

$$g_{ij} = \frac{\partial Y_r}{\partial x_i}\frac{\partial Y_r}{\partial x_j}, \tag{39.41}$$

so that the strain tensor is represented by

$$\gamma_{ij} = \frac{1}{2}(g_{ij} - G_{ij}) = \frac{1}{2}\left(\frac{\partial Y_r}{\partial x_i}\frac{\partial Y_r}{\partial x_j} - \delta_{ij}\right). \tag{39.42}$$

For thin membranes, x_3 is always a small value. Thus the derivatives of $\mathbf{r}$, may be written as

$$Y_{r,1} = y_{r,1}, \quad Y_{r,2} = y_{r,2}, \quad Y_{r,3} = \lambda_3 n_r.$$

Since all these derivatives are independent of x_3, the strain tensor, which is uniform through the thickness, is represented by

$$\gamma_{ij} = \frac{1}{2}\begin{bmatrix} y_{r,1}y_{r,1} - 1 & y_{r,1}y_{r,2} & 0 \\ y_{r,2}y_{r,1} & y_{r,2}y_{r,2} - 1 & 0 \\ 0 & 0 & \lambda_3^2 - 1 \end{bmatrix}. \tag{39.43}$$

We now assume that the material is hyperelastic of Mooney type and incompressible. This means that the strain energy density has the form (see (39.7))

$$W = C_1[(I_1 - 3) + \alpha(I_2 - 3)],$$

where C_1 and α are constant, and I_1 and I_2 are the strain invariants. If we accept the simplified version (39.43) for the strains, the invariants are

$$I_1 = y_{r,\alpha}y_{r,\alpha} + \lambda_3^2, \quad I_2 = \lambda_3^2 y_{r,\alpha}y_{r,\alpha} + \lambda_3^{-2}, \tag{39.44}$$

where $\alpha = 1, 2$. The constraint of incompressibility requires that a volume element does not change with the deformation. This implies the condition

$$h\,dx_1\,dx_2 = \lambda_3 h|\mathbf{g}_1 \times \mathbf{g}_2|dx_1\,dx_2, \tag{39.45}$$

which is an equation relating λ_3 to the gradients $y_{r,\alpha}$ of the coordinates of the deformed middle surface. The vector $\mathbf{g}_1 \times \mathbf{g}_2$ has the components

$$b_1 = y_{2,1}y_{3,2} - y_{3,1}y_{2,2},$$
$$b_2 = y_{3,1}y_{1,2} - y_{1,1}y_{3,2}, \tag{39.46}$$
$$b_3 = y_{1,1}y_{2,2} - y_{2,1}y_{1,2},$$

so that we have

$$b = b_r b_r = (y_{r,1}y_{r,1})(y_{r,2}y_{r,2}) - (y_{r,1}y_{r,2})^2.$$

Thus (39.45) determines λ_3 in terms of b:

$$\lambda_3 = \frac{1}{\sqrt{b}}. \tag{39.47}$$

Let us now suppose that the membrane is inflated against the obstacle by a uniform pressure p. The work done by this pressure when it acts on an element $dx_1\,dx_2$ of the unstrained surface is

$$p\,dV = pb_3y_3\,dx_1\,dx_2.$$

The total potential energy of the pressurized membrane can be expressed as

$$\mathscr{E} = \mathscr{W} - \mathscr{L} = \iint_A \{hC_1[(I_1 - 3) + \alpha(I_2 - 3)] - pb_3y_3\}dx_1\,dx_2, \tag{39.48}$$

where A is the rectangle defined by (39.35a) and (39.35b).

During the inflation, the membrane may come into contact with a rigid surface positioned at some distance above the undeformed middle surface of the membrane. The shape of the indentor is prescribed, but the contact region, at which the middle surface of the membrane, conforming to the shape of the indentor, is unknown. If $y_3 = f(y_1, y_2)$ is the geometric constraint on the variables y_r imposed by the obstacle, the solution must obey the restriction

$$y_3 \begin{cases} = f(y_1, y_2), & \text{in the contact region;} \\ < f(y_1, y_2), & \text{in the free region.} \end{cases} \tag{39.49}$$

Of course, $f(y_1, y_2)$ must be less than zero at C, the boundary of A. The constraint (39.49) on the solution can be converted into a single equality by means of a new function $\eta(x_1, x_2)$ defined by the following equation

$$y_3 + \eta^2 = f(y_1, y_2). \tag{39.50}$$

The function η is called a *slack* variable. The contact region is the locus along which η vanishes. The elevation y_3 of the middle surface outside the contact region is obtained from (39.50). If we consider the case of an initially flat membrane inflated against a flat wall parallel to the x_1x_2 plane and positioned at a distance d above it, constraint (39.50) becomes

$$y_3 + \eta^2 = d = \text{constant}. \tag{39.51}$$

The approximate solution of the problem, obtained by applying the Ritz method to the total energy $\mathscr{E}$, after y_3 has been expressed in terms of η by using (39.51), is then a functional $\mathscr{E}$ of the type

$$\mathscr{E} = \mathscr{E}(y_{\beta,\alpha}, \eta, \eta_{,\alpha}). \tag{39.52}$$

The admissible functions for minimizing (39.52) must conform to the geometric boundary conditions

$$\left.\begin{array}{l} y_1 = \pm a \\ y_2 = x_2 \\ \eta = \sqrt{d} \end{array}\right\} \text{ for } x_1 = \pm a, \quad \left.\begin{array}{l} y_1 = x_1 \\ y_2 = \pm b \\ \eta = \sqrt{d} \end{array}\right\} \text{ for } x_2 = \pm b. \tag{39.53}$$

The minimum of (39.52) can be sought in the following class of admissible functions:

$$\begin{aligned} y_1 &= x_1 + \sum_{i=1}^{N}\sum_{j=1}^{N} a_{ij} \sin\frac{i\pi}{a} x_1 \cos\frac{(2j-1)\pi}{2b} x_2, \\ y_2 &= x_2 + \sum_{i=1}^{N}\sum_{j=1}^{N} b_{ij} \sin\frac{i\pi}{b} x_2 \cos\frac{(2j-1)\pi}{2a} x_1, \\ \eta &= \sqrt{d} - \sum_{i=1}^{N}\sum_{j=1}^{N} C_{ij} \cos\frac{(2i-1)\pi}{2a} x_1 \cos\frac{(2j-1)\pi}{2b} x_2. \end{aligned} \tag{39.54}$$

The substitution of (39.54) into (39.52), and the computation of the double integral with respect the variables x_1 and x_2, reduces (39.52) to a function of the $3N^2$ unknown coefficients a_{ij}, b_{ij}, and C_{ij}:

$$\tilde{\mathscr{E}} = \tilde{\mathscr{E}}(a_{ij}, b_{ij}, C_{ij}). \tag{39.55}$$

Following the Ritz procedure, the unknown coefficients are determined such that, for each i and j, we require

$$\frac{\partial\tilde{\mathscr{E}}}{\partial a_{ij}} = 0, \quad \frac{\partial\tilde{\mathscr{E}}}{\partial b_{ij}} = 0, \quad \frac{\partial\tilde{\mathscr{E}}}{\partial C_{ij}} = 0. \tag{39.56}$$

This extension of the Ritz method to the problem of the unilateral contact of a membrane is open to several serious objections. First, we must prove that the nonlinear algebraic system (39.56) admits a solution. Secondly, we must choose the right solution, whenever the solution is not unique. Thirdly, we must prove that the approximate solutions converge to a limit for $N \to \infty$. Fourthly, we must verify that the limit is a minimizer for (39.48) and satisfies the constraint (39.49). Without the demonstration of these facts, the method is purely heuristic. It may work in some cases but not in others.

The correct mathematical formulation of the problem of a membrane pushed against a rigid obstacle is formidable, even under the hypothesis of small displacements and small strains, in which the equilibrium configuration of the deformed middle surface of the membrane is governed by (37.2) with the constraint $u \geqslant \psi$. The problem of the existence and uniqueness of a solution is relatively simple (Lions and Stampacchia 1967). The procedure used to solve the problem is equivalent to minimizing the strain energy

$$\mathscr{E} = \frac{1}{2}\iint_D |\nabla u|^2\, dx_1\, dx_2, \tag{39.57}$$

where u is the transverse displacement of the middle surface, in a convex closed set K of admissible functions. These functions must satisfy the boundary condition $u = 0$

on C and $u \geqslant \psi$ in D, where $\psi(x_1, x_2)$ is the Cartesian equation of the outline of the obstacle.

The functional (39.57), in the class of functions with square integrable partial derivatives, which vanish outside some compact set in D, can be written as

$$\mathscr{E} = \tfrac{1}{2}\|u\|^2_{H^1_0} \tag{39.58}$$

where $\|u\|_{H^1_0}$ is the norm of the Hilbert space $H^1_0(D)$. Solving the equilibrium problem is equivalent to finding the minimum of $\mathscr{E}$ among the functions $v \in K \subset H^1_0$. Now, let us define

$$d = \tfrac{1}{2}\inf_{v \in K} \|v\|^2_{H^1_0}, \tag{39.59}$$

and consider a minimizing sequence $\{u_n\} \in K$ such that

$$d \leqslant \|u_m\|^2_{H^1_0} < d + \varepsilon,$$
$$d \leqslant \|u_n\|^2_{H^1_0} < d + \varepsilon,$$
$$d \leqslant \left\| \frac{u_m + u_n}{2} \right\|^2_{H^1_0} < d + \varepsilon.$$

From the parallelogram law we derive

$$\frac{1}{2}\|u_m - u_n\|^2_{H^1_0} = \|u_m\|^2_{H^1_0} + \|u_n\|^2_{H^1_0} - 2\left\| \frac{u_m + u_n}{2} \right\|^2 \leqslant 2(d + \varepsilon)^2 - d^2 = 4d\varepsilon + 2\varepsilon^2. \tag{39.60}$$

This proves that $\{u_n\}$ is a Cauchy sequence in K. As K is closed, there is a $u_0 \in K$ such that we have $u_n \to u_0$.

The proof of the existence of a solution does not differ from the standard argument used for linear problems. However, matters are different insofar as the regularity of solutions is concerned. The regularity of the solution to a problem with an obstacle does not necessary increase as the data become smoother and smoother. This circumstance is conveniently expressed by saying that problems with obstacles have a "bound of regularity" independent of the smoothness of the data.[53] The problem of regularity for solutions of the Laplace operator has been studied by several authors, but the results are still only partial. The first theorem of regularity is based on the device of considering a certain nonlinear problem and showing that its solution is actually a solution of the obstacle–membrane problem, which we were considering here (Lewy and Stampacchia 1969). Let us now introduce the function $\theta(s)$ defined by

$$\theta(s) = \begin{cases} 1, & \text{for } s \leqslant 1; \\ 0, & \text{for } s > 1, \end{cases} \tag{39.61}$$

and consider the following nonlinear problem

$$\nabla^2 u = -\max(-\nabla^2 \psi, 0)\,\theta(u - \psi), \quad u \in H^1_0(D). \tag{39.62}$$

[53] These terms were introduced by Baiocchi and Capelo (1978).

We assume that D is smooth, say its boundary C is of class $\mathscr{C}^2$, and $\psi \in H^{2,p}(D)$ for some $p > 2$. This last assumption ensures the condition that $\psi \in \mathscr{C}^{1,\alpha}(D)$, where $\alpha = 1 - (2/p)$, by virtue of a Sobolev's theorem. Then the right-hand side of (39.62), being the product of $\theta\,(\in L^\infty)$ and $\max(-\nabla^2\psi, 0)$ $(\in L^p)$, is a function $\in L^p$. Hence, by the regularity results for elliptical second-order operators, we know that $u \in H^{2,p}$ or, again by Sobolev's theorem, $u \in \mathscr{C}^{1,\alpha}$. In this way we can prove that the solution to the obstacle problem is of class $\mathscr{C}^{1,\alpha}$. However, we must still prove that (39.62) is equivalent to the problem of minimizing (39.57) among the functions $u \in K$. Lewy and Stampacchia have proved the equivalence of the two problems, establishing a fruitful connection between unilateral problems and nonlinear differential equations. However, the result that, if the obstacle is smooth, the deformed surface is, at most, $\mathscr{C}^{1,\alpha}$, does not say anything about the nature of the contact region. Only a few results are known on the subject, even for the particular case when D is a circle and ψ is a smooth axisymmetric function.

40. Curved Membranes

Curved membranes are thin, two-dimensional bodies like planar membranes, the only difference being that their middle surface, instead of being a plane, is a surface endowed with curvature. Curved membranes are also characterized by the constitutive property that the stress resultants transmitted inside are tangential to the middle surface; however, in contrast to plane membranes, they can also withstand resultant compressive stresses. While a planar membrane, loaded perpendicularly to its middle surface, is unavoidably subject to tension, a curved membrane may be compressed everywhere. As many materials, such as concrete and stone, are strong in compression but weak in tension, this explains the widespread use of curved membranes as structural elements in classical architecture. In recent times, the theory of curved membranes has become important for explaining the behavior of biological tissues, such as those of cells and the walls of blood vessels.

The theory of curved membranes under large displacements and strain has been successfully developed for rotationally symmetric surfaces, because in this case the exact theory yields a single, second-order differential equation, which may be treated in full generality (Dickey 1987). If some of the terms are neglected, we obtain certain approximate theories, which have been proposed previously by other authors (Bromberg and Stoker 1945, Reissner 1958). Among axisymmetric membranes, the spherical cap and the shallow cap are configurations of special interest, because they occur in many practical circumstances.

In order to derive the exact equations governing the equilibrium of elastic caps, we introduce cylindrical coordinates r, θ, z and denote the unit vectors in the direction of the coordinate lines by $\mathbf{e}_r$, $\mathbf{e}_\theta$, and $\mathbf{k}$, respectively. In rectangular coordinates, the vectors $\mathbf{e}_r$ and $\mathbf{e}_\theta$ assume the forms

$$\mathbf{e}_r = \mathbf{i}\cos\theta + \mathbf{j}\cos\theta, \quad \mathbf{e}_\theta = -\mathbf{i}\sin\theta + \mathbf{j}\cos\theta, \tag{40.1}$$

where $\mathbf{i}$ and $\mathbf{j}$ are the unit vectors in the x and y directions. The position of a point on the undeformed surface of the membrane is given by the vector

$$\mathbf{R} = r\mathbf{e}_r + z(r)\mathbf{k}. \tag{40.2}$$

After the deformation, the point will have a new position

$$\mathbf{r} = [r + u(r)]\mathbf{e}_r + [z(r) + w(r)]\mathbf{k}, \tag{40.3}$$

where u and w are the displacements in the radial and axial directions, respectively. An element of length dS before the deformation is given by

$$(dS)^2 = d\mathbf{R} \cdot d\mathbf{R} = (1 + z'^2)(dr)^2 + r^2(d\theta)^2. \tag{40.4}$$

After the deformation, dS is transformed into a new element of length ds, where

$$(ds)^2 = d\mathbf{r} \cdot d\mathbf{r} = \frac{(1 + u')^2 + (z' + w')^2}{1 + z'^2}(1 + z'^2)(dr)^2 + \left(1 + \frac{u}{r}\right)^2 r^2(d\theta)^2. \tag{40.5}$$

The strains are determined by the equation

$$(ds)^2 - (dS)^2 = 2\varepsilon_r(1 + z'^2)(dr)^2 + 2\varepsilon_\theta r^2(d\theta)^2, \tag{40.6}$$

so that, by using (40.4) and (40.5), we obtain the expressions

$$\varepsilon_r = \frac{1}{m^2}\left(u' + z'w' + \frac{1}{2}u'^2 + \frac{1}{2}w'^2\right), \tag{40.7}$$

$$\varepsilon_\theta = \frac{u}{r} + \frac{1}{2}\frac{u^2}{r^2}, \tag{40.8}$$

where $m^2 = 1 + z'^2$.

In order to derive the equilibrium equations, we draw an element of the deformed membrane and indicate the forces acting on the positive faces (Figure 40.1). The unit normal to the face $r =$ constant is given by

$$\mathbf{n} = \frac{(1 + u')\mathbf{e}_r + (z' + w')\mathbf{k}}{[(1 + u')^2 + (z' + w')^2]^{\frac{1}{2}}},$$

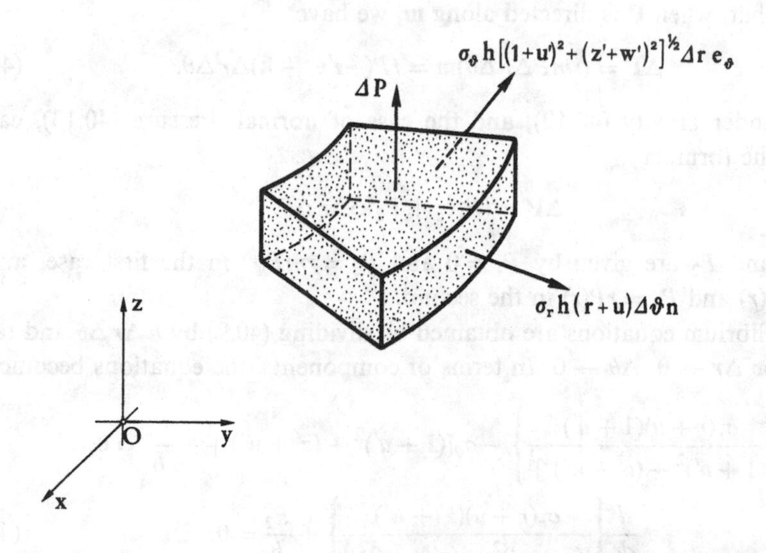

Fig. 40.1

and the unit normal to the face $\theta = \text{constant}$ is $\mathbf{e}_\theta$. The equilibrium conditions, in vector form, are

$$\left.\frac{\sigma_r h(r+u)[(1+u')\mathbf{e}_r + (z'+w')\mathbf{k}]}{[(1+u')^2 + (z'+w')^2]^{\frac{1}{2}}}\Delta\theta\right|_r^{r+\Delta r}$$
$$+ \left.\sigma_\theta h[(1+u')^2 + (z'+w')^2]^{\frac{1}{2}}\mathbf{e}_\theta\,\Delta_r\right|_\theta^{\theta+\Delta\theta} + \Delta\mathbf{P} = \mathbf{0}, \tag{40.9}$$

where $\Delta\mathbf{P}$ is the force on the undeformed area. We consider two choices of $\Delta\mathbf{P}$: in the z direction, or normal to the undeformed surface. The area of the element depicted in Figure 40.1, before the deformation has the value

$$\Delta A = \left|\frac{\partial\mathbf{R}}{\partial r} \times \frac{\partial\mathbf{R}}{\partial\theta}\right| \Delta r\,\Delta\theta.$$

From (40.2) we calculate the formulae

$$\frac{\partial\mathbf{R}}{\partial r} \times \frac{\partial\mathbf{R}}{\partial\theta} = (\mathbf{e}_r + z'\mathbf{k}) \times (r\mathbf{e}_\theta) = r(-z'\mathbf{e}_r + \mathbf{k}), \tag{40.10}$$

so that we get

$$\Delta A = r\sqrt{1+z'^2}\,\Delta r\,\Delta\theta = rm\,\Delta r\,\Delta\theta. \tag{40.11}$$

If $\Delta\mathbf{P}$ is in the z direction, we obtain

$$\Delta\mathbf{P} = (rmP\,\Delta r\,\Delta\theta)\mathbf{k}, \tag{40.12}$$

where P is the force per unit undeformed area. The unit normal to the undeformed surface is obtained from (40.10), and has the expression

$$\mathbf{m} = \frac{-z'\mathbf{e}_r + \mathbf{k}}{m}.$$

It follows that, when $\mathbf{P}$ is directed along $\mathbf{m}$, we have

$$\Delta\mathbf{P} = (rmP\Delta r\,\Delta\theta)\mathbf{m} = rP(-z'\mathbf{e}_r + \mathbf{k})\Delta r\,\Delta\theta. \tag{40.13}$$

The case under gravity (40.12), and the case of normal pressure (40.13), can be unified in the formula

$$\Delta\mathbf{P} = (P_1\mathbf{e}_z + P_2\mathbf{k})\Delta r\,\Delta\theta, \tag{40.14}$$

where P_1 and P_2 are given by $P_1 = 0$ and $P_2 = rmP(r)$ in the first case, and by $P_1 = -rz'p(r)$ and $P_2 = rP(r)$ in the second.

The equilibrium equations are obtained by dividing (40.9) by $h\,\Delta r\,\Delta\theta$ and taking the limit for $\Delta r \to 0$, $\Delta\theta \to 0$. In terms of components, the equations become

$$\frac{d}{dr}\left\{\frac{\sigma_r(r+u)(1+u')}{[(1+u')^2+(z'+w')^2]^{\frac{1}{2}}}\right\} - \sigma_\theta[(1+u')^2+(z'+w')^2]^{\frac{1}{2}} + \frac{P_1}{h} = 0,$$
$$\frac{d}{dr}\left\{\frac{\sigma_r(r+u)(z'+w')}{[(1+u')^2+(z'+w')^2]^{\frac{1}{2}}}\right\} + \frac{P_2}{h} = 0. \tag{40.15}$$

These equations must be completed by including the stress–strain law. On assuming that the material is elastic, the relation between the stress components and the strains is

$$\sigma_r = \sigma_r(\varepsilon_r, \varepsilon_\theta), \quad \sigma_\theta = \sigma_\theta(\varepsilon_r, \varepsilon_\theta). \tag{40.16}$$

If the strains are small, (40.16) may be taken in the linear version

$$\sigma_r = \frac{E}{1-\sigma^2}(\varepsilon_r + \sigma\varepsilon_\theta), \quad \sigma_\theta = \frac{E}{1-\sigma^2}(\varepsilon_\theta + \sigma\varepsilon_r). \tag{40.17}$$

The boundary conditions can assume different forms according to the way in which the boundary of the cap is constrained. For definiteness, we shall assume that they are

$$u(0) = w'(0) = w(a) = 0, \tag{40.18}$$

where a is the radius of the parallel situated at the base of the undeformed membrane. In addition, we may prescribe either the radial displacement at the boundary

$$u(a) = \mu, \tag{40.19}$$

or the radial stress

$$\sigma_r(a) = \bar{\sigma}. \tag{40.20}$$

In order to solve (40.7), (40.8), (40.15), and (40.17), we rewrite (40.7) and (40.8) in the form

$$[(1+u')^2 + (z'+w')]^{\frac{1}{2}} = m(1+2\varepsilon_r)^{\frac{1}{2}}, \tag{40.21a}$$

$$r + u = r(1+2\varepsilon_\theta)^{\frac{1}{2}}. \tag{40.21b}$$

As a consequence, the equilibrium equations (40.15) can be written as

$$\begin{gathered}\frac{d}{dr}\left[\frac{r\Sigma_r(1+u')}{m\sqrt{1+2\varepsilon_r}}\right] - m\Sigma_\theta + \frac{P_1}{h} = 0,\\ \frac{d}{dr}\left[\frac{r\Sigma_r(z'+w')}{m\sqrt{1+2\varepsilon_r}}\right] + \frac{P_2}{h} = 0,\end{gathered} \tag{40.22}$$

where Σ_r and Σ_θ have the forms

$$\Sigma_r = \sigma_r\sqrt{1+2\varepsilon_r}, \quad \Sigma_\theta = \sigma_\theta\sqrt{1+2\varepsilon_\theta}. \tag{40.23}$$

After integration, (40.22) become

$$\begin{gathered}\frac{r\Sigma_r(1+u')}{m\sqrt{1+2\varepsilon_r}} = rS - rF_1,\\ \frac{r\Sigma_r(1+w')}{m\sqrt{1+2\varepsilon_r}} = -rF_2,\end{gathered} \tag{40.24}$$

where

$$rS = \int_0^r m\Sigma_\theta \, d\tau, \tag{40.25a}$$

$$rF_1 = \frac{1}{h}\int_0^r P_1(\tau)\, d\tau, \tag{40.25b}$$

$$rF_2 = \frac{1}{h}\int_0^r P_2(\tau)\, d\tau. \tag{40.25c}$$

In (40.24), both F_1 and F_2 are known functions, because we know P_1 and P_2. Equations (40.24) imply the relation

$$r^2\Sigma_r^2 = r^2F_2^2 + r^2(S - F_1)^2. \tag{40.26}$$

So, on combining (40.26) with (40.24), we find

$$\Sigma_r = \sqrt{F_2^2 + (S - F_1)^2}, \tag{40.27a}$$

$$z' + w' = -\frac{F_2(1 + u')}{S - F_1}. \tag{40.27b}$$

From (40.21a) and (40.27b), we obtain the following results:

$$1 + u' = m\sqrt{1 + 2\varepsilon_r}\,\frac{(S - F_1)}{\Sigma_r}. \tag{40.28}$$

From (40.25a) we obtain

$$\Sigma_\theta = \frac{(rS)'}{m}, \tag{40.29}$$

and from (40.21) we derive

$$1 + u' = \frac{d}{dr}\,(r\sqrt{1 + 2\varepsilon_\theta}), \tag{40.30}$$

so that (40.28) assumes the form

$$\frac{d}{dr}\,(r\sqrt{1 + 2\varepsilon_\theta}) = m\sqrt{1 + 2\varepsilon_r}\,\frac{(S - F_1)}{\Sigma_r}. \tag{40.31}$$

The strains ε_r and ε_θ in (40.31) can be expressed in terms of Σ_r and Σ_θ by means of (40.16), written as

$$\frac{\Sigma_r}{\sqrt{1 + 2\varepsilon_r}} = \sigma_r(\varepsilon_r, \varepsilon_\theta), \tag{40.32a}$$

$$\frac{\Sigma_\theta}{\sqrt{1 + 2\varepsilon_\theta}} = \sigma_\theta(\varepsilon_r, \varepsilon_\theta). \tag{40.32b}$$

By appealing to a general property of constitutive equations for elastic materials, we assume that (40.32) have a unique solution for ε_r and ε_θ as functions of Σ_r and Σ_θ; that is,

$$\varepsilon_r = \varepsilon_r(\Sigma_r, \Sigma_\theta), \quad \varepsilon_\theta = \varepsilon_\theta(\Sigma_r, \Sigma_\theta). \tag{40.33}$$

As Σ_r and Σ_θ are known functions of S and S' by virtue of (40.27a) and (40.29), (40.31) becomes a second-order ordinary differential equation for determining S. Once S has been determined, the quantities Σ_r and Σ_θ are given by (40.27a) and (40.29) and the displacements from (40.24).

The boundary conditions on (40.31) are of two types. If the displacement is prescribed (see (40.19)) we obtain from (40.21b) the condition

$$1 + \frac{\mu}{a} = \sqrt{1 + 2\varepsilon_\theta(\Sigma_r, \Sigma_\theta)}\bigg|_{r=a}. \tag{40.34}$$

If the stress condition (40.20) is prescribed, we must have (see (40.32a))

$$\bar{\sigma} = \frac{\Sigma_r}{\sqrt{1 + 2\varepsilon_\theta(\Sigma_r, \Sigma_\theta)}}\bigg|_{r=a}. \tag{40.35}$$

Another condition is that $S(r)$ must be finite at $r = 0$.

The treatment of (40.31) may be performed in at least two ways. One is to investigate the consequences of the reduction of (40.31) when only small strains are considered (Dickey 1987). Another promising way might be to analyze the consequences of the maximum principle applied to the complete equation (40.31).

It might seem surprising that the two types of loading considered before do not consider the case in which the load is perpendicular to the deformed middle surface instead of to the undeformed one. This load condition is said to be of "hydrostatic pressure," and is perhaps more important than the others from the technical point of view, but is also more difficult to treat.

In the case of hydrostatic pressure, we again have the equilibrium equations (40.15), but the quantities P_1 and P_2 are determined from the relation

$$P_1\mathbf{e}_r + P_2\mathbf{k} = rmP\hat{\mathbf{n}}, \tag{40.36}$$

where P is the force per unit undeformed area and $\hat{\mathbf{n}}$ is the unit vector in the direction normal to the deformed surface. The unit normal $\hat{\mathbf{n}}$ is given by

$$\hat{\mathbf{n}} = \left(\frac{\partial \mathbf{r}}{\partial r} \times \frac{\partial \mathbf{r}}{\partial \theta}\right)\left|\frac{\partial \mathbf{r}}{\partial r} \times \frac{\partial \mathbf{r}}{\partial \theta}\right|^{-1}. \tag{40.37}$$

From (40.3) we calculate

$$\begin{aligned} \frac{\partial \mathbf{r}}{\partial r} &= (1 + u')\mathbf{e}_r + (z' + w')\mathbf{k}, \\ \frac{\partial \mathbf{r}}{\partial \theta} &= (r + u)\mathbf{e}_\theta. \end{aligned} \tag{40.38}$$

We may thus write (40.37) in the form

$$\hat{\mathbf{n}} = \frac{(1 + u')\mathbf{k} - (z' + w')\mathbf{e}_r}{[(1 + u')^2 + (z' + w')^2]^{\frac{1}{2}}}. \tag{40.39}$$

It then follows, after substitution of (40.39) in (40.36), that P_1 and P_2 are given by

$$P_1 = \frac{rmP(z' + w')}{[(1+u')^2 + (z'+w')^2]^{\frac{1}{2}}}, \tag{40.40}$$

$$P_2 = \frac{rmP(1+u')}{[(1+u')^2 + (z'+w')^2]^{\frac{1}{2}}}. \tag{40.41}$$

The equilibrium equations are again (40.15), but with P_1 and P_2 given by (40.40) and (40.41), respectively. Their final expression is

$$\begin{aligned} &\frac{d}{dr}\left\{\frac{(r+u)(1+u')\sigma_r}{[(1+u')^2+(z'+w')^2]^{\frac{1}{2}}}\right\} - \sigma_\theta[(1+u')^2+(z'+w')^2]^{\frac{1}{2}} \\ &-\frac{rmP(z'+w')}{[(1+u')^2+(z'+w')^2]^{\frac{1}{2}}} = 0, \end{aligned} \tag{40.42a}$$

$$\frac{d}{dr}\left\{\frac{(r+u)(z'+w')\sigma_r}{[(1+u')^2+(z'+w')^2]^{\frac{1}{2}}}\right\} + \frac{rmP(1+u')}{[(1+u')^2+(z'+w')^2]^{\frac{1}{2}}} = 0. \tag{40.42b}$$

In order to reduce these equations to a more convenient form, let us introduce the constitutive equations (the inverse of (40.16)):

$$\varepsilon_r = \varepsilon_r(\sigma_r, \sigma_\theta), \qquad \varepsilon_\theta = \varepsilon_r(\sigma_r, \sigma_\theta). \tag{40.43}$$

We assume, as before, that (40.16) have a unique inverse. The following is then a consequence of (40.21a):

$$\frac{z'+w'}{[(1+u')^2+(z'+w')^2]^{\frac{1}{2}}} = -\left[1 - \frac{(1+u')^2}{(1+u')^2+(z'+w')^2}\right]^{\frac{1}{2}} = \left\{1 - \left[\frac{(r\sqrt{1+2\varepsilon_\theta})'}{m\sqrt{1+2\varepsilon_r}}\right]^2\right\}. \tag{40.44}$$

On combining (40.30) and (40.44) with (40.42a) we find

$$\frac{d}{dr}\left[\frac{r\sqrt{1+2\varepsilon_\theta}(r\sqrt{1+2\varepsilon_\theta})'\sigma_r}{m\sqrt{1+2\varepsilon_r}}\right] - \sigma_\theta m\sqrt{1+2\varepsilon_r} + \frac{rmP}{h}\left\{1 - \left[\frac{(r\sqrt{1+2\varepsilon_\theta})'}{m\sqrt{1+2\varepsilon_r}}\right]^2\right\}^{\frac{1}{2}} = 0, \tag{40.45}$$

and using the result

$$(r\sqrt{1+2\varepsilon_\theta})' = \frac{r\varepsilon_\theta' + 2\varepsilon_\theta + 1}{\sqrt{1+2\varepsilon_\theta}},$$

(40.45) can be simplified to give

$$\begin{aligned} &\frac{d}{dr}\left[\frac{r(r\varepsilon_\theta' + 2\varepsilon_\theta + 1)\sigma_r}{m\sqrt{1+2\varepsilon_r}}\right] - \sigma_\theta m\sqrt{1+2\varepsilon_r} \\ &+\frac{rP\{m^2(1+2\varepsilon_r)(1+2\varepsilon_\theta) - (r\varepsilon_\theta' + 2\varepsilon_\theta + 1)^2\}^{\frac{1}{2}}}{h\sqrt{1+2\varepsilon_r}\sqrt{1+2\varepsilon_\theta}} = 0. \end{aligned} \tag{40.46}$$

A similar procedure may be used to rewrite (40.42b) in terms of the stresses and strains. Thus on combining (40.21), (40.30), and (40.45) with (40.42b), we find

$$-\frac{d}{dr}\left\{r\sqrt{1+2\varepsilon_\theta}\left(1-\left[\frac{(r\sqrt{1+2\varepsilon_\theta})'}{m\sqrt{1+2\varepsilon_r}}\right]^2\right)\right\}^{\frac{1}{2}}+\frac{rP(r\sqrt{1+2\varepsilon_\theta})'}{h\sqrt{1+2\varepsilon_r}}=0, \tag{40.47}$$

which can be also rewritten as

$$\frac{d}{dr}\left\{\frac{r[m^2(1+2\varepsilon_r)(1+2\varepsilon_\theta)-(r\varepsilon_\theta'+2\varepsilon_\theta+1)^2]^{\frac{1}{2}}\sigma_r}{m\sqrt{1+2\varepsilon_r}}\right\}-\frac{rP(r\varepsilon_\theta'+2\varepsilon_\theta+1)}{h\sqrt{1+2\varepsilon_r}\sqrt{1+2\varepsilon_\theta}}=0. \tag{40.48}$$

After using the constitutive equations (40.43), we find that (40.46) and (40.48) are a pair of second-order differential equations for determining σ_r and σ_θ.

Equations (40.46) and (40.48) can be studied in detail after certain simplifications have been made (Dickey 1988). If the strains are small, that is if $|\varepsilon_r| \ll 1$ and $|\varepsilon_\theta| \ll 1$, we can choose for the constitutive equations Hooke's law

$$E\varepsilon_r=\sigma_r-\sigma\sigma_\theta, \quad E\varepsilon_\theta=\sigma_\theta-\sigma\sigma_r. \tag{40.49}$$

Furthermore, if we retain in the equilibrium equations only terms linear in ε_θ and ε_r, we can modify them to form the pair of equations

$$\frac{d}{dr}\left\{\frac{r\sigma_r}{m}\right\}-m\sigma_\theta+\frac{rP}{h}\{m^2[1+2(\varepsilon_r+\varepsilon_\theta)]-[1+2(r\varepsilon_\theta'+2\varepsilon_\theta)]\}^{\frac{1}{2}}=0, \tag{40.50}$$

$$\frac{d}{dr}\left(\frac{r}{m}\{m^2[1+2(\varepsilon_r+\varepsilon_\theta)]-[1+2(r\varepsilon_\theta'+2\varepsilon_\theta)]\}^{\frac{1}{2}}\right)-\frac{rP}{h}=0. \tag{40.51}$$

Let us note that neglecting ε_r or ε_θ in relation to m^2-1 would be incorrect, as $m^2-1=(z')^2$ may itself be small. In fact we have $z'=0$ at $r=0$ if the undeformed surface is smooth.

From (40.51) we derive

$$\frac{r}{m}\{m^2(1+2\varepsilon_r)(1+2\varepsilon_\theta)-[1+2(r\varepsilon_\theta'+2\varepsilon_\theta)]\}^{\frac{1}{2}}\sigma_r=\frac{F}{h}, \tag{40.52}$$

with

$$F(r)=\int_0^r \tau P(\tau)\,d\tau.$$

On combining (40.52) and (40.50) we obtain

$$\frac{d}{dr}\left(\frac{r\sigma_r}{m}\right)-m\sigma_\theta+\frac{mPF}{h^2\sigma_r}=0. \tag{40.53}$$

If we introduce the dimensionless stresses

$$\Sigma_r=\frac{\sigma_r}{E}, \quad \Sigma_\theta=\frac{\sigma_\theta}{E},$$

Hooke's law becomes

$$\varepsilon_r=\Sigma_r-\sigma\Sigma_\theta, \quad \varepsilon_\theta=\Sigma_\theta-\sigma\Sigma_r, \tag{40.54}$$

while (40.52) and (40.53) become

$$m^2 - 1 + 2m^2(\varepsilon_r + \varepsilon_\theta) - 2r\varepsilon_\theta' - 4\varepsilon_\theta = \frac{m^2 F^2}{r^2 h^2 E^2 \Sigma_r^2}, \tag{40.55}$$

$$\left(\frac{r\Sigma_r}{m}\right)' - m\Sigma_\theta + \frac{mPF}{h^2 E^2 \Sigma_r} = 0. \tag{40.56}$$

The strains can be eliminated from (40.55) by using (40.54), so that (40.55) assumes the form

$$m^2 - 1 + [2m^2(1-\sigma) + 4\sigma]\Sigma_r + [2m^2(1-\sigma) - 4]\Sigma_\theta - 2r\Sigma_\theta' + 2\sigma r\Sigma_r = \frac{m^2 F^2}{r^2 h^2 E^2 \Sigma_r^2}. \tag{40.57}$$

Finally, after eliminating Σ_θ between (40.56) and (40.57) we get

$$m^2 - 1 + 2[m^2 + (2 - m^2)\sigma]\Sigma_r + 2(m^2 - 2 - \sigma m^2)\left[\frac{1}{m}\left(\frac{r\Sigma_r}{m}\right)' + \left(\frac{PF}{h^2 E^2 \Sigma_r^2}\right)\right]$$
$$- 2r\left\{\left[\frac{1}{m}\left(\frac{r\Sigma_r}{m}\right)'\right]' + \left(\frac{PF}{h^2 E^2 \Sigma_r}\right)'\right\} + 2\sigma r\Sigma_r' = \frac{m^2 F^2}{r^2 h^2 E^2 \Sigma_r^2}. \tag{40.58}$$

This is a single second-order differential equation for determining Σ_r. Once Σ_r is known, we find Σ_θ from (40.56), then ε_r and ε_θ from (40.54), and then the displacements u and w from the strain–displacement relations (40.21).

Equation (40.58) is also useful for estimating the consistency of the approximate theories of curved membranes. These theories are suggested by the need for simplifying (40.58) in some special situations; as, for example, when the principal radii of curvature of the cap are very slowly changing variables, so that they can be taken to be constant, or the case when the cap is shallow.

In order to illustrate this last case more precisely we write

$$z(r) = \delta\,\zeta(r), \tag{40.59}$$

with $\delta \ll 1$. Let us assume that the applied pressure is small and redefine the pressure and the radial stress by

$$P = \delta^3 p \tag{40.60}$$

$$\Sigma_r = \delta^2 S_r. \tag{40.61}$$

Equation (40.60) implies that we have

$$F = \delta^3 f = \delta^3 \int_0^r \tau p(\tau)\, d\tau. \tag{40.62}$$

Now, on substituting (40.60)–(40.62) in (40.58), and retaining the terms of lowest order in δ, we find the following equation for S_r:

$$r^2 S_r'' + 3rS_r' + \frac{f^2}{2r^2 h^2 E^2 S_r^2} = \frac{\zeta'^2}{2}. \tag{40.63}$$

In order to obtain the dimensionless circumferential stress Σ_θ, we define S_θ as

$$\Sigma_\theta = \delta^2 S_\theta, \tag{40.64}$$

and substitute this in (40.56), keeping the lowest term in δ. The result is

$$S_\theta = (rS_r)'. \tag{40.65}$$

Furthermore, we redefine the strains as

$$\varepsilon_r = \delta^2 e_r, \quad \varepsilon_\theta = \delta^2 e_\theta,$$

so that, from (40.54), we can write

$$e_r = S_r - \sigma S_\theta, \quad e_\theta = S_\theta - \sigma S_r. \tag{40.66}$$

Let us now introduce the modified displacements U and W such that we have

$$u = \delta^2 U, \quad w = \delta W, \tag{40.67}$$

and combine (40.66) with (40.7) and (40.8) to give

$$U = re_\theta, \tag{40.68}$$
$$W' = -\zeta' - [(\zeta')^2 - 2U + 2e_r]^{\frac{1}{2}}, \tag{40.69}$$

up to higher order terms in δ.

The boundary conditions associated with (40.63) are either (40.19) or (40.20). In particular, if we put

$$\mu = \delta^2 M, \tag{40.70}$$
$$\frac{\bar{\sigma}}{E} = \delta^2 S, \tag{40.71}$$

the boundary condition (40.19) becomes

$$[rS_r' + (1-\sigma)S_r]_{r=a} = \frac{M}{a}, \tag{40.72}$$

whereas (40.20) becomes

$$S_r(a) = S. \tag{40.73}$$

We also require the condition that $S_r'(0) = 0$ in both cases.

It is possible to linearize (40.63) using the following argument (Dickey 1988). We are interested in a solution of (40.63) which has the property $S_r \to 0$ for $p \to 0$. In this case, (40.63) requires the condition

$$\lim_{p\to 0} \frac{f^2}{r^2h^2E^2S_r^2} = (\zeta')^2. \tag{40.74}$$

Let us introduce a function $v(r)$ requiring $|v/\zeta'| \ll 1$ when p is sufficiently small. Then (40.74) is equivalent to

$$\frac{f}{rhES_r} = -\zeta' + v = -\zeta'\left(1 - \frac{v}{\zeta'}\right). \tag{40.75}$$

On substituting (40.74) and (40.75) in (40.63), and keeping only first-order terms, we obtain a linear equation for determining S_r:

$$r^2\left[\frac{f}{rhE\zeta'}\left(1+\frac{v}{\zeta'}\right)''+3r\left[\frac{f}{rhE\zeta'}\left(1+\frac{v}{\zeta'}\right)\right]'+\zeta'=0. \tag{40.76}$$

These equations are easily specialized to the case of the spherical cap. Let us assume that the pressure is constant, necessitating

$$f=\tfrac{1}{2}pr^2. \tag{40.77}$$

The equation of the middle surface is

$$z=\sqrt{b^2-r^2}-\sqrt{b^2-a^2}, \quad 0\leqslant r\leqslant a,$$

and hence

$$z'=\frac{\delta\left(\frac{r}{a}\right)}{\left[1-\delta^2\left(\frac{r}{a}\right)^2\right]^{\frac{1}{2}}}, \tag{40.78}$$

with $\delta=b/a\ll 1$. From (40.78), on using (40.59) and neglecting terms of higher order in δ, we derive

$$\zeta'=-\frac{r}{a}. \tag{40.79}$$

By substituting (40.77) and (40.79) in (40.76) we obtain the linear equation in $v(\rho)$ ($\rho=r/a$):

$$\rho^2\frac{d^2}{d\rho^2}\left(\frac{v}{\rho}\right)+3\rho\frac{d}{d\rho}\left(\frac{v}{\rho}\right)-\frac{2hE}{pa}\rho v=0,$$

which may also be written in the form

$$\rho^2\frac{d^2v}{d\rho^2}+\rho\frac{dv}{d\rho}-(1+\lambda^2\rho^2)v=0, \tag{40.80}$$

with $\lambda^2=(2hE/pa)$. The solution of (40.80), which is finite at $r=0$, is

$$v=AI_1(\lambda\rho), \tag{40.81}$$

where I_1 is the modified Bessel function of order one, and A is a constant to be determined by the boundary conditions (40.71) or (40.72).

41. Stability of Curved Membranes

When an elastic shell, in particular an axisymmetric cap, is subject to a uniform pressure p, and this pressure is sufficiently small, the cap assumes a unique equilibrium configuration. If, however, the pressure is progressively increased, the equilibrium state is no longer unique after a certain value of p, and the surface of the cap may suddenly switch from the state just reached to another state, sometimes very distant from the first. This phenomenon was explained intuitively by Kármán and

Tsien (1939) in their theory of snap buckling. They conjectured that the form of the response pressure–deflection curve for a spherical shallow membrane is as shown by the solid curve in Figure 41.1. In the figure, δ is the vertical deflection of the apex of the membrane, and Q_0 is a parameter proportional to the total pressure force acting on the membrane. The figure suggests that at $Q_0 = 0$ there are three equilibrium states. State (1) ($Q_0 = \delta = 0$) corresponds to the initial undeformed membrane; state (3) represents the cap "snapped through"; and the intermediate state (2) corresponds to a perfectly flat configuration of the membrane. The diagram is antisymmetric with respect to a vertical axis passing through point (2). The form of the curve, which admits three intersections with the δ axis seems to contradict the phenomenological response of shells, which exhibit a monotonically increasing $Q_0 - \delta$ diagram for small Q_0. However, in effect, there is no contradiction. Unlike membranes, shells are not perfectly flexible, so that the curve of $Q_0 - \delta$ has the form of the dashed line in Figure 41.1. For shells too, monotonicity of the curve fails, but not in the neighborhood of the origin.

In order to justify the behavior of the solid curve in Figure 41.1 we consider a meridional section of the membrane constituted by a curve C (Figure 41.2) with end points A and B (Bauer, Callegari, and Reiss 1973). The arc length along C is denoted by s, and $\theta(X)$ is the angle between the Z axis and the normal to the surface at s. We assume that $\theta(X)$ is a given differentiable function satisfying the regularity condition

$$\theta(0) = 0,$$

at the apex. Let us call $U(X)$ and $W(X)$ the tangential and the normal displacement of the middle surface. The positive directions of U and W are indicated in Figure 41.2. The membrane is subject to a pressure $p(X)$ acting normally to the middle surface. The edge point B is fixed, leading to

$$U(X_0) = W(X_0) = 0. \tag{41.1}$$

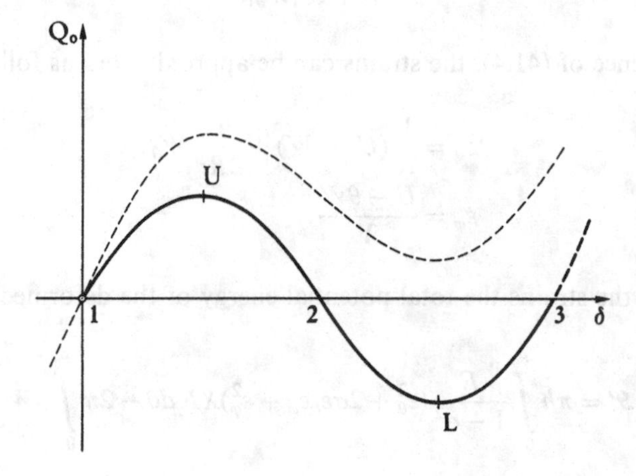

Fig. 41.1

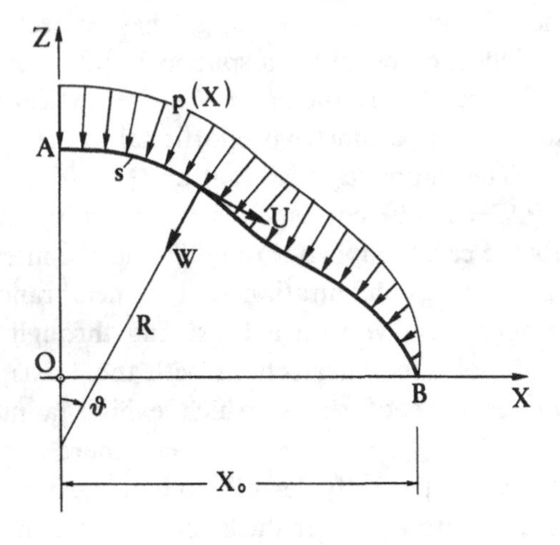

Fig. 41.2

We denote the circumferential coordinate of the membrane by ϕ. The meridional and circumferential strains ε_θ and ε_φ are

$$\varepsilon_\theta = \frac{U_\theta - W}{R}\left[1 + \frac{1}{2}\frac{(U_\theta - W)}{R}\right] + \frac{1}{2}\left(\frac{U + W_\theta}{R}\right)^2, \tag{41.2}$$

$$\varepsilon_\varphi = \frac{U\cos\theta - W\sin\theta}{X}\left[1 + \frac{1}{2}\frac{(U\cos\theta - W\sin\theta)}{X}\right], \tag{41.3}$$

where the subscript θ on U and W denotes differentiation with respect to θ, and $R(\theta)$ is the radius of curvature along the meridian.

We assume that the membrane is shallow. This implies that we may take $\sin\theta \simeq \theta$ and $\cos\theta \simeq 1$. In addition, we assume the inequality

$$|U| \ll |W_\theta|. \tag{41.4}$$

As a consequence of (41.4), the strains can be approximated as follows:

$$\varepsilon_\theta = \frac{1}{R}(U_\theta - W) + \frac{1}{2R^2}W_\theta^2, \tag{41.5a}$$

$$\varepsilon_\varphi = \frac{U - \theta W}{X}. \tag{41.5b}$$

In terms of the strains the total potential energy of the deformed membrane is

$$\mathscr{E} = \mathscr{W} - \mathscr{L} = \pi h \int_0^{\theta_0} \frac{E}{1-\sigma^2}(\varepsilon_\theta^2 + 2\sigma\varepsilon_\theta\varepsilon_\varphi + \varepsilon_\varphi^2)XR\,d\theta - 2\pi \int_0^{\theta_0} pWXR\,d\theta, \tag{41.6}$$

where θ_0 is the value of θ at $X = X_0$. Since U and W are prescribed at $X = X_0$, Euler's equations associated with the condition that $\mathscr{E}$ is stationary are

$$R\sigma_\varphi = \frac{d}{d\theta}(X\sigma_\theta), \tag{41.7}$$

$$\frac{d}{d\theta}\left(\frac{X\sigma_\theta W_\theta}{R}\right) + X\sigma_\theta + r\theta\sigma_\varphi = -\frac{XRp}{h}. \tag{41.8}$$

We eliminate σ_φ from these two equations and simplify the resulting equation by observing that, for flat membranes, we can take

$$\frac{d}{dt} = \frac{dX}{ds}\frac{ds}{d\theta}\frac{d}{dX} \simeq R\frac{d}{dX},$$

because of the definition $dX/ds = \cos\theta \simeq 1$ and $R = ds/d\theta$. On integrating the resulting equation and using the regularity of the solution at the origin, we obtain

$$\sigma_\theta\left(\frac{dW}{dX} + \theta\right) = -\frac{1}{X}\int_0^X \frac{\xi p(\xi)}{h}\, d\xi. \tag{41.9}$$

A second equation can be derived by eliminating U from (41.5). Then, on expressing the strains in terms of stresses, and using (41.7), we have

$$\frac{d}{dX}\left(X^3\frac{d\sigma_\theta}{dX}\right) = -EX\left[\theta\frac{dW}{dX} + \frac{1}{2}\left(\frac{dW}{dX}\right)^2\right]. \tag{41.10}$$

The boundary conditions associated with (41.9) and (41.10) are

$$U(X_0) = 0, \tag{41.11a}$$

$$W(X_0) = 0, \tag{41.11b}$$

and also, for symmetry, we have

$$\frac{d\sigma_\theta}{dX}(0) = 0. \tag{41.12}$$

In order to express condition (41.11a) in terms of σ_θ and W, we solve (41.5b) for U and then use the constitutive equation to give ε_φ in terms of σ_θ and σ_φ. Then (41.7) is used to eliminate σ_φ. This gives

$$U = \frac{1}{E}X\left[X\frac{d\sigma_\theta}{dX} + (1-\sigma)\sigma_\theta\right] - \theta W. \tag{41.13}$$

Condition (41.11a) can be replaced by

$$\left[X\frac{d\sigma_\theta}{dX} + (1-\sigma)\sigma_\theta\right]_{X=X_0} = 0. \tag{41.14}$$

In order to describe the solutions of (41.9) and (41.10) with the boundary conditions (41.11b)–(41.12), and (41.14), it is convenient to define the dimensionless quantities

$$x = \frac{X}{X_0}, \quad w(x) \equiv \frac{W(X)}{X_0}, \quad \Sigma_\theta(x) \equiv \frac{\sigma_\theta}{E}, \quad q(x) \equiv \frac{X_0}{h}\frac{p(X)}{E}, \tag{41.15}$$

and the new parameter

$$\tilde{\varphi}(x) \equiv \theta(X). \tag{41.16}$$

Then the boundary-value problem is reduced to

$$v\Sigma_\theta = -Q(x), \quad 0 < x < 1; \tag{41.17}$$

$$(x^3\Sigma_\theta')' = -\frac{x}{2}(v^2 - \tilde{\varphi}^2), \quad 0 < x < 1; \tag{41.18}$$

$$v = w' + \tilde{\varphi}; \tag{41.19}$$

$$\Sigma_\theta'(0) = 0, \quad \Sigma_\theta'(1) + a\Sigma_\theta(1) = 0; \tag{41.20}$$

$$w(1) = 0; \tag{41.21}$$

where $Q(x)$ and a are defined by

$$Q(x) \equiv \frac{1}{x}\int_0^x \xi q(\xi)\, d\xi, \quad a \equiv 1 - \sigma > 0. \tag{41.22}$$

The prime denotes differentiation with respect to x. When Σ_θ and w are determined, the dimensionless circumferential stress $\Sigma_\varphi \equiv \sigma_\varphi/E$ and the dimensionless tangential displacement $u \equiv U/X_0$ are evaluated from

$$\Sigma_\varphi = (x\Sigma_\theta)', \quad u = x(\Sigma_\varphi - \sigma\Sigma_\theta) - \varphi w. \tag{41.23}$$

For simplicity, we assume the condition

$$\tilde{\varphi}(x) > 0, \quad \text{for } x \neq 0, \tag{41.24}$$

which means that the cap is flat only at its cusp. Moreover, we take the value $Q(x) \equiv 0$. This hypothesis may seem too restrictive, but is sufficient to prove the conjecture by von Kármán and Tsien about the multiplicity of equilibrium states in an unloaded shallow membrane. For $Q(x) \equiv 0$ the displacement $w(x)$ is determined by combining (41.17), (41.19), and (41.21):

$$w(x) = \int_x^1 \tilde{\varphi}(\xi)\, d\xi, \tag{41.25}$$

and, consequently,

$$\delta \equiv w(0) = \int_0^1 \tilde{\varphi}(\xi)\, d\xi. \tag{41.26}$$

More precisely, we can show that, for $Q(x) \equiv 0$, the problem has exactly three solutions, given by

$$(1)\ (\Sigma_\theta)_1 = w_1 = \delta_1 = 0; \tag{41.27}$$

$$(2)\ (\Sigma_\theta)_2 = \int_0^1 G(x, \xi)\tilde{\varphi}^2(\xi)\, d\xi, \quad w_2 = \int_x^1 \tilde{\varphi}\, d\xi, \quad \delta_2 = \int_0^1 \varphi\, d\xi, \tag{41.28}$$

where $G(x, \xi)$ is Green's function defined by

$$G(x, \xi) = \frac{\xi}{2} \cdot \begin{cases} 1 - \dfrac{2}{a} - \dfrac{1}{x}, & 0 \leqslant \xi < x, \\ 1 - \dfrac{2}{a} - \dfrac{1}{\xi^2}, & x \leqslant \xi \leqslant 1; \end{cases} \tag{41.29}$$

$$\text{(3)} \quad (\Sigma_\theta)_3 = 0, \quad w_3 = 2\int_0^1 \tilde{\varphi}\, d\xi, \quad \delta_2 = 2\int_0^1 \tilde{\varphi}\, d\xi. \tag{41.30}$$

To prove the result, we observe that (41.17) with $Q(x) \equiv 0$ implies that either Σ_θ or v must vanish for each x. Furthermore, it follows from (41.18), assumption (41.24), and the continuity of Σ_θ' and Σ_θ'' that, if Σ_θ vanishes in a subinterval of [0, 1], it must vanish in the entire interval [0, 1]. Therefore, we have either $\Sigma_\theta \equiv 0$ or $v \equiv 0$ on [0, 1]. For $\Sigma_\theta \equiv 0$, (41.18) then gives

$$v^2 - \tilde{\varphi}^2 = 0,$$

and hence we have $v = +\tilde{\varphi}$ and $w' = 0$ (see (41.19)) or $v = -\tilde{\varphi}$ and $w' = -2\tilde{\varphi}$. These are just solutions (41.27) and (41.30). For $v \equiv 0$ we obtain solution (41.28) by solving the boundary-value problem (41.18), (41.20) for Σ_θ, and (41.19) and (41.21) for w.

In the special case of a shallow spherical membrane, $\tilde{\varphi}$ is given by $\tilde{\varphi} = Ax$, where A is a small constant. Then, putting $\delta = w(0)$, the three solutions are

$$(\Sigma_\theta)_1 = \delta_1 = 0;$$

$$(\Sigma_\theta)_2 = -\frac{A^2}{8\left[\left(\dfrac{1}{2} + \dfrac{1}{a}\right) - x^2\right]}, \quad \delta_2 = \frac{A}{2};$$

$$(\Sigma_\theta)_3 = 0, \quad \delta_3 = A.$$

These correspond to the three points (1)–(3) in Figure 41.1.

The problem of the shallow membrane studied by Kármán and Tsien, viewed in the light of the theory of elastic stability, is resolved in the statement that a convex membrane (see (41.24)) is never stable, because under a very low normal pressure it switches to another equilibrium configuration. Specifically, the membrane passes from the point (1) in Figure 41.1 to point (3). The particular interest in this result is that the same behavior is likely to be shown by a shallow shell with a low bending stiffness.

The curved membranes, however, show a different type of elastic instability that is unrelated to the particular shape of the middle surface but is connected with the stress–strain relation, whenever Hooke's law is invalid.

The best illustration of this phenomenon occurs when a hollow sphere of deformable but incompressible material is inflated (Carrol 1987). If we blow up a large balloon and continuously record the pressure in the balloon as a function of the volume of air supplied, we find that the pressure–volume dependence most probably differs for different materials of the sphere. For example, the pressure may change monotonically, first increasing and then decreasing, or it may increase, decrease, and then increase again, the behavior depending on the form of the strain energy function and on the initial radius and thickness of the sphere.

In order to explain these variations in behavior more precisely, we consider a radially symmetric deformation of a hollow sphere of the form

$$r^3 - r_0^3 = a^3 - a_0^3, \quad \theta = \theta_0, \quad \varphi = \varphi_0, \tag{41.31}$$

where (r, θ, φ) are spherical polar coordinates, a is the radius of the inner boundary, with the subscript 0 denoting the values of all these quantities in the reference state. Equations (41.31) ensure that the deformation is isochoric. If then b and b_0 refer correspondingly to the outer radius of the sphere in the initial condition and actual state, we put

$$\alpha = \frac{b^3}{b^3 - a^3}, \quad \alpha_0 = \frac{b^3}{b_0^3 - a_0^3}, \tag{41.32}$$

which give a measure of the "porosity" of the sphere. The radial stretch λ of the material is given by

$$\lambda = \frac{dr}{dr_0} = \frac{r_0^2}{r^2} = \left(1 - \frac{a^3 - a_0^3}{r^3}\right)^{\frac{2}{3}}, \tag{41.33}$$

while the azimuthal and circumferential stretches are both $1/\sqrt{\lambda}$. Equations (41.31)–(41.33) imply that

$$\frac{a^3}{a_0^3} = \frac{\alpha - 1}{\alpha_0 - 1}, \quad \frac{b^3}{b_0^3} = \frac{\alpha}{\alpha_0}, \quad \lambda^{\frac{3}{2}} = 1 - \frac{a_0^3(\alpha - \alpha_0)}{r^3(\alpha_0 - 1)}. \tag{41.34}$$

The equilibrium equation in the radial direction is

$$\frac{d\sigma_r}{dr} + \frac{2}{r}(\sigma_r - \sigma_\theta) = 0, \tag{41.35}$$

and the boundary conditions, in the case of internally applied pressure, are

$$\sigma_r = -P \text{ at } r = a, \quad \sigma_r = 0 \text{ at } r = b.$$

Let us assume that the material is homogeneous, isotropic, and elastic. Then the principal stress difference $\sigma_\theta - \sigma_r$ is related to λ by an equation of the form

$$\sigma_\theta - \sigma_r = \tilde{\sigma}(\lambda), \tag{41.36}$$

with λ given by (41.34). Equations (41.35) and (41.36) yield

$$P = 2\int_a^b \tilde{\sigma}(\lambda)\, \frac{dr}{r}. \tag{41.37}$$

On putting

$$x = \lambda^{\frac{3}{2}} = 1 - \frac{a^3 - a_0^3}{r^3} = 1 - \frac{a_0^3(\alpha - \alpha_0)}{r^3(\alpha_0 - 1)}, \tag{41.38}$$

we can write

$$P = \frac{2}{3}\int_{x_a}^{x_b} \tilde{\sigma}(\lambda)\,\frac{dr}{r}, \tag{41.39}$$

where

$$x_a = \frac{a_0^3}{a^3} = \frac{\alpha_0 - 1}{\alpha - 1}, \qquad x_b = \frac{b_0^3}{b^3} = \frac{\alpha_0}{\alpha}.$$

We wish to see how P depends on α. Differentiating (41.39) with respect to α gives

$$\frac{dP}{d\alpha} = \frac{2}{3(\alpha - \alpha_0)}\left\{\frac{\alpha_0 - 1}{\alpha - 1}\,\tilde{\sigma}\left[\left(\frac{\alpha_0 - 1}{\alpha - 1}\right)^{\frac{2}{3}}\right] - \frac{\alpha_0}{\alpha}\,\tilde{\sigma}\left[\left(\frac{\alpha_0}{\alpha}\right)^{\frac{2}{3}}\right]\right\}.$$

On introducing the new function $\tilde{g}$, defined by

$$\tilde{g}(x) = x\tilde{\sigma}(x^{\frac{2}{3}}), \tag{41.40}$$

the derivative $dP/d\alpha$ becomes

$$\frac{dP}{d\alpha} = \frac{2}{3(\alpha - \alpha_0)}\left[\tilde{g}\left(\frac{\alpha_0 - 1}{\alpha - 1}\right) - \tilde{g}\left(\frac{\alpha_0}{\alpha}\right)\right]. \tag{41.41}$$

The pressure is thus stationary if we have

$$\tilde{g}\left(\frac{\alpha_0 - 1}{\alpha - 1}\right) = \tilde{g}\left(\frac{\alpha_0}{\alpha}\right). \tag{41.42}$$

This condition may also be written as

$$\lambda_a^{\frac{3}{2}}\sigma_a = \lambda_b^{\frac{3}{2}}\sigma_b,$$

where λ_a, λ_b, and σ_a, and σ_b are the values of λ and $\sigma = \sigma_\theta - \sigma_r$ at $r = a$ and $r = b$. By differentiating (41.41) and using (41.42), we obtain the expression for $d^2P/d\alpha^2$ at a stationary point:

$$\left.\frac{d^2P}{d\alpha^2}\right|_{P=\text{stat.}} = \frac{2}{3(\alpha - \alpha_0)}\left[-\frac{\alpha_0 - 1}{(\alpha - 1)^2}\,\tilde{g}'\left(\frac{\alpha_0 - 1}{\alpha - 1}\right) + \frac{\alpha_0}{\alpha^2}\,\tilde{g}'\left(\frac{\alpha_0}{\alpha}\right)\right]. \tag{41.43}$$

The function $\tilde{g}$ on the interval (0, 1) determines the qualitative behavior of the pressure P. For some materials (of type A), (41.42) does not have a real root α^* with $\alpha_0 < \alpha^* < \infty$. Thus there are no stationary points of P and the pressure increases monotonically. For other materials (of type B), (41.42) has only one real root α^* with $\alpha_0 < \alpha^* < \infty$. The pressure increases monotonically to a maximum value $P^* = P(\alpha^*)$ and decreases monotonically thereafter. Finally, there is a third class of materials (of type C) for which (41.42) has two roots α^* and α^{**} with $\alpha_0 < \alpha^* < \alpha^{**} < \infty$. The pressure P increases monotonically to a local maximum $P^* = P(\alpha^*)$, decreases to a local minimum $P^{**} = P(\alpha^{**})$ and then increases monotonically.

These three types of behavior are experienced by the Mooney–Rivlin material, the strain energy function of which has the form

$$W = C_1(I_1 - 3) + C_2(I_2 - 3), \tag{41.44}$$

where C_1 and C_2 are two constants and I_1 and I_2 are the strain invariants, which, owing to symmetry, have the form

$$I_1 = \lambda^2 + \frac{2}{\lambda}, \quad I_2 = \frac{1}{\lambda^2} + 2\lambda. \tag{41.45}$$

If we put $k = C_2/C_1$, the three kinds of behavior of the function $\tilde{g}(x)$ may be classified according to a definite value of $k = k_{cr}$ of the ratio C_2/C_1. For $k \geqslant k_{cr}$ the material is of type A; for $k = 0$ it is of type B (neo-Hookean material); and for $0 < k \leqslant k_{cr}$ it is of type C.

42. Dynamic Behavior of Membranes

As stated above, no general methods for solving the problem of the equilibrium of a membrane with large displacements and strains are known. Obviously, then, it is possible to say even less about the motion of such a membrane. In the case of axisymmetric deformations, the equilibrium equations reduce to a single nonlinear second-order ordinary differential equation, and a powerful tool for treating the associated boundary-value problem is effected by using the maximum principle. However, when inertial terms are taken into account, the reduction of the problem to a single ordinary differential equation is no longer possible.

To illustrate the method by which the dynamic problem can be tackled, we consider the axisymmetric motion of a membrane stretched between two coaxial rings of different radii placed perpendicular to their common axis and at a distance ℓ apart. We assume, in particular, that the strain energy of the membrane is proportional to the surface area, as occurs in a soap film connecting the two end rings (Dickey 1966). In this case it is known (Bliss 1925) that the surface of the membrane at its position of equilibrium takes the form of the catenary of revolution

$$r = \mu \cosh\left(\frac{x}{\mu} + \lambda\right), \tag{42.1}$$

where x and r denote the distances along and from the axis of symmetry, respectively, and λ and μ are two parameters chosen so as to satisfy the boundary conditions

$$r(0) = a, \quad r(\ell) = b, \tag{42.2}$$

where a and b are the radii of the end rings. The rotationally symmetric surface the meridional section of which is the catenary (42.1) exists only for ℓ less a certain critical value, say ℓ_c. For $\ell > \ell_c$, there is no catenary of the form (42.1) connecting the two rings.

Let us assume that, at the time $t = 0$, the surface is described by $x_0 = x_0(s)$ and $r_0 = r_0(s)$, where s is the arc length of the meridional curve. The coordinates x_0, r_0 will then be related by the geometrical condition

$$x_0'^2 + r_0'^2 = 1, \tag{42.3}$$

where the prime denotes differentiation with respect to s. At the time t the coordinates of the deformed surface will be $x(s, t)$ and $r(s, t)$. The strain energy stored in this deformed state has the form

$$\mathscr{W} = 2\pi T \int_0^L r(x_s^2 + r_s^2)^{\frac{1}{2}} \, ds, \tag{42.4}$$

where T is a constant representing the tension per unit length transmitted by the surface, and L is the initial length of the meridional curve connecting the two end circles. The kinetic energy is given by

$$\mathscr{T} = 2\pi \int_0^L \frac{\rho r_0}{2} (x_t^2 + r_t^2) \, ds, \tag{42.5}$$

where ρ is the surface density of the initial state, which is assumed to be constant. Thus, if we wish to derive the equations of motion from Hamilton's principle, we must consider the functional

$$2\pi \int_0^t \int_0^L \left[\frac{1}{2} \rho r_0 (x_t^2 + r_t^2) - Tr(x_s^2 + r_s^2)^{\frac{1}{2}} \right] ds \, dt, \tag{42.6}$$

and look for its stationary value with respect to admissible variations of $x(s, t)$ and $r(s, t)$. The variational equations of motion have the form

$$\rho r_0 x_{tt} = T \frac{\partial}{\partial s} \left[\frac{r x_s}{(x_s^2 + r_s^2)^{\frac{1}{2}}} \right], \tag{42.7}$$

$$\rho r_0 r_{tt} = T \frac{\partial}{\partial s} \left[\frac{r r_s}{(x_s^2 + r_s^2)^{\frac{1}{2}}} \right] - T(x_s^2 + r_s^2)^{\frac{1}{2}}. \tag{42.8}$$

The physically interesting initial and boundary conditions are as follows. At the time $t = 0$ we know the position of any point of the surface and also the velocity, which is assumed zero:

$$x(s, 0) = x_0(s), \tag{42.9a}$$

$$r(s, 0) = r_0(s), \tag{42.9b}$$

$$x_t(s, 0) = 0, \tag{42.9c}$$

$$r_t(s, 0) = 0. \tag{42.9d}$$

We also know the positions of the end parallels at all times:

$$\begin{aligned} x(0, t) &= f_0(t), \quad x(L, t) = f_1(t), \\ r(0, t) &= a, \quad r(L, t) = b, \end{aligned} \tag{42.10}$$

where $f_0(t)$ and $f_1(t)$ are given functions.

We now try to develop a method for integrating the system (42.7) and (42.8) that is valid for small times, and requires the solution of a system of linear equations. Let us assume that the initial data are analytic in some neighborhood of $t = 0$. Let s be any point with $0 < s < L$, then, by the Cauchy–Kowalewsky theorem, we know that there is an analytic solution of system (42.7) and (42.8) in some neighborhood of $(s, 0)$. To find this solution we rewrite the system as

$$\begin{aligned} r_0 x_{tt} - \frac{\partial}{\partial s}(r_0 \psi x_s) &= 0 \\ r_0 r_{tt} - \frac{\partial}{\partial s}(r_0 \psi r_s) + \frac{T^2}{\rho^2 r_0 \psi} &= 0, \end{aligned} \tag{42.11}$$

where ψ, given by

$$\psi = \frac{Tr}{\rho r_0 (x_s^2 + r_s^2)^{\frac{1}{2}}}, \tag{42.12}$$

is an analytic function of the solution and the initial data. The function ψ represents the square of the speed of sound in the membrane. As a consequence of the analyticity of the data, we would expect to be able to write ψ as

$$\psi(s, t) = \sum_{n=0}^{\infty} \sigma_n(s) t^n, \tag{42.13}$$

convergent for small t, with $\sigma_n(s)$ analytic functions of s. Let us evaluate the σ_n terms from the initial data and the differential equations. We find that

$$\sigma_0(s) = \psi(s, 0) = \frac{Tr(s, 0)}{\rho r_0 [x_s(s, 0)^2 + r_s(s, 0)^2]^{\frac{1}{2}}} = \frac{T}{\rho}. \tag{42.14}$$

We may also show that, if (42.9b) holds, then $\sigma_1 = 0$, and (42.13) becomes

$$\psi(s, t) = \frac{T}{\rho} + \sum_{n=2}^{\infty} \sigma_n(s) t^n. \tag{42.15}$$

Therefore, for small t, we may take

$$\psi(s, t) \simeq \frac{T}{\rho}, \tag{42.16}$$

and substitute this expression for ψ in (42.11):

$$x_{tt} - \frac{T}{\rho} x_{ss} = 0, \tag{42.17a}$$

$$r_{tt} - \frac{T}{\rho} r_{ss} + \frac{T}{\rho r_0^2} r = 0 \tag{42.17b}$$

and this system can be solved explicitly. Let us consider the initial data

$$\begin{aligned} x(s, 0) &= s, \quad & r(s, 0) &= a, \\ x_t(s, 0) &= 0, \quad & r_t(s, 0) &= 0, \end{aligned} \tag{42.18}$$

and the boundary data

$$\begin{aligned} x(0, t) &= 0, \quad & x(L, t) &= L, \\ r(0, t) &= a, \quad & r(L, t) &= a. \end{aligned} \tag{42.19}$$

This means that we assume the surface to be a cylinder of radius a at time $t = 0$. This is not a position of static equilibrium, and so the membrane will be in motion after the initial instant. We solve (42.17) for this initial boundary-value problem. The solution of (42.17a) is

$$x = s. \tag{42.20}$$

In order to solve (42.17b), let us introduce

$$r(s, t) = Z(s, t) + \zeta(s) + a, \tag{42.21}$$

where

$$\zeta(s) = A^{s/a} + B^{-s/a} - a, \tag{42.22}$$

with

$$A = \frac{a(1 - e^{-L/a})}{2\sinh(L/a)}, \quad B = \frac{a(e^{L/a} - 1)}{2\sinh(L/a)}. \tag{42.23}$$

Equation (42.17b) now becomes

$$Z_{tt} - \frac{T}{\rho} Z_{ss} + \frac{T}{\rho a^2} Z = 0, \tag{42.24}$$

and Z satisfies the initial and boundary conditions

$$Z(s, 0) = -\zeta(s), \quad Z_t(s, 0) = 0, \tag{42.25}$$

$$Z(0, t) = Z(L, t) = 0. \tag{42.26}$$

We can solve (42.24) in the form of a uniformly convergent Fourier series. Then, on returning to the function $r(s, t)$, we find

$$r = Ae^{s/a} + Be^{-s/a} + \frac{4L^2}{a\pi} \sum_{n=0}^{\infty} \frac{\sin\left(\frac{2n+1}{L}\pi s\right) \cos \lambda_n t}{(2n+1)[(2n+1)^2\pi^2 + L^2/a^2]}, \tag{42.27}$$

where

$$\lambda_n = \left\{ \frac{T}{\rho} \left[\frac{(2n+1)^2\pi^2}{L^2} + \frac{1}{a^2} \right] \right\}^{\frac{1}{2}}. \tag{42.28}$$

The behavior of the solution (42.27) for different values of the time ($t = 0, 0.5, 1$) is shown in Figure 42.1. For these solutions T/ρ has been chosen as 1, and the distance between the end rings is $L = 1.3255$, which is very close to the critical distance at which there is no static solution. The dotted line in Figure 42.1 represents the equilibrium position (the catenary) which is crossed shortly after the time $t = 1$.

A special case, but of great practical importance, is that of the motion of an initially spherical membrane (Dickey 1966). In this case the initial conditions are again (42.9), but x_0 and r_0 are related by the restriction

$$x_0^2 + r_0^2 = a^2, \tag{42.29}$$

which states that the meridional curve is a circle of radius a. However, we do not prescribe end conditions like (42.10). On noting that the sphere of radius a may be parameterized as $x_0 = -a\cos(s/a)$ and $r_0 = a\sin(s/a)$, we attempt to find solutions of (42.11) of the type

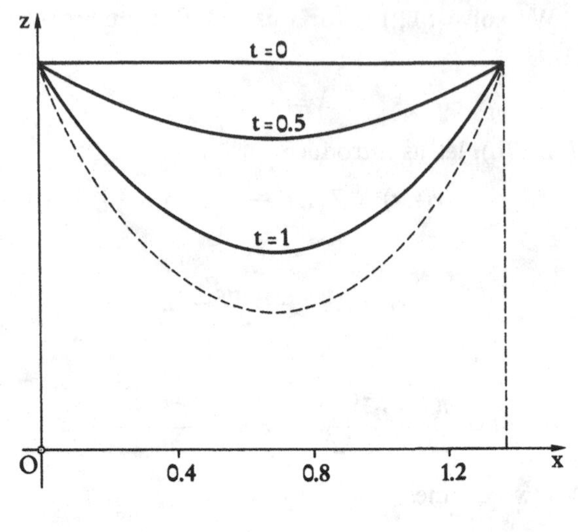

Fig. 42.1

$$x = -a\cos\left(\frac{s}{a}\right)f(t), \quad r = a\sin\left(\frac{s}{a}\right)f(t), \tag{42.30}$$

where $f(t)$ must be determined. By substituting (42.30) in (42.11) we find that both equations reduce to the same ordinary differential equation for $f(t)$:

$$\rho a^2 f'' + 2Tf = 0. \tag{42.31}$$

If we now solve (42.31) with the initial conditions $f(0) = 1$ and $f'(0) = 0$, we find the solution for the sphere to be

$$\begin{aligned} x &= -a\cos\left(\frac{s}{a}\right)\cos\left[\frac{1}{a}\left(\frac{2T}{\rho}\right)^{\frac{1}{2}} t\right], \\ r &= a\sin\left(\frac{s}{a}\right)\cos\left[\frac{1}{a}\left(\frac{2T}{\rho}\right)^{\frac{1}{2}} t\right]. \end{aligned} \tag{42.32}$$

From (42.32) we obtain the time t_c required for a soap bubble to collapse to a point:

$$t_c = \frac{\pi a}{4}\left(\frac{2\rho}{T}\right)^{\frac{1}{2}}. \tag{42.33}$$

From this we see that, for constant ρ/T, larger bubbles collapse more slowly than small bubbles.

43. Detachment and Suturing of Membranes

The prototype of a detachment problem for a membrane is the peeling of an adhesive membrane initially glued to a flat surface, which we shall call the xy plane (Burridge and Keller 1978). Let us suppose that the membrane is pulled with a force $f(x, y, t)$ normal to the plane and away from it. If the force is small and the adhesive is strong, the membrane will remain perfectly glued to the xy plane. However, as the force increases, part of the membrane may be pulled away. Let $u(x, y, t)$ be the normal displacement of the membrane in the domain $D(T)$ where it is detached. Provided that the displacement and its gradients are sufficiently small, we can assume that u satisfies the equation of motion

$$\rho u_{tt} - T\nabla^2 u = f, \text{ in } D, \tag{43.1}$$

where ρ is the constant density and T is the constant tension. At the boundary C of the detached part, we have

$$u = 0, \text{ on } C, \tag{43.2}$$

where C is, as yet, unknown. For determining C there is an additional condition, called a *breaking condition*, which is the two-dimensional counterpart of condition (35.12). This breaking condition takes the form

$$(T - \rho v^2)\frac{\partial u}{\partial n} = -F, \text{ on } C, \tag{43.3}$$

where v denotes the normal velocity of C, $\partial u/\partial n$ is the normal outer derivative of u on C, and F is the cohesive force.

The solution of the free initial boundary-value problem (43.1)–(43.3) is difficult in general, and useful results are known only for some particular cases.

As a first example, let us suppose that a concentrated steady force of magnitude f_0 is applied at the origin, so that $f(x, y) = f_0\delta(x)\delta(y)$, where $\delta(\cdot)$ is the delta function. As we may expect that the displacement u is symmetrical about the origin, we take $u = u(r)$, r being the distance of a point in the xy plane from the origin. In this case the boundary C will be a circle centered at the origin, the radius a of which is to be found. As the solution does not depend on time, from (43.1) we find

$$u(r) = -\frac{f_0}{2\pi T}\,\ell n r + \text{constant} \tag{43.4}$$

with (43.2) determining the constant to give

$$u(r) = -\frac{f_0}{2\pi T}\,\ell n\left(\frac{r}{a}\right). \tag{43.5}$$

Finally, the radius a is determined from (43.3), where again $v = 0$:

$$a = \frac{f_0}{2\pi F}.$$

The complete solution is then

$$u(r) = \frac{f_0}{2\pi F}\,\ell n\left(\frac{2\pi f r}{f_0}\right), \quad r \leqslant a. \tag{43.6}$$

As a second example (Burridge and Keller 1978), suppose that a gas at pressure p is introduced between the membrane and the plane through a small hole localized at the origin. Then the membrane will be detached from a domain D of the plane forming a lens-shaped bubble that is symmetrical about the origin. The detached surface of the membrane will be subject to a uniform pressure p and its (unknown) boundary C will be a circle of radius a. In the static case, the solution of (43.1) with $f = p =$ constant is

$$u(r) = \frac{p}{4T}(a^2 - r^2), \quad r \leqslant a. \tag{43.7}$$

The radius a is determined by condition (43.3), where again $v = 0$. The result is

$$a = \frac{2F}{p}.$$

Both these examples are a little deceptive, because the symmetry of C hides the real difficulty of the problem, which rests on the determination of C. Let us suppose that the membrane is adhering, not to the entire xy plane, but only to the half-plane $x > 0$. Assume that a concentrated force is applied at the point $(x_0, 0)$. For $x_0 > a$, the solution is again (43.5), with r measured from $(x_0, 0)$. However, if the circle $r = a$ intersects the edge $x = 0$, then a new free boundary-value problem arises, because the region of detachment is no longer a piece of circle as one might imagine.

In order to solve this problem, we start by observing that the slope of the membrane in the x direction must vanish along the edge $x = 0$. As a consequence, we may symmetrize the problem in the following way. The membrane is glued under uniform tension on the entire xy plane and two concentrated forces are applied at the points $(x_0, 0)$ and $(-x_0, 0)$ at a distance $2x_0$ apart in order to lift up the membrane. Then the displacement of uplift $u(x, t)$ of the membrane above the xy plane is a harmonic function for the region of detachment D and such that it satisfies the two boundary conditions $u = 0$ and $\partial u/\partial n = -F/T$ on C. The problem is both over-determined, and C unknown.

The structure of the solutions is determined by the parameter $\Lambda = 2\pi x_0(F/f_0)$, where f_0 is the magnitude of the two concentrated forces and F is the force at which detachment occurs (Ting 1977). For $\Lambda > 1$, the detached part consists of two isolated circles of radius $a = f_0/2\pi F$ and is centered at $(x_0, 0)$ and $(-x_0, 0)$ (Figure 43.1). For $\Lambda = 1$, the two circular boundaries just touch each other. For $\Lambda < 1$, the two boundaries merge and form a single closed boundary C. With $\Lambda < 1$, but close to one, the shape of the free boundary has approximately the form of the curve C in Figure 43.1. However, with $\Lambda \ll 1$ the free boundary C assumes the form of the dotted line drawn in the same figure. This means that the distance $|AA'|$ can be either a local minimum or a local maximum according to the value of Λ in the interval $0 < \Lambda < 1$. The transverse displacement $u(x, y)$, due to the symmetry with respect to the y axis, may be studied in the half-plane $x > 0$. The boundary condition on $x = 0$ is

$$u_x = 0. \tag{43.8}$$

Furthermore, for symmetry, we have $u_y = 0$ on $y = 0$. Thus, both partial derivatives of $u(x, y)$ vanish at the origin. The origin is a saddlepoint for $u(x, y)$, because it is a minimum with respect to x and a maximum with respect to y.

By using the method of conformal mapping (Ting 1977), it is possible to find the explicit form of the function $u(x, y)$ and the explicit equation for the curve C. The different forms of C depend on the values of the parameters f_0, F, and x_0, and it is helpful to note that the solution to this problem can also be interpreted as that of finding the planar flow in an incompressible fluid induced by two vortices of strength

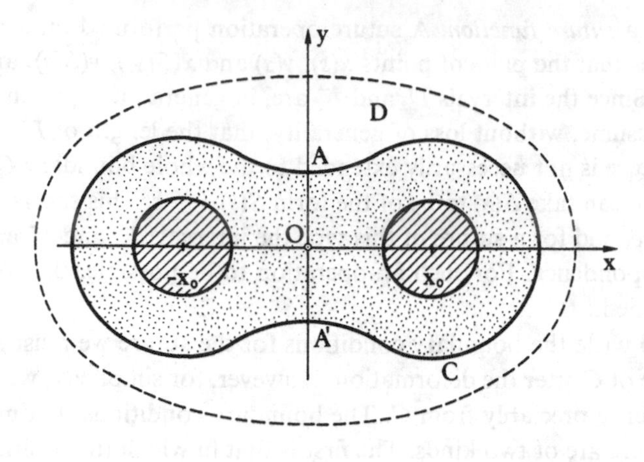

Fig. 43.1

f_0 at a distance $2x_0$ apart. In this case, F represents the constant velocity of a free stream line surrounding the two vortices.

A counterpart of the theory of the peeling of membranes is the theory of *sutures*. Suturing is the operation of closing up the opposite sides of a hole pre-existing in an indefinite planar membrane in order to reconstruct the continuity of the medium. Of course, the adjective "opposite" used above needs a more precise qualification. Let us consider a planar membrane, indefinitely extended, surrounding a hole constituted by a simple closed curve C. Let us call D the solid region exterior to the hole. The boundary C of the hole is unloaded, but the membrane may be stretched at a large distance from the hole by a uniform state at a tension described by two principal components of stress, say σ_1^∞ and σ_2^∞, both of which are non-negative in order to avoid wrinkling. The points of the membrane, in its unstretched state, can be identified by two Cartesian coordinates x, y, or, alternatively, by the complex variable $z = x + iy$. Let us assume that the points of C have the parametric representation

$$x = x(s), \quad y = y(s), \quad \text{for } -s_1 \leqslant s \leqslant s_0, \tag{43.9}$$

with

$$x(s_0) = x(-s_1), \quad y(s_0) = y(-s_1), \tag{43.10}$$

where $x(s)$ and $y(s)$ are piecewise continuously differentiable functions of s. We assume that the curve is not self-intersecting. Then, we define the intervals (Wu 1980)

$$I^+ \equiv \{C|0 < s < s_0\}, \quad I^- = \{C| -s_1 < s < 0\}, \tag{43.11}$$

and introduce a function $S(s)$ for all $s \in I^+$. The function $S(s)$ is assumed to be continuous together with its first derivative, and satisfies the conditions

$$S(s) \in I^-, \quad S'(s) \neq 0, \infty \quad \text{for all } S \in I^+;$$

$$S(0) = 0, \quad S(s_0) = -s_1. \tag{43.12}$$

$S(s)$ is called a *suture function*. A suture operation performed on a hole defined by (43.9) requires that the pairs of points $x(s)$, $y(s)$ and $x(S(s))$, $y(S(s))$ are brought into coincidence. Since the intervals I^+ and I^- are, in general, unequal in length, and we can always assume, without loss of generality, that the length of I^+ is less than that of I^-; the suture is not necessarily extended to the whole boundary C. In particular, for $s_0 = s_1$, we can take $S(s)$ to be of the form $S(s) = -s$. This special case is called a *uniform suture*, and for a suture of this type the intervals I^+ and I^- are put in a one-to-one correspondence. For such an operation to be possible, the condition $s_0 = s_1$ must be satisfied.

In order to write the boundary conditions for the suture we must know the curve C_1, the image of C after the deformation. However, for simplicity, we assume that C_1 does not differ appreciably from C. The boundary conditions that must be satisfied along the suture are of two kinds. The first is that in which the positions of the pairs of points $x(s), y(s)$ and $x(S(s)), y(S(s))$ must be the same in the final state:

$$\begin{aligned} x(s) + u(s) &= x(S(s)) + u(S(s)), \\ y(s) + v(s) &= y(S(s)) + v(S(s)), \quad \text{for } s \in I^+, \end{aligned} \tag{43.13}$$

where u and v denote the displacement components. The second condition is that the tractions on the two sides of the suture must be equal and opposite

$$t_x(s) = S'(s)t_x(S(s)), \quad t_y(s) = S'(s)t_y(S(s)), \tag{43.14}$$

for $s \in I^+$. The unit normal to the curve $x = x(s)$, $y = y(s)$ has components (ℓ, m) defined by

$$\ell = -y'(s), \quad m = x'(s).$$

Therefore the tractions $t_x(s)$ and $t_y(s)$, written in terms of the stress components, are expressed as

$$\begin{aligned} t_x(s) &= \sigma_x \ell + \tau_{xy} m = -\sigma_x y'(s) + \tau_{xy} x'(s), \\ t_y(s) &= \tau_{xy} \ell + \sigma_y m = -\tau_{xy} y'(s) + \sigma_y x'(s). \end{aligned}$$

In complex form, the traction vector becomes

$$t_x + it_y = (-\sigma_x y' + \tau_{xy} x') + i(-\tau_{xy} y' + \sigma_y x').$$

On putting $z(s) = x(s) + iy(s)$, the traction vector can be written as

$$t_x + it_y = \frac{i}{2}[(\sigma_x + \sigma_y)z' + i(\sigma_y - \sigma_x - 2i\tau_{xy})\bar{z}']. \tag{43.15}$$

In order to illustrate the consequence of a suture operation on an elastic membrane, let us consider the suturing of an elliptical hole in an infinite membrane. In this case the boundary C is an ellipse having the parametric equation

$$x(\theta) + iy(\theta) = a\left(\cos\theta + i\frac{b}{a}\sin\theta\right), \quad -\pi < \theta < \pi, \tag{43.16}$$

where a and b are the major and minor half-axes, respectively. The arc length $s(\theta)$ along C is defined by

$$ds = a\left[\sin^2\theta + \left(\frac{b}{a}\right)^2 \cos^2\theta\right]^{\frac{1}{2}} d\theta, \quad s(0) = 0.$$

The hole is to be sutured uniformly; that is, we take $S(s) = -s$. Then the membrane is stretched at infinity by principal tensile stresses σ_1^∞ and σ_2^∞, such that the angle formed by the x axis with the direction of σ_1^∞ is β.

Let us assume that the membrane is linearly elastic, characterized by a strain energy of the form

$$2W = C[I^2 - (2+c)J], \tag{43.17}$$

where I and J are the first and second variants of the strains, and C and c are two constants.

We now introduce the complex variable $z = x + iy$, and the complex differentiation

$$2(.)_z = (.)_x - i(.)_y, \quad 2(.)_{\bar{z}} = (.)_x + i(.)_y. \tag{43.18}$$

Then we represent the stresses and the positions in terms of two holomorphic functions $\Omega(z)$ and $\Psi(z)$, allowing the form

$$(x+u) + i(y+v) = \Omega(z) + \overline{\Psi(z)}, \tag{43.19}$$

where $\overline{(.)}$ denotes the complex conjugate of (.), and

$$\sigma_x + \sigma_y = 2C(2-c)[\Omega(z) + \overline{\Psi(z)}]_z, \tag{43.20}$$

$$\sigma_y + \sigma_x - 2i\tau_{xy} = -2C(2+c)[\Omega(z) + \overline{\Psi(z)}]_{\bar{z}}. \tag{43.21}$$

As the membrane is stretched at infinity, we know the behavior of (43.19) for $z \to \infty$. We may thus consider

$$(x+u) + i(y+v) = Az + \bar{B}\bar{z}, \text{ for } z \to \infty, \tag{43.22}$$

where A and B are two constants. In particular, we put $A = \bar{A}$, in order to ensure that the uniform stretch (43.22) has no rotation at infinity. The constants A and B are related to the stresses at infinity by the equations

$$A = \frac{\frac{1}{2}(\sigma_1^\infty + \sigma_2^\infty)}{C(2-c)},$$

$$B = \frac{\frac{1}{2}(\sigma_1^\infty - \sigma_2^\infty)e^{-2i\beta}}{C(2+c)}.$$

In order to determine the functions $\Omega(z)$ and $\Psi(z)$ we have the following conditions. At infinity, condition (43.22), written in terms of $\Omega(z)$ and $\Psi(z)$, becomes

$$\Omega(z) + \overline{\Psi(z)} = Az + \bar{B}\bar{z}, \text{ for } z \to \infty. \tag{43.23}$$

On the curve C, the points of the upper and lower face of the ellipse that are sutured together have coordinates z and $\bar{z}$. Therefore, conditions (43.13) can be written in the form

$$\Omega(z) - \Omega(\bar{z}) + \overline{\Psi(z)} - \overline{\Psi(\bar{z})} = 0, \text{ for } z \in C. \tag{43.24}$$

In order to write conditions (43.14) in terms of $\Omega(z)$ and $\Psi(z)$, we introduce a new function $\varphi(z)$, which has the properties

$$\left.\begin{aligned} \sigma_x + \sigma_y &= 2C(2+c)\varphi_{,z}, \\ \sigma_y - \sigma_x - 2i\tau_{xy} &= 2C(2+c)\varphi_{,\bar{z}}. \end{aligned}\right\} \tag{43.25}$$

The following can then immediately be verified

$$\varphi = \frac{2-c}{2+c}\Omega - \bar{\Psi}. \tag{43.26}$$

The traction vector can be expressed in terms of the derivative of φ:

$$\frac{d\varphi}{ds} = -\frac{i}{C(2+c)}(t_x + it_y). \tag{43.27}$$

Condition (43.14) becomes

$$\frac{d\varphi(s)}{ds} = \frac{d\varphi(S(s))}{ds}, \text{ for } s \in \mathrm{I}^+, \tag{43.28}$$

or

$$\varphi(s) = \varphi(S(s)), \text{ for } s \in I^+. \tag{43.29}$$

After using (43.26), this condition assumes the form

$$\frac{2-c}{2+c}[\Omega(z) - \Omega(\bar{z})] - \overline{\Psi(z)} + \overline{\Psi(\bar{z})} = 0, \text{ for } z \in C. \tag{43.30}$$

In order to find $\Omega(z)$ and $\Psi(z)$ from (43.23), (43.24), and (43.30), we introduce the complex ζ plane ($\zeta = \xi + i\eta$) and define the conformal mapping

$$z = m(\zeta) = \frac{R}{2}\left(\zeta + \frac{\kappa}{\zeta}\right), \text{ for } |\zeta| > 1, \tag{43.31}$$

where

$$R = a\left(1 + \frac{b}{a}\right), \quad \kappa = \left(1 - \frac{b}{a}\right) \Big/ \left(1 + \frac{b}{a}\right). \tag{43.32}$$

This function $m(\zeta)$ maps the region of the ζ plane outside the unit circle $|\zeta| = 1$ into the region of the z plane outside the ellipse (43.16). Furthermore, we define

$$\omega(\zeta) = \Omega(m(\zeta)), \quad \psi(\zeta) = \Psi(m(\zeta)), \tag{43.33}$$

which are holomorphic in $|\zeta| > 1$, and write conditions (43.23), (43.24), and (43.30) in the ζ plane as

$$\omega(\zeta) + \overline{\psi(\zeta)} = \frac{R}{2}(A\zeta + \bar{B}\bar{\zeta}), \text{ for } \zeta \to \infty; \tag{43.34}$$

$$\omega(\zeta) - \omega(\bar{\zeta}) + \overline{\psi(\zeta)} - \overline{\psi(\bar{\zeta})} = 0, \text{ for } |\zeta| = 1; \tag{43.35}$$

$$\frac{2-c}{2+c}[\omega(\zeta) - \omega(\bar{\zeta})] - \overline{\psi(\zeta)} + \overline{\psi(\bar{\zeta})} = 0, \text{ for } |\zeta| = 1; \tag{43.36}$$

where we have used the fact $\overline{m(\zeta)} = m(\bar{\zeta})$ in (43.31).

However, on the unit circle we know that $\bar{\zeta} = 1/\zeta$, and therefore it can immediately be verified that

$$\omega(\zeta) = \frac{R}{2}A\left(\zeta + \frac{1}{\zeta}\right), \quad \psi(\zeta) = \frac{R}{2}B\left(\zeta + \frac{1}{\zeta}\right), \tag{43.37}$$

represent the solution in the ζ plane.

From (43.37) we can calculate the coordinates of the points of the membrane after the deformation (see (43.19))

$$(x+u) + i(y+v) = \frac{R}{2}\left[A\left(\zeta + \frac{1}{\zeta}\right) + \bar{B}\left(\bar{\zeta} + \frac{1}{\bar{\zeta}}\right)\right]. \tag{43.38}$$

In order to find the positions of the points, which, before the suture operation, occupied the ellipse of (43.16), we introduce polar coordinates $\zeta = \rho e^{i\theta}$ and observe that ρ has the value 1 on the boundary of the unit circle. As we have $0 \leqslant \theta \leqslant \pi$, the point z describes the upper side of the suture, while the lower side is described for $-\pi \leqslant \theta \leqslant 0$. Equation (43.38) shows that the upper and lower sides of the ellipse fall, after suture, on a single rectilinear segment

$$(x+u) + i(y+v) = R(A + \bar{B})\cos\theta. \tag{43.39}$$

Hence, after suture, the hole is contracted into a unique segment, but this does not have sides parallel to the coordinate axes.

In the analysis performed so far we have neglected the most striking feature of the suture operation; that is, the large strains. However, despite the unavoidable difficulties of reformulating the boundary-value problems of suturing in nonlinear elasticity, we can, in some cases (Wu 1979), again apply the results of linear theory, with slight modifications. This can be done, for example, when the strain energy function of the material has the form (43.17), but with I and J given by

$$I = \Lambda_1 + \Lambda_2, \quad J = \Lambda_1\Lambda_2, \tag{43.40}$$

where Λ_1 and Λ_2 are the in-plane principal stretches.

44. Optimization of Membranes

The classical theory of curved membranes rests on the assumption that the form of the middle surface is known and we must determine the state of stress and displacement generated by given loads. It often happens, however, that the shape of the middle surface is not strictly fixed, but may be adjusted in order to optimize some parameter related to the solution. A typical example of a problem of this kind is the search for the meridional curve of an axisymmetric flexible container, taken to be a membrane completely filled with a heavy liquid, when we require that the membrane stresses are constant everywhere. This problem is equivalent to that of finding the shape of a liquid drop having constant surface tension, and was formulated by Laplace. Another example is that of finding the angular spread of a cylindrical dam, of constant thickness and given span, in order to minimize the total volume of material (see Figure 44.4). This problem was proposed and solved by Jorgensen (1915).

In order to formulate Laplace's problem in more precise terms, we assume that the container is completely filled through a small vertical tube at its vertex with a liquid of specific weight γ and such that the free surface of the liquid is at height h above the top of the container (Figure 44.1). Choosing the origin of cylindrical coordinates r, z at this level, the load acting on the surface in the direction of the outer normal is

$$p_n = \gamma z, \tag{44.1}$$

while the load in the direction of the tangent to the meridian is zero. On denoting the meridional stress resultant by T_φ and the hoop stress resultant by T_θ, the equilibrium equation along the normal is

$$\frac{T_\varphi}{r_1} + \frac{T_\theta}{r_2} = p_n = \gamma z, \tag{44.2}$$

where r_1 is the radius of curvature of the meridian and r_2 is the second principal radius of curvature. These radii can be expressed in terms of the single horizontal radius r by means of the relations

$$r_1 = \frac{ds}{d\varphi} = \frac{dr}{d(\sin\varphi)}, \tag{44.3a}$$

$$r_2 = \frac{r}{\sin\varphi}, \tag{44.3b}$$

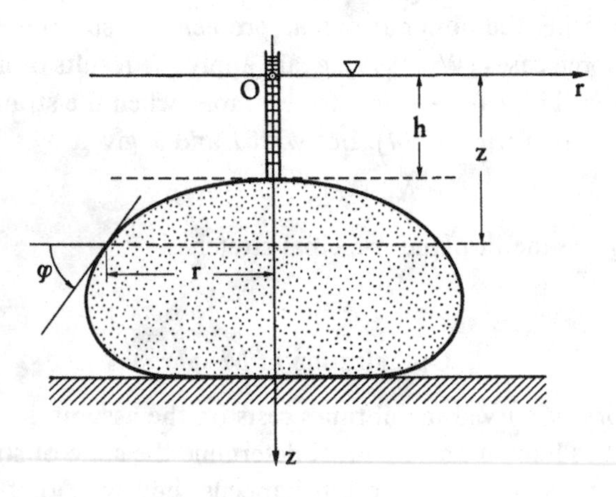

Fig. 44.1

where φ is the slope of the tangent to the meridian with respect to the r axis. As we wish to have constant membrane forces $T_\varphi = T_\theta = N_0$, (44.2), combined with (44.3), yields

$$\frac{d(\sin\varphi)}{dr} + \frac{\sin\varphi}{r} = \frac{\gamma z}{N_0}. \tag{44.4}$$

This is an ordinary differential equation, in which φ and z are the unknowns and r is the independent variable. However, we also have the geometrical relation

$$\tan\varphi = \frac{dz}{dr}, \tag{44.5}$$

so that (44.4) and (44.5) constitute a system of two first-order differential equations for determining $\varphi(r)$ and $z(r)$. If the membrane is closed, the boundary conditions are that, at the apex, where $r = 0$, the slope φ must vanish and z must be equal to h. Finding the solution for the system is usually done numerically (Tölke 1939).

An interesting variant of the above problem is the case in which the pressure, rather than varying linearly with the altitude, is constant, as occurs in a reservoir filled with gas. In this case the equilibrium equation can be solved explicitly (Tölke 1939). If p denotes the constant pressure exerted by the gas from inside, (44.4) becomes

$$\frac{d(\sin\varphi)}{dr} + \frac{\sin\varphi}{r} = \frac{p}{N_0}. \tag{44.6}$$

On using (44.3a) and multiplying both sides of (44.6) by r, we obtain

$$\sin\varphi + \frac{r}{r_1} = \frac{pr}{N_0}. \tag{44.7}$$

Let us introduce the nondimensional variables

$$r = r_0\xi, \quad z = r_0\eta, \quad r_1 = r_0\rho_1, \tag{44.8}$$

and a shape ratio ω:

$$\omega = \frac{2N_0}{pr_0} - 1. \tag{44.9}$$

We can write (44.9) in the form

$$\frac{1}{\rho_1} = \frac{2}{\omega + 1} - \frac{\sin\varphi}{\xi}, \tag{44.10}$$

and we also have

$$\tan\varphi = \frac{dz}{dr} = \frac{d\eta}{d\xi}, \tag{44.11a}$$

$$\eta = \eta_0 + \int_1^\xi \frac{\sin\varphi\, d\xi}{\sqrt{1 - \sin^2\varphi}}, \tag{44.11b}$$

where η_0 is the value of η corresponding to $\xi = 1$ (Figure 44.2).

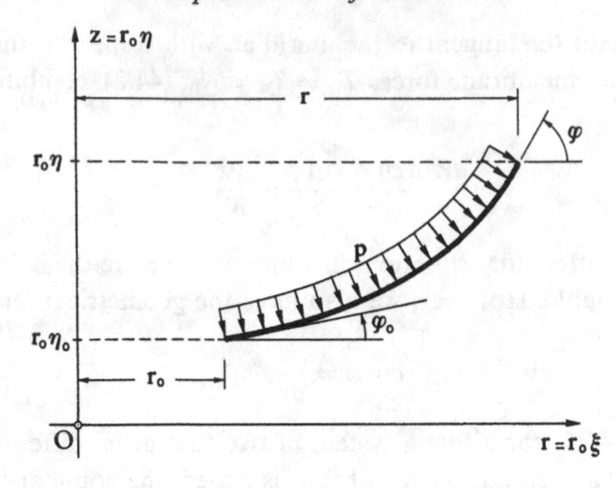

Fig. 44.2

As the vertical forces, in the direction of the z axis, are in equilibrium, we can write the following balance equation for that part of the membrane between the horizontal sections $z_0 = r_0\eta_0$ and $z = r_0\eta$:

$$N_0 2\pi r \sin\varphi - N_0 2\pi r_0 \sin\varphi_0 = \int_{z_0}^{z} 2\pi r p \, dr. \tag{44.12}$$

After integration with respect to r, (44.12) becomes

$$r\sin\varphi - r_0\sin\varphi_0 = \frac{p}{2N_0}(r^2 - r_0^2),$$

or, on using (44.8) and (44.9),

$$\sin\varphi = \frac{\sin\varphi_0}{\xi} + \frac{\xi^2 - 1}{\xi(\omega+1)}. \tag{44.13}$$

When $\sin\varphi$ is substituted in (44.11a) we obtain

$$\eta = \eta_0 + \int_1^{\xi} \frac{\xi^2 - 1 + (\omega+1)\sin\varphi_0}{\{(\omega+1)^2\xi^2 + [\xi^2 - 1 + (\omega+1)\sin\varphi_0]^2\}^{\frac{1}{2}}} \, d\xi. \tag{44.14}$$

The integral in (44.14) is reducible to a normal Legendre integral. However, without loss of generality, we may take $\eta_0 = 0$ and $\varphi_0 = \pi/2$, and discuss (44.14) in the simplified form

$$\eta = \int_1^{\xi} \frac{(\xi^2 + \omega)\, d\xi}{\sqrt{(1 - \xi^2)(\xi^2 - \omega^2)}}. \tag{44.15}$$

With the substitutions

$$\xi = \begin{cases} \sqrt{1-(1-\omega^2)\sin^2\psi}, & \text{for } \omega^2 \leqslant 1; \\ \sqrt{\omega^2-(\omega^2-1)\sin^2\psi}, & \text{for } \omega^2 > 1; \end{cases}$$

we can write (44.15) as

$$\eta = \omega \int_0^{\psi} \frac{d\psi}{\sqrt{1-(1-\omega^2)\sin^2\psi}} + \int_0^{\psi} \sqrt{1-(1-\omega^2)\sin^2\psi}\, d\psi, \quad \text{for } \omega^2 \leqslant 1;$$

$$\eta = \int_{\psi}^{\pi/2} \frac{d\psi}{\sqrt{1-\left(1-\dfrac{1}{\omega^2}\right)\sin^2\psi}} + \omega \int_{\psi}^{\pi/2} \sqrt{1-\left(1-\frac{1}{\omega^2}\right)\sin^2\psi}\, d\psi, \quad \text{for } \omega^2 > 1.$$

By introducing the moduli $\sqrt{1-\omega^2} = \kappa = \sin\theta$ and $\sqrt{1-(1/\omega^2)} = \kappa = \sin\theta$, respectively, the above integrals can be reduced to normal elliptic integrals, and ξ and η have the parametric representation

$$\left.\begin{aligned} \xi &= \sqrt{1-\sin^2\theta\sin^2\psi} \\ \eta &= F(\theta,\psi)\cos\theta + E(\theta,\psi) \end{aligned}\right\} \text{for } \omega^2 \leqslant 1, \omega = \cos\theta; \tag{44.16}$$

$$\left.\begin{aligned} \xi &= \frac{1}{\cos\theta}\sqrt{1-\sin^2\theta\sin^2\psi} \\ \eta &= \frac{1}{\cos\theta}[K(\theta)\cos\theta + E(\theta)] + \frac{1}{\cos\theta}[F(\theta,\psi)\cos\theta + E(\theta,\psi)] \end{aligned}\right\} \text{for } \omega^2 > 1, \omega = \frac{1}{\cos\theta}. \tag{44.17}$$

Having the parametric equations of the meridional curve, it is easy to draw these curves for different values of $\theta\,(0 \leqslant \theta \leqslant \pi)$; some examples are shown in Figure 44.3. For $\theta = 0$ we have a cylinder, for $\theta = \pi/2$ a sphere, and for $\theta = \pi$ the curve contracts into a point. The sphere is the only solution yielding a closed surface.

In both the examples considered above the self-weight of the membrane has been regarded as negligible in relation to the exterior loads. The opposite situation can be formulated in the same way, but the solution is only approximate (Megareus 1939). Let γ be the specific weight of the material and $h = h(s)$ be the thickness, where s denotes the arc length along the meridian measured from the vertex. On putting $m_z = \gamma h(s)$, the stress resultants T_φ and T_θ can be expressed in terms of the exterior load through the equilibrium equations:

$$T_\varphi = -\frac{1}{r\sin\varphi}\int_0^s r m_z\, ds, \tag{44.18}$$

$$T_\theta = -r m_z \cot\varphi - \frac{r T_\varphi}{r_1 \sin\varphi}, \tag{44.19}$$

where φ is the angle made by the tangent with the meridian and the z axis, and r_1 is the radius of curvature of the meridian. We wish to find the shape of the cap and the variation in thickness so that the longitudinal and circumferential stresses σ_φ and σ_θ

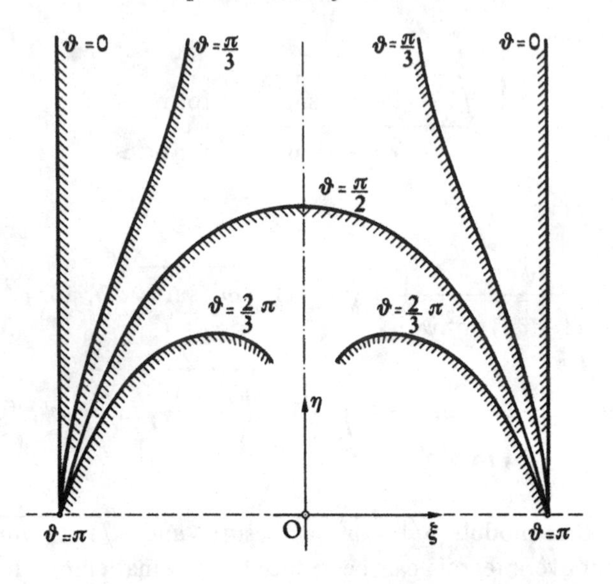

Fig. 44.3

have the same constant value $-\sigma_0$. On recalling that $r_2 = r/\sin\varphi$ is the second principal radius of curvature (see (44.3)), we derive from (44.18) and (44.19) the equations

$$\int_0^s hr\,ds = \frac{\sigma_0}{\gamma}\,hr\sin\varphi, \tag{44.20}$$

$$\cos\varphi = \frac{\sigma_0}{\gamma}\left(\frac{1}{r_1} + \frac{1}{r_2}\right), \tag{44.21}$$

where we have put $T_\varphi = T_\theta = -\sigma_0 h$. At the vertex we have $\varphi = 0$ and $r_1(0) = r_{10}$, and from (44.21) we find $r_{10} = 2\sigma_0/\gamma$. The equilibrium equations can therefore be written as

$$\int_0^s hr\,ds = \frac{r_{10}}{2}\,hr\sin\varphi, \tag{44.22}$$

$$\cos\varphi = \frac{r_{10}}{2}\left(\frac{1}{r_1} + \frac{1}{r_2}\right). \tag{44.23}$$

After differentiating (44.22) with respect to s we obtain

$$hr = \frac{r_{10}}{2}\,(h'r\sin\varphi + hr'\sin\varphi + hr\cos\varphi\varphi'), \tag{44.24}$$

and, on recalling the geometric relations

$$ds = r_1 d\varphi = \frac{dr}{\cos\varphi} = \frac{dz}{\sin\varphi},$$

we can write (44.24) as

$$hr = \frac{r_{10}}{2}\left[h'r\sin\varphi + h\cos\varphi\left(\sin\varphi + \frac{r}{r_1}\right)\right], \tag{44.25}$$

or

$$h = \frac{r_{10}}{2}\,h'\sin\varphi + \frac{r_{10}}{2}\,h\cos\varphi\left(\frac{1}{r_2} + \frac{1}{r_1}\right). \tag{44.26}$$

By using (44.23) we derive

$$h = \frac{r_{10}}{2}\,h'\sin\varphi + h\cos^2\varphi,$$

and hence we find

$$h\sin\varphi = \frac{r_{10}}{2}\,h'.$$

This equation may be also put in the form

$$\frac{dh}{h} = \frac{2}{r_{10}}\,ds\sin\varphi = \frac{2}{r_{10}}\,dz,$$

which, after integration with respect to z, yields

$$\omega = \frac{h}{h_0} = e^{2z/r_{10}}, \tag{44.27}$$

where h_0 is the thickness at $\varphi = 0$. This result can be also obtained more directly (Dašek 1937).

Equation (44.23) allows us to determine the form of the meridian for the membrane of constant strength. By using the identities

$$\frac{1}{r_2\cos\varphi} = \frac{\sin\varphi}{r\cos\varphi} = \frac{\tan\varphi}{r} = \frac{z'(r)}{r}, \tag{44.28}$$

$$\frac{1}{r_1\cos\varphi} = \frac{z''(r)}{[1 + z'^2(r)]^{\frac{3}{2}}}\sqrt{1 + \tan^2\varphi} = \frac{z''(r)}{1 + z'^2(r)}, \tag{44.29}$$

we can write (44.23) as

$$2 = \frac{z' r_{10}}{r} + \frac{r_{10} z''}{1 + z'^2}. \tag{44.30}$$

After introducing the dimensionless variables $\xi = r/r_{10}$ and $\eta = z/r_{10}$, it is easy to find the relations $\dot{\eta} = z'(r)$ and $\ddot{\eta} = r_{10} z''(r)$, where $\dot{(\,)}$ denotes the derivative with respect to ξ. Then, from (44.30) we have

$$2 = \frac{\dot{\eta}}{\xi} + \frac{\ddot{\eta}}{1 + \dot{\eta}^2},$$

or

$$\ddot{\eta} = \left(2 - \frac{\dot{\eta}}{\xi}\right)(1 + \dot{\eta}^2). \tag{44.31}$$

If the coordinates are chosen to give $\eta(0) = \dot{\eta}(0) = 0$, we can integrate (44.31) by using the power series method (Frobenius method). That is, we put

$$\eta = \eta_0 + \frac{1}{1!}\dot{\eta}_0\xi + \frac{1}{2!}\ddot{\eta}_0\xi^2 + \ldots,$$

where the first two terms on the right-hand side are zero owing to the boundary conditions. From the value of the radius of curvature at the apex

$$r_1(0) = r_{10} = \frac{(1+\dot{\eta}^2)^{\frac{3}{2}}}{\ddot{\eta}/r_{10}} = \frac{r_{10}}{\ddot{\eta}},$$

we have $\ddot{\eta}_0 = 1$.

As a consequence of the symmetry of the surface about the η axis, all terms of odd order must vanish, and therefore $\eta(\xi)$ can be represented as

$$\eta(\xi) = \frac{1}{2}\xi^2 + c_1\xi^4 + c_2\xi^6 + \ldots, \tag{44.32}$$

where the coefficients $c_1, c_2, c_3, \ldots$, are determined by substituting (44.32) in (44.31) and equating like powers of ξ to zero. The result is (Dašek 1937)

$$\eta(\xi) = \tfrac{1}{2}\xi^2 + \tfrac{1}{16}\xi^4 + \tfrac{1}{144}\xi^6 \ldots. \tag{44.33}$$

All the cases examined so far assume the presence of constant stress as a criterion for an optimal surface. This allows the problem to be reduced to a single ordinary differential equation. The problem proposed by Jorgensen (1915) is different because the shape of the optimal surface is prescribed in advance. The cylindrical dam is regarded to be sufficiently high that it can be studied as a system of independent arches subjected to the hydrostatic pressure p corresponding to the depth of the water at the level of the arch. We are given the span 2ℓ, the pressure of the water, and the constant stress σ_0 that the material can sustain (Figure 44.4). We have to determine the angle φ or, equivalently, the central radius of the arch in order to minimize the volume

$$V = 2\varphi r A, \tag{44.34}$$

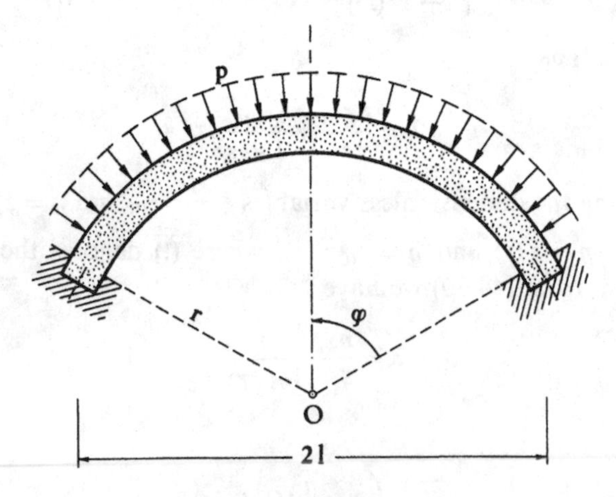

Fig. 44.4

where A is the area of the section of the arch in a radial plane passing through 0. The middle radius r is equal to $\ell/\sin\phi$, and the stress in the arch is generated by a simple circumferential normal force $T_\varphi = pr$.[54] As we have $\sigma_0 = T_0/A$, we can express the volume V in terms of the single variable φ

$$V = \frac{p\ell^2}{2\sigma_0}\frac{\varphi}{\sin^2\varphi}. \tag{44.35}$$

Stationary values of v occur when φ is a root of the equation

$$2\varphi - \tan\varphi = 0. \tag{44.36}$$

This equation admits the nontrivial solution $2\varphi_1 = 133°34'$, and it can immediately be checked that V attains a minimum at φ_1.

In other cases it is the ratio between the minor and major axes of the meridional curve (Mansfield 1981) that is required to be minimized in a membrane of revolution. These axes are defined in the following way. We choose a system of axes z and r such that the z axis coincides with the axis of revolution (Figure 44.5), and a cap, the meridional curve of which intersects the coordinate axes at two points A and B at distances a and b from the origin. We call a the major half-axis and b the minor half-axis. The membrane is pressurized from inside, and, owing to symmetry, there are only two membrane forces: T_φ acting along the meridian, and T_θ acting along the parallel. The equilibrium equations are then (44.2) and (44.12), and can be written in the form

$$\frac{T_\varphi}{r_1} + \frac{T_\theta}{r_2} = p = \text{constant}, \tag{44.37}$$

$$2\pi r T_\varphi \sin\varphi = \pi r^2 p. \tag{44.38}$$

These equations are valid in general, but in thin membranes pressurized from inside it is important that compressive stresses are avoided because they cause

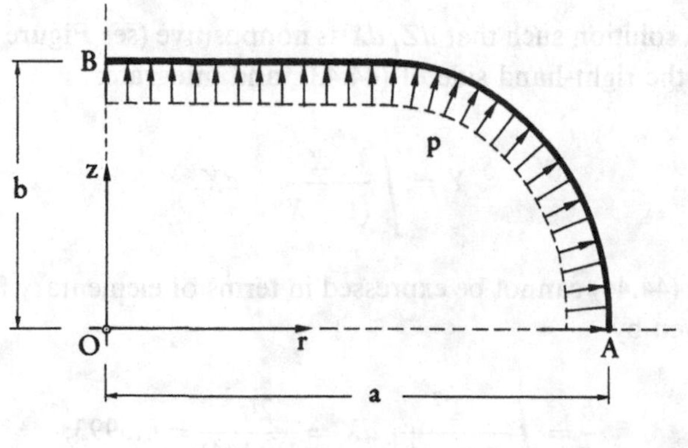

Fig. 44.5

[54] Excluding the load influence of the clamps.

wrinkling. We thus restrict our attention to those meridional curves for which, at some given place,

$$T_\theta = 0. \tag{44.39}$$

Substitution of (44.39) in (44.37) and (44.38) yields

$$r - 2r_1 \sin\varphi = 0. \tag{44.40}$$

For convenience, let us now put $r = x$, and recall the equations

$$\frac{1}{r_1} = \frac{z''}{(1+z'^2)^{\frac{3}{2}}}, \quad \tan\varphi = z', \tag{44.41}$$

where the prime denotes differentiation with respect to x. Equation (44.40) then becomes

$$xz'' - 2z'(1+z'^2) = 0.$$

If we write $t = z'$, this equation becomes

$$xt' = 2t(1+t^2),$$

which may be readily integrated to give

$$\left(\frac{x}{a}\right)^4 = \frac{t^2}{1+t^2}, \tag{44.42}$$

where a is the semimajor axis of the meridional curve. By introducing the nondimensional variables

$$X = \frac{x}{a}, \quad Z = \frac{z}{a},$$

(44.42) can be cast in the form

$$\frac{dZ}{dX} = \mp \frac{X^2}{(1-X^4)^{\frac{1}{2}}}. \tag{44.43}$$

Since we seek a solution such that dZ/dX is nonpositive (see Figure 44.5) we take the minus sign on the right-hand side of (44.43), and integrate:

$$Y = \int_X^1 \frac{X^2}{(1-X^4)^{\frac{1}{2}}}\, dX. \tag{44.44}$$

The integral in (44.44) cannot be expressed in terms of elementary functions, but the ratio b/a is given by

$$\frac{b}{a} = \int_0^1 \frac{X^2}{(1-X^4)^{\frac{1}{2}}}\, dX = \frac{2^{\frac{1}{2}}\pi^{\frac{3}{2}}}{\Gamma^2(1/4)} = 0.5993, \tag{44.45}$$

where Γ is Euler's gamma function.

In addition, the radius of curvature r_1 is given by (44.41) and (44.44):

$$r_1 = \frac{a^2}{2x}, \tag{44.46}$$

and substitution in (44.37) (with $T_\theta = 0$) yields

$$T_\varphi = \frac{pa^2}{2x}. \tag{44.47}$$

The solution is thus singular for $x \to 0$. However, this singularity implies that membrane theory is inadmissible in the neighborhood of the apex, where the curvature vanishes. In practice, a curvature at $x = 0$ is introduced by the local deformation of the membrane at the cusp, thus leading to a finite membrane force T_φ.

The theory of shape optimization is much further developed for planar membranes under small strains and displacements than for curved membranes. However, all the most important results in the subject are related to a famous conjecture of Rayleigh (1894), which states that, of all the vibrating clamped elastic membranes of constant density and fixed area, the circular membrane has the lowest fundamental frequency. Rayleigh's conjecture, which is supported by physical arguments, was proved independently by Faber (1923) and Krahn (1924), but their proofs, although valid, were incomplete because they used a system of curvilinear coordinates that is not defined everywhere, but only almost everywhere.

In order to prove Rayleigh's conjecture, we consider the eigenvalue problem

$$\begin{aligned} \nabla^2 u + \lambda u &= 0, \quad \text{in } D, \\ u &= 0, \quad \text{on } C, \end{aligned} \tag{44.48}$$

and call λ_1 the first eigenvalue and u the corresponding eigenfunction. Let us introduce a system of curvilinear coordinates at each point P in D in the following way. As we have $u > 0$ in D, we consider the level curve $u(x, y) =$ constant through P, and we direct one coordinate along this curve. The other coordinate is chosen normal to this level curve through P. It is possible to show that, if D is sufficiently regular, the gradient of u can vanish at only a finite number of points along each level curve, in this way validating the proof of Faber and Krahn.

At points at which the normal is uniquely defined we represent the element of area dA by

$$dA = dn\, d\sigma,$$

where $d\sigma$ is the arc element along the level curve and dn is the element of length along the normal. However, because for all points like P we have

$$du = \frac{\partial u}{\partial n}\, dn = -|\text{grad} u|\, dn,$$

we can write

$$dA = -\frac{d\sigma\, du}{|\text{grad} u|}, \tag{44.49}$$

or

$$A = \iint_D d\sigma\, dn = \int_0^{u_m} \left\{ \oint_{\gamma(u)} \frac{d\sigma}{|\mathrm{grad} u|} \right\} du, \tag{44.50}$$

where u_m is the maximum of u and $\gamma(u)$ is the level curve $u =$ constant. We now define

$$A(u) = \int_u^{u_m} \left\{ \oint_{\gamma(u)} \frac{d\sigma}{|\mathrm{grad} u|} \right\} du, \tag{44.51}$$

and calculate

$$-\frac{dA(u)}{du} = \oint_{\gamma(u)} \frac{d\sigma}{|\mathrm{grad} u|} \geqslant \frac{\left(\oint_{\gamma(u)} d\sigma \right)^2}{\oint_{\gamma(u)} |\mathrm{grad} u|\, d\sigma} \geqslant \frac{4\pi A(u)}{\oint_{\gamma(u)} |\mathrm{grad} u|\, d\sigma}, \tag{44.52}$$

where the last inequality is derived from the isoperimetric inequality $L^2 \geqslant 4\pi A$, L being the length of the perimeter of $\gamma(u)$.

Now the first eigenvalue of (44.48) is given by the ratio

$$\lambda_1 = \frac{\iint_D |\mathrm{grad} u|^2\, dA}{\iint_D u^2\, dA} = \frac{\int_0^{u_m} \left\{ \oint_{\gamma(u)} |\mathrm{grad} u|\, d\sigma \right\} du}{\int_0^{u_m} u^2 \left\{ \oint_{\gamma(u)} \frac{d\sigma}{|\mathrm{grad} u|} \right\} du}. \tag{44.53}$$

On inserting (44.52) in the numerator of (44.53) we obtain

$$\lambda_1 \geqslant \frac{\int_0^{u_m} 4\pi A(u) \left(-\frac{du}{dA} \right) du}{\int_0^{u_m} u^2 \left(-\frac{dA}{du} \right) du}, \tag{44.54}$$

and, on noting that $A_0 \equiv A(0)$ denotes the area of D, (44.54) becomes

$$\lambda_1 \geqslant \frac{\int_0^{A_0} 4\pi A(u) \left(\frac{du}{dA} \right)^2 dA}{\int_0^{A_0} u^2\, dA}. \tag{44.55}$$

We now set $A(u) = \pi r^2$ and use the element of area $dA = 2\pi r\, dr$. Then, by taking r as a new independent variable, we obtain

$$\lambda_1 \geqslant \frac{\int_0^{R_0} r\left(\frac{du}{dr}\right)^2 dr}{\int_0^{R_0} r u^2\, dr} \geqslant \tilde{\lambda}_1, \tag{44.56}$$

where $R_0 = (A_0/\pi)^{\frac{1}{2}}$ and $\tilde{\lambda}_1$ is the first eigenvalue of the circle of the same area. The exact value of $\tilde{\lambda}_1$ is $j_0^2 R_0^{-2}$, where j_0 is the first zero of the Bessel function $J_0(x)$.

The isoperimetric inequality (44.56) can be extended to an N-dimensional domain D. In this case, ∇^2 in (44.47) represents the N-dimensional Laplace operator and $u = u(x_1, x_2, \ldots, x_N)$. By using the same procedure as before, we can demonstrate the relation $\lambda_1 \geqslant \tilde{\lambda}_1$, where $\tilde{\lambda}_1$ is now the first eigenvalue of problem (44.48) formulated for the N-dimensional sphere, the volume of which is the same as the volume of D. The precise value of $\tilde{\lambda}_1$ is

$$\tilde{\lambda}_1 = \left(\frac{\omega_N}{N V_N}\right)^{2/N} j^2_{\frac{N-2}{2}}, \tag{44.57}$$

where ω_N denotes the surface area of the unit sphere in N-dimensions, V_N is the N-dimensional volume of D, and $j^2_{\frac{N-2}{2}}$ is the first zero of the Bessel function $J_{\frac{N-2}{2}}(x)$.

The extension of (44.56) to N dimensions may appear to be of theoretical interest only. However, on the contrary, it is useful for treating two-dimensional problems (Payne and Weinberger 1960). For instance, a four-dimensional interpretation leads to a new result for a two-dimensional problem. As the equation and the boundary condition of problem (44.48) are invariant under a translation or rotation of coordinates, we can always choose a coordinate system such that D lies in the half plane $x > 0$ (Figure 44.6), so that we can make the substitution

$$u = xV, \tag{44.58}$$

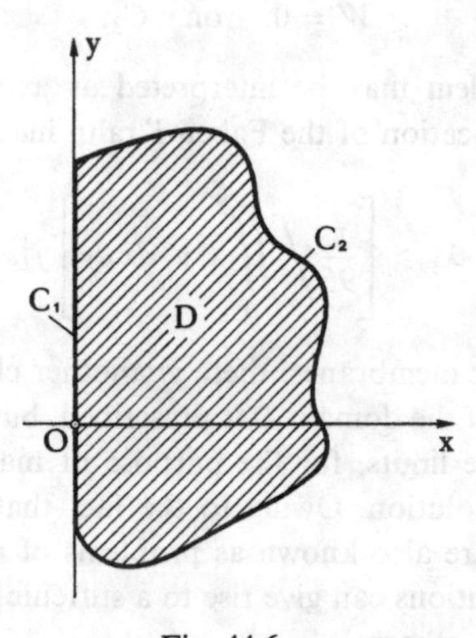

Fig. 44.6

allowing V to satisfy the equation and boundary condition

$$\frac{\partial^2 V}{\partial x^2}+\frac{\partial^2 V}{\partial y^2}+\frac{2}{x}\frac{\partial V}{\partial x}+\lambda V=0, \text{ in } D,$$

$$V=0, \text{ on } C_2.$$

If we examine this equation for V, we observe that it has the same axisymmetric form as that associated with a membrane in four dimensions; that is, V is the solution of the equation

$$\frac{\partial^2 V}{\partial x_1}+\frac{\partial^2 V}{\partial x_2}+\frac{\partial^2 V}{\partial x_3}+\frac{\partial^2 V}{\partial y^2}+\lambda V=0, \text{ in } D. \tag{44.59}$$

This depends only on the variables $x=\sqrt{x_1^2+x_2^2+x_3^2}$ and y, where D is now the four-dimensional region obtained by a rotation of D about the y axis, and the four-dimensional volume of D is

$$V_4 \propto \iint_D x^2\,dx\,dy.$$

The Faber–Krahn inequality leads to

$$\lambda_1 \geqslant \left[\frac{\pi}{8} \Big/ \iint_D x^2\,dx\,dy\right]^{\frac{1}{2}} j_1^2, \tag{44.60}$$

with equality if, and only if, D is semicircular.

Similarly, we can place D in the first quadrant and represent u as

$$u=xyW,$$

where W satisfies the equation

$$\frac{\partial^2 W}{\partial x^2}+\frac{\partial^2 W}{\partial y^2}+\frac{2}{x}\frac{\partial W}{\partial x}+\frac{2}{y}\frac{\partial W}{\partial y}+\lambda W=0, \text{ in } D,$$

$$W=0, \quad \text{on} \quad C_2,$$

(Figure 44.7). This problem may be interpreted as a membrane problem in six dimensions, and the application of the Faber–Krahn inequality yields

$$\lambda_1 \geqslant \left[\frac{\pi}{96} \Big/ \iint_D x^2 y^2\,dx\,dy\right]^{\frac{1}{3}} j_2^2. \tag{44.61}$$

In the theory of planar membranes, there is another class of optimization problems in which the shape of the domain D is prescribed, but the boundary data are to be allocated, within some limits, for the purpose of maximizing or minimizing a given functional of the solution. Owing to the fact that only the boundary data change, these problems are also known as problems of *reinforcement*, because the change in boundary conditions can give rise to a stiffening of the boundary in some parts and a weakening in others.

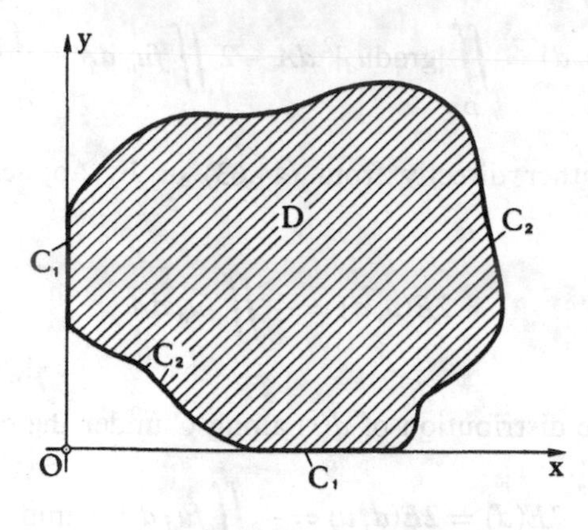

Fig. 44.7

In order to formulate a problem more precisely, let us consider (Buttazzo 1988) a regular plain domain D with boundary C, a function $f \in L^2(D)$, and a positive smooth function d defined on C. The function f represents the force per unit mass in the direction perpendicular to the plane, and the function $d(s)$ is the density of reinforcement of a point on the boundary. The membrane is not rigidly clamped at its boundary C, but the displacement u is elastically constricted by the boundary reaction of the membrane, and this reaction is proportional to d. If, for simplicity, we take $T = 1$, the equilibrium equation for the membrane, with its boundary condition, becomes

$$\begin{aligned} \nabla^2 u + f &= 0, \text{ in } D, \\ u + d\frac{\partial u}{\partial n} &= 0, \text{ on } C. \end{aligned} \tag{44.62}$$

Problem (44.62) is equivalent to the following minimum problem:

$$\min\left\{\iint_D |\text{grad}u|^2 \, dA - 2\iint_D fu \, dA + \oint_C \frac{u^2}{d} \, ds: \quad u \in H^1(D)\right\}. \tag{44.63}$$

If the function d is known, the solution u of problem (44.63) is uniquely determined. We assume, however, that d is not prescribed pointwise, but instead we are given only the total amount of material that we are allowed to distribute along the boundary; that is,

$$\oint_C d(s) \, ds = k, \tag{44.64}$$

where k is a constant. For any choice of $d(s)$ compatible with (44.64), we can solve problem (44.63) and find the corresponding solution u_d. The associated total energy is thus

$$2E(u, d) = \iint_D |\text{grad} u_d|^2 \, dA - 2 \iint_D f u_d \, dA + \oint_C \frac{u_d^2}{d} \, ds. \tag{44.65}$$

Let us note that, either directly from (44.63) or by application of Clapeyron's theorem, we have

$$2E(u, d) = -\iint_D f u_d \, dA. \tag{44.66}$$

We now look for the distribution of $d(s)$ along C under the condition

$$2E(d) = 2E(d, u) = -\iint_D f u_d \, dA = \min.$$

When the load is uniform, say $f \equiv 1$, by (44.66) the optimization problem is equivalent to that of finding the function $d(s)$ for which the mean displacement $(1/D) \iint_D u_d \, dA$ is maximum. In other words, we wish to distribute the material along C so that the overall deformability of the membrane under a uniform load is the highest possible. The solution is useful, for example, in designing a shock absorber that functions like an elastic membrane surrounded by elastic reinforcement.

To find a solution, we consider the problem

$$\min\{E(d) \colon d \in \Gamma_k\}, \tag{44.67}$$

where Γ_k is the set of functions

$$\Gamma_k \equiv \left\{ d \in L^2(C) \colon d \geqslant 0, \oint_C d(s) \, ds = k \right\}. \tag{44.68}$$

By regarding (44.63) and (44.67) as arising from a single minimization problem, we restate the problem in the following way:

$$\min\{E(u, d) \colon u \in H^1(D), \ d \in \Gamma_k\}. \tag{44.69}$$

The problem can be modified further by observing that, for every $u \in L^2(D)$, there is a solution $d_u \in \Gamma_k$ of the minimum problem

$$\min \left\{ \oint_C \frac{u^2}{d} \, ds \colon d \in \Gamma_k \right\}, \tag{44.70}$$

for which the solution is

$$d_u = k \frac{|u|}{\oint_C |u| \, ds}. \tag{44.71}$$

To prove the result, we observe that, for $u = 0$, any function d solves (44.70); if, on the other hand, u is not zero, then from Hölder's inequality we find

$$\left(\oint_C |u|\,ds\right)^2 \leqslant \oint_C \frac{u^2}{d}\,ds \oint_C d\,ds = k \oint_C \frac{u^2}{d}\,ds,$$

and hence

$$\oint_C \frac{u^2}{d_u}\,ds = \frac{1}{k}\left(\oint_C |u|\,ds\right)^2 \leqslant \oint_C \frac{u^2}{d}\,ds. \tag{44.72}$$

This proves that d_u solves (44.70). In addition, the solution is unique because the function $d \to 1/d$ is strictly convex.

Coming back to problem (44.69), we may change it into the new problem

$$\min\left\{\iint_D |\mathrm{grad}u|^2\,dA - 2\iint_D fu\,dA + \frac{1}{k}\left(\oint_C |u|\,ds\right)^2 : u \in H^1(D)\right\}, \tag{44.73}$$

the Euler–Lagrange equation of which is

$$\nabla^2 u + f = 0, \text{ in } \Omega, \tag{44.74a}$$

$$k\frac{\partial u}{\partial n} + H(u)\oint_C |u|\,ds = 0, \tag{44.74b}$$

where $H(t)$ is the multivalued function

$$H(t) = \begin{cases} \operatorname{sign} t, & \text{for } t \neq 0, \\ [-1, 1], & \text{for } t = 0. \end{cases} \tag{44.75}$$

Formulated as (44.73), the treatment of the variational problem follows the standard argument (Buttazzo 1988). Each minimizing sequence $\{u_h\}$ is bounded in $H^1(D)$, and the functional

$$G(u) = \iint_D |\mathrm{grad}u|^2\,dA - 2\iint_D fu\,dA + \frac{1}{k}\left(\oint_C |u|\,ds\right)^2 \tag{44.76}$$

is $H^1(D)$ lower semicontinuous. Then a solution exists. The uniqueness follows by contradiction, by showing that two possible different solutions, say v and w, necessarily coincide.

In order to illustrate the behavior of the solutions, let us consider the following example. Let D be the annulus

$$D \equiv \{(x, y) \in R^2 : R_1 < \sqrt{x^2 + y^2} < R_2\},$$

and take $f \equiv 1$. In this case, the optimization criterion (44.69) coincides with the problem of maximizing the mean displacement $(1/D)\iint_D u\,dA$. The solution u satisfies (44.74a) with $f \equiv 1$, and therefore it must have the form

$$u = -\frac{r^2}{4} + A \log r + B, \tag{44.77}$$

where A and B are constants. In addition, by the maximum principle, we have $u \geqslant 0$ in D, which implies the result $H(u) = 1$. It thus follows from (44.74b) that $\partial u/\partial n$ is a constant, and therefore

$$u'(R_2) + u'(R_1) = 0. \tag{44.78}$$

This equation yields $A = \frac{1}{2}R_1R_2$. On the other hand, we can write (44.74b) for one of the two boundaries, for instance, for $r = R_2$, so as to obtain the equation

$$ku'(R_2) + 2\pi[R_1u(R_1) + R_2u(R_2)] = 0, \tag{44.79}$$

whence we calculate

$$B = \frac{1}{4(R_1 + R_2)}\left[\frac{k}{\pi}(R_2 - R_1) + R_1^3 + R_2^3 - 2\pi R_1R_2(R_1 \log R_1 + R_2 \log R_2)\right]. \tag{44.80}$$

Having determined u, d_{opt} follows from the relation

$$d_{opt} = k\frac{|u_{opt}|}{\oint_C |u_{opt}|\, ds},$$

which, in the present case, yields

$$\begin{aligned} d_{ext} = d(R_2) &= \frac{1}{2(R_2 - R_1)}(4B - R_2^2 + 2R_1R_2 \log R_2), \\ d_{int} = d(R_1) &= \frac{1}{2(R_2 - R_1)}(4B - R_1^2 + 2R_1R_2 \log R_1). \end{aligned} \tag{44.81}$$

These formulae show, for $R_2 > R_1$, the result $d_{ext} < d_{int}$; that is, a larger quantity of reinforcement must be placed on the interior boundary. Solution (44.81) is, however, valid for $d_{ext} > 0$, or

$$k > \frac{\pi}{R_2 - R_1}\left[R_1(R_2^2 - R_1^2) - 2R_1^2R_2 \log\left(\frac{R_2}{R_1}\right)\right]. \tag{44.82}$$

If this is not the case, then we have $d_{ext} = 0$ and $d_{int} = k/2\pi R_1$; that is, the reinforcement must be placed only along the inner boundary.

Chapter VI
Theories of Plates

45. The Equations of Linear Theory

A planar plate, like a planar membrane, is a solid layer bounded by two parallel planes and by a cylinder that intercepts the planes at right angles. The plane midway between the faces is called the "middle plane." We can draw any cylindrical surface $\mathscr{C}$ to cut the middle plane in a curve s. The particular cylinder that coincides with the lateral surface of the plate is called the "edge." We choose rectangular axes x and y placed in the middle plane and draw the z axis at right angles to this plane, so that the upper and lower plane faces have the equations $z = h$ and $z = -h$, respectively, where $2h$ represents the thickness of the plate. We draw the outer normal $\boldsymbol{\nu}$ to s, lying in the middle plane, the unit vector $\mathbf{s}$ tangentially to s, and the unit vector $\mathbf{n}$ normal to the middle plane (Love 1927, Art. 296). These three unit vectors are oriented such that they form a right-handed orthonormal basis $(\boldsymbol{\nu}, \mathbf{s}, \mathbf{n})$.

Let δs be an arc element of the curve s, and let us suppose that the generators of $\mathscr{C}$, drawn through the extremities of δs, mark out on $\mathscr{C}$ a piece of cylindrical surface of base δs and height $2h$, called the area of this element δA. The tractions across the area δA are statically equivalent to a force applied at the centroid of δA together with a couple. We resolve, as we did for membranes, this force into three components δT, δS, and δN directed along $\boldsymbol{\nu}$, $\mathbf{s}$, and $\mathbf{n}$, respectively, and resolve the couple into three components δH, δG, and δK directed along the same vectors. We then take the limits of the ratios $\delta T/\delta s, \ldots, \delta K/\delta s$ as δs tends to zero. We postulate that the limit of $\delta K/\delta s$ is zero, but all other limits are finite. The limits of $\delta T/\delta s$, $\delta S/\delta s$, and $\delta N/\partial s$ are denoted by T, S, and N, respectively, and those of $\delta H/\delta s$, and $\delta G/\delta s$ by H and G. T is a *tension*, S and N are *shearing forces* tangential and normal to the middle plane, G is a *bending couple*, and H is a *torsional couple*. The quantities $T, \ldots, G$ represent the stress characteristics of the plate.

In particular, we can place the normal $\boldsymbol{\nu}$ parallel to the x axis and the tangent $\mathbf{s}$ parallel to the y axis. In this case all the stress characteristics are given the suffix 1 in order to signify that they act on an arc element perpendicular to x. Similarly, when the normal $\boldsymbol{\nu}$ is parallel to the y axis and $\mathbf{s}$ is parallel to the negative x axis, we give the suffix 2 to T, etc. The notion of stress characteristics in a planar plate is not necessarily related to the local stress existing in the plate, which is regarded as a three-dimensional continuum. The plate may be identified by its middle plane, and this may be conceived as a two-dimensional body that can transmit normal shearing forces and couples. However, if we wish to establish a relationship between stress characteristics and local stresses, we must define the former in terms of the latter according to formulae such as

$$T_1 = \int_{-h}^{h} \sigma_x \, dz, \tag{45.1a}$$

$$S_1 = \int_{-h}^{h} \sigma_{xy} \, dz, \tag{45.1b}$$

$$N_1 = \int_{-h}^{h} \sigma_{xz} \, dz, \tag{45.1c}$$

$$H_1 = -\int_{-h}^{h} z\sigma_{xy} \, dz, \tag{45.1d}$$

$$G_1 = \int_{-h}^{h} z\sigma_x \, dz, \tag{45.1e}$$

and

$$T_2 = \int_{-h}^{h} \sigma_y \, dz, \tag{45.2a}$$

$$S_2 = -\int_{-h}^{h} \sigma_{yx} \, dz, \tag{45.2b}$$

$$N_2 = \int_{-h}^{h} \sigma_{yz} \, dz, \tag{45.2c}$$

$$H_2 = \int_{-h}^{h} z\sigma_{yx} \, dz, \tag{45.2d}$$

$$G_2 = \int_{-h}^{h} z\sigma_y \, dz. \tag{45.2e}$$

The stress resultants are positive if they are parallel to the directions of the unit vectors **v**, **s**, and **n** (Figure 45.1(a)). These directions coincide with those of the coordinate axis on the faces $x =$ constant; however, on the faces $y =$ constant, S_2 is directed along the negative x axis. The couples are positive if their axial vectors are parallel to **v**, **s**, and **n** (Figure 45.1(b)). The characteristics G_1 and H_2 are directed along the positive axes, but H_1 and G_2 are pointing in the negative direction of the x axis. This explains the signs in (45.1) and (45.2). Let us note that, in accordance with these formulae, we have

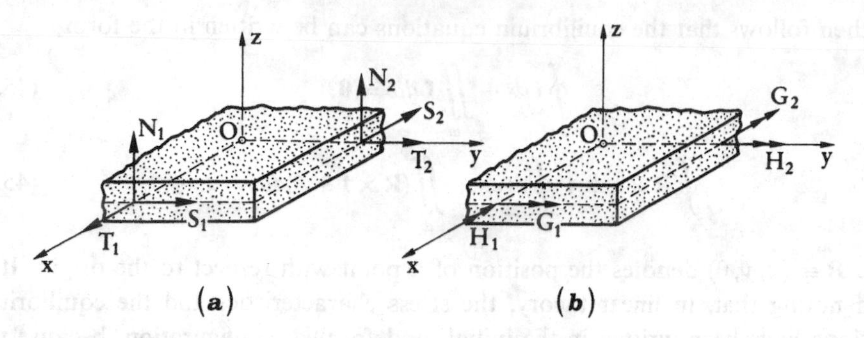

Fig. 45.1

$$S_2 = -S_1, \quad H_2 = -H_1. \tag{45.3}$$

Having defined the stress resultants and couples, we must now use a similar procedure to define the external forces. Let us again consider the cylindrical surface $\mathscr{C}$ and its intersection s with the middle plane. On the portion of plate inside $\mathscr{C}$ there are body forces and surface tractions on the faces $z = \pm h$. These forces are statically equivalent to a force applied at the centroid of the volume within $\mathscr{C}$ together with a couple. Let $\delta X'$, $\delta Y'$, and $\delta Z'$ denote the components of the force, and $\delta L'$, $\delta M'$, and $\delta N'$ denote the components of the couple along the x, y, and z axes. Let $\delta\omega$ denote the area inside the curve s (which is assumed to be simple and closed), and let us diminish $\delta\omega$ indefinitely by contracting s towards the centroid of the element of volume. We assume that the limit of the ratio $\delta N'/\delta\omega$ is zero as $\delta\omega$ tends to zero, but the limits of the other ratios are finite in general. We denote the limit of $\delta X'/\delta\omega$ by X', and the other limits by Y', Z', L', and M'. These limits represent the components of the resultant force and the resultant couple estimated per unit area of the middle plane.

We now require that the resultant force and couple of all the forces acting on the portion of plate within the cylindrical surface $\mathscr{C}$ are in equilibrium. On introducing the second-order tensors

$$\mathbf{T} \equiv \begin{bmatrix} T_1 & -S_2 & 0 \\ S_1 & T_2 & 0 \\ N_1 & N_2 & 0 \end{bmatrix}, \quad \mathbf{M} \equiv \begin{bmatrix} H_1 & -G_2 & 0 \\ G_1 & H_2 & 0 \\ 0 & 0 & 0 \end{bmatrix}, \tag{45.4}$$

we are able to find the traction and the couple acting per unit length on an element ds, the normal $\mathbf{v}$ of which has the components $(\cos\theta, \sin\theta, 0)$ along the coordinate axes. Denoting the traction by $\mathbf{t}$ and the couple by $\mathbf{m}$, we have

$$\mathbf{t} = \mathbf{T}\mathbf{v}, \quad \mathbf{m} = \mathbf{M}\mathbf{v}. \tag{45.5}$$

These formulae are the equivalent of the Cauchy stress principle in plate theory.

The external forces and couples per unit area can be represented by the vectors

$$\mathbf{f} \equiv (X', Y', Z'), \quad \boldsymbol{\ell} \equiv (L', M', 0). \tag{45.6}$$

It then follows that the equilibrium equations can be written in the form

$$\oint_s \mathbf{t}\, ds + \iint_\omega \mathbf{f}\, d\omega = \mathbf{0}, \tag{45.7}$$

$$\oint_s \mathbf{R} \times \mathbf{t}\, ds + \oint_s \mathbf{m}\, ds + \iint_\omega (\mathbf{R} \times \mathbf{f} + \boldsymbol{\ell}) d\omega = \mathbf{0}, \tag{45.8}$$

where $\mathbf{R} \equiv (x, y, 0)$ denotes the position of a point with respect to the origin. It is worth noting that, in linear theory, the stress characteristics and the equilibrium equations have been written in the initial, undeformed, configuration, because the deformed configuration, for which the equilibrium equations must be formulated, is regarded as practically the same as the initial configuration.

By substituting (45.5) in (45.7) and (45.8) we obtain

$$\oint_s \mathbf{T}\mathbf{v}\, ds + \iint_\omega \mathbf{f}\, d\omega = \mathbf{0}, \tag{45.9}$$

$$\oint_s (\mathbf{R} \times \mathbf{T}\mathbf{v} + \mathbf{M}\mathbf{v}) ds + \iint_\omega (\mathbf{R} \times \mathbf{f} + \boldsymbol{\ell}) d\omega = \mathbf{0}. \tag{45.10}$$

As $\mathbf{v} \equiv (\cos\theta, \sin\theta, 0)$ is normal to s, we can transform the line integrals into surface integrals. Then, provided that the integrands are continuous functions of their arguments, we can write (45.9) in local form:

$$\mathrm{div}\mathbf{T} + \mathbf{f} = \mathbf{0}. \tag{45.11}$$

The same argument can be applied to (45.10). Here, however, it is necessary first to use the identity (Gurtin 1972, Sect. 15)

$$\oint_s \mathbf{R} \times \mathbf{T}\mathbf{v}\, ds = \iint_\omega \mathbf{R} \times \mathrm{div}\mathbf{T}\, d\omega + 2 \iint_\omega \boldsymbol{\sigma}\, d\omega,$$

where $\boldsymbol{\sigma} \equiv \frac{1}{2}(N_2, -N_1, S_1 + S_2)$ is the axial vector corresponding to the skew part of $\mathbf{T}$. We thus obtain the result

$$\mathrm{div}\mathbf{M} + 2\boldsymbol{\sigma} + \boldsymbol{\ell} = \mathbf{0}. \tag{45.12}$$

There are only two components of (45.12), because the third is zero due to the fact that $S_1 + S_2 = 0$.

There are thus five equilibrium equations, while, if we take into account (45.3), there are eight unknowns, namely T_1, S_1, N_1, T_2, N_2, H_1, G_1, and G_2. In addition, we need to determine the shape of the middle surface after deformation. It is thus clear that the problem is highly undetermined, unless we introduce additional equations relating the various unknowns. This is done by defining the characteristics in terms of the displacement components of the middle surface. However, the possibility of giving detailed expressions for these displacements depends on having small strains and a plate thickness that is small compared with the other linear dimensions.

However, we will examine first the case in which the thickness is possibly large, but the plate is only slightly bent by couples applied at its edges. In particular, we take the case in which the edge of the plate is the rectangle of the xy plane, included inside the lines $x = \pm a$ and $y = \pm b$ parallel to the coordinate axes. The plate is bent by constant couples G_1 applied to the pair of opposite edges $x = \pm a$, and constant couples G_2 on the edges $y = \pm b$. Under these loads, we assume that the stresses in the interior of the plate are zero, except for the two components σ_x and σ_y, which are proportional to the coordinate z. These stresses have the form

$$\sigma_x = E\alpha z, \quad \sigma_y = E\beta z, \tag{45.13}$$

where α and β are constants, and E is Young's modulus. Having found the stresses, we can find the displacements by integrating the strain–displacement relations. The result is (Love 1927, Art. 90)

$$\begin{gathered} u = (\alpha - \sigma\beta)xz, \quad v = (\beta - \sigma\alpha)yz, \\ w = -\tfrac{1}{2}(\alpha - \sigma\beta)x^2 - \tfrac{1}{2}(\beta - \sigma\alpha)y^2 - \tfrac{1}{2}\sigma(\alpha + \beta)z^2. \end{gathered} \tag{45.14}$$

Under the assumption of small strains, the principal curvatures of any surface which, in the unstrained state, were parallel to the faces, are given by

$$\frac{1}{R_1} = \frac{\partial^2 w}{\partial x^2} = \sigma\beta - \alpha, \quad \frac{1}{R_2} = \frac{\partial^2 w}{\partial y^2} = \sigma\alpha - \beta, \tag{45.15}$$

and these quantities are positive if the corresponding centers of curvature lie on the side for which z is positive. From (45.15) we determine α and β in terms of the curvatures

$$\alpha = -\frac{1}{1-\sigma^2}\left(\frac{1}{R_1} + \frac{\sigma}{R_2}\right), \quad \beta = -\frac{1}{1-\sigma^2}\left(\frac{1}{R_2} + \frac{\sigma}{R_1}\right). \tag{45.16}$$

Once we have α and β, we have also the stresses from (45.13) and we are able to express G_1 and G_2 as functions of R_1 and R_2 by using (45.1e) and (45.2e). The result is

$$\begin{aligned} G_1 &= \int_{-h}^{h} z\sigma_x = -\frac{2}{3}\frac{Eh^3}{1-\sigma^2}\left(\frac{1}{R_1} + \frac{\sigma}{R_2}\right), \\ G_2 &= \int_{-h}^{h} z\sigma_y = -\frac{2}{3}\frac{Eh^3}{1-\sigma^2}\left(\frac{1}{R_2} + \frac{\sigma}{R_1}\right). \end{aligned} \tag{45.17}$$

It may appear that (45.17) are strictly related to the particular load conditions of pure bending along two orthogonal directions. However, the result also holds for more general distribution couples, provided that we interpret the preceding formulae as referring, at any point, to the pair of orthogonal axes in the middle plane, say x' and y', which, after deformation, coincide with the tangents to the lines of principal curvature of the strained middle surface, these being drawn from the strained position of the point considered. Then in the $x'y'$ system the stresses again take the form (45.13), except that they must be denoted by $\sigma'_{x'}$ and $\sigma'_{y'}$. The corresponding couples G'_1 and G'_2 are again given by formulae (45.17). Now let the direction of x' make angles φ and $\frac{1}{2}\pi - \varphi$ with the x and y axes. The stresses in the xy system are then given by

$$\begin{aligned}\sigma_x &= E\alpha z\cos^2\varphi + E\beta z\sin^2\varphi,\\ \sigma_y &= E\alpha z\sin^2\varphi + E\beta z\cos^2\varphi,\\ \sigma_{xy} &= \tfrac{1}{2}E(\alpha-\beta)z\sin 2\varphi,\end{aligned} \tag{45.18}$$

and the corresponding expressions for G_1, G_2, and H_1 are

$$\begin{aligned}G_1 &= -D\left[\frac{\cos^2\varphi}{R_1}+\frac{\sin^2\varphi}{R_2}+\sigma\left(\frac{\sin^2\varphi}{R_1}+\frac{\cos^2\varphi}{R_2}\right)\right],\\ G_2 &= -D\left[\frac{\sin^2\varphi}{R_1}+\frac{\cos^2\varphi}{R_2}+\sigma\left(\frac{\cos^2\varphi}{R_1}+\frac{\sin^2\varphi}{R_2}\right)\right],\\ H_1 &= \frac{1}{2}D(1-\sigma)\sin 2\varphi\left(\frac{1}{R_1}-\frac{1}{R_2}\right),\end{aligned} \tag{45.19}$$

where $D = \frac{2}{3}Eh^3/(1-\sigma^2)$ is the "flexural rigidity" of the plate. After introducing the curvatures

$$\begin{aligned}\kappa_1 = \frac{\cos^2\varphi}{R_1}+\frac{\sin^2\varphi}{R_2}, \qquad \kappa_2 = \frac{\sin^2\varphi}{R_1}+\frac{\cos^2\varphi}{R_2},\\ 2\tau = \sin 2\varphi\left(\frac{1}{R_1}-\frac{1}{R_2}\right),\end{aligned} \tag{45.20}$$

(45.19) become

$$G_1 = -D(\kappa_1+\sigma\kappa_2), \quad G_2 = -D(\kappa_2+\sigma\kappa_1), \quad H_1 = D(1-\sigma)\tau. \tag{45.21}$$

In deriving (45.21), the curvatures $1/R_1$ and $1/R_2$ of the mean surface represent the characteristics of the strains associated with the flexural couples. If the plate is only slightly bent, the curvature can be expressed in terms of the normal displacement $w(x, y)$ of a point on the middle surface, and written as

$$\kappa_1 = \frac{\partial^2 w}{\partial x^2}, \quad \kappa_2 = \frac{\partial^2 w}{\partial y^2}, \quad \tau = \frac{\partial^2 w}{\partial x\,\partial y}. \tag{45.22}$$

On combining (45.21) and (45.22), we find that the flexural couples can only be expressed in terms of the second derivatives of the transverse displacement of a point on the middle surface.

Let us assume that the plate is bent by transverse forces, but not by couples of the type ℓ. Then the third scalar equation (45.11) is again

$$\frac{\partial N_1}{\partial x}+\frac{\partial N_2}{\partial y} = Z' = 0, \tag{45.23}$$

but the first two (45.12) reduce to

$$\frac{\partial H_1}{\partial x}-\frac{\partial G_2}{\partial y}+N_2 = 0, \quad \frac{\partial G_1}{\partial x}-\frac{\partial H_1}{\partial y}-N_1 = 0. \tag{45.24}$$

By eliminating N_1 and N_2 from these equations, we obtain

$$\frac{\partial^2 G_1}{\partial x^2}-2\frac{\partial^2 H_1}{\partial x\,\partial y}+\frac{\partial^2 G_2}{\partial y^2}+Z' = 0,$$

and by substituting (45.21) and (45.22) in this equation we arrive at

$$D\nabla^4 w = Z', \tag{45.25}$$

where ∇^4 denotes the biharmonic operator in the xy plane. Equation (45.24) is known as the *Germain–Lagrange equation*.

In order to solve (45.25) we must prescribe the boundary conditions along the line of the edge, but the form of the boundary conditions is not obvious. If the edge is clamped, w and $\partial w/\partial \nu$ vanish; at a supported edge, w vanishes, as does the flexural couple G having the unit tangent $\mathbf{s}$ as its axis. The expression for G in terms of G_1, G_2, and H_1 is

$$G = G_1 \cos^2\theta + G_2 \sin^2\theta - H_1 \sin 2\theta, \tag{45.26}$$

or, alternatively,

$$G = -D\left[\cos^2\theta\left(\frac{\partial^2 w}{\partial x^2} + \sigma\frac{\partial^2 w}{\partial y^2}\right) + \sin^2\theta\left(\frac{\partial^2 w}{\partial y^2} + \sigma\frac{\partial^2 w}{\partial x^2}\right) + (1-\sigma)\sin 2\theta\,\frac{\partial^2 w}{\partial x\,\partial y}\right].$$

This equation can be written in a simpler form by introducing the directional derivatives along the direction $\mathbf{s}$ of the tangent to the edge line and of the normal $\boldsymbol{\nu}$:

$$\frac{\partial}{\partial s} = \cos\theta\,\frac{\partial}{\partial y} - \sin\theta\,\frac{\partial}{\partial x}, \quad \frac{\partial}{\partial \nu} = \cos\theta\,\frac{\partial}{\partial x} + \sin\theta\,\frac{\partial}{\partial y}. \tag{45.27}$$

As we have $\partial\theta/\partial\nu = 0$ and $\partial\theta/\partial s = 1/\rho'$, where ρ' is the radius of curvature of the curve, we find

$$G = -D\left[\frac{\partial^2 w}{\partial \nu^2} + \sigma\left(\frac{\partial^2 w}{\partial s^2} + \frac{1}{\rho'}\frac{\partial w}{\partial \nu}\right)\right]. \tag{45.28}$$

The expression for the boundary conditions when the line of the edge is free is more difficult to derive. In a thick plate the tractions must be specified, not simply along this line, but on the mantle of the cylindrical surface $\mathscr{C}$ drawn from the line. When the plate is thin, the actual distribution of the tractions on the cylindrical surface is of no practical importance, because, from Saint-Venant's principle, two systems of tractions that give rise to the same resultant force and couple produce the same effects at a distance from the edge, being distinguished only by the order of the thickness. In a simply bent plate the resultants of the external tractions consist of three components $\bar{N}$, $\bar{H}$, and $\bar{G}$ in the directions of the vectors $\mathbf{n}$, $\boldsymbol{\nu}$, and $\mathbf{s}$, respectively. It may seem natural that the equivalence between internal and external resultants requires that the three following equations are satisfied:

$$N = \bar{N}, \quad H = \bar{H}, \quad G = \bar{G}, \tag{45.29}$$

where $\bar{N}$, $\bar{H}$, and $\bar{G}$ are prescribed (they vanish if the boundary is free) and N, H, and G are given by (45.28) and the two corresponding equations

$$N = N_1\cos\theta + N_2\sin\theta,$$
$$H = \tfrac{1}{2}(G_1 - G_2)\sin 2\theta + H_1\cos 2\theta,$$

or, alternatively,

$$N = \cos\theta\left(\frac{\partial G_1}{\partial x} - \frac{\partial H_1}{\partial y}\right) + \sin\theta\left(\frac{\partial G_2}{\partial y} - \frac{\partial H_1}{\partial x}\right) = -D\frac{\partial}{\partial \nu}(\nabla^2 w), \tag{45.30}$$

$$H = D(1-\sigma)\left[\sin\theta\cos\theta\left(\frac{\partial^2 w}{\partial y^2} - \frac{\partial^2 w}{\partial x^2}\right) + (\cos^2\theta - \sin^2\theta)\frac{\partial^2 w}{\partial x\,\partial y}\right] = D(1-\sigma)\frac{\partial}{\partial \nu}\left(\frac{\partial v}{\partial s}\right). \tag{45.31}$$

The three equations (45.29) were adopted by Poisson, but, if prescribed separately, they are not the correct boundary conditions associated with (45.25), because a biharmonic equation requires only two independent boundary conditions, rather than three. The explanation of this apparent contradiction lies in the fact that the force N and the couple H are not independent, because a torsional couple on any finite arc of the boundary can be replaced by tractions directed at right angles to the middle plane. This result is the consequence of a theorem of statics, which states that a distribution of couples H per unit length of the edge curve, the axes of the couples being normal to this curve, is statically equivalent to a distribution of forces of amount $-(\partial H/\partial s)$, directed at right angles to the plane of the curve. The proof of the theorem is almost immediate if we consider a distribution of forces $-(\partial H/\partial s)$ along a closed curve s. The resultant of these forces is clearly zero because of the result $\oint_s(-\partial H/\partial s)ds = 0$. The moments of these forces taken about the coordinate axes are given by

$$\begin{aligned}\oint_s -y\frac{\partial H}{\partial s}\,ds &= \oint_s H\frac{\partial y}{\partial s}\,ds = \oint_s H\cos\theta\,ds,\\ \oint_s x\frac{\partial H}{\partial s}\,ds &= -\oint_s H\frac{\partial x}{\partial s}\,ds = \oint_s H\sin\theta\,ds,\end{aligned} \tag{45.32}$$

where θ is the angle that $\mathbf{v}$ forms with the x axis. The expressions on the right-hand sides of (45.32) are the components of the resultant couple of the distribution of couples H.

The result is that, instead of (45.29), the correct boundary conditions are

$$G = \bar{G}, \quad N - \frac{\partial H}{\partial s} = \bar{N} - \frac{\partial \bar{H}}{\partial s}, \tag{45.33}$$

which are known as *Kirchhoff's boundary conditions.*

Once we have solved (45.25) with its boundary conditions, we are able to determine G_1, G_2, and H_1 from (45.21), and N_1 and N_2 are then given by (45.24). In order to determine the stresses of the type σ_x, σ_y, and σ_{xy}, we use (45.18), which, after substitution of (45.16) and (45.20), yield

$$\sigma_x = \frac{3z}{2h^3}G_1, \quad \sigma_y = \frac{3z}{2h^3}G_2, \quad \sigma_{xy} = -\frac{3z}{2h^3}H_1.$$

The stresses of type σ_{zx} and σ_{zy} are not determined from the theory because it gives us only their resultants N_1 and N_2. This difficulty is overcome by assuming σ_{zx} and σ_{zy} to be quadratic functions of z that vanish at $z = \pm h$. As they satisfy (45.1c) and (45.2c), the only possible form that they can have is

$$\sigma_{zx} = \frac{3}{4h^3} N_1(h^2 - z^2), \quad \sigma_{zy} = \frac{3}{4h^3} N_2(h^2 - z^2).$$

As a brief historical sketch of the theory of planar plates, the first proposal for using a differential equation to describe the standing vibrations in a planar plate was due to Jacob Bernoulli II (Szabó 1972), who, by considering a rectangular plate as a network of orthogonal beams, established the equation (Bernoulli 1789)

$$\frac{\partial^2 w}{\partial x^2} + \frac{\partial^2 w}{\partial y^2} = \frac{w}{c^4}, \tag{45.34}$$

where c^4 is a constant. In the equilibrium state under a transverse load Z', (45.34) assumes the form

$$D\left(\frac{\partial^4 w}{\partial x^4} + \frac{\partial^4 w}{\partial y^4}\right) = Z'. \tag{45.35}$$

Compared with (45.25), this equation does not contain the term $2(\partial^4 w/\partial x^2\, \partial y^2)$, which represents the effect of torsional rigidity on the deformation. A second equation was proposed by Sophie Germain (1821):

$$\frac{\partial^6 w}{\partial x^4\, \partial y^2} + \frac{\partial^6 w}{\partial x^2\, \partial y^4} = Z', \tag{45.36}$$

but this is also incorrect. Equation (45.25) was given by Lagrange.[55] The correct deduction of the boundary conditions of a thin plate in the case of prescribed tractions is due to Kirchhoff (1850).

When the stresses σ_x and σ_y do not have the form (45.13), we say that the pure flexural deformations are accompanied by extensional deformations. The study of purely extensional deformations of a thin plate is reducible to a problem of generalized plane stress by observing that T_1, T_2, and $S_1 = -S_2$ are proportional to the averaged stresses by means of the equations

$$\bar{\sigma}_x = (2h)^{-1} \int_{-h}^{h} \sigma_x\, dz = (2h)^{-1} T_1,$$

$$\bar{\sigma}_y = (2h)^{-1} \int_{-h}^{h} \sigma_y\, dz = (2h)^{-1} T_2,$$

$$\bar{\sigma}_{xy} = (2h)^{-1} \int_{-h}^{h} \sigma_{xy}\, dz = (2h)^{-1} S_1.$$

Provided that we determine the averaged stress, we can obtain the averaged displacements $\bar{u}$ and $\bar{v}$ from the constitutive equations.

[55] Lagrange (1811) derived the correct equation after having noticed a mistake contained in an initial memoir of Sophie Germain.

The stress resultants on the edge curve s, the normal $\mathbf{v}$ of which makes angles θ and $\frac{1}{2}\pi - \theta$ with the x and y axes, are calculated from formulae such as

$$T = T_1 \cos^2\theta + T_2 \sin^2\theta + S_1 \sin 2\theta, \tag{45.37}$$

$$S = \tfrac{1}{2}(T_2 - T_1)\sin 2\theta + S_1 \cos 2\theta. \tag{45.38}$$

If the stress resultants at the boundary have prescribed values, say $\bar{T}$ and $\bar{S}$, the boundary conditions that must be satisfied by the unknown functions T_1, T_2, and S_1 are $T = \bar{T}$ and $S = \bar{S}$, where T and S are given by (45.37) and (45.38), respectively.

46. Theories of Thick Plates

The theory of thin plates is no longer applicable when the thickness of the plate is comparable to the other linear dimensions, because the detailed distribution of the displacements or tractions on the lateral surface cannot be ignored in determining the state of stress in the interior of the plate. Another case in which the approximate theory is not valid occurs when the exterior loads, defined by the limiting process leading to (45.6), are discontinuous at some points, because the state of stress and strain at these points is strongly influenced by the form of these discontinuities. For instance, under a concentrated load applied on the upper face of a plate, the assumption that the stresses have the form (45.13) is not acceptable.

The problem may be approached by considering some particular solutions of the equations of the three-dimensional elasticity in a cylindrical domain under external forces representing the most common conditions of plate loading.

Initially, we shall suppose (Michell 1900) that the plate is held by forces applied at its edge only. The faces $z = \pm h$ are then free from tractions; that is, $\sigma_z = \sigma_{zx} = \sigma_{zy} = 0$ for $z = \pm h$. The equations of equilibrium in the absence of body forces are

$$\frac{\partial \sigma_x}{\partial x} + \frac{\partial \sigma_{yx}}{\partial y} + \frac{\partial \sigma_{zx}}{\partial z} = 0, \tag{46.1a}$$

$$\frac{\partial \sigma_{xy}}{\partial x} + \frac{\partial \sigma_y}{\partial y} + \frac{\partial \sigma_{zy}}{\partial z} = 0, \tag{46.1b}$$

$$\frac{\partial \sigma_{xz}}{\partial x} + \frac{\partial \sigma_{yz}}{\partial y} + \frac{\partial \sigma_z}{\partial z} = 0. \tag{46.1c}$$

On the other hand, each stress component of the type $\sigma_x, \ldots, \sigma_{yz}, \ldots,$ must satisfy the compatibility equations

$$\begin{aligned} \nabla^2 \sigma_x &= -\frac{1}{1+\sigma}\frac{\partial^2 \Theta}{\partial x^2}, \ldots \\ \nabla^2 \sigma_{yz} &= -\frac{1}{1+\sigma}\frac{\partial^2 \Theta}{\partial y\, \partial z}, \ldots \end{aligned} \tag{46.2}$$

where $\Theta = \sigma_x + \sigma_y + \sigma_z$. In addition, we know that Θ is harmonic and each of the stress components is biharmonic.

As a consequence of the assumption that the faces $z = \pm h$ are free from tractions, it follows from (46.1c) that $\partial \sigma_z / \partial z$ vanishes at $z = \pm h$. Hence σ_z satisfies the equation

$\nabla^4\sigma_z = 0$ and the conditions $\sigma_z = 0$ and $\partial\sigma_z/\partial z = 0$ at $z = \pm h$. If the plate has no boundaries except the planes $z = \pm h$, the only possible value for σ_z would be zero, so we assume, for the moment, that $\sigma_z = 0$ applies. It then follows from $\nabla^2\Theta = 0$ and from the equation

$$\nabla^2\sigma_z = 0 = -\frac{1}{1+\sigma}\frac{\partial^2\Theta}{\partial x^2},$$

that Θ has the form $\Theta = \Theta_0 + z\Theta_1$, where Θ_0 and Θ_1 are planar harmonic functions of the only variables x and y.

The tangential stress components σ_{zx} and σ_{zy} satisfy the equations

$$\frac{\partial\sigma_{zx}}{\partial x} + \frac{\partial\sigma_{zy}}{\partial y} = 0, \quad \nabla^2\sigma_{zx} = -\frac{1}{1+\sigma}\frac{\partial\Theta_1}{\partial x}, \quad \nabla^2\sigma_{zy} = -\frac{1}{1+\sigma}\frac{\partial\Theta_1}{\partial y},$$

and the conditions $\sigma_{zx} = \sigma_{zy} = 0$ at $z = \pm h$. A particular solution of the equations is

$$\sigma_{zx} = \frac{1}{2}\frac{1}{1+\sigma}(h^2 - z^2)\frac{\partial\Theta_1}{\partial x}, \quad \sigma_{zy} = \frac{1}{2}\frac{1}{1+\sigma}(h^2 - z^2)\frac{\partial\Theta_1}{\partial y}. \tag{46.3}$$

Once we have obtained σ_{zx}, σ_{zy}, and σ_z, we can determine σ_x, σ_y, and σ_{xy} from the remaining equations of (46.1) and (46.2). The fact that the stress components satisfy equations (46.2) ensures that we can find a displacement field u, v, w that is compatible with these stresses.

The most important states of stress in a planar plate are obtained by giving Θ_0 and Θ_1 particular expressions. When σ_{zx}, σ_{zy}, and σ_z vanish throughout the plate, we say that there is a state of plane stress. This state corresponds to an expression for Θ of the form

$$\Theta = \Theta_0 + \beta z, \tag{46.4}$$

where Θ_0 is a planar harmonic function of x and y, and β is a constant. In this case the stress components σ_x, σ_y, and σ_{xy} are derived from a stress function $\chi(x, y)$ such that we have

$$\sigma_x = \frac{\partial^2\chi}{\partial y^2}, \quad \sigma_y = \frac{\partial^2\chi}{\partial x^2}, \quad \sigma_{xy} = -\frac{\partial^2\chi}{\partial x\,\partial y}, \tag{46.5}$$

and χ has the form

$$\chi = \chi_0 + z\chi_1 - \frac{1}{2}\frac{\sigma}{1+\sigma}z^2\Theta_0, \tag{46.6}$$

where

$$\nabla_1^2\chi_0 = \Theta_0, \quad \nabla_1^2\chi_1 = \beta, \quad \left(\nabla_1^2 = \frac{\partial^2}{\partial x^2} + \frac{\partial^2}{\partial y^2}\right). \tag{46.7}$$

As

$$\nabla^2\chi = \Theta_0 + z\beta = \Theta, \tag{46.8}$$

χ is a planar biharmonic functions, and the displacements can be determined by the usual methods of plane elasticity (Love 1927, Art. 300).

Let us assume instead that σ_z vanishes everywhere, whereas σ_{zx} and σ_{zy} vanish at $z = \pm h$. In this case, we take the expressions of σ_{zx} and σ_{zy} as given by (46.3). In order to determine σ_x, σ_y, and σ_{xy}, we put

$$\Theta = \Theta_0 + z\Theta_1,$$

where Θ_0 and Θ_1 are planar harmonic functions. The state of stress depending on Θ_0 is exactly of the form (45.5) and can be omitted. We therefore have the equations

$$\frac{\partial \sigma_x}{\partial x} + \frac{\partial \sigma_{yx}}{\partial y} - \frac{z}{1+\sigma}\frac{\partial \Theta_1}{\partial x} = 0, \tag{46.9a}$$

$$\frac{\partial \sigma_{xy}}{\partial x} + \frac{\partial \sigma_y}{\partial y} - \frac{z}{1+\sigma}\frac{\partial \Theta_1}{\partial y} = 0, \tag{46.9b}$$

$$\nabla^2 \sigma_x = -\frac{z}{1+\sigma}\frac{\partial^2 \Theta_1}{\partial x^2}, \tag{46.9c}$$

$$\nabla^2 \sigma_y = -\frac{z}{1+\sigma}\frac{\partial^2 \Theta_1}{\partial y^2}, \tag{46.9d}$$

$$\nabla^2 \sigma_{xy} = -\frac{z}{1+\sigma}\frac{\partial^2 \Theta_1}{\partial x\,\partial y}, \tag{46.9e}$$

$$\sigma_x + \sigma_y = z\Theta. \tag{46.9f}$$

Equations (46.9a) and (46.9b) are satisfied by putting

$$\sigma_x = \frac{z}{1+\sigma}\Theta_1 + \frac{\partial^2 \chi'}{\partial y^2}, \quad \sigma_y = \frac{z}{1+\sigma}\Theta_1 + \frac{\partial^2 \chi'}{\partial x^2}, \quad \sigma_{xy} = -\frac{\partial^2 \chi'}{\partial x\,\partial y}, \tag{46.10}$$

where χ' is a function of x, y, z. Equation (46.9f) gives

$$\nabla_1^2 \chi' = \frac{\partial^2 \chi'}{\partial x^2} + \frac{\partial^2 \chi'}{\partial y^2} = -\frac{1-\sigma}{1+\sigma} z\Theta_1, \tag{46.11}$$

and the remaining (46.9c)–(46.9e) can be written in the forms

$$\frac{\partial^2}{\partial y^2}\left(\frac{\partial^2 \chi'}{\partial z^2} - \frac{2-\sigma}{1+\sigma} z\Theta_1\right) = 0, \quad \frac{\partial^2}{\partial x^2}\left(\frac{\partial^2 \chi'}{\partial z^2} - \frac{2-\sigma}{1+\sigma} z\Theta_1\right) = 0,$$

$$\frac{\partial^2}{\partial x\,\partial y}\left(\frac{\partial^2 \chi'}{\partial z^2} - \frac{2-\sigma}{1+\sigma} z\Theta_1\right) = 0.$$

These equations show that the expression $(\partial^2 \chi'/\partial z^2) - [(2-\sigma)/(1+\sigma) z\Theta_1]$ is a linear function of x and y and may be taken to be zero without altering the values of σ_x, σ_y, or σ_{xy}. This implies that χ' must have the form

$$\chi' = z\chi_1'(x, y) + \frac{1}{6}\frac{2-\sigma}{1+\sigma} z^3 \Theta_1, \tag{46.12}$$

where χ_1' is a function of x and y only. On substituting (46.12) in (46.11), we find that χ_1' must satisfy the equation

$$\nabla_1^2 \chi_1' = -\frac{1-\sigma}{1+\sigma}\Theta_1. \tag{46.13}$$

If we substitute (46.12) in (46.10) we find that the stress components are cubic functions of z.

In order to find the displacements associated with these stresses we must use a constitutive equation of the type

$$\frac{\partial u}{\partial x}=\frac{1}{E}[\sigma_x-\sigma(\sigma_y+\sigma_z)]. \tag{46.14}$$

On recalling that we have $\sigma_z\equiv 0$, and that σ_x and σ_y are given by (46.10) with χ' of the form (46.12), we can write (46.14) as

$$\frac{\partial u}{\partial x}=-\frac{1}{E}\left[(1+\sigma)z\,\frac{\partial^2\chi_1'}{\partial x^2}+\frac{1}{6}(2-\sigma)z^3\,\frac{\partial^2\Theta_1}{\partial x^2}\right],$$

and hence, by integrating with respect to x, we obtain

$$u=-\frac{1}{E}\left[(1+\sigma)z\,\frac{\partial\chi_1'}{\partial x}+\frac{1}{6}(2-\sigma)z^3\,\frac{\partial\Theta_1}{\partial x}\right]. \tag{46.15}$$

In the same way, we can prove the result

$$v=-\frac{1}{E}\left[(1+\sigma)z\,\frac{\partial\chi_1'}{\partial y}+\frac{1}{6}(2-\sigma)z^3\,\frac{\partial\Theta_1}{\partial y}\right]. \tag{46.16}$$

To determine the component w, we use the equations

$$\frac{\partial w}{\partial x}+\frac{\partial u}{\partial z}=\frac{2(1+\sigma)}{E}\,\sigma_{zx}=\frac{1}{E}(h^2-z^2)\,\frac{\partial\Theta_1}{\partial x}, \tag{46.17}$$

$$\frac{\partial w}{\partial y}+\frac{\partial v}{\partial z}=\frac{2(1+\sigma)}{E}\,\sigma_{zy}=\frac{1}{E}(h^2-z^2)\,\frac{\partial\Theta_1}{\partial y}. \tag{46.18}$$

Here the quantities $\partial u/\partial z$ and $\partial v/\partial z$ are computable from (46.15) and (46.16). After integration of (46.17) with respect to x, or of (46.18) with respect to y, we get

$$w=\frac{1}{E}[(1+\sigma)\chi_1'+(h^2-\tfrac{1}{2}\sigma z^2)\Theta_1]. \tag{46.19}$$

The importance of the result embodied in the formulae for u, v, and w rests on the fact that it describes states of generalized plane stress in which the plane $z=0$ can undergo transverse displacements. In fact, from (46.19), we find that the normal displacement of this plane is given by

$$w=\frac{1}{E}[(1+\sigma)\chi_1'+h^2\Theta_1]. \tag{46.20}$$

As Θ_1 is harmonic and χ_1' satisfies (46.13), we have the equations

$$\nabla_1^2 w=-\frac{1-\sigma}{E}\,\Theta_1,\quad \nabla_1^4 w=0. \tag{46.21}$$

From (46.20) we obtain

$$\chi_1'=\frac{E}{1+\sigma}\,w+\frac{Eh^3}{1-\sigma^2}\,\nabla_1^2 w. \tag{46.22}$$

We already have the stresses as a function of Θ_1 and χ_1' (see (46.3) and (46.10)), and therefore we can express these stresses in terms of w and $\nabla^2 w$ only.

So far we have found the stresses in plates deformed by forces applied at the edges, but the method can be extended to plates bent by forces at right angles to the initial position of the middle plane.

When the face $z = h$ is subjected to a uniform pressure p, we have $\nabla^4\sigma_z = 0$ everywhere, $\partial\sigma_z/\partial z = 0$ at $z = \pm h$, $\sigma_z = -p$ at $z = h$, and $\sigma_z = 0$ at $z = -h$. A particular solution of the equation $\nabla^4\sigma_z = 0$ is

$$\sigma_z = \frac{1}{4}\frac{p}{h^3}(z+h)^2(z-2h) = \frac{1}{4}\frac{p}{h^3}(z^3 - 3h^2z - 2h^3), \tag{46.23}$$

and we assume that σ_z has this distribution throughout the thickness. To determine Θ, we already have $\nabla^2\Theta = 0$, and from (46.2) we derive the further equation

$$\nabla^2\sigma_z = \frac{3}{2}\frac{p}{h^3}z = -\frac{1}{1+\sigma}\frac{\partial^2\Theta}{\partial z^2}. \tag{46.24}$$

Then the most general form of Θ is

$$\Theta = -\frac{1}{4}(1+\sigma)\frac{p}{h^3}z^3 + \frac{3}{8}(1+\sigma)\frac{p}{h^3}z(x^2+y^2) + z\Theta_1 + \Theta_0,$$

where Θ_1 and Θ_0 are planar harmonic functions. We have already determined the stresses and displacements due to the terms $z\Theta_1$ and Θ_0, and therefore they may be omitted, so that Θ can take the form

$$\Theta = -\frac{1}{4}(1+\sigma)\frac{p}{h^3}z^3 + \frac{3}{8}(1+\sigma)\frac{p}{h^3}z(x^2+y^2). \tag{46.25}$$

Having determined Θ, we can find σ_{zx} and σ_{zy} from the equations

$$\frac{\partial\sigma_{zx}}{\partial x} + \frac{\partial\sigma_{zy}}{\partial y} + \frac{3}{4}\frac{p}{h^3}(z^2-h^2) = 0, \quad \nabla^2\sigma_{zx} = -\frac{3px}{4h^3}, \quad \nabla^2\sigma_{zy} = -\frac{3py}{4h^3},$$

and from the conditions that σ_{zx} and σ_{zy} vanish at $z = \pm h$. A particular solution is

$$\sigma_{zx} = \frac{3}{8}\frac{p}{h^3}(h^2-z^2)x, \quad \sigma_{zy} = \frac{3}{8}\frac{p}{h^3}(h^2-z^2)y, \tag{46.26}$$

and we take these expressions to be the values of σ_{zx} and σ_{zy}.

It is still necessary to determine σ_x, σ_y, and σ_{xy} from

$$\frac{\partial\sigma_x}{\partial x} + \frac{\partial\sigma_{xy}}{\partial y} = \frac{3}{4}\frac{p}{h^3}xz, \tag{46.27a}$$

$$\frac{\partial\sigma_{xy}}{\partial x} + \frac{\partial\sigma_y}{\partial y} = \frac{3}{4}\frac{p}{h^3}yz, \tag{46.27b}$$

$$\nabla^2\sigma_x = \nabla^2\sigma_y = -\frac{3}{4}\frac{p}{h^3}z, \tag{46.27c}$$

$$\nabla^2\sigma_{xy} = 0, \tag{46.27d}$$

$$\sigma_x + \sigma_y = \frac{1}{4}\frac{p}{h^3}\{-(2+\sigma)z^3 + 3z[\tfrac{1}{2}(1+\sigma)(x^2+y^2) + h^2] + 2h^3\}. \tag{46.27e}$$

In order to satisfy (46.27a) and (46.27b), let us take

$$\sigma_x = \frac{3}{8}\frac{p}{h^3} z(x^2+y^2) + \frac{\partial^2 \chi}{\partial y^2}, \quad \sigma_y = \frac{3}{8}\frac{p}{h^3} z(x^2+y^2) + \frac{\partial^2 \chi}{\partial x^2},$$

$$\sigma_{xy} = -\frac{\partial^2 \chi}{\partial x\, \partial y}, \tag{46.28}$$

where $\chi(x, y, z)$ must satisfy the equation (as a consequence of (46.27e))

$$\nabla_1^2 \chi = \frac{\partial^2 \chi}{\partial x^2} + \frac{\partial^2 \chi}{\partial y^2} = -\frac{2+\sigma}{4}\frac{p}{h^3} z^3 - \frac{3}{8}(1-\sigma)\frac{p}{h^3} z(x^2+y^2) + \frac{3}{4}\frac{p}{h} z + \frac{1}{2}p. \tag{46.29}$$

By combining the remaining (46.27c)–(46.27d) (46.28), we find the relations

$$\nabla^2 \sigma_x = \frac{3}{2}\frac{p}{h^3} z + \frac{\partial^2}{\partial y^2}(\nabla^2 \chi) = \frac{\partial^2}{\partial y^2}\left[\frac{9}{8}\frac{p}{h^3} z(x^2+y^2) + \nabla^2 \chi\right] = 0,$$

$$\nabla^2 \sigma_y = \frac{3}{2}\frac{p}{h^3} z + \frac{\partial^2}{\partial x^2}(\nabla^2 \chi) = \frac{\partial^2}{\partial y^2}\left[\frac{9}{8}\frac{p}{h^3} z(x^2+y^2) + \nabla^2 \chi\right] = 0,$$

$$\nabla^2 \sigma_{xy} = -\frac{\partial^2}{\partial x\, \partial y}(\nabla^2 \chi) = -\frac{\partial^2}{\partial x\, \partial y}\left[\frac{9}{8}\frac{p}{h^3} z(x^2+y^2) + \nabla^2 \chi\right] = 0,$$

which shows that

$$\frac{\partial^2 \chi}{\partial z^2} + \nabla_1^2 \chi + \frac{9}{8}\frac{p}{h^3} z(x^2+y^2)$$

must be a linear function of x and y. The function, however, can be taken to be zero without altering the stress components σ_x, σ_y, and τ_{xy}. Thus, by using (46.29) we can write the equation

$$\frac{\partial^2 \chi}{\partial z^2} - \frac{2+\sigma}{4}\frac{p}{h^3} z^3 + \frac{3}{8}(2+\sigma)\frac{p}{h^3} z(x^2+y^2) - \frac{3}{4}\frac{p}{h} z - \frac{1}{2}p = 0.$$

Integration of this equation gives the following result:

$$\chi = \frac{2+\sigma}{80}\frac{p}{h^3} z^5 - \frac{(2+\sigma)}{16}\frac{p}{h^3} z^3(x^2+y^2) - \frac{1}{8}\frac{p}{h} z^5 - \frac{1}{4}pz^2 + z\chi_1'' + \chi_0'', \tag{46.30}$$

where χ_1'' and χ_0'' are functions of x and y. These are determined by the condition that $\nabla_1^2 \chi$, as given by (46.29), is equal to $\nabla_1^2 \chi$ calculated from (46.30). We thus obtain the equations

$$\nabla_1^2 \chi_1'' = -\frac{3}{8}(1-\sigma)\frac{p}{h^3}(x^2+y^2) + \frac{3}{4}\frac{p}{h}, \quad \nabla_1^2 \chi_0'' = \frac{1}{2}p, \tag{46.31}$$

and two particular solutions of this are

$$\chi_1'' = -\frac{3}{128}(1-\sigma)\frac{p}{h^3}(x^2+y^2)^2 + \frac{3}{16}\frac{p}{h}(x^2+y^2),$$
$$\chi_0'' = \tfrac{1}{8}p(x^2+y^2). \tag{46.32}$$

It is not necessary to discuss more general solutions of (46.31), because the arbitrary planar harmonic functions that are added to (46.32) give rise to stress systems of the types discussed before.

Once we have obtained the function χ, we can calculate the detailed expressions for σ_x, σ_y, and σ_{xy}. In particular, we find that these stress components are cubic functions of z. Furthermore, we can determine the corresponding displacements and, if necessary, the stress resultants and stress couples (Love 1927, Art. 307).

It may appear that the method described so far applies only to the case of a thick plate when the faces $z = \pm h$ are free or when the face $z = -h$ is free and the face $z = h$ is subjected to a uniform pressure, but in effect we can extend the procedure to the case where the pressure on the face $z = h$ is a linear function of x and y (Michell 1900), or even when the plate is subjected to its own weight acting perpendicularly to its middle plane. This last case can be solved by superimposing two stress systems: one in which all the stress components vanish except σ_z, where $\sigma_z = g\rho(z + h)$, the z axis being drawn vertically upwards; and a second stress system, corresponding to a uniform pressure $p = 2gh\rho$ on the face $z = h$, and determined by a procedure like the one just described.

Michell's method represents an example of extraordinary ingenuity for finding the elastic solutions for a thick layer under the kind of load conditions that most frequently occur in a planar plate. The procedure is the same in each individual case. The problem is formulated in terms of stress components, associating the equilibrium equations with the compatibility equations written in terms of stresses. Then, particular expressions for σ_z, σ_{zx}, and σ_{zy} are chosen such that they satisfy the boundary conditions on the faces $z = \pm h$ and the field equations. Finally, the other stresses σ_x, σ_y, and σ_{xy} are determined by direct integration of the remaining equations. In this process, arbitrary functions are necessarily introduced, but knowledge of their detailed form is unimportant because the stresses that they add are already included in the solution.

Michell's method, however, is not systematic because certain representations of the solution for σ_x, σ_y, and σ_{xy}, such as (46.10) or (46.28), are the result of intuition rather than a general procedure for representing these solutions. An attempt to overcome this objection is to find the connections between Michell's solution and other more general formulae for three-dimensional elastic solutions (Neuber 1985, 5.8). An example of how this idea applies is represented by the relation between solutions (46.15), (46.16), and (46.19) and those elastic states known as *Boussinesq–Papkovič–Neuber solutions*. The latter are particular solutions of the equations of linear elasticity of the form (Neuber 1985, 2.2)

$$\begin{aligned} 2Gu &= -\frac{\partial F}{\partial x} + 2\alpha\Phi_1, \\ 2Gv &= -\frac{\partial F}{\partial y} + 2\alpha\Phi_2, \\ 2Gw &= -\frac{\partial F}{\partial z} + 2\alpha\Phi_3, \end{aligned} \tag{46.33}$$

where $\alpha = 2(1 - \sigma)$, Φ_1, Φ_2, and Φ_3 are three harmonic functions, and F is a biharmonic function given by

$$F = \Phi_0 + x\Phi_1 + y\Phi_2 + z\Phi_3, \tag{46.34}$$

Φ_0 being a fourth harmonic function. A simple substitution of (46.33) in the equations of elasticity proves that (46.33) is a solution. However, it is also true that every solution of the equations of elasticity can be expressed in the form (46.33).

We then ask if it is possible to find particular forms for the functions Φ_k $(k = 0, \ldots, 3)$ so as to recover (46.15), (46.16), and (46.19). For this purpose, we introduce the *planar biharmonic function* $F_0(x, y)$ and observe that the functions

$$\nabla_1^2 F_0, \quad F_0 - \frac{z^2}{2}\nabla_1^2 F_0, \quad zF_0 - \frac{z^3}{6}\nabla_1^2 F_0, \quad z\nabla_1^2 F_0, \tag{46.35}$$

are *spatial harmonic functions* of the variables x, y, and z. We then set

$$\begin{gathered}\Phi_0 = \frac{3}{2}\frac{z}{h^3}\left[\left(1 - \frac{1}{\alpha}\right)\left(F_0 - \frac{z^2}{6}\nabla_1^2 F_0\right) + \left(\frac{1}{10} + \frac{1}{10\alpha} + \frac{4}{5\alpha^2}\right)h^2\nabla_1^2 F_0\right],\\ \Phi_1 = \Phi_2 = 0,\\ \Phi_3 = \frac{3}{2}\frac{1}{\alpha h^3}\left[F_0 - \frac{z^2}{2}\nabla_1^2 F_0 + \left(\frac{1}{10} - \frac{4}{5\alpha}\right)h^2\nabla_1^2 F_0\right],\end{gathered} \tag{46.36}$$

which, by virtue of (46.35), are clearly spatial harmonic functions. Thus, from (46.34) we have

$$F(x, y, z) = \frac{3}{2}\frac{z}{h^3}\left[F_0 + \left(\frac{1}{2} + \frac{1}{\alpha}\right)\left(\frac{h^2}{5} - \frac{z^2}{3}\right)\nabla_1^2 F_0\right]. \tag{46.37}$$

The function $F_0(x, y)$ is related to $F(x, y, z)$ by the following equation:

$$\frac{1}{2h}\int_{-h}^{h} zF(x, y, z)dz = F_0(x, y), \tag{46.38}$$

as we can easily check by multiplying both sides of (46.37) by z and integrating from $-h$ to h with respect to z. Thus the biharmonic function $F_0(x, y)$ is the mean value of zF measured over the thickness.

Once we have obtained F and Φ_k $(k = 0, \ldots, 3)$ we can find the displacements from formulae (46.33):

$$\begin{aligned}2Gu &= \frac{3}{2}\frac{z}{h^3}\left[-\frac{\partial F_0}{\partial x} + \left(\frac{1}{2} + \frac{1}{\alpha}\right)\left(\frac{z^2}{3} - \frac{h^3}{5}\right)\frac{\partial}{\partial x}\nabla_1^2 F_0\right],\\ 2Gv &= \frac{3}{2}\frac{z}{h^3}\left[-\frac{\partial F_0}{\partial y} + \left(\frac{1}{2} + \frac{1}{\alpha}\right)\left(\frac{z^2}{3} - \frac{h^3}{5}\right)\frac{\partial}{\partial y}\nabla_1^2 F_0\right],\\ 2Gw &= \frac{3}{2}\frac{1}{h^3}\left\{F_0 + \left[z^2\left(\frac{1}{\alpha} - \frac{1}{2}\right) + \frac{h^2}{5}\left(\frac{1}{2} - \frac{9}{\alpha}\right)\right]\nabla_1^2 F_0\right\}.\end{aligned} \tag{46.39}$$

The stress components can be derived from (46.39) through the displacement–strain relations and the stress–strain relations. Their general form is (Neuber 1985, 5.8)

$$\sigma_x = \left(1 - \frac{\alpha}{2}\right)\nabla^2 F - \frac{\partial^2 F}{\partial x^2} + 2\alpha \frac{\partial \Phi_1}{\partial x}, \quad \ldots,$$
$$\sigma_{yz} = -\frac{\partial^2 F}{\partial y\, \partial z} + \alpha\left(\frac{\partial \Phi_2}{\partial z} + \frac{\partial \Phi_3}{\partial y}\right), \quad \ldots, \tag{46.40}$$

where the dots denote the other stress components, which can be obtained by permutating the indices. If we take F as given by (46.37) and Φ_k $(k = 1, 2, 3)$ as given by (46.36) in the formulae for stresses, we obtain

$$\begin{aligned}
\sigma_x &= \frac{\partial^2 F}{\partial y^2} - \frac{3z}{\alpha h^3}\nabla_1^2 F_0, \\
\sigma_y &= \frac{\partial^2 F}{\partial x^2} - \frac{3z}{\alpha h^3}\nabla_1^2 F_0, \\
\sigma_z &= 0, \\
\sigma_{xy} &= -\frac{\partial^2 F}{\partial x\, \partial y}, \\
\sigma_{zx} &= \frac{3}{2\alpha h^3}(z^2 - h^2)\frac{\partial}{\partial x}\nabla_1^2 F_0, \\
\sigma_{zy} &= \frac{3}{2\alpha h^3}(z^2 - h^2)\frac{\partial}{\partial y}\nabla_1^2 F_0.
\end{aligned} \tag{46.41}$$

If we compare these expressions with (46.3) and (46.10), we can see that we have recovered Michell's solution for the case of a plane stress, provided that we redefine the functions χ' and Θ_1.

The interest in the result just obtained does not rest only on the confirmation that Michell's solution may be generated by a suitable combination of general solutions of the equations of elasticity. It also provides a way of partially extending Michell's solution so as to satisfy the edge conditions that have been neglected so far. As states of displacements and stresses like (46.39) and (46.41) satisfy only the boundary conditions on the faces $z = \pm h$, the boundary conditions on the edge may be satisfied by integrating (46.41) over the thickness, and also requiring the stress resultants and stress couples on the edge line (calculated by using formulae such as (45.26), in particular the quantities G and $N - (\partial H/\partial s)$) to have given values. The solution is exact if the tractions applied at the edge are distributed in accordance with (46.41); however, if they are distributed otherwise, while continuing to have the same resultant stresses and stress couples, the solution represents the elastic state in the plate with sufficient approximation at all points that are not close to the edge. The same argument applies when, instead of tractions, displacements are prescribed on the lateral surface. In this case, we require w and $\partial w/\partial \nu$, evaluated at the edge line, to have prescribed values. It is clear that, in general, satisfying these boundary conditions does not imply that the displacements of the lateral surface of the plate are equal to those prescribed from the exterior.

Insofar as the plate is thin, satisfying the approximate boundary conditions leads to a satisfactory approximation for the state of stress and strain at all points that are sufficiently far from the edge. However, when the plate is thick, the solution is strongly influenced by the detailed distribution of exterior tractions or displacement on the lateral surface. A way of partially overcoming this difficulty is offered by the following extension of solution (46.36) (Neuber 1985, 5.8). We take

$$\Phi_0 = \frac{3}{2}\frac{z}{h^3}\left[\left(1-\frac{1}{\alpha}\right)\left(F_0 - \frac{z^2}{6}\nabla_1^2 F_0\right) + \left(\frac{1}{10}+\frac{1}{10\alpha}+\frac{4}{5\alpha^2}\right)h^2\nabla_1^2 F_0\right]$$
$$+\frac{1}{2\alpha}\left(-x\frac{\partial H}{\partial y}+y\frac{\partial H}{\partial x}\right)\sin\left(\frac{\pi z}{2h}\right),$$
$$\Phi_1 = \frac{1}{2\alpha}\frac{\partial H}{\partial y}\sin\left(\frac{\pi z}{2h}\right),\quad \Phi_2 = -\frac{1}{2\alpha}\frac{\partial H}{\partial x}\sin\left(\frac{\pi z}{2h}\right), \tag{46.42}$$
$$\Phi_3 = \frac{3}{2}\frac{1}{\alpha h^3}\left[F_0 - \frac{z^2}{2}\nabla_1^2 F_0 + \left(\frac{1}{10}-\frac{4}{5\alpha}\right)h^2\nabla_1^2 F_0\right],$$
$$F = \frac{3}{2}\frac{z}{h^3}\left[F_0 + \left(\frac{1}{2}+\frac{1}{\alpha}\right)\left(\frac{h^2}{5}-\frac{z^2}{3}\right)\nabla_1^2 F_0\right],$$

where F_0 is the same function introduced previously, and $H(x, y)$ is a function satisfying the differential equation

$$\nabla_1^2 H - \left(\frac{\pi}{2h}\right)^2 H = 0. \tag{46.43}$$

On using (46.33) again we find the following expressions for the displacements

$$2Gu = \frac{3}{2}\frac{z}{h^3}\left[-\frac{\partial F_0}{\partial x} + \left(\frac{1}{2}+\frac{1}{\alpha}\right)\left(\frac{z^2}{3}-\frac{h^3}{5}\right)\frac{\partial}{\partial x}\nabla_1^2 F_0\right] + \frac{\partial H}{\partial y}\sin\left(\frac{\pi z}{2h}\right),$$
$$2Gv = \frac{3}{2}\frac{z}{h^3}\left[-\frac{\partial F_0}{\partial y} + \left(\frac{1}{2}+\frac{1}{\alpha}\right)\left(\frac{z^2}{3}-\frac{h^3}{5}\right)\frac{\partial}{\partial x}\nabla_1^2 F_0\right] + \frac{\partial H}{\partial x}\sin\left(\frac{\pi z}{2h}\right),$$
$$2Gw = \frac{3}{2}\frac{1}{h^3}\left\{F_0 + \left[z^2\left(\frac{1}{\alpha}-\frac{1}{2}\right)+\frac{h^2}{5}\left(\frac{1}{2}-\frac{9}{\alpha}\right)\right]\nabla_1^2 F_0\right\}. \tag{46.44}$$

The stresses, derived by applying (46.40), have the forms

$$\sigma_x = \frac{\partial^2 F}{\partial y^2} - \frac{3z}{\alpha h^3}\nabla_1^2 F_0 + \frac{\partial^2 H}{\partial x\,\partial y}\sin\left(\frac{\pi z}{2h}\right),$$
$$\sigma_y = \frac{\partial^2 F}{\partial x^2} - \frac{3z}{\alpha h^3}\nabla_1^2 F_0 - \frac{\partial^2 H}{\partial x\,\partial y}\sin\left(\frac{\pi z}{2h}\right),$$
$$\sigma_z = 0, \tag{46.45}$$
$$\sigma_{xy} = -\frac{\partial^2 F}{\partial x\,\partial y} + \frac{1}{2}\left(\frac{\partial^2 H}{\partial y^2}-\frac{\partial^2 H}{\partial x^2}\right)\sin\left(\frac{\pi z}{2h}\right),$$
$$\sigma_{zx} = \frac{3}{2\alpha h^3}(z^2-h^2)\frac{\partial}{\partial x}\nabla_1^2 F_0 + \frac{\pi}{4h}\frac{\partial H}{\partial y}\cos\left(\frac{\pi z}{2h}\right),$$
$$\sigma_{zy} = \frac{3}{2\alpha h^3}(z^2-h^2)\frac{\partial}{\partial y}\nabla_1^2 F_0 - \frac{\pi}{4h}\frac{\partial H}{\partial x}\cos\left(\frac{\pi z}{2h}\right).$$

This state of stress satisfies the boundary conditions on the faces $z = \pm h$ in a manner similar to (46.41). It contains, however, the additional function $H(x, y)$, which is arbitrary, except for the constraint of being a solution of (46.43). We can thus exploit the freedom in the choice of $H(x, y)$ for the purpose of constructing a state of stress or displacement that fits the data on the lateral surface of the plate. This can be done with a higher degree of accuracy than by simply requiring the stress resultants and stress couples to balance the stress resultants and stress couples of the external tractions, or the displacements to conform to the prescribed displacements only along the edge line but not over the entire lateral surface. Of course the problem remains one of finding the correct boundary conditions that must be satisfied by an enlarged solution such as (46.45), since the choice of these conditions is not unique.

Solution (46.45) is not only interesting in that it provides an extension of Michell's theory of thick plates, but also because it justifies the subsequent attempts to refine the classical Kirchhoff's theory of thin plates by accounting for the shear strains that are neglected in this theory. In 1945, Reissner proposed a correction to Kirchhoff's theory, which allowed it to include the influence of shear strains in the deflection of the middle plane of the plate. As can be seen from the special displacement field (45.14), the u and v displacement components are related by the equations

$$u = -z\frac{\partial w}{\partial x}, \qquad v = -z\frac{\partial w}{\partial y}. \tag{46.46}$$

These relations, however, also hold in general, provided that, in the constitutive equations

$$\frac{\partial u}{\partial z} + \frac{\partial w}{\partial x} = \frac{1}{G}\sigma_{zx}, \qquad \frac{\partial v}{\partial z} + \frac{\partial w}{\partial y} = \frac{1}{G}\sigma_{zx}, \tag{46.47}$$

and that we assume $\partial w/\partial x$ and $\partial w/\partial y$ to be practically independent of z and the terms $(1/G)\sigma_{zx}$ and $(1/G)\sigma_{zy}$ are negligible. Let us instead retain the terms on the right-hand sides, and assume σ_{zx} and σ_{zy} to be quadratic functions of z, as in Kirchhoff's theory. Then (46.47) become

$$\frac{\partial u}{\partial z} + \frac{\partial w}{\partial x} = \frac{3}{4h^3}\frac{N_1}{G}(h^2 - z^2), \qquad \frac{\partial v}{\partial z} + \frac{\partial w}{\partial y} = \frac{3}{4h^3}\frac{N_2}{G}(h^2 - z^2). \tag{46.48}$$

As, by assumption, $\partial w/\partial x$ and $\partial w/\partial y$ are independent of z, integration of (46.47) with respect to z and use of the relation $G = E/[2(1+\sigma)]$ yields

$$\begin{aligned} u &= -z\frac{\partial w}{\partial x} + \frac{(1+\sigma)}{2E}\left(\frac{3z}{h} - \frac{z^3}{h^3}\right)N_1, \\ v &= -z\frac{\partial w}{\partial y} + \frac{(1+\sigma)}{2E}\left(\frac{3z}{h} - \frac{z^3}{h^3}\right)N_2. \end{aligned} \tag{46.49}$$

From the displacement components u and v, we find the strains

$$\varepsilon_x = -z\frac{\partial^2 w}{\partial x^2} + \frac{(1+\sigma)}{2E}\left(\frac{3z}{h} - \frac{z^3}{h^3}\right)\frac{\partial N_1}{\partial x},$$

$$\varepsilon_y = -z\frac{\partial^2 w}{\partial y^2} + \frac{(1+\sigma)}{2E}\left(\frac{3z}{h} - \frac{z^3}{h^3}\right)\frac{\partial N_2}{\partial y}, \tag{46.50}$$

$$\varepsilon_{xy} = -z\frac{\partial^2 w}{\partial x\,\partial y} + \frac{(1+\sigma)}{4E}\left(\frac{3z}{h} - \frac{z^3}{h^3}\right)\left(\frac{\partial N_1}{\partial y} + \frac{\partial N_2}{\partial x}\right).$$

On replacing the strains in the constitutive relations

$$E\varepsilon_x = \sigma_x - \sigma\sigma_y, \quad E\varepsilon_y = \sigma_y - \sigma\sigma_x, \quad E\varepsilon_{xy} = (1+\sigma)\sigma_{xy}, \tag{46.51}$$

and solving them with respect to σ_x, σ_y, and σ_{xy}, we obtain

$$(1-\sigma^2)\sigma_x = -Ez\left(\frac{\partial^2 w}{\partial x^2} + \sigma\frac{\partial^2 w}{\partial y^2}\right) + (1+\sigma)\left(\frac{3z}{h} - \frac{z^3}{h^3}\right)\left(\frac{\partial N_1}{\partial x} + \sigma\frac{\partial N_2}{\partial y}\right),$$

$$(1-\sigma^2)\sigma_y = -Ez\left(\frac{\partial^2 w}{\partial y^2} + \sigma\frac{\partial^2 w}{\partial x^2}\right) + (1+\sigma)\left(\frac{3z}{h} - \frac{z^3}{h^3}\right)\left(\frac{\partial N_2}{\partial y} + \sigma\frac{\partial N_1}{\partial x}\right), \tag{46.52}$$

$$(1-\sigma)\sigma_{xy} = -Ez\frac{\partial^2 w}{\partial x\,\partial y} + \frac{(1+\sigma)}{4}\left(\frac{3z}{h} - \frac{z^3}{h^3}\right)\left(\frac{\partial N_1}{\partial y} + \frac{\partial N_2}{\partial x}\right).$$

On inserting these expressions in (45.1e), (45.2d), and (45.2e), and using the equilibrium equation

$$\frac{\partial N_1}{\partial x} + \frac{\partial N_2}{\partial y} = 0, \tag{46.53}$$

we find that the couples G_1, G_2, and H_2 have the following forms:

$$G_1 = -D\left(\frac{\partial^2 w}{\partial x^2} + \sigma\frac{\partial^2 w}{\partial y^2}\right) + \frac{4}{5}h^2\frac{\partial N_1}{\partial x},$$

$$G_2 = -D\left(\frac{\partial^2 w}{\partial y^2} + \sigma\frac{\partial^2 w}{\partial x^2}\right) + \frac{4}{5}h^2\frac{\partial N_2}{\partial y}, \tag{46.54}$$

$$H_2 = -H_1 = -D(1-\sigma)\frac{\partial^2 w}{\partial x\,\partial y} + \frac{2}{5}h^2\left(\frac{\partial N_1}{\partial y} + \frac{\partial N_2}{\partial x}\right).$$

Then the equilibrium equations (45.24), again on recalling (46.53), become

$$D\frac{\partial}{\partial y}(\nabla_1^2 w) - \frac{2}{5}h^2\nabla_1^2 N_2 + N_2 = 0,$$

$$D\frac{\partial}{\partial x}(\nabla_1^2 w) - \frac{2}{5}h^2\nabla_1^2 N_1 + N_1 = 0. \tag{46.55}$$

In order to solve (46.53) and (46.55) we write

$$N_1 = -D\frac{\partial}{\partial x}(\nabla_1^2 w) + \frac{\partial H}{\partial y}, \tag{46.56a}$$

$$N_2 = -D\frac{\partial}{\partial y}(\nabla_1^2 w) - \frac{\partial H}{\partial x}, \tag{46.56b}$$

where $H(x, y)$ is a function to be determined by substituting these relations in (46.53), from which we also obtain

$$\nabla_1^4 w = 0. \tag{46.57}$$

On differentiating (46.56a) with respect to y and (46.56b) with respect to x, and subtracting, we obtain the following equation for H:

$$\nabla_1^2 H = \frac{\partial N_1}{\partial y} - \frac{\partial N_2}{\partial x} = \frac{2}{5}h^2\left[\frac{\partial}{\partial y}(\nabla_1^2 N_1) + \frac{\partial}{\partial x}(\nabla_1^2 N_2)\right] = \frac{2}{5}h^2\nabla_1^4 H, \tag{46.58}$$

which is satisfied if H is a solution of the equation

$$\nabla_1^2 H - \frac{5}{2h^2}H = 0. \tag{46.59}$$

In practice, it is more convenient to introduce, instead of w, the function

$$F_0 = D[(1-\sigma)w + \tfrac{4}{5}h^3\nabla_1^2 w], \tag{46.60a}$$

$$\nabla_1^4 F_0 = 0, \tag{46.60b}$$

so that the stress characteristics can be written in the simpler form

$$\begin{aligned}
G_1 &= \frac{\partial^2 F_0}{\partial y^2} - \frac{1}{1-\sigma}\nabla_1^2 F_0 + \frac{4}{5}h^2\frac{\partial^2 H}{\partial x\,\partial y},\\
G_2 &= \frac{\partial^2 F_0}{\partial x^2} - \frac{1}{1-\sigma}\nabla_1^2 F_0 - \frac{4}{5}h^2\frac{\partial^2 H}{\partial x\,\partial y},\\
H_2 = -H_1 &= -\frac{\partial^2 F_0}{\partial x\,\partial y} + \frac{2}{5}h^2\left(\frac{\partial^2 H}{\partial y^2} - \frac{\partial^2 H}{\partial x^2}\right),\\
N_1 &= -\frac{1}{1-\sigma}\frac{\partial}{\partial x}\nabla_1^2 F_0 + \frac{\partial H}{\partial y},\\
N_2 &= -\frac{1}{1-\sigma}\frac{\partial}{\partial y}\nabla_1^2 F_0 - \frac{\partial H}{\partial x}.
\end{aligned} \tag{46.61}$$

The equations just written were obtained by Reissner by applying a variational principle. In Reissner's theory, we have to solve a biharmonic equation similar to (46.60b) and an equation such as (46.59), which is called Helmholtz's equation. In contrast to the case in Kirchhoff's theory, which involves two boundary conditions, in Reissner's method there are three boundary conditions associated with the problem. When tractions are prescribed on the boundary, we can prescribe N, H, and G separately; alternatively, when displacements are given, we can prescribe w, $\partial w/\partial \nu$, and $\frac{\partial}{\partial z}(u\cos\theta + v\sin\theta)$ on the edge line.

Let us now return to solution (46.45), and calculate the stress characteristics by applying definitions (45.1) and (45.2). The result is

$$
\begin{aligned}
G_1 &= \frac{\partial^2 F_0}{\partial y^2} - \frac{1}{1-\sigma} \nabla_1^2 F_0 + \frac{8}{\pi^2} h^2 \frac{\partial^2 H}{\partial x \, \partial y}, \\
G_2 &= \frac{\partial^2 F_0}{\partial x^2} - \frac{1}{1-\sigma} \nabla_1^2 F_0 - \frac{8}{\pi^2} h^2 \frac{\partial^2 H}{\partial x \, \partial y}, \\
H_2 = -H_1 &= -\frac{\partial^2 F_0}{\partial x \, \partial y} + \frac{4}{\pi^2} h^2 \left(\frac{\partial^2 H}{\partial y^2} - \frac{\partial^2 H}{\partial x^2} \right), \qquad (46.62) \\
N_1 &= -\frac{1}{1-\sigma} \frac{\partial}{\partial x} \nabla_1^2 F_0 + \frac{\partial H}{\partial y}, \\
N_2 &= -\frac{1}{1-\sigma} \frac{\partial}{\partial y} \nabla_1^2 F_0 - \frac{\partial H}{\partial x}.
\end{aligned}
$$

A comparison with (46.61) shows that the solution for a thick plate, obtained from the combination of Boussinesq–Papkovič–Neuber solutions, justifies and improves Reissner's theory. The approximations made in this theory originate from the use of the factor $\frac{4}{5}$ in (46.61) instead of the more precise factor $8/\pi^2$ in (46.62). However, this difference aside, the problem is the same because it consists of finding a biharmonic function $F_0(x, y)$ and a function $H(x, y)$ that satisfies Helmholtz's equation with three boundary conditions in the edge line, using either the forces or the displacements, just as in Reissner's theory. However, solution (46.42) is open to further generalizations, because we may add other functions, such as H, to the definition of Φ_k $(k = 0, \ldots, 3)$, which enable us to satisfy the boundary conditions to a higher degree of approximation.

The effect of shear in the deflection of a moderately thick plate may be accounted for by postulating the forms of displacements inside the plate in other ways. For example, we can take (Mindlin 1951)

$$
\begin{aligned}
U(x, y, z) &= u(x, y) + z\left[\psi_x + \kappa^2 \left(\frac{z}{h}\right)^2 \left(\psi_x + \frac{\partial w}{\partial x}\right)\right], \\
V(x, y, z) &= v(x, y) + z\left[\psi_y + \kappa^2 \left(\frac{z}{h}\right)^2 \left(\psi_y + \frac{\partial w}{\partial y}\right)\right], \qquad (46.63) \\
W(x, y, z) &= w(x, y),
\end{aligned}
$$

where u, v, and w are the displacements of a point in the middle plane, ψ_x and ψ_y are the rotations of the normal about the y and x axis, respectively, and κ^2 is a positive number called the *shear factor*. From (46.63) we calculate the stresses, and hence the stress characteristics, by using (45.1) and (45.2). The expressions obtained are

$$
\begin{aligned}
G_1 &= D\left(\frac{\partial \psi_x}{\partial x} + \sigma \frac{\partial \psi_y}{\partial y}\right), \quad G_2 = D\left(\frac{\partial \psi_y}{\partial y} + \sigma \frac{\partial \psi_x}{\partial x}\right), \\
H_2 = -H_1 &= \frac{(1-\sigma)}{2} D\left(\frac{\partial \psi_x}{\partial y} + \frac{\partial \psi_y}{\partial x}\right), \qquad (46.64) \\
N_1 &= \kappa^2 G h\left(\psi_x + \frac{\partial w}{\partial x}\right), \\
N_2 &= \kappa^2 G h\left(\psi_y + \frac{\partial w}{\partial y}\right).
\end{aligned}
$$

On substituting (46.64) in the equilibrium equations we get

$$
\begin{aligned}
&\kappa^2 Gh(\nabla^2 w + \varphi) + p(x, y) = 0, \\
&\frac{D}{2}\left[(1-\sigma)\nabla^2\psi_x + (1+\sigma)\frac{\partial\varphi}{\partial x}\right] - \kappa^2 Gh\left(\psi_x + \frac{\partial w}{\partial x}\right) = 0, \\
&\frac{D}{2}\left[(1-\sigma)\nabla^2\psi_y + (1+\sigma)\frac{\partial\varphi}{\partial y}\right] - \kappa^2 Gh\left(\psi_y + \frac{\partial w}{\partial y}\right) = 0,
\end{aligned}
\tag{46.65}
$$

where $\varphi = (\partial\psi_x/\partial x) + (\partial\psi_y/\partial y)$.

As in previous, more refined, theories, in Mindlin's theory also, the number of boundary conditions naturally associated with the system (46.65) is three. For instance, in a simply supported rectangular plate with edges parallel to the coordinate axes, the conditions are:

$$
\begin{aligned}
w = \psi_y = G_1 = 0, &\quad \text{on the edges } x = \text{constant}; \\
w = \psi_x = G_2 = 0, &\quad \text{on the edges } y = \text{constant}.
\end{aligned}
\tag{46.66}
$$

The first and the last conditions in each case given in equation (46.66) are the classical boundary conditions. The second equality, in each case, states that in a simply supported plate, the filaments perpendicular to the edge do not rotate as a result of the deformation.

In order to illustrate the difference between Mindlin's solution and the classical solution, found by Navier, let us consider the following example (Levinson and Cooke 1983). Let the middle surface of the undeformed state occupy the rectangle $0 < x < a$, $0 < y < b$ of the xy plane, and let $p(x, y)$ have the form

$$
p(x, y) = p_{mn} \sin\frac{m\pi x}{a} \sin\frac{n\pi y}{b}, \tag{46.67}
$$

where m and n are integers and p_{mn} is a given constant. Then, we look at a solution of the type

$$
w = A_{mn} \sin\frac{m\pi x}{a} \sin\frac{n\pi y}{b}, \tag{46.68a}
$$

$$
\psi_x = B_{mn} \cos\frac{m\pi x}{a} \sin\frac{n\pi y}{b}, \tag{46.68b}
$$

$$
\psi_y = C_{mn} \sin\frac{m\pi x}{a} \cos\frac{n\pi y}{b}, \tag{46.68c}
$$

where A_{mn}, B_{mn}, and C_{mn} are constants to be determined. These functions satisfy the boundary conditions (46.66) along all the sides of the rectangle, and so the constants A_{mn}, B_{mn}, and C_{mn} must be determined by requiring that the field equations are also satisfied. If we substitute (46.68a) in (45.25), we obtain Navier's solution

$$
A_{mn} = \frac{p_{mn}}{D\pi^4\left[\left(\frac{m}{a}\right)^2 + \left(\frac{n}{b}\right)^2\right]^2}. \tag{46.69}
$$

If, on the other hand, we substitute (46.68) in (46.65), we obtain

$$A_{mn} = \left\{1 + \frac{D\pi^2}{\kappa^2 Gh}\left[\left(\frac{m}{a}\right)^2 + \left(\frac{n}{b}\right)^2\right]\right\} \frac{p_{mn}}{D\pi^4\left[\left(\frac{m}{a}\right)^2 + \left(\frac{n}{b}\right)^2\right]^2},$$

$$B_{mn} = -\frac{mp_{mn}}{aD\pi^3\left[\left(\frac{m}{a}\right)^2 + \left(\frac{n}{b}\right)^2\right]^2}, \tag{46.70}$$

$$C_{mn} = -\frac{np_{mn}}{bD\pi^3\left[\left(\frac{m}{a}\right)^2 + \left(\frac{n}{b}\right)^2\right]^2}.$$

A comparison between (46.69) and (46.70) shows that the correction in the deflection w generated by adopting Mindlin's theory instead of the classical theory is equivalent to an increase in the deformability of the plate.

A further generalization of (46.46), which can be used to describe the influence of more complex load conditions on the faces of a moderately thick plate, is the following (Cooke and Levinson 1983). Let us take the upper plane of the plate to lie in the xy plane and the z axis to be directed positively downwards. The displacements in the directions of the coordinate axes are assumed to have the expressions

$$u = -g(z)\frac{\partial}{\partial x}W(x, y), \quad v = -g(z)\frac{\partial}{\partial y}W(x, y), \quad w = f(z)W(x, y), \tag{46.71}$$

where $W(x, y)$ is the deflection of the upper surface, so that we have $f(0) = 1$, and $f(z)$ and $g(z)$ are functions determining the variations in the displacements through the thickness of the plate. In order to find the field equations, we derive the stresses and strains from (46.71), and write the strain energy stored in the plate

$$\mathscr{W} = \frac{1}{2}\iiint\limits_V [(\lambda + 2\mu)(f'W)^2 + \mu(f - g')^2(W_x^2 + W_y^2) + (\lambda + 2\mu)g^2(W_{xx} + W_{yy})^2$$
$$- 2\lambda f'gW(W_{xx} + W_{yy}) - 4\mu g^2(W_{xx}W_{yy} - W_{xy}^2)]dV, \tag{46.72}$$

where V is the volume of the plate, the prime denotes differentiation with respect to z, and subscripts x and y denote differentiation with respect to x and y. If h is the thickness of the plate and A its middle surface, the volume V is given by the product $A \times [0, h]$. On integrating the right-hand side of (46.72) with respect to z, we find

$$\mathscr{W} = \frac{1}{2}\iint\limits_A [k_1 W^2 + k_2(W_x^2 + W_y^2) + k_3(W_{xx} + W_{yy})^2 + k_4 W(W_{xx} + W_{yy})$$
$$+ k_5(W_{xx}W_{yy} - W_{xy}^2)]dA, \tag{46.73}$$

where it is understood that the surface integration is taken over the upper surface of the plate. The coefficients appearing in (46.72) are given by the following formulae:

$$k_1 = (\lambda + 2\mu) \int_0^h (f')^2\, dz, \quad k_2 = \mu \int_0^h (f - g')^2\, dz,$$
$$k_3 = (\lambda + 2\mu) \int_0^h g^2\, dz, \quad k_4 = -2\lambda \int_0^h f'g\, dz, \quad k_5 = -4\mu \int_0^h g^2\, dz. \tag{46.74}$$

The work done by the surface tractions can be written as

$$\mathscr{L} = \iint_A W(x, y)[p_0(x, y) - p_h(x, y)f(h)]dA + \text{edge terms}, \tag{46.75}$$

where p_0 and p_h are the pressures acting on the surfaces $z = 0$ and $z = h$, respectively. The edge boundary terms do not involve the pointwise specification of the tractions on the cylindrical mantle of the plate, but only the contracted conditions of Kirchhoff's theory.

The balance equations are the Euler equations of the variational problem

$$\delta(\mathscr{W} - \mathscr{L}) = 0, \tag{46.76}$$

the detailed forms of which are

$$k_3 \nabla^4 W + (k_4 - k_2)\nabla^2 W + k_1 W = p_0 - f(h)p_h, \tag{46.77}$$
$$I_1 f'' + (I_3 - I_2)g' - I_3 f = 0, \tag{46.78}$$
$$I_3 g'' - (I_3 - I_2)f' - I_4 g = 0, \tag{46.79}$$

where we have put

$$I_1 = (\lambda + 2\mu) \iint_A W^2\, dA, \quad I_2 = \lambda \iint_A W(W_{xx} + W_{yy})dA,$$
$$I_3 = \mu \iint_A (W_x^2 + W_y^2)dA,$$
$$I_4 = \iint_A [\lambda(W_{xx} + W_{yy})^2 + 2\mu(W_{xx}^2 + 2W_{xy}^2 + W_{yy}^2)]dA. \tag{46.80}$$

The quantities I_1, I_3, and I_4 are obviously positive definite, while nothing can be said of I_2, because λ may be either positive or negative according to the nature of the material.

The boundary conditions associated with (46.77)–(46.79) may assume different alternative forms, the most important of which are the following. At $z = 0$ and $z = h$ we must prescribe $(I_1 f' - I_2 g)$, or f, and $(I_3 g' - I_3 f)$, or g. If, as it has been assumed in the present problem, the shear stresses vanish on both the faces of the plate, then the term $(I_3 g' - I_3 f)$ must vanish at $z = 0$ and $z = h$. In addition, we have already required that $f(0) = 1$, because the upper face has been taken as the reference surface, while at $z = h$, if the plate is not geometrically constrained, we have

$$I_1 f' - I_2 g = -\iint_A p_h W\, dA. \tag{46.81}$$

As for the boundary conditions for $W(x, y)$, at the edge line of the plate we must prescribe

$$k_3\nabla^2 W + \frac{1}{2}k_4 W + \frac{1}{2}k_5 W_{ss}, \quad \text{or} \quad \tilde{\omega}_s, \tag{46.82}$$

and

$$\frac{\partial}{\partial n}\left\{\frac{1}{2}k_5 W_{ss} - \left[k_3\nabla^2 W + \left(\frac{1}{2}k_4 - k_2\right)W\right]\right\}, \quad \text{or} \quad \tilde{\omega}, \tag{46.83}$$

where n and s are normal and tangential coordinates, relative to the plate's perimeter, in the xy plane, and $\tilde{\omega}_s$ and $\tilde{\omega}$ are the mean rotations about the s axis and the mean transverse displacement, respectively. That is to say

$$\tilde{\omega}_s = -\frac{1}{2h}W_n \int_0^h (f + g')dz, \quad \tilde{\omega} = \frac{1}{h}W \int_0^h f\, dz. \tag{46.84}$$

If the edge line has a corner, that is, a point of discontinuity of the tangent, we have the condition that either $k_5\tilde{\omega}_{ns}$ or $\tilde{\omega}$ is specified. The natural boundary and corner conditions (46.82)–(46.84) may be interpreted as a prescription for the bending moment, with the conditions for the contracted shear force and the concentrated reaction force being analogous to those customarily imposed in classical plate theory. This means that, assuming a displacement field like that of (46.71) does not introduce additional degrees of freedom in the kinematics of the plate motion, beyond those already included in Kirchhoff's theory. There is, however, an important difference between the two theories. While Kirchhoff's theory is perfectly linear, system (46.77)–(46.79) is essentially nonlinear because the coefficients k_i $(i = 1, \ldots, 5)$ and I_i $(i = 1, \ldots, 4)$ are functions of the unknowns. A rigorous qualitative analysis of the well-posedness of the problem is still lacking.

47. Influence of Varying Thickness

The classical theory of thin plates and its extensions to thick plates rests heavily on the assumption that the thickness $2h$ is uniform, because only in this case can the customary hypotheses concerning the nature of the displacements and the form of certain stress components be adopted without generating incompatibilities. For instance, if the thickness were variable, the limiting process leading to the definition of the characteristics $T, \ldots, G$, discussed in Section 45, is no longer valid, because the limits might be well defined at some points but not at others. On the other hand, structural engineers have always had to recourse to plates, not only of varying thickness, but often of exceptionally variable thickness, because they can be stronger per unit weight than uniform plates. The most striking example of this is the widely used plate with densely spaced stiffeners, which almost always provides a more efficient use of material than a uniform plate. It may appear that classical theories, in particular Kirchhoff's theory, are constitutively unable to describe the bending of a thin plate with rapidly varying thickness. However, in effect, this is not so. It is possible to obtain a fourth-order equation similar to (45.25) for the transverse displacement, through an asymptotic analysis based on the three-dimensional theory of elasticity (Kohn and Vogelius 1984, 1985). The coefficients of this equation represent the constitutive law relating the bending moments to the curvatures of the central plane.

In order to illustrate the procedure, let us define the geometrical domain occupied by a plate as Ω in the x_1x_2 plane, which represents the middle plane. A function $h(x_1, x_2; \eta_1, \eta_2) \geqslant 0$, defined for $(x_1, x_2) \in \bar{\Omega}$ and $(\eta_1, \eta_2) \in \mathbb{R}^2$ represents the thickness and a real parameter a $(0 < a < \infty)$, determines the length scale of the variation in thickness. The three-dimensional undeformed configuration of the plate is

$$R(\varepsilon) = \{(x_1, x_2) \in \Omega,\ |x_3| < \varepsilon h\,(x_1, x_2; (x_1, x_2)/\varepsilon^a)\}, \tag{47.1}$$

where ε is a small parameter. We allow $h(x_1, x_2; \eta_1, \eta_2)$ to vanish for some values of $(x_1, x_2; \eta_1, \eta_2)$, which implies that the plate may have holes. We also allow h to be discontinuous at some points. However, the set for which we have $h > 0$ must be connected, with h bounded away from zero (*i.e.* $h \geqslant c > 0$) on that set for which it does not vanish. The pair of parameters (η_1, η_2) is introduced to characterize the periodic structure of h, introduced by assuming h to be a periodic function, with a period of one, of η_1 and η_2.

If $\partial_+ R(\varepsilon)$ and $\partial_- R(\varepsilon)$ are the upper and lower faces of the plate, we suppose that these faces are loaded by forces $\varepsilon^3(0, 0, f_\pm)$ per unit area of the middle plane, where $f_\pm = f_\pm(x_1, x_2; \eta_1, \eta_2)$ are defined for $(x_1, x_2) \in \Omega$, $(\eta_1, \eta_2) \in \mathbb{R}^2$, and are periodic in (η_1, η_2), with $f_+ = f_- = 0$ whenever $h = 0$. The parameter ε^3 multiplying the load has been chosen so that the vertical displacement of the plate remains bounded for $\varepsilon \to 0$. We assume that $f_\pm$ are periodic functions of η_1 and η_2 with a period of one, and we define the averaged function as

$$\begin{aligned} F(x_1, x_2) &= \frac{1}{p_1 p_2}\int_0^{p_1}\int_0^{p_2} (f_+ + f_-)\,d\eta_1\,d\eta_2 \\ &= \mathcal{M}(f_+ + f_-) \quad (p_1 = p_2 = 1). \end{aligned} \tag{47.2}$$

In the interior of $R(\varepsilon)$ the stress components must satisfy the homogeneous equations of equilibrium

$$\sigma_{ij,j} = 0, \quad i, j = 1, 2, 3, \tag{47.3}$$

whereas the boundary conditions on the faces are

$$\sigma_{ij} n_j = \begin{cases} 0, & \text{for} \quad i = 1, 2, \\ \varepsilon^3 f_\pm |n_3|, & i = 3 \quad \text{on} \quad \partial_\pm R(\varepsilon), \end{cases} \tag{47.4}$$

where n_j, the outward unit normal, is given by

$$|n_3|^{-1} n_j = \left(-\varepsilon\frac{\partial h}{\partial x_1} - \varepsilon^{1-a}\frac{\partial h}{\partial \eta_1},\ -\varepsilon\frac{\partial h}{\partial x_2} - \varepsilon^{1-a}\frac{\partial h}{\partial \eta_2},\ \pm 1\right), \tag{47.5}$$

whenever h is differentiable.

The stresses are related to the strains $\varepsilon_{ij} = \frac{1}{2}(u_{i,j} + u_{j,i})$ by the equations of linear elasticity. Let us assume, for simplicity, that the material is homogeneous and isotropic, with a Young's modulus E and a Poisson ratio σ, then the constitutive equations have the form

$$\sigma_{ij} = B_{ijk\ell}\varepsilon_{k\ell}, \tag{47.6}$$

where

$$B_{iiii} = \frac{E(1-\sigma)}{(1+\sigma)(1-2\sigma)}, \quad B_{iijj} = \frac{E\sigma}{(1+\sigma)(1-2\sigma)}, \quad (1 \neq j),$$
$$B_{ijij} = B_{ijji} = \frac{E}{2(1+\sigma)}, \quad (i \neq j), \tag{47.7}$$

while all other components are zero.

In the classical theory of thin plates the constitutive equations are preferentially written in terms of moments and curvatures rather than stresses and strains. This may be done by introducing the new elastic tensor

$$\tilde{B}_{\alpha\beta\gamma\delta} = B_{\alpha\beta\gamma\delta} - \frac{1}{B_{3333}}(B_{\alpha\beta 33}B_{\gamma\delta 33}), \quad \alpha, \beta, \gamma, \delta = 1, 2, \tag{47.8}$$

the components of which have the expressions

$$\tilde{B}_{1111} = \tilde{B}_{2222} = \frac{E}{(1-\sigma^2)}, \quad B_{1122} = B_{2211} = \frac{E\sigma}{(1-\sigma^2)},$$
$$\tilde{B}_{1212} = \tilde{B}_{1221} = \tilde{B}_{2112} = \tilde{B}_{2121} = \frac{1}{2}\frac{E}{(1+\sigma)}. \tag{47.9}$$

Together with $\tilde{B}_{\alpha\beta\gamma\delta}$, we can define the tensor

$$M_{\alpha\beta\gamma\delta} = \frac{2}{3}h^3\tilde{B}_{\alpha\beta\gamma\delta}, \tag{47.10}$$

in terms of which the relations between the flexural couples and the curvatures read

$$G_1 = -(M_{1111}w_{11} + M_{1122}w_{22}), \quad G_2 = -(M_{2211}w_{11} + M_{2222}w_{22}),$$
$$H_2 = -H_1 = -M_{1212}w_{12}. \tag{47.11}$$

The purpose of the method of averaging is to find the effective equation for the vertical displacement as ε tends to zero, by using a variational approach. We begin by postulating a representation formulae for $\mathbf{u} \equiv (u_1, u_2, u_3)$ of the type

$$\mathbf{u} \sim \mathbf{u}^* = \sum_{k=0}^{N} \varepsilon^{t_k}\mathbf{u}^{(k)}(x_1, x_2;\ (x_1, x_2)/\varepsilon^a;\ x_3/\varepsilon). \tag{47.12}$$

Each $\mathbf{u}^{(k)} = \mathbf{u}^{(k)}(x_1, x_2; \eta_1, \eta_2; \xi)$ is defined for $(x_1, x_2) \in \Omega$, $(\eta_1, \eta_2) \in \mathbb{R}^2$, $|\xi| < h$, and is periodic in η_1 and η_2. The first term in (47.12) has the form

$$\varepsilon^{t_0}\mathbf{u}^{(0)}(x_1, x_2, \eta_1, \eta_2; \xi) = (0, 0, w(x_1, x_2)), \tag{47.13}$$

where w, the limiting vertical displacement, is an as yet undetermined function of (x_1, x_2). The remaining exponents $t_k > 0$ and functions $\mathbf{u}^{(k)}$ are chosen differently depending on which condition, $a < 1$, $a = 1$, or $a > 1$, holds. In each case the result is obtained by substituting the formal expansion (47.12) in (47.3) and (47.4), collecting terms with like powers of ε, and solving successively for the functions $\mathbf{u}^{(k)}$. To obtain the fourth-order effective equation of w, we proceed as follows. We write the strain associated with $\mathbf{u}^*$ in the form

$$\varepsilon_{ij}(\mathbf{u}^*) = \varepsilon X_{ij}^{\alpha\beta}(x_1, x_2; (x_1, x_2)/\varepsilon^a; x_3/\varepsilon)w_{\alpha\beta} + O(\varepsilon), \tag{47.14}$$

where $X_{ij}^{\alpha\beta}(x_1, x_2; \eta_1, \eta_2; \xi)$ are explicit, periodic functions in η_1 and η_2, depending only on the domain $R(\varepsilon)$, and such that we have

$$X_{ij}^{\alpha\beta} = X_{ji}^{\beta\alpha}. \tag{47.15}$$

On substituting (47.14) in the expression for the total energy

$$\mathscr{W} - \mathscr{L} = \frac{1}{2}\iiint\limits_{R(\varepsilon)} B_{ijk\ell}\varepsilon_{ij}(\mathbf{u}^*)\varepsilon_{k\ell}(\mathbf{u}^*)dV - \varepsilon^3 \iint\limits_{\Omega} u_3^*(f_+ + f_-)dA, \tag{47.16}$$

integrating with respect to ξ, averaging in (η_1, η_2), and discarding terms that are of order $o(\varepsilon^3)$, we are led to consider the functional

$$\frac{1}{2}\varepsilon^3 \iint\limits_{\Omega} M_{\alpha\beta\gamma\delta}w_{\alpha\beta}w_{\gamma\delta}\,dA - \varepsilon^3 \iint\limits_{\Omega} wF\,dA, \tag{47.17}$$

where F is given by (47.2) together with

$$M_{\alpha\beta\gamma\delta}(x_1, x_2) = \mathscr{M}\left(\int_{-h}^{h} B_{ijhk}X_{ij}^{\alpha\beta}X_{hk}^{\gamma\delta}\,d\xi\right). \tag{47.18}$$

This function w must minimize (47.17); that is, it solves the equation

$$(M_{\alpha\beta\gamma\delta}w_{\gamma\delta})_{\alpha\beta} = F, \quad \text{in} \quad \Omega, \tag{47.19}$$

with some appropriate boundary conditions, which must be specified in each case. In general, we may require that the boundary is clamped, simply supported, or free, and all these conditions are expressible in terms of w and its partial derivatives up to third order, as we saw in Section 45. The boundary-value problem (47.19) is then well-posed. In order to clarify the method, we sketch briefly how it applies when the thickness of the plate varies slowly, and $h = h(x_1, x_2)$ does not depend on η_1 and η_2. In this case we give (47.12) the following particular form

$$\mathbf{u}^* = \left[-x_3w_1, \; -x_3w_2, \; w + \frac{1}{2}\frac{B_{\alpha\beta33}}{B_{3333}}(x_3)^2w_{\alpha\beta}\right], \tag{47.20}$$

and calculate the strains and isolate the terms of first order in ε. We thus find that (47.14) holds with

$$X_{ij}^{\alpha\beta}(x_1, x_2; \xi) = -\xi\delta_{ij}^{\alpha\beta} + \xi\frac{B_{\alpha\beta33}}{B_{3333}}\delta_{ij}^{33}, \tag{47.21}$$

where we have used the symbol

$$\delta_{k\ell}^{ij} = \begin{cases} 1, & \text{for } (i,j) = (k,\ell) \text{ as ordered pairs,} \\ 0, & \text{otherwise.} \end{cases} \tag{47.22}$$

It follows that we have

$$M_{\alpha\beta\gamma\delta} = \frac{2}{3} h^3(x_1, x_2) \tilde{B}_{\alpha\beta\gamma\delta}, \tag{47.23}$$

with $\tilde{B}_{\alpha\beta\gamma\delta}$ defined in (47.8). Then, recalling (47.9), substituting of (47.23) in the field equation (47.19), leads to the effective equation

$$\left[D\left(w_{\alpha\beta} + \frac{\sigma}{1-\sigma} \delta^{\alpha}_{\beta} \nabla^2 w \right) \right]_{\alpha\beta} = F, \tag{47.24}$$

with

$$D = \frac{2E}{3(1+\sigma)} h^3(x_1, x_2).$$

We thus reach the result that, if the thickness is only slowly varying, the effective equation for w is no longer biharmonic as in the classic Germain–Lagrange equation (45.25).

Let us now consider the case in which h is periodic in (η_1, η_2) with $a < 1$. The mean thickness of the plate is then much smaller than the length scale of the thickness variation. We define, for convenience, the set

$$Q(x_1, x_2) \equiv \{(\eta_1, \eta_2) \in \mathbb{R}^2 : h(x_1, x_2; \eta_1, \eta_2) > 0\}. \tag{47.25}$$

The expression for $\mathbf{u}^*$ now depends on some auxiliary functions $\varphi^{\alpha\beta}(x_1, x_2; \eta_1, \eta_2)$, periodic in (η_1, η_2), with $\varphi^{12} = \varphi^{21}$, which are chosen as to satisfy the partial differential equation

$$\frac{\partial^2}{\partial\eta_\gamma\, \partial\eta_\delta}\left(h^3 \tilde{B}_{\gamma\delta\rho\sigma} \frac{\partial^2 \varphi^{\alpha\beta}}{\partial\eta_\rho\, \partial\eta_\sigma} \right) = -\frac{\partial^2}{\partial\eta_\gamma\, \partial\eta_\delta}(h^3 \tilde{B}_{\gamma\delta\alpha\beta}), \tag{47.26}$$

in $Q(x_1, x_2)$. For $\partial Q(x_1, x) \neq \emptyset$, the boundary conditions of (47.26) are the natural ones corresponding to the minimization problem

$$\min_{\varphi^{\alpha\beta}} \mathscr{M}\left[h^3 \tilde{B}_{\gamma\delta\rho\sigma} \frac{\partial^2 \varphi^{\alpha\beta}}{\partial\eta_\gamma\, \partial\eta_\delta} \frac{\partial^2(\varphi^{\alpha\beta} + \eta_\alpha \eta_\beta)}{\partial\eta_\rho\, \partial\eta_\sigma} \right], \tag{47.27}$$

the minimum being taken among the (η_1, η_2) periodic functions. It is easy to verify that $\varphi^{\alpha\beta}$ exists and is unique in $Q(x_1, x_2)$, up to an additive function of (x_1, x_2), as (x_1, x_2) enters (47.26) and (47.27) only as a parameter.

The formula for the displacement in this case is

$$u^*_\gamma = -x_3 w_\gamma - \varepsilon^a x_3 \frac{\partial}{\partial\eta_\gamma}(\varphi^{\alpha\beta}) w_{\alpha\beta} - \varepsilon^{2a} x_3 (\varphi^{\alpha\beta} w_{\alpha\beta})_\gamma, \tag{47.28a}$$

$$u^*_3 = w + \varepsilon^{2a} \varphi^{\alpha\beta} w_{\alpha\beta} + \frac{1}{2}(x_3^2) \frac{B_{33\gamma\delta}}{B_{3333}} \frac{\partial^2}{\partial\eta_\gamma\, \partial\eta_\delta}\left(\frac{1}{2}\eta_\alpha \eta_\beta + \varphi^{\alpha\beta} \right) w_{\alpha\beta}, \tag{47.28b}$$

where $(\cdot)_\gamma$ in (47.28a) denotes differentiation which respect to x_γ. The right-hand sides of (47.28) must be evaluated at $(\eta_1, \eta_2) = (x_1, x_2)/\varepsilon$ after differentiation. Evaluating $X^{\alpha\beta}_{ij}$ gives

$$X^{\alpha\beta}_{\gamma\delta} = -\xi \frac{\partial^2}{\partial\eta_\gamma\,\partial\eta_\delta}\left(\frac{1}{2}\eta_\alpha\eta_\beta + \varphi^{\alpha\beta}\right), \quad X^{\alpha\beta}_{\gamma 3} = 0,$$
$$X^{\alpha\beta}_{33} = \xi \frac{B_{33\gamma\delta}}{B_{3333}} \frac{\partial^2}{\partial\eta_\gamma\,\partial\eta_\delta}\left(\frac{1}{2}\eta_\alpha\eta_\beta + \varphi^{\alpha\beta}\right), \tag{47.29}$$

and, on substituting these expressions in (47.18), we obtain

$$M_{\alpha\beta\delta\gamma}(x_1, x_2) = \mathscr{M}\left[\frac{2}{3}h^3\tilde{B}_{\lambda\mu\rho\sigma}\frac{\partial^2}{\partial\eta_\lambda\,\partial\eta_\mu}\left(\frac{1}{2}\eta_\alpha\eta_\beta + \varphi^{\alpha\beta}\right)\frac{\partial^2}{\partial\eta_\rho\,\partial\eta_\sigma}\left(\frac{1}{2}\eta_\gamma\eta_\delta + \varphi^{\gamma\delta}\right)\right]. \tag{47.30}$$

These formulae can be simplified by using Green's theorem, the definition of $\mathscr{M}$, and (47.26), so that we have

$$\mathscr{M}\left(h^3\tilde{B}_{\lambda\mu\rho\sigma}\frac{\partial^2\varphi^{\alpha\beta}}{\partial\eta_\lambda\,\partial\eta_\mu}\frac{\partial^2\varphi^{\gamma\delta}}{\partial\eta_\rho\,\partial\eta_\sigma}\right) = -\mathscr{M}\left(h^3\tilde{B}_{\alpha\beta\rho\sigma}\frac{\partial^2\varphi^{\gamma\delta}}{\partial\eta_\rho\,\partial\eta_\sigma}\right),$$

and (47.30) become

$$M_{\alpha\beta\gamma\delta}(x_1, x_2) = \mathscr{M}\left(\frac{2}{3}h^3\tilde{B}_{\alpha\beta\gamma\delta}\right) + \mathscr{M}\left(\frac{2}{3}h^3\tilde{B}_{\alpha\beta\rho\sigma}\frac{\partial^2\varphi^{\gamma\delta}}{\partial\eta_\rho\,\partial\eta_\sigma}\right). \tag{47.31}$$

If $h(x_1, x_2; \eta_1, \eta_2)$ does not depend on (η_1, η_2), then $\varphi^{\alpha\beta}$ is independent of (η_1, η_2), and (47.31) is identical to (47.23).

The other cases, $a = 1$ and $a > 1$, can be treated in a similar way, but the form of $\mathbf{u}^*$ must be modified (Kohn and Vogelius 1984).

The result may be illustrated by the following example. Let us consider a plate reinforced by a single family of rectangular stiffeners that can be modeled by choosing $h = h(\eta_1)$, as sketched in Figure 47.1. In this case, h is periodic with a period of one, assuming two values $h_1 < h_2$, and governed by

$$\mu_i = \text{measure}\left\{\eta_1 : |\eta_1| < \frac{1}{2}, h(\eta_1) = h_i\right\}$$

with $i = 1, 2$.

For $a < 1$ the auxiliary functions $\varphi^{\alpha\beta}$ are defined by (47.26), where the equations now become $\varphi^{12} = 0$ and

$$\frac{\partial^2\varphi^{11}}{\partial\eta_1^2} = -1 + \frac{c}{h^3}, \quad \frac{\partial^2\varphi^{22}}{\partial\eta_1^2} = \sigma\left(-1 + \frac{c}{h^3}\right),$$

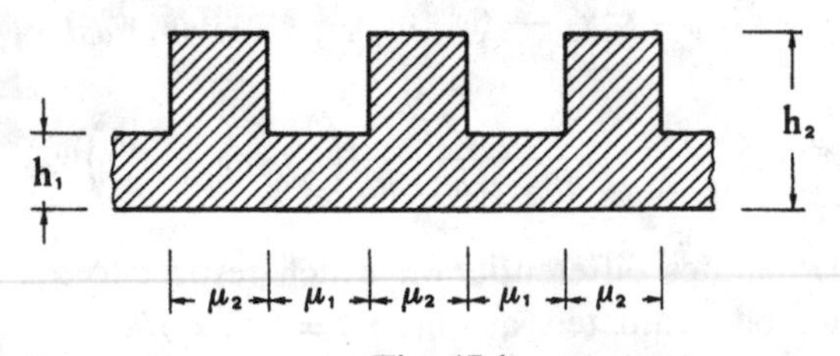

Fig. 47.1

where c is the harmonic mean of h^3:

$$c = (\mu_1 h_1^{-3} + \mu_2 h_2^{-3})^{-1}.$$

On denoting the arithmetic means m by

$$m = \mu_1 h_1^3 + \mu_2 h_2^3,$$

the nonzero components of $M_{\alpha\beta\gamma\delta}$ are

$$M_{1111} = \frac{2}{3}\frac{E}{1-\sigma^2}c, \quad M_{2222} = \frac{2}{3}Em + \frac{2}{3}\frac{E\sigma^2}{1-\sigma^2}c,$$
$$M_{1122} = M_{2211} = \frac{2}{3}\frac{E\sigma}{1-\sigma^2}c, \tag{47.32}$$
$$M_{1212} = M_{2112} = M_{1221} = M_{2121} = \frac{E}{3(1+\sigma)}m.$$

Examples of orders of magnitude of $M_{\alpha\beta\gamma\delta}$ can be given by taking $E = 1.0$ and $\sigma = 0.25$, and assuming four choices of h_1, h_2, μ_1, and μ_2, with the constraint of constant total volume. Formulae (47.32) give the results collected in Table 47.1.

A different geometrical situation, which can be treated by means of elementary considerations, is that constituted by a thin elastic plate, of uniform thickness, with periodically distributed microfissures. The novelty of the problem with respect to the case of a plate with varying thickness lies in the fact that the influence of the microfissures depends on the deformation of the plate. If the loads are such that the faces of each microcrack remain bonded together, then the plate behaves like an integer elastic solid that is able to transmit internal flexural couples applied along any direction. If, alternatively, the loads are such as to cause detachment between the faces of some fissures, a line of disconnection arises within the plate, and is characterized as being of type such that G and $N - (\partial H/\partial s)$ vanish along both sides of this line. The solution to the problem would not be particularly difficult if we knew in advance the position and the extent of the fissures which, after deformation, are open or closed. Unfortunately, however, obtaining information about the cracked region is part of the solution, the initial behavior being undetermined, except in some

Table 47.1.

$h_1 = \frac{1}{3}, h_2 = \frac{2}{3}, \mu_1 = \frac{1}{2}, \mu_2 = \frac{1}{2}$	$h_1 = \frac{1}{3}, h_2 = 1, \mu_1 = \frac{3}{4}, \mu_2 = \frac{1}{4}$
$M_{1111} = 0.047$ $M_{1122} = 0.012$ $M_{2222} = 0.114$ $M_{1212} = 0.044$	$M_{1111} = 0.035$ $M_{1122} = 0.009$ $M_{2222} = 0.187$ $M_{1212} = 0.074$
$h_1 = \frac{1}{4}, h_2 = \frac{3}{4}, \mu_1 = \frac{1}{2}, \mu_2 = \frac{1}{2}$	$h_1 = \frac{1}{4}, h_2 = \frac{5}{4}, \mu_1 = \frac{3}{4}, \mu_2 = \frac{1}{4}$
$M_{1111} = 0.021$ $M_{1122} = 0.005$ $M_{2222} = 0.147$ $M_{1212} = 0.058$	$M_{1111} = 0.015$ $M_{1122} = 0.004$ $M_{2222} = 0.334$ $M_{1212} = 0.133$

particular cases. An illustration of how the problem can be formulated is given by the following example (Lewiński and Telega 1988). Let us consider an isotropic, elastic, homogeneous plate with fissures periodically distributed along lines parallel to the x_1 axis (Figure 47.2), and let εb be the constant distance between the parallel fissures. Let κ_{11}, κ_{22}, and κ_{12} denote the curvatures of the middle surface of the plate after the deformation, and let us consider the limiting case for which $\varepsilon \to 0$. The latent fissures remain closed whenever we have $\kappa_{22} + \sigma\kappa_{11} \leqslant 0$; and the bending moment G_2 across them is such that $G_2 \geqslant 0$. For $\kappa_{22} + \sigma\kappa_{11} > 0$, we have $G_2 = 0$ because the plate cannot support such a moment across the fissure. The two-fold constitutive behavior of the plate can be described by giving $W(\kappa_{11}, \kappa_{22}, w_{12})$, the strain energy of the plate, the following form

$$W(\kappa_{11}, \kappa_{22}, \kappa_{12}) = \begin{cases} \frac{1}{2}D\left[\kappa_{11}^2 + \kappa_{22}^2 + 2\sigma\kappa_{11}\kappa_{22} + 2(1-\sigma)\kappa_{12}^2\right], & \text{wherever} \\ & \kappa_{22} + \sigma\kappa_{11} \leqslant 0, \\ \frac{1}{2}D\left[(1-\sigma^2)\kappa_{11}^2 + 2(1-\sigma)\kappa_{12}^2\right], & \text{otherwise,} \end{cases} \tag{47.33}$$

so that the moment–curvature relations become

$$G_1 = \begin{cases} -D(\kappa_{11} + \sigma\kappa_{22}), & \text{for } \kappa_{22} + \sigma\kappa_{11} \leqslant 0, \\ -D(1-\sigma^2)\kappa_{11}, & \text{otherwise;} \end{cases} \tag{47.34}$$

$$G_2 = \begin{cases} -D(\kappa_{22} + \sigma\kappa_{11}), & \text{for } \kappa_{22} + \sigma\kappa_{11} \leqslant 0, \\ 0, & \text{otherwise;} \end{cases} \tag{47.35}$$

$$H_2 = -H_1 = -D(1-\sigma)\kappa_{12}. \tag{47.36}$$

These constitutive relations indicate that the plate behaves like a monolithic plate at the point $\kappa_{22} + \sigma\kappa_{11} \leqslant 0$, and like a network of independent horizontal beams at $\kappa_{22} + \sigma\kappa_{11} > 0$. A new property of the model is that the strain energy W is convex, non-negative, and of class $\mathscr{C}^1$ in its arguments, but not strictly convex. This means that the solution of the associated boundary-value problem is not necessarily unique.

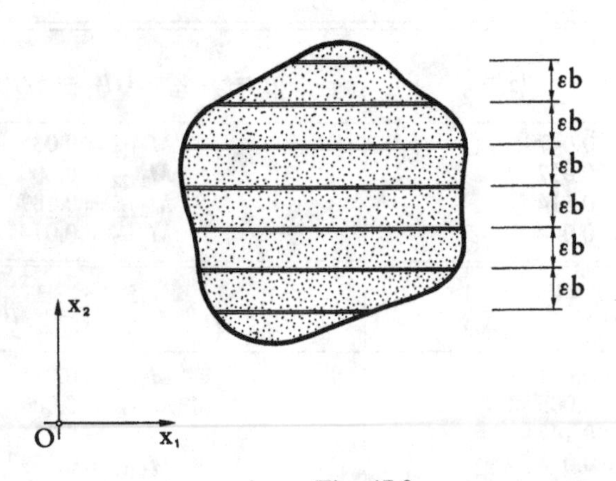

Fig. 47.2

48. The Problem of Plate Optimization

In the classical theory of plates the boundary-value problems are formulated in a standard form. We know the plane domain Ω occupied by the middle plane of the plate, the thickness, the elastic properties of the material, the loads, and the boundary conditions. We must find the vertical deflection of the middle plane and the stresses inside the plate. This formulation, however, is not necessarily the only one, nor is it the most interesting. In many practical circumstances the designer has a certain freedom in varying the thickness or the shape of the plate so as to improve some selected aspect of the mechanical performance of the entire plate, without altering the total quantity of material employed in the design. For example, we may model the plate in such a way that the maximum stress, wherever it occurs, is minimized, or the total work of the external loads is made the lowest possible. This work is also called the *compliance* of a plate, and the most common problem consists of distributing the thickness in a homogeneous plate such that it minimizes the compliance under a given system of fixed loads.

A surprising feature of this problem is that, in general, it does not have a solution, because the stiffness of a plate under prescribed loads may be indefinitely improved by placing the material along two systems of densely spaced ribs of increasing height and decreasing width, such that the total volume remains constant. As a plate with infinitely thin stiffeners is not practically realizable, it may appear that a reasonable constraint on the design variables is that of prescribing upper and lower bounds on the thickness function. However, even in this case, the optimal design problem does not have a solution, because it is possible to show that the stiffness may be improved indefinitely by distributing an infinite number of thinner and thinner stiffeners of height equal to the highest thickness allowed over the plate in an uneven pattern. The reason for this lack of solutions is that the aforementioned choice of design variables confines the solution within a class of functions that is too narrow, so that the limits of the minimizing sequences do not belong to the set of admissible functions. This difficulty can be overcome by enlarging the space of the design variables by including plates with densely spaced thin ribs. This procedure, however, is exposed to some objections of a physical nature. First of all, the designs suggested by the theory are often not practically realizable, but can only be approximated by an appropriate lumping of the stiffeners. Secondly, the wild variations in thickness may invalidate Kirchhoff's theory and a new plate theory may be needed. An alternative procedure is to restrict the design set by imposing a pointwise bound on the gradient of the thickness function (Niordson 1965). The constraint on the gradient permits a piecewise continuous distribution of the thickness, but is unfortunately inadequate for a coherent formulation of the problem within a suitable Sobolev space, because pointwise constraints must be redefined in these spaces.

Another type of local constraint on the thickness which is useful for regularizing the variational problem is the following (Cheng and Olhoff 1982). Let us consider a circular plate under a load of the type $p(r, \theta) = f(r) \cos n\theta$, where f is a given function and n is a given integer, and look for a deflection of the form $W(r, \theta) = w(r) \cos n\theta$. Let us assume that the boundary conditions are homogeneous. The plate consists of a solid part of variable thickness $h_s(r)$, $0 < h_{\min} < h_s < h_{\max}$, equipped with a system of infinitely thin stiffeners of variable concentration $\mu(r)$, placed along circumferences concentric with the axis. Each stiffener has a rectangular cross-section of height $h_{max} - h_s$, and is bisected by the middle plane (Figure 48.1). The concentration $\mu(r)$ of the stiffeners is defined by

$$\mu(r) = \lim_{\Delta r \to 0} \frac{\sum_{i=0}^{N} \Delta c_i}{\Delta r}, \quad 0 \leqslant \mu(r) \leqslant 1 \tag{48.1}$$

where Δc_i is the width of the ith stiffener of the element Δr. For $\mu(r) \equiv 0$ and $h_s = h$, the plate reduces to the solid plate of variable thickness $h(r)$ traditionally used in optimal design. Alternatively, if we have $\mu(r) \neq 0$ and $h_s \equiv h_{min}$, the model comprises another special case, namely the totally stiffened plate. There is no rigorous proof that the introduction of the two design variables $\mu(r)$ and $h_s(r)$ renders the original variational problem well-posed, but a large number of numerical results seem to confirm this conjecture (Cheng and Olhoff 1982).

We now see how, once we have chosen the formulation of the problem in terms of $\mu(r)$ and $h_s(r)$, we may derive the necessary conditions of optimality for a circular axisymmetrically loaded plate. By virtue of Clapeyron's theorem, the compliance of the plate can be written as

$$\Pi = \iint_{\Omega} \{D_{r\theta}[(1-\sigma^2)\kappa_{\theta\theta}^2 + 2(1-\sigma)\kappa_{r\theta}^2] + D_r(\kappa_{rr} + \sigma\kappa_{\theta\theta})^2\} r\, dr, \tag{48.2}$$

where κ_{rr}, $\kappa_{\theta\theta}$, and $\kappa_{r\theta}$ are the curvatures in a polar system of coordinates having the origin placed at the center, and D_r and $D_{r\theta}$ are flexural rigidities along the lines $r =$ constant and $\theta =$ constant. Note that, due to symmetry, all these quantities are functions of the single variable r. The flexural rigidities can be written in the form

$$D_r = \frac{D_{max} D_s}{\mu D_s + (1-\mu) D_{max}}, \quad D_{r\theta} = \mu D_{max} + (1-\mu) D_s, \tag{48.3}$$

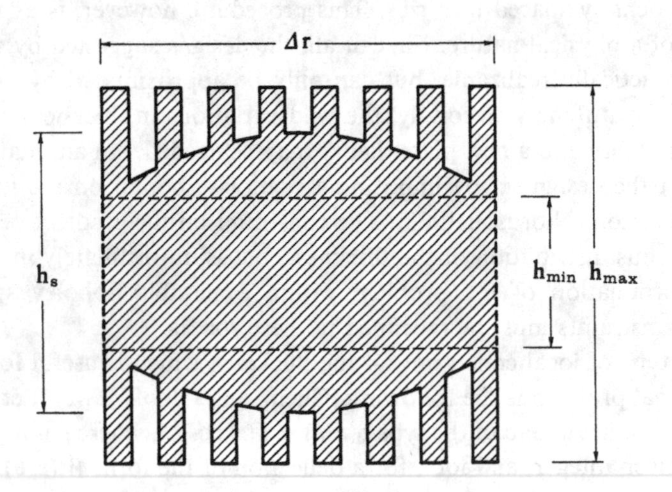

Fig. 48.1

where D_{max} and D_s are geometric quantities defined by

$$D_{max} = h_{max}^3, \quad D_s = h_s^3. \tag{48.4}$$

In order to obtain the conditions of extremum for the function (48.2), together with the constraints, we introduce the augmented functional

$$\begin{aligned}\Pi^* = \Pi - \Lambda\left\{\iint_\Omega [h_s + \mu(h_{max} - h_s)]r\,dr - 1\right\} \\ - \iint_\Omega \lambda[h_s - h_{max} + \bar{\sigma}^2]r\,dr - \iint_\Omega \beta[h_{min} - h_s + \tau^2]r\,dr \\ - \iint_\Omega \gamma[\mu - 1 + \xi^2]r\,dr - \iint_\Omega \alpha[\eta^2 - \mu]r\,dr,\end{aligned} \tag{48.5}$$

where the constraints are related by means of the Lagrangean multipliers Λ, $\lambda(r)$, $\beta(r)$, $\gamma(r)$, and $\alpha(r)$, and where the slack variables $\bar{\sigma}(r)$, $\tau(r)$, $\xi(r)$, and $\eta(r)$ are introduced to convert the inequality constraints on $h_s(r)$ and $\mu(r)$ into equality constraints.

The necessary condition for Π^* to be stationary with respect to the design variable $\mu(r)$ is then found to be

$$\begin{aligned}(D_{max} - D_s)[(1-\sigma^2)\kappa_{\theta\theta}^2 + 2(1-\sigma)\kappa_{r\theta}^2] + D_r^2\left(\frac{1}{D_s} - \frac{1}{D_{max}}\right)(\kappa_{rr} + \sigma\kappa_{\theta\theta})^2 \\ = \Lambda(h_{max} - h_s) + \gamma(r) - \alpha(r),\end{aligned} \tag{48.6}$$

and the stationarity condition with respect to $h_s(r)$ yields

$$\begin{aligned}3h_s^2(1-\mu)[(1-\sigma^2)\kappa_{\theta\theta}^2 + 2(1-\sigma)\kappa_{r\theta}^2] + \frac{3(1-\mu)}{h_s^4}D_r^2(\kappa_{rr} + \sigma\kappa_{\theta\theta})^2 \\ = \Lambda(1-\mu) + \lambda(r) - \beta(r).\end{aligned} \tag{48.7}$$

Conditions for Π^* to be stationary with respect to the Lagrangean multipliers and the slack variables lead to alternative conditions which, when combined with the constraints on h_s and μ, can be expressed as follows:

$$\gamma(r) = 0, \quad \alpha(r) \geqslant 0, \quad \text{for } \mu(r) = 0; \tag{48.8a}$$

$$\gamma(r) = 0, \quad \alpha(r) = 0, \quad \text{for } 0 < \mu(r) < 1; \tag{48.8b}$$

$$\gamma(r) \geqslant 0, \quad \alpha(r) = 0, \quad \text{for } \mu(r) = 0; \tag{48.8c}$$

and

$$\lambda(r) = 0, \quad \beta(r) \geqslant 0, \quad \text{for } h_s = h_{min}; \tag{48.9a}$$

$$\lambda(r) = 0, \quad \beta(r) = 0, \quad \text{for } h_{min} < h_s(r) < h_{max}; \tag{48.9b}$$

$$\lambda(r) \geqslant 0, \quad \beta(r) = 0, \quad \text{for } h_s = h_{max}. \tag{48.9c}$$

Equations (48.6)–(48.9) can be interpreted in different ways. Firstly, information about the problem can be obtained as follows. Let us assume that in some subregion of the plate one of the design variables $\mu(r)$ or $h_s(r)$ is unconstrained. Then either $\gamma = \alpha = 0$ or $\lambda = \beta = 0$. This implies that the gradient of the specific strain energy with respect that particular design variable should be constant. Let us instead consider a stiffened subregion characterized by

$$0 < \mu(r) < 1, \quad h_{\min} \leqslant h(r) < h_{\max}. \tag{48.10}$$

Then, in view of (48.10) and (48.8b), (48.6) reduces to

$$\frac{D_{\max} - D_s}{h_{\max} - h_s}[(1-\sigma^2)\kappa_{\theta\theta}^2 + 2(1-\sigma)\kappa_{r\theta}^2] + \frac{D_r^2}{h_{\max} - h_s}\left(\frac{1}{D_s} - \frac{1}{D_{\max}}\right)(\kappa_{rr} + \sigma\kappa_{\theta\theta})^2 = \Lambda. \tag{48.11}$$

On taking (48.10) into account together with (48.9b), we can express (48.7) as

$$3h_s^2[(1-\sigma^2)\kappa_{\theta\theta}^2 + 2(1-\sigma)\kappa_{r\theta}^2] + \frac{3}{h_s^4}D_r^2(\kappa_{rr} + \sigma\kappa_{\theta\theta})^2 \leqslant \Lambda. \tag{48.12}$$

On eliminating Λ between equation (48.11) and inequality (48.12), and expressing D_r, D_s, and $D_{\max}$ in terms of μ, h_s, and $h_{\max}$ by means of (48.4) and (48.3a), we obtain, after division by $h_{\max} - h_s$, the necessary conditions for the optimality of a stiffened subregion:

$$\begin{aligned}(2h_s + h_{\max})&[(1-\sigma^2)\kappa_{\theta\theta}^2 + 2(1-\sigma)\kappa_{r\theta}^2] \\ &\geqslant \frac{h_{\max}^3 h_s^2}{[\mu h_s^3 + (1-\mu)h_{\max}^3]^2}[h_s^2 + 2h_s h_{\max} + 3h_{\max}^2](\kappa_{rr} + \sigma\kappa_{\theta\theta})^2.\end{aligned} \tag{48.13}$$

Finally, let us consider a purely solid subregion defined by

$$\mu(r) = 0, \quad h_{\min} < h_s(r) < h_{\max}. \tag{48.14}$$

This equation, combined with (48.8a) and the fact that we have $D_r = D_s$ for $\mu = 0$, permits us to write (48.6) as

$$\frac{D_{\max} - D_s}{h_{\max} - h_s}[(1-\sigma^2)\kappa_{\theta\theta}^2 + 2(1-\sigma)\kappa_{r\theta}^2] + \frac{D_s(D_{\max} - D_s)}{D_{\max}(h_{\max} - h_s)}(\kappa_{rr} + \sigma\kappa_{\theta\theta})^2 \leqslant \Lambda \tag{48.15}$$

while the condition (48.7) reduces to

$$3h_s^2[(1-\sigma^2)\kappa_{\theta\theta}^2 + 2(1-\sigma)\kappa_{r\theta}^2] + 3h_s^2(\kappa_{rr} + \sigma\kappa_{\theta\theta})^2 - \Lambda = 0, \tag{48.16}$$

since h_s is unconstrained and we have $\mu \neq 0$. On combining the two last equations, using (48.4), and dividing by $h_{\max} - h_s$, we obtain the inequality

$$(2h_s + h_{\max})[(1-\sigma^2)\kappa_{\theta\theta}^2 + 2(1-\sigma)\kappa_{r\theta}^2] \leqslant \frac{h_s^2}{h_{\max}^3}(h_s^2 + 2h_s h_{\max} + 3h_{\max}^2)(\kappa_{rr} + \sigma\kappa_{\theta\theta})^2. \tag{48.17}$$

Conditions (48.13) and (48.17) characterize the subregions of the plate at which either stiffeners are placed or the thickness varies continuously. Unfortunately, it seems impossible to derive similar conditions for purely solid subregions where the thickness is $h_{\min}$ or $h_{\max}$.

The necessary conditions for optimality derived before may seem useless because they require the knowledge of $h_s(r)$ and of the curvatures, which are parts of the solution. They are, however, useful in the vicinity of the edge, where we know either the curvatures or the displacement from the boundary conditions. For example, at a simply supported or free edge, we have the condition $G = 0$ (see (45.28)), which is equivalent to $\kappa_{rr} + \sigma\kappa_{\theta\theta} = 0$, because singular values of G are excluded by virtue of the condition $h_s \geqslant h_{\min} > 0$. Now, in the vicinity of a simply supported or free edge we have $\kappa_{rr} + \sigma\kappa_{\theta\theta} = 0$, while $(1-\sigma^2)\kappa_{\theta\theta}^2 + 2(1-\sigma)\kappa_{r\theta}^2 > 0$ and $2h_s + h_{\max} > 0$. This implies that the condition (48.17) is not satisfied. Thus a purely solid subregion of intermediate thickness $h_{\min} < h(r) < h_{\max}$ *cannot* appear at a simply supported or free edge of an optimally designed axisymmetric plate. The plate will be either completely stiffened, or solid with maximum or minimum thickness.

At a clamped edge $r = R$ of an axisymmetric plate, we have $w(R) = w'(R) = 0$, and hence the conditions $\kappa_{\theta\theta}(R) = \kappa_{r\theta}(R) = 0$. From the results $h_s^2 + 2h_s h_{\max} + 3h_{\max}^2 > 0$ and $[\kappa_{rr}(R) + \sigma\kappa_{\theta\theta}(R)]^2 > 0$, we readily see that condition (48.13) fails to be satisfied. Thus a subregion with stiffeners *cannot* be found at the clamped edge of an optimally designed axisymmetric plate. In the vicinity of such an edge the plate will be purely solid, and its thickness will be such that we have $h_{\min} \leqslant h_s(r) \leqslant h_{\max}$.

The analysis of the necessary conditions of optimality in a circular axisymmetric plate can be extended to the dynamic case, when we wish to determine the distribution of thickness that maximizes the smallest natural frequency ω_n corresponding to the mode

$$W(r,\theta) = w(r)\cos n\theta, \tag{48.18}$$

which has a prescribed number n $(n \geqslant 0)$ of nodal diameters (Olhoff 1970, Cheng and Olhoff 1982).

In this case too, we extend the space of the design variables by introducing both the concentration $\mu(r)$ of a sequence of ribs, like those drawn in Figure 48.1, and the thickness $h_s(r)$ of a solid part of the plate. The optimization problem can be characterized mathematically as that of seeking the maximum of the quotient

$$\omega_n^2 = \left(\iint_\Omega \{D_{r\theta}(r)[(1-\sigma^2)\kappa_{\theta\theta}^2 + 2(1-\sigma)\kappa_{r\theta}^2] + D_r(\kappa_{rr} + \sigma\kappa_{\theta\theta})^2\} r\, d\theta\right) \times \left\{\iint_\Omega [h_s(r) + \mu(r)(h_{\max} - h_s(r))] w^2 r\, dr\right\}^{-1}, \tag{48.19}$$

subject to the constraints

$$\iint_\Omega [h_s(r) + \mu(r)(h_{\max} - h_s(r))] r\, dr = 1, \tag{48.20}$$

$$h_{\min} \leqslant h_s(r) \leqslant h_{\max}, \tag{48.21}$$

$$0 \leqslant \mu(r) \leqslant 1. \tag{48.22}$$

In order to derive the necessary conditions for optimality, we introduce the augmented functional

$$(\omega_n^*)^2 = \omega_n^2 - \Lambda\left\{\iint_\Omega [h_s + \mu(h_{\max} - h_s)]r\,dr - 1\right\} - \iint_\Omega \lambda[h_s - h_{\max} + \bar{\sigma}^2]r\,dr$$
$$- \iint_\Omega \beta(h_{\min} - h_s + \tau^2)r\,dr - \iint_\Omega \gamma[\mu - 1 + \xi^2]r\,dr - \iint_\Omega \alpha[\eta^2 - \mu]r\,dr, \tag{48.23}$$

where, Λ, $\lambda(r)$, $\beta(r)$, $\gamma(r)$, and $\alpha(r)$ are Lagrangean multipliers and $\bar{\sigma}(r)$, $\tau(r)$, $\xi(r)$, and $\eta(r)$ are slack variables.

The Euler–Lagrange equation determining the conditions for stationary $(\omega^*)^2$ with respect to the design variable $\mu(r)$ is now

$$(D_{\max} - D_s)[(1-\sigma^2)\kappa_{\theta\theta}^2 + 2(1-\sigma)\kappa_{r\theta}^2] + D_r^2\left(\frac{1}{D_s} - \frac{1}{D_{\max}}\right)(\kappa_{rr} + \sigma\kappa_{\theta\theta})^2$$
$$- \omega_r^2(h_{\max} - h_s)w^2 - \Lambda(h_{\max} - h_s) - \gamma(r) + \alpha(r) = 0, \tag{48.24}$$

and stationarity with respect to $h_s(r)$ leads to

$$3h_s^2(1-\mu)[(1-\sigma^2)\kappa_{\theta\theta}^2 + 2(1-\sigma)\kappa_{r\theta}^2] + \frac{3(1-\mu)}{h_s^4}D_r^2(\kappa_{rr} + \sigma\kappa_{\theta\theta})^2$$
$$- \omega_n^2(1-\mu)w^2 - \Lambda(1-\mu) - \lambda(r) + \beta(r) = 0. \tag{48.25}$$

In describing these equations we have normalized w through the condition

$$\iint_\Omega [h_s + \mu(h_{\max} - h_s)]w^2 r\,dr = 1. \tag{48.26}$$

The conditions just derived are useful for interpreting the qualitative properties of the solution in different subregions. In the presence of stiffeners we have the conditions $0 < \mu(r) < 1$ and $h_{\min} \leqslant h_s(r) < h_{\max}$, and hence the results $\lambda(r) = \gamma(r) = \alpha(r) = 0$ and $\beta(r) \geqslant 0$, as consequences of (48.8) and (48.9). Thus the optimality conditions (48.24) and (48.25) become

$$\frac{D_{\max} - D_s}{h_{\max} - H_s}[(1-\sigma^2)\kappa_{\theta\theta}^2 + 2(1-\sigma)\kappa_{r\theta}^2]$$
$$+ \frac{D_r^2}{h_{\max} - h_s}\left(\frac{1}{D_s} - \frac{1}{D_{\max}}\right)(\kappa_{rr} + \sigma\kappa_{\theta\theta})^2 - \omega_n^2 w^2 = \Lambda, \tag{48.27}$$

and

$$3h_s^2[(1-\sigma^2)\kappa_{\theta\theta}^2 + 2(1-\sigma)\kappa_{r\theta}^2] + \frac{3}{h_s^4}D_r^2(\kappa_{rr} + \sigma\kappa_{\theta\theta})^2 - \omega_n^2 w^2 \leqslant \Lambda. \tag{48.28}$$

On eliminating Λ between these two equations, the additional term $\omega_n^2 w^2$ drops out, and we arrive at the result that condition (48.13) also holds for stiffened subregions in the problem of maximizing the nth frequency. Similarly, we find that condition (48.17) also governs the optimal thickness of a solid subregion in a plate having the highest nth eigenfrequency.

We have already seen that the optimum distribution of thickness in a thin plate may be formulated in various ways, one of which is based on the choice of $h_s(r)$ and

$\mu(r)$ as design variables. Another method consists of imposing a bound on the gradient of the thickness, but the device is questionable because a pointwise bound must be reformulated when the solutions are interpreted in some weak sense. In order to answer to these objections it is necessary to find a different variational formulation for the entire problem (Bendsøe 1983, 1995).

Let us denote the plane domain occupied by the middle surface of the plate by Ω, and let us call $D_{\alpha\beta\gamma\delta}$ the fourth-order tensor representing the flexural rigidity of the plate. If $w_{\alpha\beta}$ are the curvatures, the partial differential equation that must be solved for the deflection w is

$$(D_{\alpha\beta\gamma\delta}w_{\gamma\delta})_{\alpha\beta} = p, \quad \text{in } \Omega, \tag{48.29}$$

with some boundary conditions. It is known that the boundary-value problem (48.29) can be put into another form. We introduce the Sobolev space $H^2(\Omega)$ and a subspace $V \subseteq H^2(\Omega)$ chosen in accordance with the imposed boundary conditions, and we assume that p belongs to the space $L^2(\Omega)$, and that $D_{\alpha\beta\gamma\delta}$ are elements of $L^\infty(\Omega)$. Then problem (48.29), with its boundary conditions, is equivalent to the problem of finding an element $w \in V \subseteq H^2(\Omega)$ such that

$$\iint_\Omega D_{\alpha\beta\gamma\delta}w_{\alpha\beta}v_{\gamma\delta}\,dA = a_D(w, v) = \iint_\Omega pv\,dA, \quad \forall\, v \in V. \tag{48.30}$$

The problem of minimum compliance can now, in general terms, be formulated as follows. For a given plane domain Ω, given a load $p (\in L^2)$ and boundary conditions, we want to find an admissible set of rigidities $D_{\alpha\beta\gamma\delta}$ such that

$$\pi = \iint_\Omega pw\,dA = \min, \tag{48.31}$$

where w is a solution of (48.30). In the traditional formulation of the problem of optimum design, the set of admissible rigidities is defined by putting

$$\begin{aligned} \tilde{D}_{1111} &= \tilde{D}_{2222} = \frac{2}{3}\frac{E}{(1-\sigma^2)} = k, \\ \tilde{D}_{1212} &= \tilde{D}_{1221} = \tilde{D}_{2112} = \tilde{D}_{2121} = \frac{1}{2}(1-\sigma)k, \\ \tilde{D}_{1122} &= \tilde{D}_{2211} = \sigma k, \end{aligned} \tag{48.32}$$

and considering the set

$$U_1 \equiv \left\{ h^3\tilde{D}_{\alpha\beta\gamma\delta} \,|\, h \in L^\infty(\Omega),\ \iint_\Omega h\,dA = V = \text{constant},\ 0 < h_{\min} \leqslant h \leqslant h_{\max} \right\}. \tag{48.33}$$

Here $2h$ is the variable distance between the upper and lower surfaces of the plate, which are supposed to be placed symmetrically with respect the middle plane, and the constants $h_{\min}$, $h_{\max}$, and V are chosen so that U_1 is not empty, otherwise the problem has no solution.

In order to discuss the solvability of problem (48.31), let us call F the map $h \to w_h$ and π the map $w \to \iint_\Omega pw\,dA$. The map π is linear and continuous, and hence it is weakly continuous; the map F is continuous, but nonlinear. Our goal is to find a minimum of $\pi \circ F(h)$ for $h \in U_1$. As π is weakly continuous, the minimization problem would have a solution if $F(U_1)$ were weakly closed, but this condition fails in general, and therefore we cannot conclude that there is a solution to problem (48.31).

In order to circumvent this difficulty, let us introduce the additional constraint

$$|h'(r)| \leqslant s = \text{constant}, \tag{48.34}$$

on the derivative of the thickness function, and introduce a new set of admissible functions defined by

$$U_2 \equiv \{h^3 \tilde{D}_{\alpha\beta\gamma\delta} |\, h \in \tilde{U}_2\}, \tag{48.35}$$

corresponding to a thickness function in the set

$$\tilde{U}_2 \equiv \{h \in H^1(\Omega) |\, \|h\|_{H^1} \leqslant M,\ h \geqslant h_{\min} > 0\}, \tag{48.36}$$

where the constraint $h_{\min}$ on h and M on the H^1 norm of h, namely

$$\|h\|_{H^1} = \left\{ \iint_\Omega [h^2 + (\operatorname{grad} h)^2] dA \right\}^{\frac{1}{2}} \leqslant M, \tag{48.37}$$

are chosen so that $\tilde{U}_2$ is not empty.

In the new set U_2, the problem of minimum compliance has a solution. In order to prove this result, we first note that $\tilde{U}_2$ is a bounded, closed, and convex subset of $H^1(\Omega)$, and thus is a weakly compact set of $H^1(\Omega)$. The condition $h \geqslant h_{\min} > 0$ guarantees that $F(\tilde{U}_2)$ is a bounded set of $H^2(\Omega)$ and $\pi(F(\tilde{U}_2))$ a bounded subset of $\mathbb{R}$. Now let there be a set of designs having a minimizing sequence (h_n) in $\tilde{U}_2$, that is

$$\pi(w_n) = \iint_\Omega pw_n\,dA \to \inf \pi(F(\tilde{U}_2)) > -\infty, \quad \text{for } n \to \infty, \tag{48.38}$$

where (w_n) is the sequence of responses corresponding to the sequence (h_n). As the injection $H^1(\Omega) \to L^p(\Omega)$, $p \geqslant 2$, is compact, the boundedness of $\tilde{U}_2$ ensures that there is a convergent subsequence of (h_n), so that we have

$$h_n \to h, \text{ in } L^6, \quad \text{strongly, as } n \to \infty, \tag{48.39}$$

and

$$h_n^3 \to h^3, \text{ in } L^2, \quad \text{strongly for } n \to \infty. \tag{48.40}$$

Moreover, we have $h \in \tilde{U}_2$, as $\tilde{U}_2$ is weakly compact.

The sequence (w_n) is a bounded sequence in the reflexive space $H^2(\Omega)$, so there is a subsequence with

$$w_n \to w, \quad \text{weakly, in } H^2(\Omega), \text{ as } n \to \infty. \tag{48.41}$$

This also implies the result

$$(w_n)_{\alpha\beta} \to w_{\alpha\beta}, \quad \text{weakly, in } L^2(\Omega), \text{ for } n \to \infty. \tag{48.42}$$

On combining (48.40) and (48.42), we can conclude that, in the sense of distributions, we have

$$\left[h_n^3 \tilde{D}_{\alpha\beta\gamma\delta}(w_n)_{\gamma\delta}\right]_{\alpha\beta} \to \left[h^3 \tilde{D}_{\alpha\beta\gamma\delta}(w)_{\gamma\delta}\right]_{\alpha\beta}, \quad \text{for } n \to \infty. \tag{48.43}$$

Now, for all n, we know that (w_n) is the sequence of responses corresponding to the sequence (h_n). This implies

$$\left[h_n^3 \tilde{D}_{\alpha\beta\gamma\delta}(w_n)_{\gamma\delta}\right]_{\alpha\beta} = p, \quad \text{for all } n, \tag{48.44}$$

together with the associated boundary conditions. As (48.44) holds for all n, (48.43) leads to

$$\left[h^3 \tilde{D}_{\alpha\beta\gamma\delta}(w)_{\gamma\delta}\right]_{\alpha\beta} = p, \tag{48.45}$$

with the consequence that $w \in F(\tilde{U}_2)$. In addition, we have $\pi(w_n) \to \pi(w)$ for $n \to \infty$, because π is weakly continuous. Thus, from (48.38), we have

$$\iint_\Omega pw\, dA = \inf \pi(F(\tilde{U}_2)). \tag{48.46}$$

Hence the conclusion that $h \in \tilde{U}_2$ is the solution to the minimization problem.

Once we have seen that the minimum problem is well posed, a natural question is to interpret the necessary conditions that must be satisfied by the optimal solution. Let us call $h^* \in \tilde{U}_2$ an optimal solution for the compliance $\pi(w) = \iint_\Omega pw\, dA$, with the constraint

$$a_D(w, v) = \pi(v), \quad \text{for all} \quad v \in V, \tag{48.47}$$

together with the volume constraint

$$\iint_\Omega h\, dA = V = \text{constant}, \tag{48.48}$$

the technological constraint $h \geqslant h_{\min} > 0$, and the regularity constraint (48.37). We then introduce the augmented functional

$$\pi^*(w) = \pi(w) - a_D(w, v) + \pi(v) - \Lambda\left(\iint_\Omega h\, dA - V\right) - \Phi\left[\iint_\Omega h^2\, dA + \iint_\Omega (\text{grad} h)^2\, dA - M^2 + \zeta^2\right] - \iint_\Omega \eta(h_{\min} - h + \bar{\sigma}^2)dA, \tag{48.49}$$

where v, Λ, Φ, and η are Lagrangean multipliers and ζ and $\bar{\sigma}$ are slack variables. Then the fact that π^* is stationary with respect to h yields the necessary condition

$$\iint_\Omega 3gh^2 \tilde{D}_{\alpha\beta\gamma\delta} w_{\alpha\beta} v_{\gamma\delta}\, dA - \iint_\Omega \Lambda g\, dA - 2 \iint_\Omega \Phi h g\, dA$$
$$- 2 \iint_\Omega \Phi(\mathrm{grad} h)(\mathrm{grad} g) dA + \iint_\Omega \eta g\, dA = 0, \quad \text{for all } g \in H^1(\Omega). \tag{48.50}$$

Similarly, for π^* to be stationary with respect to ζ and $\bar{\sigma}$ leads to switching conditions that can be given the form

$$\Phi \geqslant 0, \quad \text{for} \quad \|h\|_{H^1} = M; \quad \Phi = 0, \quad \text{for} \quad \|h\|_{H^1} < M. \tag{48.51}$$
$$\eta \geqslant 0, \quad \text{for} \quad h = h_{\min}; \quad \eta = 0, \quad \text{for} \quad h > h_{\min}. \tag{48.52}$$

Equation (48.50) can be rewritten as

$$3h^2 D_{\alpha\beta\gamma\delta} w_{\alpha\beta} v_{\gamma\delta} = \Lambda + 2\Phi h - 2\Phi \nabla^2 h - \eta, \quad \text{in } \Omega, \tag{48.53}$$
$$\mathrm{grad} h \cdot \mathbf{n} = 0, \quad \text{on the boundary } \partial\Omega. \tag{48.54}$$

The condition on the boundary states that the optimal thickness must be distributed at the boundary so that its gradient is perpendicular to the unit normal vector. Furthermore, on identifying the variables $g = h$ in (48.50) and $v = w$ in (48.47), we get the following formulae for the minimum $\pi(w)$ of π:

$$3\pi(w) = 3 \iint_\Omega h^3 \tilde{D}_{\alpha\beta\gamma\delta} w_{\alpha\beta} w_{\gamma\delta}\, dA$$
$$= \Lambda \iint_\Omega h\, dA + 2\Phi \iint_\Omega [h^2 + (\mathrm{grad} h)^2] dA - \iint_\Omega h\eta\, dA,$$

so that $\pi_{\min}$ is given by

$$\pi_{\min} = \frac{1}{3}\left(\Lambda V + 2\Phi M^2 - h_{\min} \iint_\Omega \eta\, dA \right), \tag{48.55}$$

where we have used (48.51) and (48.52).

Although a great deal of effort has been devoted to the problem of optimizing the thickness of a plate the shape of which is given, there are other problems suggested by applications, in which the problem must be formulated differently. In many cases the shape of the plate is not strictly prescribed, but the designer has the freedom to model the boundary, or a portion of it, in order to optimize some prescribed quantity. A typical example of this class of problem is the case of an initially rectangular plate (Figure 48.2) clamped along three consecutive sides and free along the fourth. Let us denote the clamped part of the boundary by $\partial_1\Omega$ and the free part by $\partial_2\Omega$. The plate is loaded by imposing certain deflections w and rotations $\partial u/\partial v$ along the portion $\partial_1\Omega$ of the boundary. We assume that the thickness $2h$ is constant and that the material is elastic, homogeneous, and isotropic, so that its flexural rigidity $D = \frac{2}{3}[Eh^3/(1 - \sigma^2)]$ is constant.

If $\partial_2\Omega$ is given, for example as a rectilinear segment perpendicular to the adjacent sides, the elastic deflection of any point of the plate can be found by solving the

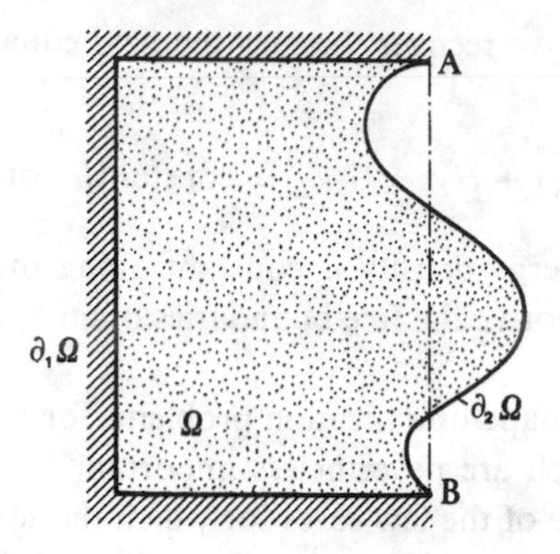

Fig. 48.2

homogeneous equation (45.25). The boundary conditions can be nonhomogeneous involving w and $\partial w/\partial \nu$ along $\partial_1\Omega$, and $G = N - (\partial H/\partial s) = 0$ along $\partial_2\Omega$. Having determined w, we can calculate the strain energy stored in the plate; that is, the value of the functional

$$\mathscr{W} = -\frac{1}{2}\iint_{\Omega}(G_1\kappa_1 - H_1\tau + H_2\tau + G_2\kappa_2)dA, \tag{48.56}$$

where the curvatures κ_1, κ_2, and τ are given by (45.22). However, frequently $\partial_2\Omega$ is not strictly fixed, and we ask whether it is possible to alter the shape of $\partial_2\Omega$ in such a way that its area is preserved; that is, to alter it such that we have the constraint

$$\iint_{\Omega} dA = V = \text{constant} \tag{48.57}$$

and, with the same end points A and B, the function $\mathscr{W}$ attains a minimum. Formulated in this way, the problem is said to be one of *shape optimization*, because the geometrical form of Ω must also be counted as unknown. No exhaustive answer to the question is available, only partial results being known, which are derived from the consideration of necessary conditions arising in the analysis of the extrema of the optimal solutions (Courant and Hilbert 1953, IV, §11). Let us call Ω the optimum domain and consider a perturbation $x + \varepsilon\xi$, $y + \varepsilon\eta$ of the independent variables, associated with a perturbation $w + \varepsilon\zeta$ of the solution. The integral (48.56), written for the perturbed domain $\Omega + \varepsilon\Omega$ can be written as

$$\begin{aligned}\mathscr{W}(w + \varepsilon\zeta) &= -\frac{1}{2}\int\int_{\Omega+\varepsilon\Omega} [(G_1 + \varepsilon G_1)(\kappa_1 + \varepsilon\kappa_1) + \ldots]dA \\ &= -\frac{1}{2}\iint_{\Omega}[(G_1 + \varepsilon G_1)(\kappa_1 + \varepsilon\kappa_1) + \ldots](1 + \varepsilon\xi_x + \varepsilon\eta_y)dA,\end{aligned} \tag{48.58}$$

where we have represented the integral over $\Omega + \varepsilon\Omega$ in terms of an integral over Ω by means of the formula for a change in the variables in double integrals. On equating the first variation of (48.58) to zero, integrating by parts, and exploiting the arbitrariness of ξ, η, and ζ we recover, besides the field equation (45.25), the other condition

$$(G_1\kappa_1 - H_1\tau + H_2\tau + G_2\kappa_2) = \text{constant}, \quad \text{on } \partial_2\Omega, \tag{48.59}$$

which states that strain energy density is constant along the free boundary of an optimal domain. In other words, the free boundaries of an optimally shaped domain are isoenergetic.

The explicit solution of shape-optimization problems for plates is known only for some particular cases, which are nevertheless important for throwing light on the properties of solutions. One of the few cases for which the solution can be explicitly found is the following (Neuber 1985). Let us consider an infinite strip bounded by two arcs of catenaries symmetrically placed with respect a longitudinal x axis, the origin O being chosen to be at the midpoint of the narrowest section (Figure 48.3). The plate is simply extended by forces of the type T_1, parallel to the x axis, applied at a long distance from the notch. The magnitude of these forces, which act theoretically at infinity, is determined by the condition that the resultant force across the section $x = 0$ has a fixed value F, determined by

$$\int_{-a}^{a} (T_1)_{x=0}\, dy = F. \tag{48.60}$$

The two boundaries of the region occupied by the plate are defined as the curves $v = \pm v_0$ of the following parametric system:

$$x = u + c \sinh u \cos v, \quad y = v + c \cosh u \sin v, \tag{48.61}$$

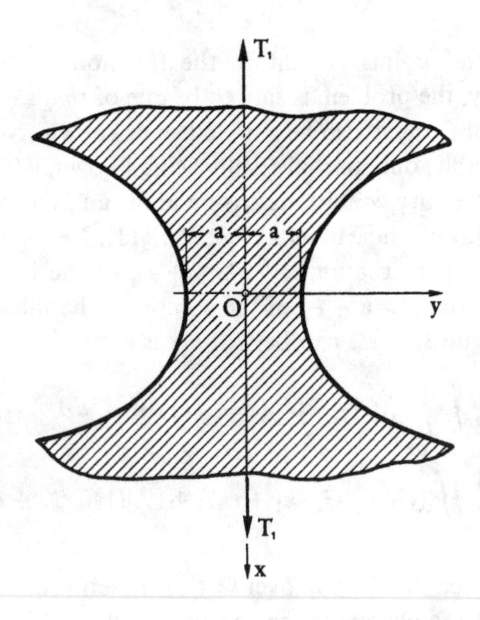

Fig. 48.3

where c is a real constant, and u and v are real parameters. The curves $u =$ constant and $v =$ constant represent two families of mutually orthogonal lines, such that the curves $v =$ constant are catenaries symmetric with respect to the x axis and the curves $u =$ constant are ovals centered symmetrically about the origin.

We assume that the normal stress σ_z vanishes throughout the plate, and that the tangential stresses σ_{zx} and σ_{zy} vanish at the faces $z = \pm h$, so that the plate can be regarded as subject to a state of generalized plane stress characterized by three averaged components of the type $\bar{\sigma}_x$, $\bar{\sigma}_y$, and $\bar{\sigma}_{xy}$, related to the stress resultants T_1, $S_1 = -S_2$, and T_2 by

$$T_1 = 2h\bar{\sigma}_x, \quad S_1 = -S_2 = 2h\bar{\sigma}_{xy}, \quad T_2 = 2h\bar{\sigma}_y. \tag{48.62}$$

Without danger of confusion, the generalized stress components are customarily denoted by σ_x, σ_y, and τ_{xy}. Determining the stress resultants is equivalent to solving a problem of generalized plane stress in the region $-v_0 < v < v_0$ of the uv plane under a given uniaxial tension at infinity. If the two curves $v = \pm v_0$ are prescribed, the problem can be solved by the known methods of plane elasticity. Here, however, v_0 is not given, but must be determined by requiring that the strain energy along the curves $\pm v_0$ is constant. As these boundaries are free from tractions, this implies that in the u, v system of coordinates the conditions

$$\sigma_v(u, \pm v_0) = \sigma_{uv}(u, \pm v_0) = 0$$

must be satisfied. Then, the strain energy is constant along the boundaries, provided that the third component $\sigma_u(u, \pm v_0)$ is a constant. Now the stresses in the u, v system can be represented in terms of two analytic functions $f(z)$ and $g(z)$ of the complex variable $z = x + iy$, which is related to the variable $w = u + iv$ by the equation

$$x + iy = z(u + iv) = z(w) = w + c \sinh w. \tag{48.63}$$

Equations (48.61) are just the real and imaginary parts of (48.63). In terms of f and g it is possible to write (Neuber 1985, Chap. 4, 16.1)

$$\sigma_u + \sigma_v = f + \bar{f}, \tag{48.64a}$$

$$\sigma_v - \sigma_u + 2i\sigma_{uv} = (\bar{z}f' + g)\frac{\dot{z}}{\dot{\bar{z}}}, \tag{48.64b}$$

where the prime denotes differentiation with respect to z and the dot denotes differentiation with respect to w.

Along an optimal free boundary $v = \pm v_0$, we have $\sigma_v = \sigma_{uv} = 0$, while σ_u is constant, being equal, say, to $2B$, where B is a real number. From (48.64) we then obtain

$$f = B, \quad g = -2B\frac{\dot{z}}{\dot{\bar{z}}}, \tag{48.65}$$

for $v = \pm v_0$ or, in complex form, for $\bar{w} = w \mp 2iv_0$. Thus, calculating $\dot{z}$ and $\dot{\bar{z}}$ from (48.65), and substituting in (48.64b), we obtain

$$g = -2B\frac{1 + c\cosh(w \pm 2iv_0)}{1 + c\cosh w}.$$

In order that g is a single-valued function, v_0 must be an integer multiple of $\pi/2$. On taking $v_0 = \pi/2$, we obtain

$$g = 2B\frac{c\cosh w - 1}{c\cosh w + 1}. \tag{48.66}$$

Having thus determined f and g, the stress components can be calculated from (48.64). Their values are

$$\begin{aligned}
\sigma_u &= \frac{2B}{h_1^2}(1 + c\cosh u\cos v + c^2\sinh^2 u\sin^2 v),\\
\sigma_v &= \frac{2B}{h_1^2}(c\cosh u\cos v + c^2\cosh^2 u\cos^2 v),\\
\tau_{uv} &= \frac{2B}{h_1^2}c^2\sinh u\cosh u\sin v\cos v,
\end{aligned} \tag{48.67}$$

where

$$h_1^2 = 1 + 2c\cosh u\cos v + c^2(\sinh^2 u + \cos^2 v). \tag{48.68}$$

In order to complete the solution, it is necessary to calculate the constant B from the condition that the total longitudinal force across the section $x = 0$ of the plate is F. From (48.60) and (48.67) we have

$$F = 2h\int_{v=-\frac{\pi}{2}}^{v=+\frac{\pi}{2}}(\sigma_u x_u + \tau_{uv}x_v)dv = 4\pi hB, \tag{48.69}$$

and hence we find B.

The conclusion is that, if we restrict the choice of the possible boundaries within the class of a pair of curves $v = \pm v_0 =$ constant, there are exactly two branches, namely $v_0 = \pm\pi/2$, for which the stress component σ_u is constant, and therefore the necessary condition of optimality is satisfied.

49. Paradoxes in the Theories of Plates

Kirchhoff's model of a plate, in the case of simply supported polygonal plates, suffers from a nonphysical phenomenon known as the plate paradox (Sapondzhyan 1952, Babuška 1961). The paradox is as follows. Consider a sequence $\{w_n\}$ of convex polygonal domains that converge to a circle. For each n, let w_n be the transverse deflection of the plate, calculated according to Kirchhoff's model, when the plate is simply supported along its boundary and subjected to a uniform load $p(x, y) \equiv 1$. Now, let w_0 be the solution to the limit problem, that is for the circle to which the polygonal domains converge. Then, for $n \to \infty$, the sequence $\{w_n\}$ converges pointwise, but the limit w_∞ is different from w_0, the error in w_∞ being approximately 40% at the center of the circle. In order to evaluate this difference, let us denote the unit circle with its center at the origin by ω_0, and let us assume that this circle is the middle surface of a thin planar plate loaded by a force $f = p/D \equiv 1$ and simply supported at the boundary $\partial\omega_0$. The deflection w_0 is the solution of the problem (see (45.28))

$$\nabla^4 w_0 = 1, \quad \text{in } \omega_0; \tag{49.1a}$$

$$w_0 = 0, \quad \text{on } \partial\omega_0; \tag{49.1b}$$

$$\sigma\nabla^2 w_0 + (1-\sigma)\frac{1}{\rho'}\frac{\partial w_0}{\partial \nu} = 0, \text{ on } \partial\omega_0. \tag{49.1c}$$

The solution of (49.1) is

$$w_0 = C_1^{(0)} + C_2^{(0)} r^2 + C_3^{(0)} r^4, \tag{49.2}$$

where $r^2 = x^2 + y^2$, $C_3^{(0)} = 1/64$, and $C_1^{(0)}$ and $C_2^{(0)}$ are determined from the boundary conditions. A simple computation gives

$$w_0(0,0) = \frac{1}{64}\frac{5+\sigma}{1+\sigma}. \tag{49.3}$$

Let us now consider a sequence of regular $(n+3)$ polygons with $\omega_n \subset \omega_{n+1}$ and $\omega_n \to \omega_0$ for $n \to \infty$. The deflection w_0 corresponding to the domain ω_∞, which is again the circle ω_0, is the solution of the boundary-value problem

$$\begin{aligned} \nabla^4 w_\infty &= 1, \quad \text{on } \omega_0; \\ w_\infty &= 0, \quad \text{on } \partial\omega_0; \\ \nabla^2 w_\infty &= 0, \quad \text{on } \partial\omega_0. \end{aligned} \tag{49.4}$$

The solution of (49.4) is

$$w_\infty = C_1^{(\infty)} + C_2^{(\infty)} r^2 + C_3^{(\infty)} r^4, \tag{49.5}$$

where $r^2 = x^2 + y^2$, $C_3^{(\infty)} = 1/64$ and $C_1^{(\infty)}$ and $C_2^{(\infty)}$ are determined from the boundary conditions. The deflection at the origin is shown to be

$$w_\infty(0,0) = \frac{3}{64}. \tag{49.6}$$

For $\sigma = 0, 3$, we have

$$\frac{w_0(0,0)}{w_\infty(0,0)} = 1.36; \tag{49.7}$$

that is, the gap between w_0 and w_∞ is 36% at the origin. This result is clearly anomalous, having the additional feature that, although in the limit the domains ω_∞ and ω_0 coincide, the limit solution does not satisfy the free-support condition on the circle.

The paradox can be resolved in one of two different ways. The first approach involves preserving Kirchhoff's theory, but making the limiting process more precise. The second is to reject Kirchhoff's theory and to investigate whether the paradox disappears in a more refined theory of plates.

The first approach is based on the idea of constructing an asymptotic expansion of the solution of (49.4) for $n \to \infty$ (Maz'ya and Nazarov 1986). In order to do this we describe the approximation of ω_0 by a polygon, and denote by $P^{(n)} = \left\{p_j^{(n)}\right\}_{j=1}^{n}$ the set of points $\partial\omega_0$ that divides this contour into n arcs of equal length. We describe a polygonal contour $\partial\omega_n$ with vertices at the point $p_j^{(n)}$ in ω_0 and denote the plate bounded within $\partial\omega_n$ by ω_n. The deflection w_n of ω_n with a freely supported boundary satisfies the boundary-value problem:

$$\begin{aligned} \nabla^4 w_n &= 1, \quad \text{in } \omega_n; \\ w_n = \nabla^2 w_n &= 0, \quad \text{on } \partial\omega_n/P^{(n)}. \end{aligned} \tag{49.8}$$

As the plate is convex, the integrals

$$\iint_{\omega_n} \left|\text{grad} w_n\right|^2 dA, \quad \iint_{\omega_n} \left|\text{grad} \nabla^2 w_n\right|^2 dA, \tag{49.9}$$

are uniformly bounded (in n). The limiting solution w_∞ and its Laplacian $\nabla^2 w_\infty$ are therefore weak limits of sequences in the Sobolev space $\overset{\circ}{W}{}_2^1(\omega_\infty)$. Hence w_∞ satisfies the boundary conditions

$$w_\infty = \nabla^2 w_\infty = 0, \text{ on } \partial\omega_0 \tag{49.10}$$

which are clearly not the conditions of free support (49.1c) on $\partial\omega_0$.

In order to solve this paradox, we consider each vertex $p_j^{(n)}$ and introduce local polar coordinates (r_i, θ_i) with the pole $p_j^{(n)}$. Let $\alpha_j(n)$ be the angle of the polygon ω_n at the vertex $p_j^{(n)}$. We then impose on w_n the additional conditions

$$\begin{aligned} M_{r_j} &= -\left[\frac{\partial^2 w_n}{\partial r_j^2} + \frac{\sigma}{r_j^2}\left(\frac{\partial^2 w_n}{\partial \theta_j^2} + r_j \frac{\partial w_n}{\partial r_j}\right)\right] \\ &= \frac{1}{4\pi n}\left[2 - \frac{\pi}{\alpha_j(n)} + \sigma\left(2 + \frac{\pi}{\alpha_j(n)}\right)\right] m\left(p_j^{(n)}\right) r_j^{-\pi/\alpha_j(n)} \cos\frac{\pi\theta_j}{\alpha_j(n)} + O\left(r_j^{\pi/\alpha_j}\right), \end{aligned} \tag{49.11}$$

where $j = 1, \ldots, n$, and m is a smooth function on $\partial\omega_0$. The relations (49.11) mean that concentrated torques M_{r_j} (or $M_{\theta_j}) = -M_{r_j} + O(r_j^{\pi/\alpha_j})$ of small magnitude are given at the vertices of ω_n. By means of a process of asymptotic expansion, it is possible to show, on choosing m correctly, that we can achieve an arbitrary distribution of moments in $\partial\omega_0$. In particular, by putting $m = -(1-\sigma)(1/\rho')(\partial\omega_0/\partial\nu)$, where $1/\rho'$ is the curvature of $\partial\omega_0$, ν is the inward normal, and w_0 is the solution of problem (49.1), we find that $\{w_n\}$ converge to w_0. In this way the paradox is resolved.

The second approach to answering the question is to regard the paradox as a consequence of the vanishing of vertical shear strains, which is implicit in Kirchhoff's theory. It is indeed possible to show that the paradox is avoided when Reissner's model or the three-dimensional theory of plates are adopted, and that it is the way in which the boundary conditions are imposed in Kirchhoff's model that causes the paradox, and not the overall assumption of vanishing shear strains (Babuška 1962, Babuška and Pitkäranta 1990).

In the three-dimensional theory of plates, the boundary conditions of simple support are typically imposed by requiring that the vertical component of the displacement, or at least its average in the vertical direction, vanishes at the edge of the plate. On the other hand, Kirchhoff's model imposes the more restrictive condition that all tangential displacements must vanish at the edge. Of course, it is also possible to impose such hard boundary conditions in the three-dimensional theory or in Reissner's model. If this is done, then the paradox occurs also in both Reissner's model and in the three-dimensional theory. The paradox, however, does not occur in these models for the case of a soft support, where only the vertical displacements are restricted at the edge of the plate. Hence we are led to the conclusion that the paradox is caused by the hard boundary conditions, which are intrinsic in Kirchhoff's model.

These results can be rendered quantitatively precise as follows. Let us take a sequence of convex (not necessarily regular) polygonal domains converging to a circle, and let $2h$ be the thickness of the sequence of plates having these polygonal domains as their middle sections. Then the relative error of Kirchhoff's solution, as compared with the three-dimensional solution with a hard support, is uniformly of order $O(h^{\frac{1}{2}})$ in the energy norm. Moreover, the relative error in the energy norm between Kirchhoff's and Reissner's solutions are of order $O(h)$ under the same assumptions. Finally, on a smooth domain, the energy norms of the three solutions are, at most, $O(h^{\frac{1}{2}})$ apart. Hence the plate paradox occurs for the case of a hard support if the thickness is sufficiently small.

The numerical calculations (Babuška and Scapolla 1989) confirm, in particular, that the error of Kirchhoff's solution with respect to the three-dimensional solution is primarily due to assuming hard boundary conditions with simply supported polygonal plates. For example, in the case of a uniformly loaded square plate the thickness of which is 1/100 the length of a side, the relative error of Kirchhoff's solution for energy norm is approximately 11% when compared with the three-dimensional solution in which the vertical component of displacement u_3 is zero on the lateral surface of the plate, and the other two conditions are natural boundary conditions describing zero components of tractions. On the other hand, if we assume hard boundary conditions in the three-dimensional model, characterized by the vanishing of all components of displacement on the lateral surface, then the relative error in energy is approximately 2% as compared to Kirchhoff's model. Moreover, the numerical results show that, on imposing various boundary conditions that are seemingly close, such as soft and hard simple supports, we may influence the solution in the entire domain, and not only in a boundary layer adjacent to the lateral surface.

50. Föppl–Kármán Theory

In the classical theory of plates, three kinds of assumption are customarily made: (1) small strains and small displacements; (2) the possibility of defining the stress characteristics and writing the equilibrium equations with reference to the planar undeformed configuration; and (3) the validity of Hooke's law. Besides the advantage of describing the problem by means of a system of linear partial differential equations, linear theory also offers the simplification that the equations governing the displacements parallel to the middle surface are separated from those describing the transverse displacement of the middle surface. This has the consequence that the two groups of equations can be treated separately.

However, the assumptions made in linear theory are often in conflict with the geometrical hypothesis that the thickness must be small compared with the other linear dimensions of the plate. Let us consider a large thin plate, fixed at its edges and subjected to pressure on its upper face. As the thickness diminishes, and the transverse displacement increases, the plate tends to behave like a curved membrane, with the consequence that the extensional strain, arising from the stretching of the initially planar middle surface, increases, and the influence of the tensile stress resultants such as T and S increases in comparison with that of the resulting shear stresses of type N. In extreme cases the transverse displacement is so large that it is not sufficient to estimate the extensional strains by using the simplified formulae of the linear theory, but it is necessary to have recourse to the general formulae valid for large strains.

The equations governing the equilibrium configurations of an initially planar plate are difficult to formulate because the relations between the stress resultants and stress couples must be written with reference to the deformed configuration, which is no longer a plane but a curved surface or a shell. The solution of these equations is possible only in some particular cases. Thus many attempts have been made to formulate theories for thin plates under large strains that are sufficiently simple to be analytically tractable and adequate enough to reveal the most relevant effects arising from large deformations.

A case of a large deformation of a plate that can be solved in full generality with surprising simplicity is that of a plate bent into a circular cylindrical form (Love 1927, Art. 335D). We suppose that an initially planar rectangular plate is bent without extension into a sector of a circular cylinder with two edges as generators, and we seek the forces that must be applied at the edges in order to hold the plate in this form. Let the planar unstrained middle surface be referred to a system of Cartesian axes x and y, such that $x = \pm a$ are the equations of the edges which, after the deformation become generators of the cylinder, and that $y = \pm b$ are the equations of the edges which become arcs of circles of radius R. The curvatures of the bent plate, in the x, y system, are

$$\kappa_1 = \frac{1}{R}, \quad \kappa_2 = \tau = 0, \tag{50.1}$$

and hence the stress couples, given by (45.21) have the forms

$$G_1 = -\frac{D}{R}, \quad G_2 = -\sigma\frac{D}{R}, \quad H_2 = -H_1 = 0, \tag{50.2}$$

while the other resultant stresses T_1, T_2, and $S_2 = -S_1$ vanish.

Bending moments of the form (50.2) satisfy the equilibrium equations at any interior point of the plate. The boundary conditions are satisfied provided that couples of magnitude $-(D/R)$ are applied at the edges $x = \pm a$ and couples of magnitude $-\sigma(D/R)$ are applied to the circular edges $y = \pm b$.

The solution clearly holds for large values of $1/R$, namely when the plate is severely bent. There is, however, a limit to the validity of this solution. The greatest extension occurs in the circumferential direction at the point of the face $z = -h$ of the plate and its value is just h/R. It is therefore necessary that this ratio is small in order that linear stress–strain relations can be employed.

When the, initially planar, plate is bent by a (not necessarily constant) pressure acting on its upper face and the transverse displacement is so large that the tensile stress resultants of type T and S are no longer negligible in comparison with N, it is possible to evaluate the transverse displacement and the stress resultants by means of a semilinear theory. This is valid only if the transverse displacement is not too large and the strains are also moderately small so that the stress–strain relations are still linear. The main advantage of this theory is that the problem can be described by a pair of nonlinear partial differential equations that, at least in principle, can be solved.

We take the unstrained middle plane of the plate to be a domain, say Ω, of the xy plane, with a regular boundary $\partial\Omega$. Let us specify the displacement of a point on Ω by the components u, v, and w, where w is the transverse displacement in the direction of the z axis, normal to the xy plane. If we assume that the displacement component w is moderately large, while u and v are small, we may define the strains in the middle surface by omitting quantities of second order in the partial derivative of u and v, but not those involving the partial derivatives of w. The strains thus become

$$\varepsilon_x = \varepsilon_1 = \frac{\partial u}{\partial x} + \frac{1}{2}\left(\frac{\partial w}{\partial x}\right)^2, \quad \varepsilon_y = \varepsilon_2 = \frac{\partial v}{\partial y} + \frac{1}{2}\left(\frac{\partial w}{\partial y}\right)^2,$$

$$2\varepsilon_{xy} = \tilde{\omega} = \frac{\partial u}{\partial y} + \frac{\partial v}{\partial x} + \frac{\partial w}{\partial x}\frac{\partial w}{\partial y}. \tag{50.3}$$

Besides the extensional and shear strains of the middle surface, it is necessary to know the curvatures of the surface. These can be expressed with sufficient approximation by using the formulae of linear theory, namely

$$\kappa_1 = \frac{\partial^2 w}{\partial x^2}, \quad \kappa_2 = \frac{\partial^2 w}{\partial y^2}, \quad \tau = \frac{\partial^2 w}{\partial x\,\partial y}. \tag{50.4}$$

Consequently the resultant stresses are given by

$$T_1 = \frac{3D}{h^2}(\varepsilon_1 + \sigma\varepsilon_2), \quad T_2 = \frac{3D}{h^2}(\varepsilon_2 + \sigma\varepsilon_1), \quad S_1 = -S_2 = \frac{3D}{2h^2}(1-\sigma)\tilde{\omega},$$

and the stress couples by

$$G_1 = -D(\kappa_1 + \sigma\kappa_2), \quad G_2 = -D(\kappa_2 + \sigma\kappa_1), \quad H_2 = -H_1 = -D(1-\sigma)\tau.$$

After the deformation, the middle surface of the plate becomes a curved surface, on which we may trace curvilinear coordinates α, β and isolate a curvilinear quadrilateral bounded by the lines α, $\alpha + d\alpha$, β, and $\beta + d\beta$. If χ' is the angle between the tangents to the curves α, β at any point, the linear element ds of a curve traced on the surface is given by

$$ds^2 = A'^2\,d\alpha^2 + 2A'B'\cos\chi'\,d\alpha\,d\beta + B'^2\,d\beta^2, \tag{50.5}$$

where $A'\,d\alpha$ is the element of arc of a curve $\beta =$ constant, and $B'\,d\beta$ is the element of arc of a curve $\alpha =$ constant. The curvatures of the surface elements can be characterized by two vectors (p_1', q_1', r_1') and (p_2', q_2', r_2') defined as follows. Let us take an orthogonal triad x', y', z' with the origin placed at the vertex of the curvilinear quadrilateral determined by the intersection of the lines $\alpha =$ constant and $\beta =$ constant. The x' axis coincides with the tangent to the $\beta =$ constant line, drawn in the sense of increasing α, and the y' axis coincides with the tangent to the surface, drawn at right angles to x'. The z' axis is perpendicular to the other axes and to the surface. As the triad is shifted by an infinitesimal amount over the surface, the directions of the axes change, and this change can be determined as an infinitesimal rigid rotation $(p_1'\,d\alpha, q_1'\,d\alpha, r_1'\,d\alpha)$ superimposed on another infinitesimal rotation $(p_2'\,d\beta, q_2'\,d\beta, r_2'\,d\beta)$. The six quantities $p_1', \dots, r_2'$ can be expressed in terms of the direction cosines of the moving triad and of their partial derivatives with respect to α and β. If, however, the lines α and β are lines of curvature on the surface, then the curvilinear quadrangle of sides $A'\,d\alpha$, $B'\,d\beta$ is a curvlinear rectangle, and the expressions for $p_1', \dots, r_2'$ simplify greatly, because in this case we have

$$\chi' = \frac{1}{2}\pi, \tag{50.6a}$$

$$p_1' = q_2' = 0, \tag{50.6b}$$

and

$$\frac{1}{R_1'} = -\frac{q_1'}{A'}, \quad \frac{1}{R_2'} = \frac{p_2'}{B'}, \tag{50.7}$$

R_1' and R_2' being the principal radii of curvature of the strained surface. The other components of rotation, r_1' and r_2', are given by

$$r_1' = -\frac{1}{B'}\frac{\partial A'}{\partial \beta}, \quad r_2' = \frac{1}{A'}\frac{\partial B'}{\partial \alpha}. \tag{50.8}$$

Formulae (50.6b)–(50.8) can easily be understood by moving the triad x', y', z' from the origin of the curvilinear rectangle where the curves $\alpha =$ constant and $\beta =$ constant intersect to the opposite vertex, that is, to the intersection of the curves $\alpha + d\alpha =$ constant and $\beta + d\beta =$ constant. The rotations are positive if taken counterclockwise with respect to each axis of the moving triad. In particular, when the strained surface can be represented with adequate approximation by the Cartesian equation $z = w(x, y)$ with respect to the fixed axes, we can take the values $\alpha = x$, $\beta = y$, and $A' = B' = 1$, so that the formulae for the rotations reduce to

$$p_1' = -q_2' = \tau = \frac{\partial^2 w}{\partial x\,\partial y}, \quad q_1' = -\frac{\partial^2 w}{\partial x^2} = -\kappa_1, \quad p_2' = \frac{\partial^2 w}{\partial y^2} = \kappa_2, \quad r_1' = r_2' = 0. \tag{50.9}$$

Let us assume that the lines $\alpha =$ constant and $\beta =$ constant are lines of curvature of the strained surface, and consider the equilibrium of the stress resultants acting on the sides of curvilinear rectangle bounded by the curves $\alpha, \beta, \alpha + d\alpha$, and $\beta + d\beta$ (Figure 50.1). The stress resultants on the side α of the rectangle give a force with the components $(-T_1 B'\,d\beta,\ -S_1 B'\,d\beta,\ -N_1 B'\,d\beta)$ along the x', y', and z' axes. The stress resultants on the side $\alpha + d\alpha$ yield forces that can be obtained from those acting on the side α, by a change of α into $\alpha + d\alpha$ and for a rotation $(p_1'\,d\alpha,\ q_1'\,d\alpha,\ r_1'\,d\alpha)$. From this the force on the side $\alpha + d\alpha$ has the expression

$$\begin{pmatrix} T_1 B'\,d\beta \\ S_1 B'\,d\beta \\ N_1 B'\,d\beta \end{pmatrix} + \frac{\partial}{\partial \alpha}\begin{pmatrix} T_1 B'\,d\beta \\ S_1 B'\,d\beta \\ N_1 B'\,d\beta \end{pmatrix} d\alpha + \begin{pmatrix} p_1'\,d\alpha \\ q_1'\,d\alpha \\ r_1'\,d\alpha \end{pmatrix} \times \begin{pmatrix} T_1 B'\,d\beta \\ S_1 B'\,d\beta \\ N_1 B'\,d\beta \end{pmatrix}. \tag{50.10}$$

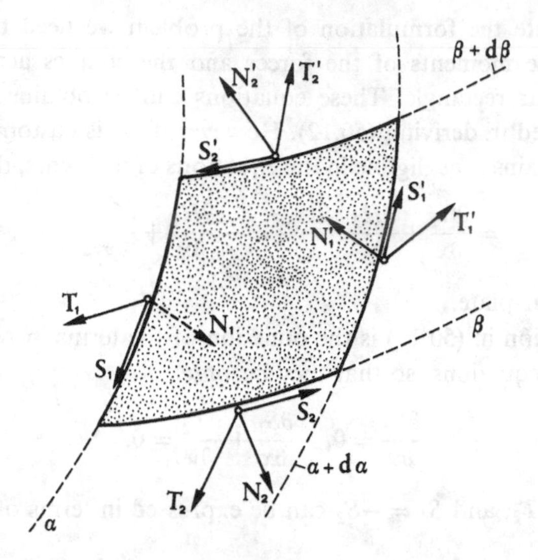

Fig. 50.1

In like manner, the force on the side β of the rectangle has the form $(S_2A'\,d\alpha,$ $-T_2A'\,d\alpha,\ -N_2A'\,d\alpha)$, and that on the side $\beta+d\beta$, resulting from the change from β into $\beta+d\beta$ by the rotation $(p_2'\,d\beta,\ q_2'\,d\beta,\ r_2'\,d\beta)$, has the form

$$\begin{pmatrix} -S_2A'\,d\alpha \\ T_2A'\,d\alpha \\ N_2A'\,d\alpha \end{pmatrix} + \frac{\partial}{\partial\beta}\begin{pmatrix} -S_2A'\,d\alpha \\ T_2A'\,d\alpha \\ N_2A'\,d\alpha \end{pmatrix} d\beta + \begin{pmatrix} p_2'\,d\beta \\ q_2'\,d\beta \\ r_2'\,d\beta \end{pmatrix} \times \begin{pmatrix} -S_2A'\,d\alpha \\ T_2A'\,d\alpha \\ N_2A'\,d\alpha \end{pmatrix} \tag{50.11}$$

If X', Y', and Z' denote the components of the external forces along the x', y', and z' axes estimated per unit area of the strained surface, it follows that the element of area $A'B'\,d\alpha\,d\beta$ is loaded by the force $\{X'A'B'\,d\alpha\,d\beta,\ Y'A'B'\,d\alpha\,d\beta,\ Z'A'B'\,d\alpha\,d\beta)$. The full equations of the equilibrium between the external and the contact forces are then given by adding the forces on the sides of the rectangle.

However, a reduced form of these equations can be used when we restrict our attention to the case in which the strained middle surface differs little from a plane, so that (x, y) replace (α, β). Furthermore, we set $A' = B' = 1$, while representing the deformed plate by the surface $z = w(x, y)$ with sufficient accuracy; in addition, the rotation elements $p_1', \ldots, r_2'$ are given by (50.9). For a normal pressure p on the face $z = -h$ the force per unit area (X', Y', Z') reduces to $(0, 0, p)$. It then follows that the equilibrium equations in the directions x', y', and z' assume the simplified form

$$\frac{\partial T_1}{\partial x} - \frac{\partial S_2}{\partial y} - N_1\frac{\partial^2 w}{\partial x^2} - N_2\frac{\partial^2 w}{\partial x\,\partial y} = 0, \tag{50.12a}$$

$$\frac{\partial S_1}{\partial x} + \frac{\partial T_2}{\partial y} - N_1\frac{\partial^2 w}{\partial x\,\partial y} - N_2\frac{\partial^2 w}{\partial y^2} = 0, \tag{50.12b}$$

$$\frac{\partial N_1}{\partial x} + \frac{\partial N_2}{\partial y} + T_1\frac{\partial^2 w}{\partial x^2} - S_2\frac{\partial^2 w}{\partial x\,\partial y} + S_1\frac{\partial^2 w}{\partial x\,\partial y} + T_2\frac{\partial^2 w}{\partial y^2} + p = 0. \tag{50.12c}$$

In order to complete the formulation of the problem we need the equilibrium equations between the moments of the forces and the couples acting across the sides of the curvilinear rectangle. These equations can be obtained by a process similar to that followed in deriving (50.12). However if, as is customarily assumed, the influence of the strains is negligible in the equations of moments, they simply give

$$N_1 = \frac{\partial G_1}{\partial x} + \frac{\partial H_2}{\partial y}, \quad N_2 = -\frac{\partial H_1}{\partial x} + \frac{\partial G_2}{\partial y},$$

as in a perfectly planar plate.

Another simplification in (50.12) is the omission of the terms in N_1 and N_2 from the first two of these equations, so that they become

$$\frac{\partial T_1}{\partial x} - \frac{\partial S_2}{\partial y} = 0, \quad \frac{\partial S_1}{\partial x} + \frac{\partial T_2}{\partial y} = 0.$$

This implies that T_1, T_2, and $S_1 = -S_2$ can be expressed in terms of a single stress function $F(x, y)$ giving

$$T_1 = \frac{\partial^2 F}{\partial y^2}, \quad T_2 = \frac{\partial^2 F}{\partial x^2}, \quad S_1 = -S_2 = -\frac{\partial^2 F}{\partial x\, \partial y}. \tag{50.13}$$

Equation (50.12c) thus becomes

$$-D\nabla_1^4 w + \frac{\partial^2 F}{\partial y^2}\frac{\partial^2 w}{\partial x^2} + \frac{\partial^2 F}{\partial x^2}\frac{\partial^2 w}{\partial y^2} - 2\frac{\partial^2 F}{\partial x\, \partial y}\frac{\partial^2 w}{\partial x\, \partial y} + p = 0. \tag{50.14}$$

The stress function $F(x, y)$ is not completely free because T_1, T_2, and S_1 are functions of the strains, and the partial derivatives of u and v must satisfy the identity

$$\frac{\partial^2}{\partial y^2}\frac{\partial u}{\partial x} + \frac{\partial^2}{\partial x}\frac{\partial v}{\partial y} = \frac{\partial^2}{\partial x\, \partial y}\left(\frac{\partial u}{\partial y} + \frac{\partial v}{\partial x}\right),$$

which is one of Beltrami's compatibility equations. Now, on using (50.3) and expressions for ε_1, ε_2, and $\tilde{\omega}$ in terms of T_1, T_2, and S_1, we arrive at the equation

$$\nabla_1^4 F + 3(1-\sigma^2)\frac{D}{h^2}\left\{\frac{\partial^2 w}{\partial x^2}\frac{\partial^2 w}{\partial y^2} - \left(\frac{\partial^2 w}{\partial x\, \partial y}\right)^2\right\} = 0, \tag{50.15}$$

which, combined with (50.14), form the so-called Föppl–Kármán system.

If we now review the deduction of (50.14) and (50.15), we immediately note that the theory is open to serious objections, because many of the simplifications introduced in these equations are not justified. First, the assumptions $A' = B' = 1$ and $\chi' = \frac{1}{2}\pi$, which imply that a rectangular element of the unstrained middle surface remains practically rectangular after the deformation, are not acceptable when the strains are large. Secondly, it is assumed that, in defining the strains, the quadratic terms in u and v are negligible and only the quadratic terms in w are to be retained. This, however, is not legitimate when the plate is severely deformed and the resultant membrane stresses of the type T_1, T_2, and $S_1 = -S_2$ become dominant. A third objection is that, in the equilibrium equations, some terms are omitted and others are retained with no precise explanation. For instance, the terms omitted in (50.12a) and (50.12b) may be of the same order of magnitude as those maintained in (50.12c). Moreover, there is no reason for neglecting the terms in $p'_1, \ldots, p'_2$ in the equations of equilibrium of moments when the same terms are retained in the equations of equilibrium between forces. However, despite these criticisms, the Föppl–Kármán equations are important for two reasons: they offer the simplest mathematical model for describing the large deformation of a thin planar plate; and the system of equations (50.14) and (50.15), although nonlinear, can be treated with a rigorous mathematical theory and, in some cases, can even be solved explicitly.

The consequences of the theory can be illustrated by the analysis of a comparatively simple example. We consider the case of a strip of infinite length bounded by two parallel straight lines with the equations $x = \pm a$. The strip is simply supported along its edges and subjected to a constant vertical force of magnitude $2P$ uniformly distributed along the straight line $x = 0$, and acting in the sense of the positive z axis. As the lines $y =$ constant deform in the same way, the displacement component v vanishes, and the other components u and w are independent of y. The constitutive equations for the stress resultants and the stress couples now become

$$T_1 = \frac{3D}{h^2}\left[\frac{du}{dx} + \frac{1}{2}\left(\frac{dw}{dx}\right)^2\right], \quad T_2 = \sigma T_1, \quad S_2 = -S_1 = 0,$$
$$G_1 = -D\frac{d^2w}{dx^2}, \quad G_2 = -\sigma G_1, \quad H_2 = -H_1 = 0, \tag{50.16}$$

and the equilibrium equations yield

$$\frac{dT_1}{dx} - N_1\frac{d^2w}{dx^2} = 0, \tag{50.17a}$$

$$\frac{dN_1}{dx} + T_1\frac{d^2w}{dx^2} = 0, \tag{50.17b}$$

$$N_1 = -D\frac{d^3w}{dx^3}, \tag{50.17c}$$

$$N_2 = 0. \tag{50.17d}$$

On combining these equations we obtain the system

$$\frac{dT_1}{dx} + D\frac{d^2w}{dx^2}\frac{d^3w}{dx^3} = 0, \tag{50.18a}$$

$$D\frac{d^4w}{dx^4} - T_1\frac{d^2w}{dx^2} = 0. \tag{50.18b}$$

As a result of the symmetry with respect to the y axis, it is sufficient to seek the solution in the interval $0 < x < a$. The boundary conditions to be satisfied are

$$u(a) = 0, \quad w(a) = 0, \quad G_1(a) = 0, \quad \frac{dw}{dx}(0) = 0, \quad N_1(0) = -P. \tag{50.19}$$

Equation (50.18a) can be integrated at once to give

$$T_1 = a_1 - \frac{1}{2}D\left(\frac{d^2w}{dx^2}\right)^2, \tag{50.20}$$

where a_1 is a constant. Since $(d^2w/dx^2)(a) = 0$, we can put $a_1 = T_1(a) = T_1^a$, which is still undetermined. Hence (50.18b) assumes the form

$$D\frac{d^4w}{dx^4} - \left[T_1^a - \frac{1}{2}D\left(\frac{d^2w}{dx^2}\right)^2\right]\frac{d^2w}{dx^2} = 0. \tag{50.21}$$

In order to integrate this equation we put $q(x) = d^2w/dx^2$ and so transform it into the equation

$$Dq''(x) - \left(T_1^a - \frac{1}{2}Dq^2\right)q = 0. \tag{50.22}$$

A first integral of (50.22) is

$$q'^2(x) - \frac{T_1^a}{D}q^2(x) + \frac{1}{4}q^4(x) = q'^2(a), \tag{50.23}$$

where $q'(a)$ is another constant to be determined. With the substitution

$$A = -2\frac{T_1^a}{D} + 2\sqrt{\left(\frac{T_1^a}{D}\right)^2 + q'^2(a)}, \quad B = 2\frac{T_1^a}{D} + 2\sqrt{\left(\frac{T_1^a}{D}\right)^2 + q'^2(a)},$$

(50.23) can be written as

$$q'^2(x) = \frac{1}{4}(A + q^2)(B - q^2),$$

whence, by a further integration, we obtain

$$x = 2\int_{q(x)}^{q(0)} \frac{d\xi}{\sqrt{(A+\xi^2)(B-\xi^2)}}. \tag{50.24}$$

Here we have chosen the positive sign in front of the integrand because, as x lies in the interval $0 < x < a$, we have $0 = q(a) < q(x) < q(0)$. Equation (50.24) evaluated at $x = a$, or $q(a) = 0$, becomes

$$a = 2\int_0^{q(0)} \frac{d\xi}{\sqrt{(A+\xi^2)(B-\xi^2)}}.$$

The substitution $\xi = -\sqrt{B}\cos\varphi$ and the value $\varphi_0 = \operatorname{arcos}[q(0)/\sqrt{B}]$ yield

$$a = \frac{2}{\sqrt{A+B}}\int_{\varphi_0}^{\frac{\pi}{2}} \frac{d\varphi}{\sqrt{1 - \frac{B}{A+B}\sin^2\varphi}} = \frac{2}{\sqrt{A+B}}F\left(\varphi_0, \sqrt{\frac{B}{\sqrt{A+B}}}\right), \tag{50.25}$$

where F denotes the elliptic function. Besides (50.25), we need a further equation to determine the constants T_1^a and q'. This condition can be simply approximated by assuming that the slope $(dw/dx)(a)$ is sufficiently small that we can take $N_1(a) \simeq -P$. This has the consequence that $Dq'(a) = -N_1(a) = P$, or $q'(a) = P/D$ and the only unknown constant T_1^a can be found from (50.25).

The general approach to the Föppl–Kármán equations can be achieved using different techniques. If the plate is circular, axisymmetrically loaded, and either simply supported or clamped at the boundary, the solution can be obtained by reducing the problem to a pair of integral equations (Dickey 1976).

As a consequence of symmetry, the functions w and F, if written in a system of r, θ polar coordinates with their origin at the center of the circle, depend on the only variable r, and the Föppl–Kármán equations can then be written in the form

$$D\nabla^4 w - \frac{1}{r}\frac{d}{dr}\left(\frac{dF}{dr}\frac{dw}{dr}\right) = p(r),$$
$$\nabla^4 F + 2hE\frac{1}{r}\frac{dw}{dr}\frac{d^2w}{dr^2} = 0, \tag{50.26}$$

where

$$\nabla^2 = \frac{1}{r}\frac{d}{dr}\left(r\frac{d}{dr}\right).$$

The boundary conditions at $r = a$, the edge of the plate, are

$$w = 0, \quad F = 0, \quad \frac{1}{r}\frac{dF}{dr} = 0, \tag{50.27}$$

which imply that the transverse displacement vanishes at the edge and no membrane forces are applied along it. In addition, if the plate is clamped, we have the condition

$$\frac{dw}{dr} = 0, \tag{50.28}$$

at $r = a$, whereas, if the plate is simply supported, the condition is

$$\frac{d^2w}{dr^2} + \frac{\sigma}{r}\frac{dw}{dr} = 0 \tag{50.29}$$

at $r = a$. Beside the conditions at $r = a$, we have additional conditions at $r = 0$, which read

$$\left|\frac{1}{r}\frac{dw}{dr}\right| < \infty, \quad \left|\frac{1}{r}\frac{dF}{dr}\right| < \infty, \quad \frac{d}{dr}(\nabla^2 w) = 0, \quad \frac{d}{dr}(\nabla^2 F) = 0. \tag{50.30}$$

In order to reduce (50.26) to a more tractable form, we put $F = 2h\varphi$ and write

$$\frac{Eh^2}{3(1-\sigma^2)}\nabla^4 w - \frac{1}{r}\frac{d}{dr}\left(\frac{d\varphi}{dr}\frac{dw}{dr}\right) = \frac{p(r)}{2h}$$
$$\frac{1}{E}\nabla^4\varphi + \frac{1}{r}\frac{dw}{dr}\frac{d^2w}{dr^2} = 0. \tag{50.31}$$

We then make the change of variables

$$\rho = \frac{r}{a}, \quad \alpha = \frac{1}{a^2 A \rho} \frac{dw}{d\rho}, \quad \beta = \frac{1}{a^2 B \rho} \frac{d\varphi}{d\rho}, \tag{50.32}$$

$$P(\rho) = \frac{C}{\rho^2} \int_0^\rho \tau p(a\tau)\, d\tau, \tag{50.33}$$

$$A = \frac{2h}{a^2 \sqrt{6(1-\sigma^2)}}, \quad B = \frac{E}{3a^2(1-\sigma^2)}, \quad C = \frac{[6(1-\sigma^2)]^{\frac{3}{2}} a^4}{8Eh^4}, \tag{50.34}$$

with the consequence that the equations can be rewritten as

$$\frac{d}{d\rho}\left(\rho^3 \frac{d\alpha}{d\rho}\right) - \rho^3 \alpha\beta = \rho^3 P(\rho), \tag{50.35a}$$

$$\frac{d}{d\rho}\left(\rho^3 \frac{d\beta}{d\rho}\right) + \rho^3 \alpha^2 = 0, \tag{50.35b}$$

together with the boundary conditions

$$\alpha'(0) = \beta'(0) = 0, \quad \beta(1) = 0, \tag{50.36}$$

with the notation $(\cdot)' = d/d\rho$. The fourth alternative condition is either

$$\alpha'(1) + (1+\sigma)\alpha(1) = 0, \tag{50.37}$$

if the plate is simply supported, or

$$\alpha(1) = 0, \tag{50.38}$$

if the plate is clamped.

However, (50.35) together with the boundary conditions (50.36)–(50.38) can be reduced to a single nonlinear integral equation. In order to achieve this, we note that the solution of the equation and boundary conditions

$$(\rho^3 v')' = \rho^3 f(\rho), \tag{50.39}$$

$$v'(0) = v(1) = 0, \tag{50.40}$$

is given by

$$v(\rho) = -\int_0^1 g(\rho, \tau)\tau^3 f(\tau)\, dr = -Gf, \tag{50.41}$$

where $g(\rho, r)$ is Green's function of (50.39) and (50.40). Similarly, the solution of (50.39) with the boundary condition

$$v'(0) = v'(1) + (1+\sigma)v(1) = 0, \tag{50.42}$$

is given by

$$v(\rho) = -\int_0^1 k(\rho, \tau)\tau^3 f(\tau)\, d\tau = -Kf, \tag{50.43}$$

where K is Green's function of (50.39) and (50.42). The detailed expressions of g and k are

$$g(\rho,\tau)=\begin{cases}\dfrac{1}{2}\left(\dfrac{1}{\rho^2}-1\right), & 0\leqslant\tau<\rho,\\[2mm] \dfrac{1}{2}\left(\dfrac{1}{\tau^2}-1\right), & \rho<\tau\leqslant 1;\end{cases}\tag{50.44}$$

$$k(\rho,\tau)=\begin{cases}\dfrac{1}{1+\sigma}+\dfrac{1}{2}\left(\dfrac{1}{\rho^2}-1\right), & 0\leqslant\tau<\rho,\\[2mm] \dfrac{1}{1+\sigma}+\dfrac{1}{2}\left(\dfrac{1}{\tau^2}-1\right), & \rho<\tau\leqslant 1.\end{cases}\tag{50.45}$$

In view of these remarks, β, the solution of (50.35b), is given by

$$\beta(\rho)=G\alpha^2.\tag{50.46}$$

Thus equation (50.35a) becomes

$$(\rho^3\alpha')'-\rho^3\alpha G\alpha^2=\rho^3P(\rho).\tag{50.47}$$

In the case of a clamped boundary, (50.47) can be rewritten as

$$\alpha(\rho)=-GP-G\big[(G\alpha^2)\alpha\big],\tag{50.48}$$

while, for a simply supported boundary, (50.47) has the form

$$\alpha(\rho)=-KP-K[(G\alpha^2)\alpha].\tag{50.49}$$

Now let us consider the linear eigenvalue problems

$$Gu=\mu u,\tag{50.50}$$

$$Kv=\gamma v.\tag{50.51}$$

Equation (50.50) has the eigenfunctions

$$u=J_1\left(\rho/\mu^{\frac{1}{2}}\right)\frac{1}{\rho},\tag{50.52}$$

for each value of μ, such that we have $J_1\left(1/\mu^{\frac{1}{2}}\right)=0$. Similarly, (50.51) has the eigenfunctions

$$v=J_1\left(\rho/\gamma^{\frac{1}{2}}\right)\frac{1}{\rho},\tag{50.53}$$

for each value of γ, satisfying

$$\left[J_1\left(\rho/\gamma^{\frac{1}{2}}\right)/\rho\right]'+(1+\sigma)J_1\left(\rho/\gamma^{\frac{1}{2}}\right)\frac{1}{\rho}=0,\tag{50.54}$$

for $\rho=1$. Both (50.50) and (50.51) have a complete set of eigenfunctions $\{u_j\}$ and $\{v_j\}$, satisfying the conditions $<u_i,u_j>=\int_0^1\rho^3u_iu_j\,d\rho=\delta_{ij}$, $<v_i,v_j>=\delta_{ij}$, and

$$\sum_{j=1}^{\infty}\mu_j^n<\infty,\quad \sum_{j=1}^{\infty}\gamma_j^n<\infty;\quad n\geqslant 1.\tag{50.55}$$

As the two equations (50.48) and (50.49) can be treated in the same way, we will discuss, for brevity, the simply supported plate; that is, equation (50.49). Let us assume for the moment that the solution can be expanded in a series of the form

$$\alpha = \sum_{j=1}^{\infty} a_j v_j(\rho). \tag{50.56}$$

This means that (50.49) becomes

$$\sum_{j=1}^{\infty} a_j v_j + KP + K\left\{\left[G\left(\sum_{j=1}^{\infty} a_j v_j\right)^2\right] \sum_{j=1}^{\infty} a_j v_j\right\} = 0. \tag{50.57}$$

If we take the inner product of (50.57) with v_ℓ, we find that the coefficients a_ℓ must satisfy the infinite system of nonlinear algebraic equations

$$\frac{1}{\gamma_\ell} a_\ell + d_\ell + \left\langle G\left(\sum_{j=1}^{\infty} a_j v_j\right)^2, \sum_{j=1}^{\infty} a_j v_j v_\ell \right\rangle = 0, \tag{50.58}$$

for $\ell = 1, 2, \ldots,$ and

$$d_\ell = \langle P, v_\ell \rangle = \int_0^1 \rho^3 P(\rho) v_\ell \, d\rho.$$

which proves the inequality

$$H^N \geqslant -\sum_{j=1}^{N} \gamma_j d_j^2 \geqslant -\gamma_1 \sum_{j=1}^{N} d_j^2 = -\gamma_1 <P, P> . \qquad (50.64)$$

Thus, for $<P, P> < \infty$, H^N is bounded from below and the bound is independent of N. As we have $H^N \to \infty$ for $\sum_{j=1}^{N} a_j^2 \to \infty$, we conclude that there is a set of values $\tilde{a}_1(N), \ldots, \tilde{a}_N(N)$ for which H^N achieves its minimum.

Inequality (50.64) also gives an upper bound for H^N. Indeed, from the inequality

$$H^N(a_1, \ldots, a_{N-1}, 0) \geqslant H^N(a_1, \ldots, a_N),$$

we have, in particular,

$$H^N(\tilde{a}_1, \ldots, \tilde{a}_{N-1}, 0) = m(N-1) \geqslant m(N) = H^N(\tilde{a}_1, \ldots, \tilde{a}_N), \qquad (50.65)$$

resulting in $m(N) \leqslant m(1)$ for all $N \geqslant 1$.

Let us now consider the $\tilde{a}_j$ terms that satisfy the equation

$$\sum_{j=1}^{N} \frac{1}{\gamma_j} \tilde{a}_j^2 + \sum_{j=1}^{N} d_j \tilde{a}_j + \langle G\tilde{\alpha}_N^2, \tilde{\alpha}_N \rangle = 0, \qquad (50.66)$$

where we have put

$$\tilde{\alpha}_N = \sum_{j=1}^{N} \tilde{a}_j v_j. \qquad (50.67)$$

From (50.66) combined with (50.61) we derive

$$\begin{aligned} \sum_{j=1}^{N} \frac{1}{\gamma_j} \tilde{a}_j^2 + 3 \sum_{j=1}^{N} d_j \tilde{a}_j &= 2 \sum_{j=1}^{N} d_j \tilde{a}_j - \langle G\tilde{\alpha}_N^2, \tilde{\alpha}_N \rangle \\ &= 2H^N(\tilde{a}_1, \ldots, \tilde{a}_N) - 2 \sum_{j=1}^{N} \frac{1}{\gamma_j} \tilde{a}_j^2 \leqslant 2m(1). \end{aligned}$$

On writing the last inequality in the form

$$\sum_{j=1}^{N} \left(\frac{\tilde{a}_j}{\gamma_j^{\frac{1}{2}}} + \frac{3}{2} \gamma_j^{\frac{1}{2}} d_j \right)^2 \leqslant 2m(1) + \frac{9}{4} \sum_{j=1}^{N} \gamma_j d_j^2, \qquad (50.68)$$

we conclude that the quantities a_j and $\gamma_j^{\frac{1}{2}}$ must lie in or on a certain sphere. From (50.68), using some inequalities derived above, we obtain

$$\begin{aligned} \sum_{j=1}^{N} \frac{\tilde{a}_j^2}{\gamma_j} &\leqslant \left\{ \frac{3}{2} \left(\sum_{j=1}^{N} \gamma_j d_j^2 \right)^{\frac{1}{2}} + \left[2m(1) + \frac{9}{4} \sum_{j=1}^{N} \gamma_j d_j^2 \right]^{\frac{1}{2}} \right\}^2 \\ &\leqslant \left\{ \frac{3}{2} (\gamma_1 <P, P>)^{\frac{1}{2}} + \left[2m(1) + \frac{9}{4} \gamma_1 <F, F> \right]^{\frac{1}{2}} \right\}^2 = M_1. \qquad (50.69) \end{aligned}$$

This means that the sum on the left-hand side is bounded independently of N. Thus, from the sequence of vectors $(\tilde{a}_1, \ldots, \tilde{a}_N, 0, \ldots)$ for $n \to \infty$, we can extract a subsequence converging to an infinite dimensional vector $[a_1^*, \ldots, a_N^*, \ldots]$ satisfying the inequality

$$\sum_{j=1}^{\infty} \frac{a_j^{*2}}{\gamma_j} \leqslant M_1.$$

The limiting function

$$\alpha^* = \sum_{j=1}^{\infty} a_j^* v_j$$

is a solution of (50.58). In order to prove this property, we use the fact that $\tilde{\alpha}_N$ is a solution of the finite system (50.60), and write

$$\left|\frac{1}{\gamma_j} a_j^* + d_j + \langle G(\alpha^*)^2, \alpha^* v_j\rangle\right| \leqslant \left|\frac{1}{\gamma_j}[a_j^* - \tilde{a}_j(N_i)]\right| + |\langle G(\alpha^*)^2, \alpha^* v_j\rangle - \langle G\tilde{\alpha}(N_i)^2, \tilde{\alpha}(N_i) v_j\rangle|. \tag{50.70}$$

The right-hand side of (50.70) tends to zero for $N \to \infty$, and thus α^* is a solution of (50.58).

The result just obtained does not tell us whether the solution is unique. However, in the absence of tangential forces at the boundary, it is possible to prove that the solution is effectively unique (Reiss 1963). This means that it is not necessary to define $(\alpha_1^*, \ldots, \alpha_N^*, \ldots)$ as the limit of subsequences, because the whole sequence of approximate solutions converges.

When the plate is not circular, but its middle surface occupies a planar region Ω, there is another method for treating the Föppl–Kármán equations, based on the application of a minimum principle (Mullin 1978). Again introducing the function $\varphi = (1/2h)F$, (50.14) and (50.15) assume the form

$$D\nabla^4 w - (\varphi_{xx} w_{xx} - 2\varphi_{xy} w_{xy} + \varphi_{yy} w_{yy}) = p, \tag{50.71a}$$

$$\nabla^4 \varphi + E(w_{xx} w_{yy} - w_{xy}^2) = 0, \tag{50.71b}$$

and we consider the case in which the boundary is clamped and no forces are applied along it, so that w and φ must satisfy the conditions

$$w = 0, \quad \frac{\partial w}{\partial \nu} = 0, \quad \varphi = 0, \quad \frac{\partial \varphi}{\partial \nu} = 0, \quad \text{on } \partial\Omega. \tag{50.72}$$

Solving (50.71) and (50.72) is equivalent to minimizing the total energy

$$\mathscr{E} = \mathscr{W} - \mathscr{L} = \frac{1}{2} DB(w) + \frac{1}{4} q_\varphi(w) - (w, p), \tag{50.73}$$

where we have

$$B = \iint_\Omega [(\nabla^2 w)^2 - 2(1-\sigma)(w_{xx}w_{yy} - w_{xy}^2)]dA, \tag{50.74}$$

$$q_\varphi = \iint_\Omega (\varphi_{xx}w_y^2 + \varphi_{yy}w_x^2 - 2\varphi_{xy}w_x w_y)dA, \tag{50.75}$$

$$(w, p) = \iint_\Omega pw\, dA, \tag{50.76}$$

in the class of functions satisfying (50.72) and the constraint (50.71b).

In order to set the minimum problem in the appropriate framework, we introduce the space $H_0^2(\Omega)$, defined as the completion of $\mathscr{C}_0^\infty$ in the norm

$$\|u\|_2^2 = \iint_\Omega u^2\, dA + \iint_\Omega (\nabla u)^2\, dA + \iint_\Omega (u_{xx}^2 + 2u_{xy}^2 + u_{yy}^2)dA, \tag{50.77}$$

and observe that $B(w)$, as defined by (50.74), satisfies the inequalities

$$\frac{1}{K}\|w\|_2 \leqslant B^{\frac{1}{2}} \leqslant K\|w\|_2.$$

Unfortunately, in the space $H_0^2(\Omega)$ the constraint (50.71b) ceases to be meaningful, because it involves the fourth-order derivatives of φ. It is thus necessary to write it in the integral form

$$\frac{1}{2}\iint_\Omega (\psi_{xx}w_y^2 + \psi_{yy}w_x^2 - 2\psi_{xy}w_x w_y)dA = \frac{1}{E}\iint_\Omega \nabla^2\varphi\nabla^2\psi\, dA, \quad \forall\, \psi \in \mathscr{C}_0^\infty, \tag{50.78}$$

or, from the definition (50.75),

$$\frac{1}{2}q_\psi(w) = \frac{1}{E}\iint_\Omega \nabla^2\varphi\nabla^2\psi\, dA, \quad \forall\, \psi \in \mathscr{C}_0^\infty. \tag{50.79}$$

Let us call C the set of pairs (w, φ) that satisfy (50.79), which is not empty because of the inclusion $(0, 0) \in C$. Since we have $\mathscr{E}(0, 0) = 0$,

$$m = \inf_C \mathscr{E}(w, \varphi) \tag{50.80}$$

satisfies the inequality

$$m \leqslant \mathscr{E}(0, 0) = 0. \tag{50.81}$$

It follows that we can find a sequence $\{(w_j, \varphi_j)\}$ in C satisfying $\mathscr{E}(w_j, \varphi_j) \leqslant 0$ and

$$\lim_{j\to\infty} \mathscr{E}(w_j, \varphi_j) = m. \tag{50.82}$$

We shall show that the sequences $\{w_j\}$ and $\{\varphi_j\}$ are bounded in $H_0^2(\Omega)$. From definition (50.73) of the energy and the inequality $\mathscr{E}(w_j, \varphi_j) \leqslant 0$, we obtain the result

$$\frac{1}{2}D\, B(w_j) + q_{\varphi_j}(w_j) \leqslant (w_j, p). \tag{50.83}$$

As both sides of (50.79) are continous linear functionals of $\psi \in H_0^2(\Omega)$, we can take $\psi = \varphi_j$ to find

$$q_{\varphi_j}(w_j) = \frac{2}{E} \iint_\Omega |\nabla^2 \varphi_j|^2 \, dA \geqslant 0, \tag{50.84}$$

and therefore we have

$$\frac{1}{2} D\, B(w_j) \leqslant (w_j, p) \leqslant \|w_j\|_{L^2} \|p\|_{L^2}, \tag{50.85}$$

where $\|w_j\|_{L^2}^2 = \iint_\Omega w_j^2 \, dA$ and $\|p\|_{L^2}^2 = \iint_\Omega p^2 \, dA$. From definition (50.77) and the following inequality for B, we find finally that

$$\|w_j\|_2 \leqslant \frac{2K}{B} \|p\|_{L^2}, \tag{50.86}$$

which proves that the sequence $\{w_j\}$ is bounded in $H_0^2(\Omega)$. In order to prove that the sequence $\{\varphi_j\}$ is also bounded in $H_0^2(\Omega)$ we observe that, by Poincaré's inequality, we have

$$\frac{1}{K_1} \|\varphi\|_2 \leqslant \|\nabla^2 \varphi\|_{L^2} \leqslant K \|\varphi\|_2, \tag{50.87}$$

thus we need only to prove that $\{\nabla^2 \varphi_j\}$ is a bounded sequence in $L^2(\Omega)$. Equation (50.79) with $\psi = \varphi_j$ yields

$$\|\nabla^2 \varphi_j\|_{L^2}^2 \leqslant \frac{E}{2} \|\varphi_j\|_2 \|w_j\|_{1,4}^2 \leqslant \frac{K_1 E}{2} \|\nabla^2 \varphi_j\|_{L^2} \|w_j\|_{1,4}^2,$$

or

$$\|\nabla^2 \varphi_j\|_{L^2} \leqslant \frac{K_1 E}{2} \|w_j\|_{1,4}^2, \tag{50.88}$$

where

$$\|u\|_{1,4}^4 = \iint_\Omega u^4 \, dA + \iint_\Omega (u_x^4 + u_y^4) dA.$$

It can be shown that we have

$$\|u\|_{1,4} \leqslant K_2 \|u\|_2, \quad \text{for all } u \in H_0^2(\Omega), \tag{50.89}$$

and that inequalities (50.86) and (50.88) yield

$$\|\nabla^2 \varphi_j\|_{L^2} \leqslant K_3 \|p\|_{L^2}^2; \tag{50.90}$$

that is, $\{\nabla^2 \varphi_j\}$ is a bounded sequence in $L^2(\Omega)$.

As a consequence of the inequalities (50.86) and (50.90), we can find a subsequence $\{(w_j, \varphi_j)\}$ such that $\{w_j\}$ converges weakly in $H_0^2(\Omega)$ to a function $\bar{w}$, and $\{\varphi_j\}$ to a function $\bar{\varphi}$. We now show the condition $(\bar{w}, \bar{\varphi}) \in C$ and

$$\mathscr{E}(\bar{w}, \bar{\varphi}) \leqslant m, \tag{50.91}$$

proving that $(\bar{w}, \bar{\varphi})$ minimizes $\mathscr{E}(w, \varphi)$ in C. Let us, in fact, consider the inequality

$$\left|q_{\varphi_j}(w_j) - q_{\bar{\varphi}}(\bar{w})\right| = \left|q_{\varphi_j}(w_j - \bar{w}, w_j + \bar{w}) + q_{\varphi_j - \bar{\varphi}}(\bar{w})\right|$$
$$\leqslant K_4 \|p\|_{L^2}^3 \|w_j - \bar{w}\|_{1,4} + \left|q_{\varphi_j - \bar{\varphi}}(\bar{w})\right|,$$

and note that the first term on the right converges to zero because $\{w_j\}$ converges weakly to $\bar{w}$ in $H_0^2(\Omega)$ and, consequently, in the norm $\|u\|_{1,4}$, as a consequence of a theorem due to Rellich. The second term tends to zero because $\{\varphi_j\}$ converges to $\bar{\varphi}$ weakly in $H_0^2(\Omega)$. Therefore we have

$$\lim_{j\to\infty} q_{\varphi_j}(w_j) = q_{\bar{\varphi}}(\bar{w}). \tag{50.92}$$

Then from (50.82) and (50.92) the following holds:

$$m = \lim_{j\to\infty} \mathscr{E}(w_j, \varphi_j) \geqslant \mathscr{E}(\bar{w}, \bar{\varphi}). \tag{50.93}$$

It remains to be shown that we have $(\bar{w}, \bar{\varphi}) \in C$; that is, that $(\bar{w}, \bar{\varphi})$ satisfies (50.79):

$$\frac{1}{2} q_\psi(\bar{w}) = \frac{1}{E}(\nabla_{\bar{\varphi}}^2, \nabla_\psi^2), \quad \forall\, \psi \in \mathscr{C}_0^\infty(\Omega).$$

This is true because the left-hand side is a continuous function of w in the norm $\|u\|_{1,4}$ and the right-hand side is a continuous function of φ in $H_0^2(\Omega)$, with $(w_j, \varphi_j) \in C$. We then have

$$\frac{1}{2} q_\psi(\bar{w}) = \lim_{j\to\infty} \frac{1}{2} q_\psi(w_j) = \lim_{j\to\infty} \frac{1}{E}(\nabla_{\varphi_j}^2, \nabla_\psi^2) = \frac{1}{E}(\nabla_{\bar{\varphi}}^2, \nabla_\psi^2),$$

which is the desired result.

Besides the existence of a solution, an important question is to establish how regular such solutions are. This is answered as follows (Berger 1967). If the load p is Hölder continuous with exponent α, then $\bar{w} \in \mathscr{C}^{4+\alpha}$ and $\bar{\varphi} \in \mathscr{C}^{6+\alpha}$. It is interesting to observe that the function $\bar{\varphi}$ is more regular than $\bar{w}$, with parity of data.

The direct derivation of the Föppl–Kármán equations is open to criticism, and several attempts have been made to justify the equations as a satisfactory systematic method of approximation relating to the three-dimensional equations of linear elasticity. In order to investigate this question (Ciarlet and Destuynder 1979, Ciarlet 1980), let us introduce a bounded open set ω of the x_1x_2 plane, with a boundary γ, and assume that γ is of class $\mathscr{C}^\infty$, ω being locally at one side of γ. We denote by $\boldsymbol{\tau} \equiv (\tau_\alpha)$ and $\boldsymbol{\nu} \equiv (\nu_\alpha)$ the unit tangent vector and the unit outer normal vector along the boundary γ, with the convention that the angle between $\boldsymbol{\nu}$ and $\boldsymbol{\tau}$ is $\pi/2$. Given a constant $\varepsilon > 0$, we put

$$\Omega^\varepsilon = \omega \times (-\varepsilon, \varepsilon), \quad \Gamma_0^\varepsilon = \gamma \times [-\varepsilon, \varepsilon],$$
$$\Gamma_+^\varepsilon = \omega \times \{\varepsilon\}, \quad \Gamma_-^\varepsilon = \omega \times \{-\varepsilon\},$$

so that the boundary of the set Ω^ε is partitioned into the lateral surface Γ_0^ε, and the upper and lower faces Γ_+^ε and Γ_-^ε, respectively.

The region Ω^ε is subjected to three kinds of applied force: body forces of density f_i^ε acting through Ω^ε; surface forces of density g_i^ε acting on the faces Γ_+^ε and Γ_-^ε; and surface forces acting on the lateral surface Γ_0^ε, for which only the resultant, obtained after integration across the thickness is known. On Γ_0^ε, besides surface forces, displacements (u_α, u_3) are prescribed. For these we require u_α to be independent of x_3, and u_3 to be zero.

We assume that the material obeys the generalized Hooke law with Young's modulus E and a Poisson ratio σ, but the strains are not necessarily infinitesimal. In this case the displacement components u_i and the stresses σ_{ij} must satisfy the following system:

$$\left.\begin{aligned} \varepsilon_{ij} &= \frac{1}{2}(u_{i,j} + u_{j,i} + u_{\ell,i}u_{\ell,j}) \\ (\sigma_{ij} + \sigma_{kj}u_{i,k})_{,j} &+ f_i^\varepsilon = 0 \end{aligned}\right\} \text{ in } \Omega^\varepsilon \tag{50.94}$$

$$\begin{aligned} &\sigma_{i3} + \sigma_{k3}u_{i,k} = \pm g_i^\varepsilon, \quad \text{on } \Gamma_\pm^\varepsilon \\ &u_3 = 0, \quad u_\alpha \text{ independent of } x_3, \text{ on } \Gamma_0^\varepsilon \end{aligned} \tag{50.95}$$

$$\frac{1}{2\varepsilon}\int_{-\varepsilon}^{+\varepsilon} (\sigma_{\alpha\beta} + \sigma_{k\beta}u_{\alpha,k})\nu_\beta \, dx_3 = h_\alpha^\varepsilon, \text{ on } \gamma. \tag{50.96}$$

Conditions (50.96) indicate that the resultant stresses parallel to the x_1x_2 plane are prescribed along γ, but we impose the additional equilibrium conditions

$$\int_\gamma h_\alpha^\varepsilon \, ds = 0, \qquad \int_\gamma (x_1h_2^\varepsilon - x_2h_1^\varepsilon)ds = 0. \tag{50.97}$$

System (50.94)–(50.96) represents the classical formulation of the elastic problem relative to the plate Ω^ε. In order to give it a weak formulation, we introduce the spaces

$$V^\varepsilon = \{v_i \in W^{1,4}(\Omega^\varepsilon), \text{ with } v_\alpha \text{ independent of } x_3 \text{ on } \Gamma_0^\varepsilon, \text{ and } v_3 = 0 \text{ on } \Gamma_0^\varepsilon\}, \tag{50.98}$$

$$\Sigma^\varepsilon = \{\sigma_{ij} \in L^2(\Omega^\varepsilon)\}. \tag{50.99}$$

The fields $u_i \in V^\varepsilon$ and $\sigma_{ij} \in \Sigma^\varepsilon$ are then solutions of the variational equations

$$\iiint_{\Omega^\varepsilon} [\sigma_{ij}\varepsilon_{ij}(v_k)]dV = \iiint_{\Omega^\varepsilon} f_i^\varepsilon v_i \, dV + \iint_{\Gamma_+^\varepsilon \cup \Gamma_-^\varepsilon} g_i^\varepsilon v_i \, dA + \int_\gamma \left\{\int_{-\varepsilon}^{\varepsilon} v_\alpha \, dt\right\} h_\alpha^\varepsilon \, ds, \qquad \forall\, v_i \in V^\varepsilon; \tag{50.100}$$

$$\iiint_{\Omega^\varepsilon} \{\varepsilon_{ij}(u_k) - \tfrac{1}{2}(u_{i,j} + u_{j,i} + u_{\ell,j}u_{\ell,i})\}\sigma_{ij} \, dV = 0, \quad \forall\, \sigma_{ij} \in \Sigma^\varepsilon. \tag{50.101}$$

The first step is to transform the problem comprised by (50.100) and (50.101) so as to formulate it in a domain that does not depend on ε. Accordingly, we make the following substitutions

$$\Omega = \omega \times (-1, 1), \quad \Gamma_0 = \gamma \times [-1, 1],$$
$$\Gamma_+ = \omega \times \{1\}, \quad \Gamma_- = \omega \times \{-1\},$$

and to each point $x_i \in \bar{\Omega}$ we associate the point $x_i^\varepsilon \in \bar{\Omega}^\varepsilon$ by means of the correspondence

$$(x_1, x_2, x_3) \in \bar{\Omega} \to x_i^\varepsilon = (x_1, x_2, \varepsilon x_3) \in \bar{\Omega}^\varepsilon.$$

With the spaces Σ^ε, V^ε defined by (50.98) and (50.99) and their functions σ_{ij} and v_i, we associate the spaces Σ and V and their functions σ_i^ε and v_i^ε defined as

$$V = \{v_i \in W^{1,4}(\Omega), \text{ with } v_\alpha \text{ independent of } x_3 \text{ on } \Gamma_0, v_3 = 0 \text{ on } \Gamma_0\} \tag{50.102}$$

$$\Sigma = L^2(\Omega), \tag{50.103}$$

$$\sigma_{\alpha\beta}(x_i^\varepsilon) = \varepsilon^2 \sigma_{\alpha\beta}^\varepsilon(x_i), \quad \sigma_{\alpha 3}(x_i^\varepsilon) = \varepsilon^3 \sigma_{\alpha 3}^\varepsilon(x_i), \tag{50.104}$$

$$\sigma_{33}(x_i^\varepsilon) = \varepsilon^4 \sigma_{33}^\varepsilon(x_i), \quad v_\alpha(x_i^\varepsilon) = \varepsilon^2 v_\alpha^\varepsilon(x_i), \quad v_3(x_i^\varepsilon) = \varepsilon v_3^\varepsilon(x_i). \tag{50.105}$$

Furthermore, for the forces, we must control the way in which they depend on ε. Accordingly, we have the requirements

$$f_\alpha^\varepsilon(x_i^\varepsilon) = O(\varepsilon^3), \quad g_\alpha^\varepsilon(x_i^\varepsilon) = O(\varepsilon^4), \tag{50.106}$$

$$f_3^\varepsilon(x_i^\varepsilon) = \varepsilon^3 f_3(x_i) + O(\varepsilon^4), \tag{50.107}$$

$$g_3^\varepsilon(x_i^\varepsilon) = \varepsilon^4 g_3(x_i) + O(\varepsilon^5), \tag{50.108}$$

$$h_\alpha^\varepsilon(y) = \varepsilon^2 h_\alpha(y) + O(\varepsilon^3), \quad \forall\, y \in \gamma, \tag{50.109}$$

where the functions f_3, g_3, and h_α are independent of ε. Formulae (50.106)–(50.109) can be justified by the criterion of uniformity with respect the powers of ε relating to some relevant integrals occurring in (50.100) and (50.101). In particular, we have

$$\iiint_{\Omega^\varepsilon} \sigma_{ij} \frac{1}{2}(v_{i,j} + v_{j,i}) dV = \varepsilon^5 \iiint_{\Omega} \sigma_{ij}^\varepsilon \frac{1}{2}(v_{i,j}^\varepsilon + v_{j,i}^\varepsilon) dV, \tag{50.110}$$

$$\iiint_{\Omega^\varepsilon} f_3^\varepsilon v_3\, dV + \iint_{\Gamma_+^\varepsilon \cup \Gamma_-^\varepsilon} g_3^\varepsilon v_3\, dA + \int_\gamma \left\{ \int_{-\varepsilon}^{\varepsilon} v_\alpha\, dt \right\} h_\alpha^\varepsilon\, ds$$
$$= \varepsilon^5 \left(\iiint_{\Omega} f_3 v_3^\varepsilon\, dV + \iint_{\Gamma_+ \cup \Gamma_-} g_3 v_\varepsilon^3\, dA + \int_\gamma \left\{ \int_{-1}^{1} v_\alpha^\varepsilon\, dt \right\} h_\alpha\, ds \right), \tag{50.111}$$

as consequences of (50.104), (50.105), and (50.107)–(50.109).

Now, after completing a purely computational process, we arrive at the following result. Let $(\sigma_{ij}^\varepsilon, u_i^\varepsilon) \in \Sigma \times V$ be constructed from a solution $(\sigma_{ij}, u_i) \in \Sigma^\varepsilon \times V^\varepsilon$ of the problem comprising (50.100) and (50.101) by means of (50.104) and (50.105), then $(\sigma_{ij}^\varepsilon, u_i^\varepsilon)$ is a solution of the variational equations

$$\mathscr{A}_0(\sigma_{ij}^\varepsilon, \tau_{ij}) + \varepsilon^2 \mathscr{A}_2(\sigma_{ij}^\varepsilon, \tau_{ij}) + \varepsilon^4 \mathscr{A}_4(\sigma_{ij}^\varepsilon, \tau_{ij}) + \mathscr{B}(\tau_{ij}, u_i^\varepsilon) + \mathscr{C}_0(\tau_{ij}, u_i^\varepsilon, u_j^\varepsilon)$$
$$+ \varepsilon^2 \mathscr{C}_2(\tau_{ij}, u_i^\varepsilon, u_j^\varepsilon) = 0, \quad \forall\, \tau_{ij} \in \Sigma, \tag{50.112}$$

$$\mathscr{B}(\sigma_{ij}^\varepsilon, v_i) + 2\mathscr{C}_0(\sigma_{ij}^\varepsilon, u_i^\varepsilon, v_j) + 2\varepsilon^2 \mathscr{C}_2(\sigma_{ij}^\varepsilon, u_i^\varepsilon, v_j) = \mathscr{F}(v_i), \quad \forall\, v_i \in V, \tag{50.113}$$

where the operators $\mathscr{A}_0, \ldots, \mathscr{F}$ are defined as follows for the arbitrary elements $\sigma_{ij}, \tau_{ij} \in \Sigma$ and $u_i, v_i \in$ V:

$$\mathscr{A}_0(\sigma_{ij}, \tau_{ij}) = \iiint\limits_{\Omega} \tau_{ij}\left(\frac{1+\sigma}{E}\sigma_{\alpha\beta} - \frac{\sigma}{E}\sigma_{\mu\mu}\delta_{\alpha\beta}\right)dV,$$

$$\mathscr{A}_2(\sigma_{ij}, \tau_{ij}) = \iiint\limits_{\Omega}\left\{2\,\frac{1+\sigma}{E}\,\sigma_{\alpha 3}\tau_{\alpha 3} - \frac{\sigma}{E}(\sigma_{33}\tau_{\mu\mu} + \tau_{33}\sigma_{\mu\mu})\right\}dV,$$

$$\mathscr{A}_4(\sigma_{ij}, \tau_{ij}) = \iiint\limits_{\Omega} \sigma_{33}\tau_{33}\, dV,$$

$$\mathscr{B}(\tau_{ij}, v_i) = -\iiint\limits_{\Omega} \tau_{ij}\frac{1}{2}(v_{i,j} + v_{j,i})dV,$$

$$\mathscr{C}_0(\tau_{ij}, u_i, u_j) = -\frac{1}{2}\iiint\limits_{\Omega} \tau_{ij}u_{3,i}v_{3,j}\, dV,$$

$$\mathscr{C}_2(\tau_{ij}, u_i, v_j) = -\frac{1}{2}\iiint\limits_{\Omega} \tau_{ij}u_{\alpha,i}v_{\alpha,j}\, dV,$$

$$\mathscr{F}(v_i) = -\left(\iiint\limits_{\Omega} f_3 v_3\, dV + \iiint\limits_{\Gamma_+\cup\Gamma_-} g_3 v_3\, dA + \int\limits_{\gamma}\left\{\int\limits_{-1}^{1} v_\alpha\, dt\right\} h_\alpha\, ds\right),$$

the functions f_3, g_3, and h_α being obtained from (50.107)–(50.109).

As the forms $\mathscr{A}_0, \ldots, \mathscr{F}$ are all independent of ε, where ε is a small parameter, a formal series of approximations of a solution $(\sigma_{ij}^\varepsilon, u_i^\varepsilon)$ of (50.112) and (50.113) can be written as

$$(\sigma_{ij}^\varepsilon, u_i^\varepsilon) = (\sigma_{ij}, u_i) + \varepsilon(\sigma_{ij}^1, u_i^1) + \varepsilon^2(\sigma_{ij}^2, u_i^2) + \ldots, \tag{50.114}$$

where the coefficients of the powers of ε on the right-hand side of (50.114) are taken to be independent of ε. On substituting this expression for $(\sigma_{ij}^\varepsilon, u_i^\varepsilon)$ in (50.112) and (50.113) and equating the terms of the same order in ε, we find that the first term (σ_{ij}, u_i) must satisfy the equations

$$\mathscr{A}_0(\sigma_{ij}, \tau_{ij}) + \mathscr{B}(\tau_{ij}, u_i) + \mathscr{C}_0(\tau_{ij}, u_i, u_j) = 0, \quad \forall\, \tau_{ij} \in \Sigma, \tag{50.115}$$

$$\mathscr{B}(\sigma_{ij}, v_i) + 2\mathscr{C}_0(\sigma_{ij}, u_i, v_j) = \mathscr{F}(v_i), \quad \forall\, v_i \in V. \tag{50.116}$$

These equations can be written in an equivalent form by observing that the six qantities τ_{ij} and the three quantities v_i can be chosen independently. The explicit form of these equations is

$$\iiint\limits_{\Omega}\left(\frac{1+\sigma}{E}\,\sigma_{\alpha\beta} - \frac{\sigma}{E}\,\sigma_{\mu\mu}\delta_{\alpha\beta}\right)\tau_{\alpha\beta}\, dV - \iiint\limits_{\Omega}\tau_{\alpha\beta}\frac{1}{2}(u_{\alpha,\beta} + u_{\beta,\alpha})dV$$

$$-\frac{1}{2}\iiint\limits_{\Omega}\tau_{\alpha\beta}u_{3,\alpha}u_{3,\beta}\, dV = 0, \quad \forall\, \tau_{\alpha\beta} \in L^2(\Omega),$$

$$\iiint\limits_{\Omega}\tau_{\alpha 3}(u_{3,\alpha} + u_{\alpha,3})dV + \iiint\limits_{\Omega}\tau_{\alpha 3}u_{3,\alpha}u_{3,3}\, dV = 0, \quad \forall\, \tau_{\alpha 3} \in L^2(\Omega),$$

$$\iiint_{\Omega} \tau_{33} u_{3,3}\, dV + \frac{1}{2} \iiint_{\Omega} \tau_{33} u_{3,3} u_{3,3}\, dV = 0, \quad \forall\, \tau_{33} \in L^2(\Omega),$$

$$\iiint_{\Omega} \sigma_{i\beta} v_{\beta,i}\, dV = \int_{\gamma} \left\{ \int_{-1}^{1} v_\alpha\, dt \right\} h_\alpha\, ds, \quad \forall\, v_\alpha \in V_\alpha, \tag{50.117}$$

$$\iiint_{\Omega} \sigma_{i3} v_{3,i}\, dV + \iiint_{\Omega} \sigma_{ij} u_{3,i} v_{3,j}\, dV$$

$$= \iiint_{\Omega} f_3 v_3\, dV + \iint_{\Gamma_+ \cup \Gamma_-} g_3 v_3\, dA, \quad \forall\, v_3 \in V_3,$$

where we have written

$$V_1 = V_2 = \{v_\alpha \in W^{1,4}(\Omega);\ v_\alpha \text{ independent of } x_3 \text{ on } \Gamma_0\}, \tag{50.118}$$

$$V_3 = \{v_3 \in W^{1,4}(\Omega);\ v_3 = 0 \text{ on } \Gamma_0\}. \tag{50.119}$$

The equations above also make sense if, instead of the spaces V_β defined in (50.118), we let the functions v_β vary, and we look for u_α, in the spaces

$$V_1' = V_2' = \{v_\alpha \in H^1(\Omega);\ v_\alpha \text{ independent of } x_3 \text{ on } \Gamma_0\}.$$

For this reason we shall consider that the problem is posed on the space $\Sigma \times V'$, with $V' = V_1' \times V_2' \times V_3$, instead of the space $\Sigma \times V$ introduced before.

Now it is possible to show (Ciarlet 1980) that (50.117) are equivalent to a two-dimensional problem, in the sense that all the unknowns u_i and σ_{ij} can be computed from the solutions (u_α^0, u_3^0) of the system

$$\frac{2E}{3(1-\sigma^2)} \nabla^4 u_3^0 - 2\sigma_{\alpha\beta}^0 u_{3,\alpha\beta}^0 = g_3(x_\alpha, +1) + g_3(x_\alpha, -1) + \int_{-1}^{1} f_3\, dt \quad \text{in } \omega,$$

$$\sigma_{\alpha\beta,\alpha}^0 = 0, \text{ in } \omega,$$

$$u_3^0 = \frac{\partial}{\partial \nu} u_3^0 = 0, \text{ on } \gamma,$$

$$\sigma_{\alpha\beta}^0 \nu_\alpha = h_\beta, \text{ on } \gamma,$$

where

$$\sigma_{\alpha\beta}^0 = \frac{E}{(1-\sigma^2)} \{(1-\sigma)\tfrac{1}{2}(u_{\alpha,\beta}^0 + u_{\beta,\alpha}^0) + \sigma \delta_{\alpha\beta} u_{\mu,\mu}\}$$
$$+ \frac{E}{2(1-\sigma^2)} \{(1-\sigma) u_{3,\alpha}^0 u_{3,\beta}^0 + \sigma u_{3,\mu}^0 u_{3,\mu}^0 \delta_{\alpha\beta}\},$$

and (u_α^0, u_3^0) is a solution with the following properties of regularity:

$$u_3^0 \in H^4(\omega), \quad u_\alpha^0 \in H^1(\omega).$$

As the system just written is exactly equivalent to the Föppl–Kármán problem, we arrive at the conclusion that the Föppl–Kármán equations can be derived from the three-dimensional theory of elasticity as the first step in an asymptotic process of approximation.

51. Berger's and Reissner's Equations

The Föppl–Kármán equations have been modified in two ways: they have been simplified by means of a suitable hypothesis about the behavior of the nonlinear terms; and a more general system has been formulated which includes all the nonlinear terms instead of only some of them. In the first approach (Berger 1955; Bucco, Jones, and Mazumdar 1978), the equations are derived from the principle of minimum energy, but by neglecting the strain energy resulting from the second invariant of the strains. The resulting equations are still nonlinear, but they can be decoupled in such a manner that they can be readily solved. In order to derive these equations we observe that the strain energy of a plate subjected to large strains, written in terms of the deflection and the strains of the middle surface Ω, is

$$\mathscr{W} = \frac{D}{2} \iint_{\Omega} \left\{ (\nabla^2 w)^2 + \frac{3}{h^2}\left(u_x + v_y + \frac{1}{2}w_x^2 + \frac{1}{2}w_y^2\right)^2 \right. \\ \left. - 2(1-\sigma)\left(\frac{3}{h^2}e_2 + w_{xx}w_{yy} - w_{xy}^2\right)\right\} dA, \tag{51.1}$$

where e_2 is the second invariant of the strains in the middle surface, that is,

$$e_2 = \varepsilon_x\varepsilon_y - \varepsilon_{xy}^2 = (u_x + \tfrac{1}{2}w_x^2)(u_y + \tfrac{1}{2}w_y^2) - \tfrac{1}{4}(u_y + v_x + w_x w_y)^2. \tag{51.2}$$

As a working hypothesis, we assume that e_2 may be neglected in the expression for $\mathscr{W}$. There is no completely satisfactory justification for this approximation, and it is only based on comparison of the resulting approximate solutions, exact solutions for the deflection being available. If we accept the aforementioned hypothesis, the strain energy $\mathscr{W}$ can be rewritten as

$$\mathscr{W} = \frac{D}{2} \iint_{\Omega} \left\{ (\nabla^2 w)^2 + \frac{3}{h^2}(u_x + v_y + \tfrac{1}{2}w_x^2 + \tfrac{1}{2}w_y^2)^2 - 2(1-\sigma)(w_{xx}w_{yy} - w_{xy}^2)\right\} dA. \tag{51.3}$$

The work done by the transverse force $q(x, y)$ is

$$\mathscr{L} = \iint_{\Omega} qw\, dA.$$

If we apply the minimum principle to the total energy $\mathscr{E} = \mathscr{W} - \mathscr{L}$, we get the necessary conditions

$$\frac{\partial}{\partial x}(u_x + v_y + \tfrac{1}{2}w_x^2 + \tfrac{1}{2}w_y^2) = \frac{\partial}{\partial x}e = 0, \tag{51.4}$$

$$\frac{\partial}{\partial y}(u_x + v_y + \tfrac{1}{2}w_x^2 + \tfrac{1}{2}w_y^2) = \frac{\partial}{\partial y}e = 0, \tag{51.5}$$

$$\nabla^4 w - \frac{3}{h^2}\left[\frac{\partial}{\partial x}(ew_x) + \frac{\partial}{\partial y}(ew_y)\right] = \frac{q}{D}, \tag{51.6}$$

from the first two of which we derive the result that e is a constant, with the value

$$e = u_x + v_y + \tfrac{1}{2}w_x^2 + \tfrac{1}{2}w_y^2 = \frac{\alpha^2 h^2}{3D}, \tag{51.7}$$

where α is a constant of integration. On substituting (51.7) in (51.6) we obtain

$$\nabla^4 w - \alpha^2 \nabla^2 w = \frac{q}{D}. \tag{51.8}$$

Owing to the presence of α the equations retain their essential nonlinearity, but they have been decoupled so that it is possible to solve (51.8) for w because it is linear in w. We then use this solution in (51.7), which is linear in u and v, to determine α. In the case in which the plate is circular, and the load and the boundary conditions are axisymmetric, (51.7) and (51.8) become

$$e = \frac{du}{dr} + \frac{u}{r} + \frac{1}{2}\left(\frac{dw}{dr}\right)^2 = \frac{\alpha^2 h^2}{12}, \tag{51.9}$$

$$\left(\frac{d^2}{dr^2} + \frac{1}{r}\frac{d}{dr}\right)\left(\frac{d^2 w}{dr^2} + \frac{1}{r}\frac{dw}{dr} - \alpha^2 w\right) = \frac{q}{D}, \tag{51.10}$$

where $u = u(r)$ denotes now the radial displacement.

As an example of the theory we consider the circular plate, which is uniformly loaded and clamped at its boundary. Then the solution to (51.10) is

$$w = AJ_0(i\alpha r) + B - q\frac{r^2}{4D\alpha^2}, \tag{51.11}$$

where J_0 is the Bessel function of the first kind and A and B are constant. On using (51.11) and the integrals of the Bessel functions, (51.9) can be integrated to give

$$u = \frac{\alpha^2 h^2}{6}r + \frac{A^2\alpha^2}{4}r[J_1^2(i\alpha r) - J_0(i\alpha r)J_2(i\alpha r)] - q\frac{Ar}{2D\alpha^2}J_2(i\alpha r) - \frac{q^2 r^3}{32D^2\alpha^2} \tag{51.12}$$

where the constant of integration has been set equal to zero, so that u is finite at the origin. The remaining constants A, B, and α are determined by the boundary conditions

$$w(R) = 0, \quad \frac{dw}{dr}(R) = 0, \quad u(R) = 0, \tag{51.13}$$

where R is the radius of the plate. From the boundary conditions for w we can express the solution in terms of the only constant α:

$$w = \frac{qR^2}{4D\alpha^2}\left\{2\frac{[J_0(i\alpha R) - J_0(i\alpha r)]}{i\alpha R J_1^2(i\alpha R)} + 1 - \left(\frac{r}{R}\right)^2\right\}, \tag{51.14}$$

$$u = \frac{\alpha^2 h^2}{6} r - \frac{q^2 R^2 r}{16 D^2 \alpha^2}\left[\frac{J_1^2(i\alpha r) - J_0(i\alpha r)J_2(i\alpha r)}{J_1^2(i\alpha R)} - \frac{4J_2(i\alpha r)}{i\alpha R J_1(i\alpha r)} + \frac{1}{2}\left(\frac{r}{R}\right)^2\right]. \tag{51.15}$$

By using the boundary condition on u to determine the constant α, we obtain

$$\left(\frac{qR^4}{2Dh}\right)^2 = \frac{\frac{1}{3}(\alpha R)^6}{\frac{3}{4} + \frac{4}{(\alpha R)^2} + \frac{J_0(i\alpha R)}{i\alpha R J_1(i\alpha R)} + \frac{1}{2}\frac{J_0^2(i\alpha R)}{J_1^2(i\alpha R)}}. \tag{51.16}$$

From this equation we find the correct value of αR to be introduced in equations (51.14) and (51.15) for the given load. If we take the limit of (51.14) as α approaches zero and use the expansions of $J_0(i\alpha R)$ and $J_1(i\alpha R)$ for small α, we get

$$\lim_{\alpha\to 0}\left(\frac{wD}{qR^4}\right) = \frac{1}{64}\left[1 - \left(\frac{r}{R}\right)^2\right]^2 \tag{51.17}$$

which is, as expected, the solution for a clamped circular plate under small deflections. Alternatively when α is large, on using the expansions for $J_0(i\alpha R)$ and $J_1(i\alpha R)$ for large α, and combining (51.14) and (51.16), we obtain

$$\lim_{\alpha\to\infty}\left[\frac{w_{\max}}{\left(\frac{qR^4}{2Eh}\right)^{\frac{1}{3}}}\right] = 0.610. \tag{51.18}$$

The analogous result for a circular membrane is (Hencky 1915)

$$\frac{w_{\max}}{\left(\frac{qR^4}{2Eh}\right)^{\frac{1}{3}}} = 0.662. \tag{51.19}$$

In the case of a plate simply supported at its edge, the radial displacement and the deflection of the edge are zero, but it is free to rotate. The boundary conditions in this case are

$$w(R) = 0, \quad u(R) = 0, \quad \frac{d^2 w}{dr^2}(R) + \frac{\sigma}{R}\frac{dw}{dr}(R) = 0. \tag{51.20}$$

By using these boundary conditions on w to solve for A and B, and substituting these values in (51.11) and (51.12), we get

$$w = \frac{qR^2}{4D\alpha^2}\left\{1 - \left(\frac{r}{R}\right)^2 - \frac{2(1+\sigma)[J_0(i\alpha R) - J_0(i\alpha r)]}{(\alpha R)^2 J_0(i\alpha R) + (1-\sigma)i\alpha R J_1(i\alpha R)}\right\}, \tag{51.21}$$

$$u = \frac{\alpha^2 h^2}{6} r + \frac{q^2 R^2 r}{16 D^2 \alpha^4}\left\{\frac{(1+\sigma)^2[J_1^2(i\alpha r) - J_0(i\alpha r)J_2(i\alpha r)]}{[\alpha R J_0(i\alpha R) - (1-\sigma)iJ_1(i\alpha R)]^2}\right.$$
$$\left. - \frac{4(1+\sigma)J_2(i\alpha r)}{\alpha R[\alpha R J_0(i\alpha R) - (1-\sigma)iJ_1(i\alpha R)]} - \frac{1}{2}\left(\frac{r}{R}\right)^2\right\}. \tag{51.22}$$

As in the case of the clamped plate, setting $u(R) = 0$ gives a relation between αR and $qR^4/2Dh$. In addition, the deflection for small values of α is given by

$$\lim_{\alpha\to 0}\left(\frac{wD}{qR^4}\right) = \frac{1}{64}\left[\frac{5+\sigma}{1+\sigma} - \left(\frac{r}{R}\right)^2\right]\left[1 - \left(\frac{r}{R}\right)^2\right], \tag{51.23}$$

which is the solution for the simply supported circular plate according to the linear theory. The deflection for large values of α is the same as in the case of the clamped plate.

In Berger's theory the simplifying assumption that the second invariant e_2 is constant may appear as a mathematical device artificially introduced to decouple the problem. Some attempts have therefore been made to justify Berger's hypothesis in the form of a systematic approximation procedure (Adrianov 1984). In order to illustrate this we write the equilibrium equation of the plate in the form of the system

$$\frac{\partial e}{\partial x} + (1-\sigma)\left\{\frac{1}{2}\frac{\partial e_{xy}}{\partial y} - \frac{\partial \varepsilon_y}{\partial x}\right\} = 0, \tag{51.24}$$

$$\frac{\partial e}{\partial y} + (1-\sigma)\left\{\frac{1}{2}\frac{\partial e_{xy}}{\partial x} - \frac{\partial \varepsilon_x}{\partial y}\right\} = 0, \tag{51.25}$$

$$D(1-\sigma^2)\nabla^4 w - 2Eh[(ew_x)_x + (ew_y)_y$$
$$+ (1-\sigma)\{(\varepsilon_y w_x)_x + (\varepsilon_x w_y)_y - \tfrac{1}{2}(\varepsilon_{xy} w_y)_x - \tfrac{1}{2}(\varepsilon_{xy} w_x)_y\}] = 0, \tag{51.26}$$

where

$$\varepsilon_x = u_x + \tfrac{1}{2}w_x^2, \quad \varepsilon_y = v_y + \tfrac{1}{2}w_y^2, \quad e_{xy} = u_y + v_x + w_x w_y, \quad e = \varepsilon_x + \varepsilon_y.$$

Appropriate boundary conditions should supplement the equilibrium equations. For instance, if the plate is clamped the conditions are

$$w = u = v = 0, \quad \frac{\partial w}{\partial v} = 0.$$

We now perform the following change of variables

$$\xi = \frac{x}{a}, \quad \eta = \frac{y}{a}, \quad (u, v, w) = \frac{1}{a}(u', v', w'), \tag{51.27}$$

where a is a characteristic length. The equilibrium equations can then be rewritten in the following form

$$\begin{aligned}
&\frac{\partial e'}{\partial \xi} + (1-\sigma)(\tfrac{1}{2} e_{\xi\eta,\eta} - \varepsilon_{\eta,\xi}) = 0, \\
&\frac{\partial e'}{\partial \eta} + (1-\sigma)(\tfrac{1}{2} e_{\xi\eta,\xi} - \varepsilon_{\xi,\eta}) = 0, \\
&\nabla^4 w' - \varepsilon^{-2}[(e' w'_\xi)_\xi + (e' w'_\eta)_\eta] \\
&+ (1-\sigma)\{(\varepsilon_\eta w'_\xi)_\xi + (e_\xi w'_\eta)_\eta - \tfrac{1}{2}(e_{\xi\eta} w'_\eta)_\xi - \tfrac{1}{2}(e_{\xi\eta} w'_\xi)_\eta\} = 0,
\end{aligned} \tag{51.28}$$

with the positions

$$\varepsilon_\xi = u'_\xi + \tfrac{1}{2} w'^2_\xi, \quad \varepsilon_\eta = v'_\eta + \tfrac{1}{2} w'^2_\eta, \quad e_{\xi\eta} = u'_\eta + v'_\xi + w'_\xi w'_\eta,$$

$$e' = \varepsilon_\xi + \varepsilon_\eta, \quad \varepsilon = \frac{h}{\sqrt{3}a}.$$

Let us represent the displacement vector in the form

$$\begin{aligned}
u' &= u'(\varepsilon^{\frac{1}{2}}\theta(\xi, \eta), \xi, \eta, \varepsilon), \\
v' &= v'(\varepsilon^{\frac{1}{2}}\theta(\xi, \eta), \xi, \eta, \varepsilon), \\
w' &= w'(\varepsilon^{\frac{1}{2}}\theta(\xi, \eta), \xi, \eta, \varepsilon),
\end{aligned} \tag{51.29}$$

where $\theta(\xi, \eta)$ is a new independent variable. As this form (51.29) requires that we have

$$\frac{\partial}{\partial \xi} = \frac{\partial}{\partial \xi} + \varepsilon^{-\frac{1}{2}} \theta_\xi \frac{\partial}{\partial \theta}, \quad \frac{\partial}{\partial \eta} = \frac{\partial}{\partial \eta} + \varepsilon^{-\frac{1}{2}} \theta_\eta \frac{\partial}{\partial \theta},$$

we write the limiting system for $\varepsilon \to 0$:

$$\frac{\partial e'}{\partial \theta} \theta_\xi + \tfrac{1}{2}(1-\sigma)(u'_{\theta\theta}\theta^2_\eta - v'_{\theta\theta}\theta_\xi\theta_\eta) = 0, \tag{51.30}$$

$$\frac{\partial e'}{\partial \theta} \theta_\eta + \tfrac{1}{2}(1-\sigma)(v'_{\theta\theta}\theta^2_\xi - u'_{\theta\theta}\theta_\xi\theta_\eta) = 0, \tag{51.31}$$

$$w'_{\theta\theta\theta\theta}(\theta^2_\xi + \theta^2_\eta)^2 - \left\{\frac{\partial e'}{\partial \theta} w'_\theta(\theta_\xi + \theta_\eta)\right\} - e' w'_{\theta\theta}(\theta^2_\xi + \theta^2_\eta) = 0. \tag{51.32}$$

Now (5.130) and (51.31), which are regarded as an algebraic system in $u'_{\theta\theta}$ and $v'_{\theta\theta}$, is compatible only for $\partial e'/\partial\theta = 0$. Then the term in the braces drops out in (51.32). Now, if we return to the variables x and y, we obtain the results $\partial e'/\partial x = \partial e'/\partial y = 0$. On taking these into account, we obtain Berger's equation from (51.32) in the initial variables.

Berger's equations yield very accurate results for the symmetrically loaded circular plate with a clamped edge. As, in this case, ε_r and ε_θ are both positive and of the same order of magnitude, it is difficult to see why Berger's hypothesis should be valid. A partial answer to the question is that the term e_2, in the expression for the strain energy (51.1), is multiplied by $(1-\sigma)$, while D is an experimentally determined quantity. Then for $\sigma \simeq 1$, the term containing e_2 is indeed negligible, although the value $\sigma = 1$ is not physically possible. However, an explanation of Berger's method may be obtained from the Föppl-Kármán equation by perturbing the solutions with respect the parameter $\mu = 1 - \sigma$ (Schmidt 1974). The total potential energy in a circular plate of radius R loaded by a symmetric force q per unit area is

$$\mathscr{E} = \mathscr{W} - \mathscr{L} = \frac{D}{2}\int_0^{2\pi}\int_0^R \left\{ \left(w_{rr}^2 + \frac{1}{r^2} w_r^2 + \frac{2\sigma}{r} w_r w_{rr} \right) + \frac{3}{h^2}[(u_r + \tfrac{1}{2}w_r^2)^2 + \frac{1}{r^2}u^2 + \frac{2\sigma}{r}u(u_r + \tfrac{1}{2}w_r^2)] \right\} r\,dr\,d\theta - \int_0^{2\pi}\int_0^R qwr\,dr\,d\theta. \tag{51.33}$$

The vanishing of the first variation of $\mathscr{E}$ and the use of the positions $\beta = w' = dw/dr$ yields the system of equations

$$\left[\frac{1}{r}(ru)'\right]' + \left(\frac{\beta^2}{r}\right)' + (1-\sigma)\frac{\beta^2}{2r} = 0 \tag{51.34}$$

$$\left[\frac{1}{r}(r\beta)'\right]' - \frac{3\beta}{h^2}\left(u' + \frac{\sigma u}{r} + \frac{\beta^2}{2}\right) = \frac{3}{Dh^2 r}\int_0^r qr\,dr, \tag{51.35}$$

where the prime denotes differentiation with respect to r. The boundary conditions associated with (51.34) and (51.35) are

$$u(R) = \beta(R) = 0, \quad \beta(0) = 0, \quad u(0) = 0. \tag{51.36}$$

In order to solve (51.34) and (51.35) by perturbation, we put

$$u = \sum_{m=0}^{\infty} \mu^m u_m(r), \quad \beta = \sum_{m=0}^{\infty} \mu^m \beta_m(r),$$

and substitute these expressions in (51.34) and (51.35), equating the terms of equal order in μ. The equations of the first order of approximation are

$$\left[\frac{1}{r}(ru_0)'\right]' + \left(\frac{\beta_0^2}{2}\right) = 0$$

$$\left[\frac{1}{r}(r\beta_0)'\right]' - \frac{3\beta_0}{h^2}\left(u_0' + \frac{u_0}{r} + \frac{\beta_0^2}{2}\right) = \frac{3}{2Dr}\int_0^r qr\,dr,$$

which coincide with Berger's equations.

When, however, the strains and the displacements in an initially planar plate are sufficiently large, it is necessary to remove some of the simplification introduced in the Föppl–Kármán theory. The correction is relatively easy when the plate is circular and the load axisymmetric (Reissner 1949). Due to the symmetry, the strained surface is a surface of revolution, which can be taken in the parametric form $r = r(\xi)$, $z = z(\xi)$, where r and z are cylindrical coordinates and ξ is a parameter by means of which the position of a point on the meridian can be specified (Figure 51.1). Any point of the surface is thus defined by ξ and the angle θ that the axial plane passing through the point makes with a fixed axial plane. On putting

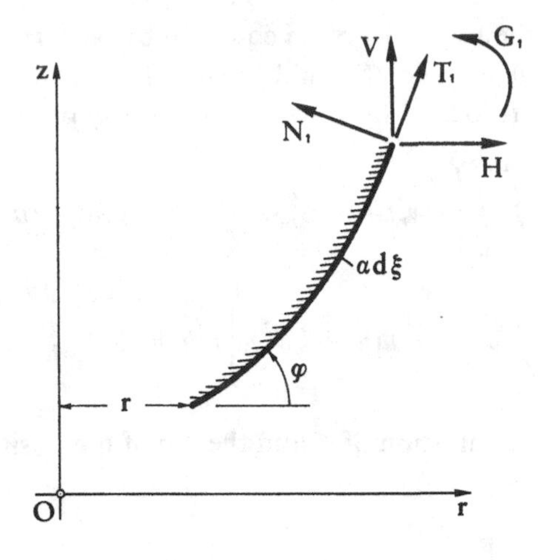

Fig. 51.1

$$\alpha^2 = r'^2 + z'^2, \tag{51.37}$$

the length of an element of arc along the meridian is $\alpha\, d\xi$. Due to symmetry, the only nonvanishing stress resultants are T_1, T_2, and N_1, while the nonvanishing resultant couples are G_1 and G_2. If we introduce the horizontal stress resultant H and the vertical resultant stress V, we have the relations

$$\begin{aligned} T_1 &= H\cos\varphi + V\sin\varphi, \\ N_1 &= -H\sin\varphi + V\cos\varphi, \end{aligned} \tag{51.38}$$

where φ is the slope of the meridian curve. Let p_H and p_V denote the components, parallel to the r and z axes of the load estimated per unit of area of the strained middle surface. Then the relevant equilibrium equations can be written as

$$(rV)' + r\alpha p_V = 0, \tag{51.39a}$$

$$(rH)' - \alpha T_2 + r\alpha p_H = 0, \tag{51.39b}$$

$$(rG_1)' - \alpha\cos\varphi\, G_2 - r\alpha N_1 = 0. \tag{51.39c}$$

Let $(\xi, 0)$ be the coordinate of a point that, after the deformation, occupies the position $[r(\xi), z(\xi)]$ so that the displacement is

$$[u(\xi), w(\xi)] = [r(\xi) - \xi, z(\xi)], \tag{51.40}$$

and the corresponding strains are given by

$$\varepsilon_1 = \frac{1}{\cos\varphi}(1 + u') - 1, \quad \varepsilon_2 = \frac{u}{\xi}. \tag{51.41}$$

The curvatures of strained surface have the expressions

$$\kappa_1 = -\varphi', \quad \kappa_2 = -\frac{\sin\varphi}{\xi}. \tag{51.42}$$

On assuming that the stress resultants and couples are related to the strains by linear constitutive equations, we have the equations

$$\begin{aligned} \varepsilon_1 &= \frac{1}{C}(T_1 - \sigma T_2), \quad \varepsilon_2 = \frac{1}{C}(T_2 - \sigma T_1), \\ \kappa_1 &= \frac{1}{D}(G_1 - \sigma G_2), \quad \kappa_2 = \frac{1}{D}(G_2 - \sigma G_1), \end{aligned} \tag{51.43}$$

with $C = 2Eh/(1-\sigma^2)$, $D = 2Eh^3/[3(1-\sigma^2)]$.

While we wish to admit large displacements of the plate, we wish at the same time to assume that the strains are not very large. In fact, only in this case can we retain the stress–strain relations (51.43). In view of this restriction we may replace $r(\xi)$ by ξ and α by 1 in the equilibrium equations and write them in the following modified form:

$$\begin{aligned} (\xi V)' + \xi p_V = 0, \quad (\xi H)' - T_2 + \xi p_H = 0, \\ (\xi G_1)' - \cos\varphi\, G_2 - \xi(-H\sin\varphi + V\cos\varphi) = 0. \end{aligned} \tag{51.44}$$

Following the analogy of what is done in the theory of small deflections, we reduce this system to two simultaneous equations for the meridian angle φ and a stress function Ψ defined by

$$\Psi = \xi H. \tag{51.45}$$

In terms of φ and Ψ we have the following expressions for the characteristics of the stress:

$$\begin{aligned} \xi V &= -\int \xi p_V \, d\xi + \text{constant}, \\ \xi T_1 &= \Psi\cos\varphi + \xi V \sin\varphi, \\ \xi N_1 &= -\Psi\sin\varphi + \xi V\cos\varphi, \\ T_2 &= \Psi' + \xi p_H, \end{aligned} \tag{51.46}$$

and

$$\begin{aligned} G_1 &= -D\left(\varphi' + \sigma\frac{\sin\varphi}{\xi}\right), \\ G_2 &= -D\left(\frac{\sin\varphi}{\xi} + \sigma\varphi'\right). \end{aligned} \tag{51.47}$$

Now, on substituting G_1 and G_2 from (51.47) in (51.39c), and introducing ε_1 and ε_2, as expressed in terms of φ and Ψ by means of (51.42) and (51.46), in the compatibility equation (as a consequence of (51.41))

$$(\xi\varepsilon_2)' - \cos\varphi\,\varepsilon_1 = \cos\varphi - 1, \tag{51.48}$$

we arrive at a system of two differential equations in φ and Ψ, the final form of which is

$$\varphi'' + \frac{1}{\xi}\varphi' - \frac{\cos\varphi\sin\varphi}{\xi^2} = \frac{1}{\xi D}(\Psi\sin\varphi - \xi V\cos\varphi), \tag{51.49}$$

$$\Psi'' + \frac{1}{\xi}\Psi' - \left(\frac{\cos^2\varphi}{\xi^2} - \frac{\sigma}{\xi}\varphi'\sin\varphi\right)\Psi = \frac{C}{\xi}(\cos\varphi - 1) - \frac{(\xi^2 p_H)'}{\xi} - \frac{\sigma\cos\varphi}{\xi^2}(\xi^2 p_H)$$
$$+ \left(\frac{\sin\varphi\cos\varphi}{\xi^2} + \frac{\sigma}{\xi}\varphi'\cos\varphi\right)\xi V + \frac{\sigma\sin\varphi}{\xi}(\xi V)'. \tag{51.50}$$

Although these equations are the result of several approximations, they are still difficult to solve. Some attempts have therefore been made to reduce the equations to a simpler form through certain plausible approximations. Equation (51.50) can be replaced by the approximate equations (Reissner 1963)

$$\xi^2\Psi'' + \xi\Psi' - \Psi = C\xi(\cos\varphi - 1) - \xi^3 p_H' - (2+\sigma)\xi^2 p_H, \tag{51.51}$$

and (51.49) can be reduced to the approximate form (Billington 1967)

$$\xi^2\varphi'' + \xi\varphi' - \varphi = \frac{\xi}{D}(\Psi\sin\varphi - \xi V\cos\varphi). \tag{51.52}$$

The solution of these new equations is still difficult, but they have a structure suited to the application of approximation procedures of integration.

Another way to investigate the properties of Reissner's equations is to see whether it is possible to derive closed-form solutions that are valid in special circumstances. This is the case, for instance (Simmonds 1983), for an annular plate subjected to vertical edge forces but no edge moments and certain radial edge displacements proportional to the bending stiffness, as specified below. In this case, Reissner's equations read

$$A(\xi^2\Psi'' + \xi\Psi' - \Psi) = \xi(\cos\varphi - 1), \tag{51.53}$$

$$D(\xi^2\varphi'' + \xi\varphi' - \sin\varphi\cos\varphi) = \xi(\Psi\sin\varphi - P\cos\varphi), \tag{51.54}$$

where $A = 1/2Eh$ and $2\pi P$ is the net vertical load applied at the outer edge $\xi = b$. For equilibrium, an opposite load is applied to the inner edge $\xi = a\,(a < b)$. The hoop strain ε_2 and the radial stress couple are given, respectively, by

$$\varepsilon_2 = \frac{1}{\xi}u = A\left[\Psi' - \sigma\frac{1}{\xi}(\Psi\cos\varphi + P\sin\varphi)\right], \tag{51.55}$$

$$G_1 = -D\left(\varphi' + \frac{\sigma}{\xi}\sin\varphi\right). \tag{51.56}$$

The boundary conditions at the edges $\xi = a$, b are $G_1 = 0$, and u takes prescribed values.

To manipulate the equations we introduce the parameters

$$\alpha = 2\left(\frac{AP}{b}\right)^{\frac{1}{2}} = 2\left(\frac{P}{2Ehb}\right)^{\frac{1}{2}},$$

$$\beta^2 = \frac{4D\alpha}{Pb} = \frac{2}{3(1-\sigma^2)}\left(\frac{2Ehb}{P}\right)^{\frac{2}{3}}\left(\frac{2h}{b}\right)^2.$$

We then set

$$\xi^2 = b^2 x, \quad \Psi = \left(\frac{P}{\alpha}\right) x^{-\frac{1}{2}} f, \quad \varphi = \alpha x^{\frac{1}{2}} g,$$

and

$$\varepsilon_2 = \left(\frac{AP}{\alpha b}\right) e, \quad u = \left(\frac{AP}{\alpha}\right) v, \quad G_1 = -\left(\frac{D\alpha}{b}\right) m.$$

Equations (51.53)–(51.56) reduce to

$$f'' = 2(\alpha x^{\frac{1}{2}})^{-2}[\cos(\alpha x^{\frac{1}{2}} g) - 1], \tag{51.57}$$

$$\beta^2\{(xg)'' + \tfrac{1}{4}x^{-1}[g - (\alpha x^{\frac{1}{2}})^{-1}\sin(\alpha x^{\frac{1}{2}} g)\cos(\alpha x^{\frac{1}{2}} g)]\} = f(\alpha x^{\frac{1}{2}})^{-1}\sin(\alpha x^{\frac{1}{2}} g) - \cos(\alpha x^{\frac{1}{2}} g), \tag{51.58}$$

$$e = x^{-\frac{1}{2}} v = 2f' - x^{-1}\{f + \sigma[f\cos(\alpha x^{\frac{1}{2}} g) + \alpha x^{\frac{1}{2}}\sin(\alpha x^{\frac{1}{2}} g)]\}, \tag{51.59}$$

$$m = 2xg' + g + \sigma(\alpha x^{\frac{1}{2}})^{-1}\sin(\alpha x^{\frac{1}{2}} g), \tag{51.60}$$

where the prime now denotes differentiation with respect to x.

In the limiting case for $\alpha \to 0$, (51.57)–(51.60) become

$$f'' = -g^2, \quad \beta^2 x(xg)'' = fg - 1, \tag{51.61a}$$

$$e = x^{-\frac{1}{2}} v = 2f' - x^{-1}(1+\sigma) f, \tag{51.61b}$$

$$m = 2xg' + (1+\sigma) g. \tag{51.61c}$$

This is just a form of the Föppl–Kármán equations in the axisymmetric case. Let us now assume that $\sigma = 1/3$. By inspection we see that

$$f = \left(\tfrac{9}{2}\right)^{\frac{1}{3}} x^{\frac{2}{3}} - \left(\tfrac{2}{9}\right)\beta^2, \quad g = \left(\tfrac{2}{9}\right)^{\frac{1}{3}} x^{-\frac{2}{3}} \tag{51.62}$$

satisfy (51.61a) and yield $m \equiv 0$, while (51.61b) reduces to

$$e = x^{-\frac{1}{2}} v = \left(\tfrac{8}{27} x\right)\beta^2. \tag{51.63}$$

Thus (51.62) is an exact solution for an annular plate under vertical loads at the edges, provided that at the edges the radial displacement have the expressions

$$v\left(\frac{a}{b}, \beta\right) = \frac{8}{27}\left(\frac{b}{a}\right)^{\frac{1}{2}} \beta^2, \quad v(1, \beta) = \frac{8}{27}\beta^2.$$

Another limiting case occurs for $\beta \to 0$, because it represents the extreme situation in which the bending stiffness is small compared with the extensional rigidity. For $\beta \to 0$ our solution reduces exactly to that for a perfectly flexible circular membrane (Schwerin 1929).

52. The Wrinkling of Plates

The most important application of the theory of moderately large strains in initially planar plates is the analysis of the stability of a plate under edge thrust parallel to its middle plane. If the thrust is not too great, the plate simply contracts in its plane and the state of stress arising in its interior is determined by the stress resultants of the type T_1, T_2, and $S_2 = -S_1$. The latter can be found by solving a problem of generalized plane stress. Often, however, the calculation of the stress resultants is immediate. Let us take, for example, a rectangular plate, the edges of which in the unstrained state are given by the equations $x = \pm a$, $y = \pm b$. Let P_1 and P_2 be the values of the thrusts along these pairs of edges. Then the stress resultants in the interior are $T_1 = -P_1$, $T_2 = -P_2$, and $S_1 = 0$, and the plate is simply contracted in its plane. If, however, P_1 and P_2 are sufficiently large the simply contracted state of equilibrium becomes unstable and the plate, under the same state of thrust, may assume a bent configuration. This configuration is characterized by a transverse displacement w, which, after omission of quantities of second order, again satisfies (50.14), where now we have

$$\frac{\partial^2 F}{\partial y^2} = -P_1, \quad \frac{\partial^2 F}{\partial x^2} = -P_2, \quad \frac{\partial^2 F}{\partial x\, \partial y} = 0.$$

If the plate is simply supported at its edges we must have

$$w = 0, \quad \frac{\partial^2 w}{\partial x^2} + \sigma \frac{\partial^2 w}{\partial y^2} = 0 \text{ at } x = \pm a, \text{ and } w = 0, \quad \frac{\partial^2 w}{\partial y^2} + \sigma \frac{\partial^2 w}{\partial x^2} = 0 \text{ at } y = \pm b.$$

The solution is of the form

$$w = W_0 \sin \frac{n\pi(x+a)}{2a} \sin \frac{m\pi(y+b)}{2b}, \tag{52.1}$$

where W_0 is a constant, and m and n are integers. However, in order that (50.14) admits nontrivial solutions, P_1 and P_2 must satisfy the equation (Bryan 1890)

$$\frac{1}{4} D\pi^2 \left(\frac{m^2}{a^2} + \frac{n^2}{b^2} \right)^2 = P_1 \frac{m^2}{a^2} + P_2 \frac{n^2}{b^2}. \tag{52.2}$$

Exactly as in the problem of the buckling of struts, the simple contracted state of equilibrium becomes critical when P_1 and P_2 reach the smallest value consistent with (52.2); for example, when P_1 and P_2 satisfy (52.2) with $m = n = 1$.

Equation (52.2) results from the application of the linearized criterion for stability. This criterion is useful for evaluating the onset of elastic instability in a plane plate compressed in its plane. However, it cannot describe the so-called postcritical deformations of a plate. The structure of the buckled states might be very different from those predicted by the linearized theory, and their analysis needs a special treatment. An unexpected feature of the nonlinear buckled configurations is that they are not necessarily symmetrical, even when the unbuckled, fundamental, state is symmetrical. A typical example is the branching of unsymmetrical equilibrium states from those which are axisymmetric in clamped circular plates subjected to a uniform edge thrust and a uniform lateral pressure (Cheo and Reiss 1973). The existence of these unsymmetric, or wrinkled, states was conjectured long ago (Friedrichs and Stoker 1942), but their detailed description is possible only by perturbation analysis.

We consider a circular plate of radius R and thickness $2h$, subject to a uniform radial edge stress T and a uniform surface pressure p. We employ polar coordinates r, θ, with respect the plate's center, and denote the displacement normal to the middle plane, and the membrane stresses by $W^*(r,\theta)$, $\sigma_r(r,\theta)$, $\sigma_\theta(r,\theta)$, and $\sigma_{r\theta}(r,\theta)$, respectively. The stress components are derived from an Airy stress function $F^*(r,\theta)$ defined by

$$\sigma_r = r^{-2}(rF_r^* + F_{\theta\theta}^*), \quad \sigma_\theta = F_{rr}^*, \quad \sigma_{r\theta}^* = -(r^{-1}F_\theta^*)_r, \tag{52.3}$$

where the subscripts denote partial differentiation with respect to r and θ.

We introduce dimensionless variables and parameters defined by the following relations:

$$x = \frac{r}{R}, \quad W(x,\theta) = CW^*(\tau,\theta)\frac{1}{2h}, \quad F(x,\theta) = \left(\frac{C^2}{4Eh^2}\right)F^*(r,\theta),$$

$$\lambda = -C^2\left(\frac{T}{E}\right)\left(\frac{R}{2h}\right)^2, \quad P = C^3\left(\frac{p}{E}\right)\left(\frac{R}{2h}\right)^4, \quad C = [12(1-\sigma^2)]^{\frac{1}{2}}. \tag{52.4}$$

The edge thrust parameter λ is positive for $T < 0$; that is, when it is compressive. In terms of the new variables, the equilibrium equations for the deformed plate assume the form

$$\nabla^4 F = -\tfrac{1}{2}[W, W];$$

$$\nabla^4 W = P + [F, W], \text{ for } x < 1,\ 0 \leqslant \theta < 2\pi; \tag{52.5}$$

$$W(1,\theta) = W_x(1,\theta) = 0, \quad F(1,\theta) = 0, \quad F_x(1,\theta) = -\lambda, \text{ for } 0 \leqslant \theta < 2\pi. \tag{52.6}$$

Here the nonlinear operator $[g, h]$ is defined, for any two functions g and h, by

$$[g,h] = \frac{1}{x}\left\{g_{xx}\left(h_x + \frac{1}{x}h_{\theta\theta}\right) + h_{xx}\left(g_x + \frac{1}{x}g_{\theta\theta}\right) - 2x\left(\frac{h_\theta}{x}\right)_x\left(\frac{g_\theta}{x}\right)_x\right\}.$$

The problem comprising (52.5) and (52.6) admits, of course, axisymmetric solutions. If $\{W_0(x), F_0(x)\}$ is any one of these solutions and we put

$$\alpha(x) = W_0'(x), \quad \gamma(x) = F_0'(x),$$

the problem can be formulated as follows:

$$L\gamma = -\frac{1}{2}\alpha^2, \quad L\alpha = \alpha\gamma + \frac{P}{2}x^2, \tag{52.7}$$

$$\alpha(0) = \gamma(0) = 0, \quad \alpha(1) = 0, \quad \gamma(1) = -\lambda, \tag{52.8}$$

where L is the differential operator

$$Lv(x) = x\left[\frac{1}{x}(xv)'\right]'.$$

For $\lambda \leqslant 0$, problem (52.7) and (52.8) has a unique solution; if λ is positive and sufficiently large, then the solutions are nonunique.

Our purpose is to analyze unsymmetric solutions that branch from a symmetric solution of problem (52.7) and (52.8). The onset of these wrinkled solutions is suggested by the following consideration. Let us define the dimensionless, axisymmetric, circumferential membrane stress

$$\tau(x) = C^2\left(\frac{R}{2h}\right)^2 \frac{\sigma\theta}{E}, \tag{52.9}$$

and investigate the axisymmetric solutions corresponding to $P = 0$ and to increasing values of λ. As λ increases, a strip with large circumferential compressive stresses develops, starting from the boundary of the plate. For sufficiently large λ, the strip behaves like a compressed ring, which may buckle unsymmetrically, generating in this way an unsymmetric wrinkling near the edge of the plate. The distribution of the dimensionless stress $\tau(x)$, for $P = 0$ and increasing values of λ is shown in Figure 52.1. The solutions correspond to buckled states arising above the lowest axisymmetric buckling load $\lambda = \lambda_1 \simeq 14.7$.

In order to determine the wrinkling loads, we express the unsymmetric solutions $\{W(x, \theta; P, \lambda), F(x, \theta; P, \lambda)\}$ of problem (52.5) and (52.6) in the form

$$\begin{aligned} W(x, \theta; P, \lambda) &= W_0(x; P, \lambda) + \varepsilon w(x, \theta; P, \lambda, \varepsilon), \\ F(x, \theta; P, \lambda) &= F_0(x; P, \lambda) + \varepsilon f(x, \theta; P, \lambda, \varepsilon), \end{aligned} \tag{52.10}$$

where $\{W_0, F_0\}$ is a solution of the axisymmetric problem and ε is a small parameter. On inserting (52.10) in (52.5) and (52.6) we obtain

$$\begin{aligned} &\nabla^4 f + [W_0, w] = -\frac{\varepsilon}{2}[w, w], \\ &\nabla^4 w - [W_0, f] - [F_0, w] = \varepsilon[f, w] \\ &w = w_x = f = f_x = 0, \ \text{for } x = 1. \end{aligned} \tag{52.11}$$

We now linearize (52.11) by setting $\varepsilon = 0$. This yields the eigenvalue problem

$$\begin{aligned} &\nabla^4 f^{(0)} + [W_0, w^{(0)}] = 0, \\ &\nabla^4 w^{(0)} - [W_0, f^{(0)}] - [F_0, w^{(0)}] = 0, \\ &w^{(0)} = w_x^{(0)} = f^{(0)} = f_x^{(0)} = 0, \ \text{for } x = 1, \end{aligned} \tag{52.12}$$

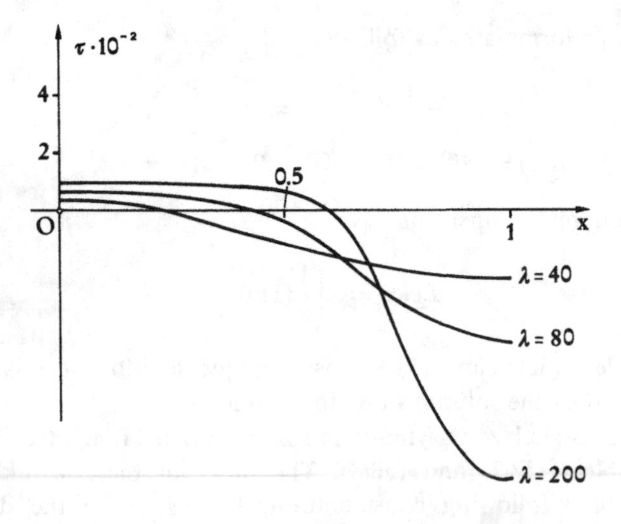

Fig. 52.1

where $w^{(0)} = w(x, \theta; P, \lambda, 0)$ and $f^{(0)} = f(x, \theta; P, \lambda, 0)$. For the combined loading problem, we wish to determine, for fixed value of λ and a specific solution of problem (52.7) and (52.8), the eigenvalues $P = P_0(\lambda)$ of (52.12). These eigenvalues are wrinkling loads. As $\{W_0(x; P, \lambda), F_0(x; P, \lambda)\}$ is a solution of the axisymmetric problem, the eigenvalue parameter appears nonlinearly in (52.12). The eigenfunctions of (52.12) are separable, and, without loss of generality, they are given by

$$\begin{aligned} f^{(0)} &= y_n(x; P, \lambda) \sin n\theta, \\ w^{(0)} &= z_n(x; P, \lambda) \sin n\theta, \end{aligned} \tag{52.13}$$

for $n = 1, 2, \ldots$. Thus $\{y_n, z_n\}$ satisfy the system

$$\begin{aligned} &L_n^2 y_n + [W_0, z_n]^0 = 0, \\ &L_n^2 z_n - [W_0, y_n]^0 - [F_0, z_n]^0 = 0, \\ &y_n = y_n' = z_n = z_n' = 0, \text{ for } x = 1, \end{aligned} \tag{52.14}$$

where $[\Phi_0, h]^0$ and the differential operators L_n are defined by

$$L_n h(x) = \frac{1}{x}(xh')' - \frac{n^2}{x^2} h \quad (n = 1, 2, \ldots),$$

$$[\Phi_0, h]^0 = \frac{1}{x}\left\{\Phi_0' h'' + \Phi_0''\left(h' - \frac{n^2}{x} h\right)\right\}.$$

As the stresses and displacements are bounded and single-valued at $x = 0$, we conclude from (52.3), (52.4), (52.10), and (52.13) that we have

$$y_n(0) = z_n(0) = 0 \quad (n = 1, 2, \ldots). \tag{52.15}$$

These equations imply that $y_n'(0)$ and $z_n'(0)$ are zero for $n = 2, 3, \ldots$, with $y_1'(0)$ and $z_1'(0)$ arbitrary. Thus the wrinkling loads $P = P_0(\lambda, n)$ are determined by solving the eigenvalue problem (52.14).

Accounting for moderately large strains in plates is necessary not only in studying buckling problems, but also for explaining some paradoxes arising in simple equilibrium problems in which the transverse load depends on the deflection (Kerr and Coffin 1990). The problem is now that of an infinite thin elastic strip, clamped along both edges, and subjected to the hydrostatic pressure of a fluid, as shown in Figure 52.2, where the free surface of the fluid coincides with the upper surface of the undeformed strip. If $w = w(x)$ is the deflection, the vertical load is

$$q(x) = \gamma w(x), \tag{52.16}$$

where γ is the specific weight of the liquids. It then follows that $w(x)$ is the solution of the boundary-value problem

$$Dw^{iv}(x) - \gamma w(x) = 0, \quad 0 < x < L; \tag{52.17}$$

$$w(0) = w(L) = 0, \quad w'(0) = w'(L) = 0. \tag{52.18}$$

The problem obviously admits the solution $w(x) \equiv 0$; there is, however, a nontrivial solution provided that we have

$$\cosh \beta L \cos \beta L = 1, \tag{52.19}$$

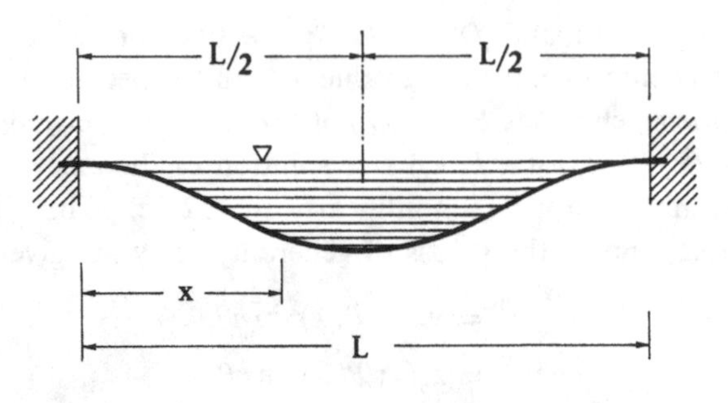

Fig. 52.2

where

$$\beta = \left(\frac{\gamma}{D}\right)^{\frac{1}{4}}. \tag{52.20}$$

Except for the meaning of γ, (52.19) is identical with the equation giving the natural frequencies of a clamped strip. The corresponding eigenvalues are $(\beta L)_1 = 4.7300$, $(\beta L)_2 = 7.8532, \ldots$. This means that, according to this analysis, the deflections are identically zero, except for those corresponding to special values of βL.

As the conclusion is not physically reasonable, it is necessary to have recourse to a nonlinear formulation. For this purpose, we exploit the symmetry of the solution and place the origin at the midpoint of the span where $x = L/2$. If the transverse deflection $w(x)$ is sufficiently large, the longitudinal strain

$$\varepsilon_x = u'(x) + \tfrac{1}{2}w'(x)^2, \tag{52.21}$$

where $u(x)$ is the axial displacement, is not negligible, and a membrane force

$$T_1 = \frac{2Eh}{(1-\sigma^2)}\,(u' + \tfrac{1}{2}w'^2) \tag{52.22}$$

acts tangentially to the deformed surface of the strip along the lines which were parallel to the x axis prior to the deformation. The approximate form of the equilibrium equation in the direction of T_1 (see (50.17a)) is $T_1' = 0$, whence, after integration with respect to x and use of (52.22), we obtain

$$\frac{2Eh}{1-\sigma^2}\,(u' + \tfrac{1}{2}w'^2) = T_1 = \text{constant}. \tag{52.23}$$

This result reduces the equilibrium equation (50.21) to

$$Dw^{iv} - T_1 w'' = \gamma w, \quad \text{for } 0 < x < \frac{L}{2}, \tag{52.24}$$

where now we have added the transverse load γw. The general solution of (52.24) is

$$w(x) = B_1 \cosh \rho x + B_2 \sinh \rho x + B_3 \cos \kappa x + B_4 \sin \kappa x, \tag{52.25}$$

where

$$\left.\begin{matrix}\rho\\ \kappa\end{matrix}\right\} = \left(\sqrt{\left(\frac{\lambda^2}{2}\right)^2 + \beta^4} \pm \frac{\lambda^2}{2}\right)^{\frac{1}{2}},$$
$$\lambda^2 = \frac{T_1}{D}, \quad \beta^4 = \frac{\gamma}{D}. \tag{52.26}$$

From the two boundary conditions at $x = 0$, namely $w'(0) = 0$ and $w'''(0) = 0$, we find the result

$$B_2 = B_4 = 0,$$

so that we have

$$w(x) = B_1 \cosh \rho x + B_3 \cos \kappa x.$$

The other boundary conditions $w(L/2) = 0$ and $w'(L/2) = 0$ yield a homogeneous system in B_1 and B_3, which admits nonzero solutions if we have

$$\kappa \cosh\left(\frac{\rho L}{2}\right) \sin\left(\frac{\kappa L}{2}\right) + \rho \sinh\left(\frac{\rho L}{2}\right) \cos\left(\frac{\kappa L}{2}\right) = 0. \tag{52.27}$$

This is the equation for determining T_1. The solution $w(x)$ can be written as

$$w(x) = B_1\left[\cosh(\rho x) - \frac{\cosh\left(\rho \frac{L}{2}\right)}{\cos\left(\kappa \frac{L}{2}\right)}\cos(\kappa x)\right]. \tag{52.28}$$

The unknown B_1 is determined from (52.23) and the boundary conditions on $u(x)$, that is $u(0) = u[L/2] = 0$. These reduce to

$$\frac{T_1 L(1-\sigma^2)}{2Eh} - \int_0^{\frac{L}{2}} w'^2\,dx = 0. \tag{52.29}$$

On substituting $w(x)$ from (52.28) in (52.29) and performing the integrations, we obtain B_1. Next, on simplifying the resulting expression in (52.28), utilizing (52.27), we write

$$w(x) = \pm\left\{\frac{2D(1-\sigma^2)[(\rho L)^2 - (\kappa L)^2]}{2Eh\Phi}\right\}^{\frac{1}{2}}\left[\cos\left(\frac{\kappa L}{2}\right)\cosh(\rho x) - \cosh\left(\frac{\rho L}{2}\right)\cos(\kappa x)\right], \tag{52.30}$$

where

$$\Phi = (\kappa L)^2 \cosh^2\left(\frac{\rho L}{2}\right) - \rho L[\rho L + 2\sinh(\rho L)]\cos^2\left(\frac{\kappa L}{2}\right). \tag{52.31}$$

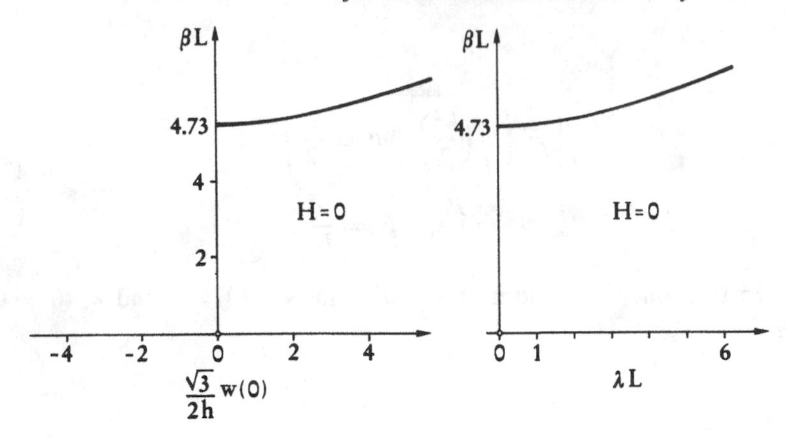

Fig. 52.3

Next, (52.30) is rewritten for $x = 0$, noting the identity $\rho^2 - \kappa^2 = \lambda^2$ and $w(x) > 0$. We thus have

$$\frac{\sqrt{3}}{2h}\, w(0) = \frac{\lambda L}{\sqrt{\Phi}}\left[\cosh\left(\frac{\rho L}{2}\right) - \cos\left(\frac{\kappa L}{2}\right)\right]. \tag{52.32}$$

This equation shows that, for $w(x) \equiv 0$, then $T_1 = 0$ and $\lambda = 0$. Therefore, for $w(0) = 0$ and $T_1 = 0$ we have $\rho = \kappa = \beta$, and (52.27) reduces to

$$\cosh\left(\frac{\beta L}{2}\right)\sin\left(\frac{\beta L}{2}\right) + \sinh\left(\frac{\beta L}{2}\right)\cos\left(\frac{\beta L}{2}\right) = 0. \tag{52.33}$$

This agrees with the result of the corresponding linear eigenvalue problem. The first root of this equation is $\beta_1 L = 4.7300$. By choosing $\beta L > 4.7300$, determining the corresponding λL values from (52.27), and noting (52.26), we can determine the deflection $w(0)$ from (52.32) for each λL, βL and the corresponding $(\rho L, \kappa L)$ pairs. The results are shown in Figure 52.3. In the range considered, the variable Φ is always greater than zero. In this case we again have the confirmation that the eigenvalues of the linearized problem permit us to determine the bifurcation points.

53. Buckled States of Plates in Nonlinear Elasticity

The most significant results in the analysis of buckling of plates are constrained by severe limitations. The only cases which can be treated with a certain generality concern circular plates subject to axisymmetric loads, and constituted by an elastic material obeying Hooke's law. The magnitude of displacements and strains does not represent too strong a restriction. In this sense the theory is constantly exposed to the objection that, in the presence of large strains, the constitutive equations, although elastic, cannot be linear. The removal of this hypothesis creates severe difficulties even in the qualitative description of the solution, whenever this is possible.

However, if we restrict ourselves to a study of the axisymmetric buckling of circular plates some results can be obtained because the equations governing the problem reduce to a system of ordinary differential equations. These equations are sufficiently general to embrace the case of nonlinearly elastic plates that can undergo changes in thickness, as well as flexure, midplane extension, and shear (Antman 1990). Despite the complexity of the resulting equations, it is possible to determine the detailed nodal properties of solutions and distinguish both bifurcating and nonbifurcating branches of these solutions.

The geometric properties of the plate before and after the deformation can be specified when we use a fixed right-handed orthonormal basis $\{\mathbf{i}, \mathbf{j}, \mathbf{k}\}$ using polar coordinates (s, φ) in the $\{\mathbf{i}, \mathbf{j}\}$ plane. We set

$$\mathbf{e}_1(\varphi) = \cos\varphi\mathbf{i} + \sin\varphi\mathbf{j}, \quad \mathbf{e}_2(\varphi) = -\sin\varphi\mathbf{i} + \cos\varphi\mathbf{j}, \quad \mathbf{e}_3 = \mathbf{k}. \tag{53.1}$$

In order to avoid the technical difficulties associated with the singularities at the origin, which arise when we work in polar coordinates, we restrict our attention to annular plates, with the intention of delaying some comments on complete plates until later. In polar coordinates, the configuration of an annular plate is determined by a pair of vectors $\mathbf{r}$ and $\mathbf{d}$ of the form

$$\mathbf{r}(s, \varphi) = \rho(s)\mathbf{e}_1(\varphi) + \zeta(s)\mathbf{k}, \quad \mathbf{d}(s, \varphi) = \delta(s)\mathbf{a}_3(s, \varphi), \tag{53.2}$$

for $a \leqslant s \leqslant 1$, and $\mathbf{a}_3$ is a unit vector normal to $\mathbf{e}_2$. We take the orthonormal triad $\mathbf{a}_1 = \mathbf{e}_2 \times \mathbf{a}_3, \mathbf{a}_2, \mathbf{a}_3$ to have the form

$$\begin{gathered} \mathbf{a}_1 = \cos\theta(s)\mathbf{e}_1(\varphi) + \sin\theta(s)\mathbf{k} \\ \mathbf{a}_2 = \mathbf{e}_2(\varphi), \quad \mathbf{a}_3 = -\sin\theta(s)\mathbf{e}_1(\varphi) + \cos\theta(s)\mathbf{k}. \end{gathered} \tag{53.3}$$

Such a configuration is determined by the four real-valued functions ρ, ζ, δ, and θ. In order to interpret these variables, we assume that the plate, in its unstressed configuration, regarded as a thin three-dimensional body, has a midplane occupying the annulus in the $\{\mathbf{i}, \mathbf{j}\}$ plane defined by the inequalities $a \leqslant s \leqslant 1$. The function $\mathbf{r}(s, \varphi)$ represents the deformed position of a material point that initially had coordinates (s, φ); the function $\mathbf{d}(s, \varphi)$ characterizes the deformed configuration of the material fiber, the reference configuration of which is the normal to the midplane drawn from the material point with coordinates s, φ. In particular, $\mathbf{a}_3$, or θ, determines the orientation and δ determines the stretch of this fiber. From the point of view of Cosserat's theory, $\mathbf{d}$ represents the *director* of the plate.

We next use the replacement

$$\frac{\partial\mathbf{r}}{\partial s}(s, \varphi) = \nu(s)\mathbf{a}_1(s, \varphi) + \eta(s)\mathbf{a}_3(s, \varphi), \tag{53.4}$$

and assume for the set of strain variables for the theory the ensemble of the following seven quantities:

$$\mathbf{w} \equiv (\tau, \nu, \eta, \sigma, \mu, \delta, \omega), \tag{53.5}$$

$$\tau(s) = \frac{1}{s}\rho(s), \quad \sigma(s) = \frac{\sin\theta(s)}{s}, \quad \mu = \theta', \quad \omega = \delta'. \tag{53.6}$$

Here, and in what follows, the prime denotes differentiation with respect to s; τ measures the azimuthal stretch, ν the stretch in the $\mathbf{a}_1$ direction, η the shear, σ the flexure about radii, μ flexure about a circular fiber, and δ the transverse stretch. Note that $\mathbf{w}$ determines a configuration to within a translation along $\mathbf{k}$ and a rotation about this axis.

The reference configuration is characterized by $\rho(s) = s$, $\theta(s) = 0$, $\delta(s) = 1$. In this configuration $\mathbf{w}$ reduces to $(1, 1, 0, 0, 0, 1, 0)$ and we shall assume that this configuration is stress free. The region occupied by the body in its reference configuration is defined by the inequalities

$$a \leqslant s \leqslant 1, \quad 0 \leqslant \varphi < 2\pi, \quad -h(s) \leqslant z \leqslant h(s), \tag{53.7}$$

where the half-thickness $h(s)$ is a positive-valued continuously differential function. The points for which $z = \pm h(s)$ form the *faces* of the plate, and those for which $s = a, 1$ form the *edges*.

We assume that, after the deformation, the material point $\mathbf{z} = s\mathbf{e}_1(\varphi) + z\mathbf{k}$ occupies the position

$$\mathbf{p} = \mathbf{r}(s, \varphi) + \beta(\delta, s, z)\mathbf{a}_3(s, \varphi), \tag{53.8}$$

where $\mathbf{r}$ and $\mathbf{d}$ are given by (53.2) and β is a continuously differentiable function of its arguments with $\beta(\sigma, s, .)$ odd, $\partial\beta/\partial z > 0$, and $\beta(1, s, z) = z$. Customarily, we take $\beta(\delta, s, z) = z\delta$.

When $\mathbf{p}$ has the form (53.8) we find

$$\begin{aligned}\frac{\partial \mathbf{p}}{\partial \mathbf{z}} =& [(\nu - \beta\mu)\mathbf{a}_1 + (\eta + \beta_s + \beta_\delta\omega)\mathbf{a}_3] \otimes \mathbf{e}_1 \\ & - (\tau - \beta\sigma)\mathbf{e}_2 \otimes \mathbf{e}_2 + \beta_z \mathbf{a}_3 \otimes \mathbf{k},\end{aligned} \tag{53.9}$$

where $\otimes$ denotes the dyadic product of vectors. From (53.9) we calculate the Cauchy–Green deformation tensor

$$\begin{aligned}\mathbf{C} = \left(\frac{\partial \mathbf{p}}{\partial \mathbf{z}}\right)^T \left(\frac{\partial \mathbf{p}}{\partial \mathbf{z}}\right) &= C_{11}\mathbf{e}_1 \otimes \mathbf{e}_1 + C_{22}\mathbf{e}_2 \otimes \mathbf{e}_2 + C_{33}\mathbf{e}_3 \otimes \mathbf{e}_3 \\ &+ C_{13}(\mathbf{e}_1 \otimes \mathbf{k} + \mathbf{k} \otimes \mathbf{e}_1),\end{aligned} \tag{53.10}$$

$$\begin{aligned} C_{11} = (\nu - \beta\mu)^2 + (\eta + \beta_s + \beta_\delta\omega)^2, \quad & C_{22} = (\tau - \beta\sigma)^2, \\ C_{33} = \beta_z^2, \quad C_{13} = (\eta + \beta_s + \beta_\delta\omega)\beta_z. & \end{aligned} \tag{53.11}$$

Let us note the null values $C_{21} = C_{23} = 0$. As we require that $\mathbf{p}$ preserve the orientation, that is, its Jacobian should be positive everywhere, we derive from (53.9) the inequalities

$$\tau > \beta(s, h(s), \delta)|\sigma|, \quad \nu > \beta(s, h(s), \delta)|\mu|, \quad \delta > 0. \tag{53.12}$$

For brevity, we denote the set of $\mathbf{w}$ terms satisfying (53.12) by $\mathscr{W}(s)$ and remark that this set is convex for fixed δ.

Besides the geometric definitions we must write the equations of equilibrium by equating to zero the resultant and the resultant moment of all the forces applied at any part of the plate. Deducing these equations is, however, not obvious because we have added the kinematic variable δ in order to take account of the transverse stretch and, consequently, an additional equilibrium equation is necessary. For this purpose let us introduce the first Piola–Kirchhoff stress tensor $\mathbf{T}(\mathbf{z})$ at the material point $\mathbf{z} = s\mathbf{e}_1 + z\mathbf{k}$. The symmetric second Piola–Kirchhoff tensor $\mathbf{S}$ is given by

$$\mathbf{T}(\mathbf{z}) = \frac{\partial \mathbf{p}}{\partial \mathbf{z}} \mathbf{S}. \tag{53.13}$$

The assumption of axisymmetry requires the conditions

$$\mathbf{e}_2 \cdot \mathbf{T}\mathbf{e}_1 = 0, \quad \mathbf{e}_2 \times \mathbf{T}\mathbf{e}_2 = \mathbf{0}, \quad \mathbf{e}_2 \cdot \mathbf{T}\mathbf{k} = 0. \tag{53.14}$$

In terms of $\mathbf{T}$ the stress resultants are defined by the formulae

$$N\mathbf{a}_1 + H\mathbf{a}_3 = \int_{-h}^{h} \mathbf{T}\mathbf{e}_1 \, dz, \quad T = \mathbf{e}_2 \cdot \int_{-h}^{h} \mathbf{T}\mathbf{e}_2 \, dz. \tag{53.15}$$

The form of (53.15) is in accord with (53.14). The quantity $N\mathbf{a}_1 + H\mathbf{a}_3$ is the resultant contact force per unit reference length of the circle $s =$ constant exerted by the material outside the cylinder of radius s and height $2h$ on the material inside this cylinder. Similarly, $T(s)\mathbf{e}_2(\varphi)$ is the resultant contact force per unit reference length of the radius $\varphi =$ constant acting across the plane $\varphi =$ constant of the cylinder. The equation of equilibrium between these stress resultants is

$$\frac{\partial}{\partial s}[s(N\mathbf{a}_1 + H\mathbf{a}_3)] - T\mathbf{e}_1 = \mathbf{0}. \tag{53.16}$$

The equation of equilibrium between the resultant couples is

$$(sM)' - \Sigma \cos\theta + s(H\nu - N\eta) = 0, \tag{53.17}$$

where

$$M = -\mathbf{e}_2 \cdot \int_{-h}^{h} (\mathbf{p} - \mathbf{r}) \times \mathbf{T}\mathbf{e}_1 \, dz, \quad \Sigma = \mathbf{a}_1 \cdot \int_{-h}^{h} (\mathbf{p} - \mathbf{r}) \times \mathbf{T}\mathbf{e}_2 \, dz. \tag{53.18}$$

In the formulae above, $M(s)\mathbf{e}_2(\varphi)$ is the bending couple; that is, the resultant contact torque about $\mathbf{e}_2$ per unit reference length of the circle $s =$ constant exerted by the material outside the cylinder of radius s and height $2h$ on the material inside this cylinder. Similarly, $\Sigma(s)\mathbf{a}_1(s, \varphi)$ is the resultant contact couple per unit reference length of the radius $\varphi =$ constant.

The last, nonclassical, equation of equilibrium reads

$$(s\Omega)' - s\Delta = 0, \tag{53.19}$$

where

$$\Omega = \mathbf{a}_3 \cdot \int_{-h}^{h} \gamma \mathbf{T}\mathbf{e}_1 \, dz, \quad \Delta = \mathbf{a}_3 \cdot \int_{-h}^{h} (\mathbf{T}\mathbf{e}_1 \gamma_s + \mathbf{T}\mathbf{k}\gamma_z) dz + \delta^{-1}(\tilde{\Sigma}\sigma + \tilde{M}\mu), \tag{53.20}$$

with

$$\delta^{-1}\tilde{\Sigma} = -\mathbf{e}_2 \cdot \int_{-h}^{h} \gamma \mathbf{T}\mathbf{e}_2 \, dz, \quad \delta^{-1}\tilde{M} = -\mathbf{a}_1 \cdot \int_{-h}^{h} \gamma \mathbf{T}\mathbf{e}_1 \, dz. \tag{53.21}$$

For $\gamma(s, z) = z$, which corresponds to the choice of $\beta = s\delta$, we then have $\tilde{\Sigma} = \Sigma$, $\tilde{M} = M$, and the expression for Δ simplifies appreciably.

As for the boundary conditions, we assume that the edge $s = 1$ of the plate is constrained to remain parallel to $\mathbf{k}$ so that we have

$$\theta(1) = 0. \tag{53.22}$$

A normal pressure of intensity λ per unit reference length is applied to the edge $s = 1$, so that we have

$$N(1) = -\lambda, \quad H_1(1) = 0. \tag{53.23}$$

We finally assume that either the thickness of the plate at this edge is fixed, giving the consequent value

$$\delta(1) = \delta_1, \tag{53.24}$$

δ_1 being a prescribed number, or, alternatively, there is no constraint preventing such a change so that we have

$$\Omega(1) = 0. \tag{53.25}$$

At the inner edge $s = a$, similar boundary conditions are prescribed, namely

$$\rho(a) = a, \quad \theta(a) = 0, \quad H(a) = 0, \quad \Omega(a) = 0. \tag{53.26}$$

Of course, other variants of these boundary conditions are possible, but we shall take those specified above for definiteness.

In order to carry out the analysis, we require a set of constitutive equations, that is, seven functions T, N, H, Σ, M, Δ, and Ω defined on $\mathscr{W}(s)$ ($s \in [a, 1]$) with the relations

$$T = T(\mathbf{w}(s), s), \quad N = N(\mathbf{w}(s), s), \quad \ldots \tag{53.27}$$

These equations can be derived from the constitutive equations for a nonlinearly elastic material, which have the frame-indifferent form

$$\mathbf{T}(\mathbf{z}) = \frac{\partial \mathbf{p}}{\partial \mathbf{z}} \mathbf{S}(\mathbf{C}, \mathbf{z}), \tag{53.28}$$

where $\mathbf{C}$ is given by (53.10). The constitutive equations in the integrated form (53.27) can be obtained by substituting (53.10) in (53.28) and then substituting the resulting expression in (53.15), (53.18), and (53.20) with $\gamma(s, z)$ chosen equal to $\beta_\delta(\delta(s), s, z)$. On setting $S^{ij} = \mathbf{e}_i \cdot \mathbf{S}\mathbf{e}_j$, and assuming that these components are independent of z, (53.27) assumes the form

$$N = \int_{-h}^{h} (\nu - \beta\mu) S^{11}\, dz, \quad M = -\int_{-h}^{h} \beta(\nu - \beta\mu) S^{11}\, dz,$$

$$H = \int_{-h}^{h} [(\eta + \beta_s + \beta_\delta\omega) S^{11} + \beta_z S^{31}] dz, \quad \Omega = \int_{-h}^{h} \beta_\delta[(\eta + \beta_s + \beta_\delta\omega) S^{11} + \beta_z S^{31}] dz,$$

$$T = \int_{-h}^{h} (\tau - \beta\sigma) S^{22}\, dz, \quad \Sigma = -\int_{-h}^{h} \beta(\tau - \beta\sigma) S^{22}\, dz, \tag{53.29}$$

$$\Delta = \int_{-h}^{h} [(\eta + \beta_s + \beta\omega) S^{13} + \beta_{\delta z} S^{33}] dz - \int_{-h}^{h} \beta_\delta[\mu(\nu - \beta\mu) S^{11} + \sigma(\tau - \beta\sigma) S^{22}] dz,$$

where the S^{ij} depend on $\mathbf{C}$ and (s, z). The constitutive equations of the three-dimensional theory are subject to certain restrictions imposed by criteria of physical admissibility. In particular, we have seen that the strong ellipticity condition is a very important inequality for ensuring the well-posedness of the boundary-value problem. The consequences of strong ellipticity in (53.29) are that the matrices

$$\frac{\partial(N, H, M, \Omega)}{\partial(\nu, \eta, \mu, \omega)}, \quad \frac{\partial(T, \Sigma)}{\partial(\tau, \sigma)}, \quad \frac{\partial \Delta}{\partial \delta}, \tag{53.30}$$

are positive-definite (Antman 1976). This condition ensures that an increase in the bending strain is accompanied by an increase in the bending couple M, and so on for the other stress- and couple-resultants. The condition of positive-definiteness of matrices (53.30) is often supplemented by the additional requirement that infinite resultants accompany extreme strains (Negrón-Marrero and Antman 1990, Antman 1995). Here, however, we prescribe a major consequence of a suitable set of such conditions, and require that, for any given numbers (N, M, H, Ω), the equations

$$N(\mathbf{w}, s) = N, \quad H(\mathbf{w}, s) = H, \quad M(\mathbf{w}, s) = M, \quad \Omega(\mathbf{w}, s) = \Omega$$

have a unique solution for (ν, η, μ, ω) as a function of $(\tau, N, H, \sigma, M, \delta, \Omega, s)$. Let us denote this solution by

$$\nu = \nu^{\#}(\tau, N, H, \sigma, M, \delta, \Omega, s), \quad \ldots, \tag{53.31}$$

where the dots stand for analogous expressions in $\eta^{\#}$, $\mu^{\#}$, and $\omega^{\#}$. The constitutive equations for T, Σ, and Δ can consequently be modified to take the form

$$T = T^{\#}(\tau, N, H, \sigma, M, \delta, \Omega, s), \quad \ldots, \tag{53.32}$$

where the dots stand for similar definitions of $\Sigma^{\#}$ and $\Delta^{\#}$. The seven functions (53.31) and (53.32) of the eight arguments $(\tau, N, H, \sigma, M, \delta, \Omega, s)$ represent an alternative, more convenient, form of the constitutive equations. We shall assume that the new constitutive functions are continuously differentiable with respect their arguments.

If we now turn again to equations (53.28), we assume that $\mathbf{S}$ satisfies the symmetry conditions:

$$S^{21} = 0 = S^{23} \text{ for } C_{21} = 0 = C_{23}, \tag{53.33}$$

$$S^{11}, S^{22}, S^{33} \text{ are even functions of } C_{13} \text{ for } C_{21} = 0 \text{ or } C_{23} = 0, \tag{53.34}$$

$$S^{13} \text{ is an odd function of } C_{13} \text{ for } C_{21} = 0 \text{ or } C_{23} = 0. \tag{53.35}$$

If conditions (53.33)–(53.35) hold, then, under the transformation

$$(\eta, \sigma, \mu) \to -(\eta, \sigma, \mu),$$

the constitutive functions H, Σ, and M defined by (53.29) change sign, while the other functions N, Ω, T, and Δ remain unchanged. The proof of this property is the result of the direct replacement of (η, σ, μ) with their negatives in (53.29). However, as a consequence of definitions (53.31) and (53.32), we can also show that, under the transformation

$$(H, \sigma, M) \to -(H, \sigma, M),$$

the constitutive functions $n^\#$, $\Sigma^\#$, and $\mu^\#$, as defined in (53.31) and (53.32), change sign, while the other constitutive functions of the same set remain unchanged. It then follows from the mean-value theorem that $\eta^\#$, $\Sigma^\#$, and $\mu^\#$ can be represented as

$$\eta^\# = \bar{\eta}_H H + \bar{\eta}_\sigma \sigma + \bar{\eta}_M M, \text{ etc.} \tag{53.36}$$

where we have

$$\bar{\eta}_H(\tau, N, H, \sigma, M, \delta, \Omega, s) = \int_0^1 \eta_H(\tau, N, tH, t\sigma, tM, \delta, \omega, s)dt, \text{ etc.} \tag{53.37}$$

The introduction of the modified constitutive equations (53.31) and (53.32) permits us to recast the boundary-value problem in a form that simplifies the ensuing analysis. We integrate (53.16) subject to the boundary condition (53.23) and use (53.31) and (53.32) to obtain

$$\begin{aligned} -N(s) &= \Lambda^\#(s) \cos\theta(s), \\ H(s) &= \Lambda^\#(s) \sin\theta(s), \end{aligned} \tag{53.38}$$

where

$$s\Lambda^\#(s) = \lambda + \int_s^1 T(\tau(t) \ldots)\, dt. \tag{53.39}$$

We next deduce from (53.17), (53.19), and (53.38) the results

$$\theta' = \mu^\#(\tau, N, \Lambda^\# \sin\theta, s^{-1} \sin\theta, M, \delta, \Omega, s), \tag{53.40}$$

$$(sM)' = \Sigma^\# \cos\theta - s\Lambda^\# \nu^\# \sin\theta + sN\eta^\#, \tag{53.41}$$

$$\delta' = \delta^\#(\tau, N, \Lambda^\# \sin\theta, s^{-1} \sin\theta, M, \delta, \Omega, s), \tag{53.42}$$

$$(s\Omega)' = s\Delta^\#, \tag{53.43}$$

where the arguments of $\Sigma^\#$, $\Delta^\#$, $\nu^\#$, and $\eta^\#$ in (53.41) and (53.43) are the same as those of $\mu^\#$ in (53.40).

Now, (53.40) and (53.41) can be modified, by using (53.36), in the form

$$\theta' = (\bar{\mu}_H \Lambda^\# + \bar{\mu}_\sigma s^{-1}) \sin\theta + \bar{\mu}_M M, \tag{53.44}$$

$$\begin{aligned}(sM)' = \{[\bar{\Sigma}_H \Lambda^\# + \bar{\Sigma}_\sigma s^{-1} - s\Lambda^\#(\bar{\eta}_H \Lambda^\# + \bar{\eta}_\sigma s^{-1})] \cos\theta - s\Lambda^\# \nu^\#\} \sin\theta \\ + [\bar{\Sigma}_M - s\Lambda^\# \bar{\eta}_M] M \cos\theta.\end{aligned} \tag{53.45}$$

The solution of our boundary-value problem for which $\theta = 0 = M$ are said to describe an *unbuckled state*. A characteristic property of the problem is that unbuckled states are typically not unique. These states are governed by the system

$$N = -\lambda - \int_s^1 T\, dt, \qquad (s\Omega)' - s\Delta = 0, \tag{53.46}$$

where the arguments of N, Ω, and Δ are $[s^{-1}\rho(s), \rho'(s), 0, 0, 0, \delta(s), \delta'(s), s]$, and those of T are the same but with s replaced by t. Without loss of generality, we suppose, for simplicity, that these constitutive functions are independent of s, that we have $\Omega = 0$ for $\eta = \sigma = \mu = \omega = 0$, and the equality $T(\tau, \nu, 0, 0, 0, \delta, 0) = N(\tau, \nu, 0, 0, 0, \delta, 0)$, as occurs in a transversely isotropic plate. After supposing that boundary condition (53.25) holds, we seek unbuckled states for which we have $\rho' = \nu_0 =$ constant and $\delta = \delta_0 =$ constant. These values must satisfy the equations

$$N(\nu_0, \nu_0, 0, 0, 0, \delta_0, 0) = -\lambda; \qquad \Delta(\nu_0, \nu_0, 0, 0, 0, \delta_0, 0) = 0. \tag{53.47}$$

If these equations admitted a unique solution, the unbuckled state would be unique. However, our monotonicity conditions are insufficient to ensure that such is the case. A sufficient condition for uniqueness would be that the matrix of partial derivatives of T, N, and Δ with respect to τ, ν, and δ is positive-definite. This positive-definiteness, however, is not a consequence of the strong ellipticity conditions, and hence the solution of (53.47) is not, in general, unique.

We pass now to an examination of the properties of buckled plates. Let us suppose that all the arguments of the barred functions in (53.44) and (53.45) are continuous in s. Equation (53.44) thus implies that θ is of class $\mathscr{C}^1$ if, and only if, (θ, M) are $\mathscr{C}^0$. Moreover, θ has a simple zero at s^* if, and only if, we have $\theta(s^*) = 0$ and $M(s^*) \neq 0$, and θ has a double zero (that is, a zero of order two or more) if, and only if, we have $\theta(s^*) = 0$ and $M(s^*) = 0$. Now let us consider the set

$$\begin{aligned}Z_k \equiv \{(\theta, M) \in \mathscr{C}^0 \colon \theta(a) = \theta(1) = 0; \theta \text{ has exactly } (k+1) \text{ zeros} \\ \text{in } [a, 1], \text{ each of which is simple}\}.\end{aligned} \tag{53.48}$$

Z_k is an open set in $\mathscr{C}^0$, and, if (θ, M) belongs to its boundary ∂Z_k, then θ has a double zero in $[a, 1]$.

Let $\mathscr{S}$ be a connected family of solutions $(\rho, N, H, \theta, M, \delta, \Omega; \lambda) \equiv (\tilde{\mathbf{u}}, \tilde{\lambda})$ with $(\tilde{\theta}, \tilde{M})$ in Z_k. Since Z_k is open in $\mathscr{C}^0$, there is a neighborhood $\mathscr{N}$ of $(\tilde{\mathbf{u}}, \tilde{\lambda})$ such that, for $(\mathbf{u}, \lambda) \in \mathscr{N} \cap \mathscr{S}$, then $(\theta, M) \in Z_k$. Let $\tilde{\mathscr{S}}$ be the largest connected component of $\mathscr{S}$ containing $(\tilde{\mathbf{u}}, \tilde{\lambda})$ and having the further property that there is no θ corresponding to one of its solutions $(\mathbf{u}, \lambda)$ having a double zero. This implies $(\theta, M) \in Z_k$ for each solution of $\tilde{\mathscr{S}}$. Now, if θ has a double zero, then the corresponding zero values for (θ, M) can be taken as initial data for (53.44) and

(53.45) and the resulting initial-value problem has the zero solution as the unique solution, by virtue of the symmetry of the coefficients of the system under the transformation $(H, \sigma, M) \to -(H, \sigma, M)$. We can restate the result by saying that, if a connected set $\mathcal{S}$ of solutions contains no unbuckled state and if has one solution with $(\tilde{\theta}, \tilde{M})$ in Z_k, then each of its solutions (θ, M) is in Z_k. Thus the number of simple zeros of Z_k is constant on such a set of solutions and characterizes it globally.

It remains for us to discover the structure of the branches of buckled states. However, this can be characterized by a theorem on the theory of bifurcation (Crandall and Rabinowitz 1971, Rabinowitz 1971) stating that, if $\bar{\lambda}$ is an eigenvalue of odd algebraic multiplicity of the linearization of the boundary-value problem about a branch of unbuckled states, and if $\bar{\mathbf{u}}(\bar{\lambda})$ is the corresponding unbuckled state, then from $(\bar{\mathbf{u}}(\bar{\lambda}), \bar{\lambda})$ a branch of buckled states bifurcates that is, either unbounded in $\mathcal{C}^0 \times \mathbb{R}$, or else returns to another such state $(\bar{\mathbf{u}}(\bar{\kappa})\bar{\kappa})$. If the eigenvalue is simple, then the bifurcating branch inherits the nodal properties of the eigenfunction.

The theory of bifurcation can be extended to anisotropic nonlinearly elastic plates. For simplicity, we assume that the material is homogeneous so that the constitutive functions are independent of s. Anisotropic plates have several applications in mechanical engineering. This occurs in cases for which we wish to describe macroscopically the constitutive response of plates reinforced by ribs placed along the radii or concentric circles. However, one of the most striking types of anisotropic plate is presented by the degenerate case of a *Taylor's plate*, which has no azimuthal tensile or flexural strength, and may thus be regarded as consisting of an infinite array of radially disposed rods. Taylor's plate is the plate-theoretic analog of a parachute. In the attempt to apply bifurcation analysis to anisotropic plates, we find (Antman and Negrón-Marrero 1987) that anisotropy generates unexpected pathologies. As in the isotropic case, the number of zeros of solutions is inherited from the eigenfunctions of the linearized problem, which correspond to simple eigenvalues, and this is preserved globally. The novelty lies in the fact that the curve $\theta = \theta(\lambda)$ may admit jumps, as shown in Figure 53.1. As λ is slowly raised through the value λ', the nontrivial solution along the branch in Figure 53.1 suddenly jumps. The effect is similar to that of snap-buckling, but the mathematical nature of the process is entirely different.

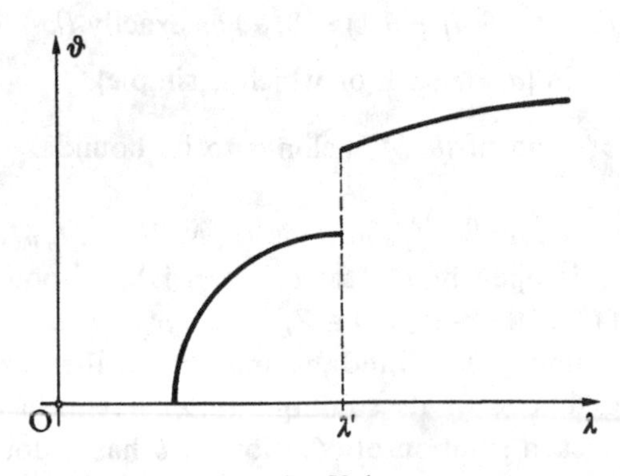

Fig. 53.1

Chapter VII

Theories of Shells

54. Geometric Definitions and Mechanical Assumptions

A curved plate or shell may be described by means of its middle surface, its edge line, and its thickness $2h$. We shall take the thickness to be constant, and consider the two surfaces of constant normal distance h from the middle surface and placed on opposite sides of it. These two surfaces constitute the *faces* of the shell. Let s denote a closed curve drawn on the strained middle surface, and consider the outer normal $\boldsymbol{\nu}$ to this curve at a point P_1, drawn in the tangent plane to the surface at P_1, and let $\mathbf{s}$ be the unit vector tangent to s, directed counterclockwise. Then let $\mathbf{n}$ denote the unit normal to the middle surface at P_1 oriented positively so that the triad $(\boldsymbol{\nu},\mathbf{s},\mathbf{n})$ is right-handed (Love 1927, Art. 328). We consider a point P_1' on the curve s, at a small distance δs from P_1, and take the two segments constituted by those parts of the normals to the middle surface at P_1 and P_1' which are included between the faces of the shell. These two pieces of normal at P_1 and P_1' mark out an element of area δA (Figure 54.1) belonging to a developable surface the generators of which are perpendicular to the middle surface. The contact tractions exerted across the area δA are made statically equivalent to a force and to a couple applied at P_1. We define the averaged components of this force and couple per unit length of s by dividing their components by δs. The limits of these averages for $\delta s \to 0$ are the stress-resultants and the stress couples transmitted across the curve s at P_1. We assume that all these

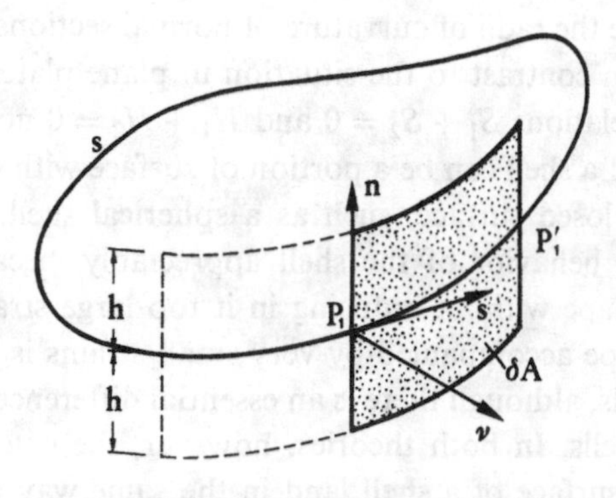

Fig. 54.1

limits exist, and that the component of the stress couple along **n** vanishes. On taking a Cartesian system x', y', z along $\boldsymbol{\nu}$, **s**, **n**, with its origin at P_1, the stress-resultants are denoted by T, S, and N, respectively, and the nonzero couple resultants by H and G. These stress- and couple-resultants may be written in terms of the stress components referred to the axes x', y', and z. On taking R' to be the radius of curvature of the normal section of the surface drawn through the tangent **s** and P_1, the relations between $T, \ldots, G$ and the stress components $\sigma'_{x'}, \sigma'_{y'}, \ldots, \sigma'_{y'z}$ are

$$T = \int_{-h}^{h} \sigma'_{x'}\left(1 - \frac{z}{R'}\right)dz, \quad S = \int_{-h}^{h} \sigma'_{x'y'}\left(1 - \frac{z}{R'}\right)dz, \quad N = \int_{-h}^{h} \sigma'_{x'z}\left(1 - \frac{z}{R'}\right)dz,$$

$$H = -\int_{-h}^{h} \sigma'_{x'y'} z\left(1 - \frac{z}{R'}\right)dz, \quad G = \int_{-h}^{h} \sigma'_{x'} z\left(1 - \frac{z}{R'}\right)dz. \tag{54.1}$$

When we take fixed axes x, y, and z on the surface, such that the x and y axes lie on the tangent plane at P_1 and z is directed along the normal **n**, the unit vectors $\boldsymbol{\nu}$ and **s** do not coincide, in general, with the x and y axes. However, in the two particular cases in which $\boldsymbol{\nu}$ is parallel to the x and y axis, (54.1) become

$$T_1 = \int_{-h}^{h} \sigma_x\left(1 - \frac{z}{R'_2}\right)dz, \quad S_1 = \int_{-h}^{h} \sigma_{xy}\left(1 - \frac{z}{R'_2}\right)dz, \quad N_1 = \int_{-h}^{h} \sigma_{xz}\left(1 - \frac{z}{R'_2}\right)dz,$$

$$H_1 = -\int_{-h}^{h} z\sigma_{xy}\left(1 - \frac{z}{R'_2}\right)dz, \quad G_1 = \int_{-h}^{h} z\sigma_x\left(1 - \frac{z}{R'_2}\right)dz, \tag{54.2}$$

and

$$T_2 = \int_{-h}^{h} \sigma_y\left(1 - \frac{z}{R'_1}\right)dz, \quad S_2 = -\int_{-h}^{h} \sigma_{xy}\left(1 - \frac{z}{R'_1}\right)dz, \quad N_2 = \int_{-h}^{h} \sigma_{yz}\left(1 - \frac{z}{R'_1}\right)dz,$$

$$H_2 = \int_{-h}^{h} z\sigma_{xy}\left(1 - \frac{z}{R'_1}\right)dz, \quad G_2 = \int_{-h}^{h} z\sigma_y\left(1 - \frac{z}{R'_1}\right)dz, \tag{54.3}$$

where R'_1 and R_2 denote the radii of curvature of normal sections parallel to the axes x and y, respectively. In contrast to the situation in plane plates, where the middle surface is planar, the relations $S_1 + S_2 = 0$ and $H_1 + H_2 = 0$ no longer hold.

The middle surface of a shell can be a portion of surface with a boundary, such as a spherical cap, or a closed surface, such as a spherical shell. The presence of a boundary modifies the behavior of the shell appreciably, because it can be bent into a very different shape without inducing in it too large strains. The possibility of large changes in shape accompanied by very small strains is a specific feature of the behavior of thin rods, although there is an essential difference between the theory of rods and that of shells. In both theories, however, the extension of any linear element of the middle surface of a shell, and in the same way the extension of the central line of a rod, must be small. Nevertheless, while a rod may assume severely bent configurations without extension of the central line, in the case of a shell this is possible only for a certain family of surfaces, namely, for those that are consistent with the unstrained middle surface. The practical consequence of this result is that a purely inextensional theory of shells is meaningful only for shells the strained middle surface of which can differ only slightly from one consistent with the unstrained middle surface. In particular, the inextensional deformation of a planar plate requires that its middle plane should deform into a developable surface.

In order to describe the possible change in shape of the shell when no restriction on the magnitude of strains is demanded, we introduce a two-dimensional bounded open connected set $\bar{\omega}$, with a Lipschitz-continuous boundary γ. Let (U^1, U^2) denote a generic point in the set $\bar{\omega}$. The middle surface of a shell, in its unstrained configuration, is the image of the set $\bar{\omega}$ through a mapping

$$\mathbf{X} = \varphi(U^\alpha) = \varphi^i \mathbf{e}_i, \tag{54.4}$$

where $(\mathbf{e}_1, \mathbf{e}_2, \mathbf{e}_3)$ is a fixed orthonormal basis, given once and for all in the Euclidean space. We assume that the mapping $\boldsymbol{\varphi}$ is injective, of class $\mathscr{C}^3$, and that the two vectors

$$\mathbf{a}_\alpha = \boldsymbol{\varphi}_{,\alpha} = (\varphi^i_{,\alpha})\mathbf{e}_i \quad (\alpha = 1, 2) \tag{54.5}$$

are linearly independent at each point U^α of $\bar{\omega}$. From this we can define the vector

$$\mathbf{a}_3 = \mathbf{a}^3 = \frac{\mathbf{a}_1 \times \mathbf{a}_2}{|\mathbf{a}_1 \times \mathbf{a}_2|}, \tag{54.6}$$

which is normal to the tangent plane to the surface at U^α, spanned by the two vectors $\mathbf{a}_\alpha$. The vectors $\mathbf{a}_\alpha$ and the vector $\mathbf{a}_3$ define a basis at the point $\boldsymbol{\varphi}(U^\alpha)$ which is called the *covariant basis* at that point.

We also introduce at each point $\boldsymbol{\varphi}(U^\alpha)$ two vectors $\mathbf{a}^\alpha$ of the tangential plane, defined by the relations

$$\mathbf{a}^\alpha \cdot \mathbf{a}_\beta = \delta^\alpha_\beta, \tag{54.7}$$

where δ^α_β is the Kronecker symbol. The vectors $\mathbf{a}^\alpha$ and the vector $\mathbf{a}^3$ define the *contravariant basis* at the point considered. Let us note that the vectors of the two bases satisfy the relations

$$\mathbf{a}_\alpha \times \mathbf{a}_\beta = \varepsilon_{\alpha\beta}\mathbf{a}^3, \quad \mathbf{a}^\alpha \times \mathbf{a}^\beta = \varepsilon^{\alpha\beta}\mathbf{a}_3, \quad \mathbf{a}_3 \times \mathbf{a}_\beta = \varepsilon_{\beta\zeta}\mathbf{a}^\zeta,$$
$$\mathbf{a}^3 \times \mathbf{a}^\beta = \varepsilon^{\beta\zeta}\mathbf{a}_\zeta, \tag{54.8}$$

where $\varepsilon_{\alpha\beta}$ and $\varepsilon^{\alpha\beta}$ are components of the Ricci tensor defined by

$$\varepsilon_{\alpha\beta} = \sqrt{a}\begin{bmatrix} 0 & 1 \\ -1 & 0 \end{bmatrix}, \quad \varepsilon^{\alpha\beta} = \frac{1}{\sqrt{a}}\begin{bmatrix} 0 & 1 \\ -1 & 0 \end{bmatrix}, \tag{54.9}$$

with $a = \det(a_{\alpha\beta})$. By assumption, a must be strictly greater than zero, and hence there is a constant a_0 such that we have

$$a(U^\alpha) \geqslant a_0 > 0, \quad \text{for all } U^\alpha \in \bar{\omega}. \tag{54.10}$$

The arc element of the middle surface is then given by

$$dS^2 = a_{\alpha\beta}\, dU^\alpha\, dU^\beta, \tag{54.11}$$

where

$$a_{\alpha\beta} = a_{\beta\alpha} = \mathbf{a}_\alpha \cdot \mathbf{a}_\beta = \boldsymbol{\varphi}_{,\alpha} \cdot \boldsymbol{\varphi}_{,\beta} \tag{54.12}$$

are the components of the *metric tensor* of the surface. The area element of the middle surface is then evaluated through the formula

$$dA = \sqrt{a}\, dU^1\, dU^2. \tag{54.13}$$

The knowledge of the metric tensor is sufficient to measure the lengths of curves on the surface. The curvatures of the surface are defined by the *curvature tensor* $b_{\alpha\beta}$ defined by

$$b_{\alpha\beta} = b_{\beta\alpha} = \mathbf{a}_3 \cdot \mathbf{a}_{\beta,\alpha} = -\mathbf{a}_\alpha \cdot \mathbf{a}_{3,\beta}, \tag{54.14}$$

where the comma denotes partial differentiation with respect to U^α on the surface. We also introduce the quantities

$$\Gamma^\rho_{\alpha\beta} = \Gamma^\rho_{\beta\alpha} = \mathbf{a}^\rho \cdot \mathbf{a}_{\beta,\alpha} = \mathbf{a}^\rho \cdot \mathbf{a}_{\alpha,\beta}, \tag{54.15}$$

which are called the *Christoffel symbols*, and are used to compute *covariant derivatives*. For instance, let $\eta_\alpha \mathbf{a}^\alpha = \eta^\beta \mathbf{a}_\beta$ be a surface vector field expressed by means of its covariant or contravariant components. Then the two functions

$$\eta_{\alpha|\beta} = \eta_{\alpha,\beta} - \Gamma^\rho_{\alpha\beta\rho}, \qquad \eta^\alpha|_\beta = \eta^\alpha{}_{,\beta} + \Gamma^\alpha_{\rho\beta}\eta^\rho, \tag{54.16}$$

are the *covariant derivatives* of the vector field. Likewise, if $T_{\alpha\beta}$ and $T^{\alpha\beta}$ denote the components of a surface tensor field, then the functions

$$\begin{aligned} T_{\alpha\beta|\rho} &= T_{\alpha\beta,\rho} - \Gamma^\delta_{\alpha\rho} T_{\delta\rho} - \Gamma^\delta_{\beta\rho} T_{\alpha\delta}, \\ T^{\alpha\beta}|_\rho &= T^{\alpha\beta}_{,\rho} + \Gamma^\alpha_{\sigma\rho} T^{\sigma\beta} + \Gamma^\beta_{\sigma\rho} T^{\alpha\sigma}, \end{aligned} \tag{54.17}$$

are the covariant derivatives of the tensor field. Let us note that the covariant derivatives $\eta_{\alpha|\beta}$ are themselves a surface tensor field, the covariant derivatives $\eta_{\rho|\alpha\beta} = (\eta_{\rho|\alpha})_{|\beta}$ of which can be computed according to (54.17). Note that the derivatives $\eta_{\rho|\alpha\beta}$ and $\eta_{\rho|\beta\alpha}$ do not commute in general, because we have

$$\eta_{\rho|\alpha\beta} - \eta_{\rho|\beta\alpha} = b_{\rho\beta} b^\sigma_\alpha \eta_\sigma - b_{\rho\alpha} b^\sigma_\beta \eta_\sigma. \tag{54.18}$$

The function $\boldsymbol{\varphi}(U^\alpha)$ defines the undeformed middle surface. When, under the action of the loads, the shell changes shape, the middle surface assumes a new configuration defined by a displacement vector $\boldsymbol{\eta} = \eta_i \mathbf{a}^i$, where η_i are the covariant components of this vector.

A nontrivial question is how to choose an appropriate measure of deformation for the strained middle surface. This measure of deformation is defined by a set of quantities, called the *strains* of the surface, but this definition is not unique because it is the result of a compromise between the criterion of simplicity and the need for considering important effects such as extension, flexure, and shear.

Among the definitions of strains, a simplified expression for them is possible when the displacements η_i are small with respect to a characteristic length of the middle surface, and their partial derivatives $\eta_{i,\alpha}$ are so small with respect to the unity that we can neglect the products of these quantities. We thus define the *linearized extensions* as

$$\gamma_{\alpha\beta} = \tfrac{1}{2}(\eta_{\alpha|\beta} + \eta_{\beta|\alpha}) - b_{\alpha\beta}\eta_3, \tag{54.19}$$

and the *linearized changes of curvature* as

$$\kappa_{\alpha\beta} = \eta_{3|\alpha\beta} + b^{\rho}_{\beta}\eta_{\rho|\alpha} + b^{\rho}_{\alpha}\eta_{\rho|\beta} + (b^{\rho}_{\beta|\alpha})\eta_{\rho} - c_{\alpha\beta}\eta_3, \tag{54.20}$$

with $c_{\alpha\beta} = c_{\beta\alpha} = b^{\rho}_{\alpha}b_{\rho\beta} = b_{\alpha\rho}b^{\rho}_{\beta}$. Note that both tensors are symmetric. On using the notation adopted in the theory of planar plates, the strain components can also be written in the forms

$$\varepsilon_1 = \gamma_{11}, \quad \varepsilon_2 = \gamma_{22}, \quad \tilde{\omega} = 2\gamma_{12}, \tag{54.21}$$

and

$$\kappa_1 = \kappa_{11}, \quad \kappa_2 = \kappa_{22}, \quad \tau = \kappa_{12}. \tag{54.22}$$

The fact that the linearized theory can be applied also allows the strained configuration of the shell to be identified with its reference configuration, which is known. The consequence is that the stress-resultants and the couple-resultants as defined by (54.3) and (54.4), can be obtained by averaging the stresses over the thickness of the undeformed configuration and equating the radii of curvature to those of the undeformed middle surface. Let us note that, having assumed the thickness $2h$ to be constant does not represent a very physical condition, because knowing the thickness of the undeformed shell tells us nothing about the thickness after deformation.

The knowledge of the strains and changes of curvature in the middle surface do not give any idea of how strains vary generally inside the shell. The law of variation of strains with z is a postulate, called the *Kirchhoff–Love hypothesis*, which states that any point situated on a normal to the middle surface remains, after the deformation has taken place, on the normal to the deformed middle surface, and that the distance between such a point and the middle surface, remains constant. The consequence of this hypothesis is that the strains in the interior of the shell have the approximate form

$$\begin{gathered}\varepsilon_x = \varepsilon_1 - \kappa_1 z, \quad \varepsilon_y = \varepsilon_2 - \kappa_2 z, \quad 2\varepsilon_{xy} = \tilde{\omega} - 2\tau z,\\ \varepsilon_{zx} = \varepsilon_{zy} = \varepsilon_z = 0.\end{gathered} \tag{54.23}$$

Although equations (54.23) are universally known as those embodying the Kirchhoff–Love hypothesis, they were not introduced by Love in his original shell theory (Love 1927, Art. 329), which rests on other, less convincing, assumptions.

To the order of approximation of the linearized theory, provided that the material is homogeneous and isotropic, the stress components of the type σ_x, σ_y, and σ_{xy} are given by the equations

$$\begin{gathered}\sigma_x = \frac{E}{(1-\sigma^2)}[\varepsilon_1 + \sigma\varepsilon_2 - z(\kappa_1 + \sigma\kappa_2)], \quad \sigma_y = \frac{E}{(1-\sigma^2)}[\varepsilon_2 + \sigma\varepsilon_1 - z(\kappa_2 + \sigma\kappa_1)],\\ \sigma_{xy} = \frac{E}{2(1+\sigma)}(\tilde{\omega} - 2\tau z).\end{gathered} \tag{54.24}$$

Hence, by substituting these results in (54.2) and (54.3) we can deduce approximate formulae for the stress resultants and stress couples:

$$T_1 = \frac{2Eh}{(1-\sigma^2)}(\varepsilon_1 + \sigma\varepsilon_2), \quad T_2 = \frac{2Eh}{(1-\sigma^2)}(\varepsilon_2 + \sigma\varepsilon_1), \quad -S_2 = S_1 = \frac{Eh}{1+\sigma}\tilde{\omega}, \tag{54.25}$$

and

$$\begin{aligned} G_1 &= -D\left[(\kappa_1 + \sigma\kappa_2) + \frac{1}{R_2'}(\varepsilon_1 + \sigma\varepsilon_2)\right], \\ G_2 &= -D\left[(\kappa_2 + \sigma\kappa_1) + \frac{1}{R_1'}(\varepsilon_2 + \sigma\varepsilon_1)\right], \\ H_1 &= D(1-\sigma)\left(\tau + \frac{\tilde{\omega}}{2R_2'}\right), \quad H_2 = -D(1-\sigma)\left(\tau + \frac{\tilde{\omega}}{2R_1'}\right), \end{aligned} \tag{54.26}$$

where R_1' and R_2' are the radii of curvature of the normal sections drawn through the x and y axes, which coincide with the tangents to β and α at their point of intersection.

The equilibrium equations are formed by equating the resultant and the resultant moment of the forces acting at a portion of the shell. These can be deduced using the method leading to the Föppl–Kármán equations, with the difference that the quantities like $p_1', \ldots, r_2'$ now have more complicated expressions. This is because we cannot identify the curvilinear coordinates α, β with the Cartesian coordinates of a plane domain $\bar{\omega}$, representing the unstrained middle surface. If, however, the approximations of linearized theory are allowed, then the rotations $p_1', \ldots, r_2'$ can be replaced by the corresponding values assumed in the unstrained middle surface, and we can drop the primes over these rotation elements. As the unstrained surface is geometrically described by its metric tensor $a_{\alpha\beta}$ and its curvature tensor $b_{\alpha\beta}$, it is convenient to know the expressions for $p_1, \ldots, r_2$ in terms of $a_{\alpha\beta}$ and $b_{\alpha\beta}$. These formulae are particularly simple when the curves $U^1 \equiv \alpha$ and $U^2 \equiv \beta$ are lines of curvature of the surface, and the tensor $a_{\alpha\beta}$ has the form (Love 1927, Art. 323)

$$a_{11} = A, \quad a_{22} = B, \quad a_{12} = a_{21} = 0, \tag{54.27}$$

and $b_{\alpha\beta}$ has the form

$$b_{11} = \frac{1}{R_1}, \quad b_{22} = \frac{1}{R_2}, \quad b_{12} = b_{21} = 0. \tag{54.28}$$

In this case we have

$$\begin{aligned} p_1 = 0, \quad q_2 = 0, \quad \frac{1}{R_1} = -\frac{q_1}{A}, \quad \frac{1}{R_2} = \frac{p_2}{B}, \\ r_1 = -\frac{1}{B}\frac{\partial A}{\partial \beta}, \quad r_2 = \frac{1}{A}\frac{\partial B}{\partial A}. \end{aligned} \tag{54.29}$$

Once we have the rotations $p_1, \ldots, r_2$, the equations of equilibrium of the forces and stress resultants acting over a curvilinear rectangle $AB\, d\alpha\, d\beta$, and resolved in the direction of a local triad, the axes of which coincide with the tangents to α and β at their point of intersection and the normal to the unstrained middle surface at this point, are

$$\begin{aligned}
&\frac{\partial}{\partial\alpha}(T_1B)-\frac{\partial}{\partial\beta}(S_2A)-(r_1S_2B+r_2T_2A)+(q_1N_1B+q_2N_2A)+ABX=0,\\
&\frac{\partial}{\partial\alpha}(S_1B)+\frac{\partial}{\partial\beta}(T_2A)-(p_1N_1B+p_2N_2A)+(r_1T_1B-r_2S_2A)+ABY=0,\quad(54.30)\\
&\frac{\partial}{\partial\alpha}(N_1B)+\frac{\partial}{\partial\beta}(N_2A)-(q_1T_1B-q_2S_2A)+(p_1S_1B+p_2T_2A)+ABZ=0.
\end{aligned}$$

The externally applied forces estimated per unit area of the unstrained middle surface are now X, Y, and Z.

The argument leading to (54.30) can also be applied by considering the equilibrium of the resultant couple acting along the edges of the curvilinear rectangle bounded by the coordinate curves α, $\alpha+d\alpha$, β, and $\beta+d\beta$ and the external couple whose components per unit area along the local axes are L, M, and 0. The couple acting on the side α has the components $(-H_1B\,d\beta, -G_1B\,d\beta, 0)$ and the couple on the side $\alpha+d\alpha$ has the components

$$\begin{bmatrix} H_1B\,d\beta\\ G_1B\,d\beta\\ 0\end{bmatrix}+\frac{\partial}{\partial\alpha}\begin{bmatrix} H_1B\,d\beta\\ G_1B\,d\beta\\ 0\end{bmatrix}+\begin{bmatrix} p_1\,d\alpha\\ q_1\,d\alpha\\ r_1\,d\alpha\end{bmatrix}\times\begin{bmatrix} H_1B\,d\beta\\ G_1B\,d\beta\\ 0\end{bmatrix}. \qquad (54.31)$$

On the side β, the components of the couple are $(G_2A\,d\alpha, -H_2A\,d\alpha, 0)$, and on the side $\beta+d\beta$ the corresponding components of the couple are

$$\begin{bmatrix} -G_2A\,d\alpha\\ H_2A\,d\alpha\\ 0\end{bmatrix}+\frac{\partial}{\partial\beta}\begin{bmatrix} -G_2A\,d\alpha\\ H_2A\,d\alpha\\ 0\end{bmatrix}+\begin{bmatrix} p_2\,d\beta\\ q_2\,d\beta\\ r_2\,d\beta\end{bmatrix}\times\begin{bmatrix} -G_2A\,d\alpha\\ H_2A\,d\alpha\\ 0\end{bmatrix}. \qquad (54.32)$$

In addition, we have the moments about the axes of the forces acting across the sides $\alpha+d\alpha$ and $\beta+d\beta$, which are $(Bd\beta N_2A\,d\alpha, -Ad\alpha N_1B\,d\beta, Ad\alpha S_1B\,d\beta + Bd\beta S_2A\,d\alpha)$. Thus, on collecting all these contributions we can write the equations of the moments in the form

$$\frac{\partial}{\partial\alpha}(H_1B)-\frac{\partial}{\partial\beta}(G_2A)-(G_1Br_1+H_2Ar_2)+(N_2+L)AB=0, \qquad (54.33a)$$

$$\frac{\partial}{\partial\alpha}(G_1B)+\frac{\partial}{\partial\beta}(H_2A)+(H_1Br_1-G_2Ar_2)-(N_1-M)AB=0, \qquad (54.33b)$$

$$G_1Bp_1+G_2Aq_2-(H_1Bq_1-H_2Ap_2)+(S_1+S_2)AB=0. \qquad (54.33c)$$

The boundary conditions associated with the equilibrium equations may assume different forms because, along the edge line of the middle surface, we can prescribe stress and couple resultants, or displacements, or a combination of the two. The most interesting among these boundary conditions are those in which forces are applied at the boundary, because account must be taken of the curvature of the surface in defining the stress and couple resultants at the edge (Love 1927, Art. 332). Let us isolate an arc element $2ds$ of the edge line s and consider the orthonormal triad ($\boldsymbol{\nu}$,**s**,**n**), having its origin at P, the midpoint of $2ds$, the $\boldsymbol{\nu}$ unit vectors directed along the principal normal to s, the unit vector **s** directed along the tangent to s, and the unit vector **n** drawn in the sense of the normal to the surface (Figure 54.2). If P′ and P″ are the end points of the arc element, the arcs P′P and PP″ have a length ds. Now the static effect of the torsional couple H acting on the arc P′P is equivalent to

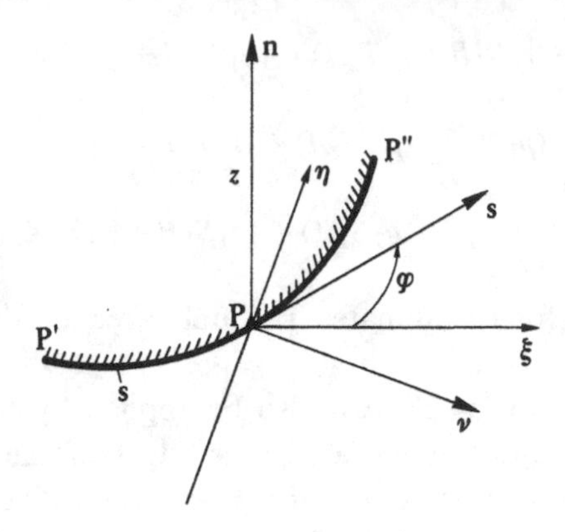

Fig. 54.2

a pair of forces, each equal to $H - dH$, applied at P′ and P, the force at P′ being parallel to the normal at P′, the force at P having the same direction and being in the positive sense of this normal. Let us now draw from P two orthogonal axes ξ and η, lying in the tangential plane to the surface and tangential to the lines of principal curvature of the surface. When referred to the axes ξ, η, and z the equation of the surface is approximately given by

$$f(\xi, \eta, z) = z - \frac{1}{2}\left(\frac{1}{R_1}\xi^2 + \frac{1}{R_2}\eta^2\right) = 0, \tag{54.34}$$

where R_1 and R_2 denote the principal radii of curvature of the surface at P. Let φ be the angle made by the ξ axis with the tangent **s** (Figure 54.2). In the ξ, η, z system, the point P′ has coordinates $(-ds \cos\varphi, -ds \sin\varphi, 0)$ and, consequently, the direction cosines of the normal at P′ (which are proportional to f_ξ, f_η, f_z) have the approximate values $((ds \cos\varphi)/R_1, (ds \sin\varphi)/R_2, 1)$. In the ξ, η, z system the direction cosines of $\boldsymbol{\nu}$ are $(\sin\varphi, -\cos\varphi, 0)$ (Figure 54.2), and the direction cosines of **s** are $(\cos\varphi, \sin\varphi, 0)$. Hence the components of the force at P, arising from the couple on P′P, along the ξ, η, z axes are $\{Hds(\cos\varphi/R_1), H - \text{dh}(\sin\varphi/R_2), H - dH\}$. The force, at P due to the couples on P′P and PP″ has components parallel to ($\boldsymbol{\nu}$,**s**,**n**) given by

$$\left\{H\, ds \sin\varphi \cos\varphi\left(\frac{1}{R_1} - \frac{1}{R_2}\right), H\, ds\left(\frac{\cos^2\varphi}{R_1} + \frac{\sin^2\varphi}{R_2}\right), -dH\right\}.$$

Hence the stress and couple resultants at the edge line s must be replaced by

$$T + \frac{1}{2}H \sin 2\varphi\left(\frac{1}{R_1} - \frac{1}{R_2}\right), \quad S + H\left(\frac{\cos^2\varphi}{R_1} + \frac{\sin^2\varphi}{R_2}\right), \quad N - \frac{\partial H}{\partial s}, \quad G. \tag{54.35}$$

Let us observe that, if the edge line s is a line of curvature, then φ is zero or $\pi/2$, and H does not contribute to T.

The equilibrium equations and the boundary conditions are not necessarily restricted to the theory of small strains and small displacements. They remain formally valid in large strains, provided that they are written for the strained middle surface.

55. Equations Deduced from Three-dimensional Theory

The classical theory of shells has a long tradition going back to Euler (1766) who was the first to propose a theory for the behavior of a shell, by dividing the shell into thin annuli each of which is regarded as a curved bar. The treatment of the problem from the point of view of the general equations of elasticity, is due to Aron (1874). The simplifications necessary to render the theory of a thin shell analogous to that of thin plates was introduced by Love (1888). An exact geometric model of thin shells based on a pure director theory was proposed by Ericksen and Truesdell (1958).

The method of Kirchhoff and Love is useful in practice because it offers a complete set of equations, which allows the solution of boundary-value problems for thin elastic shells, at least in principle, even in the case of large strains. However, the procedure by which the equations are derived is not rigorously coherent, because in some instances, such as when the stress and couple resultants are defined, it follows the three-dimensional approach. On the other hand, in other instances, in the case for which strains and changes in curvature are defined, it makes exclusive reference to the changes in the shape of the middle surface.

For the case of small strains and displacements, with linearly elastic materials, deducing the equations of shell theory from the theory of elasticity can be done in a systematic manner, by averaging the three-dimensional equations over the thickness and retaining the terms of a given power in a parameter related to the distance from the middle surface (Vlasov 1958).

We first proceed to define the strains in the interior of a shell the thickness $2h$ of which is not necessarily small. We consider the middle surface of the shell and denote by α and β a system of orthogonal curvilinear coordinates on this surface. The lines $\alpha =$ constant and $\beta =$ constant are, in addition, lines of principal curvature; that is, lines along which the radii of curvature are extrema. In the orthogonal coordinates α and β the linear element of any curve traced on the surface is given by the formula

$$dS^2 = A^2\,d\alpha^2 + B^2\,d\beta^2, \tag{55.1}$$

where A and B are, in general, functions of α and β.

Let us now draw the normal at any point of the middle surface and mark any point on this normal by its distance γ from the middle surface. The coordinates (α, β, γ) are orthogonal. The surfaces $\gamma =$ constant are equidistant from the middle surface, and the surfaces $\alpha =$ constant and $\beta =$ constant intersect along lines which are normal to the middle surface at the points $\alpha =$ constant, $\beta =$ constant. The square of the length of a linear element in the α, β, γ system can therefore be expressed as the sum

$$dS_1^2 = H_1^2\,d\alpha^2 + H_2^2\,d\beta^2 + H_3^2\,d\gamma^2, \tag{55.2}$$

where the coefficient H_3 is equal to 1, and the other coefficients given by

$$H_1 = A(1 + k_1\gamma), \quad H_2 = B(1 + k_2\gamma). \tag{55.3}$$

Here A and B are the coefficients related to dS, and $k_1 = k_1(\alpha, \beta)$ and $k_2 = k_2(\alpha, \beta)$ are the principal curvatures of the middle surface, that is, the inverse of the radii of curvature of the lines $\alpha =$ constant and $\beta =$ constant.

The coordinates α, β, γ identify any point P of the shell in its unstrained state. As a consequence of the deformation, the point (α, β, γ) assumes new coordinates $(\alpha+\xi, \beta+\eta, \gamma+\zeta)$ and occupies a position P_1. Let $u_\alpha, u_\beta, u_\gamma$ be the projections of the displacement PP_1 on the normals to the coordinates surfaces at P. If the displacement is small, so that $u_\alpha, u_\beta, u_\gamma$ and ξ, η, ζ are small quantities of the same order, we have the equations

$$\xi = \frac{u_\alpha}{H_1}, \quad \eta = \frac{u_\beta}{H_2}, \quad \zeta = \frac{u_\gamma}{H_3}. \tag{55.4}$$

From these relations we can obtain the general expressions for the linearized strains referred to the curvilinear coordinates. However, on observing that, in the present case, $H_3 = 1$, the strains have the following form (Vlasov 1958, Kap. VI):

$$\begin{aligned}
\varepsilon_\alpha &= \frac{1}{H_1} u_{\alpha,\alpha} + \frac{1}{H_1 H_2} H_{1,\beta} u_\beta + \frac{1}{H_1} H_{1,\gamma} u_\gamma, \\
\varepsilon_\beta &= \frac{1}{H_2} u_{\beta,\beta} + \frac{1}{H_1 H_2} H_{2,\alpha} u_\alpha + \frac{1}{H_2} H_{2,\gamma} u_\gamma, \\
\varepsilon_\gamma &= u_{\gamma,\gamma} \\
2\varepsilon_{\alpha\beta} &= \frac{H_1}{H_2}\left(\frac{1}{H_1} u_\alpha\right)_{,\beta} + \frac{H_2}{H_1}\left(\frac{1}{H_2} u_\beta\right)_{,\alpha}, \\
2\varepsilon_{\beta\gamma} &= H_2\left(\frac{1}{H_2} u_\beta\right)_{,\gamma} + \frac{1}{H_2} u_{\gamma,\beta}, \\
2\varepsilon_{\alpha\gamma} &= H_1\left(\frac{1}{H_1} u_\alpha\right)_{,\gamma} + \frac{1}{H_1} - u_{\gamma,\alpha}.
\end{aligned} \tag{55.5}$$

Let us now introduce the geometric assumption that a rectilinear element perpendicular to the middle surface remains rectilinear and normal to the strained middle surface and preserves its length. This hypothesis implies the conditions

$$\varepsilon_\gamma = 0 \tag{55.6a}$$

$$\varepsilon_{\alpha\gamma} = \varepsilon_{\beta\gamma} = 0. \tag{55.6b}$$

As $\varepsilon_\gamma = u_{\gamma,\gamma}$, we obtain from (55.6a) the result

$$u_\gamma = w(\alpha,\beta),$$

which states that all points placed on the same normal filament undergo the same normal displacement w. The other two equations (55.6b), written after substitution of (55.3), become

$$\begin{aligned}
\left[\frac{u_\alpha}{A(1+k_1\gamma)}\right]_{,\gamma} + \frac{1}{A^2(1+k_1\gamma)^2} w_{,\alpha} &= 0, \\
\left[\frac{u_\beta}{B(1+k_2\gamma)}\right]_{,\gamma} + \frac{1}{B^2(1+k_2\gamma)^2} w_{,\beta} &= 0.
\end{aligned}$$

On integrating each of these quantities with respect to γ, and putting $u_\alpha = u(\alpha,\beta)$ and $u_\beta = v(\alpha,\beta)$ for $\gamma = 0$, we obtain

$$u_\alpha = (1 + k_1\gamma)u - \frac{\gamma}{A}w_{,\alpha},$$
$$u_\beta = (1 + k_2\gamma)v - \frac{\gamma}{B}w_{,\beta}. \tag{55.7}$$

These equations show that the tangential displacements of any point of the shell are linear functions of the distance γ from the middle surface.

Having obtained the displacements, we can calculate the nonvanishing components of strain, namely ε_α, ε_β, and $\varepsilon_{\alpha\beta}$, from (55.7). If, however, we take account of the fact that a shell is the curvilinear sheet constituted by the points satisfying $-h \leqslant \gamma \leqslant h$, where h is small with respect to the diameter of the domain of variation in α and β, we can simplify the expressions for ε_α, ε_β, and $\varepsilon_{\alpha\beta}$ by taking

$$\frac{1}{H_1} = \frac{1}{A}(1 - k_1\gamma + k_1^2\gamma^2 + \ldots),$$
$$\frac{1}{H_2} = \frac{1}{B}(1 - k_2\gamma + k_2^2\gamma^2 + \ldots),$$
$$\frac{H_2}{H_1} = \frac{B}{A}[1 - (k_1 - k_2)(\gamma - k_1^2\gamma^2 + \ldots)], \tag{55.8}$$
$$\frac{H_2}{H_1} = \frac{A}{B}[1 + (k_1 - k_2)(\gamma - k_2^2\gamma^2 + \ldots)].$$

By substituting (55.7) and (55.8) in (55.5) and arranging the terms in increasing powers of γ, we arrive at expressions like

$$\varepsilon_\alpha = \varepsilon_1 - \kappa_1\gamma + \varphi_1\gamma^2 + \ldots,$$
$$\varepsilon_\beta = \varepsilon_2 - \kappa_2\gamma + \varphi_2\gamma^2 + \ldots, \tag{55.9}$$
$$2\varepsilon_{\alpha\beta} = \tilde{\omega} - 2\tau\gamma + \psi\gamma^2 + \ldots,$$

where the terms $\varepsilon_1, \varepsilon_2, \tilde{\omega}, \kappa_1, \kappa_2$, and τ are exactly those occurring in (54.23), and $\varphi_1 = -k_1\kappa_1, \varphi_2 = -k_2\kappa_2, \psi = \frac{1}{2}(k_1 - k_2)^2\tilde{\omega} - \frac{1}{2}(k_1 + k_2)\tau$.

We now deduce the equilibrium equations of shells from the three-dimensional theory of elasticity starting with Lamé's equations in curvilinear coordinates, written only for the components σ_α, σ_β, σ_γ, and $\sigma_{\alpha\beta}$:

$$\sigma_\alpha = (\lambda + 2\mu)\Delta - 2\mu(\varepsilon_\beta + \varepsilon_\gamma),$$
$$\sigma_\beta = (\lambda + 2\mu)\Delta - 2\mu(\varepsilon_\alpha + \varepsilon_\gamma), \tag{55.10}$$
$$\sigma_{\alpha\beta} = 2\mu\varepsilon_{\alpha\beta}, \quad \sigma_\gamma = (\lambda + 2\mu)\Delta - 2\mu(\varepsilon_\alpha + \varepsilon_\beta),$$

with $\Delta = \varepsilon_\alpha + \varepsilon_\beta + \varepsilon_\gamma$. In the present system of coordinates ($H_3 = 1$) the equilibrium equations have the form

$$(H_2\sigma_\alpha)_{,\alpha} - \sigma_\beta H_{2,\alpha} + \frac{1}{H_1}(\sigma_{\alpha\beta}H_1^2)_{,\beta} + \frac{1}{H_1}(H_1^2 H_2\sigma_{\alpha\gamma})_{,\gamma} + H_1H_2f_\alpha = 0,$$
$$(H_1\sigma_\beta)_{,\beta} - \sigma_\alpha H_{1,\beta} + \frac{1}{H_1}(\sigma_{\alpha\beta}H_2^2)_{,\alpha} + \frac{1}{H_2}(H_1H_2^2\sigma_{\beta\gamma})_{,\gamma} + H_1H_2f_\beta = 0, \tag{55.11}$$
$$(H_1H_2\sigma_\gamma)_{,\gamma} - \sigma_\alpha H_2H_{1,\gamma} - \sigma_\beta H_1H_{2,\gamma} + (H_2\sigma_{\alpha\gamma})_{,\alpha} + (H_1\sigma_{\beta\gamma})_{,\beta} + H_1H_2f_\gamma = 0,$$

where f_α is the component of the body force along the normal to the surface $\alpha =$ constant, and a similar definition holds for f_β and f_γ. Let us now substitute

(55.10) in (55.11) and express the strain components in terms of u_α, u_β, and u_γ by means of (55.5). The result of this substitution is (Vlasov 1958, Kap. IX)

$$(\lambda+2\mu)H_2\Delta_{,\alpha}-2\mu H_1\chi_{,\beta}+2\mu ABKu_\alpha-2\mu(H_2u_{\gamma,\alpha})_{,\gamma}+\frac{1}{H_1}(H_1^2H_2\sigma_{\alpha\gamma})_{,\alpha}+H_1H_2f_\alpha=0, \tag{55.12a}$$

$$(\lambda+2\mu)H_1\Delta_{,\beta}+2\mu H_2\chi_{,\alpha}+2\mu ABKu_\beta-2\mu(H_1u_{\gamma,\beta})_{,\gamma}+\frac{1}{H_2}(H_1H_2^2\sigma_{\beta\gamma})_{,\beta}+H_1H_2f_\beta=0, \tag{55.12b}$$

$$-2(\lambda+2\mu)(H+K\gamma)ABK\Delta+2\mu[(Bk_2u_\alpha)_{,\alpha}+(Ak_1u_\beta)_{,\beta}+2ABKu_\gamma]+(H_2\sigma_{\gamma\alpha})_{,\alpha}+(H_1\sigma_{\gamma\beta})_{,\beta}+(H_1H_2\sigma_\gamma)_{,\alpha}+H_1H_2f_\gamma=0, \tag{55.12c}$$

where $H=H(\alpha,\beta)$ and $K=K(\alpha,\beta)$ are the mean and the Gaussian curvatures of the middle surface, namely

$$H=\frac{1}{2}(k_1+k_2), \quad K=k_1k_2, \tag{55.13}$$

and $\chi=\frac{1}{2}(1/H_1H_2)[(H_2u_\beta)_{,\alpha}-(H_1u_\alpha)_{,\beta}]$ is the rotation of the volume element round the normal to the surface $\gamma=$ constant.

We now proceed to simplify (55.12) by putting

$$\Delta=\Delta_0+\Delta_1\gamma+\Delta_2\gamma^2+\ldots, \tag{55.14}$$

$$\chi=\chi_0+\chi_1\gamma+\chi_2\gamma^2+\ldots, \tag{55.15}$$

with $\Delta_0,\ldots,\chi_2$ being well-defined functions of α and β. We then integrate (55.12) with respect to γ from $-h$ to h, multiply (55.12a) and (55.12b) by γ, and repeat the integration with respect to γ from $-h$ to h. In this way we obtain five partial differential equations in the two independent variables α and β. These equations contain, in addition to the geometric unknowns u, v, w, Δ_0, Δ_1, Δ_2, χ_0, χ_1, and χ_2, the shear resultants

$$N_1=\frac{1}{B}\int_{-h}^{h}H_2\sigma_{\gamma\alpha}\,d\gamma, \quad N_2=\frac{1}{A}\int_{-h}^{h}H_1\sigma_{\gamma\beta}\,d\gamma. \tag{55.16}$$

By this means we obtain a system of five differential equations in 11 unknowns. On eliminating N_1 and N_2 we reduce the equations to three and the unknowns to nine. As quantities like $k_1^2h^2/3$, $k_2^2h^2/3$, and $k_1k_2h^2/3$ are small with respect to 1, we can neglect these terms in the expressions $1+(k_1^2h^2/3)$, $1+(k_2^2h^2/3)$, and $1+(k_1k_2h^2/3)$. If we collect the three differential equations just obtained with the six equations expressing $\Delta_0,\ldots,\chi_2$, we obtain nine differential equations in the nine unknowns $u,\ldots,\chi_2$.

Further in the process of averaging the equations, the known terms may be given a more familiar form. In fact the three quantities

$$X = \frac{1}{AB}\left\{\int_{-h}^{h} f_\alpha H_1 H_2\, d\gamma + H_1 H_2 \sigma_{\alpha\gamma}\Big|_{-h}^{h}\right\},$$
$$Y = \frac{1}{AB}\left\{\int_{-h}^{h} f_\beta H_1 H_2\, d\gamma + H_1 H_2 \sigma_{\beta\gamma}\Big|_{-h}^{h}\right\}, \tag{55.17}$$
$$Z = \frac{1}{AB}\left\{\int_{-h}^{h} f_\gamma H_1 H_2\, d\gamma + H_1 H_2 \sigma_{\gamma}\Big|_{-h}^{h}\right\},$$

represent the components of the external forces, per unit area of the middle surface, parallel to the coordinate lines at any point of the middle surface. In addition, the two quantities

$$L = \frac{1}{AB}\left\{\int_{-h}^{h} \gamma f_\alpha H_1 H_2\, d\gamma + \gamma H_1 H_2 \sigma_{\alpha\gamma}\Big|_{-h}^{h}\right\},$$
$$M = \frac{1}{AB}\left\{\int_{-h}^{h} \gamma f_\beta H_1 H_2\, d\gamma + \gamma H_1 H_2 \sigma_{\beta\gamma}\Big|_{-h}^{h}\right\}, \tag{55.18}$$

are the analogous components along the coordinate lines α and β of the external surface couples.

In order to formulate the boundary-value problems completely, we need to have appropriate boundary conditions. However, the general form of these boundary conditions is not known, and the boundary data are given explicitly only for some particular situations.

The derivation of the equations of the theory of shells from the three-dimensional theory of elasticity when the displacements and strains are large is much more complicated because, apart from the nonlinearity of the equation, Kirchhoff–Love's hypothesis is intimately irreconcilable with a theory of large strains. As in the linear theory, a shell is defined as the solid region contained, in its original unstrained state, between two parallel surfaces at constant distances h and $-h$ from a middle surface. For small h we attempt to describe the state of the shell approximately by functions of two independent variables (John 1971). We introduce curvilinear coordinates (U^1, U^2, U^3) and consider the body in two states. In the unstrained, natural, state the Cartesian coordinates of a particle are $X^i = X^i(U^1, U^2, U^3)$; in the deformed state the Cartesian coordinates of the same particle are $x^i = x^i(U^1, U^2, U^3)$. The linear elements in the unstrained, and strained, states are then given, respectively, by

$$dS^2 = dX^i\, dX^i = G_{ik}\, dU^i\, dU^k, \quad ds^2 = dx^i\, dx^i = g_{ik}\, dU^i\, dU^k, \tag{55.19}$$

where we have

$$G_{ik} = X^j_{,i} X^j_{,k}, \quad g_{ik} = x^j_{,i} x^j_{,k}. \tag{55.20}$$

In order to define the covariant differentiation (indicated by ";") of vector and tensor fields referred to the metric dS^2, we introduce the *Christoffel symbols* of the metric tensor G_{ik}, defined by

$$\Gamma^i_{jk} = \frac{1}{2} G^{ir}(G_{rj,k} + G_{rk,j} - G_{jk,r}), \tag{55.21}$$

where G^{ir} are the contravariant components of the tensor G_{ik}. The Christoffel symbols of ds^2 are given by

$$\bar{\Gamma}^i_{jk} = \frac{1}{2} \bar{g}^{ir}(g_{rj,k} + g_{kr,j} - g_{jk,r}), \tag{55.22}$$

where $\bar{g}^{ir}$ is the tensor reciprocal to g_{ir}, that is the tensor whose components from the matrix reciprocal to that formed by the g_{ir}. Let us note that $\bar{g}^{ir}$ does not coincide with the contravariant components

$$g^{ir} = G^{is} G^{rm} g_{sm}$$

of the tensor g_{ik}.

It is often useful to express $\bar{\Gamma}^i_{jk}$ in terms of Γ^i_{jk}. The relationship between $\bar{\Gamma}^i_{jk}$ and Γ^i_{jk} is given by the equation (John 1965)

$$\bar{\Gamma}^i_{jk} = \Gamma^i_{jk} + c^i_{jk}, \tag{55.23}$$

$$c^i_{jk} = \frac{1}{2} \bar{g}^{ir}(g_{ir;k} + g_{rk;j} - g_{jk;r}). \tag{55.24}$$

We define the *strain tensor* by

$$\varepsilon_{ik} = \frac{1}{2}(g_{ik} - G_{ik}), \tag{55.25}$$

and consider the *principal strains*, defined as the eigenvalues of the matrix

$$\varepsilon^i_k = G^{ir} \varepsilon_{rk} = \frac{1}{2}(g^i_k - \delta^i_k), \tag{55.26}$$

the elements of which are the mixed components of the strain tensor.

The material composing the shell is taken to be homogeneous, isotropic, and perfectly elastic, in the sense that it possesses a strain energy density per unit unstrained volume, which is expressed in the form

$$W = W(s_1, s_2, s_3), \tag{55.27}$$

where the s_i are symmetric functions of the eigenvalues of the strain matrix ε^i_k :

$$s_1 = \varepsilon^i_i, \quad s_2 = \varepsilon^i_j \varepsilon^j_i, \quad s_3 = \varepsilon^i_j \varepsilon^j_k \varepsilon^k_i. \tag{55.28}$$

The stress–strain relations have the form

$$t^m_i = \sqrt{\frac{G}{g}} \frac{\partial W}{\partial \varepsilon^i_m}, \tag{55.29}$$

where the t^i_k are the mixed components of the stress tensor, and

$$G = \det(G_{ik}), \quad g = \det(g_{ik}). \tag{55.30}$$

In Cartesian coordinates x^i, the components of the force acting on a surface element, which in the space of the parameters U^i has an area dA and a unit normal N_i, have the expression

$$\sqrt{g}t^{mk}x^i_{,m}N_k\,dA = \sqrt{G}G^{ms}\frac{\partial W}{\partial \varepsilon^s_k}x^i_{,m}N_k\,dA. \tag{55.31}$$

In the absence of body forces, the equations of equilibrium are

$$t^{mk}_{;k} + c^m_{ik}t^{ik} + c^k_{ik}t^{mi} = 0, \tag{55.32}$$

or, in terms of the *pseudo-stresses*, defined by

$$T^m_i = \sqrt{\frac{G}{g}}t^m_s g^s_i - W\delta^m_i = g^s_i\frac{\partial W}{\partial \varepsilon^s_m} - W\delta^m_i, \tag{55.33}$$

equations (55.32) take the simpler form

$$T^m_{i;m} = 0. \tag{55.34}$$

The equations derived so far hold for a general elastic continuum without any reference to the particular geometric form of the shell. Let us now exploit this property. The faces of the shell have the equations $U^3 = \pm h$, while the middle surface is $U^3 = 0$. For brevity, we shall write u for the coordinate U^3. The middle surface, which we shall denote by Σ_0, is locally described by its first, second, and third fundamental forms, the coefficients of which are $a_{\alpha\beta}$, $b_{\alpha\beta}$ (see (54.12) and (54.14)), and $M_{\alpha\beta} = b_{\alpha\lambda}b^\lambda_\beta$, respectively. The relations between these forms and the metric tensor G_{ik} are

$$G_{\alpha\beta} = a_{\alpha\beta} - 2ub_{\alpha\beta} + u^2M_{\alpha\beta}, \quad G_{\alpha 3} = 0, \quad G_{33} = 1. \tag{55.35}$$

We have supposed that the body is not subjected to body forces. We now add the assumption that no surface forces act on the faces, so that we have $t^{i3} = 0$ for $u = \pm h$. For the pseudo-stresses this condition is equivalent to

$$T^3_\alpha = 0, \quad T^3_3 = -W, \quad \text{for} \quad u = \pm h. \tag{55.36}$$

In order to describe the change in shape of the surface Σ_0, we introduce as dependent variables the coefficients $e_{\alpha\beta}$ and $\ell_{\alpha\beta}$ of the first and second fundamental forms of the deformed middle surface, or rather the amounts $2\varepsilon_{\alpha\beta}$ and $w_{\alpha\beta}$ by which these coefficients differ from those of Σ_0. For $u = 0$ we have

$$e_{\alpha\beta} = g_{\alpha\beta} = a_{\alpha\beta} + 2\varepsilon_{\alpha\beta}, \tag{55.37}$$

$$\begin{aligned}\ell_{\alpha\beta} &= \frac{1}{2}(\bar{g}^{33})^{-1/2}\bar{g}^{3s}(g_{\alpha s,\beta} + g_{\beta s,\alpha} - g_{\alpha\beta,s}) \\ &= (\bar{g}^3_3)^{-1/2}\left[b_{\alpha\beta} + \bar{g}^{3s}(\varepsilon_{\alpha s;\beta} + \varepsilon_{\beta s;\alpha} - \varepsilon_{\alpha\beta;s})\right].\end{aligned} \tag{55.38}$$

Let us indicate by the symbol "|" covariant differentiation with respect to the metric $a_{\alpha\beta}\,dU^\alpha\,dU^\beta$ taken on Σ_0. The changes in the coefficients of the second fundamental form due to the deformation are given by the differences

$$\begin{aligned} w_{\alpha\beta} &= \ell_{\alpha\beta} - b_{\alpha\beta} \\ &= (\bar{g}_3^3)^{-1/2} b_{\alpha\beta} - b_{\alpha\beta} + (\bar{g}_3^3)^{-1/2} \bar{g}^{3\lambda}(\varepsilon_{\alpha\lambda|\beta} + \varepsilon_{\beta\lambda|\alpha} - \varepsilon_{\alpha\beta|\lambda} - 2b_{\alpha\beta}\varepsilon_{3\lambda}) \\ &\quad + (\bar{g}_3^3)^{-1/2}(\varepsilon_{\alpha 3|\beta} + \varepsilon_{\beta 3|\alpha} - \varepsilon_{\alpha\beta;3} - 2b_{\alpha\beta}\varepsilon_{33} + b_\beta^\lambda \varepsilon_{\alpha\lambda} + b_\alpha^\lambda \varepsilon_{\beta\lambda}). \end{aligned} \tag{55.39}$$

The six quantities $\varepsilon_{\alpha\beta}$ and $w_{\alpha\beta}$ on Σ_0 are related to each other by three compatibility equations, called the *Codazzi and Gauss equations*. The geometric meaning of these equations is that the coefficients $a_{\alpha\beta}$ and $b_{\alpha\beta}$, as functions of the coordinates U^1, U^2, cannot be given arbitrarily. We therefore need three additional equations, which can be generated by averaging the equilibrium equations. For this purpose we write (55.34) in the form

$$\begin{aligned} 0 &= u^n \sqrt{\frac{G}{a}} T_{i;k}^k \\ &= \left(u^n \sqrt{\frac{G}{a}} T_i^k \right)_{,k} - u^n \sqrt{\frac{G}{a}} \left[\frac{n}{u} T_i^3 - (\log \sqrt{a})_{,\beta} T_i^\beta + \frac{1}{2} G_{\beta\gamma,i} T^{\beta\gamma} \right], \end{aligned}$$

where n is any non-negative integer and $a = \det(a_{\alpha\beta})$. On averaging these equations with respect to u for $-h < u < h$ and using the boundary conditions (55.36), we obtain a sequence of two-dimensional relations. For $i = \alpha$ we obtain the two equations

$$0 = \left(\frac{1}{2h} \int_{-h}^{h} u^n \sqrt{\frac{G}{a}} T_\beta^\alpha \, du \right)_{|\beta} + \frac{1}{2h} \int_{-h}^{h} u^n \sqrt{\frac{G}{a}} \left(-\frac{n}{u} T_\alpha^3 + u b_{\beta\gamma|\alpha} T^{\beta\gamma} - \frac{1}{2} u^2 M_{\beta\gamma|\alpha} T^{\beta\gamma} \right) du, \tag{55.40}$$

and, similarly, for $i = 3$ we obtain the equation

$$\begin{aligned} 0 &= -\frac{1}{2h} \int_{-h}^{h} \left(u^n \sqrt{\frac{G}{a}} W \right)_{,3} du + \left(\frac{1}{2h} \int_{-h}^{h} u^n \sqrt{\frac{G}{a}} T_3^\beta \, du \right)_{|\beta} \\ &\quad + \frac{1}{2h} \int_{-h}^{h} u^n \sqrt{\frac{G}{a}} \left(-\frac{n}{u} T_3^3 + b_{\beta\gamma} T^{\beta\gamma} - u M_{\beta\gamma} T^{\beta\gamma} \right) du. \end{aligned} \tag{55.41}$$

On introducing the mean curvature $H = \frac{1}{2} b_\alpha^\alpha$, and the Gaussian curvature $K = \frac{1}{2}\left(b_\alpha^\alpha b_\beta^\beta - b_\beta^\alpha b_\alpha^\beta \right)$, it is possible to write the ratio $\sqrt{G/a}$ as

$$\sqrt{\frac{G}{a}} = 1 - 2uH + u^2 K. \tag{55.42}$$

Equations (55.40) and (55.41), for general n, are equivalent to the system of equations (55.34) and the boundary conditions (55.36). If we make use only of the equations corresponding to $n = 0$, we obtain the remaining three equations necessary to determine $\varepsilon_{\alpha\beta}$ and $w_{\alpha\beta}$.

56. The Solvability of the Equations of Shell Theory

If the assumptions of the linearized theory of thin elastic shells are satisfied reasonably accurately, it is natural to try to solve them directly. However, the success of this operation is conditioned by our knowledge of how simple the equations are and whether certain necessary conditions of solvability are satisfied. For instance, a first requirement is to check whether the number of equations is equal to the number of unknowns. The equations of equilibrium (54.30) and (54.33) are a set of six equations connecting the six stress resultants $T_1, \ldots,$ and the four resultant couples $G_1, \ldots, H_2$. Four of the six stress resultants and all the resultant couples are expressed in terms of the quantities ε_1, ε_2, and $\tilde{\omega}$ and κ_1, κ_2, and τ by means of the constitutive equations (54.25) and (54.26); these quantities are expressed in terms of u, v, and w by (54.21) and (54.22). We therefore have a set of six differential equations for determining the five quantities N_1, N_2, u, v, and w.

The excess of the number of equations above the number of unknowns constitutes a serious difficulty in the solution of the problem; but the difficulty disappears if the approximate character of the equations is taken into account. The redundant equation is (54.33c), and this equation is in general incompatible with equations (54.25), (54.26). However, when more exact equations for S_1 and S_2 in terms of u, v, and w are employed, then the equation under discussion becomes an identity (Love 1927, Art. 333). For example, if we take

$$T_1 = B(\varepsilon_1 + \sigma\varepsilon_2), \quad T_2 = B(\varepsilon_2 + \sigma\varepsilon_1),$$

$$S_1 = -S_2 = B(1-\sigma)\left(\frac{\tilde{\omega}}{2} + \frac{h^3}{3}\frac{\tau}{R_1 + R_2}\right) \quad \left(B = \frac{2Eh}{1-\sigma^2}\right),$$

$$G_1 = -D(\kappa_1 + \sigma\kappa_2), \quad G_2 = -D(\kappa_2 + \sigma\kappa_1), \quad H_2 = -H_1 = -D(1-\sigma)\tau,$$

we obtain stress resultants and moments that do not contradict the sixth equation of equilibrium. There is, however, the complication that the elasticity relationships given before are not tensorial in character. However, these relations can be altered so that they acquire a tensorial form while remaining sufficiently simple (Zveriaev 1970).

Another substantial simplification in the solution of the equation of thin shells is based on the following consideration. Let us suppose that the thickness does not possess discontinuous changes, that the middle surface is continuous, as is its curvature, that the external couples (denoted by L and M) vanish, and that the loads X, Y, and Z are continuous. Then, if we solve the elastic problem, we find that the stress resultants of the type T_1, T_2, S_1, and S_2 are dominant with respect the other stress and couple resultants. In more indicative terms, we say that, provided the assumptions listed above are satisfied, a stress field of purely membrane type tends to prevail in a thin curved shell. This implies that, at a sufficient distance from the points at which the thickness, curvatures, or loads are discontinuous, the stresses in a shell can be evaluated by regarding it as a simple membrane, thus enabling flexural rigidity to be ignored. Of course, the method is inapplicable near the edges of the shell, because the boundary conditions prescribed along the edge lines are, in general, incompatible with those necessary to maintain a purely membrane state. Satisfying the boundary conditions correctly requires the application of the complete theory of shells.

However, this does not necessarily compromise the usefulness of the approximate membrane analysis, because the flexural effects are, in general, confined to a small strip adjacent to each edge line. These effects are also called *edge perturbations* and can be calculated using simplified procedures.[56]

In some circumstances the kind of approximation useful in describing the elastic response of a shell proceeds in the opposite direction. The membrane stresses and strains are regarded as negligible, while only bending effects are retained. The prevalence of flexural strains in shells is typical of many situations occurring in shell vibration analysis. This has been confirmed by a large number of experiments dating back to Rayleigh (1894), and may be understood by considering the following intuitive argument. Let us consider an open shell or bowl formed from a whole sphere with an aperture created by removing a cap cut along a small circle. When the aperture is very small, or the spherical surface is nearly complete, the vibrations approximate those of a complete spherical shell and are almost completely extensional. When, on the other hand, the aperture has a radius that is comparable with the radius of the sphere, the vibrations approximate those of a planar plate and are essentially flexural. Of course, in intermediate cases, the distinction between these two limiting behaviors is not well defined.

The general treatment of the equations of the two-dimensional linear theory may be effected without excessive conceptual difficulties by using the variational theory of boundary-value problems (Bernadou, Ciarlet, and Miara 1994). On employing the definitions of strains (54.19) and (54.20), we consider a piece of shell the unstrained middle surface of which is parametrically represented by a vector function $\boldsymbol{\varphi}(U^1, U^2)$, with $(U^1, U^2) \in \bar{\omega}$. We call γ the boundary of ω. In order to avoid unnecessary complications, we assume that the edge line of the plate is rigidly clamped and that only a surface load $\mathbf{p}$ per unit area acts on the middle surface. The unknown is the displacement $\boldsymbol{\zeta} = (\zeta_i)$, which must satisfy the boundary conditions

$$\zeta_i = \frac{\partial \zeta_3}{\partial \nu} = 0, \quad \text{on} \quad \gamma. \tag{56.1}$$

The natural Hilbert space to which the solution $\boldsymbol{\zeta} = ((\zeta_\alpha), \zeta_3)$ belongs is the product space $H^1(\omega) \times H^2(\omega)$, and in this space we define the subspace

$$V(\omega) = \left\{((\eta_\alpha), \eta_3) \in H^1(\omega) \times H^2(\omega); \quad \eta_i = \frac{\partial \eta_3}{\partial \nu} = 0, \text{ on } \gamma\right\}.$$

For any pair of vectors (η_i) and (ζ_i) the extensions and the changes of curvature $\gamma_{\alpha\beta}$ and $\kappa_{\alpha\beta}$ are well defined, as is the bilinear symmetric form

$$B(\boldsymbol{\zeta}, \boldsymbol{\eta}) = C \iint_{\omega} \left\{ \varepsilon a^{\alpha\beta\rho\sigma} \gamma_{\rho\sigma}(\boldsymbol{\eta}) \gamma_{\alpha\beta}(\boldsymbol{\zeta}) + \frac{\varepsilon^3}{3} a^{\alpha\beta\rho\sigma} \kappa_{\rho\sigma}(\boldsymbol{\eta}) \kappa_{\alpha\beta}(\boldsymbol{\zeta}) \right\} \sqrt{a}\, d\omega. \tag{56.2}$$

where C is a constant and $a^{\alpha\beta\rho\sigma}$ are elastic moduli having the form

$$a^{\alpha\beta\rho\sigma} = \frac{4\lambda\mu}{(2\mu + \lambda)} a^{\alpha\beta} a^{\rho\sigma} + 2\mu(a^{\alpha\rho} a^{\beta\sigma} + a^{\alpha\sigma} a^{\beta\rho}). \tag{56.3}$$

[56] The first of these procedures was proposed by Geckeler (1926).

In connection with the bilinear form B(ζ, η), we define the linear form

$$L(\boldsymbol{\eta}) = \iint_{\omega} \mathbf{p} \cdot \boldsymbol{\eta}\sqrt{a}\, d\omega, \quad \text{for all} \quad \boldsymbol{\eta} \in V(\omega).$$

The boundary-value problem then reduces to determine a function $\zeta \varepsilon V(\omega)$ such that we have

$$B(\boldsymbol{\zeta}, \boldsymbol{\eta}) = L(\boldsymbol{\eta}), \quad \text{for all } \boldsymbol{\eta} \in V(\omega), \tag{56.4}$$

which represents the weak formulation of the boundary-value problem.

In order to prove the existence and uniqueness of solutions to problem (56.4), we introduce the norm

$$\|\boldsymbol{\eta}\|_{H^1(\omega)\times H^2(\omega)} = \left\{ \sum_{\alpha=1}^{2} \|\eta_\alpha\|_1^2 + \|\eta_3\|_2^2 \right\}^{\frac{1}{2}}, \tag{56.5}$$

and the bilinear form $B(\boldsymbol{\zeta}, \boldsymbol{\eta})$ is continuous with respect to this norm. The form $B(\boldsymbol{\zeta}, \boldsymbol{\eta})$ is further coercive with respect the norm (56.5); that is, there is a constant $\beta > 0$ such that

$$B(\boldsymbol{\eta}, \boldsymbol{\eta}) \geqslant \beta \|\boldsymbol{\eta}\|^2_{H^1(\omega)\times H^2(\omega)}, \quad \text{for all } \boldsymbol{\eta} \in V(\omega). \tag{56.6}$$

The proof of (56.6) is not immediate, and must be obtained in two steps. First we prove that there is a constant $C_1 > 0$ such that

$$B(\boldsymbol{\eta}, \boldsymbol{\eta}) \geqslant C_1 \left\{ \sum_{\alpha,\beta=1}^{2} \|\gamma_{\alpha\beta}(\boldsymbol{\eta})\|^2_{L^2(\omega)} + \sum_{\alpha,\beta=1}^{2} \|\kappa_{\alpha\beta}(\boldsymbol{\eta})\|^2_{L^2(\omega)} \right\}, \tag{56.7}$$

for all $\boldsymbol{\eta} \in H^1(\omega) \times H^2(\omega)$. Let $(T_{\alpha\beta})$ denote an arbitrary symmetric tensor. On the one hand, we have

$$a^{\alpha\beta} a^{\rho\sigma} T_{\alpha\beta} T_{\rho\sigma} = (a^{\alpha\beta} T_{\alpha\beta})^2 \geqslant 0; \tag{56.8}$$

while, on the other, we can find a constant $C > 0$ such that

$$a^{\alpha\beta}(U^1, U^2) a^{\rho\sigma}(U^1, U^2) T_{\alpha\beta} T_{\rho\sigma} \geqslant C T_{\alpha\beta} T_{\alpha\beta}, \tag{56.9}$$

for all points $(U^1, U^2) \in \bar{\omega}$. In order to illustrate this, we write the left-hand side of (56.9) as $\boldsymbol{\theta}^T \mathbf{A} \boldsymbol{\theta}$, with

$$\mathbf{A} = \mathbf{A}^T = \begin{bmatrix} a^{11}a^{11} & 2a^{11}a^{12} & a^{12}a^{12} \\ \cdot & 2(a^{12}a^{12} + a^{11}a^{22}) & 2a^{12}a^{22} \\ \cdot & \cdot & a^{22}a^{22} \end{bmatrix}, \quad \boldsymbol{\theta} = \begin{bmatrix} T_{11} \\ T_{12} \\ T_{22} \end{bmatrix}.$$

As we have $a^{11}a^{11} > 0$, it follows that

$$\det \begin{bmatrix} a^{11}a^{11} & 2a^{11}a^{12} \\ \cdot & 2(a^{12}a^{12} + a^{11}a^{22}) \end{bmatrix} = 2\frac{a^{11}a^{11}}{a} > 0,$$

and we infer that the symmetric matrix is positive-definite at all points $(U^1, U^2) \in \bar{\omega}$, and hence the constant C exists and can be calculated.

The second step of the demonstration is to prove that, under the conditions (56.1), the semi-norm

$$|\boldsymbol{\eta}| = \left\{ \sum_{\alpha,\beta=1}^{2} \|\gamma_{\alpha\beta}(\boldsymbol{\eta})\|^2 + \sum_{\alpha,\beta=1}^{2} \|\kappa_{\alpha\beta}(\boldsymbol{\eta})\|^2 \right\}^{\frac{1}{2}}, \tag{56.10}$$

which figures in (56.7), is a norm over the subspace $V(\omega)$, and there is a constant C_3 such that

$$C_3 > 0 \quad \text{and} \quad |\boldsymbol{\eta}| \geqslant C_3 \|\boldsymbol{\eta}\|_{H^1(\omega)\times H^2(\omega)}, \qquad \text{for all } \boldsymbol{\eta} \in V(\omega). \tag{56.11}$$

The proof of this inequality is a consequence of a theorem by Rellich and of definitions (54.19) and (54.20) (Bernadou et al. 1994).

After having now proved that the bilinear form $B(\boldsymbol{\eta}, \boldsymbol{\zeta})$ is continuous and coercive in $V(\omega)$ with respect to the norm $\|\cdot\|_{H^1(\omega)\times H^2(\omega)}$, the existence and uniqueness of a solution of the variational problem (56.4) is a consequence of a lemma of Lax and Milgram.

The linearized theory of thin shells can be generalized slightly in order to allow constant shear deformations across the thickness of the shell (Naghdi 1963). These are taken in the sense that a point with a coordinate U^3 along the normal vector $\mathbf{a}^3$ is of the form $\zeta_i\mathbf{a}^i + U^3 r_\alpha \mathbf{a}^\alpha$, where ζ_i are, as before, the covariant components of the displacements of the points of the middle surface and r_α are the (linearized) covariant components of the rotation of the unit normal vector $\mathbf{a}^3$. Hence, in general, a filament perpendicular to the unstrained middle surface is no longer normal to the strained middle surface. The unknowns of the problem are, therefore, now five: the three functions ζ_i and the two functions r_α.

In order to formulate the problem of the equilibrium of an elastic thin shell satisfying the assumptions of this enlarged theory, we define the unknown as the vector

$$(\boldsymbol{\zeta}, \mathbf{r}) = ((\zeta_i), (r_\alpha)), \tag{56.12}$$

and assume that it satisfies the boundary conditions of a perfect clamp on the boundary:

$$\zeta_i = r_\alpha = 0, \quad \text{on} \quad \gamma. \tag{56.13}$$

Next, let us define the Hilbert space $H^1(\omega)$ and the subspace

$$V(\omega) = \{(\boldsymbol{\eta}, \mathbf{s}) = ((\eta_i), (s_\alpha)) \in H^1(\omega); \quad \eta_i = s_\alpha = 0, \text{ on } \gamma\}. \tag{56.14}$$

In $V(\omega)$ we define the symmetric bilinear form

$$B((\boldsymbol{\zeta}, \mathbf{r}), (\boldsymbol{\eta}, \mathbf{s})) = C \iint_\omega \Big\{ \varepsilon a^{\alpha\beta\rho\sigma} \gamma_{\rho\sigma}(\boldsymbol{\zeta}) \gamma_{\alpha\beta}(\boldsymbol{\eta}) + \frac{\varepsilon^3}{3} a^{\alpha\beta\rho\sigma} \chi_{\rho\sigma}(\boldsymbol{\zeta}, \mathbf{r}) \chi_{\alpha\beta}(\boldsymbol{\eta}, \mathbf{s}) + 8\varepsilon\mu a^{\alpha\beta} \gamma_{\alpha 3}(\boldsymbol{\zeta}, \mathbf{r}) \gamma_{\beta 3}(\boldsymbol{\eta}, \mathbf{s}) \Big\} \sqrt{a}\, d\omega, \tag{56.15}$$

where $a^{\alpha\beta\rho\sigma}$ is given by (56.3) and $\gamma_{\alpha\beta}(\boldsymbol{\eta})$ by (54.19), whereas we have put

$$\chi_{\alpha\beta}(\boldsymbol{\eta},\mathbf{s}) = \frac{1}{2}(s_{\alpha|\beta} + s_{\beta|\alpha}) - \frac{1}{2}b^{\rho}_{\alpha}(\eta_{\rho|\beta} - b_{\rho\beta}\eta_3) - \frac{1}{2}b^{\sigma}_{\beta}(\eta_{\sigma|\alpha} - b_{\sigma\alpha}\eta_3), \tag{56.16}$$

$$\gamma_{\alpha 3}(\boldsymbol{\eta},\mathbf{s}) = \frac{1}{2}(\eta_{3,\alpha} + b^{\rho}_{\alpha}\eta_{\rho} + s_{\alpha}). \tag{56.17}$$

The tensor $\chi_{\alpha\beta}(\boldsymbol{\eta},\mathbf{s})$ is the linearized change of curvature associated with an arbitrary displacement and rotational field of the middle surface and of the normal vector $\mathbf{a}_3$, respectively. The tensor $\gamma_{\alpha 3}(\boldsymbol{\eta},\mathbf{s})$ is the linearized transverse shear strain tensor.

If $\mathbf{p}$ is the vector representing the forces per unit area of the underformed middle surface, we can introduce, as before, the linear functional

$$L(\boldsymbol{\eta},\mathbf{s}) = \iint_{\omega} \mathbf{p}\cdot\boldsymbol{\eta}\sqrt{a}\,d\omega.$$

Now the unknown vector $(\boldsymbol{\zeta},\mathbf{r}) = ((\zeta_i), r_\alpha)$ solves the variational problem

$$B((\boldsymbol{\zeta},\mathbf{r}),(\boldsymbol{\eta},\mathbf{s})) = L(\boldsymbol{\eta},\mathbf{s}), \quad \text{for all } (\boldsymbol{\eta},\mathbf{s}) \in V(\omega), \tag{56.18}$$

and with this form we associate the norm

$$\|(\boldsymbol{\eta},\mathbf{s})\|_1 = \left\{\sum_{i=1}^{3}\|\eta_i\|_1^2 + \sum_{\alpha=1}^{2}\|s_\alpha\|_1^2\right\}^{\frac{1}{2}}. \tag{56.19}$$

It is not difficult to prove that the bilinear form B is coercive in the sense that there is a constant $\beta > 0$ such that

$$B((\boldsymbol{\eta},\mathbf{s}),(\boldsymbol{\eta},\mathbf{s})) \geqslant \beta\|\boldsymbol{\eta},\mathbf{s}\|_1^2, \quad \text{for all } (\boldsymbol{\eta},\mathbf{s}) \in V(\omega). \tag{56.20}$$

The form B is obviously continuous with respect the norm (56.19). Then, by again applying the Lax–Milgram lemma, we find that the problem admits one, and only one, solution.

The existence theory of linear boundary-value problems can be applied in a systematic way to linearized theories of shells. However, when strains are no longer small, a general method for approaching the problem is lacking, and only partial results are known. However, a quite complete treatment can be achieved in the case of a spherical shell under constant normal pressure, provided that only rotationally symmetric solutions are considered (Dickey 1990). In this case, the analysis can predict the onset of buckled states at a pressure significantly lower than that predicted by linear shell theory, as has been confirmed by several experimental results (Stoker 1968).

The equations describing the axisymmetric deformation of a spherical shell under normal pressure arise as a corollary of Reissner's (1949) theory. The equilibrium equations are again (51.39), but, for small strains, we can replace r with r_0 and α by α_0, where r_0 and α_0 are the values of r and α measured in the undeformed state. For a spherical shell of radius a we can set

$$r_0 = a\sin\xi, \quad z_0 = -a\cos\xi, \quad \alpha_0 = a, \quad \varphi_0 = \xi.$$

The extensional and bending strains are defined as

$$\varepsilon_1 = \frac{\cos\xi}{\cos\varphi} + \frac{u'}{a\cos\varphi} - 1, \quad \varepsilon_2 = \frac{u}{a\sin\xi},$$
$$\kappa_1 = \frac{1-\phi'}{a}, \quad \kappa_2 = \frac{1}{a}\left(1 - \frac{\sin\varphi}{\sin\xi}\right).$$

From these strains we obtain the stress and couple resultants by means of the linearized constitutive equations (51.43). Substituting these quantities into the equilibrium equations yields two differential equations similar to (51.49) and (51.50). These equations can be reduced to a pair of nonlinear ordinary differential equations of the form

$$\begin{aligned} Lq + \sigma q &= -v - \frac{1}{2}v^2 \cot\theta, \\ Lv - \sigma v &= \gamma(-Pv + q + qv\cot\theta), \end{aligned} \tag{56.21}$$

where we have

$$L = \frac{d^2}{d\theta^2} + \cot\theta\frac{d}{d\theta} - (\cot\theta)^2. \tag{56.22}$$

In (56.21), q is a quantity proportional to the radial bending moment and v is given by

$$v = u + \frac{dw}{d\theta}, \tag{56.23}$$

where u is the displacement in the θ direction and

$$w = W(\theta) - W_0, \tag{56.24}$$

W being the displacement in the negative radial direction and W_0 being the radial displacement in the unbuckled state:

$$W_0 = (1-\sigma)P. \tag{56.25}$$

The other quantities in (56.21) are

$$P = \frac{pa}{4Eh}, \quad \gamma = \frac{1-\sigma^2}{\kappa}, \quad \kappa = \frac{1}{3}\left(\frac{h}{a}\right)^2, \tag{56.26}$$

where p is the normal pressure, a is the radius of the middle surface of the shell, and $2h$ is the thickness.

The boundary conditions for (56.21) are

$$q(0) = q(\pi) = v(0) = v(\pi) = 0. \tag{56.27}$$

Once q and v have been determined, all other relevant quantities, such as radial and circumferential stresses and displacements, are determined by quadratures.

In order to solve (56.21), it is convenient to make the change of variable $x = \cos\theta$, so that the operator L in (56.21) becomes

$$L = (1-x^2)\frac{d^2}{dx^2} - 2x\frac{d}{dx} - \frac{x^2}{1-x^2}, \tag{56.28}$$

and the two equations (56.21) can be rewritten as

$$(L+\sigma)q = -v - \frac{1}{2}v^2\frac{x}{\sqrt{1-x^2}}, \tag{56.29a}$$

$$(L+\sigma)v = 2\sigma v - \gamma P v + \gamma q + \gamma q v\frac{x}{\sqrt{1-x^2}}, \tag{56.29b}$$

with the boundary conditions

$$q(-1) = q(1) = v(-1) = v(1) = 0. \tag{56.30}$$

If we introduce the integral operator G, the kernel of which is the Green's function for $-(L+\sigma)$ with the boundary conditions $u(-1) = u(1) = 0$, we can replace equation (56.29a) with the integral equation

$$q = Gv + \frac{1}{2}G\left(\frac{xv^2}{\sqrt{1-x^2}}\right). \tag{56.31}$$

On combining (56.29b) with (56.31), we find

$$\begin{aligned}(L+\sigma)v =& 2\sigma v - \gamma P v + \gamma G v + \frac{\gamma}{2}G\left(\frac{xv^2}{\sqrt{1-x^2}}\right)\\ &+ \frac{\gamma x v}{\sqrt{1-x^2}}Gv + \frac{1}{2}\frac{\gamma x v}{\sqrt{1-x^2}}G\left(\frac{xv^2}{\sqrt{1-x^2}}\right).\end{aligned} \tag{56.32}$$

Rewriting (56.32) as an integral equation yields

$$\begin{aligned}v = &-2\sigma G v + \gamma P G v - \gamma G G v - \frac{\gamma}{2}G\left(G\frac{xv^2}{\sqrt{1-x^2}}\right) - \gamma G\left(\frac{xv}{\sqrt{1-x^2}}Gv\right)\\ &- \frac{\gamma}{2}G\left(\frac{xv}{\sqrt{1-x^2}}G\frac{xv^2}{\sqrt{1-x^2}}\right) = F(v).\end{aligned} \tag{56.33}$$

In order to solve (56.33) we consider the eigenfunctions of the problem

$$\begin{aligned}LA_n + [n(n+1) - 1]A_n = 0,\\ A_n(-1) = A_n(1) = 0.\end{aligned}$$

This problem can be put in the form

$$(L+\sigma)A_n = -\lambda_n A_n, \tag{56.34}$$

where we have

$$\lambda_n = n(n+1) - (1+\sigma) \tag{56.35}$$

and the quantities A_n are normalized Legendre functions. Let us now assume for the moment that (56.33) has a smooth solution and that it can be expanded in the series

$$v = \sum_{n=1}^{\infty} a_n A_n. \tag{56.36}$$

This series would satisfy (56.33); that is,

$$\sum_{n=1}^{\infty} a_n A_n = -2\sigma \sum_{n=1}^{\infty} \frac{a_n}{\lambda_n} A_n + \gamma P \sum_{n=1}^{\infty} \frac{a_n}{\lambda_n} A_n - \gamma \sum_{n=1}^{\infty} \frac{a_n}{\lambda_n^2} A_n - \frac{\gamma}{2} G\left(G \frac{xv^2}{\sqrt{1-x^2}}\right)$$
$$- \gamma G\left(\frac{xv}{\sqrt{1-x^2}} Gv\right) - \frac{\gamma}{2} G\left(\frac{xv}{\sqrt{1-x^2}} G \frac{xv^2}{\sqrt{1-x^2}}\right). \qquad (56.37)$$

If we multiply (56.37) by one of the eigenfunctions A_n and integrate from -1 to 1 we obtain an infinite set of nonlinear algebraic equations for determining a_n:

$$a_n = -2\sigma \frac{a_n}{\lambda_n} + \lambda P \frac{a_n}{\lambda_n} - \gamma \frac{a_n}{\lambda_n^2} - \frac{\gamma}{2} \left\langle G\left(G \frac{xv^2}{\sqrt{1-x^2}}\right), A_n \right\rangle$$
$$- \gamma \left\langle G\left(\frac{xv}{\sqrt{1-x^2}} Gv\right), A_n \right\rangle - \frac{\gamma}{2} \left\langle G\left(\frac{xv}{\sqrt{1-x^2}} G \frac{xv^2}{\sqrt{1-x^2}}\right), A_n \right\rangle,$$

for $n = 1, 2, \ldots$. As G is symmetrical, we can transform this equation into the system

$$\lambda_n a_n + 2\sigma a_n - \gamma P a_n + \gamma \frac{a_n}{\lambda_n} + \frac{\gamma}{2} \left\langle G \frac{xv^2}{\sqrt{1-x^2}}, A_n \right\rangle$$
$$+ \gamma \left\langle \frac{xv}{\sqrt{1-x^2}} Gv, A_n \right\rangle + \frac{\gamma}{2} \left\langle \frac{xv}{\sqrt{1-x^2}} G \frac{xv^2}{\sqrt{1-x^2}}, A_n \right\rangle = 0, \qquad (56.38)$$

for $n = 1, 2, \ldots$.

As a first approach, we consider the approximate solution

$$v_n = \sum_{n=1}^{N} a_n A_n, \qquad (56.39)$$

and the truncated system obtained from (56.38) by replacing v with v_N and taking $n = 1, \ldots, N$. The truncated system admits the trivial solution $a_1 = a_2 = \ldots = a_N = 0$, but we wish to show that, under some circumstances, there is a nontrivial solution. We introduce the energy $E_N(a_1, \ldots, a_N)$, associated with the truncated system, defined by

$$E_N = \sum_{n=1}^{N} \left(\lambda_n + 2\sigma - \gamma P + \frac{\gamma}{\lambda_n}\right) a_n^2 + \gamma \left\langle G \frac{xv_N^2}{\sqrt{1-x^2}}, v_N \right\rangle$$
$$+ \frac{\gamma}{4} \left\langle G \frac{xv_N^2}{\sqrt{1-x^2}}, \frac{xv_N^2}{\sqrt{1-x^2}} \right\rangle. \qquad (56.40)$$

The symmetry of G implies that the approximate solution can be found by solving the equivalent system

$$\frac{1}{2} \frac{\partial E_N}{\partial a_n} = 0, \quad n = 1, \ldots, N, \qquad (56.41)$$

and we wish to prove that E_N has a minimum. For this purpose we write (56.40) in the form

$$E_N = \sum_{n=1}^{N} (\lambda_n + 2\sigma - \gamma P)a_n^2 + \frac{\gamma}{4}\left\langle G\left(\frac{xv_N^2}{\sqrt{1-x^2}} + 2v_N\right), \frac{xv_N^2}{\sqrt{1-x^2}} + 2v_N\right\rangle. \quad (56.42)$$

As G is a positive operator, (56.42) implies the inequality $E_N \geqslant 0$ if we have $(\lambda_n + 2\sigma + \gamma P) \geqslant 0$; that is (see (56.35)), for

$$P \leqslant \frac{1}{\gamma}(1+\sigma) = \frac{1}{3(1-\sigma)}\left(\frac{h}{a}\right)^2. \quad (56.43)$$

If P satisfies this inequality, the minimum of (56.42) occurs for $a_1 = \ldots = a_N = 0$; that is corresponding to the trivial solution. On the other hand, the quantities

$$P_j = \frac{1}{\gamma}(\lambda_j + 2\sigma) + \frac{1}{\lambda_j}$$

are the eigenvalues of the linearized theory. Let us assume that min P_j occurs for $j = n$. In this case we have $P_n \leqslant P_{n+1}$ and $P_n \leqslant P_{n-1}$. It follows that the necessary and sufficient condition for P_n to be the smallest eigenvalue is that the thickness parameter γ satisfies the inequality

$$\lambda_n\lambda_{n-1} \leqslant \gamma \leqslant \lambda_n\lambda_{n+1}.$$

We shall show that, for $P > P_n$ (n odd) or $P \geqslant P_n$ (n even) there are values $(a_1, \ldots, a_N)$ such that we have $E_N(a_1, \ldots, a_N) < 0$. In particular, we find the result

$$E_N(0, \ldots, a_n, \ldots, 0) = \left(\lambda_n + 2\sigma - \gamma P + \frac{\gamma}{\lambda_n}\right)a_n^2 + a_n^3\gamma\left\langle G\frac{xA_n^2}{\sqrt{1-x^2}}, A_n\right\rangle$$
$$+ \gamma\frac{a_n^4}{4}\left\langle G\frac{xA_n^2}{\sqrt{1-x^2}}, \frac{xA_n^2}{\sqrt{1-x^2}}\right\rangle, \quad (56.44)$$

and, by virtue of the definition of the normalized Legendre function

$$A_n = \sqrt{1-x^2}P_n' / \|(1-x^2)^{1/2}P_n'\|, \quad (56.45)$$

the prime denoting differentiation with respect to x, we have the result

$$\left\langle G\frac{xA_n^2}{\sqrt{1-x^2}}, A_n\right\rangle = \frac{1}{\lambda_n}\left\langle \frac{xA_n^2}{\sqrt{1-x^2}}, A_n\right\rangle = \frac{1}{\lambda_n}\int_{-1}^{1} x(1-x^2)(P_n')^3\,dx. \quad (56.46)$$

If n is odd, the integral in (56.46) vanishes because the integrand is odd. If n is even the integral in (56.46) does not vanish and is in fact positive. In this case $E_N(0, \ldots, a_n, \ldots, 0)$ becomes strictly negative even in the case $\lambda_n + 2\sigma - \gamma P + (\gamma/\lambda_n) = 0$ for an appropriate choice of a_n. Thus a sufficient condition required to guarantee that the minimum of $E_N(a_1, \ldots, a_n)$ is nonzero is

$$P \leqslant P_n = \frac{1}{\gamma}(\lambda_n + 2\sigma) + \frac{1}{\lambda_n}. \quad (56.47)$$

We shall now prove the existence of a solution to (56.41) by showing that $E_N(a_1, \dots, a_N)$ has a minimum. If this result is valid, it is a consequence of (56.43) and (56.47) that there is some value of $P = P_0$ for which we have

$$\frac{1}{\gamma}(1+\sigma) < P_0 \leqslant \frac{1}{\gamma}(\lambda_n + 2\sigma) + \frac{1}{\lambda_n} \tag{56.48}$$

(where the inequality becomes a strict equality if n is even). This condition is such that, for $P > P_0$, the minimum of E_N is negative; that is, the solution of (56.41) which minimizes the energy is nonzero. In order to prove the existence of a minimum of E_N, there is no difficulty in showing that we have $E_N \to \infty$ for $(a_1^2 + \dots + a_N^2)^{\frac{1}{2}} \to \infty$ (see (56.40)). We can also show that E_N has a lower bound which is independent of N. An immediate consequence is that E_n has a minimum that furnishes a nontrivial solution of (56.41) for $P > P_0$. Determining the lower bound for E_N is not easy, but it is possible to show that (Dickey 1990) we have

$$E_N(a_1, \dots, a_N) \geqslant -\frac{\gamma}{8}(Gg, g), \tag{56.49}$$

where g is an explicitly calculable function, continuous in $-1 \leqslant x \leqslant 1$.

Having shown that the finite system (56.41) has a nontrivial solution $(a_1, \dots, a_N)$ for $P > P_0$, we need to prove the existence of a solution to the infinite system (56.38). We have already established a lower bound for E_N. It is also easy to obtain an upper bound on E_N which is independent of N. In particular, we have

$$\begin{aligned} M(N) &= \min_{a_1, \dots, a_N} E_N(a_1, \dots, a_N) \leqslant \min_{a_1} E_N(a_1, 0, \dots, 0) \\ &= \min_{a_1} E_1(a_1) = M(1). \end{aligned} \tag{56.50}$$

We note that, for $P > P_0$ (see (56.48)), we have $M(1) < 0$. In addition, if we multiply (56.41) by a_n and sum over n, we find that $(a_1, \dots, a_N)$ satisfy the identity

$$\sum_{n=1}^{N}\left(\lambda_n + 2\sigma - \gamma P + \frac{\gamma}{\lambda_n}\right)a_n^2 + \frac{3\gamma}{2}\left\langle G\frac{xv_N^2}{\sqrt{1-x^2}}, v_N\right\rangle + \frac{\gamma}{2}\left\langle G\frac{xv_N^2}{\sqrt{1-x^2}}, \frac{xv_N^2}{\sqrt{1-x^2}}\right\rangle = 0 \tag{56.51}$$

with $v_N = \sum_{n=1}^{N} a_n A_n(x)$. Identity (56.51) can also be written as

$$\begin{aligned} \sum_{n=1}^{N}\left(\lambda_n + 2\sigma - \gamma P + \frac{\gamma}{\lambda_n}\right)a_n^2 + \frac{\gamma}{2}\left\langle G\left(\frac{xv_N^2}{\sqrt{1-x^2}} + 3v_N\right), \frac{xv_N^2}{\sqrt{1-x^2}} + 3v_N\right\rangle \\ -\frac{9\gamma}{8}\left\langle Gv_N, v_N\right\rangle = 0, \end{aligned} \tag{56.52}$$

or, on using the positivity of G, in the form

$$\sum_{n=1}^{N}\left(\lambda_n + 2\sigma - \gamma P - \frac{\gamma}{8\lambda_n}\right)a_n^2 \leqslant 0. \tag{56.53}$$

A consequence of (56.53) is the inequality

$$\sum_{n=1}^{N} n^2 a_n^2 \leqslant \gamma\left[P + \frac{1}{8(1-\sigma)}\right]\sum_{n=1}^{\infty} a_n^2. \tag{56.54}$$

It remains to bound the right-hand side of (56.54). This bound comes from (56.50) and from the fact that each E_N satisfies an inequality of the type (Dickey 1990)

$$E_N \geqslant \mu \sum_{n=1}^{N} a_n^2 - \frac{\gamma}{8}(Gg, g), \tag{56.55}$$

for $\mu > 0$ with g the function occurring in (56.49). On combining (56.55) with (56.50) we obtain

$$\sum_{n=1}^{N} a_n^2 \leqslant \frac{1}{\mu}\left[M(1) + \frac{\gamma}{8}(Gg, g)\right]. \tag{56.56}$$

The inequalities (56.54) and (56.56) imply the existence of constants K_1 and K_2, independent of N, conditioned by

$$\sum_{n=1}^{N} a_n^2 \leqslant K_1, \quad \sum_{n=1}^{N} n^2 a_n^2 \leqslant K_2. \tag{56.57}$$

Let us now define the vector

$$V_N = (a_1, \ldots, a_N, 0, 0 \ldots).$$

Inequalities (56.57) ensure that, as n tends to infinity, the vectors V_N remain in a compact set of an infinite dimensional vector space. Hence there is a subsequence $\{V_{N_i}\}$ such that we have

$$V_{N_i} \to V^* = (a_1^*, \ldots, a_n^*, \ldots), \tag{56.58}$$

with

$$\sum_{n=1}^{\infty} n^2 (a_n^*)^2 \leqslant K_2. \tag{56.59}$$

The function

$$v^* = \sum_{n=1}^{\infty} a_n^* A_n(x)$$

is continuous and the functions v_{N_i} approach v^* uniformly. It is possible to prove that the constants a_n^* solve the infinite system of algebraic equations (56.38) and v^* is a solution of (56.33).

The variational theory of boundary-value problems can be also applied to the nonlinear deformation of shallow shells, by removing the restriction of axisymmetric deformed states (Vorovich and Lebedev 1988). The displacement of the middle surface of the shell is a vector (u, v, w), where these three components depend on variables (x, y) defined in a domain Q with piecewise smooth boundary ∂Q. Here w is the displacement normal to the middle surface, and u and v are the displacements tangential to the middle surface along directions the projections of which in the x

y plane are just parallel to the coordinate axes x and y. In addition, we assume that the coordinate lines $x = \text{constant}$ and $y = \text{constant}$ on the surface are lines of principal curvature. These curvatures at any point of the surface are denoted by k_1 and k_2.

If F_i are the external loads, the equilibrium equations have the form (Vlasov 1958)

$$\begin{aligned} &D\nabla^4 w + T_1(k_1 - w_{xx}) + T_2(k_2 - w_{yy}) - 2S_1 w_{xy} + F_3 = 0, \\ &\nabla^2 u + \frac{1+\sigma}{1-\sigma}(u_y + v_x)_x + \frac{2}{1-\sigma}[(k_1 w)_x + w_x w_{xx} + \sigma(k_2 w)_x + \sigma w_y w_{xy}] \\ &\qquad + w_y w_{xy} + w_x w_{yy} + F_1 = 0, \\ &\nabla^2 v + \frac{1+\sigma}{1-\sigma}(u_y + v_x)_y + \frac{2}{1-\sigma}\left[(k_2 w)_y + w_y w_{yy} + \sigma(k_1 w)_y + \sigma w_x w_{xy}\right] \\ &\qquad + w_x w_{xy} + w_y w_{xx} + F_2 = 0. \end{aligned} \tag{56.60}$$

$$T_1 = \frac{2Eh}{(1-\sigma^2)}(\varepsilon_1 + \sigma\varepsilon_2), \quad T_2 = \frac{2Eh}{(1-\sigma^2)}(\varepsilon_2 + \sigma\varepsilon_2),$$
$$S_1 = -S_2 = \frac{Eh}{(1+\sigma)}\tilde{\omega},$$

and

$$\varepsilon_1 = u_x + k_1 w + \frac{1}{2}w_x^2, \quad \varepsilon_2 = v_y + k_2 w + \frac{1}{2}w_y^2,$$
$$\tilde{\omega} = u_y + v_x + w_x w_y.$$

In order to ensure the existence of a solution of system (56.60), some necessary conditions on the boundary data must be satisfied. We require $w(x_i, y_i) = 0$ at three points $(x_i, y_i) = 1, 2, 3$ of the domain Q that do not lie on the same straight line. Moreover, w vanishes on a part $\partial_1 Q$ of ∂Q, and the subspace of functions of class $\mathscr{C}^4$ satisfying this condition is denoted by $\mathscr{C}_1^4$. The tangential displacements u and v vanish on some part of the boundary $\partial_2 Q$ of nonzero length and they satisfy the inequality

$$\iint_Q (u^2 + v^2 + u_x^2 + u_y^2 + v_x^2 + v_y^2)\,dA \leqslant m \iint_Q (u_x^2 + v_y^2 + (u_y + v_x)^2\,dA. \tag{56.61}$$

This inequality is known as Korn's inequality. The functions (u, v) of class $\mathscr{C}^2$ satisfying (56.61) are said to belong to the set $\mathscr{C}_1^2$. On the remaining part of ∂Q the boundary conditions are natural.

Let us introduce the subspace H_1 obtained by closing the set $\mathscr{C}_1^2$, endowed with the norm

$$\|\mathbf{u}\|_{H_1}^2 = \frac{Eh}{(1-\sigma^2)}\iint_Q (e_1^2 + e_2^2 + 2\sigma e_1 e_2 + \frac{1}{2}(1-\sigma)e_{12}^2)\,dA,$$

with $e_1 = u_x$, $e_2 = v_y$, $e_{12} = u_y + v_x$. We also introduce the space H_2, obtained by closing the set of functions $\mathscr{C}_1^4$, endowed with the norm

$$\|w\|_{H_2}^2 = \frac{1}{2} D \iint_Q [(\nabla^2 w)^2 + 2(1-\sigma)(w_{xx} w_{yy} - w_{xy}^2)] dA.$$

We denote the space $H_1 \times H_2$ by H and the variation in the function f by δf. A weak solution of problem (56.60) is a vector function satisfying the integrodifferential equation

$$\iint_Q (G_1 \delta\kappa_1 + G_2 \delta\kappa_2 + 2H_2 \delta\tau + T_1 \delta\varepsilon_1 + T_2 \delta\varepsilon_2 + 2S_1 \tilde{\omega}) dA$$

$$= \iint_Q (F_1 \delta u + F_2 \delta v + F_3 \delta w) dA + \oint_{\partial Q} (f_1 \delta u + f_2 \delta v + f_3 \, dw) ds \tag{56.62}$$

where $(\delta u, \delta v, \delta w)$ is an arbitrary vector function and

$$G_1 = -D(\kappa_1 + \sigma\kappa_2), \quad G_2 = -D(\kappa_2 + \sigma\kappa_1), \quad H_2 = -D(1-\sigma)\tau,$$
$$\kappa_1 = w_{xx}, \quad \kappa_2 = w_{yy}, \quad \tau = w_{xy}.$$

In (56.62), the boundary loads f_i are not required on the part of the boundary where any of the components $(\delta u, \delta v, \delta w)$ are equal to zero. For the correct formulation of the boundary-value problem we assume the conditions $f_\alpha \in L^p(\partial Q)$ and $F_\alpha \in L^p(Q)$ $(\alpha = 1, 2)$, with $p > 1$, while f_3 and F_3 may be finite sums of δ functions and functions in ∂Q and Q, respectively.

A generalized solution of the boundary-value problem is a stationary point of the functional

$$I(\mathbf{u}) = \|w\|_{H_2}^2 + \frac{1}{2} \iint_Q (T_1 \varepsilon_1 + T_2 \varepsilon_2 + 2S_1 \tilde{\omega}) dA - \iint_Q (F_1 u + F_2 v + F_3 w) dA$$

$$- \oint_{\partial Q} (f_1 u + f_2 v + f_3 w) ds. \tag{56.63}$$

For convenience we put

$$(\mathbf{g}, \mathbf{u})_H = \iint_Q (F_1 u + F_2 v + F_3 w) dA + \oint_{\partial Q} (f_1 u + f_2 v + f_3 w) ds, \tag{56.64}$$

and give $I(\mathbf{u})$ the form

$$I(\mathbf{u}) = \|w\|_{H_2}^2 + \frac{1}{2} \iint_Q (T_1 \varepsilon_1 + T_2 \varepsilon_2 + 2S_1 \tilde{\omega}) dA - (\mathbf{g}, \mathbf{u})_H. \tag{56.65}$$

Let us now consider the functional $I(\mathbf{u})$ on an ellipsoid $T(R)$ of the space H, such that, if (u, v, w) lies on the unit sphere S, the element $(cR^2 u, cR^2 v, Rw)$ lies on the ellipsoid $T(R)$, where $c\,(> 0)$ is a certain number to be determined later. We wish to prove the result

$$I(\mathbf{u}) \to \infty, \quad \text{for } \mathbf{u} \in T(R) \text{ and } R \to \infty. \tag{56.66}$$

From a mechanical point of view, the behavior of the energy functional on $T(R)$ implies that the fraction of energy due to tangential displacements tends to become dominant as R increases. Let us divide the sphere S of the space H into two parts S_1 and S_2, and assume that the inequality

$$\|\mathbf{u}^*\|_{H_1} \geqslant \frac{1}{2}, \quad \mathbf{u}^* = (u, v) \tag{56.67}$$

is satisfied on S_1. The positive form

$$\iint_Q (T_1\varepsilon_1 + T_2\varepsilon_2 + 2S_1\tilde{\omega})\, dA \tag{56.68}$$

is homogeneous and of degree 4 on the ellipsoid $T(R)$, while the degree of homogeneity of the remaining term is not higher than 2. As we have $\|w\|^2_{H_2} \leqslant \frac{1}{2}$ on S_1, by the Sobolev embedding theorem we can state the inequality

$$\iint_Q (w_x^4 + w_y^4)\, dA \leqslant m = \text{constant}.$$

Since the integrand in (56.68) is positive in the components ε_1, ε_2, and $\tilde{\omega}$, by virtue of (56.67) we can select a constant $c > 0$ on S_1 such that

$$\iint_Q (T_{1c}\varepsilon_{1c} + T_{2c}\varepsilon_{2c} + 2S_{1c}\tilde{\omega}_c)\, dA \geqslant 1,$$

where the subscript c denotes that cu and cv are substituted in place of u and v. In order to determine c, we observe that the inequality

$$\iint_Q (T_1\varepsilon_1 + T_2\varepsilon_2 + 2S_1\tilde{\omega})\, dA \geqslant R^4, \quad \text{on} \quad T_1(R), \tag{56.69}$$

is satisfied in the image of S_1 in $T(R)$, and thereby in $T_1(R)$, for large R, and we have $I(\mathbf{u}) \geqslant \frac{1}{2}R^4$. On the part $S_2 = S/S_1$ of the unit sphere, the following inequalities hold

$$I(\mathbf{u}) \geqslant \|w\|^2_{H_2} - (\mathbf{g}, \mathbf{u})_H \geqslant \|w\|^2_{H_2} - (\mathbf{g}^*, \mathbf{u}^*)_{H_1} - \|g_3\|_{H_2}\|w\|_{H_2},$$

with

$$(\mathbf{g}^*, \mathbf{u}^*)_{H_1} = \iint_Q (F_1 u + F_2 v)\, dA + \oint_{\partial Q} (f_1 u + f_2 v)\, ds.$$

After the substitution $(u, v, w) \to (cu, cv, w)$ we obtain

$$I_c(\mathbf{u}) \geqslant \|w\|^2_{H_2} - c\|\mathbf{g}^*\|_{H_1}\|\mathbf{u}^*\|_{H_1} - \|g_3\|_{H_2}\|w\|_{H_2},$$

with the consequent inequality

$$I(\mathbf{u}) \geqslant \frac{1}{2}R^2(1 - c\|\mathbf{g}^*\|_{H_1}) - \|g_3\|_{H_2}R, \quad \text{on } T_2(R), \tag{56.70}$$

$T_2(R)$ being the image of S_2 in $T(R)$.

Now, provided that the tangential loads obey the condition

$$c\|\mathbf{g}^*\|_{H_1} \leqslant \frac{1}{2}, \tag{56.71}$$

we have $I(\mathbf{u}) \geqslant \frac{1}{5}R^2$ on $T_2(R)$ for sufficiently large R. Thus, if the tangential loads are sufficiently small so as to satisfy (56.71), there is at least one generalized solution of the equilibrium problem with finite energy; and a minimizing sequence $\{\mathbf{u}_n\}$ of $I(\mathbf{u})$ contains a subsequence converging strongly in H to a generalized solution of the problem.

57. Estimates on Solutions

One of the most striking properties of shells is that, when the thickness, the curvatures, and the loads are distributed with continuity, the state of stress is essentially of membrane type; that is, it is practically defined by stress resultants such as T_1, T_2, and $S_1 = -S_2$, while resultant couples often remain negligible. However, the influence of these flexural effects is important near the points at which the data are singular. In particular, the bending moments arising in a strip-like region close to an edge line are responsible for the high concentrations of stress that often cause the rupture of the material along a fissure running parallel to the edge line and placed at a short distance from it. The evaluation of these boundary effects can be carried out in an approximate form when the shell is a surface of revolution, symmetrically deformed, and the edge line is a parallel (Figure 57.1). By disregarding the superficial loads, which have little influence in the calculation of the edge effects, we can reduce the equation of equilibrium to a couple of ordinary differential equations for the shear force $N_1(\theta)$ and the rotation $\alpha(\theta)$ of the tangent to the meridian (Szabó 1964, §13). These equations are

$$\begin{aligned} N_1^{iv}(\theta) + 4\mu^4 N_1(\theta) = 0, \quad &\text{for} \quad 0 < \theta < \theta_0, \\ \alpha^{iv}(\theta) + 4\mu^4 \alpha(\theta) = 0, \quad &\text{for} \quad 0 < \theta < \theta_0, \end{aligned} \tag{57.1}$$

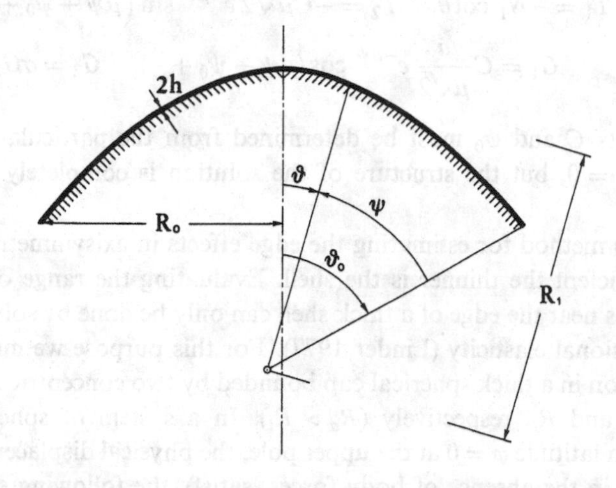

Fig. 57.1

with

$$\mu^4 = \frac{3R_1^4(1-\sigma^2)}{4h^2R_2^2}, \tag{57.2}$$

$R_1(\theta)$ and $R_2(\theta)$ being the principal radii of curvature.

Since we are interested in a solution valid only in the neighborhood of $\theta = \theta_0$, we can regard the radii of curvature R_1 and R_2 as approximately constant in the flexural region. Again, in this case, μ can be considered equal to a constant μ_0, and (57.1) as reducing to a pair of equations with constant coefficients. On introducing the angle $\psi = \theta_0 - \theta$ (see Figure 57.1) as a new independent variable, and observing that both N_1 and α must decrease with ψ, the solution of (57.1) must be of the form

$$\begin{aligned} N_1(\psi) &= Ce^{-\mu\psi}\cos(\mu\psi + \psi_0), \\ \alpha(\psi) &= \bar{C}e^{-\mu\psi}\sin(\mu\psi + \overline{\psi_0}), \end{aligned} \tag{57.3}$$

where C, $\bar{C}$, ψ_0, and $\overline{\psi_0}$ are constants of integration. The angle of rotation α and the shear force N_1 are, however, not independent of one another, and from the relationship between them we derive the conditions

$$\overline{\psi_0} = \psi_0, \quad \bar{C} = C\frac{\mu^2}{Eh},$$

so that we have

$$\alpha(\psi) = C\frac{\mu^2}{Eh}e^{-\mu\psi}\sin(\mu\psi + \psi_0). \tag{57.4}$$

Having calculated $N_1(\psi)$ and $\alpha(\psi)$ we can calculate the other stress and couple resultants. The expressions may be simplified further by replacing R_1 with the radius a of a spherical shell tangent to the given rotationally symmetric shell along the parallel $\theta = \theta_0$. This substitution is known as *Geckeler's approximation*, and allows the edge flexural effects to be evaluated through the formulae

$$\begin{aligned} T_1 &= -N_1\cot\theta, \quad T_2 = -C\mu\sqrt{2}\,e^{-\mu\psi}\sin\left(\mu\psi + \psi_0 + \frac{\pi}{4}\right), \\ G_1 &= C\frac{a}{\mu\sqrt{2}}e^{-\mu\psi}\cos\left(\mu\psi + \psi_0 + \frac{\pi}{4}\right), \quad G_2 = \sigma G_1. \end{aligned} \tag{57.5}$$

The constants C and ψ_0 must be determined from the particular constraints prescribed at $\psi = 0$, but the structure of the solution is completely characterized by (57.5).

Geckeler's method for estimating the edge effects in axisymmetric shells becomes the more efficient the thinner is the shell. Evaluating the range of propagation of flexural stress near the edge of a thick shell can only be done by solving a problem of three-dimensional elasticity (Linder 1977). For this purpose we must find the exact elastic solution in a thick spherical cap bounded by two concentric spherical surfaces of radii R_a and R_i, respectively ($R_a > R_i$). In a system of spherical coordinates (r, φ, ψ), with latitude $\varphi = 0$ at the upper pole, the physical displacement components u, v, and w, in the absence of body forces, satisfy the following system of second-order partial differential equations:

$$\nabla^2 u - \frac{2}{r^2}v_\varphi - \frac{2\cot\varphi}{r^2}v - \frac{2}{r^2\sin\varphi}v_\psi - \frac{2u}{r^2} + \frac{1}{1-2\sigma}\frac{\partial\varepsilon}{\partial r} = 0, \tag{57.6a}$$

$$\nabla^2 v + \frac{2}{r^2}u_\varphi - \frac{2\cot\varphi}{r^2\sin\varphi}w_\varphi - \frac{1}{r^2\sin^2\varphi}v + \frac{1}{1-2\sigma}\frac{1}{r}\frac{\partial\varepsilon}{\partial\varphi} = 0, \tag{57.6b}$$

$$\nabla^2 w + \frac{2}{r^2\sin\varphi}u_\psi + \frac{2\cot\varphi}{r^2\sin\varphi}v_\psi - \frac{1}{r^2\sin^2\varphi}w + \frac{1}{1-2\sigma}\frac{1}{r\sin\varphi}\frac{\partial\varepsilon}{\partial\psi} = 0, \tag{57.6c}$$

where ε is the trace of the strain tensor, that is

$$\varepsilon = u_r + \frac{2}{r}u + \frac{1}{r}v_\varphi + \cot\varphi\frac{v}{r} + \frac{1}{r\sin\varphi}w_\psi, \tag{57.7}$$

which is a solution of the Laplace equation

$$\nabla^2\varepsilon = \frac{1}{r^2}\left[\left(r^2\frac{\partial\varepsilon}{\partial r}\right)_r + \frac{1}{\sin\varphi}\left(\sin\varphi\frac{\partial\varepsilon}{\partial\varphi}\right)_\varphi + \frac{1}{\sin^2\varphi}\frac{\partial^2\varepsilon}{\partial\psi^2}\right] = 0. \tag{57.8}$$

We try to solve (57.6) by the method of separation of variables, on putting

$$\varepsilon = c_\varepsilon r^k P_k^m(\cos\varphi)\,e^{-im\psi}, \quad k = k_1 \quad \text{or} \quad k = -(k_1+1), \quad m = 0, 1, 2, \tag{57.9}$$

where $P_k^m(\cos\varphi)$ are Legendre spherical harmonics and c_ε are constants. By substituting (57.7) in (57.6a) we obtain the equation

$$u_{rr} + \frac{4}{r}u_r + \frac{1}{r^2}u_{\varphi\varphi} + \cot\varphi\frac{1}{r^2}u_\varphi + \frac{1}{r^2\sin^2\varphi}u_{\psi\psi} + \frac{2u}{r^2} - \frac{2\varepsilon}{r} + \frac{1}{1-2\sigma}\frac{\partial\varepsilon}{\partial r} = 0,$$

that is, an equation containing only u and the known function ε.

Now let us consider only solutions to the forms

$$u = U(r,\varphi)\cos m\psi, \quad v = V(r,\varphi)\cos m\psi, \quad w = W(r,\varphi)\sin m\psi,$$
$$\varepsilon = H(r,\varphi)\cos m\psi.$$

Then from (57.6b) and (57.6c) we obtain, after integration, the results

$$u = Re\left\{P_k^m\left[c_u k r^{k-1} + c_\varepsilon(2-4\sigma-k)(k+1)r^{k+1}\right]\right\}\cos m\psi,$$

$$v = Re\left\{P_k^m\left[c_u r^{k-1} - c_\varepsilon(k+5-4\sigma)r^{k+1}\right] - cr^k\frac{m}{\sin\varphi}P_k^m\right\}\cos m\psi,$$

$$w = Re\left\{-\frac{m}{\sin\varphi}P_k^m\left[c_u r^{k-1} - c_\varepsilon(k+5-4\sigma)r^{k+1}\right] + cr^k P_k^m\right\}\sin m\psi,$$

where one of the four constants of integration has been eliminated by means of (57.7). From the displacements, we derive the stress components by applying Hooke's law:

$$\sigma_r = 2\mu Re\left\{P_k^m\left[c_u r^{k-2}k(k-1) + c_\varepsilon r^k(k+1)(k+2+2\sigma-k^2)\right]\right\} \cos m\psi,$$

$$\sigma_\varphi = 2\mu Re\left\{c_u r^{k-2}\left[P_k^m\left(\frac{m^2}{\sin^2\varphi} - k^2\right) - \cot\varphi P_k^m\right] + c_\varepsilon r^k\left[P_k^m(k+1)(2+2\sigma+k^2+4k)\right.\right.$$

$$\left.\left. + (k+5-4\sigma)\left(\cot\varphi P_k^m - \frac{m^2}{\sin^2\varphi}P_k^m\right)\right] + cr^{k-1}\left[\frac{m}{\sin\varphi}(P_k^m - \cot\varphi P_k^m)\right]\right\} \cos m\psi,$$

$$\sigma_\psi = 2\mu Re\left\{c_u r^{k-2}\left[P_k^m\left(k - \frac{m^2}{\sin^2\varphi}\right) + \cot\varphi P_k^m\right] + c_\varepsilon r^k\left[P_k^m(k+1)(2+2\sigma+4\sigma k-k)\right.\right.$$

$$\left.\left. + (k+5-4\sigma)\left(\frac{m^2}{\sin^2\varphi}P_k^m - P_k^m \cot\varphi\right)\right] + cr^{k-1}\left[\frac{m}{\sin\varphi}(P_k^m - \cot\varphi P_k^m)\right]\right\} \cos m\psi,$$

$$\sigma_{r\varphi} = 2\mu Re\left\{c_u r^{k-2}P_k^m(k-1) + c_\varepsilon r^k P_k^m(1-2\sigma-k^2-2k) - \frac{c}{2}r^{k-1}P_k^m\frac{m}{\sin\varphi}(k-1)\right\}$$

$$\times \cos m\psi,$$

$$\sigma_{r\psi} = 2\mu Re\left\{c_u r^{k-2}P_k^m\frac{m}{\sin\varphi}(1-k) + c_\varepsilon r^k P_k^m\frac{m}{\sin\varphi}(k^2+2k-1+2\sigma)\right.$$

$$\left. + \frac{c}{2}r^{k-1}P_k^m(k-1)\right\} \sin m\psi,$$

$$\sigma_{\varphi\psi} = 2\mu Re\left\{c_u r^{k-2}\frac{m}{\sin\varphi}(\cot\varphi P_k^m - P_k^m) + c_\varepsilon r^k\frac{m}{\sin\varphi}(P_k^m - \cot\varphi P_k^m)(k+5-4\sigma)\right.$$

$$\left. + cr^{k-1}\left[P_k^m\left(\frac{m^2}{\sin^2\varphi} - \frac{k}{2}(k+1)\right) - \cot\varphi P_k^m\right]\right\} \sin m\psi.$$

In the particular problem of determining the pure edge effects, the faces of the shell are free of tractions. This implies that at the positions $r = R_i$ and $r = R_a$ we must have

$$\sigma_r = \sigma_{r\varphi} = \sigma_{r\psi} = 0. \tag{57.10}$$

In order to satisfy these boundary conditions we observe that, for a given m, we have two linearly independent solutions corresponding to $k = k_1$ and $k = -(k_1+1)$. We can therefore consider a linear combination of these two solutions with constants $c_\varepsilon^{(i)}$, $c_u^{(i)}$, and $c^{(i)}$ $(i = 1, 2)$, and use (57.10), after having expressed the stress components σ_r, $\sigma_{r\varphi}$, and $\sigma_{r\psi}$, as linear combinations of the two independent solutions. We thus obtain a homogeneous system of six linear equations in the constants $c_\varepsilon^{(i)}$, $c_u^{(i)}$, and $c^{(i)}$ $(i = 1, 2)$. The vanishing of the determinant of this system furnishes the eigenvalues k of the problem as functions of R_a and R_i, or, equivalently, of the ratio $(R_a + R_i)/(R_r - R_i) = R/T$. These eigenvalues are, in general, complex quantities. On using the expressions found before for the displacements and the stresses we can find the law of propagation of flexural stress from the edges; a surprising feature of the solution is that, even for very thick shells, bending moments and shear forces have a rapid decay at short distances from the edge.

Evaluating the influence of the thickness on the propagation of edge effects in axisymmetric shells can also be done by applying a variational principle, known as the *Hellinger–Toepliz–Reissner principle* (Gurtin 1972), which states that the solution of a problem of three-dimensional elasticity is also a stationary point of the functional

$$\iiint_V F\,dV - \iint_S (p_x u + p_y v + p_z w)\,dS, \tag{57.11}$$

with

$$F = \sigma_x \varepsilon_x + \sigma_y \varepsilon_y + \sigma_z \varepsilon_z + 2\sigma_{xy}\varepsilon_{xy} + 2\sigma_{yz}\varepsilon_{yz} + 2\sigma_{zx}\varepsilon_{zx} - W, \tag{57.12}$$

and

$$W = \frac{1}{2E}[\sigma_x^2 + \sigma_y^2 + \sigma_z^2 - 2\sigma(\sigma_x\sigma_y + \sigma_y\sigma_z + \sigma_z\sigma_x)] + \frac{1}{2G}(\sigma_{xy}^2 + \sigma_{yz}^2 + \sigma_{zy}^2). \tag{57.13}$$

In (57.11), p_x, p_y, and p_z are the components of the surface load, and u, v, and w are the components of the displacement in the directions x, y, and z. For axisymmetric shells the functional (57.11) can be written as (Goldberg, Korman, and Baluch 1974)

$$\iiint_V \left\{ (\sigma_\varphi \varepsilon_\varphi + \sigma_\theta \varepsilon_\theta + \sigma_z \varepsilon_z + 2\sigma_{z\varphi}\varepsilon_{z\varphi}) - \frac{1}{2E}[\sigma_\varphi^2 + \sigma_\theta^2 + \sigma_z^2 - 2\sigma(\sigma_\varphi\sigma_\theta + \sigma_\theta\sigma_z + \sigma_z\sigma_\varphi)] \right.$$
$$\left. - \frac{\sigma_{z\varphi}^2}{2G} \right\} \left(1 - \frac{z}{r_\varphi}\right)\left(1 - \frac{z}{r_\varphi}\right) r_0 r_\varphi \, d\theta\, d\varphi\, dz - \iint_S [(p^* \bar{w}_0 - q^* \bar{w}_i) + (b^* \bar{w}_i - a^* \bar{u}_0)] r_0 r_\varphi \, d\theta\, d\varphi$$
$$\tag{57.14}$$

In this formula, z denotes the distance of any point of the shell from the middle surface, φ is the angle that the vertical axis forms with the normal to the middle surface, and θ is the angle that any plane containing the axis forms with a fixed plane containing the axis. The radius of curvature of the meridian is r_φ, the other principal radius of curvature is r_θ, and $r_0 = r_\theta \sin\varphi$ is the radius of curvature of the parallel circle. The surface loads in (57.14) have the following meaning:

$$p^* = p_0\left(1 + \frac{h}{r_\varphi}\right)\left(1 + \frac{h}{r_\theta}\right),$$

$$q^* = p_i\left(1 - \frac{h}{r_\varphi}\right)\left(1 - \frac{h}{r_\theta}\right),$$

$$b^* = p^+\left(1 - \frac{h}{r_\varphi}\right)\left(1 - \frac{h}{r_\theta}\right),$$

$$a^* = p^-\left(1 + \frac{h}{r_\varphi}\right)\left(1 + \frac{h}{r_\theta}\right),$$

where p_o and p_i are the normal components of the load on the exterior and interior faces of the shell, respectively, and p^+ and p^- are the tangential components of the surface load on the same faces. The normal displacements of the outer and inner faces are denoted by $\bar{w}_\mathrm{o}$ and $\bar{w}_\mathrm{i}$, respectively, and the analogous tangential displacements are denoted by $\bar{u}_\mathrm{o}$ and $\bar{u}_\mathrm{i}$. We denote the thickness by $2h$, as usual.

Let us now express the strains in (57.14) in terms of the displacements u, v, and w, and the stresses as functions of the bending moments, the shear forces, and the membrane forces. For the tangential stresses and the normal stresses along the normal we have the following expressions

$$\sigma_{\varphi z} = \frac{1}{\left(1-\frac{z}{r_\theta}\right)}\left\{3\frac{N_1}{4h}\left[1-\left(\frac{z}{h}\right)^2\right]\left(1+\frac{z}{r_\varphi}\right)-\frac{1}{4}\left(p^+\left(1-\frac{h}{r_\theta}\right)\left[1-2\left(\frac{z}{h}\right)-3\left(\frac{z}{h}\right)^2\right]\right.\right.$$
$$\left.\left.+p^-\left(1+\frac{h}{r_\theta}\right)\left[1+2\left(\frac{z}{h}\right)-3\left(\frac{z}{h}\right)^2\right]\right)\right\}, \tag{57.15}$$

$$\sigma_{zz} = \frac{1}{\left(1-\frac{z}{r_\theta}\right)\left(1-\frac{z}{r_\varphi}\right)}\left\{-\frac{T_1}{r_\varphi}\left(\frac{h}{4r_\theta}-\frac{z}{h}+\frac{z^2}{hr_\theta}+\frac{z^3}{4h^3}\right)-\frac{T_2}{r_\theta}\left(\frac{h}{4r_\varphi}-\frac{z}{h}+\frac{z^2}{hr_\varphi}+\frac{z^3}{4h^3}\right)\right.$$
$$+G_1\left(E_\varphi+F_\varphi z+G_\varphi z^2+H_\varphi z^3\right)+G_2\left(E_\theta+F_\theta z+G_\theta z^2+H_\theta z^3\right)$$
$$\left.-\frac{3}{4}p^*\left[\frac{2}{3}-\frac{z}{h}+\frac{1}{3}\left(\frac{z}{h}\right)^3\right)-\frac{3}{4}q^*\left[\frac{2}{3}+\frac{z}{h}-\frac{1}{3}\left(\frac{z}{h}\right)^3\right]\right\}, \tag{57.16}$$

where N_1, T_1, and G_1 are stress and couple resultants defined on an unit arc $\varphi = \text{constant}$, and T_2 and G_2 are defined on an unit arc $\theta = \text{constant}$; furthermore, the quantities E_φ, F_φ, G_φ, E_θ, F_θ, and G_θ are numerical quantities depending on the shape of the middle surface. The first variation of (57.14) yields a system of differential equations reducible to a single ordinary differential equation of sixth order in the only variable φ (Goldberg, Baluch, and Tang 1971). An illustration of the differences between the results of the technical theory of thin shells of revolution (Green and Zerna 1954) and the more refined theory based on the sixth-order differential equation is given by the numerical values of the largest couple G_1 (kp/cm) acting in a hemispherical shell in which different values of the ratio $2h/r$ are prescribed, $E = 2.1 \times 10^5$ kp/cm^2, and $\sigma = 0.2$; and there is an internal radial pressure $p_i = 0.044$ kp/cm^2 (1 = psi). The hemisphere is hinged along the circle $\varphi = \pi/2$, where $r = 5$ m. The maximum values of G_1 are given in Table 57.1. From the table we can see that, for $2h/r \geqslant \frac{1}{5}$, the difference between the values predicted by the two theories is 15% or more.

A more accurate interpretation of the different approximate theories for shells rests on the following considerations (Berdichevskii and Misyura 1992). Classical shell theories agree on the conclusion that the Kirchhoff–Love hypotheses introduce in the final formula errors of order $2h/R$ in comparison with 1, where $2h$ is the

Table 57.1. *Maximum values of G_1*

	$2h/r$					
	1/5	1/7	1/9	1/11	1/13	1/15
Present theory	157.4	106.6	80.7	64.9	53.9	46.4
Technical theory	133.0	95.6	74.4	60.6	51.2	44.1

thickness; this is valid only for the stress, because an analogous statement for the displacements is incorrect.

In order to explain this loss of accuracy in the evaluation of the displacements, let us consider the problem of the deformation of an elastic isotropic cylindrical pipe under external pressure. Let the undeformed pipe occupy the region $\{V_0 | R-h \leqslant r \leqslant R+h, -\pi \leqslant \theta \leqslant \pi, 0 \leqslant |x| \leqslant L\}$ where r, θ, and x are cylindrical coordinates. We assume that the pipe slides without friction between two rigid planes $x = L$ and $x = -L$. The stress distribution is (Love 1927, Art. 100)

$$\sigma_\theta = \frac{P}{4Rh}(R+h)^2\left[1+\frac{(R-h)^2}{(R+\xi)^2}\right], \quad \xi = r - R, \tag{57.17}$$
$$\sigma_{r\theta} = \sigma_{rx} = 0.$$

In order to determine the displacement field we must fix the possible rigid motions of the shell by imposing three constraints for the displacements of the middle surface. For instance, we require that the points $[-(\pi/2), R]$ and $[(\pi/2), R]$ have zero vertical displacement, and the point $(0, R)$ has zero horizontal displacement:

$$w_\theta\left[-\frac{\pi}{2}, R\right] = w_\theta\left[\frac{\pi}{2}, R\right] = w_\theta(0, R) = 0. \tag{57.18}$$

The displacement field has only one nonzero component w_r corresponding to the stresses (57.17):

$$w_r = \frac{P(1+\sigma)}{2RhE}(R-h)^2\left[(1-2\sigma)(R+\xi)+\frac{(R+h)^2}{r+\xi}\right]. \tag{57.19}$$

In general, a tangential displacement component of the form $w_\theta = a\cos\theta - b\sin\theta + \omega r$, where a, b, and ω are constants, is associated with the component w_r (Worch 1967). However, conditions (57.18) imply the result $a = b = \omega = 0$. We denote the normal and tangential components of displacements of the middle surface by u and v:

$$u = w_r, \quad v = w_\theta \quad \text{at} \quad r = R.$$

Hence, on dropping terms of order h/R small compared to unity, (57.19) yields

$$u = \frac{PR^2(1-\sigma^2)}{2Eh} = \text{constant}, \quad v = 0. \tag{57.20}$$

Let us now consider the following problem: We cut the pipe along the plane $\theta = \pm\pi/2$ and impose the equilibrium of the upper half of the pipe. The upper semi-pipe is subjected to external pressure and contact forces applied along the cut $\theta = \pm\pi/2$ (Figure 57.2). Let us call T_1 and G_1 the stress resultant and the stress couple along the θ coordinate:

$$T_1 = \int_{-h}^{h} \sigma_\theta \, d\xi, \quad G_1 = \int_{-h}^{h} \sigma_\theta \xi \, d\xi. \tag{57.21}$$

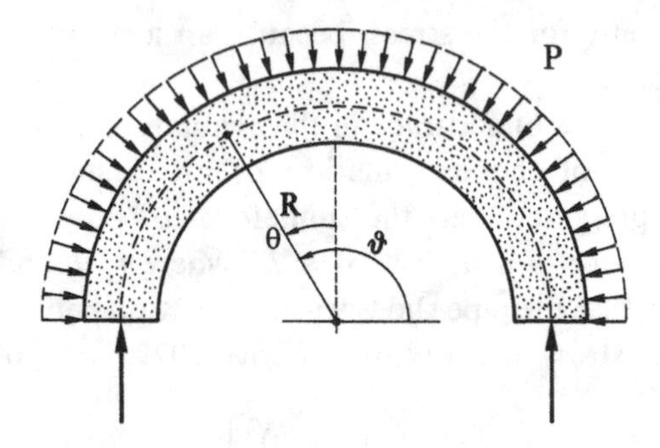

Fig. 57.2

Then the equilibrium equations of classical shell theory are

$$\frac{1}{R}T_1' + \frac{1}{R^2}G_1' = 0, \quad \frac{1}{R^2}G_1'' - \frac{T_1}{R} + P = 0, \tag{57.22}$$

where the prime denotes differentiation with respect to θ, and the boundary conditions are

$$T_1 = PR, \quad G_1 = -\frac{Ph^2}{3}, \quad G_1' = 0, \quad \text{at} \quad \theta = \pm\frac{\pi}{2}.$$

These conditions are obtained by substituting the stress distribution (57.17) in (57.21) and neglecting the terms of order $2h/R$ with respect to unity. The unique solution of equations (57.22), with the associated boundary conditions, is

$$T_1 = PR, \quad G_1 = -\frac{Ph^2}{3},$$

which does not depend on the elastic moduli.

Let us now introduce the stress–strain relations of the classical theory of shells:

$$T_1 = 4\mu h(1+\sigma_1)\gamma, \quad G_1 = -\frac{4\mu}{3}(1+\sigma_1)\rho,$$

where γ and ρ are measures of the extension and the change in curvature, given by,

$$\gamma = \frac{1}{R}(u + v'), \quad \rho = \frac{1}{R^2}(u'' - v'),$$

and where μ is the shear modulus and $\sigma_1 = \sigma/(1-\sigma)$.

In order to find the displacements we have to solve a system of differential equations with respect to the functions $u(\theta)$ and $v(\theta)$:

$$\gamma = \frac{1}{R}(u + v') = \frac{T_1}{4\mu h(1+\sigma_1)} = \frac{PR}{4\mu h(1+\sigma_1)} = \frac{PR}{2Eh}(1-\sigma^2) = \text{constant},$$

$$\rho = \frac{1}{R^2}(u'' - v') = -\frac{3G_1}{4\mu h^3(1+\sigma_1)} = \frac{P}{4\mu h(1+\sigma_1)} = \frac{P}{2Eh}(1-\sigma^2) = \text{constant},$$

with the conditions

$$v = 0 \quad \text{at} \quad \theta = -\frac{\pi}{2}, \quad 0, \quad \frac{\pi}{2}.$$

The only solution of this system of equations is

$$\begin{aligned} u &= -\frac{PR^2}{2Eh}(1-\sigma^2)\left(\frac{\pi}{2}\cos\theta - 2\right), \\ v &= \frac{R^2}{2Eh}(1-\sigma^2)\left(\frac{\pi}{2}\sin\theta - \theta\right). \end{aligned} \tag{57.23}$$

This solution differs drastically from the exact solution (57.20). The maximum error is 100%, and the example shows that the classical shell theory could give incorrect values of displacements beyond any reasonable attempt at approximation.

In order to explain this loss of accuracy, we denote by $\gamma_{\alpha\beta}$ and $\rho_{\alpha\beta}$ two measures of extension and change of curvature. Let ε_γ and ε_ρ be the amplitudes of $\gamma_{\alpha\beta}$ and $\rho_{\alpha\beta}$, defined as

$$\varepsilon_\gamma = \max_{\Omega}(\gamma_{\alpha\beta}\gamma^{\alpha\beta})^{\frac{1}{2}}, \quad \varepsilon_\rho = \max_{\Omega} h(\rho_{\alpha\beta}\rho^{\alpha\beta})^{\frac{1}{2}}, \quad \alpha, \beta = 1, 2,$$

where Ω is the domain of variation of the parameters identifying the middle surface of the shell. The extensional strains in the shell are given, in a first approximation, by the relation

$$\varepsilon_{\alpha\beta} = \gamma_{\alpha\beta} + 2h\zeta\rho_{\alpha\beta}, \tag{57.24}$$

where ζ is a dimensionless coordinate in the normal direction ($\zeta = \xi/2h$).

The classical theory allows us to calculate correct values of stresses in the first approximation, and hence also the strains $\varepsilon_{\alpha\beta}$. However, this does not mean that each term on the right-hand side of (57.24) is determined correctly. In practice, three cases may occur. When ε_γ and ε_ρ are of the same order of magnitude the classical theory predicts correct values for both $\gamma_{\alpha\beta}$ and $\rho_{\alpha\beta}$. Quite a different situation appears when we have $\varepsilon_\gamma \gg \varepsilon_\rho$. The second term on the right-hand side of (57.24) becomes a small correction term. As the classical theory guarantees only correct values of the leading terms, the bending measure found from the classical theory might be wrong, as might the displacement field. For example, from the expression of ρ given before we see that

$$\gamma = \frac{PR}{4\mu h(1+\sigma_1)} = \frac{R}{2h}2h\rho,$$

that is

$$\varepsilon_\gamma \sim \frac{R}{2h}\varepsilon_\rho \gg \varepsilon_\rho.$$

A correct displacement field requires the knowledge of all terms of order $2h/R$ in comparison with 1; that is to say, we require a more refined shell theory. The analogous situation takes place for $\varepsilon_\gamma \ll \varepsilon_\rho$. By the same reasoning, the bending measure $\rho_{\alpha\beta}$ is determined correctly, but the extension measure $\gamma_{\alpha\beta}$ can be incorrect and, more importantly, the corresponding displacements are incorrect.

The procedures for estimating the stresses and their derivatives in a thin shell can be extended to the case in which strains are large and the constitutive relations are nonlinear (John 1965, 1971). For this purpose we recall the assumptions concerning the strain energy density function W. Classically, W is a quadratic form in the ε^i_k, where ε^i_k are the mixed components of the strain:

$$\frac{1}{2}\lambda s_1^2 + \mu s_2 = \frac{E}{1+\sigma}\left(\frac{\sigma}{1-2\sigma}s_1^2 + \frac{1}{2}s_2\right). \tag{57.25}$$

More generally, W is not a quadratic form in the ε^i_k, but the terms in (57.25) merely constitute the first terms in the Taylor expansion. The ratio $W/2\mu$ is dimensionless so, in order to simplify the resulting equations, we choose a unit of force for which

$$2\mu = \frac{E}{1+\sigma} = 1,$$

and introduce the additional dimensionless parameters p and q by

$$p = \frac{\lambda}{2\mu+\lambda} = \frac{\sigma}{1-\sigma}, \quad q = 1 - p = \frac{2\mu}{2\mu+\lambda} = \frac{1-2\sigma}{1-\sigma}.$$

Now, let W have the form

$$\begin{aligned} W &= \frac{p}{2q}s_1^2 + \frac{1}{2}s_2 + \frac{1}{3}\alpha s_1^3 + \beta s_1 s_2 + \frac{1}{2}\gamma s_3 + \cdots \\ &= \frac{p}{2q}\varepsilon^i_i\varepsilon^j_j + \frac{1}{2}\varepsilon^i_j\varepsilon^j_i + \frac{1}{3}\alpha\varepsilon^i_i\varepsilon^j_j\varepsilon^k_k + \beta\varepsilon^i_i\varepsilon^j_k\varepsilon^k_j + \frac{1}{3}\gamma\varepsilon^i_j\varepsilon^j_k\varepsilon^k_i + \cdots. \end{aligned} \tag{57.26}$$

The stress–strain relations (55.33) become

$$\begin{aligned} T^m_i &= \frac{p}{q}\varepsilon^j_j\delta^m_i + \varepsilon^m_i + \left(\alpha - \frac{p}{2q}\right)\varepsilon^j_j\varepsilon^k_k\delta^m_i + \left(\beta - \frac{1}{2}\right)\varepsilon^j_k\varepsilon^k_j\delta^m_i \\ &\quad + \left(2\beta + \frac{2p}{q}\right)\varepsilon^j_j\varepsilon^m_i + (\gamma+2)\varepsilon^j_i\varepsilon^m_j + \cdots. \end{aligned} \tag{57.27}$$

The maximum strain η can be identified with the supremum in the shell of the scalar

$$\sqrt{s_2} = \sqrt{\varepsilon^i_j\varepsilon^j_i}.$$

We now assume that the coefficients $a_{\alpha\beta}$ and $a^{\alpha\beta}$ of the first fundamental form of the middle surface are bounded uniformly by a constant M. Then (see (55.35)), the G_{ik} and G^{ik} also have bounds of order M. In fact, let us put

$$\theta = \max\left(\frac{h}{D}, \sqrt{\frac{h}{R}}, \sqrt{\eta}\right), \tag{57.28}$$

where η is the maximum strain in the shell, and D and R are two lengths associated with the geometry of the undeformed shell: R is a generalized minimum radius of curvature of the middle surface, and D is the distance of any point of the shell from the edge. We assume $\theta < \frac{1}{2}$, so that we have $h/R < \frac{1}{4}$, with the consequence that we can prove the inequality

$$G > \left(1 - 2\frac{h}{R} - 2\frac{h^2}{R^2}\right)^2 a > \frac{3}{8}, \tag{57.29}$$

where $G = \det(G_{ik})$ and $a = \det(a_{\alpha\beta})$. The result is that the components of strain ε_{ik} and those of stress T_{ik} can be estimated in terms of the length of the maximum strain η. In particular, we have

$$\varepsilon_{ik} = O(\eta), \quad T_{ik} = O(\eta), \tag{57.30}$$

where O stands for a universal constant, depending only on the choice of W and on the value of M. By definition (57.28) of the dimensionless parameters θ, we have

$$\eta = O(\theta^2), \quad hb_{\alpha\beta} = O(h/R) = O(\theta^2). \tag{57.31}$$

In order to render the estimates of the strain and stresses more precise, let us introduce the following notation. If f is a scalar of the components of some tensor, and if k is some integer, the relation

$$f = O^k \tag{57.32}$$

is taken to mean that there is a constant K and an integer m such that we have

$$|f| \leqslant K\eta h^{k-m}\theta^m, \tag{57.33}$$

where the value of m is determined by the dimensions of f. In this notation two estimates $f = O^k$ and $F = O^j$ for two quantities f and F imply, by (57.30), the estimate $fF = O^{k+j+2}$ for their product. Thus the unspecified remainder terms in formulae (57.26) and (57.27) could be replaced by O^6 and O^4, respectively.

The main result in the process of estimating the covariant derivatives of ε_{ik} and T_{ik}, valid for $\theta \leqslant \theta_0$ (depending only on W), can be summarized by the formulae (John 1971)

$$\varepsilon_{i_1 i_2; j_1 \ldots j_m} = O^r, \quad T_{i_1 i_2; j_1 \ldots j_n} = O^s, \tag{57.34}$$

where r and s are determined as follows. Let a denote the total number of indices i_1 and i_2 having the value 3, and b the total by a number of indices $j_1, \ldots, j_r$ having the value 3. Then

$$\begin{aligned} &r = s = 0, && \text{for } a+b \text{ even; } r = s = -1 \text{ for } a+b \text{ odd; and } a \leqslant b;\\ &r = s = 1, && \text{for } a = 1, b = 0;\\ &s = 1, && \text{for } a = 2, b = 1;\\ &s = 2, && \text{for } a = 2, b = 0. \end{aligned} \tag{57.35}$$

All these estimates are also valid when any of the subscripts are shifted into upper positions with respect to the metric G_{ik}. In these estimates any covariant derivative "$_{;3}$" can be replaced by the partial derivative "$_{,3}$". Similarly, on Σ_0, any covariant derivative "$_{;\alpha}$" can be replaced by the covariant derivative "$|_\alpha$" along the surface. This is a consequence of the estimates for the Christoffel symbols

$$\begin{aligned} \Gamma^i_{j3} &= O(1/R) = O(\theta^2/h),\\ (\Gamma^i_{j3})_{,3} &= O(1/R) = O(\theta^2/h^2), \text{ etc.} \end{aligned}$$

On the basis of (57.35) we can estimate any polynomial in the derivatives of the ε_{ik} and T_{ik}. We find, for example,

$$\begin{gathered} g_{\alpha\beta} = G_{\alpha\beta} + O^0, \quad g_{\alpha 3} = O^1, \quad g_{33} = 1 + O^0, \\ \bar{g}^{\alpha\beta} = G^{\alpha\beta} + O^0, \quad \bar{g}^{\alpha 3} = O^1, \quad \bar{g}^{33} = 1 + O^0, \\ c^3_{\alpha\beta} = -\varepsilon_{\alpha\beta;3} + O^1, \quad c^3_{33} = \varepsilon_{33;3} + O^1, \quad c^\alpha_{\beta 3} = \varepsilon^\alpha_{\beta;3} + O^1, \\ c^\alpha_{\lambda\beta} = O^0, \quad c^3_{3\alpha} = O^0, \quad c^\alpha_{33} = O^0. \end{gathered} \tag{57.36}$$

In general, if P is a polynomial in the derivatives of ε_{ik}, T_{ik}, and G_{ik}, then from an estimate $P = O^k$ we conclude with the results $P_{;i} = O^{k-1}$ and, more precisely, $P_{;\alpha} = O^k$. These remarks apply in particular to the identities (57.27). On solving for T^α_β, ε^3_α, and ε^3_3 in terms of ε^α_β, T^3_α, and T^3_3, we find after using (57.35), that we have

$$\begin{aligned} \varepsilon^3_\alpha &= T^3_\alpha + O^3 = O^1, \\ \varepsilon^3_3 &= -p\varepsilon^\alpha_\alpha + qT^3_3 + \left[q^3\alpha + (2pq - 3p^2q)\beta - p^2q\gamma + \frac{1}{2}pq\right]\varepsilon^\alpha_\alpha\varepsilon^\beta_\beta \\ &\quad - (\beta\alpha - \frac{1}{2}q)\varepsilon^\alpha_\beta\varepsilon^\beta_\alpha + O^4 = -p\varepsilon^\alpha_\alpha + O^2, \\ T^\alpha_\beta &= p\varepsilon^\lambda_\lambda\delta^\alpha_\beta + \varepsilon^\alpha_\beta + pT^3_3\delta^\alpha_\beta + (A - \frac{1}{2}pq)\varepsilon^\lambda_\lambda\varepsilon^\mu_\mu\delta^\alpha_\beta \\ &\quad + (\beta q - \frac{1}{2}q)\varepsilon^\lambda_\mu\varepsilon^\mu_\lambda\delta^\alpha_\beta + (2\beta q + 2p)\varepsilon^\alpha_\beta\varepsilon^\lambda_\lambda + (2 + \gamma)\varepsilon^\alpha_\lambda\varepsilon^\lambda_\beta + O^4 \\ &= p\varepsilon^\lambda_\lambda\delta^\alpha_\beta + \varepsilon^\alpha_\beta + O^2, \end{aligned} \tag{57.37}$$

where $A = \alpha q^3 + 3\beta p^2 q - \gamma p^3$ is an abbreviation for a combination of elastic coefficients, and some terms could be combined by the use of the Cayley–Hamilton relation

$$\varepsilon^\alpha_\lambda\varepsilon^\lambda_\beta - \varepsilon^\alpha_\beta\varepsilon^\lambda_\lambda + (\varepsilon^\lambda_\lambda\varepsilon^\mu_\mu - \varepsilon^\lambda_\mu\varepsilon^\mu_\lambda)\delta^\alpha_\beta = 0. \tag{57.38}$$

Similarly, from (57.26) we obtain

$$\begin{aligned} W &= \frac{1}{2}p\varepsilon^\alpha_\alpha\varepsilon^\beta_\beta + \frac{1}{2}\varepsilon^\alpha_\beta\varepsilon^\beta_\alpha + T^\alpha_3 T^3_\alpha + \frac{1}{3}A\varepsilon^\alpha_\alpha\varepsilon^\beta_\beta\varepsilon^\lambda_\lambda + q\beta\varepsilon^\alpha_\alpha\varepsilon^\beta_\lambda\varepsilon^\lambda_\beta \\ &\quad + \frac{1}{3}\gamma\varepsilon^\alpha_\beta\varepsilon^\beta_\lambda\varepsilon^\lambda_\alpha + O^6. \end{aligned} \tag{57.39}$$

The estimates found before can be extended to the case of a perfectly elastic homogeneous and isotropic shell in equilibrium with applied surface, edge, and body forces. In addition, the faces of the shell in the unstrained state will be at variable distance from the middle surface (Berger 1973). If the thickness of the shell is small and does not vary too wildly, the curvature of the middle surface is not too large, and the strains are small everywhere in the shell, then certain estimates on the stresses and their derivatives are valid. In order to make the assumption rigorous, we denote the undeformed variable distance of the middle surface of the shell from each face by h, and call H the maximum and H^* the minimum of h in the region under consideration. We call R a generalized minimum radius of curvature, and D the distance of any point P of the shell from the edge. Finally, we denote an upper bound for the strain everywhere by η. Then the quantity

$$\theta = \max\left(\frac{2H}{D}, \frac{2H}{\sqrt{2H^*R}}, \frac{H}{H^*}\sqrt{\eta}\right), \tag{57.40}$$

represents a measure of the thinness, shallowness, and variation in thickness of the shell. If θ is small, a two-dimensional approximate description of the deformed shell becomes feasible. The quantity λ defined by

$$\lambda = \frac{2H}{\theta} = \min\left(\frac{D}{3}, \sqrt{2H^*R}, \frac{2H^*}{\sqrt{\eta}}\right) \tag{57.41}$$

provides a lower bound for the *wavelength* of the stresses and strains in the longitudinal directions. Here, by the wavelength of a function f, we mean a quantity proportional to $\max|f'|/|f|$.

In general, we can show that the nth-order longitudinal derivatives of the strain components are bounded by $K_n\eta/\lambda^n$ when $\theta < \theta_0$, and the constants $\theta_0, K_1, K_2, \ldots,$ depend on the elastic properties of the material alone. In addition, conditions will be imposed on the order of magnitude of the applied forces so that the internal stresses of the shell should be of similar order. If, as assumed, the maximum stress is of order ηE it can be shown that the transverse shear stresses are of order $E\eta h/\lambda$ at each point in the interior of the shell, and that the longitudinal stresses vary linearly within an error of magnitude $E\eta h^2/\lambda^2$ along a normal fiber of the shell.

The estimates derived so far are of asymptotic type; that is to say, they give the law of dependence of strains and stresses and their derivatives in terms of the parameter η that represents an upper bound for the strain everywhere in the shell. Of course, the method is exposed to the objection that the maximum strain really depends on the physically given data, in terms of geometric form of the shell and the forces applied around its edge, and it is difficult to estimate this in view of the nonlinearity of the equations and the possible occurrence of buckling instability and boundary layers. To some extent, a knowledge of η is not necessary if we have a bound for the total strain energy, as long as η is small enough to keep the linear terms dominant in the stress–strain relation. Furthermore, however, the validity of this equivalence should be proved and not simply postulated.

A different approach, which avoids these objections, is based on the possibility of estimating the error of the solution as a function of an approximate solution, which can be constructed explicitly. More precisely, we attempt to compare the solution corresponding to a two-dimensional theory of shells with that derived from the application of the three-dimensional theory (Ladezève 1976). The two-dimensional theories exploit the fact that, in the majority of load conditions, the stress field in the direction perpendicular to the middle surface is negligible. Each theory requires the determination of a field of displacements $\tilde{\mathbf{u}}$ and generalized stresses $(\tilde{\mathbf{N}}, \tilde{\mathbf{M}})$ defined on the mean surface. On the other hand, if we regard the shell as a three-dimensional continuum, the solution is constituted by a displacement field $\tilde{\mathbf{u}}_0$ and a (second) Piola–Kirchhoff stress field $\tilde{\mathbf{T}}_0$ defined on the three-dimensional domain Ω occupied by the shell-like body in its natural state. However, because these solutions are different in nature, they cannot be compared directly. We thus associate $(\tilde{\mathbf{u}}_0, \tilde{\mathbf{T}}_0)$ with a field $(\hat{\mathbf{u}}, \hat{\mathbf{T}})$ of the same mathematical nature and representative of the solution of a two-dimensional shell, in the sense that the displacements take values $\tilde{\mathbf{u}}$ on the middle surface and the stresses have resultants and moments equal to $\tilde{\mathbf{N}}$ and $\tilde{\mathbf{M}}$.

With the artifice of constructing the auxiliary state $(\hat{\mathbf{u}}, \hat{\mathbf{T}})$ we have converted, at least conceptually, the problem of interpreting a two-dimensional solution into that of comparing two three-dimensional admissible states. Comparisons of this type are classical in linear elasticity (Synge 1957), but the method of extending them to finite elasticity is not obvious. However, it is possible to make this extension without difficulty, provided that the material is hyperelastic and the strain energy function is such that, for any value of the parameter $\lambda \in [0, 1]$, the configurations $\lambda\tilde{\mathbf{u}}_0$ are stable in the sense that the second (Fréchet) derivative of the strain energy is nonnegative.

When the shell has some particular form it is possible to find more precise estimates on the solutions in the presence of large strains. This is the case for a homogeneous, elastically isotropic, semi-infinite circular tube of constant thickness $2h$ and midsurface of radius R, under axisymmetric edge loads (Horgan, Payne, and Simmonds 1990). In the classical, (first approximation) theory, the linear field equations can be reduced to the following system of coupled second-order ordinary differential equations (Libai and Simmonds 1988):

$$f'' - \varepsilon^2 f + \beta = 0, \quad \beta'' - \varepsilon^2 \beta - f = 0, \quad 0 < x < \infty, \tag{57.42}$$

where f is a dimensionless stress function, β is the angle of rotation of a generator of the midsurface of the tube, εRx is the distance along a generator from the end of the tube, and

$$\varepsilon^2 = \frac{2h}{R\sqrt{12(1-\sigma^2)}}, \tag{57.43}$$

is a small parameter.

We represent the solutions of (57.42) in the complex form

$$f + i\beta = \varphi = \varphi(0)\, e^{-px}, \quad p = (i + \varepsilon^2)^{\frac{1}{2}} = a + ib = \frac{1+i}{\sqrt{2}} + O(\varepsilon^2), \tag{57.44}$$

where $\varphi(0)$ depends on the boundary conditions at $x = 0$. From (57.44) we have the formula $|\varphi(x)| = |\varphi(0)|\, e^{-ax}$. Because we have $a = O(1)$ and $\varepsilon Rx = O(\sqrt{hR}x)$ is the distance from the edge, the linear solution exhibits a boundary layer the width of which is the geometrical mean of the thickness times its radius.

However, when the slope of the generators of the middle surface are not very small, (57.42) must be replaced by its nonlinear version, which reads

$$f'' - \varepsilon^2 f + \sin\beta = 0, \quad \beta'' - \varepsilon^2 \sin\beta - f\cos\beta = 0, \quad 0 < x < \infty, \tag{57.45}$$

with the boundary conditions that prescribe the (dimensionless) radial displacement or the shear force at the end, that is,

$$f'(0) + \sigma f(0)\sin\beta(0), \quad \text{or} \quad f(0), \tag{57.46}$$

and the bending moment or the rotation at the end of the tube, that is,

$$\beta'(0) - 2\sigma \sin^2[\beta(0)/2], \quad \text{or} \quad \beta(0). \tag{57.47}$$

Let us assume, for simplicity, that we have

$$\varphi(0) = f(0) + i\beta(0) \tag{57.48}$$

prescribed, and that

$$\int_0^\infty e^{-ax}|\varphi(x)|\,dx < \infty \quad \text{and} \quad e^{-ax}|\varphi'(x)| \to 0, \quad \text{for} \quad x \to \infty. \tag{57.49}$$

Note that the first inequality in (57.49) implies the limit

$$e^{-ax}|\varphi(x)| \to 0, \quad \text{for} \quad x \to \infty. \tag{57.50}$$

The presence of nonlinear terms in (57.45) allows for nonunique solutions. For example, if the loads at the end are zero, a semi-infinite cylinder can be in the stress-free state $f = \beta = 0$, or in the everted state $f \simeq 0$, $\beta \simeq 0$, except in a narrow zone adjacent to the edge, where f and β change rapidly to meet the stress-free boundary conditions. We also note that (57.45), (57.48), and (57.49) admit the even simpler, nondecaying constant solution

$$f = \frac{\sqrt{1-\varepsilon^8}}{\varepsilon^2}, \quad \beta = \frac{\pi}{2} + \sin^{-1}(\varepsilon^4). \tag{57.51}$$

Of course, this solution violates the assumption of small strains on which (57.45) is based. Thus, to guarantee uniqueness and decay, it is necessary to impose either bounds on the end data or an *a priori* bound on β.

The existence and the uniqueness of a solution to (57.45) and the associated boundary conditions can be proved by converting the problem into a complex-valued integral equation and showing that the contraction procedure works for sufficiently small $|\varphi(0)|$. However, it is possible to estimate the rate of decay of solutions by a direct method, based on the analysis of differential inequalities written for the energy associated with the solution.

For this purpose we construct the energy function

$$E(x) = f^2 + g(\beta), \quad 0 \leqslant x < \infty, \tag{57.52}$$

where g is an as yet unknown function. In order to guarantee that E vanishes when both f and β vanish, we shall require the condition $g(0) = 0$. We note that definition (57.52) generalizes the ordinary concept of strain energy density. In the present case the strain energy is proportional to

$$f'^2 - 2\varepsilon\sigma f' f \sin\beta + \varepsilon^2 f^2 \sin^2\beta + \beta'^2 - 4\sigma\varepsilon\beta' \sin^2(\beta/2) + 4\varepsilon^2 \sin^4(\beta/2).$$

Now, from (57.52), we find

$$E' = 2ff' + \frac{dg}{d\beta}\beta'. \tag{57.53}$$

A further differentiation of (57.53), together with (57.45) yields

$$E'' = 2f'^2 + \frac{d^2g}{d\beta^2}\beta'^2 + 2\varepsilon^2 f^2 + \varepsilon^2 \frac{dg}{d\beta}\sin\beta + f\left(\frac{dg}{d\beta}\cos\beta - 2\sin\beta\right). \tag{57.54}$$

As the last term in (57.54) is of indefinite sign, we choose g so that the term in brackets vanishes; that is, we set

$$\frac{dg}{d\beta} = 2\tan\beta,$$

and hence, from the boundary condition $g(0) = 0$, we have

$$g = \ell\text{n} \sec^2\beta. \tag{57.55}$$

Thus (57.52)–(57.54) reduce to

$$\begin{aligned} E &= f^2 + \ell\text{n} \sec^2\beta, \\ E' &= 2ff' + 2(\tan\beta)\beta', \\ E'' &= 2[f'^2 + \beta'^2 \sec^2\beta + \varepsilon^2(f^2 + \sin\beta\tan\beta)] \end{aligned} \tag{57.56}$$

for $0 \leqslant x < \infty$.

Let us introduce the hypothesis

$$-\frac{\pi}{2} < \beta(x) < \frac{\pi}{2}, \quad 0 \leqslant x < \infty, \tag{57.57}$$

this restriction being motivated by the existence of the nondecaying solution (57.51). If (57.57) holds, then we have

$$\sin\beta\tan\beta \geqslant \ell\text{n}\sec^2\beta \geqslant \sin^2\beta, \quad 0 \leqslant x < \infty, \tag{57.58}$$

and on using (57.56) we find

$$\begin{aligned} EE'' - \frac{1}{2}(E')^2 &= 2\varepsilon^2(f^2 + \ell\text{n}\sec^2\beta)(f^2 + \sin\beta\tan\beta) \\ &\quad + 2\left\{(f^2 + \ell\text{n}\sec^2\beta)(f'^2 + \beta'^2\sec^2\beta) - (ff' + \beta'\tan\beta)^2\right\}. \end{aligned} \tag{57.59}$$

By employing both inequalities of (57.58) in (54.59), we obtain

$$\begin{aligned} E'' - \frac{1}{2}(E')^2 &\geqslant 2\left\{\varepsilon^2E^2 + (f^2 + \sin^2\beta)(f'^2 + \beta'^2\sec^2\beta) - (ff' + \beta'\tan\beta)^2\right\} \\ &\geqslant 2\varepsilon^2E^2, \end{aligned} \tag{57.60}$$

where Cauchy's inequality has been used to obtain the last result. The second-order differential inequality (57.60) can be written as

$$(E^{\frac{1}{2}})'' \geqslant \varepsilon^2E^{\frac{1}{2}}, \quad 0 \leqslant x < \infty, \tag{57.61}$$

and integrated once to yield the first-order differential inequality

$$(E^{\frac{1}{2}})' + \varepsilon E^{\frac{1}{2}} \leqslant 0, \quad 0 \leqslant x < \infty, \tag{57.62}$$

provided that we have

$$\lim_{x\to\infty} e^{-\varepsilon x}(E'/2E^{\frac{1}{2}} + \varepsilon E^{\frac{1}{2}}) = 0. \tag{57.63}$$

Inequality (57.62) implies the further result

$$E'(x) \leqslant 0, \quad 0 \leqslant x < \infty, \tag{57.64}$$

so that $E(x)$ is a monotonically decreasing function of x for $E(x) > 0$. A further integration of (57.62) yields the decay estimate

$$E(x) \leqslant E(0)\, e^{-2\varepsilon x}, \quad 0 \leqslant x < \infty, \tag{57.65}$$

where

$$E(0) = f^2(0) + \ell n \sec^2 \beta(0) \tag{57.66}$$

is an explicit function of the boundary data.

However, the estimate (57.65) is not satisfactory, as the asymptotic behavior for $x \to \infty$, assumed in (57.63), is more restrictive than (57.49). Furthermore, assumption (57.57) is not natural because it imposes a restriction on $\beta(x)$ for all $x > 0$, while a condition of this type should be the result of a proof and not a postulate. But the major disadvantage of (57.65), compared to the estimates that can be derived by solving the integral equation directly, is that the decay rate of the former is much slower than in the latter. In fact, it is possible to show that, if $\varphi(x)$ is a solution of the integral equation equivalent to the nonlinear system (57.45) that decays to zero, then, for $x \to \infty$ such a solution decays neither more slowly nor more quickly than the linear solution. In other words, the nonlinear solution behaves exactly according to (57.49), and the coefficient a is, in general, much greater than 2ε. However, the energy methods have the advantage that they may be applied to more general types of shell for which an integral equation formulation might not be easy to obtain.

58. Closed Solutions for Shells

The equations of equilibrium for shells in the presence of large strains can be formulated in full generality, but integrating them is almost always impossible. We have seen that, when the shell is axisymmetric and the states of deformation are also axisymmetric, the basic equations of the problem become a system of ordinary differential equations, but the solution of this system is still difficult. In shell theory we can formulate a semilinear theory, characterized by large displacements and small strains, similar to Kirchhoff's theory of rods, but the resulting equations do not admit the elegant solutions available in the one-dimensional case.

An exact state of deformations, which change the overall geometric shape of a shell, as in the *elastica*, while the strains everywhere remain small and below the elastic limit, can be found for a helical strip subjected to an axial load and torque, as shown in Figure 58.1 (Mansfield 1973, 1980). The strip has a sufficiently slender

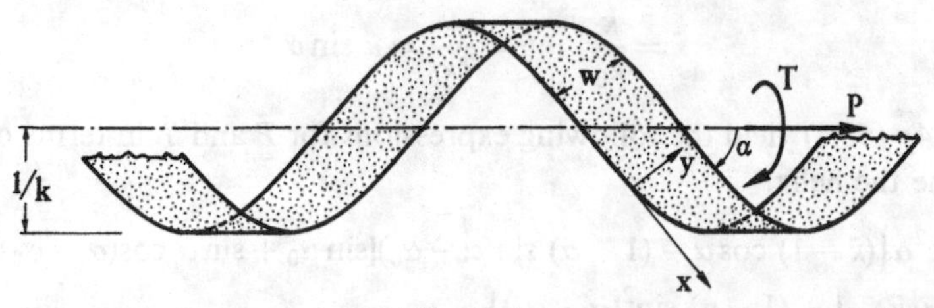

Fig. 58.1

cross-section, as in a watch or in a clock spring, so that the bending strain predominates and, as a result, the simplifying assumption of inextensionality is satisfied. Thus, in the following, attention can be confined to specific deformations, called *helical deformations*, in which the helical form of the strip is conserved under load, all changes thus being confined to the curvature and the angle of the helix.

In order to describe the deformation we introduce moving curvilinear coordinates x, y that remain embedded in the helical strip as it deforms. The x coordinate is measured along the length of the developed strip and the y coordinate is placed across the width. Now, if at any stage the helical strip lies on a cylinder with curvature κ and is inclined at an angle α to the axis of the cylinder, the curvatures referred to the x, y coordinates are given by

$$(\kappa_x, \kappa_y, \tau) = (\cos^2\alpha, \ \sin^2\alpha, \ -\sin\alpha\cos\alpha)\kappa. \tag{58.1}$$

Similar expressions hold for the curvatures $(\kappa_{x,0}, \kappa_{y,0}, \tau_0)$ in the unstressed state specified by κ_0 and α_0.

The strain energy per unit area due to inextensional deformations is given by

$$w = \frac{1}{2}D\left[(\kappa_x - \kappa_{x,0} + \kappa_y - \kappa_{y,0})^2 + 2(1-\sigma)\left\{(\tau - \tau_0)^2 - (\kappa_x - \kappa_{x,0})(\kappa_y - \kappa_{y,0})\right\}\right] \tag{58.2}$$

and hence the total strain energy in a helical strip, the developed length of which is ℓ and the width of which is w, is given by

$$\mathscr{W} = \frac{1}{2}w\ell D\left\{(\kappa - \kappa_0)^2 + 2(1-\sigma)\kappa\kappa_0 \sin^2(\alpha - \alpha_0)\right\}, \tag{58.3}$$

by virtue of (58.1).

Now, if z is the axial length of the helical strip and θ is the rotation about the axis, we have the equations

$$\begin{aligned} z &= \ell\cos\alpha, \\ \theta &= \ell\kappa\sin\alpha. \end{aligned} \tag{58.4}$$

Hence the axial load P and the torque T can be calculated from the formulae

$$P = \left(\frac{\partial\mathscr{W}}{\partial z}\right)_{\theta=\text{const.}}, \quad T = \left(\frac{\partial\mathscr{W}}{\partial\theta}\right)_{z=\text{const.}}. \tag{58.5}$$

Now let us introduce the following nondimensional terms:

$$\begin{aligned} &\tilde{P} = \frac{P}{wD\kappa_0^2}, \quad \tilde{T} = \frac{T}{wD\kappa_0}, \quad \tilde{z} = \frac{z}{\ell} = \cos\alpha, \\ &\tilde{\kappa} = \frac{\kappa}{\kappa_0}, \quad \tilde{\theta} = \frac{\theta}{\ell\kappa_0} = \tilde{\kappa}\sin\alpha \end{aligned} \tag{58.6}$$

so that (58.4)–(58.6) yield the following expressions for $\tilde{P}$ and $\tilde{T}$ in terms of $\tilde{\kappa}$ and α, which define the helix,

$$\tilde{P} = \tilde{\kappa}\operatorname{cosec}^2\alpha\left\{(\tilde{\kappa} - 1)\cos\alpha - (1-\sigma)\sin(\alpha - \alpha_0)[\sin\alpha_0 + \sin\alpha\cos(\alpha - \alpha_0)]\right\}, \tag{58.7a}$$

$$\tilde{T} = \operatorname{cosec}\alpha\left\{\tilde{\kappa} - 1 + (1-\sigma)\sin^2(\alpha - \alpha_0)\right\}. \tag{58.7b}$$

These equations constitute a formal solution to the load required for the axially symmetrical deformation of a helical strip, but certain simple subcases offer a better illustration of the results. If we assume that the torque is zero, then (58.7b) shows that we have

$$\tilde{\kappa} = 1 - (1 - \sigma) \sin^2(\alpha - \alpha_0), \tag{58.8}$$

and hence (58.7a) yields

$$\tilde{P} = -(1 - \sigma) \operatorname{cosec} \alpha \sin 2(\alpha - \alpha_0)\{1 - (1 - \sigma) \sin^2(\alpha - \alpha_0)\}. \tag{58.9}$$

Note that (58.9) shows the condition $\tilde{P}(\alpha + \pi) = -\tilde{P}(\alpha)$, so that, if $(\tilde{P}, \tilde{z})$ is an equilibrium state, so is $(-\tilde{P}, -\tilde{z})$. The curves $\tilde{P} = \tilde{P}(\tilde{z})$ are thus symmetrical with respect the origin. The case $\alpha_0 = \frac{1}{2}\pi$, which corresponds to that of a coiled spring loaded axially, is exceptional. As $|\tilde{P}|$ increases, the axial stiffness decreases and falls to zero when we have

$$\tilde{P} = \pm \frac{4}{3}\left[\frac{1 - \sigma}{3}\right]^{\frac{1}{2}}, \tag{58.10}$$

which occurs for

$$\tilde{z} = \pm\{3(1 - \sigma)\}^{-\frac{1}{2}}, \tag{58.11}$$

and $\tilde{\kappa}$ and $\tilde{\theta}$ have decreased from their initial values of unity to

$$\tilde{\kappa} = \frac{2}{3}, \quad \tilde{\theta} = \frac{2}{3}\left[\frac{2 - 3\sigma}{3(1 - \sigma)}\right]^{\frac{1}{2}}. \tag{58.12}$$

At such points the helical spring suddenly unwinds and straightens, after which the axial stiffness become theoretically infinite. However, in this case the extensional strains, ignored in the present inextensional treatment, then assume a dominant role and result in a larger but finite stiffness.

The situation complementary to that just described, occurs when the axial load vanishes, so that (58.6) and (58.7a) produce

$$\tilde{\theta} = \sin \alpha + (1 - \sigma) \tan \alpha \sin(\alpha - \alpha_0)\{\sin \alpha_0 + \sin \alpha \cos(\alpha - \alpha_0)\}, \tag{58.13}$$

and hence equation (58.7b) yields

$$\tilde{T} = (1 - \sigma) \sec \alpha \sin 2(\alpha - \alpha_0). \tag{58.14}$$

Note that we have

$$\tilde{T}(\alpha + \pi) = -\tilde{T}(\alpha), \quad \tilde{\theta}(\alpha + \pi) = -\tilde{\theta}(\alpha), \tag{58.15}$$

and hence, if $(\tilde{T}, \tilde{\theta})$ is an equilibrium state, so is $(-\tilde{T}, -\tilde{\theta})$.

Besides these two cases there are also other interesting situations. For example, if there are torsional constraints that prevent rotation under axial load, then we can put $\theta = \theta_0$, and hence from (58.4) and (58.6) we get

$$\tilde{\kappa} \sin \alpha = \sin \alpha_0,$$

so that (58.7a) yields

$$\tilde{P} = \sin\alpha_0 \operatorname{cosec}^4\alpha[(\sin\alpha_0 - \sin\alpha)\cos\alpha - (1-\sigma)\sin\alpha\sin(\alpha-\alpha_0)\times\{\sin\alpha_0 + \sin\alpha\cos(\alpha-\alpha_0)\}]. \quad (58.16)$$

The change of $\tilde{P}$ with $\tilde{z}$ shows that, for $\alpha_0 = 45°$, the axial stiffness is very small in the neighborhood of $\tilde{z} = 0$, a feature that might be used to advantage in shock-absorbing mountings.

The possibility of applying the hypothesis that the shell is flexible in bending but inextensible in stretching is allowed when the shell is initially a flat plate which deforms into a developable surface. Then, the problem of determining the shape of the plate under given loads reduces to finding the generators of the developable surface into which it deforms.

An interesting case is that of a horizontal square flat plate subjected to four equal forces at its corners, two of them upward directed at the ends of one diagonal and the other two downward directed at the ends of the other diagonal. According to the theory of small deflections, the plate undergoes a uniform twist, so that any line parallel to an edge remains straight. However, a simple experiment shows that, if the deflections are appreciable, the surface into which the plate deforms is different from that predicted by linear theory, because the surface appears to be perfectly cylindrical with the generators parallel to one or the other of the two diagonals.

The simplest problem to which the inextensional theory can be applied is that of a rectangular strip bent about the major axis of its cross-section by terminal bending moments. In the elementary infinitesimal theory of bending we find that the cross-section deforms into a circular arc, which is anticlastic with respect to the principal, longitudinal curvature. Linear theory predicts that longitudinal filaments near the long edges would be appreciably further from the axis of curvature than similar filaments near its center line. Thus, in the absence of any resultant longitudinal force in the strip, outer filaments would be stretched and inner filaments compressed. As these filaments have a curvature about the axis of curvature of the strip, the longitudinal tensions and compressions would have components forcing the outer filaments toward the axis of curvature and the inner ones away from it. Thus these forces would tend to destroy the anticlastic curvature and flatten the strip. A similar effect occurs in the flattening of bent tubes, and is known as *Brazier's effect* (Sodhi 1976).

If the middle surface of a shell, in its unstrained state, is planar or, more generally, a developable surface, the Gaussian curvature (the product of the principal curvatures) is zero. A subsequent deformation without extension preserves its value as zero at all points, and therefore the deformed middle surface is still a developable surface.

In a purely flexural deformation, the strain energy per unit area of the deformed middle surface has the form

$$W = \frac{1}{2}D[(\chi_x + \chi_y)^2 - 2(1-\sigma)(\chi_x\chi_y - \chi_{xy}^2)], \quad (58.17)$$

where χ_x, χ_y, and χ_{xy} are the changes in the curvatures of the middle surface.

The geometrical generation of a developable surface is done by considering a curve in the space, called the *edge of regression*, and the infinite tangents to this curve, called the *generators*. When the edge of regression is a point the surface is a cone or a cylinder, depending upon whether the point is positioned at a finite or an infinite distance. Let us assume that the edge of regression is a curve the vector position $\mathbf{r}(s)$ of which is a function of the arc length s, measured from a given point. Any point P on the developable surface lies on one of the generators and is determined by this generator and the distance u of P from the point M where the generator touches the curve (Figure 58.2). The position vector of the point P is then expressed by

$$\mathbf{R}(u, s) = \mathbf{r}(s) + u\mathbf{t}(s), \tag{58.18}$$

where $\mathbf{t}(s)$ is the unit vector tangential to the curve.

The coefficients of the first fundamental form of the surface are

$$g_{11} = 1 + u^2\kappa^2, \quad g_{12} = 1, \quad g_{22} = 1, \tag{58.19}$$

where κ is the principal curvature of the edge of regression. A relation between the parameters s and u, such as $f(s, u) = 0$, defines a curve on the surface, and an arc element of this curve is given by

$$d\sigma^2 = (1 + u^2\kappa^2)ds^2 + 2\,ds\,du + du^2. \tag{58.20}$$

Formula (58.20) shows that the length of the linear element on the surface does not depend on the torsion τ of the edge of regression. This property is very important because it enables the surface to be transformed into a plane without stretching, shrinking, or tearing, by simply making the torsion of the edge of regression equal to zero.

The coefficients of the second fundamental form are

$$h_{11} = u\kappa\tau, \quad h_{12} = 0 \quad h_{22} = 0, \tag{58.21}$$

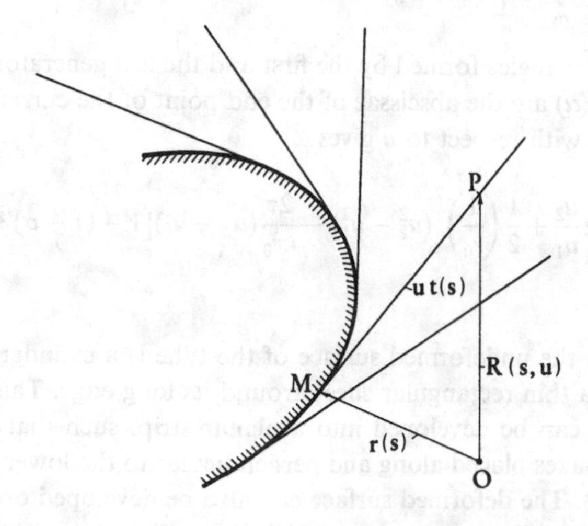

Fig. 58.2

and the principal curvatures are

$$\kappa_1 = 0, \quad \kappa_2 = \frac{\tau}{u\kappa}, \tag{58.22}$$

along the lines of curvature $s =$ constant and $s + u =$ constant.

An element of area of the surface can be expressed by

$$dA = \sqrt{g}\, du\, ds = (g_{11}g_{22} - g_{12}^2)^{\frac{1}{2}}\, du\, ds = u\kappa\, ds\, du. \tag{58.23}$$

Suppose that the undeformed surface is a thin-walled cylinder of radius r_0 so that $\kappa_x^0 = 0$, $\kappa_y^0 = 1/r_0$, and $\kappa_{xy}^0 = 0$ are the curvatures of the middle surface at any point of the undeformed state (Sodhi 1976). Let κ_x, κ_y, and κ_{xy} be the curvatures of the deformed middle surface. Then the changes in the curvatures are

$$\chi_x = \kappa_x, \quad \chi_y = \kappa_y - \frac{1}{r_0}, \quad \chi_{xy} = \kappa_{xy}. \tag{58.24}$$

The principal curvatures at a point P of the strained surface are along the perpendicular to the generator passing through that point and are equal to zero and $\tau/u\kappa$, respectively. If α is the angle that any generator forms with a fixed direction, which is taken coincident with the x axis, we can use certain geometric relations, due to Dupin and Euler, which state:

$$\begin{gathered} \kappa_x + \kappa_y = \frac{\tau}{u\kappa} = \text{mean curvature}, \\ \kappa_x\kappa_y - \kappa_{xy}^2 = 0 = \text{Gaussian curvature}, \\ \kappa_x = \frac{\tau}{u\kappa}\sin^2\alpha, \\ dA = u\kappa\, ds\, du = u\, du\, d\alpha \quad (\text{because } \kappa\, ds = d\alpha). \end{gathered} \tag{58.25}$$

On substituting the expressions for χ_x, χ_y, and χ_{xy} in (58.17), we get, after simplification,

$$\mathscr{W} = \int_{\alpha_1}^{\alpha_2}\int_{u_1}^{u_2} \frac{D}{2}\left[\left(\frac{\tau}{u\kappa} - \frac{1}{r_0}\right)^2 + 2(1-\sigma)\frac{\tau}{u\kappa}\frac{\sin^2\alpha}{r_0}\right] u\, du\, d\alpha, \tag{58.26}$$

where α_1 and α_2 are the angles formed by the first and the last generators with the x axis, and $u_1(\alpha)$ and $u_2(\alpha)$ are the abscissae of the end point of the current generator. Integration of (58.26) with respect to u gives

$$\mathscr{W} = \int_{\alpha_1}^{\alpha_2} \frac{D}{2}\left\{\left(\frac{\tau}{\kappa}\right)^2 \log\frac{u_2}{u_1} + \frac{1}{2}\left(\frac{1}{r_0}\right)^2 (u_2^2 - u_1^2) - \frac{2\tau}{\kappa r_0}(u_2 - u_1)[1 - (1-\sigma)\sin^2\alpha]\right\} d\alpha. \tag{58.27}$$

In the present case, the undeformed surface of the tube is a cylinder of radius r_0 obtained by bending a thin rectangular sheet around its long edge. This means that the undeformed tube can be developed into a planar strip, such that x and y are orthogonal Cartesian axes placed along and perpendicular to the lower long edge of the strip (Figure 58.3). The deformed surface can also be developed on to the same rectangle of the xy plane; those generators which are straight lines in the deformed surface are also straight lines in the rectangle. The deformation is such that the points $x = 0$ do not change their position. If α is the angle between a generator and the lower edge, the position of each generator is uniquely determined by the value of α and of the abscissa x of the point M$'$, where the generator meets the x axis. The generators are tangential to the edge of regression. The edge of regression is a planar curve after the development in the xy plane, and the generators are tangential to this curve. Let u be the distance of a point P of a generator from the point M at which the generator touches the edge of regression. Let u_1 and u_2 be the value of u at the points M$'$ and M$''$, respectively. Figure 58.3 shows two generators making angles α and $(\alpha + d\alpha)$ with the x axis. Let ds be the arc length of the edge of regression between two points M and Q, and let u_1 and $(u_1 + du_1)$ be the lengths of MM$'$ and QQ$'$, respectively. As the lines MM$'$ and MQ are nearly collinear, in the triangle QM$'$Q$'$ we have the geometric relation

$$u_1 = x' \sin \alpha, \tag{58.28}$$

with $x' = dx/d\alpha$.

Considering now the expression for the strain energy (58.27), we see that the limits of integration for α are $\alpha_1 = \pi/2$ and $\alpha_2 = \pi$, corresponding to the first generator at $x = 0$ and the last generator meeting the x axis at infinity. On noting the relations

$$\begin{aligned} u_2 - u_1 &= \frac{b}{\sin \alpha}, \\ \frac{u_2}{u_1} &= 1 + \frac{b}{(x' \sin^2 \alpha)}, \\ u_2^2 - u_1^2 &= \left(1 + \frac{2x'}{b} \sin^2 \alpha\right)\left(\frac{b}{\sin \alpha}\right)^2, \end{aligned} \tag{58.29}$$

where b is the width of the strip, we can express $\mathscr{W}$ as

$$\mathscr{W} = \int_{\pi/2}^{\pi} F(x', \frac{\tau}{\kappa}, \alpha) d\alpha, \tag{58.30}$$

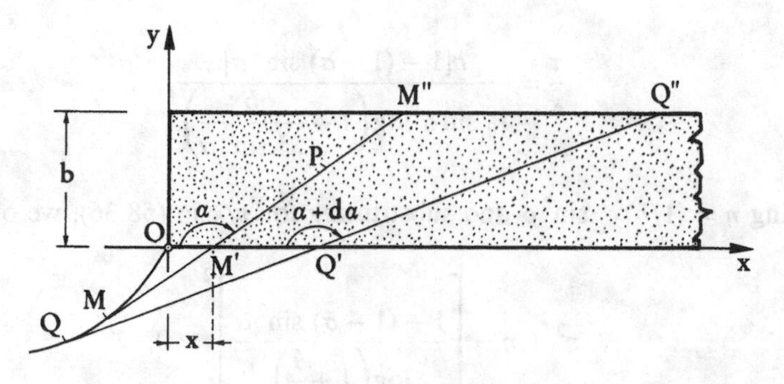

Fig. 58.3

where we have written

$$F\left(x', \frac{\tau}{\kappa}, \alpha\right) = \frac{D}{2}\left\{\left(\frac{\tau}{\kappa}\right)^2 \log\left(1 + \frac{b}{x' \sin^2 \alpha}\right) + \frac{1}{2}\frac{b^2}{r_0^2 \sin^2 \alpha}\left(1 + 2\frac{x' \sin^2 \alpha}{b}\right)\right.$$
$$\left. -2\frac{\tau}{\kappa}\frac{b}{r_0 \sin \alpha}[1 - (1 - \sigma) \sin^2 \alpha]\right\}. \tag{58.31}$$

The relation between α and x is determined from the condition that the strain energy $\mathscr{W}$ is a minimum. From this condition we obtain

$$\frac{d}{d\alpha}\left(\frac{\partial F}{\partial x'}\right) = 0 \quad \text{or} \quad \frac{\partial F}{\partial x'} = B = \text{constant}, \tag{58.32}$$

$$\frac{\partial F}{\partial\left(\frac{\tau}{\kappa}\right)} = 0, \tag{58.33}$$

and the boundary conditions

$$x = 0 \quad \text{for} \quad \alpha = \frac{\pi}{2} \text{ (geometrical boundary condition)}, \tag{58.34a}$$

$$\frac{\partial F}{\partial x'} = 0 \quad \text{for} \quad \alpha = \pi \text{ (natural boundary condition)}. \tag{58.34b}$$

The constant B in (58.32) is, from (58.34b), equal to zero. Therefore

$$\frac{\partial F}{\partial x'} = \frac{\partial F}{\partial\left(\frac{\tau}{\kappa}\right)} = 0 \tag{58.35}$$

are the differential equations relating x and α. From (58.35) we get

$$-\frac{\left(\frac{\tau}{\kappa}\right)^2}{\left(\frac{x' \sin^2 \alpha}{b} + 1\right)x'} + \frac{b}{r_0^2} = 0, \tag{58.36}$$

and

$$\frac{\tau}{\kappa} = \frac{b[1 - (1 - \sigma) \sin^2 \sigma]}{r_0 \sin \alpha \log\left(1 + \frac{b}{x' \sin^2 \alpha}\right)}. \tag{58.37}$$

On putting $\eta = (1/b)x' \sin^2 \alpha$ and substituting for τ/κ in (58.36), we obtain

$$\eta^2 + \eta - \left[\frac{1 - (1 - \sigma) \sin^2 \alpha}{\log\left(1 + \frac{1}{\eta}\right)}\right]^2 = 0,$$

or, by taking only the positive root of this equation, the required solution is

$$\eta = \frac{1}{2}\left(-1 + \left\{1 + 4\left[\frac{1-(1-\sigma)\sin^2\alpha}{\ell n\left(1+\frac{1}{\eta}\right)}\right]^2\right\}^{\frac{1}{2}}\right), \tag{58.38}$$

the negative root of the equation being meaningless. Equation (58.38) is an implicit relation from which we can obtain η as a function of α. Once we know η, then from the equation

$$\eta = \frac{x'}{b}\sin^2\alpha,$$

we can calculate x' and then

$$x(\alpha) = \int_{\pi/2}^{\alpha} x'\, d\alpha. \tag{58.39}$$

Equation (58.39) thus represents the solution to our problem.

In forming thin shells the problem arises of determining the distribution of strain necessary to convert a flat sheet of metal into a cap of positive curvature. Artisans form bowls simply by hammering a sheet of ductile metal. The compressive transverse permanent strains produce a tangential expansion, which causes the change in curvature. We consider the problem of finding the value of the tangential strain that will produce a prescribed final shape (Steele 1975). As the problem is purely kinematic, the fact that strains are permanent and not elastic is irrelevant.

We begin with an undeformed curved surface the metric of which is

$$ds^2 = a_{\alpha\beta}\, dx^\alpha\, dx^\beta, \quad \alpha, \beta = 1, 2, \tag{58.40}$$

and consider the application of an isotropic strain which changes (58.40) into

$$(ds^*)^2 = \lambda^2 a_{\alpha\beta}\, dx^\alpha\, dx^\beta, \tag{58.41}$$

where $\lambda = \lambda(x^\alpha)$ is a regular function of the curvilinear coordinates x^α. If, in particular, the coordinate lines are such that we have

$$(a_{11})^{\frac{1}{2}} = A_1, \quad (a_{22})^{\frac{1}{2}} = A_2, \quad a_{12} = 0, \tag{58.42}$$

the coefficients of the metric (58.41) become

$$A_1^* = \lambda A_1, \quad A_2^* = \lambda A_2. \tag{58.43}$$

The Gaussian curvature, which initially is

$$-K = \frac{1}{A_1 A_2}\left[\frac{\partial}{\partial x^1}\left(\frac{1}{A_1}\frac{\partial A_2}{\partial x^1}\right) + \frac{\partial}{\partial x^2}\left(\frac{1}{A_2}\frac{\partial A_1}{\partial x^2}\right)\right], \tag{58.44}$$

becomes

$$-K^* = \frac{1}{\lambda^2 A_1 A_2}\left[\frac{\partial}{\partial x^1}\left(\frac{1}{\lambda A_1}\frac{\partial \lambda A_2}{\partial x^1}\right) + \frac{\partial}{\partial x^2}\left(\frac{1}{\lambda A_2}\frac{\partial \lambda A_1}{\partial x^2}\right)\right], \tag{58.45}$$

which can be written in the form

$$-K^* = \frac{1}{\lambda^2}(-K + \nabla^2 \log \lambda), \tag{58.46}$$

where ∇^2 is the Laplace operator on the undeformed surface.

Equation (56.46) provides a nonlinear partial differential equation for the strain distribution λ, which will change the Gaussian curvature of a surface from the initial distribution K to the desired K^*. In order to illustrate the possible application of (56.46), let us assume $\varepsilon = \log \lambda$ as a measure of strain, so that (56.46) becomes

$$\nabla^2 \varepsilon + K^* e^{2\varepsilon} = K. \tag{58.47}$$

For small strains (56.47) can be linearized to give the equation

$$\nabla^2 \varepsilon + 2K^* \varepsilon = K - K^*. \tag{58.48}$$

Artisans usually leave the edges unstrained, which gives the condition $\varepsilon = 0$ along the boundary. If K is known, K^* prescribed, and the boundary curve also given, then (56.48) is a standard problem, for elliptical partial differential equations, provided that K^* is not too large, namely less than the first eigenvalue of the eigenvalue problem

$$\nabla^2 \varepsilon + 2K^* \varepsilon = 0,$$

where we have $\varepsilon = 0$ at the boundary.

As a simple example, let us consider the strain required to form a spherical cap with radius R ($K^* = R^{-2}$) from a flat circular plate of radius a ($K = 0$). Then (56.48) becomes

$$\frac{1}{r}\frac{d}{dr}\left(r\frac{d\varepsilon}{dr}\right) + \frac{2}{R^2}\varepsilon = -\frac{2}{R^2},$$

and the solution of this equation, satisfying the boundary condition $\varepsilon = 0$ at $r = a$ and bounded at the origin, is

$$\varepsilon = \frac{J_0(\sqrt{2}r/R)}{J_0(\sqrt{2}a/R)} - 1. \tag{58.49}$$

If the strain ε is small, the argument of the Bessel function must be small, so (58.49) can be replaced by the first term of its power expansion:

$$\varepsilon \simeq (a^2 - r^2)\frac{1}{2R^2}. \tag{58.50}$$

A general procedure for the solution by quadratures can be applied to infinitely long cylindrical shells that undergo planar deformations (Libai and Simmonds 1983). The analysis starts with a generalization of Kirchhoff's assumption concerning the deformation, because this takes into consideration, not only the change in curvature μ of the reference line of the cross-section of the cylinder, but also its extension λ. In addition, the elastic constitutive equation of the material is not necessarily linear.

In order to formulate the problem, let the reference line of the cylindrical shell be given by

$$x = x(s), \quad y = y(s), \tag{58.51}$$

where x, y are Cartesian coordinates referred to a fixed coordinate system in the plane, and s is the arc length. Let α be the angle of slope between the positive tangent to the curve at a point and the positive x direction such that (Figure 58.4)

$$\cos\alpha = \frac{dx}{ds}, \quad \sin\alpha = \frac{dy}{ds}, \tag{58.52}$$

and let K be the curvature of the reference line at A:

$$K = \frac{d\alpha}{ds}. \tag{58.53}$$

Under certain loads the material reference line changes its position, and the new coordinates of the point are $A_1(x_1, y_1)$, with arc length s_1, slope β, and curvature K_1.

At each point of the deformed reference line the stress resultants are a normal force N, a shearing force Q, and a bending moment M.

As measures of strain we define the extension

$$\lambda = \frac{ds_1}{ds}, \tag{58.54}$$

and the change of slope

$$\mu = \frac{d\beta}{d\alpha} = \lambda\frac{K_1}{K}. \tag{58.55}$$

Let us note that these measures of strain can be replaced by the more common measures

$$\begin{aligned} e &= \frac{ds_1 - ds}{ds} = \lambda, \\ \chi &= \frac{d(\beta-\alpha)}{ds} = K(\mu - 1). \end{aligned} \tag{58.56}$$

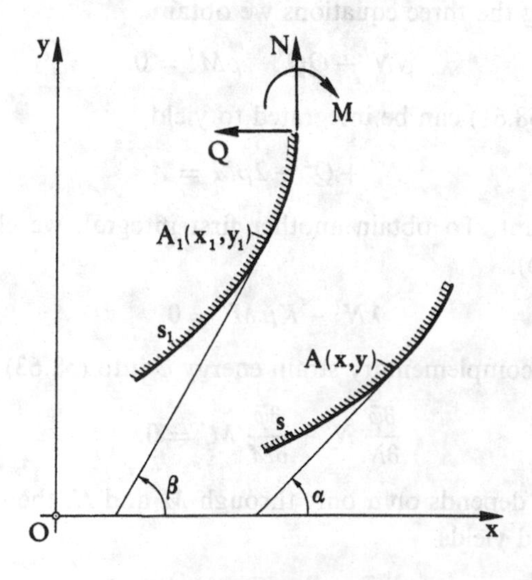

Fig. 58.4

In general, the strain e determines the normal force N, while χ determines the bending moment. In Kirchhoff's theory, there is no strain measure associated with Q.

The equilibrium equations, expressed as functions of the Lagrangean variable α, are

$$M' = \lambda \frac{Q}{K}, \tag{58.57a}$$

$$N' = \mu Q, \tag{58.57b}$$

$$Q' = \frac{\lambda \rho}{K} - \mu N, \tag{58.57c}$$

where the prime denotes differentiation with respect to α, and ρ is the exterior load, constituted by a pressure per unit length of s_1 in the direction of the negative normal to the deformed curve. For simplicity, we consider only loads of this type.

The constitutive equations are defined by a strain energy density of the form $\varphi = \varphi(\lambda, \mu)$, with

$$N = \frac{\partial \varphi}{\partial \lambda}, \quad KM = -\frac{\partial \varphi}{\partial \mu}. \tag{58.58}$$

It is often more convenient to introduce the complementary function

$$\bar{\varphi} = \lambda N - K\mu M - \varphi,$$

and write the constitutive equations in the form

$$\lambda = \frac{\partial \bar{\varphi}}{\partial N}, \quad K\mu = -\frac{\partial \bar{\varphi}}{\partial M}. \tag{58.59}$$

On using (58.55) we derive the geometric relations

$$\frac{dx_1}{ds} = \frac{\lambda}{K} \cos \beta, \quad \frac{dy_1}{ds} = \frac{\lambda}{K} \sin \beta. \tag{58.60}$$

Equations (58.54)–(58.60) constitute the field equations of the problem. To obtain a first integral of the system, we multiply (58.57a) by $-\rho$, (58.57b) by N, and (58.57c) by Q, and on adding the three equations we obtain

$$NN' + QQ' - \rho M' = 0. \tag{58.61}$$

If ρ is a constant, (58.61) can be integrated to yield

$$N^2 + Q^2 - 2\rho M = A, \tag{58.62}$$

where A is a constant. To obtain another first integral, we eliminate Q between (58.57a) and (58.57b):

$$\lambda N' - K\mu M' = 0. \tag{58.63}$$

On introducing the complementary strain energy $\bar{\varphi}$ into (58.63) we obtain

$$\frac{\partial \bar{\varphi}}{\partial N} N' + \frac{\partial \bar{\varphi}}{\partial M} M' = 0. \tag{58.64}$$

If we assume that $\bar{\varphi}$ depends on α only through M and N, then (58.64) becomes an exact differential and yields

$$\bar{\varphi} = B = \text{constant}. \tag{58.65}$$

In terms of φ, (58.65) becomes

$$\lambda\frac{\partial\varphi}{\partial\lambda}+\mu\frac{\partial\varphi}{\partial\mu}-\varphi=B; \tag{58.66}$$

and we require that φ should depend on α only through λ and $K\mu$.

Equations (58.59), (58.62), and (58.65) are four algebraic equations for the five variables N, M, Q, λ, and μ, from which it is possible to express (in principle) four terms in functions of the fifth, the latter being preferably chosen from among (M, N, Q).

In order to illustrate the procedure of integration, let us assume that the strain energy density has the form

$$\varphi=\frac{1}{2}C2h\left[\left(\lambda-\frac{1}{\lambda}\right)^2+\frac{4}{3}K^2h^2\left(\lambda-\frac{\mu}{\lambda}\right)^2\right], \tag{58.67}$$

which gives

$$N=\frac{\partial\varphi}{\partial\lambda}=2Ch\left[\lambda-\frac{1}{\lambda^3}+\frac{4}{3}K^2h^2\left(\lambda-\frac{\mu^2}{\lambda^3}\right)\right], \tag{58.68}$$

$$M=-\frac{1}{K}\frac{\partial\varphi}{\partial\mu}=\frac{1}{3}KC8h^3\left(1-\frac{\mu}{\lambda^2}\right), \tag{58.69}$$

where $2h$ is the initial (constant) thickness of the shell and C is a constant known as *Mooney's modulus*. By taking $C=E/4(1-\sigma^2)$, the formulae are applicable to shells that obey Hooke's law, provided that the strains remain small. For the energy integral, (58.66) must be used because $\bar{\varphi}$ is not conveniently available. Upon substitution of the constitutive law, (58.66) becomes

$$\lambda\frac{\partial\varphi}{\partial\lambda}+\mu\frac{\partial\varphi}{\partial\mu}-\varphi=\frac{1}{2}C2h\left[\left(\lambda^2-\frac{3}{\lambda^2}+2\right)+\frac{4}{3}(Kh)^2\left(\lambda^2-\frac{\mu^2}{\lambda^2}\right)\right]=B. \tag{58.70}$$

Elimination of μ between (58.68) and (58.70) results in a simplified energy integral, which takes the form

$$\left(\frac{N}{2Ch}\right)\lambda^3+B_1^2\lambda^2=2, \tag{58.71}$$

where $B_1=2-(2B/2Ch)$ is another constant of integration.

The formulae just derived can be made more specific in particular problems; for example, in that of a cylindrical shell subjected to two equal and opposite forces $2P$ per unit length applied to two diametrically opposite generators. The problem is well known in experimental mechanics, where elastic cylinders are used as load cells for measuring forces. Another problem is that of an open cylindrical shell the cross-section of which has the form of a quarter circle, clamped along one generator and loaded by forces along the other. The problem is statically determinate in the sense that the conditions at the second end are sufficient to calculate the constants A and B.

However, if large strains are taken into account, closed solutions in shell theory are very few. It is thus understandable that the majority of problems are treated by some approximation method. Among these methods the most natural and widely used is that of *linearization*, which consists of solving a linear problem in the neighborhood of a particular solution.

The method can be described for the case of large deflection of a shallow shell under a pressure q (Thurston 1973). The problem is described by the system of equations

$$\begin{aligned} D\nabla^4 W - L(F, W + \bar{W}) &= q, \\ \nabla^4 F + EhL(W, W + 2\bar{W}) &= 0, \end{aligned} \tag{58.72}$$

where the operator L is bilinear and has the form

$$L(S, T) = S_{,xx}T_{,yy} - 2S_{,xy}T_{,xy} + T_{,xx}S_{,yy}. \tag{58.73}$$

Here W is the component of the deflection that is normal to a reference plane, and F is a stress function. The term $\bar{W}$ defines the rise of the middle surface of the undeformed shell from the reference plane.

Equations (58.72) can be nondimensionalized to the form

$$\begin{aligned} \nabla^4 f + \frac{\lambda}{2}L(w, w) + L(\bar{w}, w) &= 0, \\ \nabla^4 w - \lambda L(f, w) - L(\bar{w}, f) &= 4\bar{q}, \end{aligned} \tag{58.74}$$

where the parameter λ is a load parameter that accounts for the possible loads acting at the edge. If a solution of (58.74) is known for $\lambda = \lambda_0$, we consider the linearized solution

$$\begin{aligned} f &= f_0(\lambda_0) + \delta f, \\ w &= w_0(\lambda_0) + \delta w, \end{aligned} \tag{58.75}$$

where δf and δw are small terms. Equations (58.74) become

$$\nabla^4(\delta f) + \lambda L(w_0, \delta w) + L(\bar{w}, \delta w) = -\frac{1}{2}(\lambda - \lambda_0)L(w_0, w_0) - \frac{1}{2}\lambda L(\delta w, \delta w), \tag{58.76a}$$

$$\nabla^4(\delta w) - \lambda L(f_0, \delta w) - \lambda L(w_0, \delta w) - L(\bar{w}, \delta f) = -(\lambda - \lambda_0)L(f_0, w_0) + \lambda L(\delta f, \delta w). \tag{58.76b}$$

The linear homogeneous equations do admit nontrivial solutions for certain eigenvalues λ_i:

$$\begin{aligned} \nabla^4 f_i + \lambda_i L(w_0, w_i) + L(\bar{w}, w_i) &= 0, \\ \nabla^4 w_i - \lambda_i L(f_0, w_i) - \lambda_i L(w_0, f_i) - L(\bar{w}, f_i) &= 0. \end{aligned} \tag{58.77}$$

When λ is far from one of these eigenvalues, the quadratic terms in $(\delta f, \delta w)$ on the right-hand side of (58.76) can be neglected, and δf and δw may be obtained as solutions of a linear problem; these solutions can possibly be calculated by iteration. When λ is near λ_i, the convergence of the iterative method breaks down and the quadratic terms in (58.76) must be retained. In this case it is convenient to introduce the expansion

$$\delta w = \sum_{i=1}^{\infty} a_i w_i, \tag{58.78}$$

where a_i are constants and w_i are the eigenfunctions corresponding to each λ_i. We then write (58.76) replacing δf by f_k, and δw by (58.78). After multiplying (58.76a) by f_k, (58.76b) by $\sum_{i=1}^{\infty} a_i w_i$, and integrating over the domain A, where the variables x and y are defined. From (58.77), we find that the coefficients a_k are solutions of the infinite set of quadratic algebraic equations

$$2(\lambda - \lambda_k)a_k = (\lambda - \lambda_0)C_k + \lambda \sum_{i=1}^{\infty} \sum_{j=1}^{\infty} a_i a_j D_{ijk}, \quad (k = 1, 2, 3, \ldots) \tag{58.79}$$

where we have

$$\begin{aligned} C_k &= \iint_A \left[N(f_0, w_0, w_k) + \frac{1}{2} N(f_k, w_0, w_0) \right] dA, \\ D_{ijk} &= \iint_A \left[N(f_i, w_j, w_k) + \frac{1}{2} N(f_k, w_i, w_j) \right] dA, \end{aligned} \tag{58.80}$$

with the obvious definitions

$$N(f_0, w_0, w_k) = L(f_0 w_0) w_k, \text{ etc.} \tag{58.81}$$

An application of this procedure is to the problem of an imperfect spherical cap under uniform pressure $\bar{q} = -1$. The equation of the undeformed middle surface of the cap is

$$\bar{w} = -\frac{1}{2}(x^2 + y^2) + \xi(x, y)$$

where $\xi(x, y)$ is a measure of the discrepancy. The nondimensional membrane solution for the perfectly spherical shell is

$$f = -(x^2 + y^2) = -r^2, \quad w = 0.$$

This suggests rewriting the system (58.74) as

$$\begin{aligned} \nabla^4 f - \nabla^2 w &= -\frac{1}{2}\lambda L(w, w) - L(\xi, w), \\ \nabla^4 w + 2\lambda \nabla^2 w + \nabla^2 f &= -4 + L(f + r^2, w) + L(\xi, f). \end{aligned} \tag{58.82}$$

Then we can relate the linearized problem with (58.82), and repeat the general argument.

59. Buckling of Shells

In the theory of large deformations of elastic shells the problem of buckling occupies a restricted role. Instead of attempting a detailed description of the relation between the applied forces and the corresponding displacements, we content ourselves with determining the possible values of the loads for which at least two equilibrium states may coexist. More precisely, when the loads are of different type, as, for example, a surface pressure acting on the upper face and edge forces applied along the boundary, they must be thought of as functions of a common parameter λ, and the displacements must be described collectively by the displacement of a particular point of the shell or, better, by one of the components of this displacement, say u. The knowledge of the function $\lambda = \lambda(u)$ may then be assumed to be a sufficiently comprehensive description of the overall elastic response of the shell under the applied loads. Any value of λ at which the number of solutions of the equilibrium problem changes is called a *bifurcation point*. If λ is made to increase slowly from zero, the smallest value of λ at which more than one equilibrium solution arises is the *buckling load*.

However, knowledge of the buckling load and, possibly, of the bifurcation points does not give any information about either the shape of the coexistent states of equilibrium or their onset. In some cases the passage from one state to another is continuous, while in others it occurs as a jump between one configuration and another; in some cases the simultaneous equilibrium configurations preserve certain common geometric properties, as, for example, the symmetries about an axis or a plane, but often these symmetries are irredemiably destroyed because the unsymmetric states reveal themselves to be more stable under the same load. The theoretical problems proposed by the mechanism of the onset of buckling and its evolution at higher levels of load are not completely understood. A surprising phenomenon is that almost none of the values for the buckling or the bifurcation points predicted by the theory agree with those measured experimentally, the experimental values being systematically lower that the theoretical ones, with serious prejudice to the safety of engineering shell-type structures. The explanation of this fact lies in the unavoidable discrepancy between practical experiments and the theoretical hypotheses upon which equations are written and calculations performed. In the theoretical analysis it is often assumed that, for example, the geometric shape of the middle surface of a shell is perfectly defined, that the nature of the boundary conditions perfectly conform to those of the theoretical model, and that the positions and the directions of the loads are defined with absolute precision. Whenever one of these assumptions is violated, or only partially satisfied, then small deviations of these data from those contemplated by the theory give rise to large perturbations in the solutions.

A complete analysis of the buckling of shells, which takes into account geometric and mechanical irregularities, can only be made in some particular cases. An example in which the calculations can be effected with success is that of a spherical cap that is imperfectly clamped at the edge due to the elastic deformation of the clamping frame. The results confirm that small changes in the edge conditions cause large changes in the buckling loads (Bauer, Keller, and Reiss 1967).

In dimensionless variables, the differential equations describing the finite axisymmetric deformations of a thin spherical cap can be written as

$$\begin{aligned} L\gamma(x) &= -\rho[\beta^2(x) + 2x\beta(x)], \\ L\beta(x) &= \rho[\beta(x)\gamma(x) + x\gamma(x) + Px^2], \end{aligned} \tag{59.1}$$

where L is a differential operator defined by

$$L\varphi(x) = x\frac{d}{dx}\left[\frac{1}{x}\frac{d}{dx}(x\varphi(x))\right].$$

The dimensionless variables are

$$x = \frac{\theta}{\Lambda}, \quad \beta(x) = \frac{1}{R\Lambda}\frac{dw}{d\theta}, \quad \rho = \frac{\Lambda^2}{C}\frac{R}{h},$$

$$P = \left(\frac{R}{h}\right)^2 \frac{p}{2EC}, \quad C^2 = \frac{2}{3(1-\sigma^2)},$$

where θ is the polar angle, 2Λ is the angle subtended to the cap at its center of curvature, $w(\theta)$ is the displacement normal to the spherical surface measured positively inward, and $\beta(x)$ is proportional to the slope of the deformed surface relative to the initial spherical surface. The quantity $\gamma(x)$ in (59.1) is a stress function defined such that the meridional and circumferential membrane stresses σ_θ and σ_φ (φ is the azimuthal angle) and the corresponding dimensionless stresses Σ_θ and Σ_φ are given by

$$\Sigma_\theta(x) = \left(\frac{R}{h}\right)\left(\frac{2}{3EC^2}\right)\sigma_\theta(\theta) = \frac{\gamma(x)}{x},$$
$$\Sigma_\varphi(x) = \left(\frac{R}{h}\right)\left(\frac{2}{3EC^2}\right)\sigma_\varphi(\theta) = \frac{d\gamma(x)}{dx}. \tag{59.2}$$

The bending stresses on the inner surface of the shell are $\sigma_\theta^0(\theta)$ and $\sigma_\theta^0(\theta)$ and the dimensionless stresses are Σ_θ^0 and Σ_φ^0. They are given in terms of the slope $\beta(x)$ by

$$\Sigma_\theta^0(x) = \left(\frac{R}{h}\right)\left(\frac{2}{3EC^2}\right)\sigma_\theta^0(\theta) = -\left(\beta' + \frac{\sigma}{x}\beta\right),$$
$$\Sigma_\varphi^0(x) = \left(\frac{R}{h}\right)\left(\frac{2}{3EC^2}\right)\sigma_\varphi^0(\theta) = -\left(\sigma\beta' + \frac{1}{x}\beta\right). \tag{59.3}$$

As the solutions are regular and symmetrical at the vertex of the cap, we have

$$\beta(0) = \gamma(0) = 0. \tag{59.4}$$

At the edge of the shell, where $x = 1$, the conditions may be of different types. We consider in particular the case for which the edge $x = 1$ is attached to a clamping frame that is perfectly rigid with respect to rotations and deflections normal to the spherical surface, but is elastic with respect to tangential deflections of the cap's edge. In this case we have

$$\beta(1) = 0, \quad \gamma'(1) - a\gamma(1) = 0 \quad (a = \sigma + 4\pi Eh\delta), \tag{59.5}$$

where δ is the coefficient of the influence of the frame due to tangential forces. On setting $\delta = 0$, we obtain the condition that the meridional displacement vanishes along the edge.

The solution of system (59.1) with the boundary conditions (59.4) and (59.5), can be obtained numerically (Bauer et al. 1967) by replacing the differential equations by difference equations that are then solved by iteration, starting by selecting P near zero and taking the first iteration of $\beta(x)$ as equal to zero everywhere. The result is that the elasticity of the frame causes a strong decrease in the buckling load, thus explaining why the experimentally determined buckling loads of a rigidly clamped cap are systematically lower than the computed values.

Imperfections arise both from the elasticity of the contraints and from the nonuniformity of loads. In order to illustrate the effect of these influences on the values of the buckling loads and on the shapes of the buckled states of the shell, it is sufficient to consider the axisymmetric deformations of a complete thin spherical shell subjected to an external pressure of the form $p(\theta) = p_0 + \tau d(\theta)$, where θ is the latitude and τ measures the deviation of the load from a uniform pressure p_0. In contrast to the case of an incomplete spherical cap, the equations of the equilibrium and buckling of a closed spherical shell can be studied in a more general form (Keller 1973).

Let (r, θ, φ) denote the spherical coordinates, with a latitude $\theta = 0$ at the north pole. The surface $r = R$ is the middle surface and the surfaces $r = R \pm h$ are the faces, so that $2h$ is the thickness. We introduce the ratio $K = \frac{1}{3}[h/R]^2$ and assume the condition $K \ll 1$. An axisymmetric but nonuniform pressure $p(\theta)$, acting in the radial direction, is considered positive when compressive.

The equations of equilibrium are again (56.21), but it is convenient to rewrite them in the form

$$\begin{aligned} Lq + \sigma q &= -\left[v + \frac{\xi}{2\sqrt{1-\xi^2}} v^2 - \tau B(\xi) \right], \\ Lv - \sigma v &= \gamma \left[q - (p + \tau C(\xi))v + \frac{\xi}{\sqrt{1-\xi^2}} qv \right], \end{aligned} \tag{59.6}$$

with $L\psi = (1-\xi^2)\psi'' - 2\xi\psi' - (\xi^2/\sqrt{1-\xi^2})\psi$. Here we have put $\xi = \cos\theta$ and $\gamma = (1-\sigma^2)/K$. The dimensionless load $P(\theta) = [p(\theta)R]/4Eh$ is assumed to have the form

$$P(\theta) = p + \tau D(\xi), \tag{59.7}$$

and then we have

$$C(\xi) = \frac{2}{1-\xi^2} \int_{\xi}^{1} \xi D(\xi)\, d\xi, \tag{59.8}$$

$$B(\xi) = -\sqrt{1-\xi^2} \frac{d}{d\xi}(2D - C) + \frac{2\xi}{\sqrt{1-\xi^2}}(D - C). \tag{59.9}$$

In order to make the decomposition (59.7) unique, we require the conditions

$$\int_{-1}^{1} \frac{D(\xi)}{\sqrt{1-\xi^2}}\, d\xi = 0, \tag{59.10}$$

$$\max_{-1 \leqslant \xi \leqslant 1} |D(\xi)| = 1, \tag{59.11}$$

$$\int_{-1}^{1} \xi D(\xi)\, d\xi = 0. \tag{59.12}$$

Thus p is the average load, and τ is the amplitude of the maximum deviation from the uniform load. Conditions (59.10) and (59.12) ensure that the additional load is self-equilibrated. As a consequence of the symmetry, the boundary conditions at the poles are

$$(q(\pm 1), v(\pm 1)) = 0. \tag{59.13}$$

In order to study the boundary-value problem, it is convenient to introduce the notation

$$\mathbf{y} = \begin{pmatrix} q \\ v \end{pmatrix}, \quad T = \begin{bmatrix} -L & 0 \\ 0 & -L \end{bmatrix}, \quad A(p, \tau) = \begin{bmatrix} \sigma & 1 \\ -\gamma & [\gamma p - \sigma + \gamma\tau C(\xi)] \end{bmatrix}$$

$$\mathbf{f}(\tau, \mathbf{y}) = \begin{bmatrix} \dfrac{\xi}{2\sqrt{1-\xi^2}} v^2 - \tau B(\xi) \\ -\dfrac{\gamma\xi}{\sqrt{1-\xi^2}} qv \end{bmatrix},$$

and to write (59.6) and (59.13) as

$$T\mathbf{y} = A(p, \tau)\mathbf{y} + \mathbf{f}(\tau, \mathbf{y}), \quad \mathbf{y}(\pm 1) = \mathbf{0}. \tag{59.14}$$

By setting $\tau = 0$ we obtain the case of a uniform load. The linearized version of this problem about the trivial solution $\mathbf{y} = 0$, is given by the system

$$T\boldsymbol{\varphi} = A(p, 0)\boldsymbol{\varphi}, \quad \boldsymbol{\varphi}(\pm 1) = \mathbf{0}. \tag{59.15}$$

The matrix $A(p, 0)$ has eigenvalues $\alpha^{\pm}$, which are the roots of

$$\alpha^2 - \gamma p\alpha + [\gamma(1 + \sigma p) - \sigma^2] = 0. \tag{59.16}$$

Thus, for $\alpha^+ \neq \alpha^-$ or, equivalently, for $p \neq (2/\gamma)(\sigma \pm \sqrt{\gamma})$, $A(p, 0)$ can be diagonalized. In the present case we can write

$$S^{-1}A(p, 0)S = \begin{bmatrix} \alpha^+ & 0 \\ 0 & \alpha^- \end{bmatrix}, \quad S = \begin{bmatrix} 1 & \dfrac{\alpha - \sigma}{\gamma} \\ \alpha - \sigma & 1 \end{bmatrix},$$

$$S^{-1} = \frac{\gamma}{\gamma - (\alpha - \sigma)^2} \begin{bmatrix} 1 & \dfrac{\sigma - \alpha}{\gamma} \\ \sigma - \alpha & 1 \end{bmatrix},$$

so that, with $\mathbf{z} = S^{-1}\boldsymbol{\varphi}$ in (59.15), we obtain

$$Lz_1 = \alpha^+ z_1, \quad z_1(\pm 1) = 0,$$
$$Lz_2 = \alpha^- z_2, \quad z_2(\pm 1) = 0.$$

The only nontrivial solutions are (to within a constant factor)

$$z_j(\xi) = P_n^{(1)}(\xi)$$

$$\text{for} \quad \alpha^{\pm} = \alpha_n = n(n+1) - 1, \quad j = 1, 2, \quad \text{and} \quad n = 1, 2, \ldots,$$

where $P_n^{(1)}(\xi)$ is the Legendre function of the first kind and order n. If α^+ or α^- is equal to α_n, then p must have the value

$$p_n = \frac{(n^2 + n - 1 + \sigma)}{\gamma} + \frac{1}{(n^2 + n - 1 - \sigma)}, \quad n = 1, 2, \ldots, \tag{59.17}$$

and for such values of p_n the eigensolutions of (59.15) are

$$\varphi^{(n)}(\xi) = S\begin{bmatrix} P_n^{(1)}(\xi) \\ 0 \end{bmatrix}, \quad \alpha^+ = \alpha_n = n(n+1) - 1, \quad n = 1, 2, \ldots. \tag{59.18}$$

If we had chosen the values $\alpha^- = \alpha_n$ and

$$\varphi^{(n)}(\xi) = S\begin{bmatrix} 0 \\ P_n^{(1)}(\xi) \end{bmatrix},$$

this would merely correspond to an interchange of z_1 and z_2; that is, to a permutation of the eigenvalues.

For $\alpha^+ = \alpha^-$, $A(p, 0)$ cannot be diagonalized and no eigensolutions exist. Thus p_n is not an eigenpressure for $p_n = (2/\gamma)(\sigma \pm \sqrt{\gamma})$.

With $p_m = p_n$ for some $m \neq n$, that is, with

$$(\alpha_n - \sigma)(\alpha_m - \sigma) = \gamma,$$

two linearly independent solutions do exist, and these are called *degenerate*.

Assuming that the eigenvalues are simple, we can rewrite the normalized eigensolutions as

$$\varphi^{(n)}(\xi) = \begin{pmatrix} a_1 \\ a_2 \end{pmatrix} P_n^{(1)}, \tag{59.19}$$

with

$$a_1(n) = \left[\frac{2n+1}{2(\alpha_n + 1)[1 + (\alpha_n - \sigma)^2]}\right]^{\frac{1}{2}},$$
$$a_2(n) = a_1(n)(\alpha_n - \sigma). \tag{59.20}$$

The eigenfunctions are normalized to give

$$\|\varphi^{(n)}\|^2 = \int_{-1}^{1} \left[\left(\varphi_1^{(n)}(\xi)\right)^2 + \left(\varphi_2^{(n)}(\xi)\right)^2\right] d\xi = 1, \tag{59.21}$$

because we have

$$\int_{-1}^{1} \left[P_n^{(1)}(\xi)\right]^2 d\xi = \frac{2(\alpha_n + 1)}{2n+1}. \tag{59.22}$$

The adjoint problem to (59.15) is

$$T\psi = A^*(p, 0)\psi, \quad \psi(\pm 1) = 0. \tag{59.23}$$

Its eigenpressures are p_n and the corresponding normalized eigensolutions have the form

$$\psi^{(n)}(\xi) = \begin{pmatrix} b_1 \\ b_2 \end{pmatrix} P_n^{(1)}(\xi), \tag{59.24}$$

with

$$b_1(n) = \left\{ \frac{2n+1}{2(\alpha_n+1)\left[1 - \frac{1}{\gamma^2}(\alpha_n - \sigma)^2\right]} \right\}^{\frac{1}{2}}, \tag{59.25}$$

$$b_2(n) = -\frac{(\alpha_n - \sigma)}{\gamma} b_1(n).$$

We now return to the original nonlinear problem (59.14) to see, according to bifurcation theory, how nontrivial solutions with $\tau = 0$ appear as the load p varies near the linearized buckling loads. Let p_n be any simple eigenpressure. In order to avoid confusion we use the notation

$$p_0 = p_n, \quad \varphi_0(\xi) = \varphi^{(n)}(\xi), \quad \psi_0(\xi) = \psi^{(n)}(\xi). \tag{59.26}$$

On setting $\tau = 0$ in (59.14), we can rewrite the problem as

$$[T - A(p_0, 0)]\mathbf{y} = [A(p, 0) - A(p_0, 0)]\mathbf{y} + \mathbf{f}(0, \mathbf{y}),$$
$$\mathbf{y}(\pm 1) = 0,$$

and try to find $\mathbf{y}(\xi)$ and p as the limit of the sequence of iterates $\{\mathbf{y}^{\nu}(\xi), p^{\nu}\}$ satisfying the recursive equations

$$[T - A(p_0, 0)]\mathbf{y}^{\nu+1} = [A(p^{\nu+1}, 0) - A(p_0, 0)]\mathbf{y}^{\nu} + \mathbf{f}(0, \mathbf{y}^{\nu}), \tag{59.27}$$
$$\mathbf{y}^{\nu+1}(\pm 1) = 0$$

$$<\psi_0, \mathbf{y}^{\nu+1} - \mathbf{y}^0> = 0, \tag{59.28}$$

where $<\varphi, \psi>$ denotes the inner product, that is

$$<\varphi, \psi> = \int_{-1}^{1} [\varphi_1(\xi)\psi_1(\xi) + \varphi_2(\xi)\psi_2(\xi)]d\xi. \tag{59.29}$$

By choosing the first iterate $\mathbf{y}^0$ as

$$\mathbf{y}^0 = \varepsilon\varphi_0(\xi), \tag{59.30}$$

for $|\varepsilon| > 0$ but sufficiently small, the iterative procedure applied to (59.27) and (59.28) converges. By the alternative theorem, in order that each iterate has a solution, it is necessary and sufficient to have the condition

$$<\psi_0, [A(p^{\nu+1}, 0) - A(p_0, 0)]\mathbf{y}^{\nu} + \mathbf{f}(0, \mathbf{y}^{\nu})> = 0. \tag{59.31}$$

From this equation, and using (59.24) and the expression for $\mathbf{f}(0, \mathbf{y}^{\nu})$, we obtain

$$p^{\nu+1}(\varepsilon) = p_0 + \frac{<v^{\nu}\left(q^{\nu} - \frac{b_1}{2\gamma b_2}v^{\nu}\right), \frac{\xi}{\sqrt{1-\xi^2}}P_n^{(1)}(\xi)>}{<v^{\nu}, P_n^{(1)}>}. \tag{59.32}$$

On using (59.30) and (59.19), with $\nu = 0$ the above yields

$$p^1(\varepsilon) = p_0 + \varepsilon\frac{3}{2}a_1(n)\frac{<(P_n^{(1)}(\xi))^2, \dfrac{\xi}{\sqrt{1-\xi^2}}P_n^{(1)}(\xi)>}{\|P_n^{(1)}\|^2}. \tag{59.33}$$

Now, because the iterative procedure converges, it is possible to show that the iterates and exact bifurcation solution $\{\mathbf{y}(\xi, \varepsilon), p(\varepsilon)\}$ are controlled by the orders of magnitude

$$\begin{aligned} |\mathbf{y}(\xi, \varepsilon) - \mathbf{y}^\nu(\xi, \varepsilon)| &= O(|\varepsilon|^{\nu+2}), \\ |p(\varepsilon) - p^{\nu+1}(\varepsilon)| &= O(|\varepsilon|^{\nu+2}). \end{aligned} \tag{59.34}$$

From (57.34), on using (59.33) it follows that a family of solutions $\{\mathbf{y}, p\}$, bifurcating from the trivial solution $\{\mathbf{0}, p_n\}$, has the form

$$\begin{aligned} \mathbf{y}_{\text{BIF}}(\xi, \varepsilon) &= \varepsilon\begin{bmatrix} a_1(n) \\ a_2(n) \end{bmatrix} P_n^{(1)}(\xi) + O(|\varepsilon|^2), \\ p_{\text{BIF}}(\varepsilon) &= p_n + \varepsilon\frac{3}{2}a_1(n)\frac{1}{\|P_n^{(1)}\|^2}\int_{-1}^{1}\frac{\xi}{\sqrt{1-\xi^2}}(P_n^{(1)})^3\, d\xi + O(|\varepsilon|^2). \end{aligned} \tag{59.35}$$

The result obviously holds for simple eigenpressures p_n of the linear theory, but we can also determine bifurcating branches corresponding to degenerate multiple eigenpressures $p_n = p_m$ for $n + m$ odd, although this is a little more complicated.

For $\tau \neq 0$ the pressure is no longer uniform, and the uniformly contracted state is not a solution. In this case the bifurcation load goes over, as τ deviates from zero, to a *branching load*, that is to say, to a load at which two distinct nontrivial branches of deformations coalesce, one of these branches being the slightly disturbed uniform contraction and the other being a perturbation of the bifurcated branch of solutions.

As for $\tau \neq 0$ we do not have a trivial solution, we seek for small $|\tau| > 0$ a small solution near a trivial solution about which the linearized problem has a nontrivial solution. In addition to (59.14), we attempt to solve the linearized problem in $\mathbf{y}$:

$$T\mathbf{\Phi} = [A(p, \tau) + \mathbf{f}_{\mathbf{y}}(\tau, \mathbf{y})]\mathbf{\Phi}, \quad \mathbf{\Phi}(\pm 1) = 0, \quad \|\mathbf{\Phi}\| \neq 0, \tag{59.36}$$

where $\mathbf{f}_{\mathbf{y}}$ is the operator

$$\mathbf{f}_{\mathbf{y}}(\tau, \mathbf{y}) = -\frac{\xi}{\sqrt{1-\xi^2}}\begin{pmatrix} 0 & -v \\ \gamma v & \gamma q \end{pmatrix}.$$

We thus apply the previous procedure and rewrite (59.14) and (59.36) as

$$[T - A(p_0, 0)]\mathbf{y} = [A(p, \tau) - A(p_0, 0)]\mathbf{y} + \mathbf{f}(\tau, \mathbf{y}), \quad \mathbf{y}(\pm 1) = 0, \tag{59.37}$$

$$[T - A(p_0, 0)]\mathbf{\Phi} = [A(p, \tau) - A(p_0, 0) + \mathbf{f}_{,\mathbf{y}}(\tau, \mathbf{y})]\mathbf{\Phi}, \quad \mathbf{\Phi}(\pm 1) = 0, \quad \|\mathbf{\Phi}\| \neq 0. \tag{59.38}$$

Here both p and τ ($\neq 0$) are to be determined with a solution to be denoted by $\{\mathbf{y}, p; \mathbf{\Phi}, \tau\}$. The iteration scheme for the iterates is

$$[T - A(p_0, 0)]\mathbf{y}^{\nu+1} = [A(p^{\nu+1}, \tau^{\nu+1}) - A(p_0, 0)]\mathbf{y}^{\nu} + \mathbf{f}(\tau^{\nu+1}, \mathbf{y}^{\nu}), \quad \mathbf{y}^{\nu+1}(\pm 1) = 0, \tag{59.39}$$

$$[T - A(p_0, 0)]\mathbf{\Phi}^{\nu+1} = [A(p^{\nu+1}, \tau^{\nu+1}) - A(p_0, 0) + \mathbf{f}_{,\mathbf{y}}(\tau^{\nu+1}, \mathbf{y})]\mathbf{\Phi}^{\nu}, \quad \mathbf{\Phi}^{\nu+1}(\pm 1) = 0. \tag{59.40}$$

These iterations are unique provided that (see (59.22)) we have

$$<\psi_0, \mathbf{y}^{\nu+1} - \mathbf{y}^0> = 0, \quad <\psi_0, \mathbf{\Phi}^{\nu+1} - \mathbf{\Phi}^0> = 0, \tag{59.41}$$

and the alternative theorem implies that (59.39) and (59.40) have solutions if, and only if,

$$\begin{aligned} &<\psi_0, [A(p^{\nu+1}, \tau^{\nu+1}) - A(p_0, 0)]\mathbf{y}^{\nu} + \mathbf{f}(\tau^{\nu+1}, \mathbf{y}^{\nu})> = 0; \\ &<\psi_0, [A(p^{\nu+1}, \tau^{\nu+1}) - A(p_0, 0) + \mathbf{f}_{,\mathbf{y}}(\tau^{\nu+1}, \mathbf{y}^{\nu})]\mathbf{\Phi}^{\nu}> = 0. \end{aligned} \tag{59.42}$$

If these relations can be solved for $p^{\nu+1}$ and $\tau^{\nu+1}$ as functionals of $\mathbf{y}^{\nu}$ and $\mathbf{\Phi}^{\nu}$, they are then used in (59.39), (59.40), and (59.41) to determine $\mathbf{y}^{\nu+1}$ and $\mathbf{\Phi}^{\nu+1}$. Again, the convergence of this procedure can be proved if we take as the initial iterate

$$\mathbf{y}^0(\xi) = \varepsilon\varphi_0(\xi), \quad \mathbf{\Phi}^0(\xi) = \varphi_0(\xi), \tag{59.43}$$

with $|\varepsilon| > 0$ sufficiently small. On putting $\nu = 0$ and using (59.43), we can show that the first iterate has the form

$$\begin{aligned} p^1(\varepsilon) = p_0 + \varepsilon 3a_1(n)\frac{1}{\|P_n^{(1)}\|^2} &<(P_n^{(1)}(\xi))^2, \frac{\xi}{\sqrt{1-\xi^2}}P_n^{(1)}(\xi)> \\ &\times \left\{1 + \frac{\alpha_n - \sigma}{2}a_2(n)\frac{<CP_n^{(1)}, P_n^{(1)}>}{<B, P_n^{(1)}>}\right\}, \end{aligned} \tag{59.44}$$

$$\tau^1(\varepsilon) = -\varepsilon^2\frac{3}{2}a_2^2(n)\frac{1}{<B(\xi), P_n^{(1)}(\xi)>} < \left(P_n^{(1)}(\xi)\right)^2, \frac{\xi}{\sqrt{1-\xi^2}}P_n^{(1)}(\xi)> . \tag{59.45}$$

Furthermore, in this case, the errors in the subsequent iterates can be bounded by formulae similar to (59.34), namely

$$\begin{aligned} |\mathbf{y}(\xi, \varepsilon) - \mathbf{y}^{\nu}(\xi, \varepsilon)| = O(|\varepsilon|^{\nu+2}), \quad |p(\varepsilon) - p^{\nu}(\varepsilon)| = O(|\varepsilon|^{\nu+1}), \\ |\mathbf{\Phi}(\xi, \varepsilon) - \mathbf{\Phi}^{\nu}(\xi, \varepsilon)| = O(|\varepsilon|^{\nu+1}), \quad |\tau(\varepsilon) - \tau^{\nu}(\varepsilon)| = O(|\varepsilon|^{\nu+2}). \end{aligned} \tag{59.46}$$

It thus follows from (59.44) and (59.45) that a family of branching solutions $\{\mathbf{y}, p; \mathbf{\Phi}, \tau\}$ has the form

$$\mathbf{y}_{BR}(\xi, \varepsilon) = \varepsilon \begin{bmatrix} a_1(n) \\ a_2(n) \end{bmatrix} P_n^{(1)}(\xi) + O(|\varepsilon|^2), \tag{59.47}$$

$$p_{BR}(\varepsilon) = p_n + \varepsilon 3a_1(n) \frac{1}{\|P_n^{(1)}\|^2} \left[\int_{-1}^{1} \frac{\xi}{\sqrt{1-\xi^2}} (P_n^{(1)}(\xi))^3 \, d\xi \right] + O(|\varepsilon|^2), \tag{59.48}$$

$$\tau_{BR}(\varepsilon) = -\varepsilon^2 \frac{3}{2} a_2^2(n) \frac{1}{\int_{-1}^{1} B(\xi) P_n^{(1)}(\xi) \, d\xi} \left[\int_{-1}^{1} \frac{\xi}{\sqrt{1-\xi^2}} (P_n^{(1)}(\xi))^3 \, d\xi \right] + O(|\varepsilon|^3), \tag{59.49}$$

$$\mathbf{\Phi}_{BR}(\xi, \varepsilon) = \begin{bmatrix} a_1(n) \\ a_2(n) \end{bmatrix} P_n^{(1)}(\xi) + O(|\varepsilon|). \tag{59.50}$$

This family exists for all $|\varepsilon| > 0$, but sufficiently small, and for each $n = 1, 2, \ldots,$ for which p_n is a simple eigenpressure of the linearized theory.

A comparison of the bifurcation load $p_{BIF}(\varepsilon)$ in (59.35) with the branching load $p_{BR}(\varepsilon)$ in (59.48) gives the result

$$(p_{BR}(\varepsilon) - p_n) = 2(p_{BIF}(\varepsilon) - p_n) + O(|\varepsilon|^2). \tag{59.51}$$

Thus, for a given amplitude ε of the component of the linearized buckling mode, the branching load $p_{BR}(\varepsilon)$ is essentially twice as far from the eigenpressure p_n as is the bifurcation load $p_{BIF}(\varepsilon)$. The complete sphere is thus sensitive to load nonuniformities in essentially the same way that it is to initial irregularities in shape. Figure 59.1 gives an idea of this effect. The straight line represents the dependence of the pressure p on a geometric parameter, called the *generalized deflection*, according to the linear theory. The two curves $p_{BIF}(\varepsilon)$ and $p_{BR}(\varepsilon)$ represent the bifurcated and branching pressures, respectively, which appear near the linearized buckling load p_n.

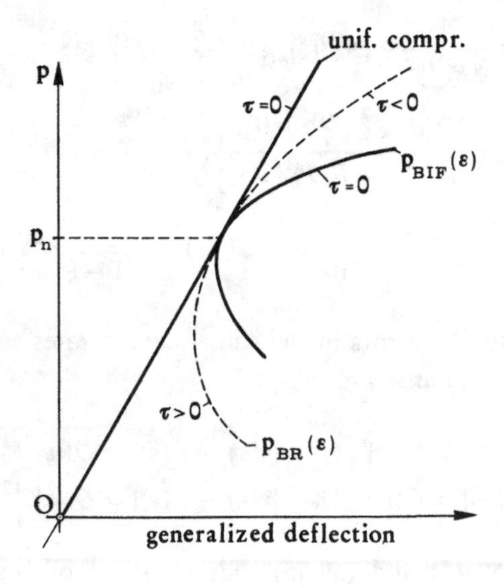

Fig. 59.1

In the analysis of the buckling of thin shells, it may be the case that some buckled states are symmetrical in the sense that the system takes up the same shape, and hence has the same energy, when modes are reversed so as to be equal and opposite in amplitude. In general, however, the final buckled state of a shell, although geometrically symmetrical with respect to some axis and symmetrically loaded with respect this axis, is not symmetrical.

The influence of the nonsymmetrical terms in the stability of a shell can be illustrated by the analysis of the buckling of a cylindrical shell when the cross-section is a thin ring of thickness $2h$, with a mean radius R, and a length 2ℓ (Hunt, Williams, and Cowell 1986) under an axial compressive force P, acting along the axis of the cylinder such that it generates a uniform compression in the fundamental state. In a possible buckled state the total energy stored by the shell is the sum of three terms:

$$\mathcal{E} = \mathcal{W}_S + \mathcal{W}_B - \mathcal{U}. \tag{59.52}$$

The first term $\mathcal{W}_S$ represents the membrane energy which can be written as

$$\mathcal{W}_S = \frac{1}{2} E2h \int_0^{2\ell} \int_0^{2\pi R} (\nabla^2 \Phi)^2 \, dx \, dy, \tag{59.53}$$

where Φ denotes a stress function such that

$$N_x = 2Eh\Phi_{yy}, \quad N_y = 2Eh\Phi_{xx}, \quad N_{xy} = -2Eh\Phi_{xy},$$

the x coordinate representing the distance of each cross-section from the bottom of the cylinder, and the y coordinate denoting the arc length of the middle circle of any cross-section, measured from a given point. The second term $\mathcal{W}_B$ is the bending energy, which has the form

$$\mathcal{W}_B = \frac{1}{2} \frac{2Eh^3}{3(1-\sigma^2)} \int_0^{2\ell} \int_0^{2\pi R} (\nabla^2 w)^2 \, dx \, dy, \tag{59.54}$$

where w is the radial displacement of the middle surface. The last term in (59.54) is the potential of the load, which is given by

$$\mathcal{U} = \frac{1}{2} 2Eh\Lambda \int_0^{2\ell} \int_0^{2\pi R} \left(\frac{\partial w}{\partial x}\right)^2 dx \, dy, \tag{59.55}$$

with $\Lambda = P/2Eh$. The two functions Φ and w are solutions of the following nonlinear differential equations:

$$\nabla^4 \Phi = \frac{1}{R} \frac{\partial^2 w}{\partial x^2} - \left[\frac{\partial^2 w}{\partial x^2} \frac{\partial^2 w}{\partial y^2} - \left(\frac{\partial^2 w}{\partial x \, \partial y} \right)^2 \right], \tag{59.56}$$

$$\frac{h^3}{3(1-\sigma^2)} \nabla^4 w = -\frac{1}{R} \frac{\partial^2 \phi}{\partial x^2} + \left[\frac{\partial^2 \phi}{\partial x^2} \frac{\partial^2 w}{\partial y^2} + \frac{\partial^2 \Phi}{\partial y^2} \frac{\partial^2 w}{\partial x^2} - 2 \frac{\partial^2 \phi}{\partial x \, \partial y} \frac{\partial^2 w}{\partial x \, \partial y} \right], \tag{59.57}$$

which are known as *Donnell's equations* for the cylindrical shell. The form of (59.56) suggests that Φ can be conceived as the sum of two contributions, $\Phi = \Phi_1 + \Phi_2$, where Φ_1 is a solution of the equation

$$\nabla^4 \Phi_1 = \frac{1}{R} \frac{\partial^2 w}{\partial x^2}, \tag{59.58}$$

and Φ_2 is a solution of

$$\nabla^4 \Phi_2 = -\left[\frac{\partial^2 w}{\partial x^2} \frac{\partial^2 w}{\partial y^2} - \left(\frac{\partial^2 w}{\partial x \, \partial y} \right)^2 \right]. \tag{59.59}$$

We now consider an approximate expression for the deflection w of the form

$$w = Q_0 R + \omega_{mn} \cos \frac{m\pi x}{2\ell} \cos \frac{ny}{R}, \tag{59.60}$$

where the first term represents a pure uniform radial dilatation and the second term is a summation over all integers $m, n \geqslant 0$, but excluding $m = n = 0$. Once we have chosen w, from (59.58) we obtain Φ_1:

$$\Phi_1 = -\frac{\Lambda y^2}{2} + \frac{1}{R} \varphi_{mn}^{(1)} \cos \frac{m\pi x}{2\ell} \cos \frac{ny}{R}, \tag{59.61}$$

with

$$\varphi_{mn}^{(1)} = -\omega_{mn} \left(\frac{m\pi}{2\ell} \right)^2 \left[\left(\frac{m\pi}{2\ell} \right)^2 + \left(\frac{n}{R} \right)^2 \right]^{-2}. \tag{59.62}$$

In a similar manner, Φ_2 may be cast in terms of w, taking care that the products of cosines are reduced to a sum of cosines. By a straightforward procedure we obtain

$$\Phi_2 = \varphi_{mn}^{(2)} \cos \frac{m\pi x}{2\ell} \cos \frac{ny}{R}, \tag{59.63}$$

where the $\varphi_{mn}^{(2)}$ are quadratic functions of the ω_{mn} terms.

After substitution of these expressions for w and Φ in the formula for the total energy (59.52), the contribution of each term can be easily discussed as a consequence of the orthogonality of the trigonometric functions. We introduce the coefficients g_{mn} defined by

$$g_{oo} = 4, \quad g_{mo} = g_{om} = 2 \text{ for } m > 0, \quad g_{mn} = 1 \text{ for } m > 0,\ n > 0. \tag{59.64}$$

After division by the factor Eh, the total potential energy can be written as

$$\mathscr{E} = \left\{ \left[\left(\frac{\varphi_{mn}^{(1)}}{R} + \varphi_{mn}^{(2)} \right)^2 + \frac{2}{3(1-\sigma^2)} \omega_{mn}^2 \right] \left[\left(\frac{m\pi}{2\ell} \right)^2 + \left(\frac{n}{R} \right)^2 \right]^2 - \Lambda \left(\frac{m\pi}{2\ell} \right)^2 \omega_{mn}^2 \right\} g_{mn} \tag{59.65}$$

where the summation convention is again employed for m and n.

On recalling that the $\varphi_{mn}^{(1)}$ and $\varphi_{mn}^{(2)}$ are, respectively, linear and quadratic in the Fourier coefficients of w, we infer that the bending energy contributes stabilizing positive quadratic terms, while the work done by the axial load contributes destabilizing negative quadratic terms. The membrane energy has three separate effects. The terms $\varphi_{mn}^{(1)}$ give rise to positive quadratic terms. Similarly, $\varphi_{mn}^{(2)}$ generate positive quadratic terms, which have more influence in the case of moderately large deflections. However, the presence of cross-products between $\phi_{mn}^{(1)}$ and $\phi_{mn}^{(2)}$, which yield cubic terms in the coefficients ω_{mn}, is responsible for the highly unstable behavior at the critical load. We thus have two opposing tendencies arising from the nonconstant part of w: the stabilizing influence of the quartic terms, which increase the critical load; and the destabilizing effect of the cubic terms.

The role played by the terms of different order is explained by choosing the following form for w:

$$w = Q_0 R + q_1 \ell \cos\frac{\pi x}{2\ell}\cos\frac{ny}{R} + q_2\ell\cos\frac{\pi x}{\ell} + q_3\ell\cos\frac{2ny}{R}, \tag{59.66}$$

which is an approximation of (59.60) truncated to the third term. The associated stress function, constructed by the superposition $\Phi = \Phi_1 + \Phi_2$, where Φ_1 and Φ_2 are solutions of (59.58) and (59.59), is

$$\begin{aligned}\Phi = &-\frac{\Lambda y^2}{2} - \left[\frac{q_1\ell}{R} + 2q_1(q_2+q_3)\left(\frac{n\ell}{R}\right)^2\right]\left\{\left(\frac{\pi}{2\ell}\right)^2\left[\left(\frac{\pi}{2\ell}\right)^2 + \left(\frac{n}{R}\right)^2\right]^{-2}\cos\frac{\pi x}{\ell}\cos\frac{ny}{R}\right. \\
&-\frac{q_1^2\ell^2}{32}\left(\frac{R\pi}{2n\ell}\right)^2\cos\frac{2ny}{R} - \left(\frac{\ell}{2\pi}\right)^2\left[\frac{4q_2\ell}{R} + \frac{q_1^2}{2}\left(\frac{n\ell}{R}\right)^2\right]\cos\frac{\pi x}{\ell} \\
&-2q_1q_2\left\{\left(\frac{n\pi}{2R}\right)^2\left[\left(\frac{3\pi}{2\ell}\right)^2 + \left(\frac{n}{R}\right)^2\right]^{-2}\cos\frac{3\pi x}{2\ell}\cos\frac{ny}{R}\right. \\
&-2q_2q_3\left\{\left(\frac{n\pi}{2R}\right)^2\left[\left(\frac{\pi}{2\ell}\right)^2 + \left(\frac{3n}{R}\right)^2\right]^{-2}\right\}\cos\frac{\pi x}{2\ell}\cos\frac{3ny}{R} \\
&-16q_2q_3\left\{\left(\frac{n\pi}{2R}\right)^2\left[\left(\frac{\pi}{\ell}\right)^2 + \left(\frac{2n}{R}\right)^2\right]^{-2}\right\}\cos\frac{\pi x}{l}\cos\frac{2ny}{R}.\end{aligned} \tag{59.67}$$

On substituting formulae (59.66) and (59.67) in (59.52), we obtain the expression

$$\begin{aligned}\mathscr{E} = &\frac{1}{2}V_{11}^0 q_1^2 + \frac{1}{2}V_{22}^0 q_2^2 + \frac{1}{2}V_{33}^0 q_3^2 - \Lambda(q_1^2 + 8q_2^2) + \frac{1}{2}V_{112}^0 q_1^2 q_2 + \frac{1}{2}V_{113} q_1^2 q_3 \\
&+ \frac{1}{24}V_{1111}^0 q_1^4 + \frac{1}{4}V_{1122}^0 q_1^2 q_2^2 + \frac{1}{4}V_{1133}^0 q_1^2 q_3^2 + \frac{1}{4}V_{2233}^0 q_2^2 q_3^2 + \frac{1}{2}V_{1123}^0 q_1^2 q_2 q_3,\end{aligned} \tag{59.68}$$

where the coefficients $V_{11}^0, \ldots$ have the forms

$$V_{11}^0 = \frac{\ell^2}{\pi R^2}\frac{8}{3(1+\beta^2)^2} + \frac{\pi^2 h^2}{3(1-\sigma^2)\ell^2}\frac{(1+\beta^2)^2}{2},$$

$$V_{22}^0 = 16\left[\frac{\ell^2}{\pi^2 R^2} + \frac{\pi^2 h^2}{3\ell^2(1-\sigma^2)}\right], \quad V_{33}^0 = \frac{16\pi^2 h^2\beta^4}{3\ell^2(1-\sigma^2)}, \quad V_{112}^0 = \beta^2\left[\frac{8}{(1+\beta^2)^2} + 1\right]\frac{\ell}{R},$$

$$V^0_{113} = \frac{8\beta^2}{(1+\beta^2)^2}\frac{\ell}{R}, \quad V^0_{1111} = \frac{3\pi^2(1+\beta^4)}{16}, \quad V^0_{1122} = 4\pi^2\beta^4\left[\frac{1}{(1+\beta^2)^2} + \frac{1}{(9+\beta^2)^2}\right],$$

$$V^0_{1133} = 4\pi^2\beta^4\left[\frac{1}{(1+\beta^2)^2} + \frac{1}{(1+9\beta^2)^2}\right], \quad V^0_{2233} = \frac{16\pi^2\beta^4}{(1+\beta^2)^2}, \quad V^0_{1123} = \frac{4\pi^2\beta^4}{(1+\beta^2)^2},$$

with

$$\beta = \frac{2n\ell}{\pi R}.$$

On differentiating with respect to q_1, q_2, and q_3, and setting the results to zero, we obtain the condition for an entremum for the total energy:

$$(V^0_{11} - 2\Lambda + V^0_{112}q_2 + V^0_{113}q_3 + \frac{1}{6}V^0_{1111}q_1^2 + \frac{1}{2}V^0_{1122}q_2^2 + \frac{1}{2}V^0_{1133}q_3^2 + V^0_{1123}q_2q_3)q_1 = 0,$$

$$(V^0_{22} - 16\Lambda)q_2 + \frac{1}{2}V^0_{112}q_1^2 + \frac{1}{2}V^0_{1122}q_1^2q_2 + \frac{1}{2}V^0_{2233}q_2q_3^2 + \frac{1}{2}V^0_{1123}q_1^2q_3 = 0,$$

$$V^0_{33}q_3 + \frac{1}{2}V^0_{113}q_1^2 + \frac{1}{2}V^0_{1133}q_1^2q_3 + \frac{1}{2}V^0_{2233}q_2^2q_3 + \frac{1}{2}V^0_{1123}q_1^2q_2 = 0. \quad (59.69)$$

In order to solve this system, we treat q_2 as known and solve the equations for the remaining three variables q_1, q_2 and Λ. Of these, the equation giving q_3 is

$$q_3 = -\frac{q_1^2(V^0_{113} + V^0_{1123}q_2)}{2V^0_{33} + V^0_{1133}q_1^2 + V^0_{2233}q_2^2}, \quad (59.70)$$

which has the unexpected feature of being independent of Λ. The first two equations then can be written in terms of just q_1 and Λ, both linear in Λ. These reduce to a cubic equation in q_1^2, which can be solved in closed form. The curve relating the load P to its axial displacement exhibits a relative minimum in the neighborhood of the origin, but not exactly at the origin. This means that the buckling load evaluated by means of nonlinear theory is less than that calculated with the linear theory. These states are called *subcritical* and are very important if we wish to estimate the effective degree of stability of a thin elastic cylinder compressed in a direction parallel to its generators.

60. Asymptotic Analysis of Shells

If we start with a set of equations that we hope will describe a certain kind of shell or emphasize certain particular phenomena in the same class of shell, we are constantly assailed by the doubt as to whether the equations we take give an adequate description of the particular effect we wish to investigate. For example, we might need to know the influence of the thickness, curvature, or large strains in specific problems. Unfortunately, no systematic method of obtaining suitably derived equations exists. If we start from the three-dimensional theory of elasticity, we are exposed to the criticism that in some cases the methods of passing to the limit as the thickness of the shell tends to zero are arbitrary. If, on the other hand, we apply the direct theory, we run the risk of having to state, without justification, that some stress or strain characteristics have more influence than others.

A partial solution to the criticisms raised above is provided by an asymptotic analysis of thin shells, which studies the convergence of the strain energy stored in a three-dimensional elastic sheet, where a parameter ε, proportional to the thickness, is made to tend to zero. Even in this analysis an element of arbitrariness remains, but at least the precise sense of how the three-dimensional solutions converge as ε tends to zero is clear (Ciarlet 1992a).

We consider, for simplicity, the case of an elastic body under small strains, so that we can apply the generalized Hooke law. Let ω be a bounded open connected subset of the two-dimensional Euclidean space with a Lipschitz continuous boundary γ, and let γ_0 be a portion of γ with length $\gamma_0 > 0$. Given a parameter $\varepsilon > 0$, we define the sets $\Omega^\varepsilon = \omega \times (-\varepsilon, \varepsilon)$, $\Gamma_0^\varepsilon = \gamma_0 \times (-\varepsilon, \varepsilon)$. The current points in $\bar{\Omega}^\varepsilon$ are denoted by (x_i^ε) $(i = 1, 2, 3)$, where $x_\alpha^\varepsilon = x_\alpha$ $(\alpha = 1, 2)$.

The middle surface of the shell is the image $S = \boldsymbol{\varphi}(\bar{\omega})$, where $\boldsymbol{\varphi} : \bar{\omega} \to \mathbb{R}^3$ is a mapping of class $\mathscr{C}^3$, which is injective, such that the two vectors $\partial_\alpha \boldsymbol{\varphi}$ are linearly independent at each point of $\bar{\omega}$. The shell is the image $\boldsymbol{\Phi}(\bar{\Omega}^\varepsilon)$ through the mapping $\boldsymbol{\Phi} : \Omega^\varepsilon \to \mathbb{R}^3$ defined by $\boldsymbol{\Phi}(x_1, x_2, x_3^\varepsilon) = \boldsymbol{\varphi}(x_1, x_2) + x_3^\varepsilon \mathbf{a}_3(x_1, x_2)$ for $(x_1, x_2, x_3^\varepsilon) \in \bar{\Omega}^\varepsilon$, where $\mathbf{a}_3$ is a unit vector perpendicular to S. The thickness of the shell is 2ε. We assume that there is an $\varepsilon_0(\boldsymbol{\varphi}) > 0$ such that for all $\varepsilon \leqslant \varepsilon_0(\boldsymbol{\varphi})$ the three vectors $\mathbf{g}_j = \partial_j^\varepsilon \boldsymbol{\Phi}$ are linearly independent and the mapping $\boldsymbol{\Phi} : \bar{\Omega}^\varepsilon \to \mathbb{R}^3$ is a diffeomorphism of class $\mathscr{C}^1$. With the covariant basis $(\mathbf{g}_j)$ we can associate the contravariant basis $(\mathbf{g}^j)$, defined by the relation $\mathbf{g}_i \cdot \mathbf{g}^j = \delta_i^j$. The products $\mathbf{g}_i \cdot \mathbf{g}_j = g_{ij}$ are the covariant components of the metric tensor and the products $\mathbf{g}^i \cdot \mathbf{g}^j = g^{ij}$ are the contravariant components of the metric tensor. The products $\Gamma_{ij}^p = \mathbf{g}^p \cdot \partial_j^\varepsilon \mathbf{g}_i$ represent the Christoffel symbols. Finally, we introduce the notations $g = \det(g_{ij})$ and $A^{ijk\ell} = \lambda g^{ij} g^{k\ell} + \mu \left(g^{ik} g^{j\ell} + g^{i\ell} g^{jk} \right)$.

In a three-dimensional problem of elasticity we wish to determine the three covariant components u_i^ε of the displacement vector $u_i^\varepsilon \mathbf{g}^i$ at each point of the shell. The unknown $\mathbf{u}^\varepsilon = (u_i^\varepsilon)$ is thus the solution of the variational problem:

$$\mathbf{u}^\varepsilon \in V(\Omega^\varepsilon) \equiv \{\mathbf{v} | \mathbf{v} \in H^1(\Omega^\varepsilon),\ \mathbf{v} = 0 \text{ on } \Gamma_0^\varepsilon\}, \tag{60.1}$$

$$\iiint_{\Omega^\varepsilon} A^{ijk\ell} \left[e_{ij}(\mathbf{u}^\varepsilon) - \Gamma_{ij}^p u_p^\varepsilon \right] \left[e_{k\ell}(\mathbf{v}) - \Gamma_{k\ell}^q v_q \right] \sqrt{g}\, dV = \iiint_{\Omega^\varepsilon} \mathbf{f}^\varepsilon \cdot \mathbf{v} \sqrt{g}\, dV \quad \forall \mathbf{v} \in V, \tag{60.2}$$

with $e_{ij}(\mathbf{v}) = \frac{1}{2}\left(\partial_j v_i + \partial_i v_j\right)$ and $\mathbf{f}^\varepsilon = \left(f^{i,\varepsilon}\right)$, where $f^{i,\varepsilon} \in L^2(\Omega^\varepsilon)$ denotes the body force.

It is possible to show that the bilinear form on the left-hand side of (60.2) is *coercive*, in the sense that there is a constant $\alpha = \alpha(\boldsymbol{\varphi}, \varepsilon)$ such that we have

$$\alpha > 0 \text{ and } \iiint_{\Omega^\varepsilon} A^{ijk\ell} \left[e_{ij}(\mathbf{v}) - \Gamma_{ij}^p v_p \right] \left[e_{k\ell}(\mathbf{v}) - \Gamma_{k\ell}^q v_q \right] \sqrt{g}\, dV > \alpha \|\mathbf{v}\|_{H^1(\Omega^\varepsilon)}^2, \tag{60.3}$$

for any $\mathbf{v} \in V$.

We now wish to study the limiting case in which the shell becomes a plate. For this purpose we introduce the family of functions representing all transformations from shell to plate:

$$\varphi_\alpha(t)(x_1, x_2) = (1-t)x_\alpha + t\varphi_\alpha(x_1, x_2), \quad \varphi_3(t)(x_1, x_2) = t\varphi_3(x_1, x_2), \tag{60.4}$$

for $0 \leqslant t \leqslant 1$, so that $\varphi(1) = \varphi$ corresponds to the given shell and $\varphi(0)$ represents a flat plate of middle surface $\bar{\omega}$.

Let us assume, for simplicity, that the functions $f^{i,\varepsilon}$ do not vary with the parameter t. For $t \in [0, 1]$, let $\mathbf{u}^\varepsilon(t) \in V$ be the solution of the variational equation (60.2) corresponding to the function $\varphi(t)$. For each $t \in [0, 1]$ the existence of a solution of (60.2) is a simple consequence of the Lax–Milgram lemma. In fact, we can write (60.2) in the form

$$a(t)(\mathbf{u}^\varepsilon(t), \mathbf{v}) = L(t)(\mathbf{v}) \quad \forall\, \mathbf{v} \in V, \tag{60.5}$$

and observe that there are two constants $\alpha(t)$ and $M(t)$ such that we have

$$a(t)(\mathbf{v}, \mathbf{v}) > \alpha(t)\|\mathbf{v}\|^2_{H^1}, \tag{60.6}$$

$$|a(t)(\mathbf{u}, \mathbf{v})| < M(t)\|\mathbf{u}\|_{H^1}\|\mathbf{v}\|_{H^1}. \tag{60.7}$$

On utilizing expression (60.2) for the bilinear form $a(t)(\mathbf{u}, \mathbf{v})$, we also find that, for each $t_0 \in [0, 1]$, there is a function $\delta : [0, +\infty] \to \mathbb{R}$ (depending on t_0) such that

$$\delta(\tau) \to 0 \text{ for } \tau \to 0^+ \text{ and } |a(t)(\mathbf{u}, \mathbf{v}) - a(t_0)(\mathbf{u}, \mathbf{v})| \leqslant \delta(|t - t_0|)\|\mathbf{u}\|_{H^1}\|\mathbf{v}\|_{H^1} \tag{60.8}$$

for any $\mathbf{u}, \mathbf{v} \in V$. This means that the form $a(t)(u, v)$ is continuous with respect to t, and, as a consequence, the solutions $\mathbf{u}^\varepsilon(t)$ of (60.5) also vary continuously with respect to t. Another consequence of (60.8) is that, as the vectors $\mathbf{g}^i$ are themselves regular functions of t, the displacement $u_i^\varepsilon \mathbf{g}^i$ is a continuous function of t.

For $t \to 0^+$, we have $\mathbf{u}^\varepsilon(t) \to \mathbf{u}^\varepsilon(0)$ in $H^1(\Omega^\varepsilon)$. We can then prove that the shell becomes a plate and $\mathbf{u}^\varepsilon(0)$ solve the three dimensional variational equation for a clamped plate

$$\iiint\limits_{\Omega^\varepsilon} B^{ijk\ell} e_{ij}[\mathbf{u}^\varepsilon(0)] e_{k\ell}(\mathbf{v})\, dV = \iiint\limits_{\Omega^\varepsilon} \mathbf{f}^\varepsilon \cdot \mathbf{v}\, dV, \tag{60.9}$$

for all $\mathbf{v} \in V$, with $B^{ijk\ell} = \lambda\delta^{ij}\delta^{k\ell} + \mu(\delta^{ij}\delta^{j\ell} + \delta^{i\ell}\delta^{jk})$. Let us take $\Omega = \omega \times (-1, 1)$ and let $\mathbf{x} = (x_1, x_2, (1/\varepsilon)x_3^\varepsilon) \in \bar{\Omega}$ be associated with $\mathbf{x}^\varepsilon = (x_1, x_2, x_3^\varepsilon) \in \bar{\Omega}^\varepsilon$. If $\mathbf{u}^\varepsilon = \mathbf{u}^\varepsilon(0)$ denotes the solution of the above plate problem, we define the scaled displacement $\mathbf{u}(\varepsilon) = (u_i(\varepsilon))$ by

$$u_\alpha^\varepsilon(\mathbf{x}^\varepsilon) = \varepsilon u_\alpha(\varepsilon)(\mathbf{x}), \quad u_3^\varepsilon(\mathbf{x}) = u_3(\varepsilon)(\mathbf{x}), \tag{60.10}$$

for all $\mathbf{x}^\varepsilon \in \bar{\Omega}^\varepsilon$, and assume that there are functions $f^i \in L^2(\Omega)$ independent of ε given by

$$f^{\alpha,\varepsilon}(\mathbf{x}^\varepsilon) = \varepsilon f^\alpha(\mathbf{x}) \quad \text{and} \quad f^{3,\varepsilon}(\mathbf{x}^\varepsilon) = \varepsilon^2 f^3(\mathbf{x}). \tag{60.11}$$

The scaled displacement $\mathbf{u}(\varepsilon)$ then converges in $H^1(\Omega)$ for $\varepsilon \to 0$, and the limit is of the form $(\zeta_1 - x_3\partial_1\zeta_3, \zeta_2 - x_3\partial_2\zeta_3, \zeta_3)$ where the functions ζ_i are independent of x_3. Furthermore, the field $(\zeta_\alpha^\varepsilon, \zeta_3^\varepsilon) \equiv (\varepsilon\zeta_\alpha, \zeta_3)$ solves the two-dimensional variational equation

$$(\zeta_\alpha^\varepsilon, \zeta_3^\varepsilon) \in V(\omega) \equiv \left\{ \eta_\alpha \in H^1(\omega), \quad \eta_3 \in H^2(\omega), \quad \eta_i = \frac{\partial \eta_3}{\partial \nu} = 0 \text{ on } \gamma_0 \right\},$$

$$\iint_\omega \left\{ \frac{\varepsilon^3}{3} b^{\alpha\beta\rho\sigma} \partial_{\alpha\beta} \zeta_3^\varepsilon \partial_{\rho\sigma} \eta_3 + \varepsilon b^{\alpha\beta\rho\sigma} e_{\alpha\beta}(\zeta_\gamma^\varepsilon) e_{\rho\sigma}(\eta_\gamma) \right\} d\omega$$

$$= \iint_\omega \left\{ \left(\int_{-\varepsilon}^{\varepsilon} f^{3,\varepsilon} \, dx_3^\varepsilon \right) \eta_3 - \left(\int_{-\varepsilon}^{\varepsilon} x_3^\varepsilon f^{3,\varepsilon} \, dx_3^\varepsilon \right) \partial_\alpha \eta_3 + \left(\int_{-\varepsilon}^{\varepsilon} f^{\alpha,\varepsilon} \, dx_3^\varepsilon \right) \eta_\alpha \right\} d\omega, \quad (60.12)$$

for all $(\eta_\alpha, \eta_3) \in V(\omega)$ with

$$b^{\alpha\beta\rho\sigma} = \frac{4\lambda\mu}{\lambda + 2\mu} \delta^{\alpha\beta} \delta^{\rho\sigma} + 2\mu(\delta^{\alpha\rho}\delta^{\beta\sigma} + \delta^{\alpha\sigma}\delta^{\beta\rho}),$$

and

$$e_{\alpha\beta}(\eta_\gamma, \eta_3) = \frac{1}{2}(\partial_\alpha \eta_\beta + \partial_\beta \eta_\alpha).$$

From (60.12) we see that, as a consequence of assumption (60.11) on the body forces, the effects of flexure and stretching appear simultaneously in the variational formulation.

In the procedure described above we have made two successive passages to the limit: first, we have studied the problem for $t \to 0$; and, second, that for $\varepsilon \to 0$. A nontrivial question is whether the result is the same when we interchange the order of the two limits. The answer is that it is not (Ciarlet 1992b).

Let ω be any open two-dimensional set, and let $S = \varphi(\omega)$ be the equation of the middle surface of a shell. The unknowns are the three covariant components $\zeta_i^\varepsilon : \bar{\omega} \to \mathbb{R}$ of the displacement $\mathbf{a}^i \zeta_i^\varepsilon$ of the points of S. We also introduce the functions

$$\gamma_{\alpha\beta}(\eta_i) = \frac{1}{2}(\partial_\alpha \eta_\beta + \partial_\beta \eta_\alpha) - \Gamma^\rho_{\alpha\beta} \eta_\rho - b_{\alpha\beta} \eta_3, \quad (60.13)$$

$$\kappa_{\alpha\beta}(\eta_i) = \partial_{\alpha\beta} \eta_3 - \Gamma^\rho_{\alpha\beta} \partial_\rho \eta_3 - c_{\alpha\beta} \eta_3 + b^\rho_\beta(\partial_\alpha \eta_\rho - \Gamma^\sigma_{\alpha\rho} \eta_\sigma)$$
$$+ b^\rho_\alpha(\partial_\beta \eta_\rho - \Gamma^\sigma_{\rho\beta} \eta_\sigma) + (\partial_\alpha b^\rho_\beta + \Gamma^\rho_{\alpha\sigma} b^\sigma_\beta - \Gamma^\sigma_{\alpha\beta} b^\rho_\sigma) \eta_\rho, \quad (60.14)$$

where $b_{\alpha\beta} = -\mathbf{a}_\alpha \cdot \partial_\beta \mathbf{a}_3$ and $c_{\alpha\beta} = b^\rho_\alpha b_{\rho\beta}$ ($b^\rho_\beta = a^{\rho\alpha} b_{\alpha\beta}$) denote the coefficients of the second and third fundamental form of S. The quantities $\gamma_{\alpha\beta}$ and $\kappa_{\alpha\beta}$ represent, respectively, the covariant components of the linearized strain tensor and of the linearized change in curvature associated with an arbitrary displacement field $\eta_i \mathbf{a}^i$ of the points of S.

In order to study the behavior of the solution $\mathbf{u}^\varepsilon$ of the three-dimensional shell problem as $\varepsilon \to 0$, we put $\Omega = \omega \times (-1, 1)$ and define the scaled unknown $\mathbf{u}(\varepsilon) : \bar{\Omega} \to \mathbb{R}^3$ by letting the components take the form

$$u_i^\varepsilon(\mathbf{x}^\varepsilon) = u_i(\varepsilon)(\mathbf{x}), \quad (60.15)$$

where $\mathbf{x} = (x_1, x_2, (1/\varepsilon)x_3^\varepsilon) \in \bar{\Omega}$ is associated with $\mathbf{x}^\varepsilon = (x_1, x_2, x_3^\varepsilon) \in \bar{\Omega}^\varepsilon$, and assume that there are functions $f^i \in L^2(\Omega)$, independent of ε such that we can write

$$f^{i,\varepsilon}(\mathbf{x}^\varepsilon) = \varepsilon^2 f^i(\mathbf{x}). \quad (60.16)$$

We now introduce the space of *inextensional displacements* (Sanchez-Palencia 1990)

$$V_0(\omega) \equiv \Big\{\eta = (\eta_i)|\eta_\alpha \in H^1(w),\ \eta_3 \in H^2(\omega),\ \eta_i = \frac{\partial \eta_3}{\partial \nu} = 0 \text{ on } \gamma_0,$$

$$\gamma_{\alpha\beta}(\eta) = 0 \text{ in } \omega\Big\}. \tag{60.17}$$

Now, for $V_0(\omega) \neq \{0\}$, the scaled unknown $\mathbf{u}(\varepsilon)$ admits an asymptotic expansion of the form

$$\mathbf{u}(\varepsilon) = \boldsymbol{\zeta} + \varepsilon \mathbf{u}^1 + \varepsilon^2 \mathbf{u}^2 + \dots. \tag{60.18}$$

Where the first term $\boldsymbol{\zeta}$ is independent of x_3, and the field $\boldsymbol{\zeta}^\varepsilon \equiv \boldsymbol{\zeta}$ solves the two-dimensional variation problem

$$\boldsymbol{\zeta}^\varepsilon \in V_0(\omega) \tag{60.19}$$

$$\iint_\omega \frac{\varepsilon^3}{3} a^{\alpha\beta\rho\sigma} \kappa_{\alpha\beta}(\boldsymbol{\zeta}^\varepsilon) \kappa_{\rho\sigma}(\eta) \sqrt{a}\, d\omega = \iint_\omega \left\{ \int_{-\varepsilon}^{+\varepsilon} \mathbf{f}^\varepsilon\, dx_3^\varepsilon \right\} \cdot \eta \sqrt{a}\, d\omega, \tag{60.20}$$

with $a^{\alpha\beta\rho\sigma} = (4\lambda\mu/\lambda + 2\mu) a^{\alpha\beta} a^{\rho\sigma} + 2\mu\big(a^{\alpha\rho} a^{\beta\sigma} + a^{\alpha\sigma} a^{\beta\rho}\big)$. Now for $V_0(\omega) = \{0\}$, the scaled unknown $\mathbf{u}(\varepsilon)$ admits the formal asymptotic expansion

$$\mathbf{u}(\varepsilon) = \varepsilon^2 \boldsymbol{\zeta} + \varepsilon^3 \mathbf{u}^3 + \varepsilon^4 \mathbf{u}^4 + \dots, \tag{60.21}$$

where the first term $\varepsilon^2 \boldsymbol{\zeta}$ is independent of x_3, and the field $\boldsymbol{\zeta}^\varepsilon \equiv \varepsilon^2 \boldsymbol{\zeta}$ solves the two-dimensional variation problem

$$\boldsymbol{\zeta}^\varepsilon \in V_1(\omega) \equiv \{\eta = (\eta_i)|\eta_\alpha \in H^1(\omega),\ \eta_3 \in L^2(\omega),\ \eta_\alpha = 0 \text{ on } \gamma_0\} \tag{60.22}$$

$$\iint_\omega \varepsilon a^{\alpha\beta\rho\sigma} \gamma_{\alpha\beta}(\boldsymbol{\zeta}^\varepsilon) \gamma_{\rho\sigma}(\eta) \sqrt{a}\, d\omega = \iint_\omega \left\{ \int_{-\varepsilon}^{\varepsilon} \mathbf{f}^\varepsilon\, dx_3^\varepsilon \right\} \cdot \eta \sqrt{a}\, d\omega, \tag{60.23}$$

for each $\eta \in V_1(\omega)$.

We thus obtain a limiting problem of pure bending for $V_0(\omega) \neq \{0\}$ or a pure membrane problem for $V_0(\omega) = \{0\}$. The reduction of the space $V_0(\omega)$ to the null element requires purely kinematic considerations, as it depends only on the geometry of the middle surface S and on the boundary conditions. An example of a case where $V_0(\omega) \neq \{0\}$ is that of a cylindrical surface clamped along one or two generators; an example of a case where $V_0(\omega) = \{0\}$ is that of a spherical cap clamped along its boundary ($\gamma_0 = \gamma$).

Let us now investigate the behavior when the form of the shell tends toward that of a plate. The midplane of the plate can be defined as the limit, for $t \to 0^+$, of a family of spherical caps, completely clamped along their common boundary, so that we have $V_0(\omega) = \{0\}$ for each $t > 0$. However, for $t = 0$, the definition (60.7) of $V_0(\omega)$ is modified to become

$$V_0(\omega) \equiv \{\eta = (0, 0, \eta_3)|\eta_3 \in H_0^2(\omega)\} \neq \{0\}. \tag{60.24}$$

In other words, a purely flexural state, such as that pertaining in a plane plate, may be reached in the limit for $t \to 0$ from thin shells governed by the equations associated with a membrane state.

Of course, there are alternative ways of defining the displacement field at the middle surface S. Let us, for example, define the space $V(\omega)$ as (Ciarlet and Lods 1995)

$$V(\omega) \equiv \left\{ \eta = (\eta_i) | \eta_\alpha \in H^1(\omega),\ \eta_3 \in H^2(\omega),\ \eta_i = \frac{\partial \eta_3}{\partial \nu} = 0 \text{ on } \gamma_0 \right\}, \tag{60.25}$$

and let ζ^ε be the solution of the variational problem

$$\iint_\omega \left\{ \frac{1}{3} \varepsilon^3 a^{\alpha\beta\rho\sigma} \kappa_{\alpha\beta}(\zeta^\varepsilon) \kappa_{\rho\sigma}(\eta) + \varepsilon a^{\alpha\beta\rho\sigma} \gamma_{\alpha\beta}(\zeta^\varepsilon) \gamma_{\rho\sigma}(\eta) \right\} \sqrt{a}\, d\omega$$

$$= \iint_\omega \mathbf{q}^\varepsilon \cdot \eta \sqrt{a}\, d\omega \quad \forall \eta \in \surd(\omega), \tag{60.26}$$

where $\mathbf{q}^\varepsilon \in L^2(\omega)$ represent the resultant on the surface S of the contravariant components of the external forces. Equation (60.26) is the mathematical formulation of a model proposed by Koiter (1966, 1970). However, the terms on the left-hand side of (60.26) are exactly the sum of the bending-dominated and membrane-dominated shell equations, and these two limiting problems correspond to two different, mutually exclusive, situations. Thus (60.26) cannot be viewed as a limit model as $\varepsilon \to 0$, but may be the correct model for moderately small value of ε.

Still following the same line of argument, a further question arising in the asymptotic analysis of shell theories is that of justifying mathematically the classical two-dimensional linear equations for shallow shells (Ciarlet and Miara 1992). More precisely, we can prove that the solution of these two-dimensional equations is the H^1 limit of the solution of the equations of the linearized three-dimensional elasticity when the thickness of the shell approaches zero. The importance of the result is that the procedure provides a rigorous criterion for deciding whether or not a shell is "shallow." In fact, the measure of shallowness is given by the mapping $\varphi(\omega) : \bar{\omega} \to \mathbb{R}$, which represents the deviation of the middle surface of the shell from a plane. Up to an additive constant, this deviation must be of the order of the thickness of the shell in order that the latter may be defined as *shallow*.

61. The Influence of Constitutive Nonlinearities

In the nonlinear theory of elastic shells it is necessary to formulate constitutive equations sufficiently general to describe the physics and sufficiently simple to ensure the success of the mathematical treatment. One of the most powerful tools for simplifying the constitutive equations is that of material symmetry, because the materials that are usually employed in the construction of shells are mechanically similar in the sense that, if we know how one part responds to all loading, we are able to infer how other parts respond. In other words, we have a way of transforming the constitutive equations governing one part to obtain those governing others.

Properties of this type are important mathematically because they permit a reduction of the constitutive equations, and are also useful in experimental mechanics because, from a limited number of experiments on a part, we are able to infer the outcome of all experiments on that part. The notion of material symmetry and exploiting its consequences is well defined in a three-dimensional continuum, and, in particular, in finite elasticity, but its extension to shell theories is not so clear because there is no generally accepted agreement about allowable transformations, as pointed out by Ericksen (1972).

An example of how a certain group of reasonable properties of symmetry can simplify the constitutive equations of elastic shells is found in the theory of hyperelastic shells, the strain energy density of which, defined on a deformable surface S, has the form

$$W = W[\mathbf{d}; \mathbf{d}_{,\alpha}; \mathbf{r}_{,\beta}; u^{\alpha}], \tag{61.1}$$

where $\mathbf{r} = \mathbf{r}(u^{\alpha}, t)$ is the position vector, u^{α} are material coordinates labeling the particles, and $\mathbf{d} = \mathbf{d}(u^{\alpha}, t)$ is a vector field defined on the surface but not tangential to it. An elastic surface characterized by a strain energy of the form (61.1) is called a *Cosserat surface*.

For theories of shells of the type considered here, we first require the *Galilean invariance*, which states that one local state $(\mathbf{r}_{,\alpha}, \mathbf{d}, \mathbf{d}_{,\alpha})$ is mechanically equivalent to another $[\bar{\mathbf{r}}_{,\alpha}, \bar{\mathbf{d}}, \bar{\mathbf{d}}_{,\alpha}]$ wherever there is a proper orthogonal transformation

$$\mathbf{Q}^{-1} = \mathbf{Q}^{\mathrm{T}}, \quad \det \mathbf{Q} = 1, \tag{61.2}$$

such that, on choosing the quantities

$$\bar{\mathbf{r}}_{,\alpha} = \mathbf{Q}\mathbf{r}_{,\alpha}, \quad \bar{\mathbf{d}} = \mathbf{Q}\mathbf{d}, \quad \bar{\mathbf{d}}_{,\alpha} = \mathbf{Q}\mathbf{d}_{,\alpha}, \tag{61.3}$$

we have

$$W(\bar{\mathbf{r}}_{,\alpha}; \bar{\mathbf{d}}; \bar{\mathbf{d}}_{,\alpha}; u^{\alpha}) = W(\mathbf{r}_{,\alpha}; \mathbf{d}; \mathbf{d}_{,\alpha}; u^{\alpha}). \tag{61.4}$$

Condition (61.4) holds for all states related by (61.3). If, however, the material enjoys same special material symmetry, we expect that two states are equivalent under transformations other than (61.3). These transformations should obviously be compatible with the equivalence established by Galilean invariance, but could alternatively be very general. In practice we introduce the somewhat artificial assumption that such transformations are linear and invertible. For simplicity, let us denote the local state by

$$(\mathbf{V}_1, \ldots, \mathbf{V}_5) = (\mathbf{r}_{,1}; \mathbf{r}_{,2}; \mathbf{d}; \mathbf{d}_{,1}; \mathbf{d}_{,2}) \tag{61.5}$$

and represent the linear transformation by

$$\hat{\mathbf{V}}_M = \mathbf{L}_M^N \mathbf{V}_N, \tag{61.6}$$

where $\mathbf{L}_1^1, \mathbf{L}_2^1, \ldots,$ are 3×3 matrices. As $\mathbf{Q}$ satisfies (61.2) the transformation $\mathbf{L}_M^N \mathbf{Q} \mathbf{V}_N$ should generate a set of vectors that is equivalent to $\hat{\mathbf{V}}_M$ in the sense of Galilean invariance. This leads to the requirement that for each $\mathbf{Q}$ there is an orthogonal transformation $\bar{\mathbf{Q}}$ such that

$$\mathbf{L}_M^N \mathbf{Q} = \bar{\mathbf{Q}} \mathbf{L}_M^N. \tag{61.7}$$

This implies that we can find scalars α_M^N and a proper orthogonal matrix $\bar{\mathbf{Q}}$ such that

$$\mathbf{L}_M^N = \alpha_M^N \hat{\mathbf{Q}}. \tag{61.8}$$

On reverting to our initial, less compact, notation, (61.6) and (61.8) take the forms

$$\begin{aligned} \hat{\mathbf{r}}_{,\alpha} &= \hat{\mathbf{Q}}(a_\alpha^\beta \mathbf{r}_{,\beta} + b_\alpha^\beta \mathbf{d}_{,\beta} + b_\alpha \mathbf{d}), \\ \hat{\mathbf{d}}_{,\alpha} &= \hat{\mathbf{Q}}(c_\alpha^\beta \mathbf{r}_{,\beta} + e_\alpha^\beta \mathbf{d}_{,\beta} + e_\alpha \mathbf{d}), \\ \hat{\mathbf{d}} &= \hat{\mathbf{Q}}(f^\alpha \mathbf{r}_{,\alpha} + g^\alpha \mathbf{d}_{,\alpha} + g\mathbf{d}), \end{aligned} \tag{61.9}$$

where a_α^β, $b_\alpha^\beta, \ldots$, are scalars restricted by the condition that the transformation should be invertible.

So far we have not exploited the property that the shell is a three-dimensional body with one dimension, the thickness, small with respect to the others. The thin dimension is the distance between the upper and lower faces, denoted by S_1 and S_2, respectively. It is commonly assumed that there is some rule by which we can set up a one-to-one correspondence between particles on S_1 and those on S_2, assigning the same surface coordinates to corresponding particles. In addition, the vector $\mathbf{d}$ is usually taken to denote the relative position vector of corresponding particles on S_1 and S_2, that is

$$\mathbf{d} = \mathbf{r}_1(u^\alpha) - \mathbf{r}_2(u^\alpha). \tag{61.10}$$

The position of the middle surface S can also be written as

$$\mathbf{r}(u^\alpha) = [1 - \lambda_1(u^\alpha)]\mathbf{r}_1(u^\alpha) + \lambda_1(u^\alpha)\mathbf{r}_2(u^\alpha), \tag{61.11}$$

with $0 < \lambda_1 < 1$. If S is effectively placed midway between S_1 and S_2 we can then take $\lambda_1 = \frac{1}{2} =$ constant. Furthermore, the position of any point of the shell can be defined by an equation of the form

$$\hat{\mathbf{r}} = \mathbf{r}(u^\alpha) + \lambda \mathbf{d}(u^\alpha), \tag{61.12}$$

with $(\lambda_1 - 1) \leqslant \lambda < \lambda_1$. If we require the mapping (61.12) to be locally invertible, we compute the Jacobian

$$\Delta(\lambda) = [\mathbf{r}_{,1} + \lambda \mathbf{d}_{,1}] \times [\mathbf{r}_{,2} + \lambda \mathbf{d}_{,2}] \cdot \mathbf{d}, \tag{61.13}$$

and impose the condition that $\Delta(\lambda)$ never vanishes. This condition characterizes the admissible states for shells and allows a further simplification of relations (61.9). More precisely, it implies the conditions

$$f^\alpha = g^\alpha = 0, \tag{61.14}$$

and that the other coefficients are expressed by

$$\begin{aligned} a_\beta^\alpha &= aH_\beta^\alpha, \quad b_\beta^\alpha = bH_\beta^\alpha, \\ c_\beta^\alpha &= cH_\beta^\alpha, \quad e_\beta^\alpha = eH_\beta^\alpha, \end{aligned} \tag{61.15}$$

where a, b, c, and e are new scalars and the H^{α}_{β} are other coefficients. The transformation (61.9) is nonsingular provided that we have

$$H(ae - bc)g \neq 0,$$
$$H = \det(H^{\alpha}_{\beta}). \tag{61.16}$$

On using this necessary condition we arrive at the result that the admissible transformation for shells must be of the form

$$\hat{\mathbf{r}}_{,\alpha} = \hat{\mathbf{Q}}H^{\beta}_{\alpha}(a\mathbf{r}_{,\beta} + b\mathbf{d}_{,\beta} + h_{\beta}\mathbf{d}),$$
$$\hat{\mathbf{d}}_{,\alpha} = \hat{\mathbf{Q}}H^{\beta}_{\alpha}(c\mathbf{r}_{,\beta} + e\mathbf{d}_{,\beta} + k_{\beta}\mathbf{d}), \tag{61.17}$$
$$\hat{d} = \hat{\mathbf{Q}}g\mathbf{d},$$

where we have put

$$b_{\alpha} = H^{\beta}_{\alpha}h_{\beta}, \quad e_{\alpha} = H^{\beta}_{\beta}k_{\beta}. \tag{61.18}$$

Once we have found an explicit form for the admissible transformations between the kinematical variables, a second step is to find the transformations induced by (61.17) in the strain energy. In order to maintain consistency with (61.4), we must assume that two transformations of the type (61.8), with the same coefficients α but different $\hat{\mathbf{Q}}$, yield the same form for W. Again, assuming a linear law, we are led to a corresponding form $\hat{W}$:

$$\hat{W} = \beta W, \tag{61.19}$$

where β is some scalar function of the α terms:

$$\beta = \beta(\alpha^{N}_{M}) \neq 0. \tag{61.20}$$

Formula (61.19) is valid in general; that is, it is valid without reverting to the reduced form (61.17). If, however, (61.17) are employed, it is possible to show that a necessary restriction on β is

$$\beta = |H|\ |g|^{q} = 1, \tag{61.21}$$

with $H \equiv \det(H^{\alpha}_{\beta})$ and q a constant. Of course, we can satisfy this relation, independently of q, by requiring the conditions

$$|H| = |g| = 1, \tag{61.22}$$

although for each q there are other transformations satisfying (61.21).

Finding the explicit form of the constitutive equations in highly deformable shells is a very difficult problem because these equations must be derived after having removed all the traditional assumptions about the structure of the deformation. In a first stage, we must abandon Kirchhoff's hypothesis; that is, that the normals to the middle surface and the distances of the points located on the normals from the middle surface maintain their directions and values, respectively. In theories based on moderate strains these assumptions must be replaced by others; for example, by introducing an averaged transverse shearing strain or an averaged change in thickness. However, both these corrections involve averaged quantities that are not incorporated in the detailed constitutive relations of the shell. Under large strains, additional effects occur and, again, treatment of these would require a yet more refined analysis. For example, under severe bending, the position of the original middle surface could undergo a large shift. As a result, this surface would move further away from being a geometrical midsurface, because points on the side under tension would tend to move toward it, whereas points under compression would tend to move away from the former mean surface. It is thus necessary to re-evaluate some of the simplifications that are commonly made in deriving the constitutive laws in terms of characteristics. In contrast to what happens in the linear case, in the non-linear materials the constitutive equations are mixed, in the sense that each stress or couple resultant depends on all the deformation variables.

The result is well illustrated by the study of the cylindrical deformation of a shell; that is, a deformation which changes a cylindrical shell into another cylindrical shell such that the deformations in the directions of the generators consist of a uniform extension (Libai and Simmonds 1981). The material is of Mooney type, with strain energy per unit undeformed volume given by

$$W = C_1(I_1 - 3) + C_2(I_2 - 3), \tag{61.23}$$

where I_1 and I_2 are the first and second invariants of the strain tensor, and C_1 and C_2 are material constants.

We first consider the deformation of a very thin cylindrical membrane of uniform thickness into another cylindrical membrane. In the undeformed configuration the position of a point is defined by a vector $\mathbf{r} \equiv (x, s, \zeta)$, where x is the distance along the generators, s is the distance along curves orthogonal to the generators, and ζ is the distance along the normals to the middle surface. The deformation satisfies the condition of plane strain such that none of the strain components are functions of x.

Under these conditions the coordinates x, s, ζ are also principal directions of strain in the deformed membrane. If λ_x, λ_s, and λ_ζ are the principal components of stretch, the linear elements dS and ds in the initial and deformed configurations are

$$(dS)^2 = dx^2 + ds^2 + d\zeta^2, \tag{61.24}$$

$$(ds)^2 = \lambda_x^2\, dx^2 + \lambda_s^2\, ds^2 + \lambda_\zeta^2\, d\zeta^2. \tag{61.25}$$

The condition of incompressibility requires that

$$\lambda_\zeta = (\lambda_x \lambda_s)^{-1}, \tag{61.26}$$

where λ_x is a constant because the state of strain is planar.

The strain energy can be written as

$$W = W[\lambda_s, \lambda_x, (\lambda_x, \lambda_s)^{-1}], \tag{61.27}$$

and the normal stresses in the deformed s direction, per unit undeformed area, are given by

$$S_{ss} = \frac{\partial W}{\partial \lambda_s}. \tag{61.28}$$

For a Mooney material, in the absence of motion in the x direction ($\lambda_x = 1$), (61.27) becomes

$$W = \frac{1}{2}C(\lambda_s^2 + \lambda_s^{-2}) - C, \tag{61.29}$$

where we have put $C = 2(C_1 + C_2)$.

The formula derived for a membrane permits the passage to the formula for a shell by considering a solid created by the superposition of thin layers in the direction ζ of the thickness. If κ and κ_1 are the curvatures of the undeformed and deformed middle surfaces, and $\lambda = \lambda_s$ ($\zeta = 0$) is the extension of the middle surface, then the linear element of any layer of the undeformed shell is given approximately by

$$dS = dS_0(1 - \kappa\zeta), \tag{61.30}$$

where dS_0 is the linear element on the middle surface and ζ is the distance of the layer from the middle surface. An analogous result holds for the linear element of any layer of the deformed shell

$$ds = ds_0(1 - \kappa_1\zeta_1), \tag{61.31}$$

leading to

$$\lambda_s(1 - \kappa\zeta) = \lambda(1 - \kappa_1\zeta_1), \tag{61.32}$$

with

$$\zeta_1 = \int_0^\zeta \lambda_\zeta \, d\zeta. \tag{61.33}$$

On differentiating (61.32) with respect to ζ, and using (61.26), we obtain

$$(1 - \kappa\zeta)\frac{\partial \lambda_s}{\partial \zeta} = \kappa(\lambda_s - \mu\lambda_s^{-1}), \tag{61.34}$$

where we have put

$$\mu = \frac{\lambda\kappa_1}{\lambda_x\kappa}. \tag{61.35}$$

The solution of the differential equation (61.34) is

$$\lambda_s = [\mu + (\lambda^2 - \mu)(1 - \kappa\zeta)^{-2}]^{\frac{1}{2}}. \tag{61.36}$$

This equation can be used to calculate the deformed coordinate ζ_1 and the thickness h_1 as functions of the undeformed thickness h, λ, and μ:

$$\zeta_1 = \int_0^\zeta (\lambda_x\lambda_s)^{-1} \, d\zeta = (\kappa\mu\lambda_x)^{-1}\left\{\lambda - [\lambda^2 - \mu + \mu(1 - \kappa\zeta)^2]^{\frac{1}{2}}\right\}, \tag{61.37}$$

$$h_1 = (\zeta_1)_{-h/2}^{h/2} = (\kappa\mu\lambda_x)^{-1}(H_1 - H_2), \tag{61.38}$$

with

$$H_{1,2} = [(\lambda^2 - \mu) + \mu(1 \pm \kappa h)^2]^{\frac{1}{2}}. \tag{61.39}$$

The distance d between the deformed and undeformed midsurfaces is given by

$$d = \frac{1}{2}\left[\zeta_1\left(\frac{h}{2}\right) + \zeta_1\left(-\frac{h}{2}\right)\right] = \frac{1}{2}(\kappa\mu\lambda_x)^{-1}(2\lambda - H_1 - H_2). \qquad (61.40)$$

For sufficiently small (κh) these quantities can be developed as power series in (κh) as follows:

$$h_1 = h(\lambda\lambda_x)^{-1}\left[1 + \frac{1}{8}\frac{\mu}{\lambda^2}\left(\frac{\mu}{\lambda^2} - 1\right)(\kappa h)^2 + O(\kappa^4 h^4)\right], \qquad (61.41)$$

$$d = \frac{1}{8}(\lambda\lambda_x)^{-1}\kappa h^2\left(\frac{\mu}{\lambda^2} - 1\right) + O(\kappa^3 h^4). \qquad (61.42)$$

Formula (61.28) gives the s component of the Piola–Kirchhoff stress; the corresponding component of the true stress, per unit deformed area, is

$$\sigma_{ss} = (\lambda_s\lambda_\zeta)^{-1} S_{ss} = \lambda_s \frac{\partial W}{\partial \lambda_s}. \qquad (61.43)$$

We now utilize the strain energy density W to find the expression for the strain energy φ per unit undeformed area of the reference surface. The force resultant N and the bending couple M, per unit undeformed width in the x direction, can be obtained by integration over the thickness. In order to discuss the result in detail we restrict the developments to the Mooney material, with the further simplification $\lambda_x = 1$. On using (61.29) and the expression (see (61.28))

$$S_{ss} = C(\lambda_s - \lambda_s^{-3}), \qquad (61.44)$$

we can calculate

$$\begin{aligned}\varphi &= \int_{-\frac{1}{2}h}^{\frac{1}{2}h} W(1 - \kappa\zeta)\,d\zeta \\ &= \frac{1}{2}Ch\left\{(\mu + \mu^{-1} - 2) + (\lambda^2 - \mu)(\kappa h)^{-1}\left[\ell n\left(\frac{1 + \frac{1}{2}\kappa h}{1 - \frac{1}{2}\kappa h}\right) - \frac{1}{\mu^2}\,\ell n\left(\frac{H_1}{H_2}\right)\right]\right\} \\ &= \frac{1}{2}Ch\left[(\lambda^2 + \lambda^{-2} - 2) + \frac{1}{12}(\kappa h)^2(1 - \mu\lambda^{-2})(\lambda^2 + 3\lambda^{-2} - 4\mu\lambda^{-4}) + O(\kappa^4 h^4)\right],\end{aligned} \qquad (61.45)$$

$$\begin{aligned}N &= \int_{-\frac{1}{2}h}^{\frac{1}{2}h} \frac{\partial W}{\partial \lambda_s}\,d\zeta \\ &= C\left\{h\mu^{-1}\left(\mu^2 - 1 + \frac{\lambda^2 - \mu}{H_1 H_2}\right) + \kappa^{-1}\sqrt{\lambda^2 - \mu}\,\ell n\left[\frac{\left(H_2 + \sqrt{\lambda^2 - \mu}\right)(1 + \frac{1}{2}\kappa h)}{\left(H_1 + \sqrt{\lambda^2 - \mu}\right)(1 - \frac{1}{2}\kappa h)}\right]\right\} \\ &= Ch\left\{(\lambda - \lambda^{-3}) + \frac{1}{24}(\kappa h)^2(1 - \mu\lambda^{-2})(2\lambda - 6\lambda^{-3} + \mu\lambda^{-1} + 15\mu\lambda^{-5}) + O(\kappa^4 h^4)\right\},\end{aligned} \qquad (61.46)$$

$$M = \int_{-\frac{1}{2}h}^{\frac{1}{2}h} \frac{\partial W}{\partial \lambda_s} \zeta \, d\zeta$$

$$= \frac{\lambda N_s}{\kappa\mu} + \frac{C}{\kappa^2 \mu}\left\{(\mu^{-1} - \mu)\kappa h - (\lambda^2 - \mu)\left[\mu^{-2}\ell\text{n}\left(\frac{H_1}{H_2}\right) + \ell\text{n}\left(\frac{1+\frac{1}{2}\kappa h}{1-\frac{1}{2}\kappa h}\right)\right]\right\}$$

$$= \frac{Ch^2}{24}[(1 - \mu\lambda^{-2})(1 + 7\lambda^{-4})\kappa h + O(\kappa^3 h^3)]. \tag{61.47}$$

In these expressions λ and μ act as variables of the deformation. The choice of these variables, although not unique, agrees with that traditionally used in shell theory. It is, however, desirable to construct the stress resultants as partial derivatives of the strain energy with respect the deformation variables. However, in the present case the derivatives of the strain energy, written in its expanded form, are

$$\frac{\partial \varphi}{\partial \lambda} = N + \underline{\frac{\kappa^2 h^2}{12\lambda^4}\mu^2 N + \frac{\mu}{\lambda}\kappa M}, \tag{61.48}$$

$$\frac{\partial \varphi}{\partial \mu} = -\underline{\left(\frac{\kappa^2 h^2}{24\lambda^3}\mu\right)N} - \kappa M, \tag{61.49}$$

which confirms that forces and moments are not the exact derivatives of the strain energy with respect to λ and μ.

A device for reducing the constitutive equations to the canonical form is that of omitting the underlined terms in (61.48) and (61.49), provided that these terms are negligible with respect the remaining terms. Another, more rigorous, procedure is to find new deformation variables in lieu of λ and μ, such that the stress resultants are the partial derivatives of the given strain energy with respect to these natural variables. Let us call these new variables α and β. We must find a continuous nonsingular transformation $\lambda(\alpha, \beta)$ and $\mu(\alpha, \beta)$ such that the equations

$$\begin{aligned} \frac{\partial}{\partial \alpha}[\varphi(\lambda, \mu)] &= N(\lambda, \mu), \\ \frac{\partial}{\partial \beta}[\varphi(\lambda, \mu)] &= -\kappa M(\lambda, \mu), \end{aligned} \tag{61.50}$$

are satisfied in the range of admissible λ and μ. In order to find the transformation we apply the definition

$$\begin{aligned} \frac{\partial \varphi}{\partial \lambda} &= \frac{\partial \varphi}{\partial \alpha}\frac{\partial \alpha}{\partial \lambda} + \frac{\partial \varphi}{\partial \beta}\frac{\partial \beta}{\partial \lambda} = N\frac{\partial \alpha}{\partial \lambda} - \kappa M\frac{\partial \beta}{\partial \lambda}, \\ \frac{\partial \varphi}{\partial \mu} &= \frac{\partial \varphi}{\partial \alpha}\frac{\partial \alpha}{\partial \mu} + \frac{\partial \varphi}{\partial \beta}\frac{\partial \beta}{\partial \mu} = N\frac{\partial \alpha}{\partial \mu} - \kappa M\frac{\partial \beta}{\partial \mu}, \end{aligned} \tag{61.51}$$

which are two partial differential equations in $\alpha(\lambda, \mu)$ and $\beta(\lambda, \mu)$. Provided that we can solve (61.50) we obtain the explicit form of the canonical variables. The conclusion is that when the deformations are large the strains and curvature changes of the reference middle surface lose their privileged status of natural variables for describing the deformation.

A general treatment of highly deformable shells which also involves nonlinear elastic constitutive equations is too difficult to give useful results. The only cases that can be dealt with with partial success are those in which the middle surface is a surface of revolution, such as a cylinder or a sphere, and the deformations are axisymmetric. Although so restricted, the theory can include shells that undergo flexure, extension, and shear (Antman 1995, Ch. X, 1, 2).

In order to define the geometry of the middle surface of an axisymmetric shell, let $\{\mathbf{i}, \mathbf{j}, \mathbf{k}\}$ be a fixed right-handed orthonormal basis for the Euclidean three-dimensional space. For each φ $(0 \leqslant \varphi \leqslant 2\pi)$ we define the vectors

$$\mathbf{e}_1(\varphi) = \mathbf{i}\cos\varphi + \mathbf{j}\sin\varphi, \quad \mathbf{e}_2(\varphi) = -\mathbf{i}\sin\varphi + \mathbf{j}\cos\varphi, \quad \mathbf{e}_3 = \mathbf{k}.$$

The actual configuration of an axisymmetric shell is determined by a pair of vector-valued functions $\mathbf{r}$ and $\mathbf{b}$ of the variable φ and of the variable s $(s_1 \leqslant s \leqslant s_2)$ of the form

$$\mathbf{r}(s, \varphi) = \rho(s)\mathbf{e}_1(s) + \zeta(s)\mathbf{k}, \quad \mathbf{b}(s, \varphi) = -\sin\theta(s)\mathbf{e}_1(\varphi) + \cos\theta(s)\mathbf{k}. \tag{61.52}$$

We also introduce

$$\mathbf{a} = \mathbf{e}_2 \times \mathbf{b} = \cos\theta(s)\mathbf{e}_1(\varphi) + \sin\theta(s)\mathbf{k}. \tag{61.53}$$

For the reference configuration, we designate the vector position of a material point on the middle surface by $\mathbf{R}(s, \varphi)$ and mark the other geometric variables with a subscript $(\cdot)_0$. For example, (61.52), referred to the reference configuration, assume the forms

$$\mathbf{R}(s, \varphi) = \rho_0(s)\mathbf{e}_1(s) + \zeta_0(s)\mathbf{k}, \quad \mathbf{b}_0(s, \varphi) = -\sin\theta_0(s)\mathbf{e}_1(\varphi) + \cos\theta_0(s)\mathbf{k}, \text{ etc.} \tag{61.54}$$

We also introduce the representations

$$\frac{\partial \mathbf{r}}{\partial s}(s, \varphi) \equiv \nu(s)\mathbf{a}(s, \varphi) + \eta(s)\mathbf{b}(s, \varphi), \tag{61.55}$$

$$\tau \equiv \frac{\rho}{\rho_0}, \quad \sigma \equiv \frac{\sin\theta - \sin\theta_0}{\rho_0}, \quad \mu \equiv \theta' - \theta_0', \tag{61.56}$$

where the prime denotes differentiation with respect to s. The strain variables are

$$\mathbf{w} \equiv (\tau, \nu, \eta, \sigma, \mu). \tag{61.57}$$

In order to ensure that these variables describe a deformation that preserves the orientation, we must require that the admissible strains satisfy a particular system of inequalities corresponding to the condition that the Jacobian of the three-dimensional deformation within the shell is positive. These inequalities are

$$\nu > h|\mu + 1|, \quad \tau > h|\sigma + 1| \tag{61.58}$$

where h is a given number in (0,1), which may be interpreted as half the thickness of the shell.

The stress characteristics are defined in the following way. Let $\mathbf{r}_1(s, \varphi)$ and $\mathbf{m}_1(s, \varphi)$ denote the resultant contact force and contact couple per unit reference length of the circle $\varphi \to \mathbf{R}(s, \varphi)$, of radius $\rho_0(s)$, that are exerted across this circular section at $\mathbf{R}(s, \varphi)$; and let $\mathbf{n}_2(s, \varphi)$ and $\mathbf{m}_2(s, \varphi)$ denote the resultant contact force and contact couple per unit reference length of the curve $s \to \mathbf{R}(s, \varphi)$ that are exerted across this section at $\mathbf{R}(s, \varphi)$. As we seek axisymmetric configurations, we require these resultants to have the form

$$\begin{aligned} \mathbf{n}_1(s, \varphi) &= \hat{N}(s)\mathbf{a}(s, \varphi) + \hat{H}(s)\mathbf{b}(s, \varphi), \\ \mathbf{n}_2(s, \varphi) &= \hat{T}(s)\mathbf{e}_2(s, \varphi), \\ \mathbf{m}_1(s, \varphi) &= -\hat{M}(s)\mathbf{e}_2(s, \varphi), \\ \mathbf{m}_2(s, \varphi) &= \hat{\Sigma}(s)\mathbf{a}(s, \varphi). \end{aligned} \tag{61.59}$$

Let us assume that the shell is subjected to an external hydrostatic pressure λ_1 per unit actual (deformed) area of the middle surface. Then, summing forces and couples on the curvilinear segment of material points with coordinates (ξ, ψ) with $s_1 < \xi < s$, $0 < \psi \leqslant \varphi$, we can write the equilibrium equations as follows:

$$\begin{aligned} &\int_0^{\varphi} \left[\hat{N}(s)\mathbf{a}(s, \psi) + \hat{H}(s)\mathbf{b}(s, \psi)\right]\rho_0(s)\, d\psi - \int_0^{\varphi} \left[\hat{N}(s_1)\mathbf{a}(s_1, \psi) + \hat{H}(s_1)\mathbf{b}(s_1, \psi)\right]\rho_0(s_1)\, d\psi \\ &+ \int_{s_1}^{s} \hat{T}(\xi)\mathbf{e}_2(\varphi)\, d\xi - \int_{s_1}^{s} \hat{T}(\xi)\mathbf{e}_2(0)\, d\xi + \lambda_1 \int_{s_1}^{s}\int_0^{\varphi} \mathbf{r}_{,\xi} \times \mathbf{r}_{,\psi}\, d\psi\, d\xi = \mathbf{0}, \end{aligned} \tag{61.60}$$

$$\begin{aligned} &\int_0^{\varphi} \left\{-\hat{M}(s)\mathbf{e}_2(\psi) + \mathbf{r}(s, \psi) \times \left[\hat{N}(s)\mathbf{a}(s, \psi) + \hat{H}(s)\mathbf{b}(s, \psi)\right]\right\}\rho_0(s)\, d\psi \\ &- \int_0^{\varphi} \left\{-\hat{M}(s_1)\mathbf{e}_2(\psi) + \mathbf{r}(s_1 \psi) \times \left[\hat{N}(s_1)\mathbf{a}(s_1, \psi) + \hat{H}(s_1)\mathbf{b}(s_1, \psi)\right]\right\}\rho_0(s_1)\, d\psi \\ &+ \int_{s_1}^{s} \left[\hat{\Sigma}(\xi)\mathbf{a}(\xi, \varphi) + \mathbf{a}(\xi, \varphi) \times \hat{T}(\xi)\mathbf{e}_2(\varphi)\right] d\xi \\ &- \int_{s_1}^{s} \left[\hat{\Sigma}(\xi)\mathbf{a}(\xi, 0) + \mathbf{a}(\xi, 0) \times \hat{T}(\xi)\mathbf{e}_2(0)\right] d\xi = \mathbf{0}. \end{aligned} \tag{61.61}$$

Note that the hydrostatic pressure appears only in (61.60). On differentiating (61.60) and (61.61) with respect to s and φ, we obtain the local form of the equilibrium equations:

$$\frac{\partial}{\partial s}\left[\rho_0(s)(\hat{N}\mathbf{a} + \hat{H}\mathbf{b})\right] - \hat{\mathbf{T}}\mathbf{e}_1 + \lambda_1(\nu\mathbf{a} + \eta\mathbf{b}) \times \rho\mathbf{e}_2 = \mathbf{0}, \tag{61.62}$$

$$\left[\rho_0(s)\hat{M}\right]' - \hat{\Sigma}\cos\theta + \rho_0\left(\nu\hat{H} - \eta\hat{N}\right) = 0. \tag{61.63}$$

In order to complete the formulation of the equilibrium problem we must add the constitutive equations. Under the hypothesis that the material is elastic, there are functions T, N, H, Σ, and M of $\mathbf{w}$ such that we can have

$$\hat{T}(s) = T(\mathbf{w}(s), s), \ldots, \tag{61.64}$$

and we require that these functions satisfy the monotonicity conditions:

$$\frac{\partial(N, H, M)}{\partial(\nu, \eta, \mu)}, \quad \frac{\partial(T, \Sigma)}{\partial(\tau, \sigma)} \text{ are positive-definite.} \tag{61.65}$$

These conditions ensure that an increase in the flexural strain μ is accompanied by a corresponding increase in the bending couple M, and so on. Functions (61.64) must, in addition, satisfy the requirement that extreme strains are enforced by extreme values of the stress resultants. Together with the monotonicity conditions these growth properties ensure that the system of nonlinear algebraic equations

$$N(\mathbf{w}, s) = n, \quad H(\mathbf{w}, s) = h, \quad M(\mathbf{w}, s) = m, \tag{61.66}$$

can be uniquely solved for (ν, η, μ) in terms of the other variables. This solution is then denoted by

$$\nu = \hat{\nu}(\tau, n, h, \sigma, m, s), \ldots. \tag{61.67}$$

In general, shells are isotropic and have a stress-free natural state. This implies that the constitutive functions (61.64) should meet the following restrictions:

$$\begin{aligned}
&\text{(a) } T, N, \Sigma, \text{ and } M \text{ are even in } \eta,\\
&\text{(b) } H \text{ is odd in } \eta,\\
&\text{(c) } T, N, \text{ and } H \text{ are unchanged under } (\sigma, \mu) \to (-\sigma, -\mu),\\
&\text{(d) } \Sigma \text{ and } M \text{ change sign under } (\sigma, \mu) \to (-\sigma, -\mu),\\
&\text{(e) } M(\mathbf{w}) = 0 = \Sigma(\mathbf{w}) \text{ for } \mu = 0 = \sigma,\\
&\text{(f) } N_\sigma(\mathbf{w}) = N_\mu(\mathbf{w}) = T_\sigma(\mathbf{w}) = T_\mu(\mathbf{w}) = 0 \text{ if } \eta = \mu = \sigma = 0,\\
&\text{(g) } N(\nu, \tau, 0, \mu, \sigma) = T(\tau, \nu, 0, \sigma, \mu),\\
&\text{(h) } M(\nu, \tau, 0, \mu, \sigma) = \Sigma(\tau, \nu, 0, \sigma, \mu),\\
&\text{(i) } N(1, 1, 0, 0, 0) = T(1, 1, 0, 0, 0) = 0.
\end{aligned} \tag{61.68}$$

All these conditions are natural, but are not sufficiently strong to reduce the problem of the equilibrium of shells to a simple formulation. In order to get some idea about the structure of solutions, it is then necessary to discuss particular cases. A useful technique is that of *asymptotic solutions* (Isaacson 1965; Needleman 1977; Antman and Calderer 1985), which consists of studying the deformed configuration of a shell under large values of the applied loads. Let us consider a closed axisymmetric shell, for instance a spheroidal shell, subjected to an internal hydrostatic pressure. Let us increase this pressure so that the enclosed volume becomes larger and larger. A natural expectation is that, as the volume tends to infinity, the shape of the shell tends to become spherical. This conjecture is effectively confirmed by theory.

In order to formulate the problem we introduce a small positive parameter ε. The large deformation of the shell is characterized by the requirement that the volume enclosed by it should be that of a sphere of radius $1/\varepsilon$, namely $4\pi/3\varepsilon^3$. We accordingly scale the strains by

$$\begin{gathered}\nu = \varepsilon^{-1}\bar{\nu}, \quad \tau = \varepsilon^{-1}\bar{\tau}, \quad \mathbf{r} = \varepsilon^{-1}\bar{\mathbf{r}}, \quad \rho = \varepsilon^{-1}\bar{\rho}, \quad \zeta = \varepsilon^{-1}\bar{\zeta},\\ \kappa = \varepsilon\bar{\kappa}.\end{gathered} \tag{61.69}$$

For large ν and τ, the leading terms in the constitutive equations for isotropic materials have the form

$$N = A(s)\nu f(\nu, \tau), \quad T = A(s)\tau f(\tau, \nu), \quad H = B(s)\eta g(\nu, \tau),$$
$$M = [D(s)\nu + C(s)\theta']f(\nu, \tau), \quad \Sigma = \left[D(s)\tau + C(s)\frac{1}{\rho_0}\sin\theta_0\right]f(\tau, \nu), \tag{61.70}$$

with $A, B, C > 0$ and $D \geqslant 0$, these inequalities being consequences of the monotonicity conditions imposed on the constitutive equations. By substituting (61.69) in (61.70) we obtain the scaled equations

$$N = \alpha(\varepsilon)\bar{N}, \quad T = \alpha(\varepsilon)\bar{T}, \quad H = \beta(\varepsilon)\alpha(\varepsilon)\bar{H}, \quad M = \gamma(\varepsilon)\alpha(\varepsilon)\bar{M},$$
$$\Sigma = \gamma(\varepsilon)\alpha(\varepsilon)\bar{\Sigma}, \tag{61.71}$$

where $\alpha(\varepsilon)$ is a positive decreasing function with $\alpha(\varepsilon) \to \infty$ as $\varepsilon \to 0$, and $\beta(\varepsilon)$ and $\gamma(\varepsilon)$ are increasing functions with $\beta(0) = 0 \leqslant \gamma(0)$. Insofar as the pressure is concerned, we introduce the scaled expression

$$p = \varepsilon^2\alpha(\varepsilon)\bar{p}. \tag{61.72}$$

Then (61.55) can be replaced by

$$\bar{\mathbf{r}}' = \bar{\nu}\mathbf{a} + \varepsilon\eta\mathbf{b}. \tag{61.73}$$

The equilibrium equations (61.62) and (61.63), written in terms of their components along the axes $\{\mathbf{i}, \mathbf{j}, \mathbf{k}\}$, when the external force is a hydrostatic pressure p, assume the form

$$(\rho_0 N)' = T\cos\theta + \rho_0\theta' H + p\rho\eta,$$
$$(\rho_0 H)' = -T\sin\theta - \rho_0\theta' N + p\rho\nu, \tag{61.74}$$
$$(\rho_0 M)' = \Sigma\cos\theta + \rho_0(\eta N - \nu H).$$

By introducing (61.71), the equilibrium equations reduce to

$$(\rho_0\bar{N})' = \bar{T}\cos\theta + \beta(\varepsilon)\rho_0\theta'\bar{H} + \varepsilon\bar{p}\rho\eta, \tag{61.75a}$$

$$\beta(\varepsilon)(\rho_0\bar{H})' = -T\sin\theta - \rho_0\theta'\bar{N} + \bar{p}\rho\bar{\nu}, \tag{61.75b}$$

$$\gamma(\varepsilon)[(\rho_0\bar{M})' - \bar{\Sigma}\cos\theta] = \rho_0[\nu\bar{N} - \varepsilon^{-1}\beta(\varepsilon)\bar{\nu}\bar{H}]. \tag{61.75c}$$

On observing that we can always change the variables so that we have $(s, \varphi) \in [-L, L] \times [0, 2\pi]$, the boundary conditions become

$$r(-L) = 0 = r(L), \quad \theta(-L) = 0, \quad \theta(L) = \pi. \tag{61.76}$$

Another form of these equations, in terms of the scaled variables, is

$$\theta(L)-\theta(-L)=\int_{-L}^{L}\theta'(s)\,ds=\pi$$

$$\bar{r}(L)-\bar{r}(-L)=\int_{-L}^{L}\bar{\mathbf{r}}'(s)\cdot\mathbf{i}\,ds=\int_{-L}^{L}(\bar{\nu}\cos\theta-\varepsilon\eta\sin\theta)ds=0, \tag{61.77}$$

$$-\int_{-L}^{L}\rho\bar{\mathbf{r}}\cdot[\bar{\nu}\mathbf{b}-\varepsilon\eta\mathbf{a}]ds=2.$$

Note that, by setting $\varepsilon = 0$, the equations obtained from (61.75a) and (61.75b) are just the equations for a membrane under hydrostatic pressure. In order to solve equations (61.75) asymptotically, we assume that there is a sequence of functions $\{\alpha_k\}$ such that each stress and couple resultant can be expressed by the expansion

$$\bar{N}(\mathbf{w},s)=\alpha(\varepsilon)\sum_{k=0}^{K}\alpha_k(\varepsilon)N^k(\bar{\mathbf{w}},s)+0(\alpha_K(\varepsilon)),\quad \ldots, \tag{61.78}$$

where the dots mean that an analogous expansion holds for the other characteristics of the stress.

Under this assumption we can prove that the solution can be represented as an asymptotic series in ε. The leading terms correspond to the state of stress existing in a hydrostatically loaded membrane. We find that, if the constitutive equations are isotropic (but not necessarily homogeneous), and if the changes in area dominate those in length in the constitutive functions T and N for large τ and ν, then the leading terms describe a spherical shell of very large radius. In particular, for large τ and ν, the isotropic strain energy function behaves as $A(\tau^2+\nu^2)^a+B(\tau\nu)^b$, where A, B, a, and b, are positive quantities.

A more detailed analysis of the equilibrium states of a nonlinearly elastic shell is possible only for very special shapes, such as, for example, when the shell is spherical and its middle surface has unit radius in its undeformed natural state. In this case we can take $s_1=0$, $s_2=\pi$ and

$$\rho_0(s)=\sin s,\quad \theta_0(s)=s. \tag{61.79}$$

In this case s is a measure of the arc length along meridians from the south pole. The boundary conditions require that the deformation is regular at the poles:

$$\rho(0)=0=\rho(\pi),\quad \eta(0)=0=\eta(\pi),\quad \theta(0)=0=\theta(\pi). \tag{61.80}$$

On using (61.79), (61.62) and (61.63) reduce to

$$[\sin s\,N(s)]'-T(s)\cos\theta(s)-\sin s\,H(s)\theta'(s)-\lambda\rho(s)\eta(s)=0, \tag{61.81a}$$
$$[\sin s\,H(s)]'+T(s)\sin\theta(s)+\sin s\,N(s)\theta'(s)+\lambda\rho(s)\nu(s)=0, \tag{61.81b}$$
$$[\sin s\,M(s)]'-\Sigma(s)\cos\theta(s)+\sin s\,[\nu(s)H(s)-\eta(s)N(s)]=0, \tag{61.81c}$$

where we have dropped the subscript 1 for λ. On combining (61.81a) with (61.81b) we obtain

$$\sin s[N(s)\sin\theta(s) + H(s)\cos\theta(s)] + \frac{\lambda}{2}\rho^2(s) = 0. \tag{61.82}$$

The material is assumed to be homogeneous and isotropic. Our constitutive assumptions ensure that (61.81) admit the trivial solution in which the shell remains spherical, unsheared, and uniformly compressed, so that the solution is

$$\begin{gathered}\nu = \tau = k = \text{constant}, \quad \eta = 0, \quad \theta(s) = s, \quad \sigma = \mu = 0,\\ N(k, k, 0, 0, 0) = -\frac{\lambda}{2}k^2.\end{gathered} \tag{61.83}$$

We are also interested in finding other solutions besides the trivial one. For this purpose we consider the linearization of (61.81) about the trivial solution. At the end of a lengthy computation we can show that these linearized equations read (Shih and Antman 1986)

$$\begin{gathered}(L\eta)(s) + A(\lambda)\eta(s)\sin s = -a(\lambda)\theta(s)\sin s,\\ (L\theta)(s) + B(\lambda)\theta(s)\sin s = -b(\lambda)\eta(s)\sin s,\\ \eta(0) = 0 = \eta(\pi), \quad \theta(0) = 0 = \theta(\pi),\end{gathered} \tag{61.84}$$

where we have put

$$\begin{gathered}(Lu)(s) = [u'(s)\sin s]' - \frac{u(s)}{\sin s},\\ A \equiv [H_\eta^0]^{-1}\left[\frac{1}{2}\lambda k^2 b + (N_\nu^0 + N_\tau^0 + \lambda k)(N_\nu^0)^{-1}(N_\tau^0 - N_\nu^0 + \lambda k + H_\eta^0)\right],\\ a \equiv (H_\eta^0)^{-1}\left[\frac{1}{2}\lambda k^2 B + (N_\nu^0 + N_\tau^0 + \lambda k)(N_\nu^0)^{-1}\left(kN_\tau^0 - kN_\nu^0 + \frac{1}{2}\lambda k^2\right)\right],\\ B \equiv 1 - \frac{M_\sigma^0}{M_\mu^0}, \quad b \equiv \left(kH_\eta^0 + \frac{1}{2}\lambda k^2\right)\frac{1}{M_\mu^0},\end{gathered} \tag{61.85}$$

where any constitutive function (such as $N, N_\nu, \ldots$), endowed with the superscript $(\cdot)^0$, corresponds to the value

$$N^0 = N[\kappa(\lambda), \kappa(\lambda), 0, 0, 0], \quad N_\nu^0 = N_\nu[\kappa(\lambda), \kappa(\lambda), 0, 0, 0], \quad \ldots. \tag{61.86}$$

Thus the linearized problem leads to a pair of coupled second-order ordinary differential equations. We now convert this system to a single equation for the complex-valued function $\varphi = \alpha\eta + \beta\theta$, where α and β are complex constants. These constants must satisfy the condition

$$(\alpha A + \beta b)\eta + (\alpha a + \beta B)\theta = C\varphi = C(\alpha\eta + \beta\theta), \tag{61.87}$$

where C is another complex constant to be determined. It follows from (61.87) that we have

$$\alpha A + \beta b = \alpha C, \quad \alpha a + \beta B = \beta C, \tag{61.88}$$

giving

$$a\alpha^2 + (B - A)\alpha\beta - b\beta^2 = 0.$$

On using (61.87), we can reduce the system (61.84) to the single Legendre equation

$$(L\varphi)(s) + C\varphi \sin s = 0, \quad \varphi(0) = 0 = \varphi(\pi). \tag{61.89}$$

This boundary-value problem has a nontrivial solution if, and only if, $C = n(n+1)$ with $n = 1, 2, \ldots$, and the corresponding eigensolutions are

$$\varphi_n(s) = P_n^1(\cos s), \tag{61.90}$$

where P_n^1 is the associated Legendre function of the first kind, of degree n and order 1. When $C = n(n+1)$, system (61.88) has a nontrivial solution if, and only if, we have

$$g(\lambda; n) \equiv [A(\lambda) - n(n+1)][B(\lambda) - n(n+1)] - a(\lambda)b(\lambda) = 0. \tag{61.91}$$

Under the reasonable assumption that we have $B < 2$, the eigenfunctions corresponding to the eigenvalue λ, which satisfy (61.89), have the form

$$[\eta(s), \theta(s)] = [[n(n+1) - B]b^{-1}, 1]P_n^1(\cos s). \tag{61.92}$$

The function $s \to P_n^1(\cos s)$ has exactly $(n+1)$ zeros in $[0, \pi]$, including those at 0 and π, each of which is simple. Hence from (61.92) we conclude that η and θ have exactly the same zeros.

Now that we have linearized the equations we can convert the nonlinear system (61.81) into an equivalent system of integral equations. It is convenient to change the unknowns by introducing new variables $(u_1, u_2, u_3) = \mathbf{u}$ by the equations

$$\begin{aligned} u_1(s)(\sin s)^{\frac{1}{2}} &= (L\rho)(s), \\ u_2(s)(\sin s)^{\frac{1}{2}} &= (\zeta' \sin s)', \\ u_3(s)(\sin s)^{\frac{1}{2}} &= L(\theta - s)(s). \end{aligned} \tag{61.93}$$

In terms of the variables (u_1, u_2, u_3), the resulting system of integral equations can be written in the form

$$\mathbf{u} = \mathbf{f}(\lambda, \mathbf{u}), \tag{61.94}$$

with $\mathbf{u} \in (\mathscr{C}^0[0, \pi])^3$ and with the further restriction

$$\int_0^\pi u_2(s)(\sin s)^{\frac{1}{2}}\, ds = 0.$$

Let us now denote by $\mathscr{E}$ the set of functions $\mathbf{u}$ satisfying the conditions above, and such that the functions $\mathbf{w}$ associated with them satisfy (61.58). It is possible to show that $\mathbf{f}$ is a compact and continuous mapping of $\mathscr{E}$ into itself, and that if $\mathbf{u}_0$ is a trivial solution, the linearization of (61.94) about $\mathbf{u}_0$, namely

$$\mathbf{v} = \mathbf{f}_\mathbf{v}[\lambda, \mathbf{u}_0(\lambda)]\mathbf{v}, \tag{61.95}$$

is equivalent to (61.84). We can now apply the global bifurcation theorem (Rabinowitz 1971) and the local bifurcation theorem (Crandall and Rabinowitz 1971). These two theorems give us the following results. Let λ_k be an eigenvalue of (61.95) of odd algebraic multiplicity. Then, bifurcating from the trivial solution pair $[\lambda_k, \mathbf{u}_0(\lambda_k)]$, there is a connected family (a branch) C_k of nontrivial solution pairs

$(\mathbf{u}, \lambda)$ of (61.94) that cannot terminate abruptly. That is to say, either C_k is unbounded in the $(\mathbf{u}, \lambda)$ space, or its closure contains another point $(\zeta, \mathbf{u}_0, (\zeta))$, where ζ satisfies (61.89) with a possibly different positive integer k. Moreover, if λ_k is simple and $\mathbf{v}_k$ is the corresponding eigenfunction, then C_k can be parameterized in the form

$$\lambda = \lambda_k + \varepsilon\mu_k(\varepsilon), \quad \mathbf{u} = \varepsilon[\mathbf{v}_k + \mathbf{w}_k(\varepsilon)], \tag{61.96}$$

where μ_k is a continuous function of ε and we have $\mathbf{w}_k(\varepsilon) \to \mathbf{0}$ for $\varepsilon \to 0$.

Another result we wish to derive is the nodal properties of the solution. Unfortunately, we have to deal with a system of ordinary differential equations instead of a single second-order differential equation as, for example, in the theory of plates. Our system can be reduced to a pair of coupled differential equations in the unknowns $\nu(s)$ and $\theta(s) - s$. The useful result emerging from the analysis of the system, however, is that there are branches that preserve the number of simultaneous zeros of $\eta(s)$ and $\theta(s) - s$, but nothing can be said about the zeros of the individual functions $\eta(s)$ and $\theta(s) - s$.

If we limit ourselves merely to the question of the existence and uniqueness of a solution of axisymmetric equilibrium states of hyperelastic shells, the fundamental result comes from the properties of the strain energy function (Cohen and Pastrone 1986). An arbitrary axisymmetric configuration of a shell can be specified by two vector equations of the forms

$$\mathbf{r} = r(s)\mathbf{e}_r(\varphi) + z(s)\mathbf{k}, \tag{61.97}$$

$$\mathbf{d} = d(s)\{\cos\psi\mathbf{e}_r(\varphi) + \sin\psi\mathbf{k}\}, \tag{61.98}$$

where $r(s) \geqslant 0$, $d(s) \geqslant 0$ ($s \in [0, \ell]$, $\varphi \in [0, 2\pi]$) are orthogonal surface coordinates (Figure 61.1). The parameter s is the arc length along the meridians of the reference configuration, and φ is the latitude. Again, $\mathbf{k}$ is the unit vector along the axis of rotation, $\mathbf{e}_r$ is the unit vector orthogonal to $\mathbf{k}$ in the plane defined by the polar angle $\varphi = 0$. In (61.98) $\mathbf{d}$ is a transverse director field, and $\mathbf{r}$ is the vector position measured from the origin O. We introduce the angle $\theta(s)$ between the tangent $\mathbf{t}$ to the meridian and the vertical, the angle $\beta(s)$ between $\mathbf{d}$ and inward normal $\boldsymbol{\nu}$ to the deformed middle surface, and the angle $\psi(s)$ between $\mathbf{d}$ and the horizontal plane. The angles ψ, θ, and β are then related by

$$\psi = \pi + \theta - \beta. \tag{61.99}$$

In the reference configuration we assume the value $\beta = \pi$. We also define the unit vector $\mathbf{e}_\varphi$ such that $\{\mathbf{e}_r, \mathbf{e}_\varphi, \mathbf{k}\}$ is an orthonormal basis and recall that the unit vector $\mathbf{t}$ is given by $\mathbf{t} = \mathbf{r}'/|\mathbf{r}'|$.

As measures of the deformation, we take the quantities

$$\sigma = \frac{r}{R} - 1, \quad \rho = \{(r')^2 + (z')^2\}^{\frac{1}{2}}, \quad \chi = \cos(\psi + \Psi) - \cos\Psi, \tag{61.100}$$

where R and Ψ are the values of r and ψ in the reference configuration.

In general, the orientation of $\mathbf{d}$ is independent of the orientation of $\boldsymbol{\nu}$; however, if we impose the constraints $\beta = \pi$ and $\mathbf{d}(s) \equiv 1$, we define a *generalized Kirchhoff shell.*

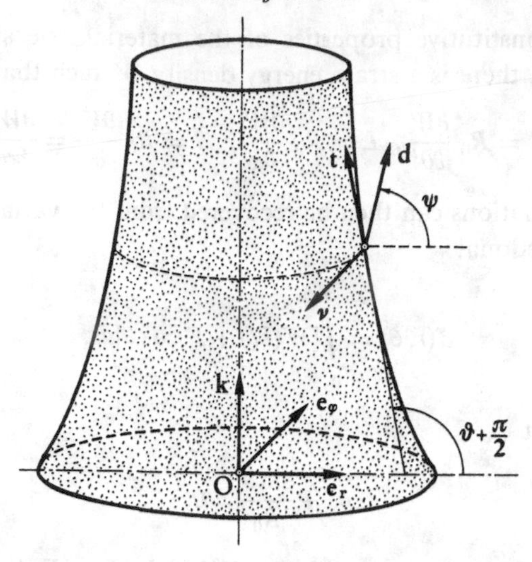

Fig. 61.1

For an axisymmetric Kirchhoff shell the deformation is defined by the vector $\varepsilon_k = \{\sigma, \rho, \chi, \theta'\}$, because we have $\theta = \psi$ from (61.99) and

$$\chi = \cos(\theta + \Theta) - \cos\Theta, \tag{61.101}$$

where Θ is the reference value of θ.

We assume, in addition, that the meridians of the shells are inextensible. This condition is necessary in many examples, because the shell is often reinforced by ribs placed along the meridians, and the effect of these ribs is a strong reduction in the deformability along the directions of the ribs. Consequently, we must impose the additional constraint

$$\mathbf{r}' \cdot \mathbf{r}' = 1 \quad \text{or} \quad \rho = 1, \tag{61.102}$$

with the result that the deformation measures reduce to $\bar{\varepsilon}_K = \{\sigma, \chi, \theta'\}$.

The stress resultants along the directions of the parallels, called *hoop stresses*, are denoted by N, and the bending moments in the same directions are denoted by L; the bending moments along the meridians are indicated by M. These stress resultants must satisfy the equilibrium equations, which is a system of differential equations relating N, L, and M, with the external load. In the present case we suppose that the load is a vertical edge force $\mathbf{F}$ per unit of edge length:

$$\mathbf{F} = -F\mathbf{k}, \quad F \geqslant 0. \tag{61.103}$$

In addition, we impose the condition that the edges of the shell corresponding to $s = 0$, $s = 1$ are constrained to lie on a cylindrical guide of radius R_0, both being free to rotate. The boundary conditions are then

$$r(0) = r(1) = 0, \tag{61.104a}$$
$$M(0) = M(1) = 0. \tag{61.104b}$$

Regarding the constitutive properties of the material, the shell is taken to be hyperelastic; that is, there is a strain energy density W such that

$$M = R_0 \frac{\partial W}{\partial \theta'}, \quad L = R_0 \frac{\partial W}{\partial \chi}, \quad N = R_0 \frac{\partial W}{\partial r} = \frac{\partial W}{\partial \sigma}. \tag{61.105}$$

The equilibrium equations can then be recovered from the variational procedure of minimizing the functional

$$\mathscr{E}(r, \theta) = \int_0^1 [W(\bar{\varepsilon}_k) + F \cos\theta] ds, \tag{61.106}$$

under the constraint

$$\sigma' + \frac{\sin\theta}{R_0} = 0, \tag{61.107}$$

and the geometric boundary conditions (61.104a). The Euler equations of this problem are

$$\begin{gathered} \left(\frac{\partial W}{\partial \theta'}\right)' + \frac{\partial W}{\partial \chi} \sin\theta - \frac{1}{R_0} \mu \cos\theta + F \sin\theta = 0, \\ \mu' - \frac{\partial W}{\partial \sigma} = 0, \\ \sigma' + \frac{\sin\theta}{R_0} = 0, \end{gathered} \tag{61.108}$$

where μ is a Lagrange multiplier. The natural boundary conditions are

$$\left(\frac{\partial W}{\partial \theta'}\right)_{s=0} = \left(\frac{\partial W}{\partial \theta'}\right)'_{s=1} = 0. \tag{61.109}$$

If W satisfies some mild conditions of growth at infinity, namely, if it is greater than linear for large values of its arguments (Antman 1971, 1995), and is strictly convex in θ' and σ, then the variational problem admits a smooth absolute minimizer. If, on the other hand, we already know a trivial solution and this solution does not furnish $\mathscr{E}(r, \theta)$ with a (relative) minimum, then the solution is not unique and nontrivial solutions must exist.

The treatment of the variational problem can be simplified by using the methods of analytical mechanics (Cohen and Pastrone 1986). Let us consider, for simplicity, the case in which the strain energy density has the form

$$W = \frac{1}{2} B(\theta')^2 + \frac{1}{2} C\sigma^2, \tag{61.110}$$

where B and C are positive constants. The Euler equations of the variational problem are

$$\begin{gathered} B\theta'' - \frac{1}{R_0} \mu \cos\theta + F \sin\theta = 0, \\ \mu' - C\sigma = 0, \quad \sigma' + \frac{\sin\theta}{R_0} = 0. \end{gathered} \tag{61.111}$$

and these equations can be derived from the Lagrangian

$$L(\theta, \theta', \sigma, \sigma') = \frac{1}{2}B(\theta')^2 + \frac{1}{2}C\sigma^2 + F\cos\theta + \mu\left(\sigma' + \frac{\sin\theta}{R_0}\right) - F.$$

On introducing the conjugate variables p and q defined by

$$p = \frac{\partial L}{\partial \theta'} = B\theta', \quad q = \frac{\partial L}{\partial \sigma'} = \mu,$$

we can write the corresponding Hamiltonian

$$H(\theta, \sigma, p, q) = \frac{1}{2}\frac{p^2}{B} - \frac{1}{2}C\sigma^2 - F\cos\theta - \frac{q\sin\theta}{R_0} - F,$$

from which the Hamiltonian equations can be readily obtained

$$\theta' = \frac{p}{B}, \quad \sigma' = -\frac{\sin\theta}{R_0}, \quad p' = -F\sin\theta + \frac{q\cos\theta}{R_0}, \quad q' = C\sigma. \tag{61.112}$$

The properties of the solutions of this system can be studied by means of the standard theory of ordinary differential equations.

The techniques of analytical mechanics can be successfully employed to solve the problem of the eversion of elastic shells. Treatment of eversion presents a paradigmatic case of multiplicity of solutions occurring in elasticity. Giving a mathematical description of the possible everted states is very difficult, except in the simple case of incompressible elastic materials. For particular types of shell, such as spherical and cylindrical forms, it is possible to prove the existence of a unique everted state under appropriate hypotheses (Szeri 1990).

In the case of a spherical shell, we represent the reference configuration of the body as the region

$$\left\{(r, \theta, \varphi) \,\middle|\, 0 < \alpha \leqslant r \leqslant 1, \quad 0 \leqslant \beta \leqslant \theta \leqslant \gamma \leqslant \pi, \quad 0 \leqslant \varphi \leqslant 2\pi\right\}, \tag{61.113}$$

where (r, θ, φ) are spherical coordinates. We choose a fixed orthonormal basis $\{\mathbf{i}, \mathbf{j}, \mathbf{k}\}$ and set (Antman 1979a)

$$\begin{aligned} \mathbf{a}_1(\theta, \varphi) &\equiv \mathbf{a}^1(\theta, \varphi) \equiv \sin\theta(\cos\varphi\,\mathbf{i} + \sin\varphi\,\mathbf{j}) + \cos\theta\,\mathbf{k}, \\ \mathbf{a}_2(\theta, \varphi) &\equiv \mathbf{a}^2(\theta, \varphi) \equiv \cos\theta(\cos\varphi\,\mathbf{i} + \sin\varphi\,\mathbf{j}) - \sin\theta\,\mathbf{k}, \\ \mathbf{a}_3(\theta, \varphi) &\equiv \mathbf{a}^3(\theta, \varphi) \equiv -\sin\varphi\,\mathbf{i} + \cos\varphi\,\mathbf{j}, \end{aligned} \tag{61.114}$$

$$\mathbf{z}(r, \theta, \varphi) = r\mathbf{a}_1(\theta, \varphi), \tag{61.115}$$

so that $\mathbf{z}$ assigns positions to the coordinates (r, θ, φ) in the undeformed configuration. For $\varepsilon = \pm 1$ we define

$$\begin{aligned} \mathbf{b}_1(\theta, \varphi) \equiv \mathbf{b}^1(\theta, \varphi) &= \mathbf{a}_1(\varepsilon\theta, \varphi) \\ &= \varepsilon\sin\theta(\cos\varphi\,\mathbf{i} + \sin\varphi\,\mathbf{j}) + \cos\theta\,\mathbf{k}, \\ \mathbf{b}_2(\theta, \varphi) \equiv \mathbf{b}^2(\theta, \varphi) &= \mathbf{a}_2(\varepsilon\theta, \varphi) \\ &= \varepsilon\cos\theta(\cos\varphi\,\mathbf{i} + \sin\varphi\,\mathbf{j}) - \sin\theta\,\mathbf{k}, \\ \mathbf{b}_3(\theta, \varphi) \equiv \mathbf{b}^3(\theta, \varphi) &= \varepsilon\mathbf{a}_3(\varepsilon\theta, \varphi) = \varepsilon(-\sin\varphi\,\mathbf{i} + \cos\varphi\,\mathbf{j}). \end{aligned} \tag{61.116}$$

The triad $\{\mathbf{b}_1, \mathbf{b}_2, \mathbf{b}_3\}$, like $\{\mathbf{a}_1, \mathbf{a}_2, \mathbf{a}_3\}$ forms an orthonormal basis. Now, let $\mathbf{p}(r, \theta, \varphi)$ denote the deformed position of the particle (r, θ, φ), and consider deformations of the type

$$\mathbf{p}(r, \theta, \varphi) = f(r)\mathbf{b}_1(\theta, \varphi), \tag{61.117}$$

which represents an inflation for $\varepsilon = 1$ and an eversion for $\varepsilon = -1$. We take here the case $\varepsilon = -1$.

The deformation gradient $\mathbf{F}$ is then

$$\mathbf{F} = \frac{df}{dr}\mathbf{b}_1 \otimes \mathbf{a}^1 + \frac{f}{r}\left(\mathbf{b}_2 \otimes \mathbf{a}^2 + \mathbf{b}_3 \otimes \mathbf{a}^3\right). \tag{61.118}$$

The quantities $u \equiv df/dr$ and $v = (1/r)f$ are the strains in the radial and in the azimuthal directions, respectively. The components of the first Piola–Kirchhoff stress tensor are defined to be $P(u, v)$ in the r direction and $Q(u, v)$ in the θ direction, where the functions P and Q are assumed to be of class $\mathscr{C}^1$ with $P(1, 1) = Q(1, 1) = 0$. This is to ensure that the reference configuration is natural.

We now suppose that the material is hyperelastic; that is, that the stress $\mathbf{S}$ can be represented as

$$\mathbf{S} = \frac{\partial W(\mathbf{F})}{\partial \mathbf{F}}.$$

However, for our purposes, it is more convenient to introduce a new function $\Phi(u, v)$ with the properties

$$P(u, v) = \frac{\partial \phi}{\partial u}, \quad Q(u, v) = \frac{\partial \phi}{\partial v}. \tag{61.119}$$

The equilibrium equation in the radial direction is then simply

$$\frac{dP}{dr} + \frac{2}{r}(P - Q) = 0, \tag{61.120}$$

and, as the surfaces of the shell are free from traction, we must have the condition

$$P(u, v) = 0, \quad \text{for} \quad r = \alpha, 1. \tag{61.121}$$

In order to complete the formulation of the problem, we need an equation relating u and v, and this can be obtained by differentiating v with respect to r to give

$$\frac{dv}{dr} = \frac{u - v}{r}. \tag{61.122}$$

We also note that for $u < 0$ and $v > 0$ the mapping $\mathbf{p}$ preserves its orientation during the deformation.

Equations (61.120) and (61.122) can be recast as a single equation in the uv plane, called the *phase plane*. By applying the chain rule to (61.120) we obtain

$$\frac{\partial P}{\partial u}\frac{\partial u}{\partial r} + \frac{\partial P}{\partial v}\frac{\partial v}{\partial r} + \frac{2}{r}(P - Q) = 0,$$

which, on making use of (61.122), yields

$$\frac{du}{dr}=\frac{2(Q-P)}{\left(\dfrac{\partial P}{\partial u}\right)r}=\left(\frac{\partial P}{\partial v}\right)\left(\frac{\partial P}{\partial u}\right)^{-1}\left(\frac{u-v}{r}\right). \tag{61.123}$$

The constitutive equations must satisfy the condition of strong ellipticity, which implies that $\partial P/\partial u$ and $\partial Q/\partial v$ are both strictly positive. If the material is hyperelastic, these conditions are equivalent to

$$\frac{\partial^2\phi}{\partial u^2}>0,\quad \frac{\partial^2\phi}{\partial v^2}>0. \tag{61.124}$$

On dividing (61.123) by (61.122), we obtain

$$\frac{du}{dv}=\frac{2(Q-P)}{\dfrac{\partial P}{\partial v}(u-v)}=\left(\frac{\partial P}{\partial v}\right)\left(\frac{\partial P}{\partial u}\right)^{-1}. \tag{61.125}$$

Before transforming (61.125), we observe that, as a consequence of (61.124) and of the natural growth conditions, we have

$$P\to\infty \text{ for } u\to 0,\quad \text{and}\quad P\to-\infty \text{ for } u\to-\infty.$$

We can invert the constitutive equations globally in the sense that, for every P and $v>0$, there is a negative u with the property

$$u=\hat{u}(P,v). \tag{61.126}$$

On multiplying (61.125) by $\partial P/\partial u$ and rearranging the terms, we have the equation

$$\frac{\partial P}{\partial u}\frac{du}{dv}+\frac{\partial P}{\partial v}=\frac{2(Q-P)}{u-v}. \tag{61.127}$$

Now, by virtue of the inversion (61.126), we can interpret the left-hand side of the last equation as the total derivative of $P=P[\hat{u}(P,v),v]\equiv\hat{P}(v)$ with respect to v. Similarly, we define $Q=Q[\hat{u}(P,v),v]\equiv\hat{Q}(v)$, so that (61.127) becomes

$$\frac{dP}{dv}=\frac{2(\hat{Q}-\hat{P})}{\hat{u}-v}, \tag{61.128}$$

with the boundary conditions $P=0$ at $r=\alpha, 1$. In addition to the growth conditions postulated for P, we also need analogous conditions for Q, namely

$$Q\to\infty \text{ for } v\to\infty\quad \text{and}\quad Q\to-\infty \text{ for } v\to 0.$$

We also require that the discriminant

$$D\equiv\frac{\partial P}{\partial u}\frac{\partial Q}{\partial v}-\frac{\partial P}{\partial v}\frac{\partial Q}{\partial u}$$

should be positive for $u<0$, $v>0$.

We now proceed to an analysis of the properties of the solutions of (61.128). The product of (61.122) and (61.128) yields

$$\frac{dP}{dv}\frac{dv}{dr}=\frac{2}{r}(\hat{Q}-\hat{P}), \tag{61.129}$$

which is either positive or negative, depending on the difference $\hat{Q} - \hat{P}$. The derivative of the right-hand side of (61.129) along a line $P =$ constant is

$$\frac{\partial}{\partial v}\{Q[\hat{u}(P, v), v] - P\}_{P=\text{const.}} = \frac{\partial Q}{\partial u}\frac{\partial \hat{u}}{\partial v} + \frac{\partial Q}{\partial v}$$
$$= \frac{\partial Q}{\partial v} - \frac{\partial Q}{\partial u}\left(\frac{\partial P}{\partial v}\right)\left(\frac{\partial P}{\partial u}\right)^{-1} = D\left(\frac{\partial P}{\partial u}\right)^{-1}. \tag{61.130}$$

The function $Q - P$ is a continuous function of v, which runs from $-\infty$ to ∞ along the line $P =$ constant as v runs from 0 to ∞. Moreover, the growth conditions on Q and P imply that $Q - P$ has a unique zero at some point v^* on each half line $v > 0$, $P =$ constant. The candidate for solution trajectories will begin on the v axis at the point $v = v_i$, move on to intersect $v^*(P)$ with zero slope, and move back to intersect the v axis at $v = v_0$. Thus at the initial point v_i we have conditions $Q > 0$ and $P = 0$, while at the end point v_0 we have the conditions $Q < 0$ and $P = 0$ (Figure 61.2). Because there are no singular points in the phase plane with $u < 0$, $v > 0$, the trajectories of (61.122) and (61.129) cannot cross; thus they are ordered by the maximum radial stress achieved at some point $(P_{\max}, v^*(P_{\max}))$. The curve $v^*(P)$ on which $(Q - P) = 0$ is the boundary in the Pv plane between the regions defined by $(Q - P) < 0$ and $(Q - P) > 0$. We should remark that, if the condition $D > 0$ is violated, then the everted states for a shell of given thickness are multiple.

Finally, we must require that the candidate trajectories intersect the physical boundaries at $r = \alpha, 1$. On integrating (61.122) we obtain

$$\int_{\alpha}^{1} \frac{dr}{r} = -\ell n\, \alpha = \int_{v(\alpha)}^{v(1)} \frac{dv}{u - v}. \tag{61.131}$$

For each $\alpha > 0$ there is only one trajectory satisfying the boundary conditions (61.121).

With a similar technique we may study the eversion of a cylindrical shell. The only complication arises from the presence of a third stretch along the axis of the cylinder. The cylindrical shell is defined by

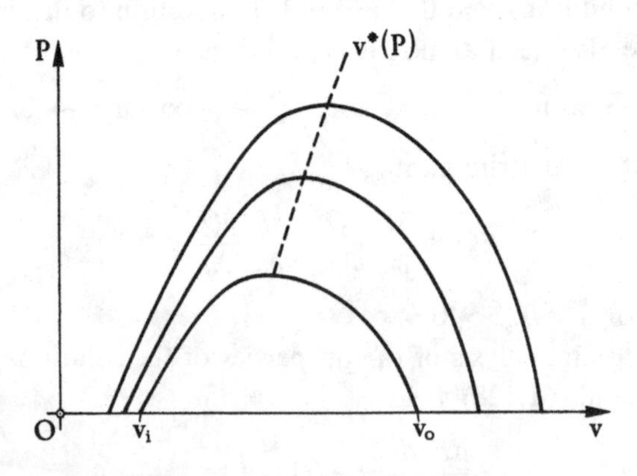

Fig. 61.2

$$\left\{(r, \theta, z) \middle| 0 < \alpha \leqslant z \leqslant 1, \quad 0 \leqslant \theta \leqslant 2\pi, \quad z_1 \leqslant z \leqslant z_2\right\}. \tag{61.132}$$

In the reference configuration we take the orthonormal basis

$$\begin{aligned} \mathbf{a}_1(\theta) &\equiv \mathbf{a}^1(\theta) \equiv \cos\theta\,\mathbf{i} + \sin\theta\,\mathbf{j}, \\ \mathbf{a}_2(\theta) &\equiv \mathbf{a}^2(\theta) \equiv -\sin\theta\,\mathbf{i} + \cos\theta\,\mathbf{j}, \\ \mathbf{a}_3 &\equiv \mathbf{a}^3 = \mathbf{k}. \end{aligned} \tag{61.133}$$

In the deformed configuration we have the orthonormal basis

$$\begin{aligned} \mathbf{b}_1(\theta) &\equiv \mathbf{b}^1(\theta) \equiv \cos\theta\,\mathbf{i} + \sin\theta\,\mathbf{j}, \\ \mathbf{b}_2(\theta) &\equiv \mathbf{b}^2(\theta) \equiv -\sin\theta\,\mathbf{i} + \cos\theta\,\mathbf{j}, \\ \mathbf{b}_3 &\equiv \mathbf{b}^3 = \mathbf{k}. \end{aligned} \tag{61.134}$$

We now consider a deformation of the form

$$\mathbf{p}(r, \theta, z) = f(r)\mathbf{b}_1(\theta) + \lambda \mathbf{b}_3,$$

where λ is a fixed number. The deformation gradient is

$$\mathbf{F} = \frac{df}{dr}\mathbf{b}_1 \otimes \mathbf{a}^1 + \frac{f}{r}\mathbf{b}_2 \otimes \mathbf{a}^2 + \lambda \mathbf{b}_3 \otimes \mathbf{a}^3,$$

with the determinant $[(1/r)f'(r)f\lambda] > 0$, which preserves the orientation. This implies the condition $f'(r) < 0$ and $\lambda < 0$ in the case of eversion. On introducing the phase variables $u \equiv df/dr$ and $v = (1/r)f$, the radial stress is $P(u, v, \lambda)$ and the hoop stress is $Q(u, v)$. The equilibrium equation is

$$\frac{dP}{dr} + \frac{1}{r}(P - Q) = 0, \tag{61.135}$$

with the boundary conditions $P = 0$ for $r = \alpha$, 1, and the further condition that the flat end faces of the shell are constrained to lie against two lubricated plates perpendicular to the axis.

As for spherical shells, we can prove that there is only one trajectory which meets the boundary conditions for each value of α and a fixed negative λ.

62. Dynamical Problems for Shells

The dynamics of thin shells introduces a problem of considerable practical interest, both for the purely mechanical analysis of the stresses generated in the interior of a shell during the motion, and for the acoustic problem of establishing the nature of the sound emanated by an instrument having the shape of a shell, such as a pipe or a bell. The problem of the vibration of a bell was first approached by Euler (1766), who considered the bell as a shell divided into thin annuli, each of which behaves as a curved bar. The method was improved by J. Bernoulli (the younger) (1789), who regarded the shell as a double sheet of curved bars intersecting each other at right angles. The theory of vibrations of cylindrical and spherical shells, regarded as two-dimensional continua, was developed essentially by Rayleigh in various memoirs and collected in his treatise (Rayleigh 1894). Rayleigh's theory is exposed to some criticism, because it is strictly applicable to very thin shells and not to shells of finite thickness. When the thickness is not infinitely small, it may become necessary to take into account certain *local perturbations* that arise in the neighborhood of the edges, and which radically alter the state of stress at large distances from the edges.

One of the assumptions made in the dynamics of shells is that the state of the strain has the same form as that determined by solving the equations of equilibrium. This assumption, however, is not completely acceptable. A common feature of the strain distributions assumed in the shell theories is that the stress component, say σ_z, perpendicular to the middle surface, vanishes everywhere. More precisely, if the shell is a plate σ_z vanishes to a second approximation, while a curved shell σ_z vanishes to a first approximation. However, when the plate or shell is in rapid vibration, the inertial forces also acting perpendicular to the middle surface generate stresses of the type σ_z, which are not negligible. The problem of determining the influence of the stress σ_z in the motion of a plane plate was approached by Rayleigh (1894), who arrived at the conclusion that there are classes of vibrations in which σ_z vanishes throughout the plate, and other classes in which σ_z can be expanded in increasing powers of h (the half-thickness) and z, and the expansions do not contain terms lower than fourth-order.

In the analysis of the deformation of a thin shell, an important question is to decide whether or not the middle surface undergoes extension. In the former case, the deformation is called *extensional* and the potential energy is proportional to the thickness $2h$ of the shell. As the kinetic energy associated with the motion is also proportional to h, the frequencies of vibration are in this case independent of h. When the flexural deformations are taken into account, the potential energy is proportional to the cube of the thickness, and the eigenfrequencies are proportional to h^2. In general, both terms occur, and it is only in the limit that one of the two types of motion prevails. However, in intermediate cases, neglecting one of the two motions may be arbitrary. That the term in h^3 should be negligible in the case of extensional vibration seems reasonable enough, but the reason for omitting the term in h while retaining that in h^3 is not so clear. The logic invoked in making this choice rests on the fact that the natural frequencies of vibration of a shell are determined by using the variational principle of minimizing the ratio between the strain energy and the kinetic energy. The strain energy will, in general, include an extensional part, proportional to h, and a bending part, proportional to h^3. As this energy is to be as small as possible, the actual displacement will be such as to give preference to the bending form, if a displacement of this kind is consistent with the edge conditions. In the case of small motions, for which the linearized theory of elasticity applies, the possibility of a separation between the extension and the flexural vibration considerably simplifies the solution of the equations.

The dynamics of nonlinear elastic shells is a very difficult problem that can be solved only for some particular cases. In contrast to the equilibrium problem, in which the *a priori* restrictions on the constitutive equations allow us to characterize the qualitative properties of the solution, in the problem of the motion the constitutive restrictions are not themselves sufficient to ensure that a well-posed formulation of the corresponding differential problem.

Another difficulty peculiar to the nonlinear elastodynamic problem is that, due to the large displacements, different parts of the shell might collide with each other. As these collisions are not smooth, shock waves could be generated from the regions of mutual contact, with the result of transforming instantaneously the boundary and initial conditions of the system.

A sufficiently general treatment of the dynamical problem is possible only in the presence of strong symmetry, as in the case of the radial motion of a spherical shell under a constant pressure difference between the inner and outer surfaces, and under zero body forces (Calderer 1983).

The material composing the shell must also be homogeneous, isotropic, hyperelastic, and incompressible. This last property might seem an artificial device introduced to simplify the problem; however, on the contrary, the treatment of an incompressible material is more difficult than that of an ordinary compressible one. This is because, in the former case, the concavity hypothesis made on the strain energy function $W(\mathbf{F})$ is not admissible, and therefore the so called *concavity method* (Knops, Levine, and Payne 1974) is no longer applicable. In the particular case of the radial motion of an incompressible spherical or cylindrical shell, the behavior of solutions can be determined by a direct analysis of the differential problem (Calderer 1983).

In order to establish the equations of motion, we introduce spherical coordinates and define the reference configuration of the shell as the domain

$$\Omega = \left\{(R, \Theta, \Phi) \;\middle|\; R_1 < R < R_2\right\}, \tag{62.1}$$

where R_1 and R_2 are the inner and outer radii, respectively. Let the position of the particle, labeled by (R, Θ, Φ) in the reference configuration, be given by

$$\begin{aligned} r^3(t) &= R^3 + r_1^3(t) - R_1^3, \\ \theta(t) &= \Theta, \quad \varphi(t) = \Phi. \end{aligned} \tag{62.2}$$

From (62.2) we can calculate the physical components of the deformation gradient $\mathbf{F}$:

$$\mathbf{F} = \operatorname{diag}\left(\frac{R^2}{r^2}, \frac{r}{R}, \frac{r}{R}\right), \tag{62.3}$$

and note the constraint $\det(\mathbf{F}) = 1$. The right and left Cauchy–Green tensors $\mathbf{C}$ and $\mathbf{B}$ are

$$\mathbf{C} = \mathbf{B} = \operatorname{diag}\left(\frac{R^4}{r^4}, \frac{r^2}{R^2}, \frac{r^2}{R^2}\right). \tag{62.4}$$

The principal stretches λ_1, λ_2, and λ_3 of the deformation have the forms

$$\lambda_1 = \lambda = \frac{R^2}{r^2}, \quad \lambda_2 = \lambda^{-\frac{1}{2}} = \lambda_3 = \frac{r}{R}, \tag{62.5}$$

and the principal invariants of $\mathbf{B}$ are

$$I_1 = \lambda^2 + \frac{2}{\lambda}, \quad I_2 = \frac{1}{\lambda^2} + 2\lambda, \quad I_3 = 1. \tag{62.6}$$

The strain energy function can be expressed as a function of I_1, I_2, or, alternatively, of λ_1, λ_2, and λ_3:

$$W(\mathbf{F}) = W(I_1, I_2) = \sigma(\lambda_1, \lambda_2, \lambda_3). \tag{62.7}$$

Because the material is incompressible, the Cauchy stress tensor has the form

$$\mathbf{T} = -p\mathbf{I} + \psi_1 \mathbf{B} + \psi_{-1}\mathbf{B}^{-1}, \tag{62.8}$$

where

$$\psi_1 = 2\frac{\partial W}{\partial I_1}, \quad \psi_{-1} = 2\frac{\partial W}{\partial I_2}, \tag{62.9}$$

and p is an undetermined hydrostatic pressure. On substituting (62.4) in (62.8) we obtain

$$\mathbf{T} = \operatorname{diag}\left(-p + \psi_1 \frac{R^4}{r^4} + \psi_{-1}\frac{r^4}{R^4},\ -p + \psi_1\frac{r^2}{R^2} + \psi_{-1}\frac{R^2}{r^2},\ -p + \psi_1\frac{r^2}{R^2} + \psi_{-1}\frac{R^2}{r^2}\right).$$

The spherical components of the acceleration are

$$a_r = \ddot{r}, \quad a_\theta = \alpha_\varphi = 0. \tag{62.10}$$

The balance equations in spherical coordinates reduce to

$$\frac{\partial p}{\partial \theta} = 0, \quad \frac{\partial p}{\partial \varphi} = 0$$

$$\begin{aligned} \frac{\partial T_r}{\partial r} + \frac{2}{r}(T_r - T_\theta) = {} & \frac{\partial}{\partial r}\left(-p + \psi_1\frac{R^4}{r^4} + \psi_{-1}\frac{r^4}{R^4}\right) \\ & + \frac{2}{r}\left[\psi_1\left(\frac{R^4}{r^4} - \frac{r^2}{R^2}\right) + \psi_{-1}\left(\frac{r^4}{R^4} - \frac{R^2}{r^2}\right)\right] = \rho\ddot{r}. \end{aligned} \tag{62.11}$$

From the first two equations we find $p = p(r, t)$, and on differentiating (62.2) with respect to t we obtain

$$\ddot{r} = -\frac{\partial A}{\partial r}, \tag{62.12}$$

with

$$A = \frac{1}{r}\left(2r_1\dot{r}_1^2 + r_1^2\ddot{r}_1\right) - \frac{1}{2}\frac{1}{r}r_1^4\dot{r}_1^2. \tag{62.13}$$

As far as the boundary conditions are concerned, we assume that the outer face is free from traction and the inner one is subjected to a constant pressure $p > 0$ measured per unit area in the present configuration. Hence we have

$$T_r(r_2, t) = 0, \quad T_r(r_1, t) = -p. \tag{62.14}$$

We integrate (62.11) with respect to r and use (62.12) and (62.13) to obtain

$$-\rho(A_1 - A_2) + 2\int_{r_2}^{r_1}\left(\frac{r^2}{R^2} - \frac{R^4}{r^4}\right)\left(\psi_1 - \frac{r^2}{R^2}\psi_{-1}\right)\frac{dr}{r} = -p, \tag{62.15}$$

with

$$A_1 = r_1\ddot{r}_1 + \frac{3}{2}\dot{r}_1^2,$$
$$A_2 = \frac{1}{r_2}\left(2r_1\dot{r}_1^2 + r_1^2\ddot{r}_1\right) - \frac{1}{2}r_2^{-4}r_1^4\dot{r}_1^2. \tag{62.16}$$

On setting

$$x(t) = \frac{r_1(t)}{R_1} > 0, \quad \xi(R,t) = \frac{r^3}{R^3}, \quad \delta = \frac{R_2^3}{R_1^3} - 1 > 0, \tag{62.17}$$

we can write

$$\frac{r_2^3}{R_2^3} = \frac{\delta + x^3}{1+\delta}, \quad \frac{r_2^3}{r_1^3} = 1 + \frac{\delta}{x^3}, \quad \xi(R,t) = 1 + \frac{R_1^3}{R^3}(x^3 - 1),$$
$$\lambda = \xi^{-\frac{2}{3}}. \tag{62.18}$$

By substituting (62.16)–(62.18) in (62.15), we obtain

$$2x\ddot{x}\left[1 - \left(1 + \frac{\delta}{x^3}\right)^{-\frac{1}{3}}\right] + \dot{x}^2\left[3 - 3\left(1 + \frac{\delta}{x^3}\right)^{-\frac{1}{3}} + \frac{\delta}{x^3}\left(1 + \frac{\delta}{x^3}\right)^{-\frac{4}{3}}\right] + g(x^3,\delta) = P,$$
$$x > 0, \tag{62.19}$$

with

$$P = \frac{2p}{\rho R_1^2} \quad \text{and} \quad g(x^3,\delta) = \frac{4}{3\rho R_1^2}\int_{\frac{x^3+\delta}{1+\delta}}^{x^3} \xi^{\frac{-7}{3}}(1+\xi)\left(\psi_1 - \psi_{-1}\xi^{\frac{2}{3}}\right)d\xi. \tag{62.20}$$

We now consider the density ρ as a constant. Then, P is also a constant and, therefore, (62.19) is autonomous. By introducing the one-dimensional strain energy $W(\xi) = W[I_1(\xi), I_2(\xi)]$, or $\sigma(\lambda) = W(\xi(\lambda))$, we can differentiate $W(\xi)$ using (62.9) and can write (62.20) as

$$g(x^3,\delta) = \frac{2}{\rho R_1^2}\int_{\frac{x^3+\delta}{1+\delta}}^{x^3} \frac{W'(\xi)}{\xi - 1}\,d\xi. \tag{62.21}$$

We now look for solutions of (62.19) that satisfy the initial conditions $x(0) = x_0$ and $\dot{x}(0) = \dot{x}_0$, where x_0 and $\dot{x}_0$ are prescribed ($x_0 > 0$).

In order to study the behavior of the solutions, we apply the energy method. Multiplying (62.19) by $x^2\dot{x}$ and integrating with respect to t, we obtain

$$\left[1 - \left(1 + \frac{\delta}{x^3}\right)^{-\frac{1}{3}}\right]x^3\dot{x}^2 + G(x^3,\delta) - \frac{1}{3}Px^3 = E, \tag{62.22}$$

with

$$G(x^3, \delta) = \frac{1}{3}\int_1^{x^3} g(\lambda, \delta)\, d\lambda,$$

where E is the total energy, which depends on $(x_0, \dot{x}_0)$. Let us observe that $G(x^3, \delta)$ is a function of x^3, δ being a parameter. The quantity

$$V(x, \delta) = G(x^3, \delta) - \frac{1}{3}Px^3 \tag{62.23}$$

represents the potential energy.

Now make the following assumptions:

(1) $g \in \mathscr{C}^2(0, \infty)$;
(2) $g(x^3, \delta) > 0$ for $x > 1$, and $g(x^3, \delta) < 0$ for $x < 1$;
(3) There are two constants $D \geqslant C > 0$ and a real number $\alpha > 0$, for x large, such that we have $Cx^{\alpha-3} \leqslant g(x^3, \delta) \leqslant Dx^{\alpha-3}$;
(4) There are two constants $M \geqslant N > 0$ and $\beta > 0$ for x small, satisfying $-Mx^{-\beta} \leqslant g(x^3, \delta) \leqslant -Nx^{-\beta}$.

On putting $g'(\xi, \delta) = (\partial/\partial\xi)g(\xi, \delta)$, we can first characterize the equilibrium configurations. Looking at (62.19), we can immediately see that the solutions are defined by the roots of $g(x^3, \delta) = P$, and that they satisfy the inequality $\bar{x} > 1$, as follows from hypothesis (2). In addition, we can state that, for $g'(\bar{x}^3, \delta) > 0$, the solution $(\bar{x}, 0)$ is a center and $V(\bar{x}, \delta)$ is a relatively minimum; if $g'(\bar{x}^3, \delta) < 0$ then $(\bar{x}, 0)$ is a saddlepoint and $V(\bar{x}, \delta)$ is a relative maximum. The results follows from the fact that the eigenvalues of the linearization of (62.19) about $(\bar{x}, 0)$ are

$$\lambda = -\mu = \left\{ \frac{-3\bar{x}g'(\bar{x}^3, \delta)}{2\left[1 - \left(1 + \dfrac{\delta}{\bar{x}^3}\right)^{-\frac{1}{3}}\right]} \right\}^{\frac{1}{2}}, \tag{62.24}$$

and the equilibrium solution $(\bar{x}, \delta)$ is a center if the eigenvalues of the linearized equation are pure imaginary, equal, and with opposite sign, whereas $(\bar{x}, \delta)$ is a saddlepoint if the eigenvalues are real, with the same absolute value, and of opposite sign. Let us note that for $g'(\bar{x}^3, \delta) = 0$, then $\lambda = \mu = 0$ and the linearization method fails. In this case we have $P = P_{\mathrm{cr}}$, where P_{cr} is a point of bifurcation.

The asymptotic behavior of $V(x, \delta)$, for x large, is as follows. Let us assume that condition (3) above is satisfied with $\alpha > 3$. Then, for a sufficiently large fixed $\alpha > 0$ and large x, we can write

$$\int_1^{x^3} g(\xi, \delta)\, d\xi = \int_1^{a^3} g(\xi, \delta)\, d\xi + \int_{a^3}^{x^3} g(\xi, \delta)\, d\xi$$

$$\leqslant \int_1^{a^3} g(\xi, \delta)\, d\xi + \frac{3D}{\alpha}(x^\alpha - a^\alpha). \tag{62.25}$$

On using (62.23), we have

$$V(x,\delta) \leqslant \frac{(D+1)}{\alpha}x^{\alpha} - \frac{P}{3}x^3.$$

Therefore, for $\alpha > 3$, we obtain

$$\frac{C}{2\alpha}x^{\alpha} < V(x,\delta) < \frac{(D+1)}{\alpha}x^{\alpha}. \tag{62.26}$$

If, on the other hand, (3) holds for $\alpha = 3$, then, for x large, and a given $P > 0$, we can prove, by using the same technique, the result

$$\frac{1}{6}(C-P)x^3 < V(x,\delta) < (D-P)x^3. \tag{62.27}$$

Finally, if (3) holds for $\alpha < 3$ we have

$$\lim_{x\to\infty} \frac{V(x,\delta)}{\frac{P}{3}x^3} = 1. \tag{62.28}$$

The asymptotic behavior of $V(x,\delta)$, for $x > 0$ small, is controlled by a similar result. If (4) holds for $\beta > 3$, we consider a sufficiently small fixed $a > 0$, and, for a sufficiently small $x > 0$, and apply (4) to the following inequality:

$$\int_1^{x^3} g(\xi,\delta)\,d\xi = \int_1^{a^3} g(\xi,\delta)\,d\xi + \int_{a^3}^{x^3} g(\xi,\delta)\,d\xi$$

$$\geqslant \int_1^{a^3} g(\xi,\delta)\,d\xi + \frac{3N}{3-\beta}(a^{-\beta+3} - x^{-\beta+3}). \tag{62.29}$$

On using (62.23) we can write

$$\frac{M+1}{(\beta-3)}\frac{1}{x^{\beta-3}} \geqslant V(x,\delta) \geqslant \frac{N}{2(\beta-3)}\frac{1}{x^{\beta}-3} > 0. \tag{62.30}$$

From (62.29) and (2), we derive the inequalities

$$\infty > \int_0^1 -g(\xi,\delta) > 0,$$

so that, for $\beta > 3$, we find that

$$V_0 = \lim_{x\to 0^+} V(a,\delta)$$

exists, is finite, and positive. For $\beta = 3$ we have only the growth condition

$$\lim_{x\to 0^+} \frac{V(x,\delta)}{-\log x} = 1. \tag{62.31}$$

References

Adrianov, I. V. (1984). On the theory of Berger's plates. *Prikl. Matem. Mekhan.* **47**: 142–144.

Amerio, L. and Prouse, G. (1975). Study of the motion of a string vibrating against an obstacle. *Rend. Mat. (2), Ser. VI* **8**: 837–859.

Antman, S. S. (1968). General solutions for plane extensible elastica having nonlinear stress–strain laws. *Q. Appl. Math.* **XXVI**: 35–47.

Antman, S. (1969a). Equilibrium states of nonlinearly elastic rods. In: *Bifurcation Theory and Nonlinear Eigenvalue Problems* (Eds J. B. Keller and S. Antman). New York: Benjamin.

Antman, S. (1969b). Appendix: The case $n = 1$. In: *Bifurcation Theory and Nonlinear Eigenvalue Problems* (Eds J. B. Keller and S. Antman). New York: Benjamin.

Antman, S. (1971). Existence and nonuniqueness of axisymmetric equilibrium states of nonlinearly elastic shells. *Arch. Rat. Mech. An.* **40**: 329–371.

Antman, S. S. (1972). The theory of rods. In: *Handbuch der Physik*, Vol. VIa/2. Berlin: Springer.

Antman, S. (1974). Kirchhoff's problem for nonlinearly elastic rods. *Q. Appl. Math.* **XXXII**: 221–240.

Antman, S. (1976). Ordinary differential equations of one-dimensional nonlinear elasticity. I: Foundations of the theories of nonlinearly elastic rods and shells. *Arch. Rat. Mech. Anal.* **67**: 307–351.

Antman, S. (1979a). The eversion of thick spherical shells. *Arch. Rat. Mech. Anal.* **70**: 113–123.

Antman, S. S. (1979b). Multiple equilibrium states for nonlinear elastic strings. *SIAM J. Appl. Math.* **37**: 588–604.

Antman, S. S. (1988a). A zero-dimensional shock. *Q. Appl. Math.* **XLVI**: 569–581.

Antman, S. S. (1988b). The paradoxical asymptotic status of massless springs. *SIAM Journ. Appl. Math.* **48**: 1319–1334.

Antman, S. (1990). Global properties of buckled states of plates that can suffer thickness changes. *Arch. Rat. Mech. Anal.* **110**: 103–117.

Antman, S. S. (1995). *Nonlinear Problems of Elasticity*. New York: Springer.

Antman, S. S. and Adler, C. L. (1987). Design of material properties that yield a prescribed buckling response. *J. Appl. Mech.* **54**: 263–268.

Antman, S. and Calderer, M. C. (1985). Asymptotic shapes of inflated spherical nonlinearly elastic shells. *Math. Proc. Camb. Phil. Soc.* **97**: 541–549.

Antman, S. and Marlow, R. S. (1992). Transcritical buckling of columns. *Z. Angew. Math. Phys.* **43**: 7–27.

Antman, S. S. and Négron-Marrero, P. V. (1987). The remarkable nature of radially symmetric equilibrium states of aeolotropic nonlinearly elastic bodies. *J. Elasticity* **18**: 131–164.

Antman, S. and Pierce, J. F. (1990). The intricate global structure of buckled states of compressible columns. *SIAM J. Appl. Math.* **50**: 395–419.

Aris, R. (1978). *Mathematical Modelling Techniques*. San Francisco: Pitman.

Aron, H. (1874). Das Gleichgewicht und die Bewegung einer unendlich dünnen, beliebig gekrümmten elastischen Schale. *Crelle J. Math.* **LXXVIII**: 136–174.

Babuška, I. (1961). Stability of the domain under perturbation of the boundary in fundamental problems of the theory of P. D. E. 1, 2. *Czech. Math. J.* **11**: 75–105, 165–203.

Babuška, I. (1962). The stability of domains and the question of the formulation of the plate problem. *Apl. Mat.* **7**: 463–467.

Babuška, I. and Pitkäranta, J. (1990). The plate paradox for hard and soft simple support. *SIAM J. Math. Anal.* **21**: 551–576.

Babuška, I. and Scapolla, T. (1989). Benchmark computation and performance evaluation for a rhombic plate bending problem. *Int. J. Num. Methods Eng.* **28**: 155–181.

Baiocchi, C. and Capelo, A. (1978). *Disequazioni Variazionali e Quasivariazionali, Applicazioni a Problemi di Frontiera Libera.* Bologna: Pitagora.

Ball, J. M. (1977). Convexity conditions and existence theorems in non-linear elasticity. *Arch. Rat. Mech. Anal.* **63**: 337–403.

Barnes, D. C. (1988). The shape of the strongest column is arbitrarily close to the shape of the weakest column. *Q. Appl. Math.* **46**: 605–609.

Batra, R. C. (1972). On nonclassical boundary conditions. *Arch. Rat. Mech. Anal.* **48**: 163–191.

Bauchau, O. A. and Hong, C. H. (1988). Nonlinear composite beam theory. *J. Appl. Mech.* **55**: 156–163.

Bauer, F., Keller, H. B. and Reiss, R. L. (1967). Boundary imperfections in the buckling of spherical caps. *SIAM J. Appl. Math.* **15**: 273–283.

Bauer, L., Callegari, A. J. and Reiss, E. L. (1973). On the collapse of shallow elastic membranes. In: *Nonlinear Elasticity* (Ed. R. W. Dickey), pp. 1–34. New York: Academic Press.

Beck, M. (1952). Die Knicklast des einseitig–eingespannten, tangential gedrückten Stabes. *Z. Angew. Math. Phys.* **3**: 225–228, 476–477.

Beck, M. (1955). Knickung gerader Stäbe durch Drück und conservative Torsion. *Ing. Arch.* **23**: 231–253.

Bendsøe, M. P. (1983). On obtaining a solution to optimization problems for solid, elastic plates by restriction of the design space. *J. Struct. Mech.* **11**: 501–521.

Bendsøe, M. P. (1995). *Optimization of Structural Topology, Shape and Material.* Berlin: Springer.

Benson, R. C. (1981). The deformation of a thin incomplete ring in a frictional channel. *J. Appl. Mech.* **48**: 895–899.

Berdichevskii, V. L. (1980). On the energy of an elastic rod. *Prikl. Matem. Mekh.* **45**(4): 704–718.

Berdichevskii, V. L. and Misyura, V. (1992). Effect of accuracy loss in classical shell theory. *J. Appl. Mech.* **59**: 217–223.

Berdichevskii, V. L. and Starosel'skii, L. A. (1983). On the theory of curvilinear Timoshenko-type rods. *Prikl. Matem. Mekhan.* **47**(6): 1015–1024.

Berdichevskii, V. L. and Starosel'skii, L. A. (1985). Bending, extension, and torsion of naturally twisted rods. *Prikl. Matem. Mekham.* **49**(6): 746–755.

Berdichevskii, V. L. and Truskinovskii, L. (1985). Energy structure and localization. In: *Local Effects in the Analysis of Structures* (Ed. P. Ladezève). Amsterdam: Elsevier.

Berger, H. M. (1955). A new approach to the analysis of large deflections of plates. *J. Appl. Mech.* **22**: 465–472.

Berger, M. S. (1969). A bifurcation theory for nonlinear elliptic partial differential equations and related systems. In: *Bifurcation Theory and Nonlinear Eigenvalue Problems* (Eds J. B. Keller and S. Antman). New York: Benjamin.

Berger, M. S. (1967). On von Kármán equations and the buckling of a thin elastic plate I. *Commun. Pure. Appl. Math.* **XX**: 687–718.

Berger, N. (1973). Estimates for stress derivatives and error in interior equations for shells of variable thickness with applied forces. *SIAM J. Appl. Math.* **24**: 97–120.

Bernadou, M., Ciarlet, P. G. and Miara, B. (1994). Existence theorems for two-dimensional shell theories. *J. Elasticity* **34**: 111–138.

Bernoulli, D. (1733). Theoremata de Oscillationibus Corporum Filo Flexili Connexorum et Catenae Verticaliter Suspensae. *Commun. Acad. Sci. Petrop.* **VI** (1732–1733): 108–122.

Bernoulli, Ja. (1694). Curvatura laminae elasticae. Acta Eruditorum Lipsiae. In: *Opera* **1**: 576–600.

Bernoulli, Ja. (1705). Véritable hypothèse de la résistance des solides avec la démonstration de la courbure des corps qui font ressort. In: *Opera Omnia* **Genf 1744**: 976.
Bernoulli II, J. (1789). Essay théorique sur les vibrations des plaques élastiques, rectangulaires et libres. *Nova Acta Petropolitanae* **V**: 197.
Betti, E. (1872). Teoria dell'elasticità. *Nuovo Cim.* **7–8**: 5–21, 69–97, 158–180; **9**: 34–43; **10** (1873): 58–84.
Bickley, W. G. (1934). The heavy elastica. *Phil. Mag.* **17**(VII): 603–622.
Bieberbach, L. (1926). Ueber Tchebychefsche Netze auf Flächen Negativer Krümmung. *Sit. Preuss Akad. Wiss. (Phys. Math.)* **23**: 294–321.
Billington, D. P. (1967). Note on finite symmetrical deflections in thin shells of revolution. *J. Appl. Mech.* **34**: 763.
Bisshopp, K. E. and Drucker, D. C. (1945). Large deflections of cantilever beams. *Q. Appl. Math.* 3: 272–275.
Bliss, G. (1925). *Calculus of Variation*. La Salle: Open Court.
Boussinesq, J. (1885). *Application des Potentiels à l'Etude de l'Equilibre et des Mouvement des Solides Elastiques*. Paris: Gautier-Villars.
Bromberg, E. and Stoker, J. J. (1945). Non-linear theory of curved elastic sheets. *Q. Appl. Math.* **III**: 246–265.
Bryan, G. H. (1890). On the stability of a plane plate under thrust in its own plane, with applications to the "buckling" of the sides of a ship. *Proc. London Math. Soc.* **XXII**: 54–67.
Bucco, D., Jones, R. and Mazumdar, J. (1978). The dynamic analysis of shallow spherical shells. *J. Appl. Mech.* **45**: 690–691.
Burridge, R. and Keller, J. B. (1978). Peeling, slipping and cracking – some one-dimensional free-boundary problems in mechanics. *SIAM Rev.* **20**: 31–61.
Buttazzo, G. (1988). Thin insulating layers: the optimization point of view. In: *Proceedings of Material Instabilities in Continuum Mechanics and Related Mathematical Problems* (Ed. J. Ball), pp. 11–19. Oxford: Oxford University Press.
Byrd, P. F. and Friedman, M. D. (1971). *Handbook for Elliptic Integrals for Engineers and Physicists*. Berlin: Springer.
Calderer, C. (1983). The dynamical behavior of nonlinear elastic spherical shells. *J. Elasticity* **13**: 17–46.
Callegari, A. J. and Keller, J. B. (1974). Contact of inflated membranes with rigid surfaces. *J. Appl. Mech.* **41**: 189–191.
Callegari, A. J. and Reiss, E. L. (1970). Non-linear boundary value problems for the circular membrane. *Arch. Rat. Mech. Anal.* **31**: 390–400.
Carrier, G. F. (1945). On the nonlinear vibration problem of the elastic string. *Q. Appl. Math.* **3**: 157–165.
Carrier, G. F. (1949). The spaghetti problem. *Am. Math. Monthly* **56**: 669–672.
Carrol, M. M. (1987). Pressure maximum behavior in inflation of incompressible elastic hollow spheres and cylinders. *Q. Appl. Math.* **XLV**: 141–154.
Cauchy, A. L. (1828). De la pression ou tension dans un système de points matériels. *Exer. Math.* **III**, 213.
Cauchy, A. L. (1829). Sur l'équilibre et le mouvement intérieur des corps considérés comme des masses continues. *Exer. Math.* **IV**: 293–319; **IX**: 343–369.
Celigoj, C. (1979). Geometrisch quadratische Biegetorsionstheorie gerader, elastischer Stäbe. *Ing. Archiv.* **48**: 113–119.
Cheng, K.-T. and Olhoff, N. (1982). An investigation concerning design of solid elastic plates. *Int. J. Solids Struct.* **17**: 305–323.
Cheo, L. S. and Reiss, E. L. (1973). Unsymmetrical wrinkling of circular plates. *Q. Appl. Math.* **XXXI**: 75–91.
Chree, C. (1886). Longitudinal vibrations of a circular bar. *Q. J. Math.* **XXI**: 287–298.
Christensen, R. M., Schmidt, R., Levinson, M. and Leko, T. (1977). Discussion on the paper by Nicholson, J. W. and Simmonds, J. G. [1977]. *J. Appl. Mech.* **44**: 797–799.
Chu, C. (1951). The effect of initial twist on the torsional rigidity of thin prismatical bars and tubular members. *Proc. First U.S. Nat. Cong. Appl. Mech*, 265–269. Ann Arbor: Edwards.
Chwalla, E. (1939). Die Kipp-Stabilität gerader Träger mit doppelt-symmetrischem I-Querschnitt. *Forschungshefte aus dem Gebiete des Stahlbaues*, H.2. Berlin: Springer.

Chwalla, E. (1943). Einige Ergebnisse der Theorie des aussermittig gedrückten Stabes mit dünnwandigem, offenem Querschnitt. *Forschungshefte aus dem Gebiete des Stahlbaues*, H.6, pp. 12–21. Berlin: Springer.
Ciarlet, P. G. (1980). A justification of the von Kármán equations. *Arch. Rat. Mech. Anal.* **73**: 349–389.
Ciarlet, P. G. (1988). *Mathematical Elasticity*, Vol. 1: *Three-Dimensional Elasticity*. Amsterdam: North Holland.
Ciarlet, P. G. (1992a). Echange de limites en théorie asymptotique de coques. II. En premier lieu, l'épaisseur tend vers zéro. *C. R. Acad. Sci. Paris* **315**(I): 227–233.
Ciarlet, P. G. (1992b). Echange de limites en théorie asymptotique de coques. I. En premier lieu, la coque devient une plaque. *C. R. Acad. Sci. Paris* **315**(I): 107–111.
Ciarlet, P. G. and Destuynder, P. (1979). A justification of a nonlinear model in plate theory. *Comput. Meth. Appl. Mech. Eng.* **17/18**: 227–258.
Ciarlet, P. G. and Lods, V. (1995). Asymptotic analysis of linearly elastic shells. III. Justification of Koiter's shell equations. *Publ. Lab. Anal. Num. Univ. P. M. Curie, 95023*, pp. 1–12.
Ciarlet, P. G. and Miara, B. (1992). On the ellipticity of linear shell models. *Z. Angew. Math. Phys.* **43**: 243–253.
Ciarlet, P. G. and Nečas, J. (1987). Injectivity and self-contact in nonlinear elasticity. *Arch. Rat. Mech. Anal.* **87**, 171–188.
Cimetière, A., Geymonat, P., Le Dret, H., Raoult, A. and Tutek, Z. (1988). Asymptotic theory and analysis for displacements and stress distribution in nonlinear elastic straight slender rods. *J. Elasticity* **19**: 111–161.
Clebsch, A. (1862). *Theorie der Elastizität der festern Körpers*. Leipzig: Teubner.
Cohen, H. and Pastrone, F. (1986). Axisymmetric equilibrium states of non-linear elastic cylindrical shells. *Int. J. Non-Linear Mech.* **21**: 37–50.
Cooke, D. W. and Levinson, M. (1983). Thick rectangular plates. II. The generalized Levy solutions. *Int. J. Mech. Sci.* **25**: 207–215.
Cosserat, E. and Cosserat, F. (1898). Sur les équations de la théorie de l'élasticité. *C. R. Acad. Sci. Paris* **126**: 1129–1132.
Cosserat, E. and Cosserat, F. (1901). Sur la déformation infiniment petite d'une envelope sphérique élastique. *C. R. Acad. Sci. Paris* **133**: 326–329.
Cosserat, E. and Cosserat, F. (1907). Sur la statique de la ligne déformable. *C. R. Acad. Sci. Paris* **145**: 1409–1412.
Cosserat, E. and Cosserat, F. (1909). *Théorie des Corps Déformables*. Paris: Hermann.
Costello, G. A. and Phillips, J. W. (1973). Contact stresses in thin twisted rods. *J. Appl. Mech.* **40**: 629–630.
Costello, G. A. and Phillips, J. W. (1974). A more exact theory for twisted wire cables. *J. Eng. Mech. Div. ASCE.* **100**: 1096–1099.
Coulomb, C. A. (1787). Recherches théoriques et expérimentales sur la force de torsion, et sur l'élasticité des fils de métal. *Histoire Acad. Paris* **1784**: 229–269.
Courant, R. and Hilbert, D. (1953). *Methods of Mathematical Physics*, Vol. 1. New York: Interscience.
Cowper, G. R. (1966). The shear coefficient in Timoshenko's beam theory. *J. Appl. Mech.* **33**: 335–340.
Crandall, M. G. and Rabinowitz, P. H. (1971). Bifurcation from simple eigenvalues. *J. Funct. Anal.* **8**: 321–340.
d'Alembert, J. le R. (1747). Recherches sur la courbe que forme une corde tendue mise en vibration. *Hist. Acad. Sci. Berlin (1749)* **III**: 214–219.
Dašek, V. (1937). Zur Berechnung von Behälterböden gleicher Festigkeit. *Beton Eisen.* **36**: 54.
Day, W. A. (1981). Generalized torsion: The solution of a problem of Truesdell's. *Arch. Rat. Mech. Anal.* **76**: 283–288.
Day, W. A. (1987). Adiabatic invariants for strings and membranes subjected to slowly-varying tension. *Q. Appl. Math.* **XLV**: 349–360.
Day, W. B. (1980). Buckling of a column with nonlinear restraints and random initial displacement. *J. Appl. Mech.* **47**: 204–205.
De Giorgi, E. (1957). Sulla differenziabilità e l'analiticità delle estremali degli integrali multipli regolari. *Mem. Acc. Sci. Torino (Cl. Sc.)* **3**: 25–43.

Denman, H. H. and Schmidt, R. (1968). Chebyshev approximation applied to large deflections of elastica. *Ind. Math.* **18**: 63–74.
Dickey, R. W. (1966). Dynamic behaviour of soap films. *Q. Appl. Math.* **XXIV**: 97–106.
Dickey, R. W. (1967). The plane circular elastic surface under normal pressure. *Arch. Rat. Mech. Anal.* **26**: 219–236.
Dickey, R. W. (1969). The nonlinear string under a vertical load. *SIAM J. Appl. Math.* **17**: 172–178.
Dickey, R. W. (1970a). Infinite systems of nonlinear oscillations equations with linear damping. *SIAM J. Appl. Math.* **19**: 208–214.
Dickey, R. W. (1970b). Free vibrations and dynamic buckling of the extensible beam. *J. Math. Anal. Appl.* **29**: 443–454.
Dickey, R. W. (1976). Nonlinear bending of circular plates. *SIAM J. Appl. Math.* **30**: 1–9.
Dickey, R. W. (1983). The nonlinear circular membrane under a vertical force. *Q. Appl. Math.* **XLI**: 331–338.
Dickey, R. W. (1987). Membrane caps. *Q. Appl. Math.* **XLV**, 697–712.
Dickey, R. W. (1988). Membrane caps under hydrostatic pressure. *Q. Appl. Math.* **XLVI**: 95–104.
Dickey, R. W. (1990). Minimum energy solution for the spherical shell. *Q. Appl. Math.* **XLVIII**: 321–339.
Dökmeci, M. C. (1972). A general theory of elastic beams. *Int. J. Solids Struct.* **8**: 1205–1222.
Downs, B. (1979). The effect of substantial pretwist on the stiffness properties of thin beams of cambered sections. *J. Appl. Mech.* **46**: 341–344.
Drucker, D. C. and Tachau, H. (1945). A new design criterion for wire rope. *J. Appl. Mech.* **12**: A33–38.
Duhem, P. (1893). Le potentiel thermodynamique et la pression hydrostatique. *Ann. Ecole Norm.* **10**: 187–230.
Duncan, W. G. (1937). Galerkin method in mechanics and differential equations. *Aer. Res. Commun., Rep. Mem. 1718*.
Duva, J. M. and Simmonds, J. G. (1990). Elementary, static beam theory is as accurate as you please. *J. Appl. Mech.* **57**: 134–137.
Eisenhart, L. P. (1947). *An Introduction to Differential Geometry with the use of the Tensor Calculus*. Princeton, NJ: Princeton University Press.
Eliseyev, V. V. (1988). The non-linear dynamics of elastic rods. *Prikl. Matem. Mekhan.* **52**: 493–498.
Ericksen, J. L. and Truesdell, C. (1958). Exact theory of stress and strain in rods and shells. *Arch. Rat. Mech. Anal.* **1**: 295–323.
Ericksen, J. L. (1960). Tensor fields. In: *Handbuch der Physik*, Vol. III/1. Berlin: Springer.
Ericksen, J. L. (1970). Simpler static problems in nonlinear theories of rods. *Int. J. Solids Struct.* **6**: 371–377.
Ericksen, J. L. (1972). Symmetry transformations for thin elastic shells. *Arch. Rat. Mech. Anal.* **47**: 1–14.
Ericksen, J. L. (1973). Loading devices and stability of equilibrium. In: *Nonlinear Elasticity* (Ed. R. W. Dickey), 161–174. New York: Academic Press.
Ericksen, J. L. (1975). Equilibrium of bars. *J. Elasticity.* **5**: 191–201.
Ericksen, J. L. (1977). Bending a prism to helical form. *Arch. Rat. Mech. Anal.* **66**: 1–18.
Ericksen, J. L. (1980). On the status of St-Venant's solutions as minimizers of energy. *Int. J. Solids Struct.* **16**: 195–198.
Euler, L. (1744). Additamentum I (de curvis elasticis), methodus invieniendi lineas curvas maximi minimive proprietate gaudentes. In: *Opera Omnia I*, **24**: 231–297.
Euler, L. (1757). Sur la force des colonnes. *Mém Acad. Berlin* **XIII**: 252–282.
Euler, L. (1766). Tentamen de sono campanarum. *Novi Commentarii Acad. Petrop.* **X**: 261.
Euler, L. (1775). De pressione funium tensorum in corpora subjecta eorumque motu a frictione impedito . . . . *Novi Com. Acad. Sci. Petrop.* **XX**: 304–326, 327–342 (1776).
Faber, G. (1923). Beweis dass unter aller homagen Membramen von gleichen Fläche und gleicher Spannung die Kreisförmige den tiefsten Grundton gibt. *Sitz. Bayer Akad. Wiss.* 169–172.
Farshad, M. (1976). Configurations of momentless extensible rods. *Int. J. Mech. Sci.* **19**: 113–119.

Feng, W. W. and Yang, W. H. (1973). On the contact problem of an inflated spherical nonlinear membrane. *J. Appl. Mech.* **40**: 209–214.
Feng, W. W., Tielking, J. T. and Huang, P. (1974). The inflation and contact constraint of a rectangular Mooney membrane. *J. Appl. Mech.* **41**: 979–984.
Fillunger, P. (1930). Ueber die Spannungen im Mittelschnitt eines Eisbahnzughakens. *Z. Angew. Math. Mech.* **10**: 3, 218–227.
Fischer, D. H. (1969). *Historians' Fallacies.* New York: Harper & Row.
Föppl, A. (1907). *Vorlesungen ueber technische Mechanik*, Bd. 5. Leipzig: Teubner.
Fraser, W. B. and Budiansky, B. (1969). The buckling of a column with random initial deflection. *J. Appl. Mech.* **36**: 233–240.
Friedrichs, K. O. and Stoker, J. J. (1942). Buckling of a circular plate beyond the critical thrust. *J. Appl. Mech.* **9**: A7–A14.
Galilei, G. (1638). *Discorsi e Dimostrazioni Matematiche su due Nuove Scienze.* Leiden: Plantijn.
Geckeler, J. (1926). Ueber die Festigkeit Achsensymmetrischer Schalen. *Forsch. Arb. Ingwes.* **276**: 21.
Geckeler, J. W. (1928). Elastostatik. In: *Handbuch der Physik*, Vol. VI, Chap. 3. Berlin: Springer.
Germain, S. (1821). *Recherches sur la théorie des surfaces élastiques.* Paris: Courcier.
Goldberg, J. E., Baluch, M. H. and Tang, Y. K. (1971). Analysis of moderately thick shells of revolution. *Proc. Conf. Struct. Mech. Reactor Technology*, Berlin: W. Ernst.
Goldberg, J. E., Korman, T. and Baluch, M. (1974). Bemerkungen ueber die Gültigkeitsgrenze der Theorie der dünnen Schalen. *Beton Stahlbetonbau.* **69**: 171–173.
Golubitsky, M. and Schaeffer, D. G. (1985). *Singularities and Groups in Bifurcation Theory*, Vol. II. Berlin: Springer.
Gordienko, B. A. (1979). Theory of spatially curvilinear elastic beams. *Prikl. Matem. Mekh.* **43**(2): 374–380.
Green, A. E. and Laws, N. (1966). A general theory of rods. *Proc. R. Soc. London, Ser. A* **293**: 145–156.
Green, A. E. and Zerna, W. (1954). *Theoretical Elasticity.* Oxford: Clarendon Press.
Green, G. (1839). On the laws of the reflexion and refraction of light at the common surface of two non-crystallized media. *Trans. Cambridge Phil. Soc.* **VII**: 1–24.
Green, G. (1842). On the propagation of light in crystallized media. *Trans. Cambridge Phil. Soc.* **VII**: 121–140.
Greenhill, A. G. (1881). Determination of the greatest height consistent with stability that a vertical pole or mast can be made, and of the greatest height to which a tree of given proportion can grow. *Cambridge Phil. Soc. Proc.* **IV**: 65–73.
Gurtin, M. E. (1972). The linear theory of elasticity. In: *Handbuch der Physik*. Vol. VIa/2, pp. 1–295. Berlin: Springer.
Hall, H. M. (1951). Stresses in small wire ropes. *Wire Wire Products* **26**: 257–259.
Hammersley, J. M. (1973). Maxims for manipulators. *Bull. IMA* **9**: 276; **10**: 368.
Healey, T. J. (1990). Stability and bifurcation of rotating nonlinearly elastic loops. *Q. Appl. Math.* **XLVIII**: 679–698.
Hencky, H. (1915). Ueber den Spannungszustand in kreisrunden Platten mit verschwindender Biegungssteifigkeit. *Z. Math. Phys.* **63**: 311–317.
Hooke, R. (1675). The true mathematical and mechanical form of all manner of arches for building. In: *Posthumous Works of Robert Hooke, M. D., S.R.S.*, p. 1705, R. Walker. London: J. Martyn.
Horgan, C. O., Payne, L. E. and Simmonds, J. G. (1990). Existence, uniqueness and decay estimates for solutions in the nonlinear theory of elastic, edge-loaded, circular tubes. *Q. Appl. Math.* **XLVIII**: 341–359.
Hertz, H. (1881). Ueber die Berührung fester elastischer Körper. *J. R. Ang. Math. Band* **XCII**: 156–171.
Hruska, F. H. (1951). Calculation of stresses in wire rope. *Wire Wire Products* **26**: 766–767, 799–801.
Hsu, S. B. and Wang, S. F. (1988). Analysis of large deformation of a heavy cantilever. *SIAM J. Math. Anal.* **19**: 854–866.
Huang, N. C. (1979). Finite extension of an elastic strand with a central core. *J. Appl. Mech.* **45**: 852–858.

Huddleston, J. V. and Dowd, J. P. (1979). A nonlinear analysis of extensible tie-rods. *J. Struct. Div. ASCE.* **105**: 2456–2460.
Hunt, G. W., Williams, K. A. J. and Cowell, R. G. (1986). Hidden symmetry concepts in the elastic buckling of axially-loaded cylinders. *Int. J. Solids Struct.* **22**: 1501–1515.
Irvine, H. M. (1980). Energy relations for a suspended cable. *Q. J. Mech. Appl. Math.* **XXXIII**: 227–234.
Irvine, H. M. (1981). *Cable Structure*. Cambridge, MA: Harvard University Press.
Isaacson, E. (1965). The shape of a balloon. *Commun. Pure Appl. Math.* **XVIII**: 163–166.
Jahnke, E., Emde, F. and Lösch, F. (1966). *Tafeln höherer Funktionen*. Stuttgart: Teubner.
John, F. (1965). Estimates for the derivatives of the stress in a thin shell and interior shell equations. *Commun. Pure Appl. Math.* **XVIII**: 235–267.
John, F. (1971). Refined interior equations for thin elastic shells. *Commun. Pure Appl. Math.* **XXIV**: 583–615.
Jorgensen, L. (1915). The constant angle arch dam. *Trans. ASCE* **78**: 685–695.
Kármán, Th. v. (1910). Festigkeitsprobleme im Maschinenbau. In: *Encyclopädie der Math. Wiss.*, Vol. IV/4, pp. 601–694. Leipzig: Teubner.
Kármán, Th. v. and Tsien, H.-S. (1939). The buckling of spherical shells by external pressure. *J. Aero. Sci.* **7**: 43–50.
Kauderer, H. (1958). *Nichtlineare Mechanik*. Berlin: Springer.
Keller, H. B. (1973). Buckling of complete spherical shells under slightly nonuniform loads. In: *Nonlinear Elasticity* (Ed. R. W. Dickey), pp. 299–254. New York: Academic Press.
Keller, J. B. (1960). The shape of the strongest column. *Arch. Rat. Mech. Anal.* **5**: 275–285.
Kelvin, Lord W. T. (1848). Note on the integration of the equations of equilibrium of an elastic solid. *Cambridge Dublin Math. J.* **III**: 87–89.
Kelvin, Lord W. T. and Tait, P. G. (1867). *Natural Philosophy*. Oxford: Oxford University Press.
Kerr, A. D. and Coffin, D. W. (1990). On membrane and plate problems for which the linear theories are not admissible. *J. Appl. Mech.* **57**: 128–133.
Kirchhoff, G. (1850). Ueber das Gleichgewicht und die Bewegung einer elastischen Scheibe. *Crelles J.* **40**: 51–88.
Kirchhoff, G. (1859). Ueber das Gleichgewicht und die Bewegung eines unendlich dünnen elastischen Stabes. *J. Reine Angew. Math. (Crelle)* **56**: 285–343.
Kirchhoff, G. (1876). *Vorlesungen ueber Mathematische Physik, Mechanik*, Vol. 28. Leipzig: Teubner.
Kneser, A. (1900). *Lehrbuch der Variationsrechnung*. Braunschweig: Vieweg.
Knops, R. J., Levine, H. A. and Payne, L. E. (1974). Non-existence, instability, and growth theorems for solutions of a class of abstract nonlinear equations with applications to nonlinear elastodynamics. *Arch. Rat. Mech. Anal.* **55**: 52–72.
Kohn, R. V. and Vogelius, Y. (1984). A new model for thin plates with rapidly varying thickness. *Int. J. Solids Struct.* **20**: 333–350.
Kohn, R. V. and Vogelius, M. (1985). A new model for thin plates with rapidly varying thickness. II: a convergence proof. *Q. Appl. Math.* **XLIII**: 1–22.
Koiter, W. T. (1966). On the nonlinear theory of thin elastic shells. *Proc. Kon. Ned. Akad. Wetensch.* **B69**: 1–54.
Koiter, W. T. (1970). On the foundations of the linear theory of thin elastic shells. *Proc. Kon. Ned. Akad. Wetensch.* **B73**: 169–195.
Kolosov, G. V. (1909). On an application of complex function theory to a plane problem in the mathematical theory of elasticity. Dissertation, Dorpat (Yuriev) University [in Russian].
Kolosov, G. V. (1914). Applications of the complex variable to the theory of elasticity. *Z. Math. Phys.* **62**: 383–409.
Korn, A. (1907). Sur un problème fondamental dans la théorie de l'élasticité. *C. R. Acad. Sci. Paris*, **145**, 165–169.
Korn, A. (1908). Solution générale du problème d'équilibre dans la théorie de l'élasticité, dans le cas ou les éfforts sont données à la surface. *Ann. Fac. Sci. Toulouse* **10**(2): 165–269.
Kotchine, N. E. (1926). Sur la théorie des ondes de choc dans un fluide. *Rend. Circ. Mat. Palermo* **50**: 305–344.

Krahn, E. (1924). Ueber eine von Rayleigh formulierte Minimaleigenschaft des Kreises. *Math. Ann.* **94**: 97–100.

Krein, M. G. (1955). On certain problems on the maximum and minimum of characteristic values and on Lyapounov stability. *Math. Soc. Transl., Ser.* 2: 163–187.

Kunoh, T. and Leech, C. M. (1985). Curvature effects on contact position of wire strands. *Int. J. Mech. Sci.* **27**: 465–470.

Ladezève, P. (1976). Critères de validité de la théorie non-linéaire des coques minces élastiques. *J. Struct. Mech.* **4**: 327–348.

Lagrange, J. L. (1759). Recherches sur la nature et la propagation du son. *Misc. Taurin* **1**(3): I–X, 1–112.

Lagrange, J. L. (1811). Posthumous news. *Ann. Chim. Phys.* **30**: 149; 207 (1828).

Lamb, H. (1909). On the flexure of a narrow beam. *Atti Cong. Naz. Matem, Roma* **3**: 12–32.

Landau, L. D. and Lifschitz, E. M. (1971). *The Classical Theory of Fields*, 3rd edn, Oxford: Pergamon.

Lauricella, G. (1906). Sull'integrazione delle equazioni dell'equilibrio dei corpi elastici isotropi. *Att. Acc. Lincei Rend.* **15**: 426–432.

Levinson, M. and Cooke, D. W. (1983). Thick rectangular plates. I. The generalized Navier solution. *Int. J. Mech. Sci.* **25**: 199–205.

Lewiński, T. and Telega, J. J. (1988). Asymptotic method of homogenization of fissured elastic plates. *J. Elasticity* **19**: 37–62.

Lewy, H. and Stampacchia, G. (1969). On the regularity of the solutions of a variational inequality. *Commun. Pure Appl. Math.* **XXII**: 153–188.

Libai, A. and Simmonds, J. S. (1981). Large-strain constitutive laws for the cylindrical deformation of shells. *Int. J. Non-Linear Mech.* **16**: 91–103.

Libai, A. and Simmonds, J. G. (1983). Highly non-linear cylindrical deformations of rings and shells. *Int. J. Non-Linear Mech.* **18**: 181–197.

Libai, A. and Simmonds, J. G. (1988). *The Nonlinear Theory of Elastic Shells: One Space Dimension*. Boston: Academic Press.

Linder, F. (1977). Das Randstörungs Problem der dicken Kalottenschale. *Z. Angew. Math. Mech.* **57**: 101–105.

Lions, J. L. and Magenes, E. (1968). *Problèmes aux limites non homogènes et applications*. Paris: Dunod.

Lions, J. L. and Stampacchia, G. (1967). Variational inequalities. *Commun. Pure Appl. Math.* **XX**: 493–519.

Littlewood, J. E. (1963). Lorentz's pendulum problem. *Ann. Phys.* **21**: 232–242.

Lottati, I. and Elishakoff, I. (1987). On a new destabilization phenomenon: effect of rotatory damping. *Ing. Arch.* **57**: 413–419.

Love, A. E. H. (1888). The small free vibrations and deformation of a thin elastic shell. *Phil. Trans. R. Soc., Ser. A.* **179**: 491–546.

Love, A. E. H. (1927). *A Treatise on the Mathematical Theory of Elasticity*, 4th edn. Cambridge: Cambridge University Press.

Maddocks, J. H. and Keller, J. B. (1987). Ropes in equilibrium. *SIAM J. Appl. Math.* **47**(6): 1185–1200.

Maisonnève, O. (1971). Sur le principe de Saint-Venant. Thesis, University of Poitiers, France.

Mansfield, E. H. (1973). Large-deflection torsion and flexure of initially curved strips. *Proc. R. Soc., London, A Ser.* **334**: 279–298.

Mansfield, E. H. (1980). On finite inextensional deformation of a helical strip. *Int. J. Non-Linear Mech.* **15**: 459–467.

Mansfield, E. H. (1981). An optimum surface of revolution for pressurised shells. *Int. J. Mech. Sci.* **23**: 57–62.

Mansfield, L. and Simmonds, J. G. (1987). The reverse of spaghetti problem; drooping motion of an elastica issuing from a horizontal guide. *J. Appl. Mech.* **54**: 147–150.

Maz'ya, V. G. and Nazarov, S. A. (1986). Paradoxes of limit passage in solutions of boundary value problems involving the approximation of smooth domains by polygonal domains. *Math. USSR.-Izv.* **29**: 511–533.

Megareus, G. (1939). Die Kuppel gleicher Festigkeit. *Bauing.* **17/18**: 232–234.

Michell, A. G. H. (1904). The limits of economy of material in frame-structures. *Phil. Mag., Ser.* **VIII**: 589–597.

Michell, J. H. (1900). On the direct determination of stress in an elastic solid, with applications to the theory of plates. *Proc. London Math. Soc.* **31**, 100–124.

Michell, J. H. (1901). The theory of uniformly loaded beams. *Q. J. Math.* **XXXII**: 28–42.

Mindlin, R. D. (1951). Effect of rotatory inertia and shear on flexural vibrations of isotropic, elastic, plates. *J. Appl. Mech.* **18**: 31–38.

Mindlin, R. D. and Herrmann, G. (1952). A one-dimensional theory of compressed waves in an elastic rod. *Proc. First Nat. Cong. Appl. Mech.*, Chicago, 1951. New York: ASME.

Müller, I. (1972). *Thermodynamik*. Düsseldorf: Bertelsmann.

Mullin, M. (1978). The energy approach to the Kármán–Föppl equations. *SIAM J. Math. Anal.* **9**: 151–156.

Nageswara, B. R. and Venkateswara, G. R. (1986). Large deflections of cantilever beams with end rotational load. *Z. Angew. Math. Mech.* **66**: 507–509.

Naghdi, P. M. (1963). Foundations of elastic shell theory. In: *Progress in Solid Mechanics*, 1–90. Amsterdam: North-Holland.

Nanson, E. J. (1878). Note on hydrodynamics. *Mess. Math.* **7**: 182–185.

Narashima, R. (1968). Non-linear vibration of an elastic string. *J. Sound Vib.* **8**: 131–146.

Navier, C. L. M. H. (1827). Mémoire sur les lois de l'équilibre et du mouvement des corps solides élastiques (1821). *Mém. Acad. Sci. Inst. France* **VII**: 375–393.

Needleman, A. (1977). Inflation of spherical rubber balloons. *Int. J. Solids Struct.* **13**: 409–421.

Negrón-Marrero, P. V. and Antman, S. (1990). Singular global bifurcation problems for the buckling of anisotropic plates. *Proc. R. Soc., London, Ser. A.* **427**: 95–137.

Nemat-Nasser, S. (1970). Thermoelastic stability under general loads. *Appl. Mech. Rev.* **23**: 615–624.

Neuber, H. (1934). Ein neuer Ansatz zur Lösung räumlicher Probleme der Elastizitätstheorie. *Z. Angew. Math. Mech.* **14**, 203–212.

Neuber, H. (1985). *Kerbspannungslehre*. Berlin: Springer.

Neumann, F. (1885). *Vorlesungen ueber die Theorie der Elastizität der festen Körper und des Lichtäthers*. Leipzig: Teubner.

Nicholson, J. W. and Simmonds, J. G. (1977). Timoshenko beam theory is not always more accurate than elementary beam theory. *J. Appl. Mech.* **44**: 337–338.

Nicolai, E. L. (1928). On the stability of the rectilinear form of equilibrium of a bar in compression and torsion. *Izv. Leningr. Politekh. Inst.* **31**: 357–387.

Nicolai, E. L. (1930). Ueber den Einfluss der Torsion auf die Stabilität rotierender Wellen. In: *Proc. 3rd Int. Congr. Appl. Mech.* **2**, 101–110. Stockholm: Tryckerier.

Niordson, F. I. (1965). On the optimal design of a vibrating beam. *Q. Appl. Math.* **XXIII**: 47–53.

Noll, W. (1958). A mathematical theory of the mechanical behavior of continuous media. *Arch. Rat. Mech. Anal.* **2**: 197–226.

Noll, W. and Virga, E. G. (1988). Fit regions and functions of bounded variation. *Arch. Rat. Mech. Anal.* **102**: 1–21.

Novozilov, V. V. (1948). *Foundations of the Nonlinear Theory of Elasticity*. Moscow: Gostekhizdat. [English transl. F. Bagemihl, H. Komm, and W. Seidel, Rochester, NY: Graylock (1953)].

Olhoff, N. (1970). Optimal design of vibrating circular plates. *Int. Solids Struct.* **6**: 139–156.

Olhoff, N. and Rasmussen, S. H. (1977). On single and bimodal optimum buckling loads of clamped columns. *Int. J. Solids Struct.* **13**: 605–614.

Ono, K. (1979). Lateral motion of an axially moving string on a cylindrical guide surface. *J. Appl. Mech.* **46**: 905–912.

Papkovič, P. F. (1932). Expressions générales des composantes des tensions, ne renfermant comme fonctions arbitraires que des fonctions harmoniques. *C. R. Acad. Sci. Paris* **195**: 754–756.

Parker, D. F. (1979). The role of Saint-Venant's solutions in rod and beam theories. *J. Appl. Mech.* **46**: 861–866.

Payne, L. E. and Weinberger, H. F. (1960). A Faber–Krahn inequality for wedgelike domains. *J. Math. Phys.* **39**: 182–188.

Peirce, C. S. (1931). *Collected Papers of Charles Sanders Peirce* (Eds C. Hartshorne and P. Weiss). Cambridge, MA: Harvard University Press.
Phillips, J. W. and Costello, G. A. (1973). Contact stresses in twisted wire cables. *J. Eng. Mech. Div. ASCE* **99**: 331–341.
Phillips, J. W. and Costello, G. A. (1979). General axial response of stranded wire helical springs. *Int. J. Non-Linear Mech.* **14**: 247–257.
Phillips, J. W. and Costello, G. A. (1985). Analysis of wire ropes with internal wire rope cores. *J. Appl. Mech.* **52**: 510–516.
Pipkin, A. C. (1980). Some developments in the theory of inextensible networks. *Q. Appl. Math.* **XXXVIII**: 343–355.
Pipkin, A. C. (1982). Inextensible networks with slack. *Q. Appl. Math.* **XL**: 63–71.
Pipkin, A. C. (1984). Equations of Tchebychev nets. *Arch. Rat. Mech. Anal.* **85**: 81–97.
Pipkin, A. C. (1986). Energy minimization for nets with slack. *Q. Appl. Math.* **XLIV**: 249–253.
Plaut, R. H. (1971). A new destabilization phenomenon in nonconservative systems. *Z. Angew. Math. Mech.* **51**: 319–321.
Pochhammer, L. (1876). Beitrag zur Theorie der Biegung des Kreiszylinders. *Crelle J. Math.* **LXXXI**: 33–61.
Prager, W. and Synge, J. L. (1947). Approximation in elasticity based on the concept of function space. *Q. Appl. Math.* **V**: 241–269.
Prandtl, L. (1899). Kipperscheinungen. Dissertation, University of Munich, Germany.
Protter, M. H. and Weinberger, H. F. (1967). *Maximum Principles in Differential Equations.* Englewood Cliffs, NJ: Prentice-Hall.
Rabinowitz, P. H. (1971). Some global results for nonlinear eigenvalue problems. *J. Funct. Anal.* **7**: 487–513.
Rayleigh, J. W. S. (1894). *The Theory of Sound.* London: Macmillan.
Reiss, E. L. (1963). A uniqueness theorem for the nonlinear axisymmetric bending of circular plates. *AIAA J.* **1**: 2650–2652.
Reiss, E. L. (1980). A new asymptotic method for jump phenomena. *SIAM J. Appl. Math.* **39**: 440–455.
Reiss, E. L. and Matkowsky, B. J. (1971). Nonlinear dynamic buckling of a compressed elastic column. *Q. Appl. Math.* **XXIX**: 245–259.
Reissner, E. (1945). The effect of transverse-shear deformation on bending of elastic plates. *J. Appl. Mech.* **12(A)**: 68–67.
Reissner, E. (1949). On axisymmetrical deformations of thin shells of revolution. *Proc. Symp. Appl. Math.* **1**: 27–52.
Reissner, E. (1958). Rotationally symmetric problems in the theory of thin elastic shells. In: *Third U.S. Nat. Congr. Appl. Mech.*, p. 51. Ann Arbor: Edwards.
Reissner, E. (1963). On the equations for finite symmetrical deflections of thin shells of revolution. *Prog. Appl. Mech.*, Prager Ann. Vol., 171–178.
Rivlin, R. S. (1955). Plane strain of a net formed by inextensible cords. *J. Rat. Mech. Anal.* **4**: 951–974.
Rivlin, R. S. (1959). The deformation of a membrane formed by inextensible cords. *Arch. Rat. Mech. Anal.* **2**: 447–476.
Routh, E. J. (1891). *A Treatise on Analytical Statics*, Vol. 1. Cambridge: Cambridge University Press.
Rosen, A. (1980). The effect of initial twist on the torsional rigidity of beams – another point of view. *J. Appl. Math.* **47**: 389–392.
Rosen, A. (1983). Theoretical and experimental investigation of the nonlinear torsion and extension of initially twisted bars. *J. Appl. Mech.* **50**: 321–326.
Rosen, A. and Friedmann, P. (1979). The nonlinear behavior of elastic slender straight beams undergoing small strains and moderate rotations. *J. Appl. Mech.* **46**: 161–168.
Russel, B. (1903). *The Principles of Mathematics.* Cambridge: Cambridge University Press.
Rychter, Z. (1988a). A simple and accurate beam theory. *Acta Mech.* **75**: 57–62.
Rychter, Z. (1988b). An engineering theory for beam bending. *Ing. Arch.* **58**: 25–34.
Saint-Venant, B. de (1855). Mémoire sur la torsion des prismes. *Mém. Savants étrangers* **XIV**: 233–560.
Sanchez-Palencia, E. (1990). Passage à la limite de l'élasticité tridimensionelle à la théorie asymptotique des coques minces. *C. R. Acad. Sci. Paris* **331**(II): 909–916.

Sapondzhyan, O. M. (1952). Bending of a simply supported polygonal plate. *Izv. Akad. Nauk. SSR.T.V.* **2**: 29–46 [in Russian].
Sawyers, K. N. and Rivlin, R. S. (1974). The Throusers' test for rupture. *Eng. Fract. Mech.* **6**: 557–562.
Schmidt, R. (1974). On Berger's method in the nonlinear theory of plates. *J. Appl. Mech.* **41**: 521–522.
Schmidt, R. and Da Peppo, D. A. (1971). Approximate analysis of large deflections of beams. *Z. Angew. Math. Mech.* **51**: 333–334.
Schwerin, E. (1929). Ueber Spannungen und Formänderungen Kreisförmiger Membrane. *Z. Techn. Phys.* **12**: 651–659.
Servant, M. (1902). Sur l'habillage des surfaces. *C. R. Acad. Sci.* **135**: 575–577.
Servant, M. (1903). Sur l'habillage des surfaces. *C. R. Acad. Sci.* **137**: 112–115.
Sewell, M. J. (1967). On configuration-dependent loading. *Arch. Rat. Mech. Anal.* **23**: 327–351.
Shearer, M. (1985). Elementary wave solutions of the equations describing the motion of an elastic string. *SIAM J. Math. Anal.* **16**: 447–459.
Shih, K.-G. and Antman, S. (1986). Qualitative properties of large buckled states of spherical shells. *Arch. Rat. Mech. Anal.* **93**: 357–384.
Signorini, A. (1930). Sulle deformazioni termoelastiche finite. *Proc. 3rd Inst. Cong. Appl. Mech.* **2**: 80–89. Stockholm: Tryckerier.
Signorini, A. (1949). Transformazioni termoelastiche finite. *Ann. Mat. Pura Appl.* **39**: 147–201.
Simmonds, J. G. (1983). Closed-form, axisymmetric solution of the von Kármán plate equations for Poisson's ratio one-third. *J. Appl. Mech.* **50**: 897–898.
Simpson, A. and Tabarrok, B. (1976). On the equilibrium configuration of a chain subjected to uniform fluid flow in a horizontal plane. *Int. J. Mech. Sci.* **18**: 91–94.
Sodhi, D. S. (1976). Inextensional deformation of thin developable shells with edge constraints. *J. Appl. Mech.* **43**: 69–74.
Sommerfeld, A. (1944). *Vorlesungen ueber Theoretische Physik*, Vol. 1: *Mechanik*. Leipzig: Akademie Verlag.
Steele, C. R. (1975). Forming of thin shells. *J. Appl. Mech.* **42**: 884.
Stephen, N. G. (1980). Timoshenko's shear coefficient from a beam subjected to gravity loading. *J. Appl. Mech.* **47**: 121–127.
Stephen, N. G. (1981). Considerations on second order beam theories. *Int. J. Solids Struct.* **17**: 325–333.
Sternberg, E. and Knowles, J. K. (1966). Minimum energy characterizations of Saint-Venant's solution to the relaxed Saint-Venant problem. *Arch. Rat. Mech. Anal.* **21**: 87–107.
Stevin, S. (1586). *De Beghinselen der Weeghconst*. Leyden: Plantijn.
Stoker, J. J. (1950). *Nonlinear Vibrations in Mechanical and Electrical Systems*. New York: Interscience.
Stoker, J. J. (1968). *Nonlinear Elasticity*. New York: Gordon & Breach.
Stokes, G. G. (1845). On the theories of the internal friction of fluids in motion, and of equilibrium and motion of elastic solids. *Cambridge Phil. Trans.* **8**: 287–319.
Synge, J. L. (1957). *The Hypercircle in Mathematical Physics*. Cambridge: Cambridge University Press.
Szabó, I. (1963). *Einführung in die Technische Mechanik*. Berlin: Springer.
Szabó, I. (1964). *Höhere Technische Mechanik*. Berlin: Springer.
Szabó, I. (1972). Die Geschichte der Plattentheorie. *Bautechnik* **49**: 1–8.
Szeri, A. J. (1990). On the everted state of spherical and cylindrical shells. *Q. Appl. Math.* **XLVIII**: 49–58.
Tadjbakhsh, I. and Keller, J. B. (1962). Strongest columns and isoperimetric inequalities for eigenvalues. *J. Appl. Mech.* **29**: 157–164.
Tadjbakhsh, I. (1969). Buckled states of elastic rings. In: *Bifurcation Theory and Nonlinear Eigenvalue Problems* (Eds J. B. Keller and S. Antman). New York: Benjamin.
Tarnai, T. (1980). Destabilizing effect of additional restraint on elastic bar structures. *Int. J. Mech. Sci.* **32**: 379–390.
Tarski, A. (1953). A general method in proofs of undecidability. In: *Undecidable Theories* (Eds. A. Tarski, A. Mostowski, and R. R. Robinson). Amsterdam: North Holland.

Tchebychev, P. L. (1878). Sur la coupe des vêtements. *Ass. Franç. pour l'Avanc. des Sci. Cong. Paris*, 154–155.

Tchebychev, P. L. (1907). Sur la coupe des vêtements. *Oeuvres* II, Ac. Sc., St. Petersbourg, 708.

Thevendran, V. and Wang, C. M. (1985). Minimum weight design of cables with supports at different levels. *Int. J. Mech. Sci.* **27**: 519–529.

Thurston, G. A. (1973). On the interrelation between stability and computation. In: *Nonlinear Elasticity* (Ed. R. W. Dickey), pp. 252–287. New York: Academic Press.

Tielking, J. T. and Feng, W. W. (1974). The application of the minimum potential energy principle to nonlinear axisymmetric membrane problems. *Appl. Mech.* **41**: 491–497.

Timoshenko, S. P. (1921). On the correction for shear of the differential equations for transverse vibrations of prismatic bars. *Phil. Mag.* **41**: 744–746.

Ting, L. (1977). Interaction of two vortices in a fluid with lower stagnation pressure. *Q. Appl. Math.* **XXXV**: 353–363.

Todhunter, I. and Pearson, K. (1886). *A History of the Theory of Elasticity*, Vols I and II (1893). Cambridge: Cambridge University Press.

Tölke, F. (1939). Ueber Rotationsschalen gleicher Festigkeit für konstanten Innen- oder Aussendrück. *Z. Angew. Math. Mech.* **19**: 338–343.

Trosch, A. (1952). Stabilitätsprobleme bei tordierten Stäben und Wellen. *Ing. Arch.* **20**, 265–277.

Truesdell, C. (1959). The rational mechanics of materials – past, present, future. *Appl. Mech. Rev.* **12**: 75–80.

Truesdell, C. (1960). The rational mechanics of flexible or elastic bodies, 1638–1788. In: *L. Euleri Opera Omnia*, Vol. 11b. Zurich: Orell-Füssli.

Truesdell, C. (1966). *The Elements of Continuum Mechanics*. Berlin: Springer.

Truesdell, C. (1978). Some challenges offered to analysis by rational thermomechanics. In: *Contemporary Developments in Continuum Mechanics and Partial Differential Equations* (Eds G. M. de la Penha and L. A. Madeiras), pp. 495–603. Amsterdam: North Holland.

Truesdell, C. (1980). *The Tragicomical History of Thermodynamics 1822–1854*. Berlin: Springer.

Truesdell, C. (1991). *A First Course in Rational Continuum Mechanics*, Vol. 1. New York: Academic Press.

Truesdell, C. and Noll, W. (1965). The non-linear field theories of mechanics. In: *Handbuch der Physik*, Vol. III/3. Berlin: Springer.

Truesdell, C. and Toupin, R. A. (1960). The classical field theories. In: *Handbuch der Physik*. Vol. III/1. Berlin/Göttingen/Heidelberg: Springer.

van der Heijden, A. V., Koiter, W. T., Reissner, E. and Levine, H. S. (1977). Discussion on the paper by Nicholson, J. W. and Simmonds, J. G. [1977]. *J. Appl. Mech.* **44**: 357–360.

Verma, G. R. and Keller, J. B. (1984). Hanging rope of minimum elongation. *SIAM Rev.* **26**: 569–571.

Vielsack, P. (1975). Lineare Stabilitätstheorie elastischer Stäbe nach der zweiten Näherung. *Ing. Archiv.* **44**: 143–152.

Vlasov, V. S. (1958). *Allgemeine Schalentheorie und ihre Anwendung in der Technik* [Trans. from the 1949 Russian edition). Berlin: Akademie.

Voigt, W. (1887). Theoretische Studien ueber die Elastizitätsverhältnisse der Kristalle. *Abh. Ges. Wiss. Göttingen* **34**: 1–100.

Vorovich, I. I. and Lebedev, L. P. (1988). On the solvability of non-linear shallow shell equilibrium problems. *Prikl. Matem. Mekhan.* **52**: 635–641.

Wang, C.-Y. (1981a). Large deflections of an inclined cantilever with an end-load. *Int. J. Non-Linear Mech.* **16**: 155–164.

Wang, C.-Y. (1981b). Large deformations of a heavy cantilever. *Q. Appl. Math.* **XXXIX**: 261–273.

Wang. C.-Y. (1982). Rotation of a free elastic rod. *J. Appl. Mech.* **49**: 225–227.

Wang. C.-Y, and Watson, L. T. (1980). Theory of the constant force spring. *J. Appl. Mech.* **47**: 956–958.

Watson, L. T. and Wang. C.-Y. (1983). The equilibrium states of a heavy rotating column. *Int. J. Solids Struct.* **19**: 653–658.

Weinberger, H. F. (1965). *A First Course in Partial Differential Equations*. Waltham: Blaisdell.

Wempner, G. (1981). *Mechanics of Solids with Application to Thin Bodies*. Alphen aan den Rijn: Sijthoff and Noordhoff.

Wilson, J. F. and Snyder, J. H. (1988). The elastica with end-load flip-over. *J. Appl. Mech.* **55**: 845–848.

Worch, G. (1967). Elastische Scheiben. In: *Beton-Kalender*, Vol. II, pp. 1–127. Berlin: Ernst.

Wu, C. H. (1979). Large finite strains membrane problems. *Q. Appl. Math.* **XXXVI**: 347–359.

Wu, C. H. (1980). Sutures in stretched membranes. *Q. Appl. Math.* **XXXVIII**: 109–119.

Yabuta, T., Yoshizawa, N. and Kojima, N. (1982). Cable kink analysis: cable loop stability under tension. *J. Appl. Mech.* **49**: 584–588.

Yang, W. H. and Hsu, K. H. (1971). Indentation of a circular membrane. *J. Appl. Mech.* **38**: 227–231.

Zachmann, D. W. (1979). Nonlinear analysis of a twisted axially loaded elastic rod. *Q. Appl. Math.* **XXXVII**: 67–72.

Ziegler, H. (1952). Die Stabilitätskriterien der Elastomechanik. *Ing. Arch.* **20**: 49–56.

Zveriaev, E. M. (1970). On elasticity relationships in the linear theory of thin elastic shells. *Prikl. Mat. Mech.* **34**: 1136–1138.

Index of Authors Cited

Index